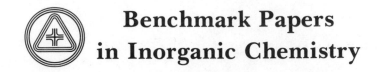

Benchmark Papers
in Inorganic Chemistry

Series Editor: Harry H. Sisler
University of Florida

Published Volumes and Volumes in Preparation

METAL–AMMONIA SOLUTIONS
William L. Jolly
COMPOUNDS CONTAINING PHOSPHORUS–PHOSPHORUS BONDS
Alan H. Cowley
CHLORAMINATION REACTIONS
Stephen E. Frazier
HARD AND SOFT ACIDS AND BASES
Ralph G. Pearson
SYMMETRY IN CHEMICAL THEORY
John P. Fackler, Jr.
LIGAND FIELD THEORY
R. Carl Stoufer
COMPOUNDS CONTAINING ARSENIC NITROGEN AND ANTIMONY
NITROGEN BONDS
Larry K. Krannich

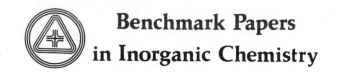

Benchmark Papers
in Inorganic Chemistry

————— A *BENCHMARK* TM Books Series —————

HARD AND SOFT
ACIDS AND BASES

Edited by
RALPH G. PEARSON
Northwestern University

Dowden, Hutchinson & Ross, Inc.
Stroudsburg, Pennsylvania

546.24
P31h
89851
Sept 1974

Library of Congress Cataloging in Publication Data

Pearson, Ralph G   comp.
    Hard and soft acids and bases.

    (Benchmark papers in inorganic chemistry)
    1. Acids--Addresses, essays, lectures.
2. Bases (Chemistry)--Addresses, essays, lectures.
I. Title.
QD477.P39   546'.24   72-93262
ISBN 0-87933-021-X

Acknowledgments
and Permissions

The following papers have been reprinted with the permission of the authors and copyright owners.

The American Chemical Society—*Journal of the American Chemical Society*
"Correlation of the Relative Rates and Equilibria with a Double Basicity Scale"
"The Factors Determining Nucleophilic Reactivities"
"Hard and Soft Acids and Bases"
"Chemical Reactivity and the Concept of Charge- and Frontier-Controlled Reactions"
"A Four-Parameter Equation for Predicting Enthalpies of Adduct Formation"
"Cation-Induced Linkage Isomerism of the Thiocyanatopentacyanocobaltate(III) Complex"
"Nucleophilic Constants and Substrate Discrimination Factors for Substitution Reactions of Platinum(II) Complexes"
"Ligand Substitution Catalysis *via* Hard Acid–Base Interaction"
"Formation of Metal Ion–Nitrene Complexes"
"Homolytic and Ionic Mechanisms in the Ligand-Transfer Oxidation of Alkyl Radicals by Copper(II) Halides and Pseudohalides"
"Application of the Principle of Hard and Soft Acids and Bases to Organic Chemistry"
"Nucleophilic Reactivity Constants toward Methyl Iodide and *trans*-$[Pt(py)_2Cl_2]$"
"The Relative Nucleophilicity of Some Common Nucleophiles toward Sulfinyl Sulfur. The Nucleophile-Catalyzed Hydrolysis of Aryl Sulfinyl Sulfones"
"The Relative Nucleophilicity of Some Common Nucleophiles toward Sulfonyl Sulfur. The Nucleophile-Catalyzed Hydrolysis and Other Nucleophilic Substitution Reactions of Aryl α-Disulfones"

The American Chemical Society—*Inorganic Chemistry*
" 'Symbiotic' Ligands, Hard and Soft Central Atoms"
"Use of the Edwards Equation to Determine Hardness of Acids"
"Inorganic Linkage Isomerism of the Thiocyanate Ion"
"Relative Stabilities of Some Halide Complexes of Rhodium and Iridium"
"Monohalide Displacements of *trans*-$[Pt(P(C_2H_5)_3)_2Cl_2]$ in Dipolar Aprotic Solvents"
"Reactivity Patterns in Inner- and Outer-Sphere Reductions of Halogenopentaamminecobalt(III) Complexes"

The American Chemical Society—*Journal of Organic Chemistry*
"Symbiotic Effects in Nucleophilic Displacement Reactions on Carbon"

The American Chemical Society—*Journal of Chemical Education*
"Hard and Soft Acids and Bases, HSAB, Parts I, II, Fundamental Principles, Underlying Theories"

Academic Press—*Advances in Inorganic Chemistry and Radiochemistry*
"The General, Selective, and Specific Formation of Complexes by Metallic Cations"

The Chemical Society, London—*Quarterly Reviews of the Chemical Society*
"The Relative Affinities of Ligand Atoms for Acceptor Molecules and Ions"

The Chemical Society, London—*Chemical Communications*
"Failure of Pauling's Bond Energy Equation"

"Failure of the Principle of Hard and Soft Acids and Bases to Explain the Amount of Cyclization of Various Hex-5-enyl Derivatives During Acetolysis"

The Chemical Society, London—*Journal of the Chemical Society*
"Metal-β-Diketone Complexes. Part II. Carbon-Bonded Platinum(II) Complexes of Trifluoroacetylacetone and Benzoylacetone"

American Association for the Advancement of Science—*Science*
"Acids and Bases"

The Royal Institute of Chemistry and the Chemical Society—*Chemistry in Britain*
"Hard and Soft Acids and Bases"

Springer-Verlag, New York—*Structure and Bonding, Volume 1*
"The Classification of Acceptors and Donors in Inorganic Reactions"
"Factors Contributing to (b)-Behaviour in Acceptors"
"Electric Polarizability, Innocent Ligands and Spectroscopic Oxidation States"

Springer-Verlag, New York—*Structure and Bonding, Volume 3*
"Relations between Softness, Covalent Bonding, Ionicity and Electric Polarizability"

Springer-Verlag, New York—*Structure and Bonding, Volume 5*
"Thermodynamics of Complex Formation between Hard and Soft Acceptors and Donors"

Butterworth and Company, Publishers, Ltd. and the International Union of Pure and Applied Chemistry—*Pure and Applied Chemistry*
"Electrostatic and Non-electrostatic Contributions to Ion Association in Solution"

Microforms International Marketing Corporation—*Journal of Inorganic and Nuclear Chemistry*
"Electronegativity, Acids, and Bases—I. Hard and Soft Acids and Bases and Pauling's Electronegativity Equation"
"Electronegativity, Acids, and Bases—II. Size Effects and the Principle of Hard and Soft Acids and Bases"

The Chemical Society of Japan—*Bulletin of the Chemical Society of Japan*
"Evaluation of Softness from the Stability Constants of Metal-Ion Complexes"

North-Holland Publishing Company—*Chemical Physics Letters*
"Scales of Softness for Acceptors and Donors"

Elsevier Publishing Company—*Coordination Chemical Review*
"Stabilization of Metal Complexes by Large Counter-Ions"
"The Concept of Hard and Soft Acids and Bases and Nucleophilic Displacement Reactions"

Elsevier Publishing Company—*Journal of Electroanalytical Chemistry*
"Chemical Softness and Specific Adsorption at Electrodes"

Pergamon Press—*Journal of Inorganic and Nuclear Chemistry*
"The Hard and Soft Acid–Base Principle and Metal Ion Assisted Ligand Substitution Processes"

U. Croatto—*Inorganica Chemica Acta, Reviews*
"The Effects of Ligand and Metal Hardness on the Formation of Hydrido- and Organo-Metallic Complexes by the β-Interaction"

Verlag Chemie, GmbH, Weinheim—*Angewandte Chemie, International Edition*
"The Concept of Hard and Soft Acids and Bases as Applied to Multi-Center Chemical Reactions"
"Nucleophilic Reactivity of the Thiophosphoryl Group"

The Croatian Chemical Society—*Croatica Chemica Acta*
"Structural Factors Involved in Ionic Adsorption"

Società Chimica Italiana—*La Chemica e l'Industria*
"General Features of Homogeneous Catalysis with Transition Metals"

Akademische Verlagsgesellschaft—*Zeitschrift für Physikalische Chemie*
"Adsorptive and Catalytic Properties of Chromia"

Series Editor's Preface

Throughout the last century the study of acid–base phenomena has consistently generated and maintained a high degree of interest on the part of professional chemists as well as chemistry students. Chemical and physical phenomena explainable in terms of acid–base theories have proved to be almost unlimited in mumber. The broadening of the concepts of acids and bases has contributed in a major way to the understanding of the physical and biological worlds, has lent an underlying simplicity to chemical theory, and has helped enormously in the achievement of a workable intellectual model of the physical world. In this volume we have an editor, Dr. Ralph Pearson, who has himself contributed effectively to the development of acid–base chemistry through the creation of the concept of "hard" and "soft" acids and bases, selecting from the literature those papers which are Benchmarks in the development of that concept, in its application to the chemical problems of recent years, in the refinement of the concept, and in the delineation of its limitations. In his commentary, Dr. Pearson shares his personal history of the development of this concept and, as a by-product, allows us to sample some of the human side of science. I am sure that the reader will find this volume interesting and pleasant to read as well as helpful in the understanding of an important field of chemistry.

Harry H. Sisler

Preface

This book attempts to perform two functions. The first is to tell something of the history of hard and soft acids and bases from the personal side of the author. Many readers may be interested in how a somewhat controversial scientific concept evolves, and how it is received by the scientific community.

The second, more important, function is to show by these selections from the literature how the HSAB principle has been used in many areas of chemistry, both inorganic and organic. The reader may then see how HSAB can be useful in his own work, whether it be learning chemistry or practicing it.

The articles chosen for inclusion were selected from literally hundreds that referenced HSAB in one way or another. Of course in most of these, HSAB was only a small part of the content. The selected articles are among those where the concept played a major role.

In addition, these are articles which (a) are historically important; (b) are good examples of application to a given field; or (c) offer a chance to make some qualifying or explanatory remarks about the application of HSAB.

My special thanks go to the original authors who have kindly allowed me to use their material. I regret that space limitations made it necessary to leave out a large number of equally fine articles.

My thanks also go to Miss Susan E. Wolf, who aided greatly in the preparation of this volume.

Ralph G. Pearson
Evanston, Illinois
February, 1973

Contents

Part V

Part VI

Part VII

Contents by Author

Introduction

The origin of the word acid is from the Latin *acidus* meaning sour. The word base comes eventually from the Latin *bassus* meaning low. The alchemists used Platonic concepts to divide metals into two classes. Those which resisted corrosion were rich in the element of nobility, and were called noble metals. Those which tarnished or rusted readily contained the element of baseness, and were called base metals.

The more reactive metals, of course, produced alkaline solutions under the influence of air and water. The word alkali is of Arabic origin, its meaning being "the ashes." Wood ashes are indeed a source of alkali, mainly sodium and potassium carbonates. In English we recognize this by our use of the name potassium (from "pot-ash").

The alchemists identified acids and bases in a purely operational manner, a procedure in the best scientific tradition. Acids were substances which tasted sour, turned litmus red, dissolved chalk with effervescence, and so on. Bases tasted bitter, turned litmus blue, and had other distinct properties. However, the most important characteristic of a base was that it could neutralize the properties of an acid. Thus, from the very beginning, acids and bases were linked together in a complementary fashion.

The next stage of development came at the end of the eighteenth century and the beginning of the nineteenth. With a modern understanding of the chemical elements, it was natural to try to identify acids and bases as compounds containing certain elements. Lavoisier regarded oxygen as the "acidifying principle." This view was questioned by Davy, who pointed to the importance of hydrogen.

In the decade between 1880 and 1890, Arrhenius and Ostwald showed that acidic properties were found in compounds which gave rise to hydrogen ions in aqueous solution. Similarly, basic properties were related to the formation of hydroxide ions in solution. Neutralization was neatly explained in terms of the favorable reaction $H^+ + OH^- \rightarrow H_2O$. The Law of Mass Action was applied to ionization equilibria and a satisfactory quantitative account was given for a number of phenomena.

The definition of acids and bases as substances that produce hydrogen ions and hydroxide ions, respectively, in aqueous solution was accepted for many years, and is still in some use today. Nevertheless, certain difficulties existed. For example, it was necessary to invent hypothetical substances such as NH_4OH to account for the basicity of ammonia. Solvents other than water were troublesome. The work of Franklin and others showed that in liquid ammonia, the ions that took the place of H^+ and OH^- were NH_4^+ and NH_2^-. In other protic solvents, similar ions could be postulated. The hydrogen ion in water began to be considered as H_3O^+.

It is of interest to note that Arrhenius' concept of weak and strong acids met considerable opposition. It was argued that a so-called weak acid required just as much base to neutralize it as did a strong acid. Hence there was no essential difference. This objection was obviously not valid in the light of Arrhenius' definition of a weak acid (only partially ionized) and the Law of Mass Action.

By a curious coincidence, 1923 saw the birth of two new definitions of acids and bases—those of J. N. Brønsted and G. N. Lewis. These new definitions are the ones in common use today. They exist side by side in a competitive, and still unresolved, situation.

According to Brønsted, "an acid is a species tending to lose a proton and a base is a species tending to gain a proton." The fundamental acid–base reaction is

$$A \rightleftharpoons B + H^+ \tag{1}$$

The symbol H^+ represents the proton and not the solvated hydrogen ion. Reaction (1) is hypothetical and the actual reactions that occur are

$$A_1 + B_2 \rightleftharpoons A_2 + B_1 \tag{2}$$

in which the proton is transferred from one base to another. Note that acids and bases always occur in conjugate pairs differing by a proton, as in reaction (1). The Brønsted definition is a powerful and general one for all processes involving the proton.

The definition of Lewis is based in part on a return to empirical procedures, and in part on the theory of chemical bonding. The argument is first made that substances with similar properties should be put in the same category. If one of these substances, say HCl, is an acid, then the related substances, such as $SnCl_4$ or BF_3, are also acids. A base would be any substance that could neutralize the properties of an acid.

On the theoretical side, a base is defined by Lewis as any substance that can donate a pair of electrons to form a coordinate bond. An acid is any substance that can accept the pair of electrons. The fundamental reaction becomes

$$A + :B \rightleftharpoons A : B \tag{3}$$
$$\text{Acid} + \text{base} \rightleftharpoons \text{complex} \tag{4}$$

A Lewis base is essentially the same as a Brønsted base, where the proton takes the role of A in equation (3). However, Lewis acids are far more numerous than Brønsted acids. The latter, in fact, would be classified as acid–base complexes in the Lewis scheme.

It is unfortunate that these different definitions are in competitive use. In Europe it is common to call A and B in equation (3) electron acceptors and electron donors, respectively. This obscures the fact that a donor is also clearly a base, but otherwise has much to recommend it.

N. Bjerrum has made the suggestion that A should be called an antibase, to emphasize its characteristic property of neutralizing bases. Since antiacid is a common medical term for a base, we could then rewrite equation (4) to read

$$\text{antibase} + \text{antiacid} \rightleftharpoons \text{anticomplex?} \tag{5}$$

which does not seem to be helpful.

The situation is also complicated by the well-entrenched practice of calling a base a nucleophile, and of calling a Lewis acid an electrophile when discussing kinetic rather than thermodynamic behavior. This usage has some definite advantages and probably should be retained.

Strictly speaking, one should always say Brønsted acid or Lewis acid to clarify one's meaning. The term generalized acid for Lewis acid would also be acceptable. Unfortunately in common practice the identifying prefix is often dropped. In the pages that follow, Lewis acid will almost always be meant, even when the term used is simply "acid."

I next wish to discuss the development of the concept of hard and soft acids and bases, and some of the history of this concept in the form of my personal experiences. I make no claims for the superiority of Lewis' definitions over those of Brønsted. Clearly the material to be covered requires the use of a definition which is highly generalized. From the practical point of view, it would be just as useful to speak of hard and soft donors and acceptors. As an American, it was natural for me to think of Lewis acids and bases, though there was no conscious intent to make the eagle scream.

I first developed a strong interest in generalized acid–base behavior in 1961, when John Edwards of Brown University spent a sabbatical leave at Northwestern University. I found Edwards to be very stimulating and together we wrote a paper on nucleophilic reactivity which turned out to be of considerable interest and utility. In it we showed that electrophiles could be divided into categories, depending on whether they were sensitive to the proton basicity or to the polarizability of various nucleophiles. The alpha effect was also identified and named.

The following year, I decided to do the same thing for equilibrium data that Edwards and I had done for rate data. That is, I would categorize Lewis acids in equation (3) in terms of whether they were sensitive to the proton basicity or the polarizability of the base. The criterion was to be the stability of the acid–base complex formed. By using the Lewis definitions, it was obvious that all organic and inorganic compounds could be considered as acid–base complexes.

The result of this effort was the first paper on hard and soft acids and bases in 1963. The classification that I adopted was strongly influenced by early papers by Schwarzenbach in Switzerland and by Chatt, Ahrland, and Davies in England. Both Chatt and Schwarzenbach had been thinking about the preferences of various metal ions for various ligands and had independently come to very similar conclusions. This was the well-known classification of metal ions into two classes.

Historically, such a classification goes back to the days of Berzelius. In his *Nouveau Système de Minéralogie,* Berzelius pointed out that certain metals, such as mercury, silver, and lead, were commonly found as their sulfide ores, whereas other metals were found as oxide, carbonate, or sulfate ores. In the case of both Chatt, who used (a) and (b), and Schwarzenbach, who used (A) and (B), the labels for the two groups of metal ions came from the A and B subgroups of the Periodic Table. Thus calcium ion was class (a) and Group IIA and cadmium ion was class (b) and group IIB.

Unfortunately, another trans-Atlantic conflict had arisen. In 1932, H. G. Deming in America had reversed the A and B labels for all groups except I and II. Thus thallium is in IIIA rather than IIIB. Accordingly, the logic of the class (a) or class (b) label was somewhat weakened.

I had already put many Lewis acids and bases into one of two categories for each, and in fact, I originally thought to call them class (a) or class (b) acids, and class (a) or class (b) bases. However, I often felt the need to use comparative terms, implying a gradual transition rather than a sharp demarcation between classes. I had also spent much time trying to decide on the fundamental properties which led to class (a) or (b) behavior.

At this point I took what may well have been a fatal step, I decided to call my categories *hard* and *soft,* rather than (a) and (b). These labels were simple and descriptive, and seemed to fit the properties of the various acids and bases, and allowed for the easy use of both the comparative and the superlative.

I had first heard Saul Winstein use the colorful expression "soft, mushy base" to describe the iodide ion. Later I heard Daryle Busch call the hydroxide ion a "hard, tight base." Sir Robert Robinson, in some of his papers published during the thirties, would speak of an atom such as sulfur being softer than an atom such as nitrogen. The repulsive part of potential energy curves between atoms or molecules are called hard or soft, depending on how steeply they rise. In spite of these precedents, it turned out that many scientists do not like the names hard and soft acids and bases. Hard and soft obviously are four-letter words and should not be used in serious scientific discussion, though they might be accepted in informal conversation. Even those who use the terms have some reservations. It is usual to put quotes around "hard" and "soft" when displayed in print. When spoken aloud, it is conventional to smile.

Later I contemplated whether the classic approach would have been better. The Greek word for hard is *scleros,* and for soft, *malacos.* The Latin words are *durus* and *mollis.* Thus I could have stated the Principle of Hard and Soft Acids and Bases in the following way: malacotic donors prefer malacotic acceptors, and sclerotic donors prefer sclerotic acceptors. On the whole, I prefer the approach I used. The abbreviation HSAB is convenient (pronounced hassab). The alternative SHAB, often used in England, is not preferred, for obvious reasons.

In German, HSAB becomes harte und weiche Sauren und Basen; in Spanish, ácidos y bases, fuertes y débiles, or ABFD. The French have a more difficult time; they must say les acides durs et moux, et les bases dures et molles. Similarly in Italian, acidi duri e molle, e basi dure e molle.

In 1965, I had an opportunity to hear all of these phrases when the American Cyanamid's European Research Institute in Geneva decided to sponsor an international symposium on HSAB. The leaders at CERI were R. F. Hudson, C. K. Jørgensen, and G. Klopman. H. H. Schmidtke, F. Calderazzo, and K. Noack were also involved. There were some 70 persons attending the symposium and 12 speakers, most of whom elected to speak in their own languages.

Papers of particular importance to the HSAB concept were presented by Ahrland, Hudson, Jørgensen, Schwarzenbach, and Brian Saville. The latter introduced his useful rules for selection of electrophilic and nucleophilic catalysts. Ahrland showed the dominant role played by the d electrons in determining the properties of transition metal ions.

Klixbull Jørgensen had earlier made a most important observation concerning the effect of other attached groups on the properties of a central atom. He was the first to use the word symbiosis for this effect. John Burmeister has pointed out that this is not the correct word. Symbiosis in biological systems refers to *unlike* organisms living together in a cooperative fashion. Jørgensen wished to stress the flocking together of *like* bases on a central atom. However, the name seems well established in acid–base terminology.

In Geneva, Jørgensen discussed his new concept of innocent and noninnocent ligands. His remark that soft bases were likely to be noninnocent later led Henry Freiser to remark that chemists should perhaps speak of oxidation and seduction as important classes of reactions.

The meeting was greatly enlivened by the presence of Fajans, who criticized the HSAB concept. His early work explaining the difference between NaCl and AgI in terms of polarization, or covalency effects, should certainly be recognized. His suggestion that chemical bonding be interpreted by the quanticule theory has not met with much enthusiasm.

CERI was closed down by American Cyanamid in 1968 (an event which I trust was unrelated to its sponsorship of the HSAB conference) and the leading scientists left to become professors at universities in Europe or in the United States.

In 1967, a second international symposium was held in London. The organizer was M. J. Frazer, then head of the chemistry department at Northern Polytechnic. About 250 persons attended and some 20 papers were given. Compared to Geneva, it was gratifying to find that everyone seemed to know something about HSAB, and all of the papers had at least some connection with the subject. To my relief, R. J. P. Williams, who had previously castigated me at the Chemical Society meeting in Oxford, did not attend. His place in the program was taken by L. M. Venanzi, who concentrated on denouncing π-bonding in phosphine complexes.

By 1969, I had given literally hundreds of lectures on hard and soft acids and bases, including an oft-repeated one-day short course for the American Chemical Society. By this time I was heartily sick of the subject and resolved to give no more lectures. I also intended never to write on the subject again. Fortunately by this time I had developed a new interest, symmetry rule for chemical reactions.

In spite of the above resolution, I persuaded myself that it would be useful to take responsibility for the present collection of reprints, and to make a few final remarks on HSAB as I now see it. At the least, a few common misconceptions might be cleared up for some readers.

First of all, what is HSAB? It is supposed to be a unifying concept which makes easier the remembering of a vast body of chemical facts and which enables predictions of a limited nature to be made. The essence is contained in the Principle of Hard and Soft Acids and Bases, which states that hard acids prefer to combine with hard bases and soft acids prefer to combine with soft bases (hard and soft being defined in a qualitative way).

The principle is supposed to be concise statement of fact, something in the nature of a law of chemistry. Yet, it would be presumptuous to call it a law because it is very imprecise. It is definitely not a theory, though many people habitually call it that. The reasoning seems to be that its validity is dubious. Hence it is like theories, which are also dubious, and not like laws or facts, which are assumed to be true. I would say that the HSAB Principle is not a theory because it does not attempt to *explain* anything. To say that mercuric ion forms a strong complex with a phosphine because both are soft is not an explanation. The theory underlying HSAB is in fact the whole theory of chemical bonding, as R. P. Bell pointed out at the London meeting.

As to how accurately it mirrors the facts, much depends on how accurately one would like the facts to be known. Certainly it is not quantitative, nor is it likely ever to be. Even in a qualitative sense, there are many exceptions, or apparent exceptions. Many of these arise from a misconception, for which I am responsible. In my desire to make a statement which was at the same time both very concise and widely applicable, I managed to convince many people that hard acids would form stable compounds *only* with hard bases, and correspondingly with soft acids and bases.

This is clearly not the case experimentally, nor did I mean to create this impression. My consistent goal in writing and lecturing has been to point out that two parameters, at least, are necessary to describe the bonding properties of a Lewis acid or base. I chose to call these properties strength and softness or hardness. It was assumed from the beginning that Lewis acids and bases were intrinsically of widely differing strengths. The concept of hardness and softness was then used to explain the varying orders of acid or base strength that were found, depending on the reference used. The HSAB Principle perhaps should be modified to read "complexes of hard acids and bases or of soft acids and bases have an added stabilization."

Another common misuse of HSAB has been the many attempts to explain rather small differences in chemical behavior. I believe it is pointless to try to estimate whether piperidine is harder or softer than triethylamine, for example. It is even doubtful whether piperidine and pyridine can be differentiated in a consistent manner. The most important factor which determines the softness of a base is the nature of the donor atom. It is very clear that electronegativity is the dominant feature, bases with the more electronegative donor atoms being harder.

It would be quite possible to use scales of electronegativity to rate bases in an order of hardness or softness. Varying oxidation state and the nature of attached groups would vary the electronegativity in a rather predictable manner. It is worth noting that Pauling believes that electronegativity should be used only for some average valence state of an element, and that corrections for oxidation state, etc., weaken its usefulness.

In the case of the Lewis acids, it is not so clear that electronegativity is a factor. Generally speaking, the least electronegative metals form ions that are hard, the more electronegative metals form ions that are soft. Yet increasing the oxidation state, which surely increases the electronegativity, usually makes the Lewis acid harder.

My belief is that the HSAB concept, with all its imperfections, is still very useful. I believe that it should be introduced first in high school chemistry. In college it should be used in the general chemistry course and in the first organic course as well. It is in these descriptive courses, assuming that they still contain some descriptive material, that such a simple concept is most helpful. It becomes a device for organizing quantities of material which, to the student at least, are vast.

The use of the HSAB approach also has the advantage of making familiar to the student the generalized acid–base concept. Unfortunately, at the present time the important ideas of Lewis are barely mentioned in most elementary texts. I, and many others, consider the generalized acid–base approach the best way of organizing the greater part of chemistry.

Coordination chemists naturally use such an approach. It is not difficult for other inorganic chemists to see that a generalized acid–base picture is useful. Organic chemists may be more resistant to the notion, but it is clear that any modern organic chemist is organizing his information in this way, perhaps not consciously. Of course, one-electron oxidation–reduction reactions and free radical reactions normally require further classification.

At the research level, it seems to be true that many chemists have found the HSAB Principle to be a useful first approximation. This is particularly so for those chemists who actually run reactions and try to isolate products. Scientists who are interested in precise measurements are not likely to find HSAB useful, or even acceptable.

The belief that a concept is not valid unless it can be accurately defined and measured is firmly rooted in science. As George Hammond has pointed out, such statements tend to become shibboleths. Certainly concepts such as electronegativity and solvent polarity have been useful in chemistry, though no one agrees on their definitions.

The statement that a Lewis acid is hard is somewhat akin to saying that an object is red. Certainly in the latter case it is more scientific to measure an absorption spectrum. But for many purposes it is quite adequate to know about the redness.

As to what the future holds for HSAB, no one can say. It seems inevitable that the entire concept will be replaced by statements that are less ambiguous and more felicitously phrased. Nevertheless, these statements will still have to accomplish the same mission of condensing a maximum amount of chemical information into a minimum number of words.

I would hope that the general procedure of looking at chemistry in terms of generalized acid –base interactions will be preserved. This aspect, due of course to the genius of G. N. Lewis, has still not been very widely adopted. As a unifying concept, it can do much to simplify the teaching of chemistry, the remembering of chemical facts, and the interpretation of data.

Part I

Editor's Comments on Papers 1–4

We start with a group of papers published prior to 1963, when the first paper on hard and soft acids and bases appeared. These earlier papers were a source of ideas and inspiration for which full acknowledgment is gratefully given.

Chronologically, the first paper is the one by Edwards in which he introduces the four-parameter equation, later called the oxibase scale. This was by no means the first attempt to improve the two-parameter Brønsted relationship. Swain and Scott had presented the equation

$$\log (k/k_0) = sn + s'e$$

where n is a nucleophilicity factor and E an electrophilicity factor (Swain and Scott, 1953). However, e is related to the solvent as a rule, and the equation was really based on the push–pull concept of nucleophilic substitution.

Thus the Edwards equation was the first to emphasize *two* properties each of the nucleophile and of the electrophile. Also it had the virtue of trying to correlate rate and equilibrium data with numbers obtainable from completely independent information. The redox potentials selected as important for characterizing bases are clearly related to the later concept of softness. A soft base had a large value of E_n.

A follow-up paper by Edwards (Edwards, 1956) discussed the possible use of polarizability as a parameter, instead of redox potential. This works well in practice, but the form of the equation was clearly unsound from a theoretical point of view.

Both Schwarzenbach and the trio of Ahrland, Chatt, and Davies deserve credit for the classification of metal ions into (a) and (b) categories. The first paper was by Schwarzenbach in 1956 (Schwarzenbach, 1956), but was not very detailed. The 1958 paper of Ahrland *et al.* reprinted here was much more detailed and also much more influential; Schwarzenbach's paper was in German and in a journal not widely read.

Later, in 1961, Schwarzenbach presented his views more completely in a chapter written for *Advances in Inorganic and Radiochemistry*. This is an excellent discussion of the many factors influencing the stabilities of metal complexes. I have chosen to use only a portion of this chapter. It is the section which clearly gives Schwarzenbach's approach to class A and B metal ions.

The last paper is the one by Edwards and Pearson on nucleophilic reactivity. While not quantitative, it was based on the idea of a four-parameter equation. It stressed the point that for certain systems, one pair of parameters was dominant, and for other systems a second pair of parameters was dominant. Thus there was a classification of electrophiles and nucleophiles into two categories. It was clear to both Edwards and myself that this could be extended to thermodynamic data as well.

References

Edwards, J. O., *J. Amer. Chem. Soc.* **78**, 1819 (1956).
Schwarzenbach, G., *Experientia*, Suppl. V, 162 (1956).
Swain, C. G., and C. B. Scott, *J. Amer. Chem. Soc.* **75**, 141 (1953).

[Reprinted from the Journal of the American Chemical Society, 76, 1540 (1954).]

[CONTRIBUTION FROM THE METCALF CHEMICAL LABORATORIES OF BROWN UNIVERSITY]

Correlation of Relative Rates and Equilibria with a Double Basicity Scale

BY JOHN O. EDWARDS

RECEIVED OCTOBER 29, 1953

1

A new equation, which is a combination of a nucleophilic scale and a basicity scale, is presented for the correlation of the reactions of electron donors. A new nucleophilic scale for donors, based on electrode potentials, is devised. Data used to test the equation and the scale include rates of displacement reactions of carbon, oxygen, hydrogen and sulfur, and equilibrium constants for complex ion associations, solubility products and iodine and sulfur displacements. The results are good for most of the correlations, and are especially encouraging for those cases, such as complex ion constants, which have been treated heretofore only qualitatively. Advantages and consequences of a double basicity scale for electron donors are discussed briefly.

It has been known for some time that rates of nucleophilic displacements on alkyl carbon atoms do not follow the normal basicities (to protons) of the entering electron donors.[1] Recently, a linear free energy relationship, based on the displacement rates of methyl bromide as standard substrate, was employed by Swain and Scott[2] (hereafter S.S.) to correlate many rate data. The rates of nucleophilic displacements in aromatic compounds follow similar patterns as those in aliphatic compounds but there appears to be a greater dependence on the basicity of the entering donor.[3] Similar difficulties involving lack of correlation of basicity to protons with basicity to Lewis acids (such as aqueous cations[4,5]) and to formally positive sulfur compounds[6,7] are also well established. Although there has been no quantitative treatment of these data to date, Foss[6,7] pointed out that there appears to be a relationship between nucleophilic character and electrode potentials.

In this study, the correlation of many rate data (of the nucleophilic displacement type) and equilibrium data (involving some degree of covalent bond formation) has been attempted. The equation

$$\log\left(\frac{K}{K_0}\right) = \alpha E_n + \beta H \tag{1}$$

where K/K_0 is a relative (to water) rate or equilibrium constant,[8] E_n is a nucleophilic constant characteristic of an electron donor, H is the relative basicity of the donor to protons, and α and β are substrate constants, has been used for these correlations. The calculations have been made by using a least squares analysis of the observed data to determine α and β. In many cases, K_0 is known, but in others it is necessary to treat K_0 as an additional parameter.

The H Scale.—For one standard reaction of donors, the basicity to protons is used. The normal pK_a values of the conjugate acids in aqueous solution are employed; they are changed, however,

by the addition of the constant 1.74 which is the correction for the pK_a of H_3O^+. By definition

$$H = pKa + 1.74$$

Many of the needed pK_a values are either unknown or poorly known. This is true for HCl, HBr, HI, HSCN and others; estimated values are employed in these cases. In Table I, H values for the donors (for which the symbol N is used) are presented; those values which are estimates are placed in parentheses.

TABLE I

ELECTRODE POTENTIALS AND DONOR CONSTANTS

N	E^{0a}	E_n	H
NO_3^-	......	0.29[e]	(0.40)
SO_4^-	−2.01	.59	3.74
$ClCH_2COO^-$	......	.79[f]	4.54
CH_3COO^-	......	.95[g]	6.46
C_5H_5N	......	1.20[h]	7.04
Cl^-	−1.3595	1.24	(−3.00)
$C_6H_5O^-$	......	1.46[i]	11.74
Br^-	−1.087	1.51	(−6.00)
N_3^-	......	1.58[j]	6.46
OH^-	−0.95	1.65	17.48
NO_2^-	− .87	1.73	5.09
$C_6H_5NH_2$	......	1.78[j]	6.28
SCN^-	− .77	1.83	(1.00)
NH_3	− .76	1.84	11.22
$(CH_3O)_2POS^-$	− .56[b]	2.04	(4.00)
$C_2H_5SO_2S^-$	− .54[b]	2.06	(−5.00)
I^-	− .5355	2.06	(−9.00)
$(C_2H_5O)_2POS^-$	− .53[b]	2.07	(4.00)
$CH_3C_6H_4SO_2S^-$	− .49[b]	2.11	(−6.00)
$SC(NH_2)_2$	− .42[c]	2.18	0.80
$S_2O_3^-$	− .08	2.52	3.60
SO_3^-	− .03	2.57	9.00
CN^-	.19[d]	2.79	10.88
S^-	.48	3.08	14.66

[a] Data from ref. 9, except where indicated otherwise. [b] Ref. 7. [c] P. W. Preisler and L. Berger, THIS JOURNAL 69, 322 (1947). [d] Assuming the free energy change for the reaction $C_2N_2(g) \rightleftarrows C_2N_2(aq)$ is negligible. [e] Calculated from $Hg(NO_3)_2$ complex; ref. 5. [f] Calculated from Ag-$(ClCH_2COO)_2^-$ complex. [g] Calculated from epichlorohydrin rate. [h] Calculated from mustard cation rate. [i] Calculated from ICH_2COO^- rate. [j] Calculated from methyl bromide rate.

The E_n Scale.—A search was made for a standard state for the nucleophilic character of a donor which would be more fundamental than rates of reaction with an arbitrary substrate. While nucleophilic character is strongly linked to electron polarizability, it is difficult to relate the measured

(1) (a) P. D. Bartlett and G. Small, THIS JOURNAL, 72, 4867 (1950); (b) C. K. Ingold, "Structure and Mechanism in Organic Chemistry," Cornell Univ. Press, Ithaca, N. Y., 1953, Chapter VII, pp. 306–418; (c) E. D. Hughes, Quart. Revs., 5, 245 (1951).

(2) C. G. Swain and C. B. Scott, THIS JOURNAL, 75, 141 (1953).

(3) (a) J. F. Bunnett and R. E. Zahler, Chem. Revs., 49, 273 (1951); (b) Ref. 1b, Chapter XV, pp. 797–815.

(4) E. C. Lingafelter, THIS JOURNAL, 63, 1999 (1941).

(5) J. Bjerrum, Chem. Revs., 46, 381 (1950).

(6) O. Foss, Kgl. Norske Vid. Selsk. Skrifter, 2 (1945).

(7) (a) O. Foss, Acta Chem. Scand., 1, 8 (1947); (b) 1, 307 (1947); (c) 3, 1385 (1949).

(8) Wherever possible, data for room temperature (18–25°) and aqueous solution will be used.

polarizability of a donor particle to the nucleophilic strength of a specific part of this donor.

Inspection of electrode potentials given by Latimer[9] for oxidative dimerizations of the type

$$2I^- \rightleftharpoons I_2 + 2e^-$$

showed that these potentials become more positive in the same order as the n values of S.S. become larger. That the relationship is a linear one may be seen in Fig. 1; in this figure the data for the six donors for which there are reliable data are presented.

As all of the rate and equilibrium constants correlated in this study are values relative to the water values, the potential for the couple

$$2H_2O \rightleftharpoons H_4O_2^{+2} + 2e^-$$

is needed. However, it is not reported[9] and there does not appear to be any way to calculate it from the present thermodynamic data. It was, therefore, necessary to calculate a value for this couple by means of a least squares correlation of the data shown in Fig. 1. From this calculation the potential value -2.60 is obtained for the above couple.[10]

From the value of the potential of the couple and the values of the oxidation potentials of the donors, a nucleophilic scale defined by the equation

$$E_n = E° + 2.60$$

is set up. In Table I, E_n values for all of the donors to be correlated are presented. In the absence of electrode potential data, E_n values were obtained by secondary standardizations; for example, the E_n value for $ClCH_2COO^-$ was obtained using the known formation constant for the complex Ag-$(ClCH_2COO)_2^-$, the known H value for $ClCH_2$-COO^- and the values of α and β found for other complexes containing a silver ion and two donors.

Displacement Rates on Carbon.—S.S. correlated the rates of nucleophilic displacements on carbon with eight substrates. For six of these substrates, the experimental data can be reasonably well correlated using only the αE_n term. For mustard cation and β-propiolactone, however, the results show a skewness which can best be interpreted as the need for the βH term. For example with OH^- attacking the mustard cation, the former calculated rate[2] was over forty-fold lower than the observed rate.

For comparison of the present correlation with that of S.S., log (K/K_0) data are presented in Table II. The deviations of the calculated from the observed values using equation 1 are all less than 0.30, i.e., a factor of 2, and the average deviation for the 13 pieces of data calculated using equation 1 is 0.14. Concerning the need for the βH term in these correlations, it is pertinent to point out that the calculated values using the equation of S.S.

(9) W. Latimer, "Oxidation Potentials," 2nd Ed., Prentice–Hall, Inc., New York, N. Y., 1952.

(10) This value is quite reasonable for it is more negative than -1.77 which is the potential for the couple

$$2H_2O \rightleftharpoons H_2O_2 + 2H^+ + 2e^-$$

and the equilibrium constant for the reaction

$$H_4O_2^{+2} \rightleftharpoons H_2O_2 + 2H^+$$

is overwhelmingly in favor of the right side.

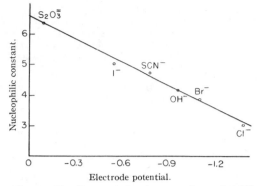

Electrode potential.

Fig. 1.—The linear relation between the nucleophilic constant of Swain and Scott and the electrode potential.

are low for OH^-, CH_3COO^-, $C_6H_5NH_2$ and $S_2O_3^=$, are high for Br^- and I^-, and are close to the observed values for SCN^- and Cl^-. Only in one case (Cl^- with mustard cation) is the deviation significantly inverted to the direction one would expect because of the lack of a βH term.

TABLE II

CORRELATIONS OF TWO DISPLACEMENTS ON CARBON

N	β-Propiolactone Obsd.[a]	Calcd.[b]	S.S.[c]	Mustard cation Obsd.[d]	Calcd.[e]	S.S.[c]
CH_3COO^-	2.49	2.35	2.10	2.72	2.80	2.58
Cl^-	2.26	2.27	2.34	3.04	2.81	2.89
Br^-	2.77	2.60	3.00	..	..	..
OH^-	6.08[f]	..	..	5.62	5.33	3.99
$C_6H_5NH_2$	..	..	..	4.60	4.82	4.26
SCN^-	3.58	3.74	3.67	4.54	4.54	4.53
I^-	3.48	3.51	3.87	4.54	4.37	4.79
$S_2O_3^=$	5.28	5.30	4.90	6.15	6.43	6.04

[a] Data of ref. 1a. [b] Calculated using $\alpha = 2.00$ and $\beta = 0.069$. [c] Calculated by S.S. [d] Data of ref. 11 and 2. [e] Calculated using $\alpha = 2.45$ and $\beta = 0.074$. [f] This rate is not comparable with the others since OH^- reacts with β-lactones in a manner different from that of other donors.

There are other pieces of data from the studies on mustard cation[11] which can be correlated by equation 1. These are presented in Table III. The results are poorer, especially for thiourea. Since thiourea is about twenty times more nucleophilic to alkyl bromides than is pyridine,[12] it is difficult to understand why the two donors should show quite similar reactivities to mustard cation.

TABLE III

OTHER DISPLACEMENTS ON MUSTARD CATION

N	Obsd.[a]	Calcd.[b]
SO_4^-	2.41	1.72
C_6H_5N	3.45[c]	..
$SC(NH_2)_2$	3.90	5.39
$CH_3C_6H_4SO_2S^-$	4.43	4.73
$(C_2H_5O)_2POS^-$	5.34	5.36

[a] Ref. 11. [b] Calculated using $\alpha = 2.446$ and $\beta = 0.0741$. [c] Used as secondary standard.

(11) A. G. Ogston, E. R. Holiday, J. St. L. Philpot and L. A. Stocken, Trans. Faraday Soc., **44**, 45 (1948).

(12) R. G. Pearson, S. H. Langer, F. V. Williams and W. J. McGuire, This Journal, **74**, 5130 (1952).

In Table IV, the results for ten series of displacements on carbon are presented. Of the 59 pieces of data, four were used for secondary standards and for only two did the calculated rate differ from the observed rate by more than an order of magnitude. One of these two cases is shown in Table III; the other is with CN^- replacing I^- in ICH_2COO^- with the calculated rate being high by almost two orders of magnitude.

TABLE IV
SUBSTRATE CONSTANTS FOR DISPLACEMENT ON CARBON

Substrate	M^a	α	β
Ethyl tosylate[b]	5	1.68	0.014
Benzyl chloride[c]	3	3.53	− .128
β-Propiolactone[d]	7	2.00	.069
Epichlorohydrin[e]	6	2.46	.036
Glycidol[e]	6	2.52	.000
Mustard cation[f]	12	2.45	.074
Methyl bromide[g]	6	2.50	.006
Benzoyl chloride[h]	4	3.56	.008
Diazoacetone[i]	4	2.37	.191
Iodoacetate anion[j]	6[k]	2.59	− .052

[a] Number of bases correlated including H_2O. The relative rates are given along with experimental conditions in Table I of ref. 2, except for diazoacetone and iodoacetate ion. [b] H. R. McCleary and L. P. Hammett, THIS JOURNAL, 63, 2254 (1941). [c] G. W. Beste and L. P. Hammett, ibid., 62, 2481 (1940). [d] Ref. 1a. [e] J. N. Brönsted, M. Kilpatrick and M. Kilpatrick, ibid., 51, 428 (1929). [f] Ref. 2 and 11. [g] Ref. 2; E. A. Moelwyn-Hughes, Trans. Faraday Soc., 45, 167 (1949); A. Slator and D. F. Twiss, J. Chem. Soc., 95, 93 (1909). [h] Ref. 2. [i] C. E. McCauley and C. V. King, THIS JOURNAL, 74, 6221 (1952), in water at 25°. [j] H. J. Backer and W. H. van Mels, Rec. trav. chim., 49, 177, 363, 457 (1930); C. Wagner, Z. physik. Chem., A115, 121 (1925); in water at 25°. [k] Since the K_0 value is not known, it was necessary to calculate it in the least squares analysis. The calculated log K_0 value is −6.25. The rate with CN^- was not used in the least squares as it is inconsistent with the other data.

In their review, Bunnett and Zahler[3a] point out that there have been no experiments designed to elucidate the order of nucleophilic strength of donors toward aromatic carbon displacements. They do, however, give preliminary data in their Tables 31 and 32 which indicate that the basicity to protons is important in the nucleophilic character of the entering group. The approximate order of strength found (with 1-chloro-2,4-dinitrobenzene as substrate under a variety of conditions; see ref. 3) is

$$SO_3^- \sim OH^- > C_6H_5O^- > C_6H_5NH_2 > NH_3 > I^- > Br^-$$

The halogens are much less reactive than are the donors which have higher H values. It is apparent that displacements on aromatic carbons cannot be correlated with displacements on aliphatic carbons by means of a single scale of base strength; rather a double scale such as equation 1 will be required.

Displacement Rates on Hydrogen.—In only one case, the mutarotation of glucose, is there an adequate amount of data to test equation 1. These data are presented in Table IVA; the observed relative rates are compared with calculated values from equation 1 in column three and with values calculated by the Brönsted equation[13] (modified to give relative rates) in the last column. The new

(13) J. N. Brönsted, Chem. Revs., 5, 231 (1928).

calculation gives excellent results (average deviation = 0.08).

Although the data on the enolization of acetone do not include as many donors, it is important to note that the deviations from Brönsted's equation are in the opposite direction to those for the mutarotation of glucose. Acetate ion is high relative to H_2O and OH^{-14} which indicates the need for a positive value of α. For this reason, poor results are obtained if the acetone and glucose data are plotted in accordance with the ideas of Pfluger.[15] Further, the fact that deviations are in both directions from the proton basicity scale indicates that statistical factors, and perhaps steric factors, are secondary in comparison to some more fundamental cause of deviation with these two substrates.

TABLE IVA
RELATIVE RATES IN MUTAROTATION OF GLUCOSE

N	Obsd.[a]	Calcd. (1)[b]	Brönsted[c]
SO_4^-	1.61	1.52	1.56
$ClCH_2COO^-$	1.74	1.75	1.90
CH_3COO^-	2.43	2.60	2.70
C_5H_5N	2.92	2.82	2.94
NH_3	4.47	4.53	4.69
OH^-	7.60	7.56	7.31

[a] J. N. Brönsted and E. A. Guggenheim, THIS JOURNAL, 49, 2554 (1927); and G. Kilde and W. F. K. Wynne-Jones, Trans. Faraday Soc., 49, 243 (1953). [b] Calculated using $\alpha = -0.407$ and $\beta = 0.4705$. [c] Calculated using $\beta = 0.418$.

Displacement Rates on Oxygen.—Reactions of hydrogen peroxide with electron donors have rate laws which indicate that the mechanism involves displacement on oxygen.[16] In Table IVB, the correlations for two groups of rate constants of H_2O_2 reactions are presented. The agreement of calculated and observed values is satisfactory for the average deviation is 0.26 and the maximum deviation is 0.46. It is apparent, moreover, that these rates would be poorly correlated by either the E_n scale or the H scale alone.

TABLE IVB
OXIDATIONS BY HYDROGEN PEROXIDE[a]

Base	Rate law I Obsd.	Rate law I Calcd.[b]	Rate law II Obsd.	Rate law II Calcd.[c]
Cl^- [d]	−6.96	−7.33	−4.30	−4.54
Br^- [d]	−4.64	−4.43	−1.85	−2.31
I^- [e]	−0.16	0.22	1.02	1.40
$S_2O_3^-$ [f]	−1.61	−1.84	0.22	0.27
CN^- [g]	−3.00	−3.01	....	....

[a] For details concerning the mechanisms, etc., see ref. 16. [b] Calculated using $\alpha = 6.31$, $\beta = -0.394$ and log $K_0 = -16.33$. [c] Calculated using $\alpha = 5.20$, $\beta = -0.279$ and log $K_0 = -11.83$. [d] A. Mohammed and H. A. Liebhafsky, THIS JOURNAL, 56, 1680 (1934). [e] H. A. Liebhafsky and A. Mohammed, ibid., 55, 3977 (1933). [f] E. Abel, Monatsh., 28, 1239 (1907). [g] O. Masson, J. Chem. Soc., 91, 1449 (1907).

Since the data available for the reactions of H_2O_2 with donors do not include the water rate, it is necessary to employ equation 1 in the long form

(14) (a) H. M. Dawson and E. Spivey, J. Chem. Soc., 2180 (1930); (b) C. G. Swain, THIS JOURNAL, 72, 4578 (1950); (c) R. P. Bell and P. Jones, J. Chem. Soc., 88 (1953).
(15) H. L. Pfluger, THIS JOURNAL, 60, 1513 (1932).
(16) J. O. Edwards, J. Phys. Chem., 56, 279 (1952).

and to solve this equation using K_0 as an additional parameter.

There is another test which can be used to check the validity of these correlations. This test concerns the reaction of H_2O_2 with either H_2O or OH^- to give oxygen atom exchange. Using the correlation for rate law I, it is calculated that the second-order rate constant for the reaction

$$HOOH + \overset{*}{O}H^- \rightleftarrows HO\overset{*}{O}H + OH^-$$

should be about 2×10^{-13} liter/mole/sec. Using the correlation for rate law II, it is calculated that the first-order rate constant for the reaction

$$HOOH + H\overset{*}{O}H \rightleftarrows HO\overset{*}{O}H + HOH$$

is about 1×10^{-10} per sec. in normal acid. These calculations predict that the rate of oxygen isotope exchange of H_2O_2 with H_2O is extremely slow in aqueous solution at room temperature. It is reassuring to find that the exchange of labeled oxygen atoms between H_2O_2 and H_2O has never been observed.[17]

Although the lack of observed isotope exchange is not a quantitative test for the validity of $\log K_0$ (the "calculated water rate") the predictions based on the calculated rate constants are certainly borne out qualitatively.

Ross[18] found that H_2O_2 oxidized $(HOC_2H_4)_2S$ to the corresponding sulfoxide with the same type of rate behavior as had been found for the oxidations of $S_2O_3^-$ and the halide ions.[16] Unfortunately, neither E_n nor H is known for this compound but it certainly should be strongly nucleophilic while being a weak base to protons. The nature of the product also indicates that displacement on oxygen in H_2O_2 has transpired.

Displacement Rates on Chalcogens.—Although there are few quantitative data available, it seems probable that the rates of displacements on the chalcogens can be correlated using equation 1.

It was found[2] that the rates of displacement of Cl^- from benzenesulfonyl chloride in 50% water–50% acetone at 0.5° decrease in the order

$$OH^- > C_6H_5NH_2 > H_2O$$

Using equation 1 and the observed rates, the values calculated for α and β are 2.56 and 0.094, respectively.

Foss[7c] has found that catalysis of the decomposition of monotelluropentathionate ion by bases decreases in the order

$$OH^- > S_2O_3^- > I^- > H_2O$$

and that CH_3COO^- probably is a catalyst while Cl^- does not appear to be one. In the decompositions of $S_5O_6^-$ and $SeS_4O_6^-$, OH^- is a better catalyst than $S_2O_3^-$. Evidence that the decomposition is initiated by a displacement reaction is discussed by Foss[7c]; the fact[19] that the reaction of $S_5O_6^-$ and

OH^- follows bimolecular kinetics is in agreement with a displacement mechanism in the rate-determining step. In order to correlate these data, both αE_n and βH terms probably are necessary.

Bases which react with polythionates and related compounds but do not appear to cause decomposition are CN^-, $SO_3^=$, $S^=$, $C_5H_{10}NH$, mercaptide ions, xanthate ions and dithiocarbamate ions; the products of these reactions indicate that displacements on sulfur have occurred.[6,7] The reactions of $S_4O_6^-$ and $S_5O_6^-$ with CN^- and SO_3^- have been subjected to kinetic study[20–22]; all four reactions are bimolecular, all are rapid and in each case $S_5O_6^-$ reacts faster than $S_4O_6^-$. In the case of the SO_3^- reactions, tracer studies[23] have proved that these reactions are indeed displacements. The recent tracer studies of the reaction of S^- with $S_4O_6^-$[24] are also in agreement with a displacement mechanism.

One can conclude from the data available for displacements on chalcogens that the donors which are most active usually are both strongly nucleophilic and strongly basic to protons. Alternatively, any correlations of chalcogen reactions will probably require positive values for both α and β.

Complex Ion Equilibria.—Using equation 1 and the values of E_n and H from Table I, the correlation of equilibrium, as well as rate, data has been attempted. The results of fourteen correlations of complex ion formation constants are presented in Table V; in Table VI, the values of α and β (obtained by least squares analysis of the observed constants) used to calculate the formation constants are given. As both of the standard states employed in this study are systems which involve covalent bonding to a fair degree, only those complex ions for which there is quite a bit of covalent character can be correlated by equation I.

Since these formation constants are really relative constants to water (the cations are undoubtedly hydrated), the value of $\log K_0$ for equation 1 used for these correlations was 1.74 times the coördinated number of the cation for the particular series being investigated. For example, in the correlation of Hg^{+2} complexes, the values 1.74, 3.48 and 6.96 were used for HgN^{+2}, HgN_2^{+2} and HgN_4^{+2}, respectively. It is assumed that each entering donor replaces a water molecule; this is very likely a valid assumption.

The observed values in Table V for AgN_2^+, HgN_2^{+2}, CdN_4^{+2}, ZnN_4^{+2}, CuN_2^+ and CuN_4^{+2} come, with a few exceptions, from the review by Bjerrum.[5] A larger number of other values have been taken from this review[5] and from the book by Latimer.[9] Listings of Hg^{+2} complexes with the halides are given by Sillen,[25] of SO_4^- complexes by Whiteker and Davidson,[26] of Fe^{+3} complexes by

(17) (a) E. R. S. Winter and H. V. A. Briscoe, THIS JOURNAL, **73**, 496 (1951); (b) P. Baertschi, *Experientia*, **7**, 215 (1951); (c) J. Halperin and H. Taube, THIS JOURNAL, **74**, 380 (1952); (d) M. Dole, G. Muchow, DeF. P. Rudd and C. Comte, *J. Chem. Phys.*, **20**, 961 (1952).

(18) S. D. Ross, THIS JOURNAL, **68**, 1484 (1946).

(19) J. A. Christiansen, W. Drost-Hansen and A. E. Nielsen, *Acta Chem. Scand.*, **6**, 333 (1952).

(20) F. Ishikawa, *Z. physik. Chem.*, **130**, 73 (1930).

(21) F. Foerster and K. Centner, *Z. anorg. Chem.*, **157**, 45 (1926).

(22) B. Foresti, *Z. anorg. allgem. Chem.*, **217**, 33 (1934).

(23) J. A. Christiansen and W. Drost-Hansen, *Nature*, **164**, 759 (1949).

(24) (a) H. B. v. d. Heijde and A. H. W. Aten, Jr., THIS JOURNAL, **74**, 3706 (1952); (b) H. B. v. d. Heijde, *Rec. trav. chim.*, **72**, 510 (1953).

(25) L. G. Sillen, *Acta Chem. Scand.*, **3**, 539 (1949).

(26) R. A. Whiteker and N. Davidson, THIS JOURNAL, **75**, 3081 (1953).

TABLE V
COMPLEX ION FORMATION CONSTANTS (LOG K_f)

N	AgN+ Obsd.	Calcd.	AgN2+ Obsd.	Calcd.	HgN2+2 Obsd.	Calcd.	CdN4+2 Obsd.	Calcd.	ZnN4+2 Obsd.	Calcd.	HgN4+2 Obsd.	Calcd.	HgN+2 Obsd.	Calcd.
NO2-	..	...	(−3.9)	−1.5	0.0	*	(−5.4)	−4.8	.....	.....	..	..	...	..
SO4-	0.2	−0.2	0.2	−0.1	2.3	3.8	(2.2)	−1.9	(2.2)	−1.1	..	..	...	..
ClCH2COO-	.6a	.1	.5a	*	....	...	.....	.....	.....	.....	..	..	...	..
CH2COO-	.7a	.7	.6a	1.8	....	..	1.8	1.3	(−0.2)	2.8	..	..	...	..
C5H5N	2.0	1.4	4.2	3.5	10.3	11.3	1.8	3.2	1.9d	4.7	11.1	11.9	5.1	5.2
Cl-	3.1b	2.3	5.1b	6.0	13.2	12.9	2.2	0.9	−1.0	−1.6	15.0	14.4	6.7	6.8
Br-	..	...	(9.0)	8.7	17.3	16.6	2.7c	2.0	−2.6	−1.9	21.0	19.6	9.0	9.0
OH-	2.3	2.0	3.6	4.3	22.7	16.1	9.8	9.0	15.4	14.2	..	..	(7.9)	6.6
C6H5NH2	1.4	3.2	3.2	7.8	....	..	.....	...	.....	.....	..	..	...	..
SCN-	..	...	7.6	9.4	(16.9)	20.0	2.6	6.0	(−4.2)	4.5	19.2	23.6	...	..
NH3	3.4	3.1	7.2	7.1	17.5	19.2	7.4	8.7	9.4	11.2	19.3	21.9	8.8	8.7
I-	..	...	(14.0)	13.3	23.8	24.0	5.6c	5.1	(−5.4)	−0.6	29.8	29.4	12.8	13.0
S2O3-	..	...	13.0	13.7	...	..	7.4	11.5	(−0.2)	10.3	..	..	...	..
SO3-	..	...	8.5	12.8	....	..	.....	.....	.....	.....	..	..	...	..
CN-	..	...	18.7	14.0	(34.9)	31.5	18.6	15.3	19.0	16.7	41.6	38.0	...	..

N	CdN+2 Obsd.	Calcd.	CuN4+2 Obsd.	Calcd.	CuN2+ Obsd.	Calcd.	PbN+2 Obsd.	Calcd.	FeN+3 Obsd.	Calcd.	AuN4+3 Obsd.	Calcd.	InN+3 Obsd.	Calcd.
NO2-	...	...	.....	...	....	..	0.4f	−0.8	0.0	−0.8	..	..	...	..
SO4-	0.8	−0.2	(2.2)	−0.6	....	..	.....	.....	2.0	1.8	..	..	...	..
CH2COO-	..	...	3.4	3.6	....	..	2.1g	2.7	.....	.....	..	..	...	..
C5H5N	..	...	5.8	5.4	....	..	.....	.....	.....	.....	..	..	...	..
Cl-	1.2c	0.8	−4.6	−4.1	5.5	5.0	1.1	1.0	0.6	−0.3	(25.3)	24.4	2.4h	1.7
Br-	1.5c	1.1	(−6.6)	−5.7	5.9	6.2	1.1	0.9	−0.3	−1.3	(32.2)	30.5	2.2h	1.7
OH-	..	...	16.2	17.5	...	..	7.8	7.6	11.3	12.2	(40.9)	41.1	10.2h	10.2
SCN-	1.4e	2.3	.....	...	....	..	.....	.....	3.0	3.4	(42.6)	40.9	...	..
NH3	2.7	3.1	13.0	12.4	10.9	13.6	.....	...	.....	...	..	..	...	..
I-	2.0c	2.1	.....	...	8.9	9.5	1.5	1.7	.....	.....	..	..	2.0h	2.6
CN-	5.5	5.1	.....	...	(22.9)	20.7	...		...		..	..	...	..

a F. H. MacDougall and L. E. Topol, *J. Phys. Chem.*, 56, 1090 (1952). b E. Berne and I. Leden, *Svensk Kem. Tidskrift*, 65, 88 (1953). c P. M. Strocchi and D. N. Hume, A.C.S. Meeting at Los Angeles, Cal., March, 1953. d C. J. Nyman, THIS JOURNAL, 75, 3575 (1953). e I. Leden, *Z. physik. Chem.*, [A] 188, 160 (1941). f H. M. Hershenson, M. E. Smith and D. N. Hume, THIS JOURNAL, 75, 507 (1953). g S. M. Edmonds and N. Birnbaum, *ibid.*, 62, 2367 (1940). h L. G. Hepler and Z. Z. Hugus, Jr., *ibid.*, 74, 6115 (1952).

TABLE VI
COMPLEX ION SUBSTRATE CONSTANTS

Complexa	Mb	α	β	Ip e	Ionic radiusf
CuN2+	5	7.55	0.284	7.736	0.96
CuN4+2	7	4.69	.958	20.287	0.80
AgN+	8	3.08	− .078	7.575	1.13
AgN2+	15c	7.14	− .226	7.575	1.13
AuN4+3	4	26.02	.294	....	..
ZnN4+2	11	5.94	.650	17.960	0.83
CdN+2	7	2.18	.071	16.9052	1.03
CdN4+2	12	6.98	.255	16.9052	1.03
HgN+2	6	6.57	− .137	18.752	1.12
HgN2+2	10c,d	12.92	− .099	18.752	1.12
HgN4+2	7	16.84	− .187	18.752	1.12
InN+3	4	3.61	.342	28.04	0.92
PbN+2	6	2.83	.270	15.03	1.32
FeN+3	6	2.52	.557	43.43	0.67

a Data for TlN4+3, PbN4+2, AuN2+, SnN4+2 and FeN6+3 are given by Bjerrumb but they are not deemed sufficiently accurate to justify correlation. b Number of donors correlated, not including H2O. c One value used for a secondary standardization. d OH− value not used to calculate α and β. e Ionization potential, ref. 9, pp. 15–16. f Ref. 5.

Rabinowitch and Stockmayer,[27] and of various complexes by C. L. van Panthaleon van Eck.[28] Equilibrium constants obtained from other sources are referenced in Table V. Those observed values which are estimates are placed in parentheses. All formation constants are given as log values with only one figure after the decimal point.

There are 108 pieces of data given in Table V; of these two were used for secondary standardiza-

(27) E. Rabinowitch and W. Stockmayer, THIS JOURNAL, 64, 335 (1942).

(28) van Panthaleon van Eck, *Rec. trav. chim.*, 72, 50, 529 (1953).

tions and 20 are estimates of complexity constants from indirect measurements. Of the 86 other data, 59 or over two-thirds of the observed and the calculated values agree within an order of magnitude, and only for 13 are the deviations greater than two orders of magnitude. Since the observed log values range from −4.6 for CuCl4−2 to +41.6 for Hg-(CN)4−2, the agreement certainly is encouraging.

It should be mentioned at this point that errors in complexity constants are liable to be much greater than in rate constants. An excellent discussion of certain of the difficulties which arise in the evaluation of complexity constants has been given by Young and Jones.[29] For this reason, and since the values of α are exceptionally large (it would require an error of only 0.06 in E_n to cause an order of magnitude error in K_f for a HgN4+2 complex), large discrepancies between calculated and observed values are to be expected in complex ion correlations.

The values of α probably should correlate with the ionization potentials of the metals.[28,30] It would be expected that the value of α would increase as the ionization potential of the metal increased. For related metals in Table V, this is generally true; the exceptions noted are similar to those observed by van Panthaleon van Eck.[28] The values of β should correlate with ionic radii and with charge. As the radius gets larger in a particular family of the Periodic Chart, it is expected that β should get smaller since the electrostatic attraction of metal ion to ligand should decrease.

(29) T. F. Young and A. C. Jones, Section on "Solutions of Electrolytes" in *Ann. Rev. Phys. Chem.*, 3, 275 (1952).

(30) A summary and a list of references which deal with correlation of complex stabilities and ionization potentials may be found in ref. 28.

TABLE VII
SOLUBILITY PRODUCT CORRELATIONS ($-\text{Log } K_{sp}$)[a]

X⁻[b]	AgX Obsd.	AgX Calcd.[c]	CuX Obsd.	CuX Calcd.[d]	Hg₂X₂ Obsd.	Hg₂X₂ Calcd.[e]	TlX Obsd.	TlX Calcd.[f]	PbX₂ Obsd.	PbX₂ Calcd.[g]
CH₃COO⁻	2.6	3.9	..	..	9.4	12.3	Sol.	0.8	V.Sol.	7.4
Cl⁻	9.6	9.2	6.5	7.6	16.9	16.6	3.7	3.6	4.8	5.4
Br⁻	12.3	12.0	8.2	8.5	21.3	20.3	5.4	5.0	5.3	5.7
N₃⁻	8.6	8.0	..	..	(12)	20.7	(4)	2.2	(12)	10.7
OH⁻	[h]	..	[h]	..	[h]		0.1	−0.1	14.8	15.2
SCN⁻	12.0	11.6	13.4	13.1	19.5	24.3	3.2	4.1	(8)	10.0
I⁻	16.1	16.6	12.0	11.3	27.5	27.7	7.1	6.9	8.1	7.4
CN⁻	13.8	14.3	Ins.	23.1	39.3	36.7	(1)	4.1	V.S. Sol.	18.7

[a] Sol. = soluble, V.Sol. = very soluble, V.S.Sol. = very slightly soluble, Ins. = insoluble. Values in parentheses are estimates; they plus the CH₃COO⁻ values have not been used to calculate values of α and β. [b] X⁻ is used to symbolize a monovalent anion. [c] Calculated using $\alpha = 6.52$ and $\beta = -0.354$. [d] Calculated using $\alpha = 6.97$ and $\beta = 0.335$. [e] Calculated using $\alpha = 13.28$ and $\beta = 0.040$. [f] Calculated using $\alpha = 2.36$ and $\beta = -0.231$. [g] Calculated using $\alpha = 5.24$ and $\beta = 0.373$. [h] No OH⁻ precipitate known.

Such is the case, as may be seen by comparing β values for ZnN₄⁺², CdN₄⁺² and HgN₄⁺², etc. As the charge on the metal ion gets larger, the value of β probably should increase. Since the charge affects the radius and since β is a function of radius, it is difficult to say with certainty whether the charge is exerting any direct effect on β.

Solubility Product Equilibria.—Quantitative correlation of solubility product equilibria using equation 1 would, at first sight, seem very improbable. Correlative attempts are, however, worthy of further consideration since it has been found in recent studies[31] that the solubility of a salt is sometimes related to the stability of the corresponding complex.

In Table VII, five correlations of solubility product data are presented along with the values of α and β which were employed to obtain the calculated values. The estimated values (given in parentheses) and the values of $-\log K_{sp}$ for acetates were not used in the calculation of α and β. Also it was assumed that $\log K_0$ of equation I was zero.

The agreement between the observed values and the calculated values is surprisingly good when one considers that the forces acting in a crystal are in many ways different from those acting in homogeneous liquid solution. There is evidence, furthermore, that this correlation of solubility product data is not fortuitous; this evidence is found in the values of α and β. Comparing values for the precipitates AgX and CuX, the value of α for AgX is slightly lower than that of CuX; this is exactly the same trend as is found for the complexes AgN₂⁺ and CuN₂⁺ and as is noted for the ionization potentials. The values of β also are close to those found for AgN₂⁺ and CuN₂⁺. Similar qualitative agreement is found for PbX₂ precipitates and PbN⁺² complexes.

As the other seven anions being considered are roughly symmetrical and can take part in bonding at both ends of an axis, it is not surprising that the acetate salts are more soluble than would be predicted by the values of α and β given in Table VII.

Most of the observed values of solubility products were obtained from the book by Latimer.[9] The mercurous halide values are those given by Sillen,[25] and the approximate values were estimated from solubility data as given by Seidell.[32]

(31) Cf. R. J. P. Williams, J. Chem. Soc., 3770 (1952).
(32) A. Seidell, "Solubilities of Inorganic and Metal Organic Compounds," 3rd Ed., D. Van Nostrand Co., New York, N. Y., 1940.

Equilibria Involving Iodine.—In Table VIII some results for correlation of the reactions

$$\text{I}_2(\text{aq.}) + \text{X}^- \rightleftharpoons \text{I}_2\text{X}^-(\text{aq.})$$

are presented. Since iodine is known to be present in aqueous solution as a water complex, equation 1 was employed for the calibrations in the same way as for complex ions. Use of the value 1.74 for $\log K_0$ gave the results to be found in the column marked $\log K_1$ of the table. Although the worst deviation is only 0.44, this deviation is large in respect to the value of α and the calculated values differ from the observed values in a manner which does not seem to be random. Using the value of 3.48 for $\log K_0$, the results presented in the last column of the table were obtained. The calculated values are in excellent agreement with the observed values, for the largest deviation is only 0.06.

TABLE VIII
CORRELATION OF IODINE COMPLEXES

X⁻	log K Obsd.	log K₁ Calcd.[d]	log K₂ Calcd.[e]
Cl⁻	0.32[a]	0.76	0.29
Br⁻	1.08[b]	1.36	1.11
SCN⁻	2.04[c]	1.82	2.09
I⁻	2.85[d]	2.50	2.79

[a] Ref. 35. [b] F.-H. Lee and K.-H. Lee, J. Chinese Chem. Soc., 4, 126 (1936). [c] R. O. Griffith and A. McKeown, Trans. Faraday Soc., 31, 868 (1935). [d] Calculated using $\alpha = 1.96$, $\beta = -0.023$ and $\log K_0 = -\log [\text{H}_2\text{O}]$. [e] Calculated using $\alpha = 3.044$ and $\log K_0 = -2 \log [\text{H}_2\text{O}]$.

It is surprising to find that these equilibria appear to involve a loss of two water molecules from the iodine molecule when it is being complexed by an anion. Using equation 1 without the βH term, a least squares analysis of the observed constants gave a value of 3.52 for $\log K_0$, in excellent agreement with the value 3.48 which one would expect if two water molecules are freed. As the writer does not know of any data in the literature that can substantiate this rather surprising conclusion, quantitative significance should only be assigned with caution to this $\log K_0$ value[33] at the present time.

In contrast to the above series of equilibrium

(33) Since no similar difficulty arose in the assignment of $\log K_0$ values for the complex ion correlations of Table V and since the calculated values based on the $\log K_0$ value of 3.48 are in such good agreement with the observed values, it is doubtful if this anomaly can be blamed either on the use of 1.74 for $\log$ [H₂O] or on experimental errors in evaluating the equilibrium constants.

constants, a definite need for the βH term is found for reactions of the type

$$I_2 + N \rightleftarrows IN^+ + I^-$$

From the data of Bell and Gelles,[34] it is possible to calculate relative equilibrium constants. For N equals H_2O, I^- and OH^-, these are 2.2×10^{-13}, 1 and 30, respectively. The data, when substituted into equation 1, give the values $\alpha = 6.85$ and $\beta = 0.162$. From the work of Awtrey and Connick[35] one can estimate a lower limit for $N = S_2O_3^=$. This estimate, which is that the relative equilibrium constant (see above) is $\geq 10^6$, is in fair agreement with the calculated relative equilibrium constant of 2×10^5. Although the quantitative calculations on this reaction are not convincing, it is certain that correlation of the data requires a significantly large and positive value of β as well as a large value of α.

Equilibria Involving Sulfur Compounds.—Although the available data for equilibria involving displacement on sulfur are qualitative in nature, there is much information that is interesting and pertinent. Foss[6,7] places considerable emphasis on the importance of electrode potentials of the donors in sulfur reactions and on the displacement nature of these reactions. He has, however, found it necessary to divide up his donors into two classes, the "thio anions" such as $S_2O_3^=$ and SCN^- and the "anthio anions" such as $SO_3^=$ and CN^-.

The thio anions can be arranged in order of increasing displacement ability. From studies of displacement reactions of compounds containing formally positive sulfur atoms, Foss[6,7] has found that the order of increasing strength of thio anions is $Cl^- < SCN^- < (CH_3O)_2POS^- < C_2H_5SO_2S^- < p\text{-}CH_3C_6H_4SO_2S^- < S_2O_3^= <$ thiocarbonyl anions and mercaptides.[36] For example: (a) from disulfur dithiocyanate $S_2(SCN)_2$, $CH_3C_6H_4SO_2S^-$ displaces SCN^- to give $S_2(SO_2SC_6H_4CH_3)_2$; and (b) ethyl xanthate (a thiocarbonyl anion) will displace a thiosulfate ion from tetrathionate ion in accordance with the equation

$$O_3S_2 \cdot S_2O_3^= + S_2COC_2H_5^- \rightleftarrows O_3S_2 \cdot S_2COC_2H_5^- + S_2O_3^=$$

Comparison of this series of displacement strengths with the known electrode potentials shows good agreement.

For the anthio anions, Foss[6,7] observed that the order of increasing strength was

$$CH_3C_6H_4SO_2^- < C_2H_5SO_2^- < SO_3^= < CN^- \sim (CH_3O)_2PO^-$$

Although this order is not conclusive by itself since the E_n values for three of these anions are unknown, it is interesting to find that $SO_3^=$ and CN^- are able to displace the stronger (more nucleophilic) thio anions such as mercaptide ions and xanthate ions in some cases. As these anthio anions are stronger bases to protons than are thio anions, it is probable that the difference between Foss' two classes lies in the need of a positive βH term as well as a large αE_n term in correlation of sulfur equilibria.

Other data such as: (a) the dissolution of sulfur in solutions containing amines, $S^=$, $SO_3^=$, CN^- and OH^-, but not in solutions containing SCN^-, $S_2O_3^=$ and the halide ions; and (b) the observation[6] that $CH_3C_6H_4SO_2S^-$ is a stronger thio anion than $C_2H_5SO_2S^-$ while, for the corresponding anthio anions, $C_2H_5SO_2^-$ is stronger than $CH_3C_6H_4SO_2^-$, also indicate that correlations of displacement equilibria of sulfur compounds require both αE_n and βH terms.

Miscellaneous Correlations.—In the previous sections, rate constants of displacement reactions and equilibrium constants for allied reactions have been discussed in terms of equation 1. In this section, data of various sorts which cannot be classified in either of the above groups will be correlated.

A.—Jette and West[37] found that the abilities of anions to quench the fluorescence of quinine bisulfate follow the order

$$I^- > SCN^- > Br^- > Cl^- > C_2O_4^- > CH_3COO^- > SO_4^- > NO_3^- > F^-$$

Ignoring $C_2O_4^-$ for which there is no E_n value available this order is identical to the order of E_n values.[38] The quenching of fluorescence of disodium fluorescein (Uranin), uranyl salts, and other compounds was found to follow the same general order.[37,39] Further it is reported[39] that $S_2O_3^=$ and $SO_3^=$ are quenching agents for the fluorescence of these compounds.

A study of the quenching of the fluorescence of uranin by various amines showed "that the basicity of the nitrogen atom (i.e., the availability of the unshared pair of electrons for bond formation) is not important in the quenching reaction."[40] For example, with three amine donors of known E_n value, the order was found to be

$$C_6H_5NH_2 > C_5H_5N > NH_3$$

Although the above data on quenching of fluorescence indicate that the nucleophilic strength of the quencher is of primary importance, it should be mentioned that the fluorescence of certain other compounds is quenched by an entirely different group of anions which include IO_3^-, BrO_3^-, NO_3^-, $S_4O_6^=$ and $AsO_4^=$.[39]

B.—In the course of their studies on the mechanism of displacement reactions in coördination compounds, Basolo, Pearson and co-workers[41] found that the rates of reaction of entering donors with cis-$[Co(en)_2NO_2H_2O]^{+2}$ follow the order

$$NO_2^- > N_3^- > SCN^- > H_2O$$

in aqueous solution at 35°. It seems probable that these data could be satisfactorily correlated using equation 1.

C.—In a recent study of the differential capacity of the electrical double layer at a mercury solution interface,[42] it was found that the anion present in

(34) R. P. Bell and E. Gelles, *J. Chem. Soc.*, 2734 (1951).

(35) A. D. Awtrey and R. E. Connick, THIS JOURNAL, **73**, 1341 (1951).

(36) Foss[6,7] estimated that the E^0 values for the various thiocarbonyl anions and mercaptides are about $+0.3$, thus the E_n values should be about 2.9.

(37) E. Jette and W. West, *Proc. Roy. Soc. (London)*, **A121**, 299 (1928).

(38) From the data of Latimer,[9] the E_n value for F^- is calculated to be -0.27.

(39) G. K. Rollefson and R. W. Stoughton, THIS JOURNAL, **63**, 1517 (1941).

(40) J. C. Rowell and V. K. LaMer, *ibid.*, **73**, 1630 (1951).

(41) (a) F. Basolo, *Chem. Revs.*, **52**, 459 (1953); (b) R. G. Pearson, personal communication.

(42) D. C. Grahame, M. A. Poth and L. J. Cummings, THIS JOURNAL, **74**, 4422 (1952).

solution exerted a marked effect. The order of increasing capacity was found to be

$$ClO_4^- < NO_2^- < F^- < SO_4^- < CH_3COO^- < HCO_3^- < Cl^- < CO_3^- < OH^- < CNS^- < Br^- < I^-$$

The investigators[42] pointed out that their capacities correlate well with the solubilities of Hg_2^{+2} salts (cf., section on solubility products).

Discussion

In this article, a new equation for the correlation of the reactions of electron donors has been presented, and a new nucleophilic scale, based on electrode potentials, has been defined. A large amount of quantitative and qualitative data has been given in support of the equation and scale. The results are generally good, but in several correlations (e.g., ZnN_4^{+2} complexes in Table V) the agreement between calculated and observed values leaves much to be desired. It seems quite conclusive, however, that further efforts in the fields of correlation of reactions of electron donors (and acceptors, too) will require an equation of the general type of equation I and a set of constants akin to those of Table I.

The values of E_n and H for some of the donors probably will require revision as better data become available; it may even be necessary to assign separate E_n values for different types of correlations. For example, the E_n value for CN^- gives low calculated results with cation complexes; yet there are indications that high results will be obtained with displacements on carbon when this E_n value is employed.[43] Similar difficulties arise with NH_3; while the results for this donor are good with cation complexes and displacements on hydrogen, data on displacements in aromatic compounds[3a] and on quenching of fluorescence[40] indicate that the ammonia E_n value is too large for these cases. All of the results for SO_4^- indicate that the present E_n value is low. Since the E_0 value assigned to the couple

$$2SO_4^- \rightleftharpoons S_2O_8^- + 2e^-$$

is not accurately known,[9] it is difficult to find the correct reason for these discrepancies with SO_4^-. In order to be as general and as fundamental as possible, no attempt was made to obtain the best possible correlation by trying different standard substrates. For this reason and so that the important discrepancies may be recognized as such, the calculated values are tabulated in their presently imperfect form.

Although the new equation and its nucleophilic scale are not final and despite the fact that cases which cannot be correlated will arise, there is much to be said for this treatment of rate and equilibrium constants. In order: (a) the new scale for the nucleophilic character of a donor is a thermodynamic scale and is related to electrode potentials which can be measured; (b) the worst discrepancies in the previous correlations of displacement on carbon[2] have been eradicated; (c) the data available indicate that this correlation equation can be extended to displacement rates on other atoms such as oxygen and the chalcogens; it is also possible that the deviations from Brönsted's equation may be explained, in part,[44] by equation 1; (d) contrary to previous feelings[2] that the word nucleophilic should be used only in discussion of rate and that the word basic refers to equilibrium conditions, it is obvious that almost any correlation made will require a greater or lesser contribution of both basic and nucleophilic character to the covalent bonds; (e) the equation can be applied to equilibrium constants for complex ions of certain types, solubility products (albeit poorly), iodine displacements and sulfur displacements. This writer does not know of any previous correlation of complex ion constants or solubility product constants in which the influence of the donors was studied.

One of the most important conclusions which has come out of this study is that Lewis acids can now be quantitatively correlated along with the proton. In Table IX, the substrate constants (α and β) for ten acids for which there are quantitative data on the reaction

$$acid + donor \rightleftharpoons salt$$

in aqueous solution at 25° are given. Because of its very small diameter, and its charge, the proton has a low value of α and a high value of β. The differences between the proton and the other cations probably are not of kind but, rather, of degree.

Table IX
Substrate Constants for Various Acids[a]

Acid	α	β	Acid	α	β
H^+	0.00	1.000	Hg^{+2}	6.57	−0.137
Ag^+	3.08	−0.078	Pb^{+2}	2.83	.270
Cu^{+2}	2.04	.275	In^{+3}	3.61	.342
Zn^{+2}	1.32	.206	Fe^{+3}	2.52	.557
Cd^{+2}	2.18	.071	I_2	3.04	.000

[a] The data for CuN^{+2} (5 donors) and ZnN^{+2} (6 donors) have not been presented in detail. All calculated values of the 11 constants agree with the observed values within an order of magnitude.

Acknowledgment.—For their comments and advice, the author wishes to thank Professor R. G. Pearson, of Northwestern University, and the members of this department, especially Drs. J. S. Belew, R. H. Cole and G. V. R. Mattock. The financial aid of a Brown University Faculty Summer Research Grant made this work possible.

Providence, Rhode Island

(43) In the displacements of I^- from ICH_2COO^-, CN^- is less active than $S_2O_3^-$ or SO_3^-, and the calculated rate is too large by almost two orders of magnitude.

(44) Although the nucleophilic strength of the electron donor may be one of the prime reasons for deviations from the Brönsted equation, steric factors,[45] etc., can also cause significant deviations.

(45) R. G. Pearson and F. V. Williams, This Journal, **75**, 3073 (1933).

Reprinted from *Advances in Inorganic and Radiochemistry*, **3**, 257; 265–271 (1961)

THE GENERAL, SELECTIVE, AND SPECIFIC FORMATION OF COMPLEXES BY METALLIC CATIONS

G. Schwarzenbach

Laboratorium für anorganische Chemie, Eidgenössische Technische Hochschule, Zurich, Switzerland

2

I. Introduction

The overwhelming majority of reactions used to detect or determine metallic ions in aqueous solution involve the replacement of water molecules in the hydration shell by other ligands. Reactions of this type include not only complex formation in homogeneous solution but also the formation of precipitates. The solution remains homogeneous if the reagent added as a ligand occupies one or several coordination positions on one and the same individual cation (unidentate ligands, e.g., NH_3, $CH_3CO_2^-$; multidentate chelating ligands, e.g., en, gl$^-$, ox$^=$).* Precipitation occurs if the ligand links two or several metallic ions together (bridging ligands, e.g., $H_2N—NH_2$, OH^-, $CO_3^=$), so that the infinite network of a coagulate or an ordered crystal lattice results. Transition from the simple metal ion to precipitation naturally proceeds through polynuclear species of low molecular weight, though these intermediate stages are almost always difficult to detect and cannot usually be isolated (*63*) (e.g., $Be_3(OH)_3^{3+}$, $Bi_6(OH)_{12}^{6+}$). An attempt is made in this article to survey the course of such reactions in order to point out some empirical regularities, without, however, giving more than an indication of their theoretical interpretation.

* en = ethylenediamine; gl = glycinate anion; ox = oxalate anion.

V. Ligands and Ligand Atoms

The nature of the atom which binds the ligand to the metal largely determines the ligand behavior. Only a few ligand atoms will be considered here (the halogens, oxygen, sulfur, nitrogen, and carbon). There is very little quantitative information on complex formation which entails attachment of selenium, tellurium, phosphorus, or arsenic to a metal.

(a) The halogens are able to coordinate only as the simplest anions F^-, Cl^-, Br^-, and I^-. The mononuclear fluoro, chloro, bromo, and iodo complexes, or sparingly soluble halides, result.

(b) The most important ligand containing oxygen is H_2O, the aquo complexes of which serve as a starting point since water is the solvent normally used. Mononuclear hydroxo complexes are first produced by deprotonation, but the formation of polynuclear species and precipitation almost always follows (63); in spite of this, K_1 for OH^- as a ligand is known for most cations. The ion O_2^- cannot be studied as a ligand in aqueous solution. Examples of other inorganic oxygen donors are: NO_2^-, NO_3^-, CO_3^-, PO_4^{3-}, and polyphosphates; $S_2O_3^-$, SO_3^-, SO_4^- and ClO_4^-. Organic complexing agents are able to offer the metal ion oxygen in the form of ether $R-O-R$, and alcohol or phenol, and ketonic or carboxylic oxygen.

(c) Sulfur donors are: HS^- and S^-, SO_3^-, $S_2O_3^-$, SCN^-, thioethers $R \cdot S \cdot R$, mercaptans and their anions $R \cdot SH$ and $R \cdot S^-$, thioketones and the mono- and dithiocarboxylate groups

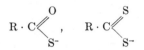

(d) Nitrogen is coordinated by addition of NH_3. Deprotonated ammonia occurs as a ligand for mercury (35, 36, 47). The ion NO_2^- usually adds on to metal ions through nitrogen and the same is also often the case with SCN^- (60, 66). Organically bound nitrogen coordinates in complex formation with primary, secondary, and tertiary amines, RNH_2, R_2NH, R_3N; Schiff's bases, $>C=NR$; carboxylic amides, $R \cdot CO \cdot NH_2$; nitroso groups $R \cdot N=O$; oxime groups $C=N \cdot OH$, and azo groups.

(e) The only complex-forming group with carbon as the ligand atom that has been studied in aqueous solution is CN^-.

VI. A- and B-Metal Cations

(a) Cations with a rare-gas configuration (d^0 cations), i.e., Be^{++}, Mg^{++}, Ca^{++}, Sr^{++}, Ba^{++}, Al^{3+}, Sc^{3+}, Y^{3+}, lanthanides^{3+}, Ti^{4+}, Zr^{4+}, Hf^{4+}, Th^{4+},

Nb^{5+}, and Ta^{5+}, constitute a related group from the point of view of complex chemistry and will be treated together here as A-metal cations. Their characteristic is that in aqueous solution they are able to form complexes only with F$^-$ and oxygen as donor atoms.

Insoluble fluorides, which are often soluble in an excess of fluoride to form mononuclear complexes, are known to every analyst. It cannot be proved, however, that the chloride ion or any other of the heavy halide anions can be added to the A-metal cations. Possibly loose chloro complexes occur in quite strong hydrochloric acid with the A ions that have a valence greater than three, but they decompose at once on diluting the solution. In contrast to this, the A metals react with all oxygen donors. All of them which are polyvalent naturally form hydroxo complexes which, apart from those of the heavy alkaline earths, go over into sparingly soluble hydroxide precipitates through intermediate polynuclear species.

Difficultly soluble precipitates are likewise formed with CO$_3$$^=$ and PO$_4$$^{3-}$. All the important organic complexing agents for these A cations are also oxygen donors (e.g., tartrates, citrates, the enolates of β-diketones and the anions of aminopolycarboxylic acids, which present several oxygen ligand atoms in addition to one or two nitrogen atoms). Sulfur, on the other hand, is not added to the A cations. Addition of ammonium or alkali sulfide gives slightly soluble hydroxides. In addition, no reaction occurs with sulfur donors such as dithiocarbamates and xanthates, and there is no color reaction with dithizone.

Ammonia also precipitates hydroxides and no ammine complexes result apart from the loose complexes which have been found for alkali earth metals (10). Cyanide also precipitates hydroxides because the ligand takes a proton from the solvent and leaves OH$^-$ to react with the metal ion.

The stability of the complex increases rapidly with increase in charge on the metal ion and, with a series of cations of the same valence, those with the smallest radius form the most stable complexes. The validity of this rule can be recognized readily from values of the solubility products of fluorides and hydroxides, as well as from K_1 values for mononuclear fluoro and hydroxo complexes. It is also valid in general for chelate complexes of A cations, such as complex oxalates, tartrates, and citrates. Deviations from the dependence on the cation radius arise if the chelating species possesses more than 3–4 ligand atoms. With EDTA complexes, for example, where the anion has six ligand atoms, the following stability series are obtained:

$$Be^{++} < Mg^{++} < Ca^{++} > Sr^{++} > Ba^{++} \quad \text{and} \quad Al^{3+} < Sc^{3+} > Y^{3+} > La^{3+}.$$

Here the stabilities of complexes of the smallest cations are smaller than those which follow them because the coordination number is often only four (Be^{++}), and the small cation, unlike the larger, cannot, on steric grounds, make use of all six bulky ligand groups in chelation.

Considering the dependence of complex stability on the ligand structure in the case of fluorine- and oxygen-containing ligands, it is at once apparent that anionic ligands give much more stable complexes than uncharged ligands. Oxygen in ethers, ketones, and alcoholic OH groups is able to compete as a ligand with the solvent H_2O, and to attach itself to the metal ions if it forms part of a chelate ligand which also contains more firmly bound anionic oxygen atoms. For a series of anionic ligands the following rule holds: the stability of the complex increases rapidly with the basicity of the ligand oxygen (i.e., its ability to add on a proton), so that we obtain the following stability sequence: OH^- > phenolate > carboxylate > F^-. For the series of isoelectric oxo anions the charge decreases simultaneously with the basicity, so that the following stability sequence is explicable: $CO_3^=$ ≫ NO_3^-; PO_4^{3-} ≫ $SO_4^=$ ≫ ClO_4^-. This is manifest in the smaller solubility of carbonates compared with nitrates and in that of phosphates compared with sulfates and perchlorates. Some data are available for the K_1 of mononuclear complexes, namely $MgCO_3$, $10^{2.2}$ (45); $MgNO_3^+$, $10^{-0.1}$; $CaPO_4^-$, 10^5 (22); $CaSO_4$, 10^2. The value for the species $CaClO_4^+$ has not been determined.

(b) Cations with an outer shell of 18 electrons (d^{10}) will be referred to as B cations. They are Cu^+, Ag^+, Zn^{++}, Cd^{++}, Hg^{++}, Ga^{3+}, In^{3+}, Tl^{3+} and Sn^{4+}.

Comparing first of all the three univalent ions of the noble metals with the alkali ions, a radically different behavior is found: slightly soluble chlorides, bromides, iodides, and sulfides are in equilibrium with the corresponding mononuclear complexes, all of which are very stable, and very stable ammine and cyano complexes are also formed. It is at once clear that in these cases factors of a different sort from those for the A cations must be operating. That simple coulombic forces are not decisive is seen from the fact that the largest of the three ions gives more stable complexes than the smallest, for the stability sequence is, in general, Cu(I) < Ag(I) < Au(I). It was observed quite early that this sort of complex formation is shown particularly by cations of noble metals and, later, it was found that the first or second ionization potential of the metal from which the cation is derived is a good criterion for assessing its tendency to complex formation (14, 21, 28, 29, 37), for this quantity in general parallels the stability of the complex measured by $\log K_1$ or $\log \beta_n/n$.

As far as the influence of the ligand goes, it is found for the three univalent noble metal cations that there is an increase in the stability of their complexes in the following sequences: F^- < Cl^- < Br^- < I^-; OH^- ≪ SH^- ≪ $S^=$; F^- < OH^- < NH_3 < CN^-. To establish this sequence

one can compare either solubilities or the stability constants of the mononuclear complexes. Table II, for example, shows the values of log K_1 for

TABLE II
STABILITY CONSTANTS OF SILVER COMPLEXES (20–25°)

Complex log K_1 or $\frac{1}{2}$ log β_2	AgF	AgCl	AgBr	AgI	Ag(OH)	Ag(SH)	AgS⁻	AgNH₃	Ag(CN)₂
	−0.3 (10)	3.0 (10)	4.3 (10)	8.1 (10)	2.0 (4, 5)	13.6 (70)	20.3 (70)	3.1 (10)	10.5 (10)

mononuclear silver complexes. Most of the ligands are singly charged
anions, but the uncharged ammonia molecule is also included. On examining the values, one has the impression that the charge does not play a decisive role in determining stability. Thus the ammonia molecule forms a
complex which is appreciably more stable than that from the smaller OH^-
and F^- anions. It seems, rather, that the electronegativity of the atom of
the ligand element is important, for these values fall in the following
sequence:

$$F > O > N > Cl > Br > I \sim C \sim S;$$

this is also the order in which the log K_1 values in Table II increase.

If different ligands having the same donor atom are compared, one
again finds an increase in complex stability with increase in basicity. The
thioether grouping can be added only to the cations of very noble metals
and is greatly inferior to the much more basic mercaptide group. The silver
complexes of various amines have been widely studied and it has been
found that in general those with the greatest pK form the most stable
complexes and that log K_1 even increases linearly with pK (8, 11, 33). In
this case, however, only aliphatic amines, or aromatic and pyridine bases,
should be compared with one another.

We will now consider the coordination tendency of the polyvalent B
cations by considering a series of isoelectronic ions such as Ag^+, Cd^{++}, In^{3+}
and Sn^{4+}. The quite unexpected observation is then made that the stability
of the complexes with less electronegative ligand atoms falls with increasing
charge instead of increasing, as would be anticipated. The value of log K_1
for the ammonia complexes is 3.1 and 2.7, respectively, for Ag^+ and Cd^{++},
and indium and tin behave toward ammonia similarly to A cations in that
the hydroxide precipitates without forming an ammine complex. A fall in
stability in passing from Cu^+ to Zn^{++}, Ag^+ to Cd^{++}, or Au^+ to Hg^{++} is
also found invariably for anionic ligands of low electronegativity. It is
less pronounced in going from II- to III-valent B cations, for the stability
often rises somewhat and, in other cases, drops. The following data give
log K_1 values for the 1:1 complexes shown (10): AgCl, 3.0; CdCl⁺, 1.6;

InCl^{++}, 2.4; AgBr, 4.3; CdBr^{+}, 2; InBr^{++}, 2; AgI, 8.1; CdI, 2.2; InI^{++}, 1.6; AgCN, 10.0; CdCN^{+}, 5.4; AuCN, 20; HgCN^{+}, 18; TlCN^{++}, 10.

On the other hand, with fluoride as a ligand the usual A cation behavior is found, i.e., a rise with increasing charge on the metal ion (AgF, -0.3; CdF^{+}, 0.6; InF^{++}, 4; SnF^{3+}, 5.) This is also the case for mononuclear hydroxo complexes (AgOH, 2; CdOH^{+}, 4–5; InOH^{++}, 10). The solubility products drop correspondingly in the following sequence, which is also that for A cations with increasing charge:

$$AgOH < Cd(OH)_2 < In(OH)_3 < Sn(OH)_4.$$

The few constants known for complexes with organic oxygen donors (e.g., oxalates) also show a rise in stability with increasing cation charge (*10*).

VII. Electrovalent and Nonelectrovalent Interaction

It is not intended in this article to go into the nature of the forces between the metal ion and ligand. Consideration of the known facts, however, leads us to distinguish between two sorts of associative forces. Coulombic forces are certainly present, and, even if it is impossible to predict quantitatively the free energy of association due to them (*24*), it is nevertheless clear that simple electrostatic forces must operate in such a way that the stability of the complex rises with the valence of the metal ion, and falls with its radius. We find that this relationship holds without exception for the A metal ions, the rare-gas structure of which would, indeed, lead us to expect that their behavior would be particularly simple. The interaction for A cations may therefore perhaps be referred to as "electrovalent," implying that simple electrostatic considerations can be used in discussing the forces between cation and ligand. If this is so, it follows from the observed facts for A cations that for metal ions in aqueous solutions classic coulombic forces allow only fluoride ions and the various oxygen donors to form complexes, with replacement of molecules in the hydration shell. The heavier halogen ions, Cl$^-$, Br$^-$, and I$^-$, the ions HS$^-$ and S$^=$, ammonia, cyanide ion, and the many other organic ligands with N and S are clearly not able, on the basis of coulombic forces, to compete as ligands in aqueous solution with the water molecule, or with OH$^-$, the product of its deprotonation. For none of the A cations, whatever the radius or charge, is it possible to establish with certainty an association with Cl$^-$, Br$^-$, I$^-$, S, N, or C.

Apart from coulombic forces, there must be other types of interaction which are capable of interpretation only in terms of quantum mechanics. These will be referred to as nonelectrovalent forces and the term will include everything not due to classical electrostatic action, including the possibility of crystal field stabilization (*7*). Nonelectrovalent behavior is encountered in its purest form for noble-metal cations of low charge. In the rough picture of bond formation by the sharing of an electron pair between the central atom and the ligand, it is significant that the stability of the complex is found to increase both with the tendency of the cation to take up electrons (i.e., with increasing ionization potential of the metal involved), and with the tendency of the ligand atom to give up electrons (i.e., with decreasing electronegativity of the nonmetal).

In studying ion association, nonelectrovalent behavior is never encountered in its pure form since coulombic forces must always be operative. For B cations we always have to deal with a superposition of electrovalent and nonelectrovalent interaction. Electrovalent interaction must, however, increase with increasing charge and decreasing radius of the cation and, at the same time, it becomes increasingly difficult for the ligands Cl^-, Br^-, I^-, SH^-, CN^-, NH_3, etc. to compete successfully with H_2O or OH^-. The stability relationships for the complexes of isoelectronic B cations of increasing charge which have been mentioned above are understandable, at least qualitatively, on this basis. The nonelectrovalent interaction with ligands of B cations of low charge changes gradually to electrovalent interaction with increasing charge.

VIII. Transition Metal Cations

These are ions with between 0 and 10 d-electrons. Those of the first long period with a valence of two have been particularly fully studied. In the process the Irving-Williams order was discovered (28, 29), according to which the stability of complexes increases in the series

$$Mn^{++} < Fe^{++} < Co^{++} < Ni^{++} < Cu^{++} > Zn^{++}$$

up to copper and then falls in passing to zinc. This rule is valid for almost every ligand. Taking the value of β_i as a criterion, exceptions occur when $i > 4$, for then those cations which usually have a coordination number of only 4 are at a disadvantage compared with those which have a coordination number of 6. There are also polydentate chelate ligands whose structures do not fit sterically into the quasi-square coordination geometry assumed by Cu^{++}, so that octahedral Ni^{++} then forms a more stable complex than Cu^{++}. Finally, Fe^{++} gives complexes of a special sort with CN^-, phenanthroline, and certain other heterocyclic bases, which are characterized by their diamagnetism and deep color. In such cases these cations no longer fit normally into the Irving-Williams series.

The otherwise good validity of the Irving-Williams rule is readily understood in terms of the difference between electrovalent and nonelectrovalent interaction. Radii of transition-metal cations decrease somewhat from Mn^{++} to Cu^{++}, while Zn^{++} has a somewhat greater size. Moreover, ionization potentials of the metals from Mn to Cu increase and fall again with zinc. We would therefore expect that both the electrovalent and the non-electrovalent behavior of the transition cations would change in accordance with the rule. Since the large changes in ionization potential are much more effective in their action than small changes in ionic radii, nonelectrovalent interaction must undergo a much more marked change than electrovalent interaction. In fact the change in stability for complexes with ligands of low electronegativity is much more marked than for complexes with metal-oxygen bonds. For example log K_1 increases in the case of ethylenediamine complexes from 2.7 for Mn^{++} to 10.8 for Cu^{++} and then sinks again to 5.9 for Zn^{++}, while all oxalate complexes of the series have log K_1 values between 4 and 5. We can say that the electrovalent behavior of the transition ions of like charge remains almost constant, whereas their nonelectrovalent behavior changes markedly. The Mn^{++} ion behaves not very differently from a bivalent A cation, while Cu^{++} shows a markedly more pronounced nonelectrovalent behavior than the d^{10} cation Zn^{2+}.

In recent years it has become customary to explain the Irving-Williams series with the aid of crystal field theory (9, 40). The ions Mn^{++} and Zn^{++} cannot, as d^5 and d^{10} cations, show any crystal field stabilization, while this will increase steadily in the series from d^6 to d^8. This results in the sequence of stability which actually is observed for ligands producing a stronger ligand field than water. The question remains unanswered as to why ligands with N, S, and C produce a particularly strong ligand field when they are quite unable to compete with water in the case of A cations.

Few quantitative measurements are available on the complex-forming behavior of transition cations of the second and third long periods. The nonelectrovalent interaction is, however, certainly the same as for the corresponding cations in the first transition series with the same number of d electrons. It increases again in passing to the right and is particularly large for d^7, d^8, and d^9, i.e., for the cations of Rh, Pd, Ir, Pt, as well as for Ag(II) and Au(III). These groups of metals form inert complexes, however, and all reactions occur slowly; this is very useful in preparative studies, but it makes equilibrium measurements difficult or impossible.

3

THE RELATIVE AFFINITIES OF LIGAND ATOMS FOR ACCEPTOR MOLECULES AND IONS

By Sten Ahrland, J. Chatt, and N. R. Davies

(Akers Research Laboratories, Imperial Chemical Industries Limited, Welwyn)

In 1941 Sidgwick [1] summarised the few available data on the relative affinities of the commoner ligand atoms for various acceptor molecules and ions. Since then the experimental material has increased enormously, and so it seems profitable to attempt a revised and extended correlation involving all the ligand atoms except hydrogen. Admittedly, the quantitative data concerning the heavy donor atoms of Groups V and VI are still sparse, but together with semiquantitative and qualitative evidence there are sufficient to provide a fairly coherent picture. On the other hand, in the case of Group VII where the affinities of the simple halide ions for metal ions can usually be measured conveniently in aqueous solution, the number of quantitative data is now considerable.[*][2]

The Relative Co-ordinating Affinities of Ligand Atoms from the Same Group.—There is no uniform pattern of relative co-ordinating affinities of all ligand atoms for all acceptor molecules and ions, not even when only simple unidentate ligands of closely analogous structures are considered, *e.g.*, the alkyl derivatives PR_3, R_2S, etc. Rather, their relative affinities depend on the acceptor concerned. Thus towards trimethyl gallium [3] the relative tendencies of the alkyls of co-ordinating atoms from Groups V and VI to form complexes under comparable conditions are $N > P > As > Sb$ and $O > S < Se > Te$, but towards platinum(II) the order appears to be $N \ll P > As > Sb$ [4] and $O \ll S \gg Se < Te$,[5] and towards silver $N \ll P > As$ [6] and $O \ll S < Se < Te$.[3, 7] Other similarly diverse examples could be given.

In spite of this lack of uniformity two regular features have emerged: (1) There is in general a very great difference between the co-ordinating affinities of the first and the second element from each of the three Groups of ligand atoms in the Periodic Table, *i.e.*, between N and P, O and S, F and Cl. (2) There are two classes of acceptor: (*a*) those which form their

[1] Sidgwick, *J.*, 1941, 433.

[2] " Stability Constants ", Part II, by J. Bjerrum, G. Schwarzenbach, and L. G. Sillén, The Chemical Society, London, 1958, pp. 88—127.

[3] Coates, *J.*, 1951, 2003.

[4] Chatt, *ibid.*, p. 652.

[5] Cf. Chatt and Venanzi, *J.*, 1955, 2787.

[6] Ahrland, Chatt, Davies, and Williams, *J.*, 1958, 276.

[7] *Idem, ibid.*, p. 264.

[*] Equilibrium data have been chosen as a measure of stability because they are the only quantitative data widely available. Heats and entropies are still sparse.

most stable complexes with the first ligand atom of each Group, *i.e.*, with N, O, and F, and (*b*) those which form their most stable complexes with the second or a subsequent ligand atom.

In this context each oxidation state and perhaps even each magnetic state must be regarded as a different acceptor ; *e.g.*, copper(I) is markedly different from copper(II) in its complex chemistry, and the different oxidation states of iron also behave very differently towards one and the same donor.

Most metals in their common valency states belong to class (*a*). It also contains the hydrogen ion, and therefore the affinities of ligands for class (*a*) acceptors tend to run roughly parallel to their basicities, except when steric and other factors intervene. The acceptors of class (*b*), which are less numerous, are almost all derived from a number of neighbouring elements occupying an area of more or less triangular shape in the Periodic Table. The base of this triangle stretches, in the sixth period, from about tungsten to polonium and its apex is at copper, copper(I) being a definite class (*b*) acceptor and copper(II) on the border between the classes.

The most pronounced class (*b*) acceptors are formed by elements in the central part of this area. These are the metals which form stable olefin complexes, *viz.*, Cu(I), Rh(I), Pd(II), Ag(I), Pt(II), and Hg(II).[8] Evidently the co-ordination of carbon as a ligand atom, *e.g.*, in ligands such as C_2H_4, CO, and CNR, occurs only to acceptors of a pronounced class (*b*) character. Class (*b*) character appears to depend on the availability of electrons from the lower *d*-orbitals of the metal for dative π-bonding.[9]

The type (*a*) or type (*b*) character of many acceptors is so well defined that they can be classified beyond any doubt even from purely qualitative observations of the stabilities of their complexes, but this is not the case in the border region. On the contrary the boundary is somewhat diffuse, mainly because the various oxidation states of the border elements have different characters. But even when the oxidation state is specified the boundary still remains slightly diffuse, depending in its detail on the specific Group of ligand atoms under consideration. The transition between classes (*a*) and (*b*) within the Periodic Table is thus a more or less gradual one, as visualised in Table 1, where a tentative classification of the acceptor atoms in their normal valency states is given. The border region around the core of pronounced class (*b*) acceptors is fairly extensive. In this region, exact quantitative determinations are necessary to determine the character of a certain acceptor towards a given Group of ligand atoms.

The border is most clearly marked to the right of Group VIII because multiple valency is not common in this part of the Periodic Table and the number of acceptor atoms to be considered is small, also the transition in the character of the acceptors is rapid (compare Ag^+ and Cd^{2+} in Tables 2 and 5). The border almost certainly runs between copper and zinc, and between cadmium and indium. Thallium, lead, bismuth, and tellurium

[8] Leden and Chatt, *J.*, 1955, 2936.

[9] Chatt, *Nature*, 1950, **165,** 637 ; Craig, Maccoll, Nyholm, Orgel, and Sutton, *J.*, 1954, 332 ; Jaffé, *J. Phys. Chem.*, 1954, **58,** 185.

are all border-line elements, and according to their complex halides [10-14] just on the side of class (*b*).   Copper(I) is in class (*b*), but cobalt(II), nickel(II), and copper(II) are on the border, having a weak class (*a*) character towards halide ions,[15-17] and a weak class (*b*) character towards Group VI [18] and perhaps Group V ligand atoms.

TABLE 1.   *Classification of acceptor atoms in their normal-valent states*

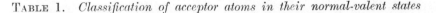

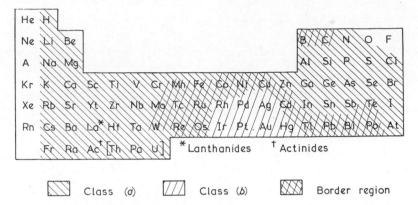

It is surprising that the halide ions follow so closely the pattern of stabilities set by the uncharged atoms of Groups V and VI.   Since they are negatively charged, and neutralisation of charge occurs when they form complexes with positive ions, the changes in hydration during the formation of complexes should introduce an important perturbing factor.   Nevertheless the pattern of halide-complex stabilities appears to follow in its main features that deduced from complex formation by ligands containing uncharged ligand atoms from Groups V and VI.

To the right of the copper Group of acceptors the higher-valent states have a greater class (*b*) character than the lower-valent states of the same element.   This is to be expected because the electrons which enter the metal during its reduction do not enter the *d*-orbitals, which are filled, but the *s*-orbitals where they will screen the *d*-orbitals making them less available for dative π-bonding.   Thallium(I) and thallium(III) in their complex halides provide a case in point.   The few data, which relate only to halide ions,[10, 11] indicate that both valency states are class (*b*) acceptors and that thallium(III) has by far the stronger class (*b*) character.

[10] Nilsson, *Arkiv Kemi*, 1957, **10**, 363.

[11] Benoit, *Bull. Soc. chim. France*, 1949, 518 ; Grenthe and Norén, personal communication.

[12] Biggs, Parton, and Robinson, *J. Amer. Chem. Soc.*, 1955, **77**, 5844 ; Karlsson, personal communication.

[13] Ahrland and Grenthe, *Acta Chem. Scand.*, 1957, **11**, 1111.

[14] Aynsley and Campbell, *J.*, 1957, 832.   [15] Carleson and Irving, *J.*, 1954, 4390.

[16] Ahrland, *Acta Chem. Scand.*, 1956, **10**, 723.

[17] Ahrland and Rosengren, *ibid.*, p. 727.

[18] Tichane and Bennett, *J. Amer. Chem. Soc.*, 1957, **79**, 1293.

To the left of the nickel Group of metals the border is not so well defined. Multiple valency is common and many elements occur in both classes according to their oxidation states. In this region the higher valencies tend to class (a) character and the low valencies to class (b) character, because the d-electrons are more available for dative π-bonding the lower the oxidation state. Thus iron(III) in its complexes of high magnetic moment forms very stable complex fluorides and belongs to class (a) [15] but iron(0) undoubtedly belongs to class (b). All transition metals in their zero-valent states appear to belong to class (b). It may be that the metal ions to the left of the nickel Group will best be separated, class (a) from class (b), by a surface through a three-dimensional arrangement of acceptor atoms which are plotted with their Group numbers as abscissæ, their Period numbers as co-ordinates, and their valencies vertically, so that all under the surface would have class (b) character and all above it class (a).

Boron is a particularly interesting example of a border element because it does not lie near the typical class (b) acceptors associated with Group VIII. Perhaps, together with carbon as acceptor, it may be regarded as a low island of class (b) character in a sea of class (a) elements.

Data relating to the complex compounds of boron with the heavier ligand atoms are relatively common. These indicate that boron lies on both sides of the border according to its environment in the acceptor molecule and the Group of ligand atoms under comparison. Thus boron trifluoride as an acceptor molecule has class (a) character and borine has mild class (b) character with all ligand atoms from Groups V and VI in their methyl derivatives. In similar circumstances trimethylborine appears to have mild class (a) character with Group V and mild class (b) character with Group VI.[19] It seems surprising that boron with no filled d-orbitals can have any class (b) character at all, and it is probable that the bonding electrons in borine and to a smaller extent in trimethylborine are so weakly held that hyperconjugation occurs, providing weak dative π-bonding to suitable ligand atoms, as Graham and Stone[19] have discussed. Such an explanation is consistent with the existence of the weak carbonyl and phosphorus trifluoride complexes of borine.[20]

In a border-line case such as boron, it can be misleading to take the equilibrium constants K (connected with ΔG) as a measure of the strength of the acceptor-to-donor bond because entropy effects can now make a significant contribution to ΔG. Thus the dissociation constants of the complexes Me_3B,NMe_3 and Me_3B,PMe_3 at 100° are 0·472 and 0·128 respectively,[21] and it appears that the amine complex is the least stable, so trimethylborine is a class (b) acceptor. However, the greater dissociation of the amine complex is due to its greater entropy of dissociation ($\Delta S = 45·7$) than that of the phosphine complex ($\Delta S = 40·0$). The heat of formation, ΔH, of the amine complex (17·6 kcal./mole) is actually greater than that of the phosphine complex (16·5 kcal./mole), indicating that in these analogous

[19] Graham and Stone, *J. Inorg. Nuclear Chem.*, 1956, **3**, 164 ; and references therein.
[20] Parry and Bissot, *J. Amer. Chem. Soc.*, 1956, **78**, 1524.
[21] Brown, *J.*, 1956, 1248.

complexes the B–N bond is stronger than the B–P bond and that trimethyl-borine is just on the side of the class (a) acceptors.

We shall now consider the relative co-ordinating affinities of the ligand atoms in each Group. Those of Group V are the simplest, followed by Group VII, but Group VI is very complicated and more data are necessary before any general plan will emerge. In this correlation every attempt has been made to keep the comparison as free from extraneous factors as possible. Most weight has been given to unidentate ligands imposing a minimum of steric strain on the system and having ligand atoms as free as possible from conjugation. Both steric and conjugative effects can cause major anomalies. For example, it would be wrong to compare NPh_3 and PPh_3 as typical amines and phosphines. We might wrongly conclude from the existence of PPh_3HI and non-existence of NPh_3HI that the relative affinities of phos-phorus(III) and nitrogen(III) for the proton are in the sequence P > N and so that the hydrogen ion has class (b) character. Here the nitrogen and the phosphorus atom are not under similar conditions. The compatibility of orbital size between carbon and nitrogen is such that the nitrogen atom is much more strongly conjugated than the phosphorus atom to the benzene nuclei and so in effect loses its lone pair of electrons. It is for this reason that the measurements in Table 2 do not determine whether cadmium is in class (a) or (b). There is no doubt in the case of silver where the aromatic phosphine, relatively inert amongst phosphines, co-ordinates more strongly than any amine, aliphatic or aromatic.

The Relative Co-ordinating Affinities of Ligand Atoms from Group V.— These ligand atoms in their simplest stable derivatives, the trialkyls, show a simple uniformity of behaviour. With class (a) acceptors their co-ordinating affinities lie in diminishing sequence N ≫ P > As > Sb > Bi, with class (b) in the sequence N ≪ P > As > Sb > Bi. No examples are known from qualitative or quantitative work where any other se-quence than P > As > Sb > Bi has even been suggested, and no com-pound containing bismuth as ligand atom is known. The qualitative and semiquantitative data relating to this Group are summarised in Table 2. Nearly all these are concerned with Group III acceptor atoms and vapour-phase measurements. The examples, except palladium and silver, are all of class (a) or have only very weak class (b) character, but qualitatively there is no doubt that Pt(II), Cu(I), Au(I), and Au(III) all have very strong class (b) character, even stronger than Ag(I).

With class (b) acceptors nitrogen with its lower affinity than phosphorus can be placed in the P > As > Sb > Bi sequence, and the position of nitrogen in this sequence will serve as a measure of class (b) character. Thus in the case of BH_3, tertiary alkylamines have affinities lying between those of the tertiary phosphines and tertiary arsines, P > N > As, connoting definite but weak class (b) character. In the case of Ag(I) the amines undoubtedly lie below the corresponding arsines in affinity, P > As > N. This appears to be true of all except border-line class (b) acceptors, so that the tertiary alkylamines and tertiary alkylstibines compete for third place in the sequence.

32

Explanatory Note to Tables 2, 4, and 5

Where the first stability constant ($\log K_1$) is given the measurements were made in aqueous solution. Other measurements or ligand displacement reactions were carried out in the vapour phase. Semiquantitative measurements of relative affinities are indicated by symbols and signs, *e.g.*, N > P, etc. Quantitative data are represented by numerical values of ΔH (= heat of dissociation (kcal./mole)) or $1/K_p$ (t°) [= association constant (atm.$^{-1}$) at t° c]. The constants arranged horizontally in the Tables are comparable. No account is taken of slight differences in temperature or ionic strengths. V. = very.

Ligands are indicated by superscript letters as follows:

a Methyl derivatives, NMe$_3$, Me$_2$O, and MeF and analogues, as appropriate to the reference.

b 3-NH$_2$·C$_6$H$_4$·SO$_3^-$	*g* O(CH$_2$·CH$_2$·NH$_2$)$_2$	*l* 4-PhSe·C$_6$H$_4$·SO$_3^-$
c 3-PPh$_2$·C$_6$H$_4$·SO$_3^-$	*h* S(CH$_2$·CH$_2$·NH$_2$)$_2$	*m* NH(CH$_2$·CO$_2^-$)$_2$
d As(3-C$_6$H$_4$·SO$_3^-$)$_3$	*i* 4-MeO·C$_6$H$_4$·SO$_3^-$	*n* NH(CH$_2$·CH$_2$·NH$_2$)$_2$
e O(CH$_2$·CO$_2^-$)$_2$	*j* 4-EtS·C$_6$H$_4$·SO$_3^-$	*p* 4-NMe$_2$·C$_6$H$_4$·SO$_3^-$
f S(CH$_2$·CO$_2^-$)$_2$	*k* 4-PhS·C$_6$H$_4$·SO$_3^-$	

q K = Equilibrium constant in *isooctane* (2 : 2 : 4-trimethylpentane) of the reaction :
(Oct^{n}NH$_2$)$_2$PdCl$_2$ + PBu$^n_3$ $\rightleftharpoons$ PBu$^n_3$,Oct^{n}NH$_2$,PdCl$_2$ + Oct^{n}NH$_2$.

TABLE 2. *Relative affinities of ligand atoms from Group* V

Acceptors		Relative affinities							Ref.
Class (a)		N	>	P	>	As	>	Sb	
BeMe$_2$		N	>	P	>	[As]^a			1
BF$_3$		N	>	P	>	[As]^a			2
BMe$_3$	ΔH	17·6		16·5	>	As^a			3, 4
AlMe$_3$		N	>	P^a					5
GaMe$_3$	$1/K_p$(100°)	V. large		33 (130°)		0·75		V. small^a	6
InMe$_3$	ΔH	19·9		17·1	>	As^a			7
Class (b)		N	<	P	>	As	>	Sb	
Pd(II)	$\log K$		3·5^q						8
Ag$^+$	$\log K_1$	1·23^b		8·15^c		5·36^d			9
Cd^{2+}	$\log K_1$	0·26^b		0·9^c					10
BH$_3$		N	<	P	$\geqslant$	As^a			2
TlMe$_3$		N	$\approx$	P	$\geqslant$	As^a			7

References : [1] Coates and Huck, *J.*, 1952, 4501. [2] Graham and Stone, *J. Inorg. Nuclear Chem.*, 1956, **3**, 164 ; and references therein. [3] Brown, *J.*, 1956, 1248. [4] Brown, quoted by Chatt, *J.*, 1951, 652. [5] Davidson and Brown, *J. Amer. Chem. Soc.*, 1942, **64**, 316. [6] Coates, *J.*, 1951, 2003. [7] Coates and Whitcombe, *J.*, 1956, 3351. [8] Meddings and Burkin, *ibid.*, p. 1115. [9] Ahrland, Chatt, Davies, and Williams, *J.*, 1958, 276. [10] *Idem, ibid.*, p. 1403.

TABLE 3. *Relative affinities of ligand atoms from Group* VII *

(All figures are log K_1)

Acceptors	Relative affinities			
Class (a)	F >	Cl >	Br >	I
H^+	3·17†	− 7†	− 9†	− 9·5†
Be^{2+}	*ca.* 5	Small		
Mg^{2+}	1·82†	V. small		
Sc^{3+}	7·08†	Small		
La^{3+}	3·56†	− 0·12		
Ce^{3+}	3·99†		0·38†	No complex
Zr^{4+}	9·80†	0·30		
Th^{4+}	8·65†	1·38†		
VO^{2+}	3·30‡	0·04‡		
Cr^{3+}	5·20†	0·60†		
UO_2^{2+}	4·54	− 0·10	− 0·30	
Pu^{4+}	6·77	− 0·24		
Fe^{3+}	6·04†	1·41†	0·49†	
Ni^{2+}	0·66	Small		
Cu^{2+}	1·23†	0·05†	− 0·03†	
Zn^{2+}	0·77	− 0·19	− 0·60	< − 1·3
Ga^{3+}	5·86†	− 0·6†		
In^{3+}	3·78	2·36	2·01	1·64
Sn^{2+}	3·95	1·15	0·73	
Class (b)	F <	Cl <	Br <	I
$Pt(\text{II})$	< 1§	2·52§	3·04§	4·60§
Ag^+	0·36†	3·04†	4·38†	8·13
Cd^{2+}	0·57	1·59	1·76	2·08
Hg^{2+}	1·03	6·74	8·94	12·87
Tl^+	0·10†	0·68†	0·93†	
		in 4N-NaClO$_4$ $\rightarrow$	0·32	0·72
Tl^{3+}		8·1†	9·7†	
Pb^{2+}	< 0·3	0·96	1·11	1·26
$Te(\text{IV})$	F <	Cl <	Br <	I

* Compiled from " Stability Constants, Part II ", by J. Bjerrum, G. Schwarzenbach, and L. G. Sillén, The Chemical Society, London, 1958, where many more data and original references are given.

† Corrected to zero ionic strength.

‡ Ahrland and Norén, *Acta Chem. Scand.*, 1958, **12** (in the press).

§ log K_1 of the reaction : *trans*-[C$_2$H$_4$,H$_2$O,PtCl$_2$] + X$^-$ $\rightleftharpoons$ *trans*-[C$_2$H$_4$PtX,Cl$_2$] + H$_2$O at 25° and an ionic strength of 0·2 (HClO$_4$) (Leden and Chatt, *J.*, 1955, 2936).

TABLE 4. *Relative affinities of ligand atoms from Group* VI

Acceptors		O		S		Se		Te	Ref.
Class (a)		O	>	S		Se		Te	
$BeMe_2$		O	>	$[S]^a$					1
Ca^{2+}	$\log K_1$	3.4^e		1.4^f					2
BF_3		O	>	S	>	Se^a			3, 4
$AlMe_3$		O	>	S	>	Se	>	Te^a	5, 6
$GaMe_3$	$1/K_p(100°)$	0.93		≈ 0.4		0.66		$\approx 0.4^a$	5
Zn^{2+}	$\log K_1$	3.6^e		3.0^f					2
Class (b)		O	<	S		Se		Te	
Co^{2+}	$\log K_1$	2.7^e		3.4^f					2
Ni^{2+}	$\log K_1$	2.8^e		4.1^f					2
	$\log K_1$	5.5^g		7.3^h					7, 8
Cu^{2+}	$\log K_1$	3.9^e		4.5^f					2
	$\log K_1$	8.6^g		9.1^h					8
Ag^+	$\log K_1$	-0.12^i		2.6^j					9
	$\log K_1$			1.7^k		2.6^l			9
				S	<	Se	<	Te^a	5
BH_3		O	<	S	>	Se^a			3
BMe_3		O	<	S	>	Se^a			3
$TlMe_3$		O	<	S^a					10

Footnotes to Table 4 :

References : [1] Coates and Huck, *J.*, 1952, 4501.   [2] Tichane and Bennett, *J. Amer. Chem. Soc.* 1957, **79**, 1293.   [3] Graham and Stone, *J. Inorg. Nuclear Chem.*, 1956, **3**, 164 ; and references therein.   [4] Brown and Adams, *J. Amer. Chem. Soc.*, 1942, **64**, 2557.   [5] Coates, *J.*, 1951, 2003.   [6] Davidson and Brown, *J. Amer. Chem. Soc.*, 1942, **64**, 316.   [7] Gonick, Fernelius, and Douglas, *ibid.*, 1954, **76**, 4671.   [8] Lotz, Ph.D. Thesis, Penn. State Univ., 1954.   [9] Ahrland, Chatt, Davies, and Williams, *J.*, 1958, 264.   [10] Coates and Whitcombe, *J.*, 1956, 3351.

Footnotes to Table 5 :

References : [1] Coates and Huck, *J.*, 1952, 4501.   [2] Tichane and Bennett, *J. Amer. Chem. Soc.*, 1957, **79**, 1293.   [3] Chaberek and Martell, *ibid.*, 1952, **74**, 5052.   [4] Lotz, Ph.D. Thesis, Penn. State Univ., 1954.   [5] Jonassen, Hurst, LeBlanc, and Meibohm, *J. Phys. Chem.*, 1952, **56**, 16 ; and references therein.   [6] Ahrland, Chatt, Davies, and Williams, *J.*, 1958, 276.   [7] *Idem, ibid.*, p. 264.   [8] Prue and Schwarzenbach, *Helv. Chim. Acta*, 1950, **33**, 985.   [9] Graham and Stone, *J. Inorg. Nuclear Chem.*, 1956, **3**, 164; and references therein.   [10] Brown and Adams, *J. Amer. Chem. Soc.*, 1942, **64**, 2557.   [11] Brown, Bartholomay, and Taylor, *ibid.*, 1944, **66**, 435.   [12] Schlesinger and Burg, *Chem. Rev.*, 1942, **31**, 1.   [13] Davidson and Brown, *J. Amer. Chem. Soc.*, 1942, **64**, 316.   [14] Coates, *J.*, 1951, 2003.   [15] Coates and Whitcombe, *J.*, 1956, 3351.   [16] Ahrland, Chatt, Davies, and Williams, *J.*, 1958, 1403.   [17] Brown, *J.*, 1956, 1248.   [18] Stone and Burg, *J. Amer. Chem. Soc.*, 1954, **76**, 386.

TABLE 5. *Relative affinities of ligand atoms from the same Period*

Acceptors		Relative affinities			Ref.
Period II		N	>	O	
BeMe$_2$		N	>	O^a	1
Co^{2+}	log K_1	7·0^m		2·7^e	2, 3
Ni^{2+}	log K_1	8·2^m		2·8^e	2
	log K_1	10·8^n		5·5^g	4, 5
Cu^{2+}	log K_1	10·4^m		3·9^e	2
	log K_1	16·1^n		8·6^g	4, 5
Ag$^+$	log K_1	0·76^p		− 0·12^i	6, 7
Zn^{2+}	log K_1	7·0^m		3·6^e	2, 3
Hg^{2+}	log K_1	23^n		15·8^g	4, 8
BF$_3$	1/K_p	V. large (177°)^a		5·8 (99°)^a	9, 10
BMe$_3$	1/K_p	2·12 (100°)^a		V. small (− 78°)^a	9, 11
BH$_3$	1/K_p	V. large (125°)^a		V. small (− 78·5°)^a	12
AlMe$_3$		N	>	O^a	13
GaMe$_3$	1/K_p	V. large (100°)^a		0·93 (100°)^a	14
InMe$_3$		N	>	O^a	15
TlMe$_3$		N	>	O^a	15
		N	<	O	
Ca^{2+}	log K_1	2·7^m		3·4^e	2
Period III		P	>	S	>	Cl	Ref.
BeMe$_2$		P	>	S^a		1
Ag$^+$	log K_1	8·15^c		1·7^k	6, 7
Cd^{2+}	log K_1	≈ 0·9^c		0·67^k	16
½B$_2$H$_6$	1/K_p	V. large (200°)^a		1·56* (60°)^a	9
BMe$_3$	1/K_p	7·8 (100°)^a		V. small (25°)^a	9, 17
AlMe$_3$		P	>	S	>	Cl^a	13
GaMe$_3$	1/K_p	33 (130°)^a		≈ 0·4 (100°)^a	14
InMe$_3$		P	>	S^a		15
Period IV		As	>	Se	
Ag$^+$	log K_1	5·36^d		2·6^l	6, 7
BH$_3$	1/K_p	V. large (80°)^a		V. small (6°)^a	9, 18
GaMe$_3$	1/K_p (100°)	0·75^a		0·66^a	14
		As	<	Se	
TlMe$_3$		As	<	Se^a	15
Period V		Sb	<	Te	
GaMe$_3$	1/K_p (100°)	V. small		≈ 0·4^a	14

* Atm.$^{-\frac{1}{2}}$.

The Relative Co-ordinating Affinities of Ligand Atoms from Group VII.—
In this Group the simple alkyls, RX, form very few metal complexes
analogous to those formed by the trialkyls of Group V and dialkyls of
Group VI ligand atoms. The only strongly complexing series of ligands are
the halide ions. They are comparatively small and carry a negative charge,
and so they are not strictly comparable with the ligands in Tables 2 and 4
containing uncharged co-ordinating atoms from Groups V and VI. The
effects of hydration will be more marked. Nevertheless two regular
sequences of stabilities of complex halides have been noted.[22-24, 15, 16]
With acceptor ions of class (a) in aqueous solution the affinities of the halide
ions lie in the sequence $F \gg Cl > Br > I$, and with class (b) in the sequence
$F \ll Cl < Br < I$. This differs from Group V in the occurrence of opposing
sequences $Cl > Br > I$ and $Cl < Br < I$. It seems reasonable to sup-
pose that some acceptor near the border between class (a) and class (b) might
form a series of halide complexes of almost identical stabilities and then a
mixed sequence could occur, e.g., $F > Cl < Br > I$ or $F < Cl > Br < I$.
A vast number of halide complexes have been investigated quantitatively,
but only Bi^{3+} shows any tendency to give a mixed sequence.[13] A selection
from the large number of data compiled by Bjerrum, Schwarzenbach, and
Sillén,[2] together with a few recent additions, is given in Table 3. The
values have been chosen to give the best relative affinities of each metal ion
for the different halide ions and not of the various metal ions for the same
halide ion.

The Relative Co-ordinating Affinities of Ligand Atoms from Group VI.—
With metal ions of strong class (a) character the sequence $O \gg S > Se > Te$
would be expected since the corresponding sequence is found in both Groups
V and VII and this is confirmed with $AlMe_3$ as acceptor (Table 4). With
acceptors of class (b) almost any sequence of S, Se, and Te might occur
since the stabilities of complexes formed by the corresponding elements in
Groups V and VII run in opposing sequences, $P > As > Sb$ and $Cl < Br < I$.
In fact almost every sequence seems to occur with acceptor atoms of mild
class (a) to strong class (b) character. These are listed in Table 4. Many
more data are needed before any rules can be formulated for this Group of
ligand atoms, but it seems likely that the data relating to this Group, where
sequences depend so much on subtle differences between the acceptor
atoms, will be more valuable than any other in classifying the various
acceptor molecules and ions according to their co-ordination chemistry.

**The Relative Co-ordinating Affinities of Ligand Atoms from the Same
Period.**—There are few comparable quantitative data relating to ligand
atoms in similar circumstances in the same Period, and these are concerned
almost entirely with nitrogen and oxygen in Period II (Table 5). There
seems little doubt that with all acceptor atoms except the most electropositive the affinities fall, $N > O > F$, in a series of similar ligands, e.g.,
NR_3, R_2O, and RF. Such a sequence is to be expected wherever covalent
donor bonding makes a significant contribution to the strength of the

[22] Leden, Diss., Lund, 1943, p. 27.   [23] J. Bjerrum, *Chem. Rev.*, 1950, **46**, 381.
[24] Ahrland and Larsson, *Acta Chem. Scand.*, 1954, **8**, 354.

co-ordinate bond, because these ligand atoms are all very electronegative and the order is that of rapidly increasing electronegativity and so of decreasing availability of the lone pair of electrons. Further, in combination with class (b) acceptors with their filled d-orbitals, there will be increasing lone-pair repulsion between the electrons in the d-orbitals and the non-bonding lone pairs in the p-orbitals of these small ligand atoms, as the lone pairs become more numerous along any series of analogous ligands containing nitrogen, oxygen, and fluorine as co-ordinating atoms. Thus we would expect to find a very rapid decrease in co-ordinating affinity $N \gg O \gg F$ for strong class (b) acceptors, a more gradual decrease for border-line acceptors, a slow decrease for the weak class (a), and for the very strong class (a), *i.e.*, the ions of the most electropositive metals, where the bonding is mainly electrostatic, even a mild reversal of the order to $F > O > N$. The few data available fit with this suggested pattern of co-ordination (Table 5). The expected reversal has been observed with Ca^{2+}.

In Period III, P, S, and Cl, similar arguments hold, but these atoms are more readily polarisable, and their electronegativities are lower and more nearly equal; also because they have vacant 3d-orbitals, dative π-bonding becomes possible and lone-pair repulsions should be less. We might therefore expect a similar pattern to that of Period II, but with smaller differences.

In Periods IV and V these trends continue so that the differences between the ligand atoms become small and reversals in the order of the stabilities of their complexes become more probable. In Period V the reversal As < Se appears to occur except with acceptors of moderate to strong class (b) character, and in Period VI the reversed order Sb < Te appears to be the rule even with such strong class (b) acceptors as platinum [5] and palladium.[25] It is worth noting, in considering the fifth Period, that organic iodides, RI, show some tendency to form complexes, especially to silver(I) [26] which also forms surprisingly strong complexes with dialkyl tellurides.[3]

This Review has concerned itself mainly with the relative affinities of ligand atoms for acceptor atoms under the conditions where complex compounds are formed reversibly, as in the vapour phase or in solution. No attempt has been made to interpret the facts in any except a rudimentary manner, because without more data of bond strengths and heats and entropies of complex formation the interpretation would be largely speculative. Many factors influence the stabilities of complexes but these relative affinities which determine what may be called the compatibility of ligand and acceptor atoms is one of the most important. The regularities which we see emerging from the still very inadequate quantitative data may be summarised as follows :

Acceptor ions and molecules are of two well-defined types : class (a), which form their most stable complexes with the first ligand atom in each Group ; class (b), which form their most stable complexes with the second or a subsequent ligand atom of each Group. In each Group of ligand atoms

[25] Chatt and Venanzi, *J.*, 1957, 2351.
[26] Andrews and Keefer, *J. Amer. Chem. Soc.*, 1950, **72**, 3113.

the co-ordinating affinities are generally related as follows : Group V, with class (a) acceptors, N ≫ P > As > Sb > Bi ; class (b), N ≪ P > As > Sb > Bi ; Group VI, with class (a), O ≫ S and S, Se, and Te variable ; class (b), O ≪ S, and S, Se, and Te variable ; Group VII (halide ions), with class (a), F ≫ Cl > Br > I ; class (b), F ≪ Cl < Br < I. When we consider the ligand atoms in Periods, the comparable data are too few to allow definite generalisation. Many more data, especially relating to the heavier ligand atoms, are necessary before any finality will be reached, but tentatively it appears that the affinities run as follows : Period II, with all except the ions of very electropositive metals (e.g., Ca), N > O > F in a similar series of ligands, e.g., NR_3, R_2O, and RF ; Period III, similarly, P > S > Cl ; Period IV, As > Se > Br and As < Se < Br, perhaps about equally distributed amongst acceptors ; Period V, Sb < Te > I with most acceptors, Sb > Te > I rarely, if at all.

Reprinted from *Journal of the American Chemical Society*, **84**, 16–24 (1961)
Copyright 1961 by the American Chemical Society

4

[CONTRIBUTION FROM THE CHEMISTRY DEPARTMENTS OF BROWN UNIVERSITY, PROVIDENCE, RHODE ISLAND, AND
NORTHWESTERN UNIVERSITY, CHICAGO, ILLINOIS]

The Factors Determining Nucleophilic Reactivities

BY JOHN O. EDWARDS[1] AND RALPH G. PEARSON
RECEIVED MAY 17, 1961

Three important factors determining the reactivity of nucleophilic reagents are considered. These are basicity, polarizability and the presence of unshared pairs of electrons on the atom adjacent to the nucleophilic atom, the alpha effect. The theoretical bases for these three factors are discussed. Experimental data for a number of substrates are given which make it clear that the reactivities of some substrates depend almost entirely on basicity of the nucleophile, and some substrate reactivities depend entirely on the polarizability. Substrates which resemble the proton in having a high positive charge and a low number of electrons in the outer orbitals of the central atom depend on basicity. Substrates with a low positive charge and with many electrons in the outer orbitals of the central atom depend on polarizability. The alpha effect appears to be general for all substrates.

There is now available in the literature a large amount of data on the rates of the generalized bimolecular substitution reaction

(1) Metcalf Research Laboratory, Brown University, Providence 12, Rhode Island.

$$N + SX \longrightarrow NS + X \qquad (1)$$

Here N is a nucleophilic reagent (ligand in inorganic chemistry) and SX is a substrate containing a replaceable group X and an electrophilic atom S. Other groups, of course, may be bound to S. The

40

nucleophilic reactivity of N is measured by the rate of reaction 1 for a given substrate.[2a] Often the relative rate with respect to a standard nucleophile such as water is used as a measure of reactivity. It is well known that the substrate, SX, determines not only the magnitudes of the rates but also the order for a series of nucleophiles. Information is available in cases where S is a carbon atom, either tetrahedral or trigonal, phosphorus, oxygen, boron, nitrogen or sulfur, and where S is a metal atom or ion in a complex. With such a wide range of information at hand, it seems desirable to discuss the following points in as fundamental a manner as possible: what factors make N a good nucleophilic reagent in general, and what specific factors in SX will tend to produce a certain order of reactivity among the various N groups.

To limit the problem somewhat, certain factors will be deliberately omitted from discussion. These include solvation effects, for it is known that different orders of reactivity can be found in different solvents.[2b] Also steric factors, such as strain in the transition state for reaction 1, will not be mentioned further nor will features such as hydrogen bonding or cyclic structures for the transition state. This means that entropies of activation will not be considered and the emphasis will be on the electronic factors which determine the energies of the ground states of N and SX and the activated complex N—S—X.

General Nucleophilic Reactivity.—Several approaches might be used to set up a normal scale of reactivity. For example, a standard substrate might be used as in the work of Swain and Scott.[2a] Or a correlation of rate data with quite independent properties of the nucleophile could be attempted as in the equation of Edwards,[3] wherein the reactivity of N is correlated with its ability to be oxidized (electrode potential) and to take up a proton (basicity). A correlation[4] of nucleophilicity with basicity and polarizability of the form

$$\log (k/k_0) = \alpha P + \beta H \qquad (2)$$

where (k/k_0) is the rate relative to water, P is defined as $\log (R_N/R_{H_2O})$ with R standing for molar refractivity,[5] and H is a function of basicity $(H = pK_a + 1.74)$, has been given. The coefficients α and β are determined by experiment for each substrate, and with suitable choice of values the equation can be made to fit a large amount of rate data. Recently it has been shown[6] that a certain group of nucleophiles seem to react at rates invariably higher than can be accounted for by (2). These nucleophiles can be represented by the formula YN, where N is the nucleophilic atom and Y is an electronegative atom which contains one or more pairs of unshared electrons. Examples would be NH_2OH, ClO^- and $R_2C{=}NO^-$.

The three properties of basicity, polarizability and unshared pairs on the neighboring atom are sufficient for a fundamental discussion of nucleophilic reactivity. Each of these properties now will be considered in detail and the mechanism of their contribution to the stability of the transition state in a substitution reaction discussed.

Basicity.—The relationship of basicity to nucleophilic character is implicit in the fact that substitution reactions are generalized acid–base reactions. Reaction 1 where S is a proton is one example of the more general class and hence N is always a base. Basicity is measured in terms of an equilibrium constant for a reaction like (3) or (4). Ideally the gas phase proton affinities would be most desirable to avoid solvation difficulties.[7]

$$N(g) + H(g)^+ \rightleftharpoons NH(g)^+ \qquad (3)$$

In the absence of values for most proton affinities, the base constant in water is ordinarily used.

$$N(aq) + H(aq)^+ \rightleftharpoons NH(aq)^+ \qquad (4)$$

The large number of such constants and their precision make them valuable for correlation of rate data. The rate constants of reactions such as (3) and (4) are sometimes available, but they are no more fundamental than the equilibrium constants.

In the case of basicity to the proton, N is clearly interacting with a positive center. In substitution reactions, N is interacting with the atom S most directly. In the ground state SX, it is conceivable that S has a net negative charge. However in the transition state for reaction 1 it may be assumed that S has developed a positive charge of some magnitude since the leaving of X would always remove negative charge from S. We now seek the relationship between the charge on S in the transition state, the basicity of N to the proton and the rate of the substitution reaction.

First it is necessary to examine the process of binding a proton to a base in more detail. Consider the ground state of N, the base. It is characterized by a certain distribution of nuclei and electrons with a wave function, ϕ_0, the square of which represents the spatial density of the electron cloud. The proton is now added to this system. It will seek out the position in the molecule which has the greatest negative potential. This negative potential will be partly the result of the original charge distribution and partly the result of the redistribution caused by the presence of the proton. An important point, however, is this: for the electronegative atoms with which nucleophilicity is concerned, the perturbation of the original charge cloud by proton is not great.

For example, a number of calculations of the proton affinities of simple molecules and ions have been made by quantum mechanical perturbation methods recently.[8] Surprisingly good results can

(2) (a) C. G. Swain and C. B. Scott, *J. Am. Chem. Soc.*, **75**, 141 (1953). (b) For examples see E. A. S. Cavell, *J. Chem. Soc.*, **4217** (1958); S. Winstein, *et al.*, *Tetrahedron Letters*, No. **9**, 24 (1960); R. G. Pearson and D. C. Vogelsong, *J. Am. Chem. Soc.*, **80**, 1048 (1958); J. Miller and A. J. Parker, *ibid.*, **83**, 117 (1961).

(3) J. O. Edwards, *J. Am. Chem. Soc.*, **76**, 1540 (1954).

(4) J. O. Edwards, *ibid.*, **78**, 1819 (1956).

(5) There is an inconsistency in this equation in that theory predicts the form $P = R_N - R_{H_2O}$. The logarithm is better empirically, probably because only a fraction of the total refraction is concerned with the nucleophilic center.

(6) W. P. Jencks and J. Carriuolo, *J. Am. Chem. Soc.*, **82**, 1778 (1960).

(7) R. G. Pearson and D. C. Vogelsong, *ibid.*, **80**, 1038 (1958).

(8) H. Hartmann, *et al.*, *Z. Naturforsch.*, **2a**, 489 (1947); *Z. physik. Chem.*, **19**, 29 (1959); *ibid.*, **22**, 305 (1959); R. Gaspar, *et al.*, *Acta Phys. Acad. Sci. Hung.*, **7**, 151, 44 (1957); *Ann. Physik*, **2**, 208 (1958); *Acta Phys. Acad. Sci. Hung.*, **10**, 149 (1959); A. F. Saturno and R. G. Parr, *J. Chem. Phys.*, **33**, 22 (1960).

be derived by simply burying the proton in the charge cloud of the base without any change in the electron wave function.[9] Better results can be obtained if the wave function ϕ_0 is modified to move the electron cloud towards the proton. There are two ways of producing this electronic distortion. One is to add to ϕ_0 a wave function ϕ_H which is centered on the proton. This is the usual linear combination of atomic orbitals method as applied to polar. covalent bonds. The other method is to add to ϕ_0 one or more terms which represent additions to the wave function still centered on the nucleophilic atom of N but distorted toward the proton. As an example, in HF only wave functions centered on fluorine are used. The reason for the success of this method is that the final wave function for HF is not greatly different from that for F⁻.[10]

One concludes from the above that basicity is determined chiefly by the original distribution of charge and, to a lesser degree, by the redistribution of charge caused by the proton. A high negative potential can be caused by a large negative charge on N over-all, by an electronegative atom which concentrates much of the charge near it, and/or by a favorable arrangement of the other nuclei and electrons. This last point is illustrated by the presence of base-strengthening or acid-weakening dipoles in the molecule N.

As a simple example of some of the principles mentioned, fluoride ion is much more basic than iodide ion in aqueous solution. In the gas phase fluoride ion is more basic by some 50 kcal./mole, the difference between the proton affinities. The total charge is the same but is concentrated near the nucleus in the case of F⁻ and widely dispersed in the case of I⁻. A high negative potential is produced near the fluorine nucleus by this tight charge cloud. The diffuse charge cloud of iodine produces a less negative potential. The redistribution of charge caused by the proton is more extensive for the less electronegative, more polarizable iodide ion. However the energy gain from redistribution fails to compensate for the lower energy due to the diffuse, original charge distribution.

The effect of favorable arrangement of nuclei is shown by the fact that the basicities of the ions F⁻, OH⁻, NH₂⁻ and CH₃⁻ increase markedly in the order given.

With this concept of the mechanism for basicity in hand, an answer can be given to the question of the relation between the charge on the substrate atom S, the basicity of N and the rate of reaction. A high positive charge on S in the transition state can lead to a strong interaction with the high negative potential of a basic reagent N. This will lower the energy of the activated complex and cause a high rate of reaction. Thus basicity will be an increasingly important factor in rate of substitution as the positive charge on the electrophilic atom in the substrate increases. The co-

efficient β in equation 2 thus will increase as the charge on the substrate increases.

There is an important restriction to the conclusion derived in the above paragraph. In order for a strong lowering of the energy by electrostatic effects to occur, it is necessary that either the positive charge be situated in the region of negative potential or the negative charge be situated in the region of positive potential. The proton is unique in that it always can be placed in the most favorable region without restrictions. Any other atom S will be seriously restricted by the additional electrons that it has. Repulsion between these electrons and the electrons of N will raise the energy rapidly as S and N are brought together. It should be noted that this repulsion is due to the operation of the Pauli exclusion principle and is far greater than simple electrostatic repulsion.

The importance of this repulsion is summarized by pointing out that the equilibrium bonding distance of the proton to the first row atoms contained in N is about 1.0 Å., whereas the transition state distance between S and the basic atom of N usually is estimated to be of the order of 1.5 to 2.0 Å. The greater separation is partly compensated for by some of the electronic charge of N drifting closer to S. However, as pointed out, this has had the effect of partly destroying the basicity of N.

Polarizability.—It is known that polarizable molecules and ions such as thiourea, iodide ion and unsaturated systems are more nucleophilic than their basicities would warrant. Indeed, often such species have negligible basicity. The reason for the beneficial effect of high polarizability on rate has not been well understood. Two factors have been considered by various writers. One is that polarization of the bonding electrons in the direction from N toward S occurs. This permits better electrostatic interaction without bringing in Pauli exclusion effects due to the rest of the N molecule as explained above. The other factor considered[2a] is the polarization of non-bonding electrons on N away from S. This has the desirable effect of reducing electrostatic repulsions between N and the leaving group X. However, since it also has the effect of reducing the electrostatic attraction between N and S, it is not clear that the over-all balance is a favorable one. A more important consideration is that such removal of non-bonding electrons away from the S–N bonding region diminishes repulsions due to Pauli exclusion. This makes closer approach of S to N possible.

Quantum mechanical calculations of polarizability involve putting the atom or molecule in a weak, homogeneous electric field and carrying out a second-order perturbation calculation of the energy.[11] The lowering of the energy in the field is proportional to the polarizability. The procedure is to mix in to the ground state wave function ϕ_0 excited wave functions which combine with

(9) J R Platt, *J Chem Phys.*, **18**, 932 (1950); H. C. Longuet-Higgins, *J. Inorg. Nuclear Chem.*, **1**, 60 (1955); also R. Gaspar, *et al.*, ref. 8

(10) There are examples of bases in which the electronic distribution is markedly changed on adding a proton. The pseudo-acids, such as nitromethane, obviously form such anions.

(11) For an elementary discussion see K. S. Pitzer, "Quantum Chemistry," Prentice–Hall, New York, N. Y., 1953, p. 69; for recent calculations see A. Dalgarno and D. Parkinson, *Proc. Roy. Soc. (London),* **A250**, 422 (1959), and R. M. Sternheimer, *Poly. Rev.,* **115**, 1198 (1959).

ϕ_0 to give a resultant which corresponds to a shift of the electronic charge distribution toward more positive potential. For example, an atomic p orbital can be mixed with an s ground state orbital to form a hybrid with the center of the electron cloud moved away from the nucleus. An atomic d orbital can be mixed with a ground state p orbital, and so on.

The conclusions from the above calculations are that *high polarizability results from the existence of low-lying excited states which, when mixed with the ground state, produce polarity.* Only the electrons in the highest energy atomic or molecular orbitals of the ground state are affected. Weakly held electrons are most easily distorted because the energy required to excite them is small. It may be noted that basicity and polarizability generally do not go together since they depend on quite different factors. The factors are really somewhat incompatible as shown by the different properties of F^- and I^-; however, it is possible to have both factors in the same molecule as typified by the sulfide ion S^-.

For the purpose of understanding nucleophilic reactivity, the important result from the above discussion is that certain highly reactive nucleophiles are characterized by having empty orbitals available which are relatively low in energy. These empty orbitals can be used to accommodate some of the electrons of the molecule N in the transition state. The additional possibility exists that in some cases these empty orbitals can be used to hold some of the electrons on the substrate S. Consider the case of an iodide ion attacking a substrate in which the electrophilic atom is oxygen (as in peroxide). There will be Pauli repulsion between the non-bonded p electrons on oxygen and on iodine. By forming a p–d hybrid on iodine, two new orbitals will be formed, one oriented away from oxygen and holding an oxygen electron pair. This is illustrated in Fig. 1. The net result is a lower energy than if both pairs of electrons tried to occupy the same region in space.

The high polarizability of unsaturated systems is evidenced by the exaltation of the molar refraction for such compounds. This can be explained by stating that the anti-bonding molecular orbital of the π-system is mixed in with the bonding orbital. This creates a polar structure, e.g.,

$$\text{\\C=C/} \leftrightarrow \text{\\C-C/}^{-\ +}$$

. In a crude way it can be said that a partly empty atomic orbital has been created. It should be noted that there are definite directional properties for polarization in molecular systems. This can affect that possibility of using such empty, excited orbitals in the transition state.

The efficiency of high polarizability in the nucleophile in lowering the energy of the activated complex will be a function of the substrate SX. It will always produce some lowering of the energy because of the flexibility it gives to the system. In the case of bonding to the proton, it is not very effective as has already been discussed. Polarizability will be of the greatest help in the case of a substrate which has many electrons in the outer

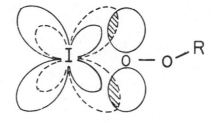

<u>p</u>d hybrids

Fig. 1.—Polarization of pd orbital hybridization in the reaction of iodide ion with a peroxide. The solid line hybrid is filled and the dotted line hybrid is empty.

orbitals of S, particularly if these orbitals project well out from the atom S and form an obstruction to the close approach of N.

The Unshared Pair on the Adjacent Atom. The Alpha Effect.—If basicity is used as a criterion, a certain group of nucleophiles is found to react more rapidly than expected with a number of substrates. The nucleophilic reagents include hydroxylamine, hydrazine, hydroxamic acids, N-hydroxyphthalimide, isonitrosoacetone, the anions of peroxides and hydrogen peroxide, hypochlorite ion, oxime anions and others. The substrates which show the effect include activated esters,[6] peroxides,[12] nitriles,[13] tetrahedral phosphorus,[14] activated double bonds[15] and, possibly, acetaldehyde hydrate[16] (in a proton transfer reaction). The one common feature that can be distinguished in these nucleophiles is the presence of an electronegative atom containing one or more pairs of unshared electrons adjacent to the nucleophilic atom. For reasons that will be brought out in the discussion that follows, it is proposed that the excess reactivity shown by this class of reagents be called the "alpha effect." The reference is to the pairs of electrons on the alpha atom.

The mechanism by which these alpha electrons can influence the rate has been discussed, but it cannot be said that it is understood.[6] A reasonable explanation can be given by considering the limiting case of a nucleophilic substitution. Imagine a pair of electrons leaving the nucleophile for a substrate a large distance away. This would resemble the ionization of a halide ion from an organic halide to form a carbonium ion. By analogy any factor

$$N^m \longrightarrow N^{m+2} + 2e^- \qquad (5)$$
$$RX \longrightarrow R^+ + X^- \qquad (6)$$

which would stabilize the carbonium ion R^+ should also stabilize the denuded nucleophile N^{m+2}. This would include the presence of an unshared pair of

(12) D. L. Ball and J. O. Edwards, *J. Am. Chem. Soc.*, **78**, 1125 (1956).

(13) K. B. Wiberg, *ibid.*, **77**, 2519 (1955).

(14) L. Larsson, *Acta Chem. Scand.*, **12**, 723 (1958); A. L. Green, G. L. Sainsbury, B. Saville and M. Stansfield, *J. Chem. Soc.*, 1583 (1958); J. Epstein, M. M. Demek and D. H. Rosenblatt, *J. Org. Chem.*, **21**, 796 (1956).

(15) C. A. Bunton and C. J. Minkoff, *J. Chem. Soc.*, 665 (1949); H. O. House and R. S. Ro, *J. Am. Chem. Soc.*, **80**, 2428 (1958); H. E. Zimmerman, L. Singer and B. S. Thyagarajan, *ibid.*, **81**, 108 (1959).

(16) R. P. Bell, *J. Phys. Chem.*, **55**, 885 (1951).

electrons on the adjacent atom. An α-halo ether would be an example of the stabilization for a carbonium ion.

$$R\text{-}O\text{-}CH_2\text{-}Cl \longrightarrow R\overset{+}{-}O{=}CH_2 + Cl^- \quad (7)$$

So, in a similar fashion, one can write

$$Cl\text{-}O^- \longrightarrow \overset{+}{Cl}{=}O + 2e \quad (8)$$

To the extent to which the transition state represents some removal of an electron pair from the reactive atom, the π-bonding shown in 8 should make some contribution to the stability of the system. In the same way, removal of chloride ion in the transition state for 7 is not complete, but the enhancement in rate by the oxygen atom, compared to a methylene group, is many powers of ten.[17]

Since excess reactivity for these special reagents such as NH_2OH is referred to basicity as a standard, the possibility of stabilization of the conjugate acid, such as NH_3OH^+ or $HOCl$, must also be considered. Because the proton polarizes some of the electrons toward itself, some effect of π-bonding must be expected. To have enhanced nucleophilic reactivity, it is necessary that removal of sigma electrons be more complete in the activated complex of the nucleophilic reactions than in the normal state of the acid. In view of earlier remarks on the small perturbation of the electron cloud of the base due to the proton, this probably is always the case. The available rate data do not indicate any substrate for which this special group of nucleophiles does not show enhanced reactivity. Some important substrates, such as saturated carbon, have not as yet been investigated quantitatively.

A further prediction from the theory advanced above is that other carbonium ion stabilizing factors should create better nucleophiles.[18,19] Such factors could be alkyl and aryl substitution and unsaturation. To be effective such groups must be on the alpha atom; however, in these cases there are complications by other phenomena. Increased steric strain and significant electronic arrangement on proton addition to the nucleophile are two possible factors which would render the above prediction invalid.

Orders of Nucleophilic Character.—In a displacement reaction, the order of nucleophilic strength is a marked function of the nature of the substrate. It is the purpose of this section to discuss the orders of nucleophilic strength for a number of substrates. In most cases, we shall not give numerical data, often because of their incomplete nature; references from which the results were obtained will be given, however. The data presented in this section will be discussed

later in relation to the conclusions of the previous section.

Hydrogen.—In general, rates of nucleophilic displacement on hydrogen (as given by general base catalyzed reactions) follow the equilibrium basicity scale moderately well. There are, however, significant exceptions some of which are discussed by Bell.[16,20,21] For example, hydroxide ion often is kinetically less reactive than one would expect from a Brønsted plot,[20] while oximate ions react more rapidly than would be expected[16]; the latter nucleophile is one which has a free electron pair on the alpha atom. The adverse effect that strong electron delocalization in anions can have on rates is pointed out by Bell[16,20]; this is particularly noteworthy in the case of the anions of pseudo-acids like nitromethane.

Carbonyl Carbon.—Jencks and Carriuolo[6] discuss the reaction of p-nitrophenyl acetate with a large number of nucleophiles. To a large extent, the nucleophilic strength correlates with basicity, although there are some deviations[6,22] as may be seen in Table I. The influence of a spare pair

TABLE I

RATES OF NUCLEOPHILIC REPLACEMENTS[a]

Nucleophile	pK_{HA}	Substrates Carbonyl carbon[b]	Tetrahedral phosphorus[c]
HOO^-	11.5	2×10^5	1.0×10^5
Acetoximate	12.4	3.6×10^3	....
Salicylaldoximate	9.2	3.2×10^3	1.5×10^3
OH^-	15.7	9×10^2	1.6×10^3
$C_6H_5O^-$	10.0	1×10^2	34
NH_2OH	6	1×10^2	1.3
OCl^-	7.2	1.6×10^3	7×10^2
CO_3^-	10.4	1.0	75
NH_3	9.2	16	....
CN^-	10.4	11	....
$C_6H_5S^-$	6.4	....	7.4×10^{-3}
$C_6H_5NH_2$	4.6	1.5×10^{-2}	....
C_5H_5N	5.4	0.10	....
NO_2^-	3.4	1.3×10^{-3}	....
$CH_3CO_2^-$	4.8	5×10^{-4}	....
F^-	3.1	1×10^{-3}	Very reactive[d]
$S_2O_3^-$	1.9	1×10^{-3}	Unreactive
H_2O	-1.7	6×10^{-7}	1×10^{-6}

[a] Rate constant units are l. mole^{-1} min.$^{-1}$. [b] p-Nitrophenyl acetate as substrate (ref. 6). [c] Isopropoxy-methylphosphoryl fluoride (Sarin) as substrate (ref. 24). [d] Estimated from other similar substrates.

of electrons on the alpha atom is shown by perhydroxyl ion, hypochlorite ion and others. It was also found that polarizable, non-basic nucleophiles such as iodide ion and thiourea are not reactive to this ester. The results obtained are explained in terms of a tetrahedral intermediate[22] which may go on to product or revert to starting material, but there can be no question that nucleophilic character in this case is primarily dependent on

(17) P. Ballinger, P. B. D. de la Mare, G. Kohnstam and B. Prestt, *J. Chem. Soc.*, 3641 (1955).

(18) By similar reasoning one also could predict that any factor which stabilizes free radicals would also create stronger nucleophilic reagents. This stems from the relationship between nucleophiles and reducing agents (C. K. Ingold, "Structure and Mechanism in Organic Chemistry," Cornell University Press, Ithaca, N. Y., 1953, Chapter 5).

(19) R. G. Pearson and F. V. Williams, *J. Am. Chem. Soc.*, **76**, 258 (1954); R. P. Bell and A. F. Trotman-Dickenson, *J. Chem. Soc.*, 1286 (1949).

(20) R. P. Bell, "Acid–Base Catalysis," Clarendon Press, Oxford, 1941, page 92.

(21) R. P. Bell, "The Proton in Chemistry," Cornell University Press, Ithaca, N. Y., 1959.

(22) (a) M. L. Bender, *Chem. Revs.*, **60**, 53 (1960); especially pages 62–64; (b) M. L. Bender and W. A. Glasson, *J. Am. Chem. Soc.*, **81**, 1590 (1959).

basicity. Bruice and Lapinski[23] have found Brønsted slopes of about 0.8 in the reaction of p-nitrophenyl acetate with several series of nucleophiles, even though each series fell on a different line. The high value of the slope implies that basicity is an important factor.

This conclusion is not limited to activated esters nor to esters only. Data in the literature indicate that ordinary esters show an even greater dependence on basicity.[22] Acyl halides and acid anhydrides also show a high sensitivity to basicity in their reactions with nucleophiles.[22a]

Tetravalent Phosphorus.—The literature data on displacements in neutral four-coördinate phosphorus compounds are scattered, thus it is difficult to prepare a list of nucleophiles in order of relative strength. Some numerical data are presented in Table I,[14,24] and the order in water appears to be $OOH^- > OH^- \simeq OCl^- > NH_2OH > NO_2^- > N_3^- > H_2O$. In ethanol, the order $F^- > C_2H_5O^- > C_6H_5O^-$ was found.[25] Sulfur nucleophiles such as $S_2O_3^=$ and $C_6H_5S^-$ do not seem to be particularly reactive. The conclusion that nucleophilic strength to four-coördinate phosphorus primarily follows basicity seems certain. The reactivity of fluoride ion is surprisingly high, however, as are the reactivities of nucleophiles with unshared electrons on the alpha atom.

Individual phosphorus compounds show considerable variation in the extent to which basicity plays a role in nucleophilic strength. The slopes of plots of log k_2 against pK_a (of conjugate acid of nucleophile) are 0.9, 0.7 and 0.5 for the substrates Sarin (isopropoxy-methyl-phosphoryl fluoride), T-EPP (tetraethyl pyrophosphate) and Tabun (dimethylamide-ethoxy-phosphoryl cyanide), respectively, all with substituted hydroxamic acids.[14]

This selectivity order does not result from a reactivity order of Tabun > TEPP > Sarin. Similarly it has been shown[24b] that in compounds of the type R_3PX, the importance of basicity in the nucleophile decreases as alkyl R is converted to alkoxyl OR. Such variations are understood in terms of increasing π-bonding (donation of electrons from oxygen in OR to phosphorus) which cuts down on the positive charge on the phosphorus atom. Nevertheless, even in strongly π-bonded systems, there is no good evidence that polarizable, but non-basic, nucleophiles become effective.

Trigonal Boron.—Not many data are available. In a series of reactions of R_2BX compounds with various reagents, it was found[26a] that the rate order was $OH^- > OR^- > NH_3 > R_2NH \simeq SR^-$. Thus basicity rather than polarizability seems the important factor.

Tetrahedral Boron.—The breaking of the boron–nitrogen bond in H_3NBF_3 has been found to be catalyzed by anions.[26b] The order of nucleophilic strength appears to be $OH^- > F^- > H_2O$ with Cl^- showing no influence. Although the amount of data is limited, we conclude that nucleophilic attack on tetrahedral boron is primarily a function of basicity.

Tetrahedral Sulfur.—There are no data available from which a quantitative scale of nucleophilic character can be derived. It is possible, however, to gain some idea of nucleophilic character from available data on competitive reactions.[27–29] Bunnett and Bassett[27] treated p-nitrophenyl p-toluenesulfonate with various nucleophiles and isolated the products in high yield. The most striking result obtained was that very basic nucleophiles preferred to attack tetrahedral sulfur whereas more polarizable (albeit still somewhat basic) nucleophiles attacked the aromatic carbon atom. Similar results had been found in the reaction of neopentyl p-toluenesulfonate,[28] wherein the competition for the nucleophile is between saturated carbon and tetrahedral sulfur. From such data, we conclude that the rough order of nucleophile strength in attack on tetrahedral sulfur is $OH^- \simeq CH_3O^- > C_6H_5O^- > RNH_2 >$ Piperidine $> C_6H_5$-$NH_2 > C_6H_5S^-$. Thus basicity is of prime importance for nucleophilic character in this case, although there may well be some dependence on polarizability also.

Bivalent Sulfur.—Parker and Kharasch[30] in their review on the breaking of the sulfur–sulfur bond have compiled lists from which orders of nucleophilic strength may be obtained. Since many of the data are qualitative in nature and since both rate and equilibrium data are considered, the order given below is at best a rough one. The order is $RS^- > R_3P > C_6H_5S^- \simeq CN^- > SO_3^= > OH^- > S_2O_3^= > SC(NH_2)_2 > SCN^- > Br^- > Cl^-$. From this order it would seem that nucleophilic attack on bivalent sulfur requires both polarizability and basicity.

There is evidence that polarizability is more important than basicity in at least one case, sulfur in the form of S_8 and S_6 species in solution. Here it has been shown by Bartlett[31] that triphenylphosphine, HS^- and HSO_3^- are powerful nucleophiles for sulfur, whereas tertiary amines are not.

Aromatic Carbon.—From the studies of Bunnett[27,32] and Huisgen[33] it is possible to list the nucleophiles in order of relative strength. The order observed is $C_6H_5S^- \simeq CH_3O^- > C_5H_{10}NH > C_6H_5O^- > N_2H_4 > OH^- > C_6H_5NH_2 > Cl^- > CH_3OH$. Also the order $C_6H_5NH_2 > NH_3 > I^- > Br^-$ is known. Apparently nucleophilic attack requires both polarizability and basicity. The discrimination between nucleophiles by the substrate is very large; for example, methoxide ion is 10^4 times as reactive as aniline which, in turn, is 10^9 times as reactive as the solvent methanol.

(23) T. C. Bruice and R. Lapinski, *J. Am. Chem. Soc.*, **80**, 2265 (1958).

(24) (a) L. Larsson, *Svensk Kem. Tidskr.*, **70**, 405 (1959); (b) G. Aksnes, *Acta Chem. Scand.*, **14**, 1515 (1960); (c) J. Epstein, private communication.

(25) I. Dostrovsky and M. Halmann, *J. Chem. Soc.*, 502, 508, 511, 516 (1953).

(26) (a) D. W. Aubrey and M. F. Lappert, *Proc. Chem. Soc. (London)*, 148 (1960); (b) L. G. Ryss and S. L. Idel, *Russ. J. of Phys. Chem.*, **33**, 374 (1959).

(27) J. F. Bunnett and J. Y. Bassett, *J. Am. Chem. Soc.*, **81**, 2104 (1959).

(28) F. G. Bordwell, B. M. Pitt and M. Knell, *ibid.*, **73**, 5004 (1951).

(29) R. L. Burwell, Jr., *ibid.*, **74**, 1462 (1952).

(30) A. J. Parker and N. Kharasch, *Chem. Revs.*, **59**, 583 (1959).

(31) P. D. Bartlett, E. F. Cox and R. E. Davis, *J. Am. Chem. Soc.*, **83**, 103 (1961); P. D. Bartlett, A. K. Colter, R. E. Davis and W. R. Roderick, *ibid.*, **83**, 109 (1961).

(32) J. F. Bunnett and R. E. Zahler, *Chem. Revs.*, **49**, 273 (1951).

(33) J. Sauer and R. Huisgen, *Angew. Chem.*, **72**, 294 (1960).

Saturated Carbon.—The reaction of nucleophiles with tetrahedral carbon atoms has been studied often and, at least in major detail, is well understood. The order of nucleophilic strength is $C_4H_9S^- > C_6H_5S^- > S_2O_3^- > SC(NH_2)_2 > I^- > CN^- > SCN^- > OH^- > N_3^- > Br^- > C_6H_5O^- > Cl^- > C_5H_5N > CH_3CO_2^- > H_2O$; these data are collected from several sources[2a,3,4,34] and refer primarily to reactions in aqueous solution at room temperature. This order follows electrode potentials for oxidation of nucleophiles[2b] surprisingly closely. In Table II, calculated rate constants for displacement on saturated carbon by a variety of nucleophiles are presented.

TABLE II
RATES OF NUCLEOPHILIC REPLACEMENTS[a]

Nucleophile	Saturated[b] carbon	Peroxide[c] oxygen	Platinum (II)[d]
SO_3^-	2.3×10^{-4}	2×10^{-1}	......
S_2O_3^-	1.7×10^{-4}	2.5×10^{-2}	......
SC(NH_2)_2	2.5×10^{-5}	Very fast	8×10^{-1}
I^-	1.2×10^{-5}	6.9×10^{-1}	2×10^{-1}
CN^-	1×10^{-5}	1.0×10^{-3}	......
SCN^-	3.2×10^{-6}	5.2×10^{-4}	4×10^{-1}
NO_2^-	1.8×10^{-6}	5×10^{-7}	4×10^{-3}
OH^-	1.2×10^{-6}	[e]	[e]
N_3^-	8×10^{-7}	....	8×10^{-3}
Br^-	5×10^{-7}	2.3×10^{-5}	$(5 \times 10^{-3})^f$
NH_3	2.2×10^{-7}	....	$(8 \times 10^{-4})^f$
Cl^-	1.1×10^{-7}	1.1×10^{-7}	9×10^{-4}
C_5H_5N	9×10^{-8}	....	3×10^{-3}
H_2O	1×10^{-10}	[e]	5×10^{-7}

[a] Rate constant units are l. mole^{-1} sec^{-1}. [b] Substrate is hypothetical methyl compound whose rate with Cl$^-$ is same as peroxide rate with Cl$^-$. [c] Substrate is hydrogen peroxide (ref. 41). [d] Substrate is Pt(dien)Br$^+$ (where dien is diethylenetriamine) ref. 42. [e] Rate too low to measure. [f] Estimated from other similar substrates.

The observed order can be considered to be made up of a combination of polarizability and basicity factors with the former being more significant. The importance of polarizability is demonstrated by the fact that malonate ion reacts several times more rapidly with ethyl bromide than does ethoxide ion, even though ethoxide is 500 times more basic.[35] The high rate of reaction of the anions of pseudo acids in general is noteworthy in view of the slowness of proton transfers involving these anions. Similarly tri-substituted phosphines are more reactive in displacements on carbon than are trisubstituted amines, even though the amines are more basic.[36]

Trivalent Nitrogen.—Compounds of the type NH_2X will react with nucleophiles by displacement at nitrogen[37,38] when X is a good leaving group such as Cl$^-$ or SO$_4^-$. Some of the nucleophiles are NH_3, RNH_2, R_3N, R_3As, R_2S, OH^- and $C_2H_5O^-$. The phosphines react moderately rapidly,[39]

the amines at a measurable rate[40] and the oxygen bases somewhat more slowly. Hence polarizability is important and basicity probably less so.

Bivalent Oxygen.—Data for nucleophilic displacements on oxygen in peroxide have been collected.[41] The order observed with hydrogen peroxide is presented in Table II. Also known are the facts that (a) trialkyl phosphines react rapidly even at low temperature, (b) olefins and organic sulfides react as nucleophiles and (c) oxygen anions (except peroxides which have unshared electrons on the alpha atom) are extremely poor nucleophiles. The observations lead to the conclusion that nucleophilic displacements on oxygen follow the polarizability scale closely, with little if any contribution from basicity. Explanations for the observed order of nucleophile reactivity have recently been given.[41]

Platinum(II) Compounds.—In a recent study, Gray[42] found that nucleophiles will react with many planar platinum (II) compounds by what is most certainly a nucleophilic displacement mechanism. Quantitative data are shown in Table II. The order is similar to that observed with peroxide oxygen, and this similarity is confirmed by the facts that olefins and phosphines are good nucleophiles while hydroxide ion and ethoxide ion are very poor nucleophiles.[43] Accordingly polarizability is dominant for platinum(II) and basicity seems to play no important role.

The nature of the platinum complex does determine nucleophilic order to some extent, however. In the case of PtCl$_4^-$ and PtNH$_3$Cl$_3^-$, nitrite ion is over 1000 times as reactive as chloride ion.[44] For Pt(dien)Cl$^+$, as shown in Table II, the reactivity ratio is about 5. From this, and other observations, it is likely that polarizability enhances nucleophilic character toward platinum-(II) to a greater extent as the positive charge on platinum diminishes. Ignoring this effect, the general order of reactivity toward platinum(II) seems to be $R_3P \simeq$ thiourea $\simeq SCN^- \simeq I^- > N_3^- > NO_2^- >$ pyridine $>$ aniline $>$ olefin $\simeq NH_3 \simeq Br^- > Cl^- >$ glycine $\simeq OH^- \simeq H_2O \simeq F^-$.

Halogen Compounds.—It is rather surprising to find that nucleophilic displacements on fluorine are possible, but such is apparently the case in the compound perchloryl fluoride FClO$_3$. Enolate ions, nitronate ions, enamines, vinyl ethers and related organic compounds (all of which have a canonical form with a free electron pair on carbon) attack the fluorine in perchloryl fluoride to give products containing carbon–fluorine bonds.[45] Bases such as ethoxide ion and ammonia attack the chlo-

(34) A. Streitwieser, *Chem. Revs.*, **56**, 571 (1956); *cf.*, page 582.

(35) R. G. Pearson, *J. Am. Chem. Soc.*, **71**, 2212 (1949).

(36) W. A. Henderson, Jr., and C. A. Streuli, *ibid.*, **82**, 5791 (1960); W. A. Henderson, Jr., and S. A. Buckler, *ibid.*, **82**, 5794 (1960).

(37) H. H. Sisler, 135th Meeting of ACS at Boston, Mass., April, 1959, paper 184, Division of Organic Chemistry.

(38) H. H. Sisler, 138th Meeting of ACS at New York, N. Y., September, 1960, paper 83, Division of Inorganic Chemistry.

(39) H. H. Sisler, A. Sarkis, H. S Ahuja, R. J. Grage and N. L. Smith, *J. Am. Chem. Soc.*, **81**, 2982 (1959).

(40) F. N. Collier, Jr., H. H. Sisler, J. G. Calvert and F. R. Hurley, *ibid.*, **81**, 6177 (1959).

(41) (a) J. O. Edwards, paper given at the Peroxide Reaction Mechanisms Conference at Brown University, Providence, R. I., June, 1960; (b) M. C. R. Symons, *Chem. and Ind. (London)*, **48**, 1480 (1960).

(42) H. B. Gray, Ph.D. thesis, Northwestern University, Evanston, Ill., 1960.

(43) D. Banerjea, F. Basolo and R. G. Pearson, *J. Am. Chem. Soc.*, **79**, 4055 (1957).

(44) H. B. Gray and R. Olcott, unpublished results.

(45) F. L. Scott, R. E. Oesterling, E. A. Tyczkowski and C. E. Inman, *Chem. and Ind. (London)*, 528 (1960); and other papers referenced therein.

rine to form substituted perchlorates, but there is no evidence found for these basic nucleophiles attacking the fluorine atom. We conclude that attack on covalent fluorine requires a highly polarizable nucleophile and basicity is of no importance. Conversely attack on tetrahedral chlorine seems to require basicity.

Discussion

The three sources of high nucleophilicity mentioned in the first section are basicity, polarizability and the alpha effect. They seem to be essentially independent of each other. As far as can be ascertained at present, the alpha effect is common to all substrates, though it is expected that the magnitude of the effect will be found to vary over a wide range eventually. From the examples of the previous section, it is seen that different substrates show marked differences with respect to susceptibility toward basicity and polarizability. Stated another way, the values of α and β in equation 2 show large variations with the substrate.

The particular cases of the substrates of carbonyl carbon, tetrahedral phosphorus, tetrahedral sulfur, trigonal boron and the proton seem to depend almost entirely on basicity. The examples of oxygen, bivalent sulfur, fluorine and platinum seem to depend almost entirely on polarizability. Aliphatic tetrahedral carbon, aromatic carbon and trivalent nitrogen depend on both factors with polarizability rather dominant.

The theoretical discussion of the first section enables these variations to be understood in a very satisfactory way. To the extent that the interaction between S and N in the transition state of reaction 1 resembles the interaction of a proton with N, we should find basicity important. The characteristics of the proton are a high positive charge and an absence of outer electrons. It is just those substrates which most nearly fulfill these requirements for which basicity is indeed dominant. In the case of ester hydrolysis, it is

structure $\overset{+}{\underset{}{C}}\overset{-}{-O}$ which is available for the transi-

tion state. This presents a more positive carbon to the incoming nucleophile and, equally important, one less pair of electrons attached to carbon, in comparison to a substrate such as an alkyl halide. In the compounds of tetrahedral phosphorus and sulfur, there is both a high positive charge and a set of empty d orbitals of some stability on the central atom. As π-bonding from oxygen to the central atom reduces the positive charge and fills up these orbitals, so the susceptibility to basicity decreases as expected. Trivalent boron is positive in nature and has an empty p orbital as well. The function of the empty orbitals in all these cases is two-fold: to reduce the number of repelling electron pairs on the substrate atom and to provide a positive site for the acceptance of electrons from N.

The cases where polarizability is the chief factor show a common pattern. The central element is electronegative, often negatively charged in the ground state and has a number of outer orbital electrons. The first two factors cause basicity

to play a minor role and the last factor insures that polarizability will be important. Recall that the important property of a highly polarizable nucleophile is that it can provide a low-energy, empty orbital to accommodate electrons from the substrate. In the cases of oxygen, bivalent sulfur and fluorine, the central atom S has a full set of p orbital electrons. In the case of platinum(II), there is a full set of d orbital electrons except for one vacant d orbital in the plane of the complex. The eight d electrons project in all directions from which a nucleophile could reasonably approach the central atom.

The dual nature of a polarizable nucleophile in both donating electrons to the substrate atom in a sigma bond and accepting electrons from the substrate atom in a π-bond has led to the suggestion that such nucleophiles be called biphilic reagents.[46] There is much evidence of fragmentary nature to indicate that biphilicity is important for complexes of the transition metal ions which are of relatively low positive charge and which have many d orbital electrons. For example, in the metal carbonyls it is only a biphilic reagent such as carbon monoxide itself, trialkyl phosphine and the like, which can easily displace an attached carbon monoxide molecule. This may be due to thermodynamic factors, but, in a few cases, it has been shown that simple nucleophilic displacements occur for these systems.[47] Hydroxide ion is often surprisingly poor as a reagent. For example, in the case of cis and trans $Rh(en)_2Cl_2{}^+$, the rate of reaction with hydroxide ion is hardly greater than the rate of reaction with water.[48] This is unexpected in view of the high positive charge for rhodium(III). The data for platinum(II) cited earlier again show hydroxide ion to be a poor reagent. For metal atoms with no outer d electrons and with a high charge, hydroxide ion becomes a good reagent again. This is true for chromium(VI), silicon(IV), germanium(IV) and tin(IV).

A biphilic reagent is not the same thing as an ambident reagent.[49] The latter has two different nucleophilic sites within it, such as the oxygen atom and the nitrogen atom in the nitrite ion. In any given transition state only one site is involved. Generally in an ambident reagent, one nucleophilic atom is more basic and one atom is more polarizable or presents a structure with an empty orbital. As explained by Kornblum and his co-workers,[49] if the substrate SX resembles a proton, the more basic atom will be the reactive one. This would be the case if a carbonium ion-like intermediate was formed. The more polarizable site would be the reactive one in the case of a substrate which favored polarizability over basicity, such as primary alkyl halide.

The reactions of alkyl halides and similar compounds require both basicity and polarizability, with the latter more important. This results naturally from the structure of the transition state

(46) R. G. Pearson, H. B. Gray and F. Basolo, *J. Am. Chem. Soc.*, **82**, 787 (1960).

(47) F. Basolo and A. Wojcicki, *ibid.*, **83**, 520, 525 (1961).

(48) S. A. Johnson, private communication.

(49) N. Kornblum, R. A. Smiley, R. K. Blackwood and D. C. Iffland, *J. Am. Chem. Soc.*, **77**, 6269 (1955).

in a displacement reaction. The central carbon atom is somewhat positive, but not greatly so. The orbitals surrounding carbon are all filled, but the electrons in them are somewhat removed from the critical region by bonding to other atoms or groups. The general situation is clearly intermediate between that of a proton as the substrate and an oxygen atom of a peroxide as a substrate, but rather closer to the oxygen atom case. As one goes from tetrahedral carbon compounds, R_3CX, across the periodic table to R_2NX, ROX and finally FX, one expects polarizability to become more important and basicity less important, as found. The fact that tetrahedral boron compounds, R_3BX, depend more on basicity than does R_3CX, is also expected.

It might also be predicted that as the groups surrounding carbon in R_3CX promote a mechanism with more carbonium ion-like character, the dependence on basicity should increase. This would follow from the increased positive charge on carbon in the transition state. It must be remembered, however, that groups that favor an S_N1 mechanism do so by processes which remove positive charge from the central carbon atom. This greatly reduces the expected effect, as was pointed out by Swain and Scott.[2a] The experimental facts are not quite enough at present to make the situation clear.

Acknowledgments.—The authors are grateful to their colleagues, especially Myron L. Bender, John F. Neumer, William T. King and William P. Jencks, for their suggestions and comments. J.O.E. wishes to thank Brown University for sabbatical leave and Northwestern University for kind hospitality during the leave.

Part II

Editor's Comments on Papers 5–9

In this part several papers which put forward the fundamental ideas of HSAB are presented in chronological order. Unfortunately the papers are necessarily repetitious to a considerable degree. Still it may be of interest to read them in order, with the purpose of tracing changes in the concept. My own ideas changed somewhat between 1963 and 1968. It was also necessary to change the emphasis in presentation to clarify certain areas of confusion that had arisen.

I include Klixbull Jørgensen's paper on symbiosis, since it certainly forms a part of the fundamental principles. While I had noted the moderating effect of other attached groups in 1963, I had not realized that a simple, logical statement could be made which summarized these effects as Jørgensen has done in this reprint.

There is a rather important error in Table 2 of article 9. A value of $\log K_{eq} \approx 30$ is given for $(CH_3)_2Hg$, suggesting that this molecule is extremely stable. This number is incorrect, being based on an incorrect value for the heat of formation. A better value gives $\log K_{eq} \simeq -10$, which shows that $(CH_3)_2Hg$ is *not* very stable.

Such a result seems to violate both the HSAB principle and the principle of symbiosis. Nevertheless it is correct and can be shown to be part of a very consistent pattern of behavior. There exists a class of compounds for which the expected symbiotic effect is overwhelmed by a *trans* effect which works in exact opposition.

Several examples will show what is happening. Compounds such as $Pt(CH_3)_3$-$(H_2O)_3{}^+$, $Au(CH_3)_2(H_2O)_2{}^+$, $Co(CN)_3(H_2O)_3$ always exist in the form in which the soft ligands are *cis* to each other and have a *trans* water molecule as a partner (Tobias, 1970; Krishnamurthy *et al.,* 1967), e.g.:

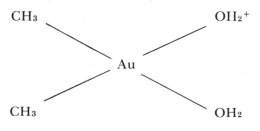

Dimethylsulfoxide forms a complex with Pd(II) in which two molecules are S-bonded and two are O-bonded (Wayland and Schramm, 1969; 1972):

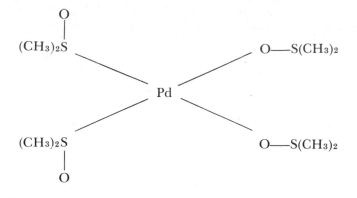

The two S-bonded molecules are *cis* to each other. Stability constants for *bis* complexes of Cu(I), Ag(I), Hg(II), Cd(II) show uniformly that *bis* phosphine complexes are less stable relative to *mono* phosphine complexes than *bis* amine complexes are relative to *mono* amine complexes (Meier, 1967).

Transition metal hydrides of the general formula $ML_l(CO)_m H_n X_o$ almost always have the hydride ligands *cis* to each other. If X is a halogen, or an O or N donor ligand, it almost always appears *trans* to either H or CO, which avoid being *trans* to each other (Jesson, 1971). Similar rules apply for the alkyl derivatives of the heavier transition metals.

Chatt and Heaton (Chatt and Heaton, 1968) were the first ones to give an explanation for the kind of phenomena mentioned above. They pointed out "that groups of high *trans* effect, as ethylene in $Pt(C_2H_4)Cl_3^-$, render the position in mutual *trans* position more susceptible to bonding by what are now called hard bases." They also pointed out that such a result was expected from the general theory of the *trans* effect (Langford and Gray, 1965).

Since the ligands of high *trans* effect are always soft, we can restate the rule of Chatt and Heaton as follows: *two soft ligands in a mutual trans position will have a destabilizing effect on each other when attached to soft metal atoms.* Assuming that covalent bonding is important for soft bases, two such *trans* ligands will always be in competition with each other, either for orbitals, or for electrons. The competition will be most serious for soft metal atoms, where covalency is greatest. Hard metal ions, where the bonding is more ionic, will show little of this antisymbiotic *trans* effect.

We can now see that $(CH_3)_2Hg$ should indeed be quite unstable, since it contains two CH_3^- *trans* to each other. This does not require that the Hg—CH_3 coordinate bond be unusually weak so much as it requires equilibria such as

$$Hg(CH_3)_2 + Hg^{2+} \rightleftharpoons 2CH_3Hg^+$$

to lie well to the right.

The symbiotic effect will still be present in the *cis* position of square planar complexes. Symbiosis will be most effective in tetrahedral complexes, such as those of boron, and in organic compounds. Octahedral complexes will still exhibit symbiotic stabilization for *cis* ligands, but not for *trans*.

References

Chatt, J., and B. T. Heaton, *J. Chem. Soc. A*, 2745 (1968).

Jesson, J. P., in "Transition Metal Hydrides" (E. L. Muetterties, Ed.), Marcel Dekker, Inc. New York, 1971, chap. 4.

Krishnamurthy, R., W. B. Schaap, and J. R. Perumareddi, *Inorg. Chem.* **6**, 1338 (1967).

Langford, C. H., and H. B. Gray, "Ligand Substitution Processes," W. A. Benjamin, Inc., New York, 1965, p. 25.

Meier, M., Ph.D. Dissertation, ETH, Zurich, 1967.

Tobias, R. S., *Inorgan. Chem.* **9**, 512 (1970).

Wayland, B. B., and R. F. Schramm, *Inorg. Chem.* **8**, 971 (1969); **11**, 1280 (1972).

[Reprinted from the Journal of the American Chemical Society, **85**, 3533 (1963).]

5

PHYSICAL AND INORGANIC CHEMISTRY

[Contribution from the Department of Chemistry, Northwestern University, Evanston, Ill.]

Hard and Soft Acids and Bases

By Ralph G. Pearson

Received June 14, 1963

A number of Lewis acids of diverse types are classified as (a) or (b) following the criterion of Ahrland, Chatt, and Davies. Other, auxiliary criteria are proposed. Class (a) acids prefer to bind to "hard" or nonpolarizable bases. Class (b) acids prefer to bind to "soft" or polarizable bases. Since class (a) acids are themselves "hard" and since class (b) acids are "soft" a simple, useful rule is proposed: hard acids bind strongly to hard bases and soft acids bind strongly to soft bases. The explanations for such behavior include: (1) various degrees of ionic and covalent σ-binding; (2) π-bonding; (3) electron correlation phenomena; (4) solvation effects.

In a recent publication[1] the rate data for the generalized nucleophilic displacement reaction were reviewed and analyzed.

$$N + S-X \longrightarrow N-S + X \qquad (1)$$

Here N is a nucleophilic reagent (ligand, Lewis base) and S–X is a substrate containing a replaceable group X (also a base) and an electrophilic atom (Lewis acid) S. Other groups may also be bound to S. It was found that rates for certain substrates, S–X, were influenced chiefly by the basicity (toward the proton) of N, and other substrates had rates which depended chiefly on the polarizability (reducing power, degree of unsaturation) of N.

In this paper the equilibrium constants of eq. 1 will be considered, instead of the rates.

$$N \text{ (base)} + S-X \text{ (acid–base} \rightleftarrows N-S \text{ (acid–base} + X \text{ (base)}$$
$$\text{complex)} \qquad \text{complex)} \qquad (2)$$

Thus the relative strengths of a series of bases, N, will be compared for various acids, S. The reference base X will be constant for each comparison. In solution X may simply be the solvent, and in the gas phase X may be completely absent. Thus the discussion of equilibrium constants is concerned only with the stability of acid–base adduct N–S and the stability of the free (or solvated) base N. The nature of N–S may be that of a stable organic or inorganic molecule, a complex ion, or a charge transfer complex. In all cases it will be assumed that N is acting in part as an electron donor and S as an electron acceptor so that a coordinate, covalent bond between N and S is formed. Other types of interaction, sometimes stronger, sometimes weaker, may occur. These will be discussed later.

In terms of equilibria, rather than rates, it again turns out that various substrate acids fall into two categories: those that bind strongly to bases which bind strongly to the proton, that is, basic in the usual sense; those that bind strongly to highly polarizable or unsaturated bases, which often have negligible proton basicity. Division into these two categories is not absolute and intermediate cases occur, but the classification is reasonably sharp and appears to be quite useful. It will be convenient to divide bases into

two categories, those that are polarizable, or "soft," and those that are nonpolarizable, or "hard."[2] Now it is possible for a base to be both soft and strongly binding toward the proton, for example, sulfide ion. Still it will be true that hardness is associated with good proton binding. For example, for the bases in which the coordinating atom is from groups V, VI, and VII (the great majority of all bases), the atoms F, O, and N are the hardest in each group and also most basic to the proton. The reason for this has been discussed in reference 1. The atoms in each group become progressively softer with increasing atomic weight. They bind protons less effectively, but increase their ability to coordinate with certain other Lewis acids.

For the special case of metal ions as acids, Ahrland, Chatt, and Davies[3] made a very important and useful classification. All metal ions were divided into two classes depending on whether they formed their most stable complexes with the first ligand atom of each group, class (a), or whether they formed their most stable complexes with the second or a subsequent member of each group, class (b).[4] Thus the following sequences of complex ion stability are very often found

(a) $N \gg P > As > Sb > Bi$
(b) $N \ll P > As > Sb > Bi$
(a) $O \gg S > Se > Te$
(b) $O \ll S \sim Se \sim Te$
(a) $F \gg Cl > Br > I$
(b) $F < Cl < Br \ll I$

The classification is very consistent in that a metal ion of class (b) by its behavior to the halides, for example, will also be class (b) with respect to groups V and VI also.

Note that nothing is said concerning relative stabilities of group V ligands *vs.* group VI, for example, for a given metal ion. For a typical class (b) metal ion the order of decreasing stability of complexes for different ligand atoms is generally found to be $C \sim S >$

(1) J. O. Edwards and R. G. Pearson, *J. Am. Chem. Soc.*, **84**, 16 (1962)

(2) The descriptive adjectives "hard" and "soft" were suggested by Professor D. H. Busch of Ohio State University.

(3) S. Ahrland, J. Chatt, and N. R. Davies, *Quart. Rev.* (London), **12**, 265 (1958).

(4) The terms (a) and (b) appear to have no significance except that most class (b) metal ions belong to B subgroups of the periodic table Class (a) metal ions belong to both A and B subgroups.

3533

I > Br > Cl ∼ N > O > F, which is the same as that of increasing electronegativity and of increasing hardness. For a class (a) metal ion a strong, but not complete, inversion of this order occurs.[5] The inversion can be strong enough so that for some class (a) metal ions only O and F complexes can be obtained in aqueous solution. The failure to get complete inversion of the order is, as mentioned before, that some soft bases are still strong proton acceptors. The proton is, in fact, the most typical class (a) ion and other class (a) metals will bind strongly to ligands that are basic to the proton, whether they are hard or soft. Class (b) metal ions will bind to all soft bases whether they are good proton bases or not.

It seems clear that Chatt's (a) and (b) metal ions are the exact analogs of Edwards' and Pearson's substrates which are sensitive to proton basicity in the nucleophile (class (a)) and to polarizability in the nucleophile (class (b)).[1] Thus we could say that a substrate like a phosphate ester is a class (a) electrophilic reagent, or more properly that the phosphorus atom in the ester is a class (a) electrophilic center. The oxygen atom of peroxides is a class (b) electrophilic center. Nucleophiles also can be classified as hard (nonpolarizable) or soft (polarizable). Furthermore we can now examine equilibrium data for other Lewis acids than metal ions and classify them as (a) or (b) in type. Finally, the interesting question of why two contrasting kinds of behavior should exist will be examined.

Classification of Lewis Acids as Class (a) or (b).— Table I contains a listing of all generalized acids for which sufficient information could be found in the literature to enable a choice between class (a) or class (b) to be made. A few borderline cases are also given. The listing of reference 3 for the metal ions is left essentially unchanged. In classifying other Lewis acids, the criterion of Ahrland, Chatt, and Davies[3] was used whenever possible, that is, to compare the stabilities of F vs. I, O vs. S, and N vs. P type complexes. When this was not possible, two other criteria were used. One is that class (b) acids will complex readily with a variety of soft bases that are of negligible proton basicity. These include CO, olefins, aromatic hydrocarbons, and the like. The other auxiliary criterion is that if a given acid depends strongly on basicity and little on polarizability as far as *rates* of nucleophilic displacements are concerned, then it will depend even less on polarizability as far as equilibrium binding to bases is concerned. Such an acid will therefore be in class (a).

The justification of this rule comes partly from theory and partly from experimental facts. In a transition state there is an increased coordination number for displacement type reactions and an increased transfer of negative charge to the acid atom S in S–X. The theories to be described later all predict increased class (b) behavior for one or both of these reasons. For experimental proof we can cite data on tetrahedral carbon (see below) and metal ions such as platinum(II) or rhodium(III). For the metal ions, for example, hydroxide ion is a poor nucleophilic reagent but a strongly binding ligand at equilibrium.

Some other generalizations can serve both as precautionary remarks in the use of Table I, and as aids in the prediction of the class (a) or (b) character of new acids. As Chatt, *et al.*, observed[3] the class of a given element is not constant, but varies with oxidation state. A safe rule is that class (a) character increases with increasing positive oxidation state and *vice versa* for class (b) behavior. The acid class of a given element in a fixed oxidation state is also affected by the other groups attached to it, not counting the base to

which it coordinates. Groups which transfer negative charge to the central atom will increase the class (b) character of that atom since such transfer of charge is equivalent to a reduction of the oxidation state. The groups which most easily transfer negative charge will be the soft bases, particularly if negatively charged. Thus, hydride ion, which is highly polarizable,[6] alkide ions, and sulfide ion will be very effective.

For bases which are ions, there will be a strong solvent dependence for their strength of binding. This will be of different magnitudes for hard and soft ions and hence inversions in the binding order of the halide ions, for example, can occur with changes in solvent. Conclusions as to class (a) or (b) character using as references ionic bases are hence a function of the environment. This topic will be discussed in more detail later. Fortunately for neutral bases the nature of the solvent, or even its complete absence, seems to have little effect on class (a) or (b) behavior.

The data from which Table I is constructed are of diverse kinds, most being true equilibrium data, some being heat data only, and, in a few cases, merely observations that certain reactions occur easily or that certain compounds are stable. Accordingly, the evidence for the new examples not listed in reference 3 will be given in some detail.

TABLE I
CLASSIFICATION OF LEWIS ACIDS

Class (a) or hard	Class (b) or soft
H⁺, Li⁺, Na⁺, K⁺	Cu⁺, Ag⁺, Au⁺, Tl⁺, Hg⁺,
Be²⁺, Mg²⁺, Ca²⁺, Sr²⁺, Sn²⁺	Cs⁺
Al³⁺, Sc³⁺, Ga³⁺, In³⁺, La³⁺	Pd²⁺, Cd²⁺, Pt²⁺, Hg²⁺,
Cr³⁺, Co³⁺, Fe³⁺, As³⁺, Ir³⁺	CH₃Hg⁺
Si⁴⁺, Ti⁴⁺, Zr⁴⁺, Th⁴⁺, Pu⁴⁺,	Tl³⁺, Tl(CH₃)₃, BH₃
VO²⁺	RS⁺, RSe⁺, RTe⁺
UO₂²⁺, (CH₃)₂Sn²⁺	I⁺, Br⁺, HO⁺, RO⁺
BeMe₂, BF₃, BCl₃, B(OR)₃	I₂, Br₂, ICN, etc.
Al(CH₃)₃, Ga(CH₃)₃, In-	Trinitrobenzene, etc.
(CH₃)₃	Chloranil, quinones, etc.
RPO₂⁺, ROPO₂⁺, SO₃	Tetracyanoethylene, etc.
RSO₂⁺, ROSO₂⁺, SO₃	O, Cl, Br, I, R₃C(?)
I⁷⁺, I⁵⁺, Cl⁷⁺	M⁰ (metal atoms)
R₃C⁺, RCO⁺, CO₂, NC⁺	Bulk metals
HX (hydrogen bonding molecules)	

Borderline
Fe²⁺, Co²⁺, Ni²⁺, Cu²⁺,
Zn²⁺, Pb²⁺
B(CH₃)₃, SO₂, NO⁺

Use of Rate Data.—Assignments of class (a) character have been made on the basis of kinetic data for the acids RCO⁺ (carbonyl carbon as in esters, acyl halides), RPO₂⁺ and ROPR₂⁺ (tetrahedral phosphorus), RSO₂⁺ and ROSO₂⁺ (tetrahedral sulfur), I (V) and Cl (VII) (tetrahedral halogen), and Si⁴⁺. These substrates all react rapidly with strong bases and are little influenced by polarizability in the nucleophile.[7] Hence, as explained above, this strongly indicates that basicity will be the dominant factor to an even greater degree at equilibrium. This conclusion is certainly supported by the compounds formed by these acids in which oxygen atom donors are the most numerous by far.

Thermal data for SO₃, which is closely related to the sulfate and sulfonate esters, also support class (a) assignment. The heats of reaction, in the mixed liquids, are given as[8]

(5) G. Schwarzenbach, *Advan. Inorg. Chem. Radiochem.*, **3** (1961).

(6) M. G. Veselov and L. N. Labzovskii, *Vestnik Leningrad Univ.*, **15**, 5 (1960); *Chem. Abstr.*, **55**, 1671 (1961).

(7) Reference 3 and R. G. Pearson, D. N. Edgington, and F. Basolo, *J. Am. Chem. Soc.*, **84**, 3233 (1962).

(8) A. A. Woolf, *J. Inorg. Nucl. Chem.*, **14**, 21 (1960).

$$SO_2 + HF \longrightarrow HSO_2F \quad \Delta H = -20.9 \text{ kcal.}$$

$$SO_2 + HCl \longrightarrow HSO_2Cl \quad \Delta H = -6.0 \text{ kcal.}$$

$$SO_2 + HBr \longrightarrow \text{ no reaction}$$

Equilibria in Solution.—Ir^{3+} is put in class (a) on the basis of equilibrium data[9] for the hydrolysis of Ir-$(NH_3)_5X^{2+}$. The acid OH^+ is put in class (b) on the basis of the equilibria[10]

$$HOCl(aq) + Br^+(aq) \longrightarrow HOBr(aq) + Cl^-(aq)$$
$$\Delta G^\circ = -7.6 \text{ kcal.} \quad (3)$$

$$HOCl(aq) + I^-(aq) \longrightarrow HOI(aq) + Cl^-(aq)$$
$$\Delta G^\circ = -23.4 \text{ kcal.} \quad (4)$$

In a similar way the class (b) nature of the oxygen atom as a Lewis acid is shown by[10]

$$OCl^-(aq) + Br^-(aq) \longrightarrow OBr^-(aq) + Cl^-(aq)$$
$$\Delta G^\circ = -5.9 \text{ kcal.} \quad (5)$$

$$OCl^-(aq) + I^-(aq) \longrightarrow OI^-(aq) + Cl^-(aq)$$
$$\Delta G^\circ = -18.6 \text{ kcal.} \quad (6)$$

This is supported by bond energy data in the gas phase[11]

$$(n\text{-}C_4H_9)_3PO(g) \longrightarrow (n\text{-}C_4H_9)_3P(g) + O(g)$$
$$\Delta E = 138 \text{ kcal.} \quad (7)$$

$$Cl_3PO(g) \longrightarrow PCl_3(g) + O(g) \quad \Delta E = 119 \text{ kcal.} \quad (8)$$

$$NH_3O(g) \longrightarrow NH_3(g) + O(g) \quad \Delta E < 44 \text{ kcal.} \quad (9)$$

The last figure comes from the fact that the heat of formation of the unknown NH_3O must be more positive than that of its stable isomer NH_2OH. The greater strength of the PO bond compared to NO is also shown by the ease of reactions of the following type from a preparative standpoint.[12]

$$R_3NO + R'_3P \longrightarrow R_3N + R'_3PO \quad (10)$$

The expected class (b) nature of CH_3Hg^+ is shown by equilibrium studies in water.[13] In similar studies $(CH_3)_2Sn^{2+}$ is found to be class (a).[14] It is somewhat surprising to note that Lindquist[15] reports that the neutral molecules $SnCl_4$ and $SbCl_5$ form thio-adducts which are sometimes more stable and sometimes less stable than the corresponding oxo-adducts. It is expected that Sn^{4+} and even $SnCl_2^{2+}$ would be definitely class (a).

Class (b) behavior for RS^+, RSe^+, and RTe^+ is indicated by the extensive surveys of Parker and Kharasch and Pryor.[16] These studies are only semi-quantitative for the most part but they do indicate which bases are strong enough to cleave disulfides and related compounds in times long enough so that equilibrium may be assumed. The conclusion is that bases in which S or P is the active atom are much more effective than bases in which O or N donors are involved. The relatively slow reaction with hydroxide ion probably occurs chiefly because RSOH is unstable and disproportionates to RSO_3H and RSSR.

Parker has also discussed[17] the important case of basicity toward ordinary tetrahedral carbon compounds, that is, the acid strength of carbonium ions, R_3C^+.

The point was made above that polarizability is much less important in the equilibrium situation than for rates. Conversely, basicity toward the proton becomes of much greater importance. For anionic bases, strong solvent effects exist as expected. In water solution the rate data for the forward and reverse reactions[18] can be used to calculate the equilibrium constant for

$$CH_3I(aq) + F^-(aq) \underset{k_2}{\overset{k_1}{\rightleftarrows}} CH_3F(aq) + I^-(aq) \quad (11)$$

At 70° the value of K_{eq} is 5, which is class (a) behavior, but just barely. The heat of reaction is 2.0 kcal. endothermic so that entropy effects account for the stability of CH_3F compared to CH_3I in this solvent. Data from the same source[18] show that CH_3I is slightly more stable than CH_3Br, which is class (b) behavior.

Bunnett, et al.,[19] have equilibrium data on the reaction

$$RSH + OH^- \longrightarrow ROH + SH^- \quad (12)$$

which show $K_{eq} > 10^3$. Here R^+ is a rather complex tertiary carbonium ion and a mixed solvent, acetone–water, was used. Conversely, Miller[20] has examples of reactions in ethanol which show the exactly opposite behavior for CH_3^+.

$$CH_3SP + OP^- \longrightarrow CH_3OP + SP^- \quad K_{eq} < 10^{-4} \quad (13)$$

Here SP^- is a complex thiophosphoric ester. The over-all conclusion from the limited data in solution is that R_3C^+ is a borderline case between (a) and (b). However, some data in the gas phase to be given next show more clearly class (a) behavior.

Gas Phase Equilibria.—Heats of formation and heats of evaporation allow the ΔH or ΔE of a number of gas phase reactions of interest to be calculated. These are of the type

$$CH_3I(g) + F^-(g) \longrightarrow CH_3F(g) + I^-(g) \quad (14)$$

for example. In such reactions the changes in entropy will be small and ΔH can be used to discuss equilibrium constants. Table II gives data for a number of reactions of interest. It can be seen that the acids shown, and all Lewis acids, would be class (a) if judged by their affinity for halide ions in the gas phase.[21] Such a classification would be of little value, however, since the reactions are purely hypothetical.

As mentioned, the equilibria in aqueous solution are actually used as a basis. The great effect of water on reaction 14, which is 57 kcal. exothermic in the gas, is shown by returning to reaction 11 in water which is 2 kcal. endothermic. The difference is chiefly due to the heats of hydration of the ions. The negative hydration heats for the halide ions are 61, 74, 85, and 117 kcal./mole for I^-, Br^-, Cl^-, F^-, respectively.[22] The difference of 56 kcal. between I^- and F^- account almost exactly for the great change of the heat of reaction. The heats of hydration of neutral molecules are very much smaller, to begin with, and the difference between CH_3I and CH_3F would be even less, about 3 kcal. from the data.

The effect of water solvent is thus to lower the basicity of small (hard) anions with respect to related large (soft) anions. The same deactivation is found for OH^- compared to SH^-. For neutral bases, the influence of the solvent is small. The negative heats of hydration

(9) A. B. Lamb and L. T. Fairhall, *J. Am. Chem. Soc.*, **45**, 378 (1923).

(10) National Bureau of Standards Circular No. 500 "Selected Values of Chemical Thermodynamic Properties," 1952; I. E. Flis, K. P. Mishchenko, and N. V. Pakhomova, *Zh. Neorgan. Khim.*, **3**, 1781 (1958).

(11) T. L. Cottrell, "The Strengths of Chemical Bonds," Academic Press, Inc., New York, N. Y., 1954, pp. 192, 218, 253; C. L. Charwick and H. A. Skinner, *J. Chem. Soc.*, 1401 (1956).

(12) L. Horner and H. Hoffmann, *Angew. Chem.*, **68**, 480 (1956); E. Howard, Jr., and W. F. Olszewski, *J. Am. Chem. Soc.*, **81**, 1483 (1959).

(13) M. Schellenberg and G. Schwarzenbach, Proc. of Seventh Intern. Conf. on Coord. Chem., Stockholm, June, 1962, paper 4A6.

(14) M. Yasuda and R. S. Tobias, *Inorg. Chem.*, **2**, 207 (1963).

(15) I. Lindquist, "Inorganic Adduct Molecules of Oxo-Compounds," Academic Press, Inc., New York, N. Y., 1963, p. 108.

(16) A. J. Parker and N. Kharasch, *J. Am. Chem. Soc.*, **82**, 3071 (1960); A. J. Parker, *Acta Chem. Scand.*, **16**, 855 (1962); see also W. A. Pryor, "Mechanisms of Sulfur Reactions," McGraw-Hill Book Co., Inc., New York, N. Y., 1962, p. 60.

(17) A. J. Parker, *Proc. Chem. Soc.*, 371 (1962).

(18) R. H. Bathgate and E. A. Moelwyn-Hughes, *J. Chem. Soc.*, 3642 (1959).

(19) J. F. Bunnett, C. F. Hauser, and K. V. Nahabedian, *Proc. Chem. Soc.*, 305 (1961).

(20) B. Miller, *ibid.*, 303 (1962).

(21) F. Basolo and R. G. Pearson, "Mechanisms of Inorganic Reactions," John Wiley and Sons, Inc., New York, N. Y., 1958, p. 179.

(22) Reference 21, p. 67.

TABLE II
HEATS OF GAS PHASE REACTIONS

$$MX(g) + Y^-(g) \longrightarrow MY(g) + X^-(g)$$

Re-actants	Products	ΔH, kcal./mole[a]	Re-actants	Products	ΔH, kcal./mole[a]
HI/F$^-$	HF/I$^-$	-63	AlI$_3$/F$^-$	AlF$_3$/I$^-$	-72×3
LiI/F$^-$	LiF/I$^-$	-39	AsI$_3$/F$^-$	AsF$_3$/I$^-$	-68×3
CsI/F$^-$	CsF/I$^-$	-27	HI/Cl$^-$	HCl/I$^-$	-27
HgI/F$^-$	HgF/I$^-$	-12	LiI/Cl$^-$	LiCl/I$^-$	-18
I$_2$/F$^-$	IF/I$^-$	-23	AgI/Cl$^-$	AgCl/I$^-$	-5
IBr/F$^-$	BrF/I$^-$	-11	I$_2$/Cl$^-$	ICl/I$^-$	-2
CH$_3$I/F$^-$	CH$_3$F/I$^-$	-57	CH$_3$I/Cl$^-$	CH$_3$Cl/I$^-$	-25
CNI/F$^-$	CNF/I$^-$	-51	CH$_3$Br/F$^-$	CH$_3$F/Br$^-$	-45
NOI/F$^-$	NOF/I$^-$	-33	COBr$_2$/F$^-$	COF$_2$/Br$^-$	-61×2
HgI$_2$/F$^-$	HgF$_2$/I$^-$	-24×2			

[a] Thermodynamic data at 25° from Lewis and Randall, "Thermodynamics," 2nd Ed. revised by L. Brewer and K. S. Pitzer, McGraw-Hill Book Co., Inc., New York, N. Y., 1961, pp. 679–683; JANAF Interim Thermodynamic Tables, the Dow Chemical Co., ARPA Program, 1960; W. H. Evans, T. R. Munson, and D. D. Wagman, *J. Research Natl. Bur. Standards*, **55**, 147 (1955). The heats of formation used for the gaseous halide ions were -65.1, -58.8, -59.3, and -58.0 kcal./mole for F$^-$, Cl$^-$, Br$^-$, and I$^-$, respectively.

for H_2O, H_2S, and H_2Se and 9.8, 4.6, and 9.3 kcal./mole, for example.

Solvents other than water would give effects in the same direction, but less in magnitude as a rule. Hydrogen-bonding solvents (protonic) would be most like water. Aprotic solvents, especially if highly polarizable, will produce a much smaller differentiation between fluoride ion and iodide ion, for example. The heats of reaction would resemble those in water more than in the gas phase, from the general rule that any solvent is much better than none as far as ions are concerned.[23] In summary, solvents tend to bring out class (b) character for acids compared to the gas phase. Water does this more than other common solvents. The effect is far greater for anionic bases.

Table II shows large negative values of ΔH for reactions of the type shown for class (a) acids and smaller values for typical class (b) acids. This enables us to classify As^{3+}, NC$^+$, and R$_3$C$^+$ as class (a), I$^+$, Br$^+$, and Cs$^+$ as class (b), and NO$^+$ as intermediate.

Hydrogen Bonding.—Considering the hydrogen-bonding interaction as acid–base in nature

$$Y + HX \longrightarrow Y{-}HX$$

then the acids HX shows all the expected behavior for class (a). The interaction is strong when Y is F and not I, O and not S, N and not P. Some earlier confusion in the literature has been removed by a recent study[24] in which it was found that hydrogen bonding of neutral bases to a reference phenol decreased in strength in the order RF > RCl > RBr > RI and R$_2$O > R$_2$S > R$_2$Se. It is of interest to note that the theories of acid classes to be given in the Discussion state that the hydrogen bond is chiefly electrostatic in nature, if class (a) behavior is found.

Charge Transfer Complexes.—The typical charge transfer complex is formed as a result of an acid–base reaction.[25] The electron donor is a base and the acceptor is a Lewis acid. A number of acceptors are listed in Table I as class (b) acids. The list includes I$_2$, Br$_2$, ICN, tetracyanoethylene, trinitrobenzene, chloranil, quinone. Obviously a number of similar molecules could be added. There is a large body of equilibrium data for such acids.[26] The behavior is strongly

class (b) in type. Thioethers are preferred as bases to ethers, alkyl iodides to alkyl fluorides; aromatic hydrocarbons form quite stable complexes, as do other molecules with negligible proton basicities.

Halogen Atoms and Free Radicals.—A number of reports have appeared in the literature which indicate that free halogen atoms are stabilized by aromatic solvents and not by solvents containing oxygen or nitrogen atoms.[27] The data on recombination of iodine atoms requiring a third body is now interpreted[28] as involving the sequence

$$I + M \longrightarrow IM \tag{15}$$
$$IM + I \longrightarrow I_2 + M \tag{16}$$

The efficiency of M, the third body, increases with the critical temperature of M. The critical temperature depends largely on London forces which, in turn, depend on polarizability. Furthermore, aromatic molecules and alkyl iodides are unusually effective. It has been suggested[29] that the ready reaction of free radicals with sulfur- and phosphorus-bearing molecules involves complex formation prior to reaction. A complex between the isooctyl radical and methyl bromide appears to be formed.[29] All of this shows class (b) behavior certainly for electrophilic radicals such as Cl, Br, and I, and possibly for simple aliphatic radicals as well.

Metal Atoms and Metal Surfaces.—From the effect of changing the positive oxidation state of metals, it can be predicted that metal atoms at zero oxidation state will always be class (b) acids.[3] Complexes with neutral metal atoms typically contain the soft bases characteristic of class (b). These include CO, P, and As ligands, olefins, aromatics, isonitriles, and heterocyclic chelate amines.[30] It seems reasonable that metal atoms at the surface of a bulk metal would have the same properties. Certainly CO and unsaturated molecules are strongly adsorbed. The strong absorption of basic molecules on a metal surface is usually considered an electron donation process from the base to the metal, *i.e.*, an acid–base reaction.[31] Bases containing P, As, Sb, S, Se, and Te in low oxidation states are poisons in heterogeneous catalysis involving metallic catalysts. Strong oxygen- and nitrogen-containing bases are not poisons.[32] This agrees with the expectation if the metal is a class (b) acid since poisons are bases held so firmly that the active sites are blocked off to weaker bases, according to the usual view.

Discussion

The common features of the two classes of Lewis acids are easily discernible from Table I. The features which bring out class (a) behavior are small size and high positive oxidation state. Class (b) behavior is associated with a low or zero oxidation state and/or with large size. Both metals and nonmetals can be either (a) or (b) type acids depending on their charge and size. Since the features which promote class (a) behavior are those which lead to low polarizability, and those which create type (b) behavior lead to high polarizability, it is convenient to call class (a) acids "hard" acids and class (b) acids "soft" acids. We

(23) R. G. Pearson, *J. Chem. Phys.*, **20**, 1478 (1952).

(24) R. West, D. L. Powell, L. S. Whatley, M. K. T. Lee, and P. von R. Schleyer, *J. Am. Chem. Soc.*, **84**, 3221 (1962).

(25) See R. S. Mulliken: (a) *J. Phys. Chem.*, **56**, 801 (1952); (b) *J. Am. Chem. Soc.*, **74**, 811 (1952).

(26) See G. Briegleb, "Elektronen-Donator-Acceptor-Komplexe," Springer-Verlag, Berlin, 1961; J. D. McCullough and I. C. Zimmerman, *J. Phys. Chem.*, **65**, 888 (1961).

(27) G. A. Russell, *J. Am. Chem. Soc.*, **79**, 2977 (1957); S. J. Rand and R. I. Strong, *ibid.*, **82**, 5 (1960); T. A. Gover and G. Porter, *Proc. Roy. Soc.* (London), **A262**, 476 (1961).

(28) G. Porter and J. A. Smith, *ibid.*, **A261**, 28 (1961); M. I. Christie, *J. Am. Chem. Soc.*, **84**, 4066 (1962).

(29) M. Szwarc in "The Transition State," Special Publication No. 16, The Chemical Society, London, 1962, p. 103.

(30) J. Chatt, *J. Inorg. Nucl. Chem.*, **8**, 515 (1958).

(31) See R. L. Burwell, Jr., in "Actes au Deuxieme Congres International de Catalyze," Paris, 1960, p. 1005.

(32) G. C. Bond, "Catalysis by Metals," Academic Press, Inc., New York, N. Y., 1962.

then have the useful generalization that hard acids prefer to associate with hard bases, and soft acids prefer soft bases.

Polarizability is simply a convenient property to use as a classification. It may well be that other properties which are roughly proportional to polarizability are more responsible for the typical behavior of the two classes of acids. For example, a low ionization potential is usually linked to a large polarizability and a high ionization potential to a low polarizability. Hence ionization potential, or the related electronegativity, might be the important property. Unsaturation, with the possibility of acceptor π-bonding in the acid–base complex, and ease of reduction, favoring strong electron transfer to the acid, are also associated with high polarizability. For example, Edwards[33] has developed a fairly successful equation for predicting nucleophilic reactivity using proton basicity and oxidation–reduction potential as the two parameters on which reactivity depends. Later it was shown that a polarizability term could take the place of the oxidation–reduction term with about equal success.[34] To help in deciding which properties are of importance, it is necessary to examine the theories which have been advanced to account for the facts on which Table I is based. Different investigators, looking at different aspects, have come up with several explanations. These may be called: (1) the ionic–covalent theory; (2) the π-bonding theory; (3) electron correlation theory; (4) solvation theory.

To anticipate the analysis of these different views, there seems to be no reason to doubt that all of the factors involved in the above theories are of importance in explaining the behavior of acids. Different examples will depend more or less strongly on the several factors.

The Ionic–Covalent Theory.—This is the oldest and usually the most obvious explanation.[35] The class (a) acids are assumed to bind bases with primarily ionic forces and the class (b) acids hold bases by covalent bonds. High positive charge and small size would favor strong ionic bonding and bases of large negative charge and small size would be held most strongly. Mulliken[25] has developed a theory of covalent bonding suitable for discussing soft bases and soft acids. Bonding will be strong if the electron affinity of the acid is large and the ionization potential of the base is low.[36] Softness in both the acid and base means that the repulsive part of the potential energy curve rises less sharply than for hard acids and bases. Thus closer approach is possible and better overlap of the wave functions used in covalent bonding.

Mulliken's treatment is intended chiefly for charge transfer complexes which involve type (b) acids. It is not applicable to type (a) acids where, as we have seen, the base of *highest* ionization potential, for example F$^-$, is bound most strongly. In the theory of covalent bonding, it is generally considered necessary that both bonded atoms be of similar electronegativity to have strong covalent bonding.[37] That is, the coulomb integrals on both bonded atoms should be similar, and the sizes of the bonding atomic orbitals should be similar to get good overlap. These considerations show that hard acids will prefer hard bases even when considerable covalency exists. Soft bases will mismatch with hard acids for good covalency, and ionic bonding will also

be weak because of the small charge or large size of the base.

The π-Bonding Theory.—Chatt[38] has made important contributions to the theory of Lewis acids, applied chiefly to metallic complexes. The important feature of class (b) acids in his view is considered to be the presence of loosely held outer d-orbital electrons which can form π-bonds by donation to suitable ligands. Such ligands would be those in which empty d-orbitals are available on the basic atom, such as P, As, S, I. Also unsaturated ligands such as CO and isonitriles would also be able to accept metal electrons by the use of empty, but not too unstable, molecular orbitals. Class (a) acids would have tightly held outer electrons, but also there would be empty orbitals available, not too high in energy, on the metal ion. Basic atoms such as O and F particularly could form π-bonds in the opposite sense, by donating electrons from the ligand to the empty orbitals of the metal. With class (b) acids, there would be a repulsive interaction between the two sets of filled orbitals on metal and O and F ligands.

With some imagination, this model can be generalized to fit most of the entries in Table I. The soft acids are potential d- or p-electron donors *via* π-bonds. The hard acids are potential π-bond acceptors. Such effects are, of course, in addition to σ-bonding interactions. The hydrogen-bonding molecules and carbonium ions, R_3C^+, in class (a) do not seem to fit in with π-bonding ideas. At least one class (b) acid, Tl^{3+}, has such a high ionization potential that it is hard to imagine it as an electron donor.

Electron Correlation Effects.—Pitzer[39] has suggested that London, or van der Waals, dispersion forces between atoms or groups in the same molecule may lead to an appreciable stabilization of the molecule. Such London forces depend on the product of the polarizabilities of the interacting groups and vary inversely with the sixth power of the distance between them.[40] They are large when both groups are highly polarizable. Similarly, Bunnett[41] has noted that reaction rates are usually fast when a nucleophile of high polarizability reacts with a substrate carrying a highly polarizable substituent near the reaction site. This was attributed to stabilization of the transition state by London forces.

Even for bonded atoms it may be argued that the electron correlations responsible for London forces will operate for the nonbonding electrons.[39] It has been calculated that some 11 kcal. of the bromine–bromine bond energy may be due to London forces.[42] It then seems plausible to generalize and state that additional stability due to London forces will always exist in a complex formed between a polarizable acid and a polarizable base. In this way the affinity of soft acids for soft bases can be accounted for.

Mulliken[43] has given a different explanation for the extra stability of the bromine–bromine and iodine–iodine bonds. It is assumed that d$_\pi$–p$_\pi$ orbital hybridization occurs so that both the π_u bonding molecular orbitals and π_g antibonding orbitals contain some admixed d-character. This has the twofold effect of strengthening the bonding orbital by increasing overlap and weakening the antibonding orbital by decreasing

(33) J. O. Edwards, *J. Am. Chem. Soc.*, **76**, 1540 (1954).

(34) J. O. Edwards, *ibid.*, **78**, 1819 (1956).

(35) See A. A. Grinberg, "An Introduction to the Chemistry of Complex Compounds," translated by J. R. Leach, Pergamon Press, London, 1962, Chapter 7; G. Schwarzenbach, ref. 5; R. J. P. Williams, *Proc. Chem. Soc.*, 20 (1960).

(36) J. Weiss, *J. Chem. Soc.*, 245 (1942).

(37) C. A. Coulson, *Proc. Phil. Soc.*, **33**, 111 (1937).

(38) References 3 and 28; *Nature*, **165**, 859 (1950); **177**, 852 (1956); J. Chatt, L. A. Duncanson, and L. M. Venanzi, *J. Chem. Soc.*, 4456 (1955).

(39) K. S. Pitzer, *J. Chem. Phys.*, **23**, 1735 (1955); K. S. Pitzer and E. Catalano, *J. Am. Chem. Soc.*, **78**, 4844 (1956).

(40) J. C. Slater and J. G. Kirkwood, *Phys. Rev.*, **37**, 682 (1931).

(41) J. F. Bunnett, *J. Am. Chem. Soc.*, **79**, 5969 (1957); J. F. Bunnett and J. D. Reinheimer, *ibid.*, **84**, 3284 (1962).

(42) G. L. Caldow and C. A. Coulson, *Trans. Faraday Soc.*, **58**, 633 (1962).

(43) R. S. Mulliken, *J. Am. Chem. Soc.*, **77**, 884 (1955).

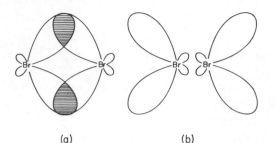

(a)   (b)

Fig. 1.—Atomic orbital hybrids for (a) bonding and (b) antibonding molecular orbitals.

overlap. Figure 1 shows schematically and exaggeratedly the d_π–p_π hybrids for both situations. Pearson and Edwards[1] proposed essentially the same mechanism to account for the high rate of reaction of polarizable nucleophiles with polarizable substrates.[44]

There is considerable similarity to the proposals of London forces and of orbital hybridization. They both represent electron correlation phenomena. The basic cause is different in the two cases, however. London correlation occurs because of the electrostatic repulsion of electrons for each other. The proposed π-orbital hybridization occurs largely because of nonbonded repulsion effects arising from the Pauli exclusion principle. It would appear that the latter would be more important for interactions between bonded atoms and the former for more remote interactions. Thus for the interaction of a soft acid and a soft base, orbital hybridization should usually be more important than stabilization due to van der Waals forces.

Mulliken's theory is the same as Chatt's π-bonding theory as far as the π_u bonding orbital is concerned. The new feature is the stabilization due to the π_g molecular orbital. As Mulliken points out[43] this effect can be more important than the more usual π-bonding. The reason is that the antibonding orbital is more antibonding than the bonding orbital is bonding, if overlap is included. For soft–soft systems, where there is considerable mutual penetration of charge clouds, this amelioration of repulsion due to the Pauli principle would be great. Unshared pairs of π-electrons would be affected more than electrons used for bonding purposes (compare I^- and $(C_2H_5)_3P$). An MO calculation, necessarily very approximate, gives some idea of the energies involved.

Consider the π-interactions of a system consisting of p_π and d_π atomic orbitals on an atom such as iodine and a p'_π atomic orbital on an atom such as oxygen. For simplicity let the p_π coulomb integrals be equal on both atoms, say q. The coulomb integral of the d_π orbital will be much nearer zero and may be set equal to zero for simplicity. The p_π–p'_π exchange integral will be β and assume the d_π–p'_π integral between the two atoms is also β, though it might be $\beta/2$, for example, without changing the argument. The d_π–p_π exchange for the same atom will be zero in the one-electron approximation.

The MO's found on solving the secular equation

$$\begin{array}{c|ccc} & p & p' & d \\ \hline p & q-E & \beta & 0 \\ p' & \beta & q-E & \beta \\ d & 0 & \beta & -E \end{array} = 0$$

are

$$\phi_1 = \frac{(\psi_p + \lambda\psi_d) + (1+\lambda)\psi_{p'}}{2+2\lambda} \tag{17}$$

$$\phi_2 = \frac{(\psi_p - \lambda\psi_d) - (1-\lambda)\psi_{p'}}{2-2\lambda} \tag{18}$$

$$\phi_3 = (\psi_d - \lambda\psi_{p'}) \tag{19}$$

The mixing parameter λ is equal to β/q which is small, say 0.10, and terms in λ^2 have been omitted; ϕ_1 and ϕ_2 resemble the bonding and antibonding orbitals, respectively, shown in Fig. 1. The corresponding energies, omitting terms in λ^2, are

$$E_1 = q + \beta + \beta^2/q \tag{20}$$

$$E_2 = q - \beta + \beta^2/q \tag{21}$$

$$E_3 = -2\beta^2/q \tag{22}$$

The net stabilization for four electrons, two in the ϕ_1, or π_u, orbital and two in the ϕ_2, or π_g, orbital will be $4\beta^2/q$.

If only the usual π-bonding had been considered, the net stabilization would have been equal to $2\beta^2/q$, so in this case the two effects are equal in magnitude. Putting in the overlap integral does not change the above calculations in the first approximation, and the uncertainties involved prevent a more detailed calculation. Mulliken has made some further estimates.[43]

It will be noted that the effect of π-bonding depends directly on the square of the exchange integral β^2 and inversely on the excitation energy q, between the stable p-orbital and the unstable d-orbital. This explains why the first row elements cannot benefit from such π-bonding, even though empty d- and p-orbitals exist also at an energy near zero.[45] It also explains why the lowest empty d-orbital is most suitable for π-bonding even though many other states of near equal energy exist. It is expected that the overlap, and hence β, would be best for this d-orbital.

Polarizability also depends inversely on the excitation energy, q, to the excited levels.[46] This phenomenon is not restricted to the first empty d-orbital but includes all excited states. It can be seen that, generally speaking, the metallic cations of class (b) are of high polarizability, not only because of large size, but also because of easily excited outer d-orbital electrons. While some class (b) acids, such as the oxygen atom, do not appear polarizable compared to some class (a) acids such as $Al(CH_3)_3$, one must bear in mind that the acid site in the latter case is really a modified Al^{3+} ion. Furthermore, the unshared electron pairs on the oxygen atom will benefit more from correlation effects than the shared pairs of electrons in the Al–C bonds.

The Solvation Theory.—Parker[47] in particular has stressed the effect of solvents on reducing the basicity of small anions and hence causing large anions to appear abnormally strong. The implication that such solvation is the common explanation for the strong binding, or high rates of reaction, of polarizable bases seems to be incorrect. As was pointed out above, differences in solvation energies between neutral molecules such as ROH and RSH are very much smaller than the differences for the corresponding anions. Furthermore, much of the data on which Table I is based comes from the gas phase or from nonpolar solvents. Also solvation effects alone would not cause a division into two distinct classes of acids as are found.

What solvation does do is to generally destroy class (a) character and enhance class (b) character. The

(44) Other explanations[1] for such high reactivity in terms of polarization of the σ-electrons seem to be equivalent to the covalent bonding discussed above.

(45) W. Klemperer, J. Am. Chem. Soc., **83**, 3910 (1961); A. F. Saturno and J. F. Eastham, ibid., **84**, 1313 (1962).

(46) See H. Eyring, G. E. Kimball, and I. Walter, "Quantum Chemistry," John Wiley and Sons, Inc., New York, N. Y., 1944, p. 121.

(47) A. J. Parker, J. Chem. Soc., 1328 (1961); Quart. Rev. (London), **14**, 163 (1962); J. Miller and A. J. Parker, J. Am. Chem. Soc., **83**, 117 (1961).

magnitude of the class (a) character in the gas phase will determine if a solvent can cause inversion (compare Table II).[48] It is clear then that solvation effects, while of great importance, particularly for ions, do not explain why some acids prefer hard bases and some acids prefer soft bases. The explanation for this must come from interactions existing in the acid–base complex. Such interactions include ionic–covalent σ-bonding, π-bonding, and electron correlation effects, all of which seem to play a role in determining class (a) and (b), or hard and soft, character.

Some Consequences of Hard–Soft Classification of Acids and Bases.—The simple rule that hard acids prefer to bind hard bases and soft acids prefer soft bases permits a useful systematization of a large amount of chemical information. Some illustrations of the usefulness of the rule will now be given.

Stabilization of Metal–Metal Bonds.—It was shown above that zero-valent metals and bulk metals are class (b) acids. Metals can also act as Lewis bases, since donation of electrons is more characteristic of them than the acceptance of electrons. They will only be good bases in a zero or low valent state. Thus they will be soft bases. To form a compound with a stable metal to metal bond, then, requires that *both* metal atoms be in a low or zero oxidation state. Typical examples would be Hg_2^{2+} and $Mn_2(CO)_{10}$. To stabilize the metal–metal bond other ligands attached to the metal should be typical soft bases such as CO, R_3P, and alkide ions, R^-.[49] This has the dual effect of stabilizing each metal atom in the low valent condition and of increasing softness by increasing the electron density on the metal atoms.

Metal–metal bonding appears in an interesting way in heterogeneous catalysis. Metal ions of class (b) are poisons for metal surfaces, while metal ions of class (a) are not.[32] This can be explained by the softness of the metal considered as a base. The softness of the metal as an acid (Table I) explains why phosphines, sulfides, etc., are poisons.

Classification of Solvents as Hard or Soft.—Solute–solvent interactions may often be considered as acid–base interactions of varying degree. Considering solvents as acids, HF, H_2O, and hydroxylic solvents will be hard solvents. They will strongly solvate hard bases such as F^-, OH^-, and other oxygen anions. A variety of dipolar, aprotic solvents such as dimethyl sulfoxide, sulfolane, dimethylformamide, nitroparaffins, and acetone will be soft acid solvents. These solvents will have a mild preference for solvating large anions which function as soft bases.[50] The class (a), or hard,

solvents will tend to level basicity while class (b) or soft, solvents will not. Hence the high reactivity of OH^- and OR^- noted in class (b) solvents.[50]

Solvents can be classified as hard or soft by virtue of their basic properties as well, and this will influence their interaction with cations. Even neutral solutes will be affected to a lesser degree. The obvious rule is that hard solvents dissolve hard solutes well and soft solvents dissolve soft solutes well.[51]

Stabilization of Valence States and Ligands.—It is well known in coordination chemistry that ligands of large size, low charge, and low electronegativity are good for stabilizing metal ions in low valence states.[52] For metal ions in high positive oxidation states, the fluoride ion and oxide ion are the best stabilizing groups.[52] Obviously these are examples of preferential soft–soft and hard–hard interactions. A somewhat less obvious corollary has to do with the preparation of certain classes of compounds containing unstable ligands such as H^- and R^-.

It has been found by Chatt[53] that such complexes for transition metals are stabilized by the presence of typical soft ligands. Chatt has emphasized the high ligand field strength of such ligands as a factor in their stabilizing ability. Equally important is the concept that such ligands keep the metal in the class (b) condition necessary for it to combine effectively with the highly polarizable hydride and alkide ions.

Formation of Unexpected Complexes.—The common use of water, a hard solvent, and of H^+, a hard acid, as a reference for basicity, justify the statement that soft acids form unexpected complexes, particularly in aqueous solutions. Since soft bases often do not bind the proton at all in water (H_3O^+ being formed instead), the fact that they are bases is often forgotten. The complexes formed with suitable soft acids, even in aqueous solutions, are then considered rather abnormal. Examples would be I_3^-, I_2SCN^-, $Ag(C_2H_4)^+$, $PtCl_3$-$C_2H_4^{2-}$, and charge-transfer complexes in general. It is true that increasing familiarity with such complexes, and increasing sophistication, cause them to appear less surprising now than a few years ago. It is probable that the entire area of molecule–molecule and ion–molecule interactions can be examined with profit as examples of generalized acid–base phenomena. If so, the concept of hard and soft acids and bases should prove useful.

Acknowledgment.—This work was supported in part by the U. S. Atomic Energy Commission under Contract At(11-1)-1087.

(48) A. J. Poë and M. S. Vaidya, *J. Chem. Soc.*, 1023 (1961).

(49) Compare R. S. Nyholm, E. Coffey, and J. Lewis, paper 1H3, Proc. of Seventh Conf. on Coord. Chem., Stockholm, June, 1962. Other factors influencing metal–metal bonds are discussed in this paper and in R. S. Nyholm, *Proc. Chem. Soc.*, 273 (1961).

(50) (a) A. J. Parker, *Quart. Rev.* (London), **16**, 163 (1962); (b) see also N. Kornblum, *et al.*, *J. Am. Chem. Soc.*, **85**, 1141, 1148 (1963).

(51) Reference 50a gives a number of examples.

(52) Reference 28 and W. Klemm, *J. Inorg. Nucl. Chem.*, **8**, 532 (1958); also the succeeding papers by various authors.

(53) J. Chatt, Tilden Lecture, *Proc. Chem. Soc.*, 318 (1962); see also G. Calvin, G. E. Coates, and P. S. Dixon, *Chem. Ind.* (London), 1628 (1959).

6

"Symbiotic" Ligands, Hard and Soft Central Atoms

Sir:

Pearson[1] recently published a fascinating paper entitled "Hard and Soft Acids and Bases." However, it is not certain that the central atoms in definite oxidation numbers can be arranged according to a single parameter, going from extreme Chatt–Ahrland (a) "hard acid" to (b) "soft acid." Thus, the (b) class seems to involve three rather disparate categories[2]: unusually low oxidation numbers; certain high oxidation numbers; and the s^2-family Sn(II), Sb(III), Tl(I), Pb(II), and Bi(III) showing (b) characteristics toward heavy halides and chalcogenides but definite (a) aversion against σ-bonded cyanides and amines. Many chemists would not admit that an element with highly varying oxidation number z, say manganese, would show (b) character as well for high z (+4 or +7) as for low z (−1 or +1), the maximum (a) tendencies occurring at an intermediate value of z, here +2. The M(a)–NCS and M(b)–SCN bonds formed by the ambidentate ligand thiocyanate are frequently used as a test case,[3] and it would not be surprising if Fe(II) was found to bind N and Fe(III) S, showing a maximum (a) tendency for z = +2. By the same token, the molybdenum(III) complex[4] $Mo(NCS)_6^{3-}$ is (a) and the dark red Mo(V) thiocyanates possibly belong to class (b). The question really is whether the covalent bonding becomes so much stronger for increasing z-values that the actual, fractional charge of the central atom rather decreases, promoting (b) behavior again. There is some evidence from the nephelauxetic effect[5] that this phenomenon occurs.

Closely connected with Pearson's ideas is the question why soft (b) ligands flock together in the same complexes, a true inorganic symbiosis. Bjerrum[6] emphasized that mixed complexes MX_aY_b of moderately different pairs of ligands, say water and ammonia, or fluoride and water, usually tend to be even more frequent than suggested by the statistical contribution to the higher entropy. Most exceptions can be explained away by high-spin–low-spin variations, making the intermediate complexes less favored, though, as Kida[7] pointed out, certain facts about cyanides are difficult to understand. The symbiosis is clear in cobalt(III) complexes, where $Co(NH_3)_5X^{2+}$ is far better bound for X = F than I, showing (a) characteristics, whereas $Co(CN)_5X^{3-}$ is most stable with X = I and not even known for X = F. The very soft base H^- also forms $Co(CN)_5H^{3-}$. We may remember Pauling's effect that, under equal circumstances, the characteristic coordination number of (b) ligands tends to be lower than of (a) ligands (*cf.* $CoCl_4^{2-}$ and $Co(H_2O)_6^{2+}$). This is not so much a question of larger atomic size as of an appropriate amount of electronic density being donated from a lower number of soft ligands.

The author and his colleague, Dr. H.-H. Schmidtke, attempt now to elucidate the symbiotic tendency by various MO calculations. Since the optical electronegativity[8] of the central atom nd shell increases monotonically with z, the empty $(n + 1)s$ orbital may be more important. The nonmonotonic behavior as a function of z might be connected with the empty $(n + 1)s$ orbital producing strong characteristics of a soft acid for low z, whereas the somewhat less pronounced (b) properties for very high z are caused by the near coincidence of the energies of a partly filled nd shell with the filled orbitals of the ligands. In the intermediate z range, covalent bonding is less conspicuous because both $(n + 1)s$ and nd have much higher energy than the ligand orbitals.

(1) R. G. Pearson, *J. Am. Chem. Soc.*, **85**, 3533 (1963).
(2) C. K. Jørgensen, "Inorganic Complexes," Academic Press, London, 1963.
(3) A. Wojcicki and M. F. Farona, *Inorg. Chem.*, **3**, 151 (1964).
(4) J. Lewis, R. S. Nyholm, and P. W. Smith, *J. Chem. Soc.*, 4590 (1961).
(5) C. K. Jørgensen, *Progr. Inorg. Chem.*, **4**, 73 (1962).
(6) J. Bjerrum, *Chem. Rev.*, **46**, 381 (1950).
(7) S. Kida, *Bull. Chem. Soc. Japan*, **34**, 962 (1961).
(8) C. K. Jørgensen, "Orbitals in Atoms and Molecules," Academic Press, London, 1962.

Cyanamid European Research Institute
Cologny (Geneva), Switzerland

C. Klixbull Jørgensen

Received March 23, 1964

Reprinted from *Science*, **151**, 172–177 (1966)
Copyright 1966 by the American Association for the Advancement of Science

7

Acids and Bases

Hard acids prefer to associate with hard bases,
and soft acids prefer to associate with soft bases.

Ralph G. Pearson

The most important of all classes of chemical reactions is the generalized acid-base reaction (*1*):

$$A + :B \rightleftharpoons A:B \qquad (1)$$

A is a Lewis acid, or electron acceptor, and : B is a Lewis base, or electron donor; A : B is the complex formed between them by partial donation of electrons from : B to A. Examples of such complexes include coordination compounds and complex ions in which A is a metal atom or ion, most ordinary inorganic and organic molecules, charge-transfer complexes, hydrogen-bonded complexes, and complexes between free radicals (which act as acids) and various bases. When A is a metal ion, the base B is called a ligand. When the rates of reaction 1 are being discussed, A is called an electrophilic reagent and B is called a nucleophilic reagent.

Indeed one can see that very much of chemistry is included under the heading of acid-base interactions. Any

The author is professor of chemistry at Northwestern University, Evanston, Illinois.

generalizations that can be made about the equilibrium constants for reaction 1, or the stability of the acid-base complex, A : B, will have wide applicability. The special case where A is a metal ion has been extensively studied, and many equilibrium constants for reaction 1 are known (*2*). Actually what is usually known is the equilibrium constants for the competition reaction

$$A:B' + A':B \rightleftharpoons A:B + A':B' \qquad (2)$$

where A′ and : B′ are the common reference acid and base, H_2O.

Several earlier workers, especially Fajans (*3*) and J. Bjerrum (*4*), had noted that the metal ions fall into two categories according to the kinds of bases they prefer to coordinate with. Schwarzenbach (*5*) divided the metal ions into two classes, A and B. The most typical metal ions of class A were those of the representative elements having no *d*-orbital electrons. The class B metal ions had 8 to 10 outer *d* electrons, occurring near the end of a transition series.

The overall order of stability of

class-B metal ions for various bases falls in the approximate sequence

$$S \sim C > I > Br > Cl > N > O > F$$

where the atom shown is the donor atom of the base. For class A metal ions this order is strongly inverted. Hence stable complexes in water solution can only be formed with oxygen donors and F− in many cases. For class A ions the stability of the complexes increases with increasing positive charge: $Al^{3+} > Mg^{2+} > Na^+$. For class B ions, the reverse is true, at least for the best donor atoms in the series. $Ag^+ > Cd^{2+} > Au^{3+} > Sn^{4+}$.

Chatt, Ahrland, and Davies made a very useful advance (*6*) when they classified metal ions according to whether they form their most stable complexes with the first ligand atom of each group, class (a), or with the second or a subsequent member of each group, class (b). The following sequences of complex-ion stability are then found:

(a) N >> P > As > Sb > Bi
(b) N << P > As > Sb > Bi
(a) O >> S > Se > Te
(b) O << S ~ Se ~ Te
(a) F > Cl > Br > I
(b) F < Cl < Br < I

Chatt, Ahrland, and Davies' class (a) metal ions are the same as Schwarzenbach's class A, and their class (b) metal ions are the same as his class B. To avoid confusion with symbols used for Lewis acid and Lewis base, I use (a) and (b) from here on.

The rules of Ahrland, Chatt, and Davies can also be used to classify other kinds of generalized Lewis acids (*7*). Where the necessary equilibrium data are not available, other criteria may be used. One is that class (b)

1

acids complex readily with a variety of bases of negligible proton basicity. These include carbon monoxide, olefins, and aromatic hydrocarbons. Rate data may also be used. Table 1 shows a variety of Lewis acids for which the available data allow an assignment of class (a) or class (b) behavior. Some borderline cases are also shown.

Hard and Soft Acids and Bases

In order to better characterize the behavior of class (a) and class (b) acids, let us define a soft base as one in which the donor atom is of high polarizability and of low electronegativity and is easily oxidized or is associated with empty, low-lying orbitals (8). These terms are not independent, for they describe in different ways a base in which the donor electrons are not held tightly but are easily distorted or removed. They are not exactly equivalent, however. Hard bases have the opposite properties. The donor atom is of low polarizability and high electronegativity, is hard to reduce, and is associated with empty orbitals of high energy and hence inaccessible.

Table 2 lists a variety of bases in order of increasing softness, at least as judged by a criterion of reaction rate with a typical class (b) acid, $Pt(C_5H_5N)_2Cl_2$. Other hard bases are oxygen donors such as acetate, sulfate, and phosphate ions. Other soft bases are carbon monoxide, aromatic hydrocarbons, olefins, alkyl isocyanides, alkide ions, and hydride ions.

We now see that class (a) acids prefer to bind to hard bases (O over S, N over P, F over I) and class (b) acids prefer to bind to the softer, more polarizable bases. The latter acids also form complexes with a variety of soft bases that class (a) acids generally ignore, at least in aqueous solution.

If we examine the properties of class (a) acids, we find that they can be either metals or nonmetals, as can class (b) acids. The distinguishing features of class (a) are small size, high positive oxidation state, and the absence of any outer electrons which are easily excited to higher states. These are all properties which lead to low polarizability, and we may call such acids hard acids. Class (b) acids have one or more of the following properties: low or zero positive charge, large size, and several easily excited outer electrons. For metals these outer electrons are d-orbital electrons. All of these properties lead to high polarizability, and class (b) acids may be called soft acids.

We can now state a useful, general principle: *Hard acids prefer to associate with hard bases and soft acids prefer to associate with soft bases.*

This rule must not be taken to mean more than it says. For example, it certainly does not say that soft acids do not ever complex with hard bases, or that hard acids do not form stable complexes with any soft bases. Some hard bases such as OH^- form stable complexes in water with most positively charged acids, hard or soft. Since H^+ is the prototype hard acid, any base which binds strongly to the proton will bind to other hard acids as well. In the case of soft bases, these will be negatively charged bases such as H^-, R^-, and S^{2-}.

Perhaps the best way to illustrate what this general principle does tell us is by an example (9). Let us compare CH_3Hg^+, a typical soft acid, and H^+, a hard acid. Both form stable complexes with OH^-, a hard base, and with S^{2-}, a soft base. However, the stability constants are such that the competition reaction

$$H^+ + CH_3HgOH \rightleftharpoons H_2O + CH_3Hg^+ \quad (3)$$

has an equilibrium constant of $10^{6.3}$. The competition reaction

$$H^+ + CH_3HgS^- \rightleftharpoons HS^- + CH_3Hg^+ \quad (4)$$

on the other hand, has an equilibrium constant of $10^{-8.4}$. The preferences of the proton for the hard base and of CH_3Hg^+ for the soft base are dramatically demonstrated.

The hardness of an acid is a function of the oxidation state of the acceptor atom, usually increasing as this number becomes more positive. Also

the hardness of a given acceptor atom is a function of the other groups attached to it. Thus, BF_3 is a class (a) acid, but BH_3 is a typical class (b) acid, forming complexes such as BH_3CO with the soft base carbon monoxide. The group $Co(NH_3)_5{}^{3+}$ shows class (a) behavior, but $Co(CN)_5{}^{2-}$ shows class (b) behavior. In these cases the boron atom and the cobalt atom have a formal oxidation state of $3+$. The rule that is obeyed is that soft bases, when coordinated to an acid, tend to make it softer. The effect is one of reducing the positive charge on the acceptor atom. Thus, the actual positive charge on the boron atom is probably much less in BH_3 than it is in BF_3.

The above rule is the basis for what Jørgensen calls symbiotic behavior (10). For a given metal ion, or other acid center, hard ligands will tend to flock together or soft ligands will flock together. For instance, stable $Co(NH_3)_5F^{2+}$ and $Co(CN)_5I^{3-}$, and unstable $Co(NH_3)_5I^{2+}$ and $Co(CN)_5F^{3-}$, illustrate the symbiotic principle. In organic chemistry we find that compounds with several oxygen or fluorine atoms attached to the same saturated carbon atom are unusually stable. Hine has explained this by double bond–no bond resonance (11).

$$O-C-O \leftrightarrow \overset{+}{O}=C\ O^- \quad (5)$$

This explanation is a combination of two of the theories of hard and soft behaviors discussed below.

Applications of the Principle

The rule that soft acids prefer soft bases and hard acids prefer hard bases is only qualitative, or, at best, semiquantitative. However, it does permit

Table 1. Classification of Lewis acids.*

Hard [class (a)]	Soft [class (b)]
H^+, Li^+, Na^+, K^+	Cu^+, Ag^+, Au^+, Tl^+, Hg^+, Cs^+
Be^{2+}, Mg^{2+}, Ca^{2+}, Sr^{2+}, Mn^{2+}	Pd^{2+}, Cd^{2+}, Pt^{2+}, Hg^{2+}, CH_3Hg^+
Al^{3+}, Sc^{3+}, Ga^{3+}, In^{3+}, La^{3+}	Tl^{3+}, Au^{3+}, Te^{4+}, Pt^{4+}
Cr^{3+}, Co^{3+}, Fe^{3+}, As^{3+}, Ce^{3+}	$Tl(CH_3)_3$, BH_3, $CO(CN)_5{}^{2-}$
Si^{4+}, Ti^{4+}, Zr^{4+}, Th^{4+}, Pu^{4+}	RS^+, RSe^+, RTe^+
Ce^{4+}, Ge^{4+}, VO^{2+}	I^+, Br^+, HO^+, RO^+
$UO_2{}^{2+}$, $(CH_3)_2Sn^{2+}$	I_2, Br_2, ICN, etc.
$BeMe_2$, BF_3, BCl_3, $B(OR)_3$	Trinitrobenzene, etc.
$Al(CH_3)_3$, $Ga(CH_3)_3$, $In(CH_3)_3$, AlH_3	Chloranil, quinones, etc.
$RPO_2{}^+$, $ROPO_2{}^+$	Tetracyanoethylene, etc.
$ROS_2{}^+$, $ROSO_2{}^+$, SO_3	O, Cl, Br, I, N
I^{7+}, I^{5+}, Cl^{7+}, Cr^{6+}, Se^{6+}	M^n (metal atoms)
RCO^+, CO_2, NC^+	Bulk metals
HX (hydrogen bonding molecules)	

* The following are in a borderline class between (a) and (b): Fe^{2+}, Co^{2+}, Ni^{2+}, Cu^{2+}, Zn^{2+}, Pb^{2+}, Sn^{2+}, Sb^{3+}, Bi^{3+}, Rh^{3+}, Ir^{3+}, $B(CH_3)_3$, SO_2, NO^+, Ru^{2+}, Os^{2+}, R_3C^+.

2

correlation and better understanding of a very large amount of chemical information.

One application is the explanation of the stabilities of various compounds and complexes. For example, the acylium ion, RCO^+, is listed as a hard acid in Table 1. This means that CH_3COF is more stable than CH_3COI, and that CH_3COOR is more stable than CH_3COSR. But RS^+, the sulfenyl group, is listed as a soft acid. This means that RSI is more stable than RSF, and RSSR is more stable than RSOR.

The case of carbonium ions, R_3C^+, is of special interest. It is put in the borderline category in Table 1. For example, CH_3F is slightly more stable than CH_3I, but CH_3SCH_3 is more stable than CH_3OCH_3 by a factor of 10^3. We may expect that various substituents on tetrahedral carbon will cause changes so that a carbonium ion may be either class (a) or class (b). Hine has recently discussed the subject of carbon basicity in terms of the equilibrium constant for the reaction

$$R_3COH + B^- \rightleftharpoons R_3CB + OH^- \quad (6)$$

for a number of carbonium and acylium ions (11).

Hydrogen-bonded complexes are class (a), so that stronger bonds are formed to N, O, and F donors than to P, S, and I donors. Charge-transfer complexes between acids such as I_2, $(NC)_2C=C(CN)_2$, quinones, and various bases show that the acids are in class (b). More stable complexes are formed to thio ethers than to ethers, for example. Aromatic molecules are good bases for these acids.

We can also consider the stabilization of a given element in a certain oxidation state. If we wish to have a low, or zero, oxidation state for a metal, it is necessary to surround it with soft bases, or soft ligands. This follows because the metal would be a soft acid if zero-valent. Suitable ligands would be carbon monoxide, phosphines, isocyanides, and the like. An element in a high oxidation state, Fe(V) or Cr(VI), would be stabilized best by hard ligands such as F^- or O^{2-}.

In a similar way, we may wish to prepare organic compounds of a metal which ordinarily does not form stable alkyl derivatives, such as one of the transition metals. The reasoning would be that R^- is a soft base; therefore, the metal should be in a low valence

state, or soft condition. This would require stabilizing it with a number of other soft ligands. Therefore, we expect $CH_3Mn(CO)_5$, but not $CH_3Mn(H_2O)_5$, to be stable.

The formation of compounds containing metal-metal bonds is of great current interest (12). Some help in understanding what systems will be stable comes from the new principle. A metal atom can be a Lewis acid, as already indicated. It can also be a Lewis base, since a metal is defined chemically as an electron donor. Hence, we may regard a metal-metal bond formally as an example of an acid-base complex,

$$L-M + :M'-L \rightarrow L-M:M'-L \quad (7)$$

where L stands for other ligands attached to the metal atom.

Since a metal atom must be in a low oxidation state to be a good electron donor, we see that M' should be a soft metal atom. From our rule, M must also be a soft metal atom, to act as a soft acid. Accordingly, M and M' should be of zero or low valence, and the groups L should be soft bases to stabilize the metals in this condition. Most of the known metal-metal bonds satisfy these conditions (13):

$$Mn_2(CO)_{10}, Fe_2(CO)_8^{2-}, [MoC_5H_5(CO)_3]_2$$
$$Pt(NH_3)_4PtCl_4, Ph_3PAuCo(CO)_4,$$
$$Pt(SnCl_3)_5^{3-}$$

A few examples are known of a hard metal forming a metal-metal bond (14). In these cases the second metal, which acts as the base, is also relatively hard.

The idea that a metal atom in the zero oxidation state is both a soft acid and a soft base can be used to explain surface reactions of bulk metals. Soft bases such as carbon monoxide and olefins are strongly adsorbed on surfaces of the transition metals. Bases containing P, As, Sb, Se, and Te in low oxidation states are the typical poisons in heterogeneous catalysis by metals. These soft bases are strongly adsorbed, blocking off the active sites. Strong bases containing oxygen and nitrogen are not poisons. Also metal ions of class (b) are poisons, whereas metal ions of class (a) are not. This shows the soft-base character of the free metal. Heterogeneous catalysis on metals is generally viewed today as the formation of unstable organometallic compounds and hydrides on the metal surface (15).

There is much evidence that free radicals and atoms behave as Lewis acids in that they form complexes with bases prior to reaction 7. The most stable complexes are formed with molecules of high polarizability, or soft bases. The reactivity of chlorine atoms, for example, shows that aromatic solvents stabilize them so they are less reactive, but oxygen and nitrogen donors do not. This is the expected behavior if chlorine is a soft acid. A recent study (16) shows that nitrogen atoms attack sulfur atoms in the molecules S_2Cl_2, H_2S, CS_2, COS, S_8 and SCl_2, and do not react with SO_2, $SOCl_2$, and SO_3. Thus the nitrogen atom acts as a soft Lewis acid, attacking only soft basic atoms.

Solubility may often be discussed in terms of acid-base interactions between solvent and solute molecules. Each solvent can be classified as hard or soft, though a separate classification for its acid function and its basic function may be necessary. A useful rule is that hard solutes dissolve well in hard solvents and soft solutes dissolve well in soft solvents (17). This is a rephrasing of the old adage *similia similibus dissoluntur*.

Water is a hard solvent in both its acid and its basic functions. It will solvate strongly such bases as OH^-, F^-, and other oxygen anions. Dipolar, aprotic solvents such as dimethyl sulfoxide, sulfolane, dimethylformamide, nitroparaffins, and acetone will be much softer. Hard solvents will level basicity while soft solvents will not. For example, CH_3O^- is some 10^9 times more reactive in dimethylsulfoxide than in methanol as a basic catalyst for a proton removal reaction (18). Class (a) characteristics will always be partly destroyed, and class (b) characteristics enhanced, by hard solvents, in comparison with soft solvents, and by any solvent, in comparison with the gas phase (7).

In precipitating salts, chemists have long known that a large cation is precipitated best by a large anion, while a small cation needs a small anion. This is, of course, a lattice energy effect, but it also is an example of the softness-hardness principle. Consider the solid-state reaction

$$LiI + CsF \rightarrow LiF + CsI \quad (8)$$

which is exothermic to the extent of 33 kcal/mole. Thus, the hard Li^+ prefers the hard F^- and the much softer Cs^+ is left with the soft I^-.

3

The standard method for increasing solubility in hydrocarbon solvents, or lipid solubility, is to add long alkyl chains to the molecule. It is also possible to decrease water solubility and increase solubility in soft solvents by replacing oxygen atoms in the solute molecule with softer sulfur or selenium atoms (19).

Application to Kinetic Behavior

Instead of equilibrium data for acid-base reactions, we may look at rate data. These will usually be in the form of second-order rate constants for the nucleophilic displacement reaction

$$A:B' + B \xrightarrow{k} A:B + B' \quad (9)$$

though a certain amount of information on the electrophilic displacement reaction is also available.

$$A:B + A' \xrightarrow{k} A':B + A \quad (10)$$

If the vast amount of data on reaction 9 is analyzed (20), it is found that for some substrates, A : B′, the rates are sensitive chiefly to the ordinary proton basicity of the nucleophile B. Other substrates are sensitive chiefly to the polarizability of B. The properties of the acid site (or electrophilic center) of A : B′ determine which type of behavior is found. If the properties of the acid center are those that make it a hard acid, then basicity is the dominant factor. If the acid site is a soft center, then polarizability is the important factor in the rates.

Accordingly, we may call the P(V) atom of a phosphate ester a hard electrophilic center. Displacements on P(V) will be fast for OH⁻ and F⁻ and other good bases toward the proton. A Pt(II) complex will have a soft electrophilic center. Fast substitution reactions will occur with phosphines, olefins, and iodide ion (see Table 2). Hard reagents will be ineffective.

Experimentally it turns out to be true that polarizability is always more important for rates than for equilibria. Any central atom which is known from rate data to be a hard electrophile will certainly be a hard acid by equilibrium standards. A borderline hard acid, such as the methyl carbonium ion, will become a rather soft electrophilic center. Thus, for nucleophilic displacements on tetrahedral carbon, polarizability and basicity of the nucleophile will both be

Table 2. Softness parameter of some bases.[*]

Base	Constant[†]
H_2O	0
OH⁻, OCH₃⁻, F⁻	<1
Cl⁻	1.65
NH_3	1.67
C_5H_5N	1.74
NO_2^-	1.83
N_3^-	2.19
NH_2OH	2.46
H_2N-NH_2	2.47
C_6H_5SH	2.75
Br⁻	2.79
I⁻	4.03
SCN⁻	4.26
SO_3^{2-}	4.40
$(C_6H_5)_3Sb$	5.26[‡]
$(C_6H_5)_3As$	5.36[‡]
SeCN⁻	5.71
$C_6H_5S^-$	5.78
$S=C(NH_2)_2$	5.78
$S_2O_3^{2-}$	5.95
$(C_6H_5)_3P$	7.51[‡]

* For another scale, see 29.   † From log k for rates of reaction with trans-Pt $(C_5H_5N)_2Cl_2$ (see 30).   ‡ See 31.

important, with polarizability the dominant factor.

For the displacement reactions

$$CH_3I + RO^- \rightarrow CH_3OR + I^- \quad (11)$$

$$CH_3I + RS^- \rightarrow CH_3SR + I^- \quad (12)$$

the rate constant for reaction 12 will exceed that of reaction 11 by a factor of 100. For displacement reactions on $(R'O)_2POX$, OR⁻ will be a better nucleophile than SR⁻ by a factor of about 30 (21).

$$\underset{(SR^-)}{(R'O)_2P-X} + OR^- \rightarrow \underset{(SR)}{(R'O)_2P-OR} + X^- \quad (13)$$

The rates of proton transfer reaction usually follow the Brønsted law (22).

$$HA + B \xrightarrow{k} A^- + BH^+ \quad (14)$$

The rates increase with increasing (ordinary) acid strength of HA and increasing base strength of B. Linear relationships are found between log k and pK_a or pK_b. However, if either B or A⁻ is a soft base, then lower rate constants are found than are predicted from the Brønsted relation (23). Proton transfers to and from soft bases are abnormally slow (provided the equilibrium constant of reaction 14 is not overwhelmingly large). An extreme example is the slow protonation of carbanions.

$$R_3C^- + HA \rightarrow R_3CH + A^- \quad (15)$$

Saville (24) has used the softness-hardness principle as a guide to select-

ing electrophilic catalysts for nucleophilic displacement reactions.

$$B: + A:B' + A' \rightarrow B:A + A':B' \quad (16)$$

The rules are simply that if B′ is a hard base, then a hard acid A′ is used as a catalyst; if B′ is a soft base, then A′ should be a soft acid. The nucleophile B should also match the characteristic of the electrophilic center A. For example,

$$\underset{}{F^- + (RO)_2 \overset{\overset{O}{\|}}{P} SEt + Ag^+ \rightarrow (RO)_2 \overset{\overset{O}{\|}}{P} F + AgSEt} \quad (17)$$

$$I^- + H_2O_2 + H^+ \rightarrow IOH + H_2O \quad (18)$$

The best possibilities for catalysis occur when the substrate, A : B′, has a mismatch of a hard acid with a soft base, or vice versa. In reaction 17 the hard base F⁻ is used to attack the hard phosphorus atom. Catalysis is provided by the soft Ag⁺, which helps pull off the soft SEt⁻ group. In reaction 18 we have a nucleophilic attack by the soft I⁻ on the soft electrophilic oxygen atom of OH⁺. Catalysis is brought about by coordinating the hard hydrogen ion to the hard nucleophilic oxygen of OH⁻. Note that the same atom, in this case the oxygen of H_2O_2, can simultaneously be a soft acid and a hard base center. Electron donation puts greater stress on polarizability than does electron acceptance.

The position of attack by an ambient nucleophile can usually be predicted by the use of the softness-hardness principle (21). An ambient ion has two possible donor atoms, such as nitrogen or oxygen in NO_2^-. Usually one donor atom is softer than the other—for example, sulfur in NCS⁻ or in $(RO)_2POS$, and nitrogen in NO_2^-. In these cases the hardness or softness of the electrophile determines the point of attack. A hard center like Fe(III) in $Fe(H_2O)_6^{3+}$ is attacked by nitrogen to form $Fe(H_2O)_5NCS^{2+}$, whereas the softer center Co(III) in $Co(CN)_5H_2O^{2-}$ forms $Co(CN)_5SCN^{3-}$.

Theory Underlying
Hard and Soft Behavior

What are the reasons for the preferential hard-hard and soft-soft acid-base interactions? Actually, no one factor seems universally responsible, and several different theories have been proposed by investigators looking at different aspects of acid-base behavior. It should be emphasized that the answer must lie in interactions occurring in the acid-base complex itself. Solvation effects, while important for ionic acids and bases, will not cause a separation into class (a) and class (b), each with its characteristic behavior.

The oldest and most obvious explanation may be called the ionic-covalent theory. It goes back to the ideas of Grimm and Sommerfeld for explaining the differences in properties of AgI and NaCl. Hard acids are assumed to bind bases primarily by ionic forces. High positive charge and small size would favor such ionic bonding. Bases of large negative charge and small size would be held most tightly—for example, OH^- and F^-. Soft acids bind bases primarily by covalent bonds. For good covalent bonding, the two bonded atoms should be of similar size and similar electronegativity. For many soft acids ionic bonding would be weak or nonexistent because of the low charge or the absence of charge. It should be pointed out that a very hard center, such as I(VII) in periodate or Mn (VII) in MnO_4^-, will certainly have much covalent character in its bonds, so that the actual charge is reduced much below +7. Nevertheless, there will be a strong residual polarity.

The π-bonding theory of Chatt (25) seems particularly appropriate for metal ions, but it can be applied to many of the other entries in Table 1 as well. According to Chatt the important feature of class (b) acids is considered to be the presence of loosely held outer d-orbital electrons which can form π-bonds by donation to suitable ligands. Such ligands would be those in which empty d orbitals are available on the basic atom, such as phosphorus, arsenic, sulfur, or iodine. Also, unsaturated ligands such as carbon monoxide and isonitriles would be able to accept metal electrons by means of empty, but not too unstable, molecular orbitals. Class (a) acids would have tightly held outer electrons, but also there would be empty orbitals available, not too high in energy, on the metal ion. Basic atoms such as oxygen and fluorine in particular could form π-bonds in the opposite sense, by donating electrons from the ligand to the empty orbitals of the metal. With class (b) acids, there would be a repulsive interaction between the two sets of filled orbitals on metal and oxygen and fluorine ligands. Figure 1 shows schematically a p orbital on the

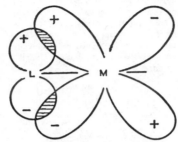

Fig. 1. A p atomic orbital on a ligand atom and a d orbital on a metal atom suitable for π-bonding. The d orbital is filled and the p orbital is empty for a soft acid–soft base combination. The d orbital is empty and the p orbital is filled for a hard acid–hard base combination. The plus and minus signs refer to the mathematical sign of the orbital.

ligand and a d orbital on the metal which are suitable for forming π-bonds.

The importance of the d electrons in metals for determining class (a) or class (b) behavior is very marked. In fact no class (b) metal ion containing less than five d electrons is known. A neutral metal atom, such as potassium or calcium, while soft in some respects, will still not show many of the typical reactions of class (b) metals, such as formation of normal carbonyl or olefin complexes. A decrease in the shielding of the d shell by removal of outer electrons will sometimes enhance (b) character (26). Thus, Tl(III) is softer than Tl(I) in spite of its greater positive charge. Also Sn(IV) and As(V) seem to show more (b) behavior than Sn(II) or As(III).

Occasionally metal ions show soft behavior even when they are of high positive charge—for example, Pt(IV) in Table 1. This behavior seems paradoxical. However, an explanation may be offered for these cases in terms of the actual charge on the metal ion rather than the formal charge (10). Thus, there is evidence that platinum in PtI_6^{2-} has a charge near zero, and certainly not plus four. It is characteristic of cases where a high oxidation state leads to soft behavior that all the ligands are soft. Covalent bonding will then cause a large transfer of charge from ligands to metal and extensive charge neutralization on the metal ion. Thus, we have the complexes $Mo(NCS)_6^{3-}$, with Mo(III) acting as a hard acid, and $Mo(SCN)_6^-$, with Mo(V) acting as a soft acid.

In the first case we have metal-nitrogen bonding and in the second case, metal-sulfur bonding.

Pitzer has suggested (27) that London, or van der Waals, disperson forces between atoms or groups in the same molecule may lead to an appreciable stabilization of the molecule. Such London forces depend on the product of the polarizabilities of the interacting groups and vary inversely with the sixth power of the distance between them. These forces are large when both groups are highly polarizable. It seems plausible to generalize and state that additional stability due to London forces will always exist in a complex formed between a polarizable acid and a polarizable base. In this way the affinity of soft acids for soft bases can be partly accounted for.

Mulliken has given a different explanation for the extra stability of the bonds between large atoms—for example, two iodine atoms (28). It is assumed that d- p-orbital hybridization occurs, so that both the π-bonding molecular orbitals and the π^*-antibonding orbitals contain some admixed d-character. This has the twofold effect of strengthening the bonding orbital by increasing overlap and weakening the antibonding orbital by decreasing overlap.

Mulliken's theory is the same as Chatt's π-bonding theory as far as the π-bonding orbital is concerned. The new feature is the stabilization due to the antibonding molecular orbital. As Mulliken points out, this effect can be more important than the more usual π-bonding. The reason is that the antibonding orbital is more antibonding than the bonding orbital is bonding, if overlap is included. For soft-soft systems, where there is considerable mutual penetration of charge clouds, this amelioration of repulsion due to the Pauli principle would be great.

One final point in connection with theory is illustrated by reaction 8. Clearly it is the strong binding in LiF, compared to all other factors, which drives the equilibrium to the right. Since most acid-base reactions are actually double exchanges, or competitions (see reaction 2), it is the strongest bonding, A : B, which dominates. The weakest bonding, A' : B', may simply be dragged along as a necessary consequence. The strongest bonding will usually be between the hard acid and the hard base.

5

65

Biological Applications
of the Principle

One wonders, naturally, what the applications of the concept of hard and soft acids and bases may be in biochemistry and biology. To the extent that biochemistry can be considered to involve simple chemical reactions, many of the applications discussed above can be taken over directly. Unfortunately, most reactions of biological systems are characterized not only by complexity but also by specificity. This means that it is unlikely that general rules will be of much value.

Nevertheless, a few generalizations can be drawn. If the lists of hard and soft acids and bases of Tables 1 and 2 are examined, it is seen that hard acids and bases are usually the normal, abundant components of biological systems. Thus, an organism will tolerate most hard acids or bases, unless some specific reaction occurs. One may say, *"La vie est dure."*

Contrariwise, most soft acids and bases are poisons to living organisms. While specific effects are no doubt common, the general effect must be poisoning by the formation of complexes with the soft bases and acids that are present, in small amounts, in the organism—the heavier metals and sulfide groups, for example. It is of interest to find that the same substances that are poisons in heterogeneous catalysis are poisons for living things.

References and Notes

1. Introductory material on the generalized acid-base concept will be found in W. Luder and S. Zuffanti, *The Electronic Theory of Acids and Bases* (Wiley, New York, 1946); R. P. Bell, *Acids and Bases* (Methuen, New York, 1952); C. A. Vanderwerf, *Acids, Bases and the Chemistry of the Covalent Bond* (Reinhold, New York, 1961).
2. See *Stability Constants*, L. G. Sillen and A. E. Martell, Eds. (Chemical Society, London, ed. 2, 1964).
3. K. Fajans and G. Joos, *Z. Physik* **23**, 1 (1924); K. Fajans, *Naturwissenschaften* **11**, 165 (1923).
4. J. Bjerrum, *Chem. Rev.* **46**, 381 (1950).
5. G. Schwarzenbach, *Experentia Suppl.* **5**, 162 (1956); *Advan. Inorg. Chem. Radiochem.* **3** (1961).
6. S. Ahrland, J. Chatt, N. R. Davies, *Quart. Rev. London* **12**, 265 (1958).
7. R. G. Pearson, *J. Am. Chem. Soc.* **85**, 3533 (1963).
8. Unsaturation always produces such orbitals. They are the π^* antibonding orbitals.
9. G. Schwarzenbach, paper presented at a symposium on SHAB, Cyanamid European Research Institute, Geneva, May 1965.
10. C. R. Jørgensen, *Inorg. Chem.* **3**, 1201 (1964).
11. J. Hine and R. D. Weimar, *J. Am. Chem. Soc.* **87**, 3387 (1965).
12. For reviews see D. P. Craig and R. S. Nyholm, in *Chelating Agents and Metal Chelates*, F. P. Dwyer and D. P. Mellor, Eds. (Academic Press, New York, 1964), and J. Lewis, *Plenary Lectures, 8th International Conference on Coordination Chemistry* (Butterworths, London, 1965).
13. Cases where metal atoms are held together by other atoms, but where short metal-metal distances suggest some bonding, must be considered separately. An example would be the unit $Mo_6Cl_8^{4+}$.
14. D. F. Shriver, *J. Am. Chem. Soc.* **85**, 3509 (1963).
15. See R. L. Burwell and J. B. Peri, *Ann. Rev. Phys. Chem.* **15**, 131 (1964); W. O. Ord, F. J. McQuillin, P. L. Simpson, *J. Chem. Soc.* **1963**, 5996 (1963).
16. J. L. Smith and W. L. Jolly, *Inorg. Chem.* **4**, 1006 (1965).
17. See A. J. Parker, *Quart. Rev. London* **16**, 163 (1962).
18. D. J. Cram, B. Rickborn, C. A. Kingsbury, P. Haberfield, *J. Am. Chem. Soc.* **83**, 3678 (1961).
19. H. G. Mautner and E. M. Clayton, *ibid.* **81**, 6270 (1959).
20. J. O. Edwards and R. G. Pearson, *ibid.* **84**, 16 (1962).
21. R. F. Hudson, paper presented at a symposium on SHAB, Cynamid European Research Institute, Geneva, May 1965.
22. See R. P. Bell, *The Proton in Chemistry* (Cornell Univ. Press, Ithaca, N.Y., 1959), chap. 10.
23. M. Eigen, *Angew. Chem. Intern. Ed. Engl.* **3**, 1 (1964).
24. B. Saville, paper presented at a symposium on SHAB, Cyanamid European Research Institutes, Geneva, May 1965.
25. J. Chatt, *Nature* **177**, 852 (1956); *J. Inorg. Nucl. Chem.* **8**, 515 (1958).
26. S. Ahrland, paper presented at a symposium on SHAB, Cyanamid European Research Institute, Geneva, May 1965.
27. K. S. Pitzer, *J. Chem. Phys.* **23**, 1735 (1955).
28. R. S. Mulliken, *J. Am. Chem. Soc.* **77**, 884 (1955).
29. R. S. Drago and B. B. Wayland [*J. Am. Chem. Soc.* **87**, 3571 (1965)] have another set of parameters which characterize bases. The ratio C_B/B_B from their paper may be used as a measure of softness.
30. U. Bellucco, L. Cattalini, F. Basolo, R. G. Pearson, A. Turco, *J. Am. Chem. Soc.* **87**, 241 (1965).
31. H. Sobel, unpublished results.

Reprinted from *Chemistry in Britain*, **3**, 103–107 (1967)
Copyright 1967 by the Royal Institute of Chemistry and the Chemical Society, London

Hard and Soft Acids and Bases

Ralph G Pearson

The ultimate goal of science is to make order out of chaos, that is, to find the underlying coherent patterns which exist in the bewildering multitude of events which occur in nature. One of the most important areas in which order is sought for is that of chemical reactions. The chemist is aware of a very large number of chemical processes of apparently diverse types. Much progress has been made in systematizing chemical reactions but much still remains to be done.

Professor Pearson of the Department of Chemistry, Northwestern University, Evanston, Illinois, is the leading exponent of the concept of 'hard and soft acids and bases'—the subject of a symposium this month at Northern Polytechnic, London (Chem. in Britain, *1966*, **2**, 318).

THE NEW FIELDS of physical-organic chemistry, starting during the thirties, and physical-inorganic chemistry, starting some dozen years later, have concentrated on understanding reaction mechanisms, the details of how chemical reactions occur. In this way it has been possible to break down complicated chemical reactions into a series of simpler ones called elementary processes. An elementary process is a reaction undergone by a single molecule (a unimolecular process), two molecules (a bimolecular process) or, rarely, by three molecules (a termolecular process).

Viewed in terms of the elementary processes, almost all chemical reactions can be considered to be in one of two main categories: generalized acid–base reactions and oxidation–reduction reactions. The meaning of the term generalized acid–base reactions follows from the definitions of acids and bases due to G. N. Lewis in 1923.[1] The definitions emphasize the ability of a base to partially donate a pair of electrons to form a coordinate, or dative, covalent bond, and the ability of an acid to partially accept a pair of electrons from the base. A base is then an atom, molecule or ion which has at least one pair of valence electrons which is not already being shared in a covalent bond. An acid is similarly a unit in which at least one atom has a vacant orbital in which a pair of electrons can be accommodated. The typical acid–base reaction is then

$$A + :B \longrightarrow A:B \quad .. \quad .. \quad (1)$$

The species $A:B$ may be called a *coordination compound*, an adduct or an *acid–base complex*. A wide variety of acid–base complexes exists under different names.

Since the proton is an acid as defined by Lewis, there is no real difference between a base according to Brønsted and a base according to Lewis. However, an acid as defined by Brønsted will be an acid–base complex according to Lewis. It has been suggested that Lewis acids be called electron acceptors and Lewis bases be called electron donors, to preserve the concept that acids are related to the hydrogen ion. The current tendency seems to be to call proton acids, Brønsted acids or just acids, and to call the wider class of acids, Lewis acids or generalized acids.

Acid-base reactions

Probably the most important of all class of chemical reaction is the generalized acid–base reaction of equation (1). The easiest way to appreciate this is to consider the different kinds of acid–base complexes, $A:B$, that may be formed. For example, all metal atoms or ions are Lewis acids. They are usually found coordinated to several Lewis bases simultaneously since they are polyvalent. The bases in this special case are called ligands. The combination may be electrically charged, in which case we have a complex ion formed. An example is $FeCl_4^-$. The combination may be electrically neutral, in which case a normal inorganic molecule such as $SnCl_4$ is formed.

All cations are Lewis acids and all anions are bases, hence all salts are automatically acid–base complexes. $MgCl_2$ in the solid state consists of the acids Mg^{2+} coordinated to six neighbouring Cl^- ligands. In solution the magnesium ion is coordinated to the base water forming the solvated ion, $Mg(H_2O)_6^{2+}$. The chloride ion also forms an acid–base complex with water, $Cl(H_2O)_n^-$ and in this case the number of water molecules, n, is not known but is probably four.

Most inorganic compounds, as solids, liquids, gases or in solution, are examples of acid–base complexes and the same thing can be said for organic compounds. One can

103

mentally dissect the organic molecule into two fragments, one of which is a Lewis acid and the other a base. For example, ethyl alcohol can be thought of as composed of the ethyl carbonium ion, $C_2H_5^+$, and the hydroxide ion OH^-. All carbonium ions are Lewis acids and the hydroxide ion, of course, is a base.

Similarly, ethyl acetate can be thought of as a complex of the acylium ion, CH_3CO^+, which is an acid, and the base ethoxide ion, $C_2H_5O^-$. Even a hydrocarbon can be thought of as an acid such as H^+ and a base carbanion. Thus methane can be considered to be H^+ and CH_3^- combined. It should be noted that a certain group of atoms is often designated as an acid or base even if it has no stable existence. A carbonium ion such as CH_3^+ is considered to be Lewis acid because its Lewis structure

$$H$$
$$H:C^+$$
$$\overset{..}{H}$$

shows that it can accept a pair of electrons from a base.

Other kinds of acid–base complexes are the co-called charge transfer complexes which are responsible for the colours produced when many substances are mixed. An example is iodine and the intense brown colour it gives in solvents which are bases, such as water or alcohols. Finally, free atoms and radicals act as Lewis acids and form complexes with a variety of bases. These complexes of free radicals cannot be isolated but they have a very great effect on the reactivity of the radicals.

Hard and soft

It is apparent that it is possible, in principle at least, to view the greater part of chemistry as examples of interaction of generalized acids and bases. This in turn means that any rules that can be developed concerning the stability of complexes $A:B$ in reaction (1) will have very wide application and can be useful in many areas. Recently a rule has been suggested, 'The Principle of Hard and Soft Acids and Bases', which does seem to have value in understanding a wide variety of chemical phenomena.[2]

Let us define a soft base as one in which the valence electrons are easily distorted—polarized—or removed. A hard base has the opposite properties, holding on to its valence electrons much more tightly. We also define a hard acid as one of small size, high positive charge and with no valence electrons that are easily distorted or removed. Strictly speaking it is the acceptor atom of the acid which has those properties. A soft acid is one in which the acceptor atom is of large size, small or zero positive charge, or has several valence electrons which are easily distorted or removed.

A general principle can now be stated: Hard acids prefer to coordinate with hard bases and soft acids prefer to coordinate with soft bases.

Perhaps the first to notice an example of the hard and soft principle was Berzelius who commented that some metals occurred as ores which were oxides or carbonates

and other metals occurred as sulphide ores. The metal ions which are hard acids such as magnesium, aluminium and calcium, occurred as oxides and carbonates, which are hard bases. Soft metal ions are commonly found as their sulphides, the sulphide ion being a soft base. Examples are copper, mercury and lead.

Table I shows a number of representative Lewis acids which are classified as being either hard or soft. Table II shows a number of bases classified in the same way. Of course, in both cases there is a continuous range of softness or hardness, so that intermediate cases exist as well. Unfortunately, at the present time there is no quantitative scale of softness that can be written down. Several physical properties such as polarizability, electronegativity and oxidation potential are obviously related to softness but none of them is suitable as an exact measure.

Let us look at the problem of the stability of compounds $A:B$ in a somewhat different way, in terms of the acid strength or base strength of A and B. To define strength in a simple way, consider the acid–base substitution reactions

$$A' + A:B \longrightarrow A':B + A \qquad .. \quad (2)$$

$$B' + A:B \longrightarrow A:B' + B \qquad .. \quad (3)$$

TABLE I

CLASSIFICATION OF LEWIS ACIDS

Hard	Soft
H^+, Li^+, Na^+, K^+	Cu^+, Ag^+, Au^+, Tl^+, Hg^+
Be^{2+}, Mg^{2+}, Ca^{2+}, Sr^{2+}, Mn^{2+}	Pd^{2+}, Cd^{2+}, Pt^{2+}, Hg^{2+},
	CH_3Hg^+, $Co(CN)_5^{2-}$, Pt^{4+},
	Te^{4+}
Al^{3+}, Sc^{3+}, Ga^{3+}, In^{3+}, La^{3+}	Tl^{3+}, $Tl(CH_3)_3$, BH_3, $Ga(CH_3)_3$,
N^{3+}, Gd^{3+}, Lu^{3+}	$GaCl_3$, GaI_3, $InCl_3$
Cr^{3+}, Co^{3+}, Fe^{3+}, As^{3+}	RS^+, RSe^+, RTe^+
Si^{4+}, Ti^{4+}, Zr^{4+}, Th^{4+}, U^{4+}	I^+, Br^+, HO^+, RO^+
Pu^{4+}, Ce^{3+}, Hf^{4+}	
UO_2^{2+}, $(CH_3)_2Sn^{2+}$, VO^{2+},	I_2, Br_2, ICN, etc.
MoO^{3+}	
$BeMe_2$, BF_3, $B(OR)_3$	Trinitrobenzene, etc.
$Al(CH_3)_3$, $AlCl_3$, AlH_3	Chloranil, quinones, etc.
RPO_2^+, $ROPO_2^+$	Tetracyanoethylene, etc.
RSO_2^+, $ROSO_2^+$, SO_3	O, Cl, Br, I, N
I^{7+}, I^{5+}, Cl^{7+}, Cr^{6+}	M^0 (metal atoms)
RCO^+, CO_2, NC^+	Bulk metals
HX (hydrogen bonding molecules)	CH_2, carbenes

Borderline
Fe^{2+}, Co^{2+}, Ni^{2+}, Cu^{2+}, Zn^{2+}, Pb^{2+}, Sn^{2+}, Sb^{3+}, Bi^{3+}, Rh^{3+}, Ir^{3+}, $B(CH_3)_3$, SO_2, NO^+, Ru^{2+}, Os^{2+}, R_3C^+, $C_6H_5^+$, GaH_3

TABLE II

CLASSIFICATION OF BASES

The symbol R stands for an alkyl group such as CH_3 or C_2H_5

Hard	Soft
H_2O, OH^-, F^-	R_2S, RSH, RS^-
$CH_3CO_2^-$, PO_4^{3-}, SO_4^{2-}	I^-, SCN^-, $S_2O_3^{2-}$
Cl^-, CO_3^{2-}, ClO_4^-, NO_3^-	R_3P, R_3As, $(RO)_3P$
ROH, RO^-, R_2O	CN^-, RNC, CO
NH_3, RNH_2, N_2H_4	C_2H_4, C_6H_6
	H^-, R^-

Borderline
$C_6H_5NH_2$, C_5H_5N, N_3^-, Br^-, NO_2^-, SO_3^{2-}

104

68

If the reactions go as indicated, it means that A' is a stronger Lewis acid than A, and that B' is a stronger base than B. If it were possible to put all Lewis acids into an order of decreasing strength, and the same for all bases, then it would be possible to predict the stabilities of all possible acid–base complexes. That is, we could predict what chemical reactions would occur under various conditions, what compounds would be stable, *etc.*

We would then expect the equilibrium constant of equation (1) to be given by

$$\log K = S_A S_B \quad \quad .. \quad \quad .. \quad (4)$$

where S_A is a strength factor for the acid and S_B is a strength factor for the base. It would then only be necessary to determine S_A and S_B values in a given environment once for all Lewis acids and bases.

Unfortunately, it is not possible to write down any universal order of acid or base strength. If a series of different bases, B', are used in testing against a fixed reference $A:B$, then the order of base strength that one gets is very much a function of the nature of the reference acid, A. When A is the chromic ion, Cr^{3+}, one gets a different order from the case when A is the ferrous ion, Fe^{2+}. In fact, the order is different for ferric ion and for ferrous ion so that a change in the oxidation state of the reference acid can have an effect on relative base strengths.

As might be expected, the order of base strengths depends very much on whether A is a hard or soft acid centre. If A is soft then, as Schwarzenbach has pointed out,[3] the order of decreasing base strength for various donor atoms is roughly

$$S > C \sim P > I > Br > Cl > N > O > F$$

For hard acids, A, the order is strongly inverted so that the most stable bonds are formed when the donor atom of the base B is O, F or N. Only weak bonds are formed to most S, P and I donors. The orders are not invariant but depend on each individual case to some degree.

Nevertheless, it can be seen that the soft acids coordinate best to soft donor atoms, *i.e.* to soft bases. The hard acids coordinate best to hard donor atoms such as we have in OH^-, F^- and NH_3. Chatt, Ahrland and Davies pointed out a more reliable rule[4]: hard acids* coordinate best to the lightest atom of a family of elements in the periodic table. Soft acids* coordinate best to one of the heavier atoms of the same family. Thus we have the order of stabilities of complexes $A:B$:

hard A	soft A
$N \gg P > As > Sb > Bi$	$N \ll P > As > Sb > Bi$
$O \geqslant S > Se > Te$	$O \ll S \sim Se \sim Te$
$F > Cl > Br > I$	$F < Cl < Br < I$

It can be seen that hard acids bind best to the least polarizable (hardest) atom of a family, whilst soft acids bind to a more polarizable (softer) atom of the family of

* Many people prefer the terminology, class (a) acceptor instead of hard acid and class (b) acceptor instead of soft acid. This terminology, while having historical precedence,[3,4] does not emphasize the properties which make for class (a) or (b) behaviour.

elements. However, soft acids do not form their most stable compounds with the most polarizable (softest) atom. The reason is that bases of some of these very soft atoms, such as the stibines R_3Sb, are very weak bases towards all acids. What the Principle of Hard and Soft Acids does say in such a case is that the complexes of R_3Sb with soft acids will generally be more stable than those with hard acids.

The implication is that equation (4) would be modified to include a term which shows a special affinity of certain acids and bases for each other. One possibility would be

$$\log K = S_A S_B + \sigma_A \sigma_B \quad .. \quad \quad .. \quad (5)$$

Here σ_A is a softness factor for the acid and σ_B a softness factor for the base. For hard acids and bases, the σ values might be negative, so that their product is positive. For soft acids and bases, both σ values would be positive.

Equation (5) allows for the fact that a very strong acid such as the proton, even though hard, might form a more stable complex with a soft base, such as R_3Sb, than does a soft, but weak, acid such as a quinone. Also we can see that a quite stable compound could be formed between a strong acid and a strong base even if there is a mismatch between their σ values.

Applications of the principle

The principle at present is largely qualitative in nature. Nevertheless, it can be used to help remember and to explain a very large. number of apparently unrelated phenomena. It also has considerable predictive powers. For example, the acid RS^+ is a soft acid, since the positive charge is not large and the sulphur atom, which is the acceptor atom, has two pairs of valence electrons which are not tightly held. The fact that sulphenyl iodide, RSI, is stable and that sulphenyl fluoride, RSF, is unstable can be readily understood since the iodide ion is a soft base because of its large size, whereas the fluoride ion is a hard base being very small.

While other explanations for the relative properties of RSI and RSF can be put foward, the above explanation has the advantage of being general. It can be used to explain why the complex ions AgI_2^- and I_3^- are stable but AgF_2^- and I_2F^- are not. Ag^+ and I_2 are soft acids. Furthermore, the fact that disulphides, RSSR, are stable whereas the corresponding compounds, RSOR, are unknown, can be explained in the same way. RS^+, being a soft acid, forms a stable complex with RS^-, which is soft. The complex of RS^+ with the hard base RO^- is not as stable.

The hardness of any element obviously increases with its positive oxidation state. To stabilize an element, metal or non-metal, in a high oxidation state, it should be surrounded with hard bases such as O^{2-}, OH^- or F^-. To stabilize an element in a low oxidation state, it should be coordinated with soft bases such as carbon monoxide, phosphines, cyanides and arsines.

A kind of symbiosis exists, as was pointed out by C. K. Jørgensen, for Lewis acids which can coordinate with several Lewis bases simultaneously.[5] Hard bases tend to flock together and soft bases tend to flock together on a given acid centre. For example, BF_3 is a typical hard acid, but BH_3 is a soft acid. In both cases the boron has an oxidation state of B^{3+}. However, the soft hydride ions, being easily polarized, lose a portion of their negative charge to the boron atom. Thus the boron atom has an effective charge of much less than $3+$. The fluoride ion, not being polarizable, holds on to its electrons and hence the effective charge on boron is close to $3+$. BH_3CO is a stable complex and BF_3CO is not known. Since carbon monoxide, CO, is a typical soft base, we see that the soft H^- groups have made it possible for boron to also hold the soft CO group. BF_3, on the other hand, forms stable complexes such as BF_3NR_3 or BF_3OR_2, where R_2O is an ether. The hard bases, F^- and R_2O, are compatible.

Organic bases in which the donor atom is carbon are soft. This is also true of bases such as olefins and aromatic hydrocarbons. Stable complexes with metal ions are best formed if the metal ion is soft. The typical soft acids such as Pt^{2+} and Hg^{2+} have long been known to form complexes with these organic bases. In order to make organometallic compounds of the important transition elements such as chromium, manganese, iron and nickel, it is necessary that these elements be in a low or zero oxidation state so they will be soft. This in turn means that the other ligands attached to the metal must be soft to preserve it in the low oxidation state and to enhance the symbiotic effect. A realization of this situation has helped lead to the synthesis of hundreds of compounds with organic ligands such as CH_3^-, $C_5H_5^-$, C_2H_4, C_6H_6, CO or hydrides H^-, attached to the transition metals.

Since metal atoms in a zero oxidation state are soft, bulk metals should also act as soft acids. Soft bases such as carbon monoxide and olefins are strongly adsorbed on surfaces of the transition metals. These metals are important heterogeneous catalysts.

Bases containing P, As, Sb, Se and Te in low oxidation states are the typical poisons in heterogeneous catalysis by metals. These soft bases are strongly adsorbed, blocking off the active sites. Strong oxygen and nitrogen containing bases are not poisons. Soft metal ions are poisons, whereas hard metal ions are not, showing the soft base character of the free metal. That is, a metal atom can also act as an electron donor and be a base as well as an acid. Heterogeneous catalysis on metals is generally viewed today as the formation of unstable organometallic compounds and hydrides on the metal surface.

There is much evidence which shows that free radicals and atoms behave as Lewis acids which form complexes with bases prior to reaction. The most stable complexes are formed with donors of high polarizability, or soft bases. The reactivity of Cl atoms, for example, shows that aromatic solvents stabilize them so they are less reactive, but oxygen and nitrogen donors do not. This is the expected behaviour if Cl is a soft acid. Also O and N atoms act as soft Lewis acids, attacking only soft basic atoms.

Solubility can be discussed in terms of acid–base interactions between solvent and solute molecules. Each solvent can be classified as hard or soft, though a separate classification for its acid function and its basic function may be necessary. A useful rule is that hard solutes dissolve well in hard solvents and soft solutes dissolve well in soft solvents. Water is a very hard solvent, both as an acid and as a base. It strongly solvates small anions such as OH^- and F^-, and small highly charged cations. Solvents which do not contain acidic hydrogen atoms for forming hydrogen bonds will usually be softer. Examples are dimethyl sulphoxide, dimethylformamide and acetone. Solvents such as benzene are the softest of all.

It is of interest to consider the oxidation potentials of the metals in terms of whether the ions which they form are hard or soft. The process involved is

$$M_{(solid)} \longrightarrow M^+_{(aqueous\ solution)} + electron \quad .. \quad (6)$$

If the metal ion, M^+, is hard, it will be strongly solvated. The binding in the metallic state, which is largely covalent, will be weak compared to the solvation energy, which is largely ionic. Hence the reaction goes well to the right as shown and the metal has a high oxidation potential and is reactive. If the metal ion M^+ is soft, then it will be weakly solvated in water relative to its binding in the metallic state. These metals will have a low oxidation potential and will be inert. Examples are gold, silver, mercury and platinum.

Carbene, CH_2, is a Lewis acid since it has only six valence electrons, like the oxygen atom. Now surely CH_2 is softer than CH_3^+, since the latter differs by one proton which should have a 'hardening' effect. We can now predict that ylids such as

$$CH_2\text{-}SR_2 \text{ or } CH_2\text{-}\underset{\underset{O}{\parallel}}{S}\text{-}R_2$$

will be much more stable than the corresponding oxygen ylids $CH_2\text{-}OR_2$. Also the ylid $CH_2\text{-}PR_3$ will be more stable than $CH_2\text{-}NR_3$ and CH_2I^- will be more stable than CH_2F^-.

It can be seen that an entire new approach to the stability of organic molecules is possible by the application of the HSAB principle. For example, the symbiotic effect plays an important role. Hard oxygen or fluorine atoms tend to cluster together on a central carbon, or soft hydride ions, or carbanions, tend to cluster together on a central carbon. These facts had been pointed out earlier by J. Hine.[7]

Reaction rates predicted

The hard–soft principle can also be used to predict something about the rates of chemical reactions in many cases. Very common types of chemical reaction are the acid substitution reaction (2) and the base substitution reaction (3) or the electrophilic and nucleophilic substitution reactions, as they are called. In both cases the important factor which determines the speed of reaction is the hard or soft character of the various acid and base centres. For example, in reaction (3) if A is a hard centre, then hard bases, B', will react fast. If A is a soft centre, then soft bases, B', will react fast.

An important example is furnished by substitution reactions of organic molecules, such as methyl chloride CH_3Cl. The methyl carbonium ion, CH_3^+, is a moderately soft acid centre, as mentioned. Hence soft bases such as RS^-, R_3P, $S_2O_3^{2-}$ and I^- react very rapidly. Hard bases react slowly, unless like OH^-, they are very strong bases towards all acids. In this case hydroxide ion reacts at a moderate rate.

The carbonyl group in compounds like esters and ketones is a hard acid centre. That is, CH_3CO^+ is in the hard category of Table I. It is found that hard bases attack the carbonyl group rapidly, as in ester hydrolysis, whereas soft bases react very slowly. Again, a very strong base, such as CH_3^-, will react rapidly even though it is soft. Soft bases which are weak, such as I^-, $S_2O_3^-$ and R_3As will have a negligible reactivity, though they are very effective towards methyl chloride.

R. F. Hudson has used the HSAB principle extensively to correlate rates of nucleophilic substitution reactions.[8] In particular he has shown how predictions can be made as to the point of attack in molecules with multiple electrophilic or nucleophilic centres. Since even so simple a molecule as isopropyl bromide has three different acidic sites, or electrophilic centres, such predictions can be of great value. The three acid sites in i–C_3H_7Br are the methyl protons, the CH carbon atom and the bromine atom, regarded as Br^+. The behaviour of i–C_3H_7Br towards a hard base such as OH^- is quite different from that towards a soft base, such as malonic ester anion, even though OH^- and $(C_2H_5OOC)_2CH^-$ are bases of very similar strength.

B. Saville has shown how it is possible to select catalysts which speed up the rates of reactions (2) and (3).[9] The trick in (2) is to use a basic catalyst which binds A effectively, and in (3) to add an acid catalyst which binds B effectively. The choice of a hard or soft acid or base catalyst is then dictated by the hard or soft nature of A or B. Note that the catalyst is added in addition to the reagents A' or B'.

Underlying theories

An interesting question to ask is: Why does the Principle of Hard and Soft Acids seem to work? It should be emphasized that the principle is supposed to be a statement of scientific facts and not a theory in itself. That is, it is of the nature of a law, but unfortunately not a quantitative law. This means that there must be an underlying theory, or theories, which explain the facts. It seems certain that the theory will not be simple, since to explain the stability of the acid–base complex $A:B$ must require a consideration of all of the effects which determine the strengths of chemical bonds generally. All of the theory of chemical bonding must be used.

Some general points do seem to emerge. It seems apparent that most combinations of hard acids and hard bases are held together by ionic, or polar, bonds. Most combinations of soft acids with soft bases are held together by largely covalent bonds. Combinations of hard and soft acids and bases suffer from a mismatch of their respective bonding tendencies.

Also the phenomenon of π bonding enters. The electron density of the bond is not concentrated along the line joining the two bonded nuclei, but is above and below this line. It is true that most hard acids are π bond acceptors. That is, they have empty outer orbitals which can accept π electrons from a base. Most soft acids are π bond donors. They have filled outer orbitals which can donate π electrons. Hard bases have filled outer orbitals which can donate electrons, via a π bond, to a hard acid. Soft bases have certain empty outer orbitals which can accept π electrons from a soft acid. Again a hard-soft combination of either kind is mismatched.

Finally, any two large atoms close to each other will have a stabilizing effect because of the existence of van der Waals's forces. These are always stronger for large, polarizable atoms than for small atoms and probably play some part in stabilizing soft acid–base combinations.

It is of interest that R. S. Drago[10] has developed an equation very similar to (5) for correlating the heats of forming various acid–base complexes not involving metal ions. The interpretation given by Drago of the first term on the right hand side of equation (5) is that it represents ionic interaction. The second term represents covalent interactions. It seems likely that if the transition metal ions were included, that a third term representing π bonding interactions might also be required.[11]

Conclusion

Whatever the explanations, it appears that the Principle of Hard and Soft Acids and Bases does describe a wide range of chemical phenomena in a qualitative if not quantitative way. It has usefulness in helping to correlate and remember large amounts of data, and it has useful predictive power. It is not infallible, since many apparent discrepancies and exceptions exist. These exceptions usually are an indication that some special factor exists in these examples. In such cases the principle can still be of value by calling attention to the need for further consideration.

REFERENCES

1. Introductory material on generalized acids and bases may be found in W. Luder and S. Zuffanti, *The Electronic Theory of Acids and Bases*, New York: John Wiley and Sons, Inc., 1946; R. P. Bell, *Acids and Bases*, London: Methuen, 1952.
2. R. G. Pearson, *J. Amer. Chem. Soc.*, 1963, **85**, 3533; *Science*, 1966, **151**, 172.
3. G. Schwarzenbach, *Experientia*, Supplement, 1956, **5**, 162.
4. S. Ahrland, J. Chatt and N. R. Davies, *Quart. Rev.*, 1958, **12**, 265.
5. C. K. Jørgensen, *Inorg. Chem.*, 1964, **3**, 1201.
6. R. G. Pearson and J. Songstad, *J. Amer. Chem. Soc.*, in press.
7. J. Hine, *ibid.*, 1963, **85**, 3239.
8. R. F. Hudson, *Structure and Mechanism in Organophosphorus Chemistry*, New York: Academic Press, 1963; *Coordination Chemistry Reviews*, 1966, **1**, 89.
9. B. Saville, *Chem. Eng. News*, 1965, **43**, 100; *Angew. Chem.*, in press.
10. R. S. Drago and B. B. Wayland, *J. Amer. Chem. Soc.*, 1965, **87**, 3571.
11. M. P. Johnson, D. G. Shriver and S. A. Shriver, *ibid.*, 1966, **88**, 1588.

107

Hard and Soft Acids and Bases, HSAB, Part I

Fundamental principles

Ralph G. Pearson
Northwestern University
Evanston, Illinois 60201

According to G. N. Lewis a base is an atom, molecule, or ion which has at least one pair of valence electrons which are not already being shared in a covalent bond. An acid is similarly a unit in which at least one atom has a vacant orbital in which a pair of electrons can be accommodated. The typical acid-base reaction is

$$A + :B \rightarrow A:B \qquad (1)$$

The species A:B may be called a coordination compound, an adduct, or an acid-base complex. In fact, a wide variety of acid-base complexes exist under different names. The species A is usually called a Lewis acid, or a generalized acid, to avoid confusion with Brønsted acids. A Lewis base, B, is identical with a Brønsted base.

Sidgwick suggested the terms electron acceptor, in place of Lewis acid, and electron donor, in place of base. These terms are widely used, particularly in Europe. Also certain types of weak generalized acid-base interactions are almost always discussed under the heading of donor-acceptor complexes. The disadvantage is that a different term, electron donor, is used to describe substances which are generally and conveniently called bases. It is also true that special names are sometimes used for special categories. The use of the term ligand in place of base when A in eqn. (1) is a metal ion is firmly established. Also in speaking of rates of reaction it is usual to call A an electrophile and to call B a nucleophile.[1]

Probably the most important class of chemical reaction is the generalized acid-base reaction of eqn. (1). The easiest way to appreciate this is to consider the different kinds of acid-base complexes, A:B, that may be formed. For example, all metal atoms or ions are Lewis acids. They are usually found coordinated to several bases or ligands simultaneously since they are polyvalent. The combination may be electrically charged, in which case we have a complex ion formed. Also the combination may be electrically neutral in which case a normal inorganic molecule such as $SnCl_4$ is formed.

Most cations are Lewis acids and most anions are bases. Hence salts are automatically acid-base complexes. $MgCl_2$ in the solid state consists of the acid Mg^{2+} coordinated to six neighboring Cl^- ligands. In dilute solution the magnesium ion is coordinated to the

base water forming the solvated ion. The chloride ion also forms an acid-base complex, *via* hydrogen bonding in which water molecules are Lewis acids.

Inorganic compounds, as solids, liquids, gases, or in solution, are examples of acid-base complexes. The same thing can be said for organic compounds. The method here is to mentally dissect the organic molecule into two fragments, one of which is a Lewis acid and the other a base. For example, ethyl alcohol can be thought of as composed of the ethyl carbonium ion, $C_2H_5^+$, and the hydroxide ion, OH^-. All carbonium ions are Lewis acids and the hydroxide ion is the base.

Similarly, ethyl acetate, can be thought of as a complex of the acylium ion, CH_3CO^+, which is an acid, and the base ethoxide ion, $C_2H_5O^-$. Even a hydrocarbon can be broken down (conceptually) into an acid such as H^+ and a carbanion, which is a base. Thus methane can be considered to be H^+ and CH_3^- combined. It is equally true that CH_4 can be viewed as a combination of CH_3^+ and H^-. Such ambiguity is, in fact, universal among both organic and inorganic molecules. While at first confusing, it turns out to be absolutely necessary to explain the variety of reactions undergone by these molecules. That is, methane sometimes reacts as if it were splitting into CH_3^+ and H^-, and sometimes as if it were splitting into CH_3^- and H^+. In addition, there are reactions in which it behaves as $CH_3\cdot$ and $H\cdot$. These are redox, or free radical reactions, however, and do not concern us here.

In the case of ethyl acetate, the molecule can be viewed as an acid-base complex as explained above. It is also true that it is a Lewis acid *and* a Lewis base. Ethyl acetate acts as a Lewis base when it forms complexes through one of its oxygen atoms to the proton, or other Lewis acids. It acts as a Lewis acid when it adds bases, such as the hydroxide ion, to its carbonyl group. Such acid-base processes are important in the acid and base catalyzed hydrolyses of esters.

It should be noted that a certain group of atoms is often designated as an acid or base even if it has no stable existence. A carbonium ion such as CH_3^+ is considered to be a Lewis acid because its structure shows that it can accept a pair of electrons from a base. The breaking down of a molecule, such as CH_3Cl, into a methyl carbonium ion and a chloride ion is a purely conceptual process and has nothing to do with the stability of CH_3^+. The point is that most reactions of methyl chloride can be classified as being exchanges of the chloride ion by other bases, or of the methyl cation by other Lewis acids. Just as the proton does not exist free under ordinary circumstances, so it is likely that CH_3^+ does not ordinarily exist as a free species.

EDITOR'S NOTE: This is a two part article. The second part, which discusses the theorys underlying the HSAB principle, will appear in the October issue.
[1] HAYEK, E., *Osters. Chem.-Zent.*, **63**, 109 (1962), has a general discussion of the problem of acid-base terminology.

[Reprinted from Journal of Chemical Education, Vol. 45, Pages 581–587, September, Pages 643–648, October, 1968.]

This brings up the point that eqn. (1) is oversimplified. What actually occurs in most cases is the exchange reaction

$$A:B' + A':B \rightleftharpoons A:B + A':B' \qquad (2)$$

In solution A' and B' are often solvent molecules. Thus, as already mentioned in the case of ions dissolved in water, most solute-solvent interactions can be classified as generalized acid-base reactions. A polar molecule will always have an electron rich, or basic site, and an electron poor, or acid site.

Other kinds of acid-base complexes are the so-called charge transfer complexes which are responsible for the colors produced when many substances are mixed. An example is iodine and the intense brown color it gives in solvents which are bases, such as water, alcohols, or ethers. Many charge transfer complexes are formed between Lewis acids which are unsaturated molecules with electron-withdrawing substituents, such as tetracyanoethylene, and unsaturated molecules with electron donating substituents, such as toluene. These systems are called π-acids and π-bases, respectively. Finally, free atoms and radicals containing electronegative elements act as Lewis acids and form complexes with a variety of bases. These complexes of free radicals cannot be isolated but they have a very great effect on the reactivity of the radicals.

It is apparent that it is possible, in principle at least, to view the greater part of chemistry as examples of interaction of generalized acids and bases. This in turn means that any rules that can be developed concerning the stability of complexes A:B in reaction (1) will have very wide application and can be useful in many areas. Recently a rule has been suggested, "The Principle of Hard and Soft Acids and Bases," or HSAB principle, which does seem to have value in understanding a wide variety of chemical phenomena (1).

Acid and Base Strength

The concept of acid or base strength comes in at this point. In a qualitative way, what is meant by generalized acid or base strength is simply that a strong acid, A, and a strong base, B, will form a stable complex, A:B. A weaker acid and base will form a less stable complex. Operationally we may define generalized acid and base strength by competition experiments. Consider the acid-base substitution reactions

$$A' + A:B \rightarrow A':B + A \qquad (3)$$

$$B' + A:B \rightarrow A:B' + B \qquad (4)$$

If the reactions go as indicated, it means that A' is a stronger Lewis acid than A, and that B' is a stronger base than B. If it were possible to put all Lewis acids into an order of decreasing strength, and the same for all bases, then it would be possible to predict the stabilities of all possible acid-base complexes. That is, we could predict what chemical reactions would occur under various conditions, what compounds would be stable, etc.

We might then expect the equilibrium constant of eqn. (1) to be given by an equation such as

$$\log K = S_A S_B \qquad (5)$$

where S_A is a strength factor for the acid and S_B is a strength factor for the base. Equation (5) is not the only equation that might result, but it would be the simplest one that would correctly predict the direction of displacement reactions such as (3) and (4). Of course, S_A and S_B would be functions of the environment and the temperature. The "intrinsic" strengths would presumably refer to gas phase reactions. There would then be strong solvent corrections, but even so an equation such as eqn. (5) would be most useful since a series of S_A and S_B values would need to be determined once and for all in water at 25°C, and the stabilities of possible complexes would be known in that medium. S_A and S_B values for 100 acids and 100 bases would predict the stabilities of 10,000 complexes, for example.[2]

Unfortunately it is not possible to write down any universal order of acid or base strength (2). If a series of different bases, B', are used in testing against a fixed reference A:B, then the order of base strength that one gets is very much a function of the nature of the reference acid, A. When A is the chromic ion, one gets a different order from the case when A is the cuprous ion. In fact, the order is different for cupric ion and for cuprous ion so that even a change in oxidation state of the reference acid can have an effect on relative base strengths.

We are forced to the conclusion that there is no straight-forward way of evaluating base strengths or acid strengths even if we were to agree that only gas phase data for reaction (1) be used. Of course, a useful scale of base strengths towards the special Lewis acid, the proton, in aqueous solution does exist. It must always be remembered that this scale is not valid for all Lewis acids. It can be useful in a general sense, however, in telling us that some bases such as ClO_4^-, are very weak, and other bases, such as H^-, are very strong. Such a rough ordering will usually be valid if two bases of widely different strengths are being compared.

A comparable scale to order Lewis acids in aqueous solution has not been devised, partly because many Lewis acids are decomposed by water. For many others, a scale in which the hydroxide ion is a reference base could be used. In aqueous solution we could write both the reactions

$$H^+(aq) + B(aq) \rightleftharpoons BH^+(aq) \qquad (6)$$

$$OH^-(aq) + A(aq) \rightleftharpoons AOH^-(aq) \qquad (7)$$

Any arbitrary value of S_A could be assigned to the proton. Let us pick 9.0, for example. This would lead to values of S_B ranging from about 5 for a strong base such as CH_3^- to -1 for a weak base such as I^-. The value of S_B for the hydroxide ion would then be $(15.74/9) = 1.75$. The number 15.74 is the $\log_{10}$ of the equilibrium constant for eqn. (6) when B is the hydroxide ion $(55.5/1.0 \times 10^{-14})$.

This value for S_B of OH^- then leads to values of S_A of 5.9 and 8.6 for the aqueous Hg^{2+} and CH_3^+, respectively. These figures come simply from the acid ionization constants of $Hg(H_2O)_2^{2+}$ and $CH_3OH_2^+$ applied to reaction (7). Again, such numbers are useful

---

[2] One might ask at this point why standard free energies of formation are not used as the fundamental properties. The answer is that these involve prior knowledge of the stability of A:B. We are seeking a method that predicts the properties of A:B from a knowledge of A and B only, in a given medium.

in telling us that H^+ and CH_3^+ are probably stronger acids than Hg^{2+}. However, they cannot be combined with the values of S_B previously obtained to accurately predict the stabilities of $CH_3HgH_2O^+$, $HgI(H_2O)^+$, CH_3I, or C_2H_6 in aqueous solution. They may however, give some rough idea of stabilities. Furthermore, there will be some Lewis acids sufficiently like the proton so that a Brønsted relationship exists between the equilibrium constants, K_A, for the reaction

$$A(aq) + B(aq) \rightleftharpoons A:B(aq) \qquad (8)$$

and the equilibrium constant for eqn. (6), K_a,

$$K_A = GK_a{}^\alpha \qquad (9)$$

In such cases eqn. (5) will be valid since eqn. (9) is simply another form of the equation with α equal to $S_A/S_H{}^+$.

Usually the simple eqn. (5) will not be adequate, and it would be logical to replace it with a more complex equation involving more parameters. That is, instead of having only one parameter, S, characteristic of each acid and each base, it will be necessary to have at least two. Such an equation (not the only one possible) would be

$$\log K = S_A S_B + \sigma_A \sigma_B \qquad (10)$$

Now σ_A and σ_B are parameters for each acid and each base which measure some different characteristic from that of strength. We will call them "softness" parameters for reasons that will become clearer later on.

Equations of the form of eqn. (10) have often been used to represent rate or equilibrium data. Such four-parameter equations are the next necessary stage after it is found that two-parameter equations (linear free energy relationships) (3) have suggested an empirical equation. For example, Drago and Wayland (3) have suggested an empirical equation

$$-\Delta H = E_A E_B + C_A C_B \qquad (11)$$

which accurately reproduces heats of reaction for certain complex forming reactions in nonpolar solvents. These are reactions between neutral bases and weak Lewis acids such as I_2 or phenol, which forms hydrogen bonds.

The E parameters in eqn. (11), vary with the polarity of the acids and bases in such a way as to suggest that their product represents electrostatic bonding. Similarly, the C parameters vary in such a way as to suggest that their product represents covalent bonding between A and B. The heat of reaction, ΔH, is closely related to the free energy changes $\Delta G°$, and hence to the log of the equilibrium constant.

Another four-parameter equation is the celebrated Edwards equation (4)

$$\log_{10}(K/K_0) = \alpha E_n + \beta H \qquad (12)$$

E_n is a redox factor defined by $E_n = E° + 2.60$, where $E°$ is the standard oxidation potential for the process.

$$2\,B^- \rightleftharpoons B_2 + 2e^- \qquad (13)$$

H is a proton basicity factor defined by $H = 1.74 + pK_a$. Both definitions are arranged so that $H = 0$ and $E_n = 0$ for water at 25°C. K_0 is therefore the constant when the base, B, is water. Equation (14) may conveniently be called the oxibase scale (5). It quite successfully correlates a large amount of rate and equilibrium data.

Table 1. A Comparison of α and β Values for the Edwards Equation (4)

Lewis Acid	α	β
Hg^{2+}	5.786	−0.031
Cu^+	4.060	0.143
Ag^+	2.812	0.171
Pb^{2+}	1.771	0.110
Cd^{2+}	2.132	0.171
Cu^{2+}	2.259	0.233
Mn^{2+}	1.438	0.166
Au^{3+}	2.442	0.353
Mg^{2+}	1.402	0.243
Zn^{2+}	1.367	0.252
Ga^{3+}	3.795	0.767
Fe^{3+}	1.939	0.523
Ba^{2+}	1.786	0.411
Ca^{2+}	1.073	0.327
Al^{3+}	−0.749	1.339
H^+	0.000	1.000
I_2	3.04	0.000

Data from Yingst and McDaniel (6) except for I_2, Edwards (4).

The way in which α and β in eqn. (12) depend upon the Lewis acid substrate is quite revealing. Table 1 shows a compilation of values of α and β for the formation of a number of metal complexes with various bases (6). It can be seen that β is large for Lewis acids with a high positive charge and small size, and small for Lewis acids of low charge and large size. In other words β varies just exactly as S_A is expected to vary in eqn. (5). The term that it is multiplied by, H, is simply another way of expressing S_B. We can accordingly identify the product βH with $S_A S_B$. Then αE_n must be identified with $\sigma_A \sigma_B$ which means that σ_B is large for bases that are easily oxidized, such as I^-, and small, or negative, for bases that are hard to oxidize, such as F^-. We can also see that α, or σ_A, is large for Lewis acids of large size, low positive charge, and containing unshared electrons in p or d orbitals in the valence shell, such as Ag^+. Also σ_A is small for Lewis acids of the opposite characteristics, such as Mg^{2+}.

While the Edwards equation is of the form of eqn. (5), it is not the only equation that might be used, even in aqueous solution. For example, an entirely empirical equation, such as eqn. (11), might be developed. In fact, one can see that if eqn. (5), or eqn. (10), or eqn. (11) were valid, that equations of the form

$$\log(K/K_0) = \gamma \log(K'/K_0') + \delta \log(K''/K_0'') \qquad (14)$$

should generally exist, where K, K', and K'' are any series of related equilibrium constants. The constants K' might be taken as the H values of the Edwards equation, and K'' might be values for a typical Lewis acid with the opposite properties to the proton in Table 1, such as Hg^{2+}.

An even better standard is the methylmercury(I) cation, CH_3Hg^+, for which a large amount of equilibrium data in aqueous solution has been accumulated by Schwarzenbach.

$$CH_3Hg^+ (aq) + B(aq) \rightleftharpoons CH_3HgB^+(aq) \qquad (15)$$

Like the proton, CH_3Hg^+ has the advantage of having a coordination number of one, which simplifies the equilibria involved.

Table 2. Equilibrium Constant in H_2O at 25°C for the Reaction $CH_3Hg(H_2O)^+ + BH^+ \rightleftharpoons CH_3HgB^+ + H_3O^+$

B	$\log K_{eq}$	B	$\log K_{eq}$
F^-	-1.35	SO_3^{2-}	1.3
Cl^-	12.25	$S_2O_3^{2-}$	9.0
Br^-	15.62	N_3^-	1.3^a
I^-	18.1	NH_3	-1.8
OH^-	-6.3	$NH_2C_6H_4SO_3^-(p)$	-0.5
HPO_4^{2-}	-1.8	$NH_2CH_2CH_2NH_2$	-1.8
HPO_3^{2-}	-1.6	$(C_6H_5)_2PC_6H_4SO_3^-(p)$	9.0
S^{2-}	7.0	$P(C_2H_5)_3$	6.2
$HOCH_2CH_2S^-$	6.6	CN^-	5.0
SCN^-	6.7	CH_3^-	$\approx 30^b$

^{a} See Musgrave and Keller, *Inorganic Chem.*, **4**, 1793 (1965).
^{b} Estimated by assuming that K_{eq} for $CH_3HgCl + CH_4 \rightleftharpoons$ $(CH_3)_2Hg + HCl$ is the same in the gas phase and in water. Remaining data from G. Schwarzenbach and M. Schellenberg (7).

With two reference acids of different properties we can test various bases to see if they prefer to bind to H^+ or CH_3Hg^+. Table 2 shows some data (7) for the equilibrium constant for the exchange reaction

$$BH^+ + CH_3Hg(H_2O)^+ \rightleftharpoons CH_3HgB^+ + H_3O^+ \quad (16)$$

The important feature which we note is that bases in which the donor atom is N, O, or F prefer to coordinate to the proton. Bases in which the donor atom is P, S, I, Br, Cl, or C prefer to coordinate to mercury.

The donor atoms in the first group are those which are of high electronegativity, of low polarizability, and hard to oxidize. Let us call bases containing these donors "hard" bases, to emphasize the fact that they hold on to their electrons tightly. The donor atoms of the other bases are of low electronegativity, of high polarizability, and easy to oxidize. Let us call bases containing these donor atoms "soft" bases, a term which graphically describes the looseness with which they hold their valence electrons.

The Principle of Hard and Soft Acids and Bases

We can now classify every conceivable base into one or the other of three categories, hard, soft, or borderline. Table 3 shows such a classification. The borderline category takes into account such factors as the presence of unsaturation in some nitrogen donors, which should loosen up the valence electrons. It also recognizes that chloride ion is less soft than bromide ion, which in turn is less soft than iodide ion. An important feature

Table 3. Classification of Bases

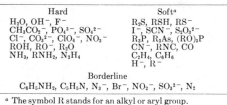

Hard	Soft^a
H_2O, OH^-, F^-	R_2S, RSH, RS^-
$CH_3CO_2^-$, PO_4^{3-}, SO_4^{2-}	I^-, SCN^-, $S_2O_3^{2-}$
Cl^-, CO_3^{2-}, ClO_4^-, NO_3^-	R_3P, R_3As, $(RO)_3P$
ROH, RO^-, R_2O	CN^-, RNC, CO
NH_3, RNH_2, N_2H_4	C_2H_4, C_6H_6
	H^-, R^-

Borderline
$C_6H_5NH_2$, C_5H_5N, N_3^-, Br^-, NO_2^-, SO_3^{2-}, N_2

^{a} The symbol R stands for an alkyl or aryl group.

is that Table 3 could be constructed in two independent ways: by considering the properties of the donor atom (easily oxidized, polarizable, etc.), or by estimating the equilibrium constant of reaction (16). The results would be the same, with a few exceptions. For example, Cl^- has an unusually high affinity for mercury ion, as do Br^- and I^-, and the values of K_{eq} in Table 2 are somewhat anomalous for this reason. Perhaps Cl^- should be considered a borderline base.

We can now proceed to an equivalent classification of Lewis acids into three categories, including borderline cases. We simply ask the question whether the acid is like the proton in preferring the hard bases of Table 3, or like CH_3Hg^+ in preferring the soft bases. Operationally this is best done by using the criteria of Schwarzenbach (8) and of Ahrland, Chatt, and Davies (9). For complexes with different donor atoms, the following sequences of stabilities are found

Class (a) acids (hard) $\begin{cases} N \gg P > As > Sb \\ O \gg S > Se > Te \\ F > Cl > Br > I \end{cases}$

Class (b) acids (soft) $\begin{cases} N \ll P > As > Sb \\ O \ll S < Se \sim Te \\ F < Cl < Br < I \end{cases}$

If we consider any Lewis acid, such as Cu^+, NO^+, or I_2, we simply examine the literature to see if a complex such as CuI or CuF is more stable, or if $Cu(PR_3)_2^+$ is more stable than $Cu(NH_3)_2^+$.

The term stable is ambiguous, as ordinarily used, and a strict definition would refer to the equilibrium constants for reactions in water such as

$$CuF(aq) + I^-(aq) \rightleftharpoons CuI(aq) + F^- \quad (17)$$

$$Cu(PR_3)^+(aq) + NH_3(aq) \rightleftharpoons Cu(NH_3)^+(aq) + PR_3(aq) \quad (18)$$

Oftentimes the data is incomplete and a variety of interpolations must be made to draw a conclusion (1). Nevertheless, it is usually possible to conclude that a Lewis acid prefers either the hard bases of Table 3, or the soft bases.

When this is done for a large number of Lewis acids, the results are as shown in Table 4. The entries in the

Table 4. Classification of Lewis Acids

Hard	Soft
H^+, Li^+, Na^+, K^+	Cu^+, Ag^+, Au^+, Tl^+, Hg^+
Be^{2+}, Mg^{2+}, Ca^{2+}, Sr^{2+}, Mn^{2+}	Pd^{2+}, Cd^{2+}, Pt^{2+}, Hg^{2+}, CH_3Hg^+, $Co(CN)_5^{2-}$, Pt^{4+}, Te^{4+}
Al^{3+}, Sc^{3+}, Ga^{3+}, In^{3+}, La^{3+}	Tl^{3+}, $Tl(CH_3)_3$, BH_3, $Ga(CH_3)_3$
N^{3+}, Cl^{3+}, Gd^{3+}, Lu^{3+} Cr^{3+}, Co^{3+}, Fe^{3+}, As^{3+}, CH_3Sn^{3+}	$GaCl_3$, GaI_3, $InCl_3$ RS^+, RSe^+, RTe^+
Si^{4+}, Ti^{4+}, Zr^{4+}, Th^{4+}, U^{4+} Pu^{4+}, Ce^{3+}, Hf^{4+}, WO^{4+}, Sn^{4+}	I^+, Br^+, HO^+, RO^+
UO_2^{2+}, $(CH_3)_2Sn^{2+}$, VO^{2+}, MoO^{3+}	I_2, Br_2, ICN, etc.
$BeMe_2$, BF_3, $B(OR)_3$	trinitrobenzene, etc.
$Al(CH_3)_3$, $AlCl_3$, AlH_3	chloranil, quinones, etc.
RPO_2^+, $ROPO_2^+$	tetracyanoethylene, etc.
RSO_2^+, $ROSO_2^+$, SO_3	O, Cl, Br, I, N, $RO\cdot$, $RO_2\cdot$
I^{7+}, I^{5+}, Cl^{7+}, Cr^{6+}	M° (metal atoms)
RCO^+, CO_2, NC^+	bulk metals
HX (hydrogen bonding molecules)	CH_2, carbenes

Borderline
Fe^{2+}, Co^{2+}, Ni^{2+}, Cu^{2+}, Zn^{2+}, Pb^{2+}, Sn^{2+}, Sb^{3+}, Bi^{3+} Rh^{3+}, Ir^{3+}, $B(CH_3)_3$, SO_2, NO^+, Ru^{2+}, Os^{2+}, R_3C^+, $C_6H_5^+$, GaH_3

left hand column are class (a) acids, and those on the right are class (b) acids. However, instead of following this method of naming, which has historical precedence, a different system of naming has some advantages.

If we examine the class (a) Lewis acids, we find that the acceptor atoms are small in size, of high positive charge, and do not contain unshared pairs of electrons in their valence shell (not all of these properties need be possessed by any one acid). Now these are all properties which lead to high electronegativity and low polarizability. It seems appropriate to call such acids "hard." The class (b) Lewis acids, generally speaking, have acceptor atoms large in size, of low positive charge, and containing unshared pairs of electrons (p or d electrons) in their valence shell. These properties lead to high polarizability and low electronegativity. Again it seems reasonable to call these Lewis acids "soft."

A comparison of Tables 1 and 4 shows that Lewis acids with large α and β values are all in the soft group and those with small α and large β values are in the hard group. Thus we can say that soft acids form stable complexes with bases that are highly polarizable and are good reducing agents, and not necessarily good bases towards the proton. Hard acids, of which the proton itself is typical, will usually form stable complexes with bases that are good bases towards the proton. Polarizability, or the reducing properties of the base, play a minor role.

If we arrange the donor atoms of the most common bases in an order of increasing electronegativity, we will have As, P < C, Se, S, I < Br < N, Cl < O < F. Soft Lewis acids will form more stable complexes with left hand members of this series, and hard Lewis acids will form more stable complexes with the right hand members of the series. For example, a rare earth ion, La^{3+}, will form complexes only with N, O, and F donors, or with hard bases.

If one accepts the system of naming used in Tables 3 and 4, a very simple rule can now be stated concerning the stability of acid-base complexes. The rule is that: *Hard acids prefer to bind to hard bases and soft acids prefer to bind to soft bases.*

The rule is a concise statement which sums up the experimental information used to compile Tables 3 and 4. It is a statement of fact and is not to be regarded as a theory or as a hypothesis. Such generalized statements covering many facts were often called laws. However, modern scientific practice is to reserve the name "law" only for those generalizations which are capable of rather precise mathematical formulation. For this reason it seems best to call the above rule "The Principle of Hard and Soft Acids and Bases." Eventually, it may become possible to make quantitative predictions based on equations similar to eqn. (10).

Unfortunately eqn. (10) cannot be expected to be exact, or even nearly exact. It is too simple to represent the complexity of changes that occur when electron-donating groups combine with electron-accepting groups. It is certainly much better than eqn. (5) since it has more parameters in it. It is as good as eqns. (11) and (12), since it is simply the general prototype of any four parameter equation. Yet eqns. (11) and (12), while very good over a limited range, cannot reproduce data over a very wide range of Lewis acids and bases. Of course, in any case, different values of all the parameters of these equations would be needed, if one changed the solvent or the temperature.

In the case of the HSAB principle we have a simple, but imprecise, law with a very wide range of applicability. In spite of the lack of precision the rule does appear to have considerable utility. It can be used in prediction; perhaps more important, it can be extremely helpful in correlating the vast amount of chemical information which we already have at hand.

Estimations of Strength and of Softness

What has been suggested in the previous section is that two properties of an acid and a base are needed to make an estimate of the stability of the complex which they might form. One property is what we may call the intrinsic strength (S_A or S_B in eqn. (10)); the other is the hardness or softness (σ_A or σ_B in eqn. (10)). While various arbitrary scales of strength and softness can be devised, such as pK_a for strength and $pK_{CH_3Hg^+}$ for softness, it seems best to leave them undefined operationally at present, and to use qualitative definitions based on the properties of the acids and bases. That is, we need to have methods of estimating the strength and the softness of an acid or base which depend on a knowledge of their chemical compositions and electronic structures.

In fact we have long had such rules for estimating Lewis acid or base strength. We know that for cations increased charge and decreased radius make for strong acids. For anions increased charge and decreased radius also increase base strength. Thus O^{2-} is a stronger base than OH$^-$, and stronger than Se^{2-}. The ions Al^{3+}, AlCl^{2+}, AlCl$_2^+$, and AlCl$_3$ will have steadily decreasing intrinsic strengths. Neutral acids and bases will have strengths proportional to the local dipoles at the acceptor or donor atom sites. More remote substituents also have rather predictable effects on acid or base strength in terms of electron withdrawing properties that they may have, or the local dipoles that they create (10).

Of course, in chemistry we rarely are concerned with the properties of a bare ion, such as Al^{3+}. Instead we deal with such ions in various environments. This alters the nature and strength of the Lewis acid involved. We are more likely to need to know that the Lewis acids Al(H$_2$O)$_5^{3+}$, Al(H$_2$O)$_4$Cl^{2+}, and Al(H$_2$O)$_3$Cl$_2^+$ decrease steadily in acid strength. The coordination number (number of base molecules or ligands attached to the central acceptor molecule) is reduced by one in these examples to show the unit which is the Lewis acid. In Table 4 are listed a number of ions. These are always meant to be the aquated ions less one molecule of water, Ni(H$_2$O)$_5^{2+}$, B(H$_2$O)$_3^{3+}$, etc. Such aquo ions are very much weaker acids than the rather hypothetical bare ions would be.[3]

---

[3] Solvation effects on base strengths are also very important. For example, CH$_3^-$ is a much stronger base than OH$^-$ in aqueous solution. In the gas phase, however, the base strengths are virtually identical, that is, the intrinsic strengths of CH$_3^-$ and OH$^-$ are the same. Strong solvation of OH$^-$ by water accounts for the difference in solution. JOLLY, W. L., J. CHEM. EDUC., **44**, 304 (1967), has a discussion of solvation effects.

Just as we can make reasonable guesses about intrinsic acid and base strength, so we can make estimates of hardness and softness. This was, in fact, done for bases in constructing Table 3 by a simple examination of the nature of the donor atom (electronegative and nonpolarizable like F, or polarizable and not electronegative like I). We can also assign increasing softness within related series without much ambiguity. Thus SbR_3, AsR_3, PR_3, and NR_3 should be of decreasing softness, as are CH_3^-, NH_2^-, OH^-, and F^-. The effect of oxidation state on a given donor element is predictable; sulfur(IV) in SO_3^{2-} should be a harder base than sulfur($-II$) in S^{2-}.

It is somewhat suprising that there does not seem to be much difference in hardness between H_2O, $OH,^-$ and O^{2-}. All three are very hard bases by any criteria and any difference between them is masked by other effects. We would have expected the polarizability (and hence the softness) to increase in the order H_2O $<OH^-<O^{2-}$. Also it is not easy to decide if there is any difference in hardness between various oxygen donors such as CH_3COO^-, SO_4^{2-}, PO_4^-, etc.

For the Lewis acids, the important properties that determine softness are size, charge or oxidation state, electronic structure, and the other attached groups. Both metals and nonmetals can be acceptor atoms in acid-base complexes. For elements of variable valence there is usually a smooth increase in hardness as the oxidation state increases. Thus $Ni(0)$ (in $Ni(CO)_4$, for example) is soft, $Ni(II)$ is borderline, and $Ni(IV)$ is hard. The sulfur atom of the sulfenyl group RS^+ is a soft Lewis acid whereas the sulfur atom of the sulfonyl group RSO_2^+ is hard. The formal oxidation state changes from plus two to plus six. Other examples can be found in Table 4.

Exceptions do occur at the end of the transition series. It is certain that $Tl(III)$ is softer than $Tl(I)$ and it is likely that $Hg(II)$ is softer than $Hg(I)$. The evidence is incomplete, but it is possible that $Pb(IV)$ is softer than $Pb(II)$. These cases all involve the inert pair of electrons in the $5s$ or $6s$ orbitals. It seems that the presence of electrons in these particular orbitals decreases softness by a shielding effect on the outer d electrons.

The importance of the d electrons for metal ions is very great. As Ahrland has pointed out (11) no really good class (b), or soft, acceptor among the metals exsists which does not have at least a half-filled outer d shell. This accounts for another anomaly. In going across a transition series, e.g., from Ca to Zn, the ionization potentials of the atoms increase because of increasing nuclear charge. One would interpret this as meaning that the elements become more electronegative, that is, harder, as one goes across from Ca to Zn (12). In fact, chemically the elements become *softer*. This is a consequence of the increasing number of d electrons, a factor which outweighs increasing electronegativity.

Fortunately, for the representative elements, and for the nonmetals in particular, this complication does not arise and softness seems to be a predictable function of oxidation state. Of equal importance is the nature of the other groups attached to the acceptor atom. We see in Table 4 that BF_3 is a hard acid, and BH_3 is a soft acid. Experimentally it is found that $BF_3 \cdot OR_2$

is more stable than $BF_3 \cdot SR_2$, whereas for BH_3 the reverse is true. Borine will even form a carbonyl, BH_3CO. Formation of complexes with carbon monoxide or olefins is a good test for soft behavior.

In both BF_3 and BH_3 the boron is formally in a plus three oxidation state, yet quite different behavior is noted. The presence of hard fluoride ions in BF_3 makes it easy to add other hard bases. The presence of soft hydride ions in BH_3 makes it easy to add other soft bases. This important effect was particularly commented on by Jørgensen who coined the name "symbiosis" to describe it (13). Soft bases tend to group together on a given central atom and hard ligands tend to group together ("birds of a feather flock together"). The mutual stabilizing effect is called symbiosis.[4]

The explanation for symbiosis is rather easily seen. The hard F^- ligands form a complex which is largely ionic. Hence the boron atom in BF_3 is nearly B^{3+} and is hard. The soft H^- ligands donate negative charge extensively to the central boron, by covalency or by simple polarization. As a result the boron atom in BH_3 is almost neutral and naturally becomes soft.

The conclusion that it is the actual charge on the central atom, rather than the formal charge, which determines softness seems perfectly logical. While it complicates the assignment of hardness or softness in some cases, it helps explain many otherwise puzzling phenomena (14). For example, the existence of ions such as AsS_4^{3-} and $Mo(SCN)_6^-$ can be rationalized in spite of the high formal oxidation state of the central atom. In the latter case, the thiocyanate ion is believed to be S-bonded rather than N-bonded.

The mode of bonding of the thiocyanate ion is often used as a test of hard or soft behavior. The sulfur end is assumed to be much softer than the nitrogen end, and hence to prefer soft Lewis acids. An interesting test of this assumption has been made (15). A study was made of the complexing of both an alkyl thiocyanate, RSCN, and an alkyl isothiocyanate, RNCS, with the soft Lewis acid, iodine, and the hard Lewis acid, phenol. The thiocyanate, RSCN, with a free nitrogen end to act as a donor, formed a more stable complex with phenol (hydrogen-bonding) than with I_2 (charge-transfer). With the isothiocyanate, exactly the opposite results were found. However, the differences were very small and the criterion must be used with caution.

Note that saying that the group $Mo(SCN)_5$ is a soft Lewis acid is not the same as saying that $Mo(V)$ is a soft acid. In fact, $MoO(NCS)_5^{2-}$ turns out to be N-bonded (16). Similarly, $Rh(NH_3)_5^{3+}$ is a borderline case whereas $Rh(SCN)_5^{2-}$ is a soft acid (17). As Jørgensen (18) points out, it is likely that most elements in a very high oxidation state could form complexes containing a maximum number of soft groups, if it were not for spontaneous oxidation-reduction (MnS_4^-,

---

[4] Symbiosis is counter-balanced by other factors. Otherwise all equilibria of the type

$$MX_n + MY_n \rightleftharpoons 2MX_mY_m$$

would lie to the left. In fact, they can go in either direction. The other factor is probably that of intrinsic strength which will favor the mixed species if X and Y differ markedly in base strength.

for instance). However, mixed species such as MnO_2-S_2^- should be unstable in any case because they lack symbiotic stabilization.

Literature Cited

(1) Pearson, R. G., *J. Am. Chem. Soc.*, **85**, 3533 (1963); *Science*, **151**, 172 (1966); *Chemistry in Britain*, **3**, 103 (1967).

(2) Lewis, G. N., "Valence and the Structure of Atoms and Molecules," The Chemical Catalog Co., New York, N. Y., **1923**; Lewis, G. N., *J. Franklin Institute*, **226**, 293 (1938).

(3) Drago, R. S., and Wayland, B. B., *J. Am. Chem. Soc.*, **87**, 3571 (1965).

(4) Edwards, J. O., *J. Am. Chem. Soc.*, **76**, 1540 (1954).

(5) Davis, R. E., *J. Am. Chem. Soc*, **87**, 3010 (1965).

(6) Yingst, A., and McDaniel, D. H., *Inorganic Chem.*, **6**, 1067 (1967).

(7) Schwarzenbach, G., and Shellenberg, M., *Helv. Chim. Acta*, **48**, 28 (1965).

(8) Schwarzenbach, G., *Experentia Suppl.*, **5**, 162 (1956).

(9) Ahrland, S., Chatt, J., and Davies, N. R., *Quart. Revs.*, (London) **12**, 265 (1958).

(10) Bell, R. P., "The Proton in Chemistry," Cornell University Press, Ithaca, **1959**, Chapter 7.

(11) Ahrland, S., *Structure and Bonding*, **1**, 207, (1966).

(12) Williams, R. J. P., and Hale, J. D., *Structure and Bonding*, **1**, 249 (1966).

(13) Jørgensen, C. K., *Inorg. Chem.*, **3**, 1201 (1964).

(14) Basolo, F., and Burmeister, J., *Inorg. Chem.*, **3**, 1587 (1964).

(15) Wayland, B. B., and Gold, R. H., *Inorg. Chem.*, **5**, 154 (1966).

(16) Mitchell, P. C. H., *Quart. Rev.* (London), **20**, 103 (1966).

(17) Schmidtke, H. H., *J. Am. Chem. Soc.*, **87**, 2522 (1965).

(18) Jorgensen, C. K., *Structures and Bonding*, **1**, 234 (1966).

Reprinted from *Journal of Chemical Education*, **45**, 643–648 (October, 1968)

9b

Hard and Soft Acids and Bases, HSAB, Part II

Underlying theories

Ralph G. Pearson
Northwestern University
Evanston, Illinois 60201

It must be emphasized again that the HSAB principle is intended to be phenomenological in nature. This means that there must be underlying theoretical reasons which explain the chemical facts which the principle summarizes. It seems certain that there will be no one simple theory. To explain the stability of acid-base complexes, such as A:B, will require a consideration of all the factors which determine the strength of chemical bonds.

Any explanation must eventually lie in the interactions occurring in A:B itself. Solvation effects, while important, will not in themselves cause a separation of Lewis acids and bases into two classes, each with its characteristic behavior. Of course a major part of solvent-solute interaction is itself an acid-base type of reaction (*19*). With regard to the bonding in A:B, several pertinent theories have been put forward by various workers interested in special aspects of acid-base complexation.

The oldest and most obvious explanation may be called the ionic-covalent theory. It goes back to the ideas of Grimm and Sommerfeld for explaining the differences in properties of AgI and NaCl. Hard acids are assumed to bind bases primarily by ionic forces. High positive charge and small size would favor such ionic bonding. Bases of large negative charge and small size would be held most tightly—for example, OH^- and F^-. Soft acids bind bases primarily by covalent bonds. For good covalent bonding, the two bonded atoms should be of similar size and similar electronegativity. For many soft acids ionic bonding would be weak or nonexistent because of the low charge or the absence of charge. It should be pointed out that a very hard center, such as I(VII) in periodate or Mn-(VII) in MnO_4^-, will certainly have much covalent character in its bonds, so that the actual charge is reduced much below $+7$. Nevertheless, there will be a strong residual polarity.

The π-bonding theory of Chatt (*20*) seems particularly appropriate for metal ions, but it can be applied to many of the other entries in Table 4 as well. According to Chatt the important feature of class (b) acids is considered to be the presence of loosely held outer d-orbital electrons which can form π bonds by donation to suitable ligands. Such ligands would be those in which empty d orbitals are available on the basic atom, such as

phosphorus, arsenic, sulfur, or iodine. Also, unsaturated ligands such as carbon monoxide and isonitriles would be able to accept metal electrons by means of empty, but not too unstable, molecular orbitals. Class (a) acids would have tightly held outer electrons, but also there would be empty orbitals available, not too high in energy, on the metal ion. Basic atoms, such as oxygen and fluorine in particular, could form π bonds in the opposite sense, by donating electrons from the ligand to the empty orbitals of the metal. With class (b) acids, there would be a repulsive interaction between the two sets of filled orbitals on metal and oxygen and fluorine ligands. Figure 1 shows schematically a p orbital on the ligand and a d orbital on the metal atom which are suitable for forming π bonds.

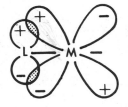

Figure 1. A *p*-atomic orbital on a ligand atom and *d* orbital on a metal atom suitable for π-bonding. The *d* orbital is filled and the *p* orbital is empty for a soft acid-soft base combination. The *d* orbital is empty and the *p* orbital is filled for a hard acid-hard base combination. The plus and minus signs refer to the mathematical sign of the orbital.

Pitzer (*21*) has suggested that London, or van der Waals, dispersion energies between atoms or groups in the same molecule may lead to an appreciable stabilization of the molecule. Such London forces depend on the product of the polarizabilities of the interacting groups and vary inversely with the sixth power of the distance between them. These forces are large when both groups are highly polarizable. It seems plausible to generalize and state that additional stability due to London forces will always exist in a complex formed between a polarizable acid and a polarizable base. In this way the affinity of soft acids for soft bases can be partly accounted for.

Mulliken (*22*) has given a different explanation for the extra stability of the bonds between large atoms—for example, two iodine atoms. It is assumed that d-p-orbital hybridization occurs, so that both the π-bonding molecular orbitals and the π^*-antibonding orbitals contain some admixed d character. This has the two-fold effect of strengthening the bonding orbital by increasing overlap and weakening the antibonding orbital by decreasing overlap.

The first part of this article appeared on p. 581 of the September issue of THIS JOURNAL and discussed the fundamental principles of the law of Hard and Soft Acids and Bases. Numbers of equations, footnotes, and references follow consecutively those in Part I.

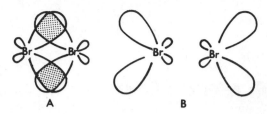

Figure 2. Atomic orbital hybrids for (a) bonding and (b) antibonding molecular orbitals. These atomic hybrids are formed by combining a 4 p and a 5 d orbital on each bromine atom. The hybrids are then combined to form the molecular orbitals.

Figure 2 shows the appearance of the hybrid orbitals on two bromine atoms. These are now added and subtracted in the usual way to form bonding and anti-bonding molecular orbitals. The bonding orbital will clearly have a greater overlap than if it were formed by adding a p atomic orbital from each bromine atom. Hence it will be more bonding. The anti-bonding molecular orbital will overlap less than if it were formed by substracting two p atomic orbitals. Hence it will be less anti-bonding.

Mulliken's theory is the same as Chatt's π-bonding theory as far as the π-bonding orbital is concerned. The new feature is the stabilization due to the antibonding molecular orbital. As Mulliken points out, this effect can be more important than the more usual π-bonding. The reason is that the antibonding orbital is more antibonding than the bonding orbital is bonding, if overlap is included. For soft-soft systems, where there is considerable mutual penetration of charge clouds, this amelioration of repulsion due to the Pauli principle would be great.

Klopman (23) has developed an elegant theory based on a quantum mechanical perturbation theory. Though applied initially to chemical reactivity, it can apply equally well to the stability of compounds. The method emphasizes the importance of charge and frontier-controlled effects. The frontier orbitals are the highest occupied orbitals of the donor atom, or base, and the lowest empty orbitals of the acceptor atom, or acid. When the difference in energy of these orbitals is large, very little electron transfer occurs and a charge-controlled interaction results. The complex is held together by ionic forces primarily.

When the frontier orbitals are of similar energy, there is strong electron transfer from the donor to the acceptor. This is a frontier-controlled interaction, and the binding forces are primarily covalent. Hard-hard interactions turn out to be charge-controlled and soft-soft interactions are frontier-controlled. By considering ionization potentials, electron affinities, ion sizes, and hydration energies, Klopman has succeeded in calculating a set of characteristic numbers, $E^{\ddagger}$, for many cations and anions.

These numbers, Table 5, show an astonishingly good correlation with the known hard or soft behavior of each of the ions as a Lewis acid or base. The only exception is H^+, which turns out to be a borderline case by calculation, but experimentally is very hard. Probably it is a special case because of its small size. Tl^{3+} is predicted to be softer than Tl^+, as is known to be true experimentally.

Table 5. Calculated Softness Character (Empty Frontier Orbital Energy) of Cations and Donors[a]

Ion	Orbital energy (eV)	De-solvation[b] energy (eV)	$E^{\ddagger}_m$ (eV)	
Al^{3+}	26.04	32.05	6.01	
La^{3+}	17.24	21.75	4.51	
Ti^{4+}	39.46	43.81	4.35	
Be^{2+}	15.98	19.73	3.75	
Mg^{2+}	13.18	15.60	2.42	
Ca^{2+}	10.43	12.76	2.33	
Fe^{3+}	26.97	29.19	2.22	Hard
Sr^{2+}	9.69	11.90	2.21	
Cr^{3+}	27.33	29.39	2.06	
Ba^{2+}	8.80	10.69	1.89	
Ga^{3+}	28.15	29.60	1.45	
Cr^{2+}	13.08	13.99	0.91	
Fe^{2+}	14.11	14.80	0.69	
Li^+	4.25	4.74	0.49	
H^+	10.38	10.8	0.42	Borderline
Ni^{2+}	15.00	15.29	0.29	
Na^+	3.97	3.97	0	
Cu^{2+}	15.44	14.99	−0.55	
Tl^+	5.08	3.20	−1.88	
Cd^{2+}	14.93	12.89	−2.04	
Cu^+	6.29	3.99	−2.30	Soft
Ag^+	6.23	3.41	−2.82	
Tl^{3+}	27.45	24.08	−3.37	
Au^+	7.59	3.24	−4.35	
Hg^{2+}	16.67	12.03	−4.64	
			$E^{\ddagger}_n$	
F^-	6.96	5.22	−12.18	
H_2O	15.8	(−5.07)[c]	−(10.73)	
OH^-	5.38	5.07	−10.45	Hard
Cl^-	6.02	3.92	−9.94	
Br^-	5.58	3.64	−9.22	
CN^-	6.05	2.73	−8.78	
SH^-	4.73	3.86	−8.59	
I^-	5.02	3.29	−8.31	Soft
H^-	3.96	3.41	−7.37	

[a] KLOPMAN (23).
[b] Refers to aqueous solution.
[c] This value is negative, as it would be in general for neutral ligands, because the solvation increases rather than decreases during the removal of the first electron. The numerical value has been put equal to the value for OH^- in absence of more reliable data.

The numbers, $E^{\ddagger}$, consist of two parts: the energies of the frontier orbitals themselves, in an average bonding condition, and the changes in solvation energy that accompany electron transfer, or covalent bond formation. It is the desolvation effect that makes Al^{3+} hard, for example, since it loses much solvation energy on electron transfer. All cations would become softer in less polar solvents. Extrapolation to the gas phase would, in fact, seem to make the hardest cations in solution become the softest! In the same way, the softest anions in solution seem to become the hardest in the gas phase. This suggests that it is not reasonable to extrapolate the interpretations from solution into the gas.

It should be remembered that much of the data on which Table 4 (Part I) is based was obtained from studies in the gas phase, or in solvents of very low polarity. Thus the characteristic behavior of hard and soft Lewis acids exists even in the absence of solvation effects. For example, the reaction

$$CaF_2(g) + HgI_2(g) \rightleftharpoons CaI_2(g) + HgF_2(g) \qquad (19)$$

is endothermic by about 50 kcal. The hard calcium ion prefers the hard fluoride ion, and the soft mercury ion prefers the soft iodide ion, just as they would in solution.

When the electron donor and electron acceptor are brought together (in solution) to form a complex, the change in energy may be calculated by Klopman's method. The calculation does not involve multiplying together $E^{\ddagger}_m$ and $E^{\ddagger}_n$. Instead their difference becomes important, as well as the magnitude of the exchange integral between the frontier orbitals. This must be estimated in some way.

The most stable combinations are found for large positive values of $E^{\ddagger}_m$ with large negative values of $E^{\ddagger}_n$ (hard-hard combination), or for large negative values of $E^{\ddagger}_m$ with small negative values of $E^{\ddagger}_n$ (soft-soft combinations). This explains the HSAB principle. It is also noteworthy that the theory predicts that complexes formed by hard cations and hard anions exist because of a favorable entropy term, and in spite of unfavorable enthalpy change. Complexes of soft cations and anions exist because of a favorable enthalpy change. This is exactly what is observed in aqueous solution (24).

The generally good agreement between Klopman's approach and the experimental properties of the various ions does suggest that the simple explanation based on hard-hard binding being electrostatic and soft-soft binding being covalent, is a good one. There is no reason to doubt, however, that π-bonding and electron correlation in different parts of the molecule can be more or less important in various cases. The electron correlation would include both London dispersion and Mulliken's hybridization effect.

It is just because so many phenomena can influence the strength of binding that it is not likely that one scale of intrinsic acid-base strength, or of hardness-softness, can exist. It has been a great temptation to try to equate softness with some easily identified physical property, such as ionization potential, redox potential, or polarizability. All of these give roughly the same order, but not exactly the same. None is suitable as an exact measure (18). The convenient term *micropolarizability* may sometimes be used in place of softness to indicate that deformability of an atom, or group of atoms, at bonding distances is the important property.

Some Applications of the HSAB Principle

In conclusion we may say that in the broadest sense the HSAB principle is to be regarded as an experimental one. Its use does not depend upon any particular theory, though several aspects of the theory of bonding may be applicable. No doubt the future will bring many changes in our ideas as to why HOI is stable compared to HOF, whereas the reverse is true for HF compared to HI. While the explanations will change, the chemical facts will remain. It is these facts that principle deals with.

In spite of several efforts, it does not seem possible to write down quantitative definitions of hardness or softness at this time. Perhaps it is not even desirable, lest too much flexibility be lost. The situation is somewhat reminiscent of the use of the terms "electronegativity" and "solvent polarity." Here also no precise definitions exist or, rather, many workers have established their own definitions. The several definitions, while conflicting in detail, usually conform to the same general pattern.

The looseness of meaning in the terms hard and soft does create some pitfalls in the application of the HSAB principle. Problems do arise particularly in discussing the "stability" of a chemical compound in terms of the HSAB principle. A great deal of confusion can result when the term stable is applied to a chemical compound. One must specify whether it is thermodynamic or kinetic stability which is meant, stability to heat, to hydrolysis, etc. The situation is even worse when a rule such as the principle of hard and soft acids is used. The rule implies that there is an extra stabilization of complexes formed from a hard acid and a hard base, or a soft acid and a soft base. It is still quite possible for a compound formed from a hard acid and a soft base to be more stable than one made from a better matched pair. All that is needed is that the first acid and base both be quite strong, say H^+ and H^- combined to form H_2.

A safer use of the rule is to use it in a comparative sense, to say that one compound is more stable than another. This is really only straightforward if the two compounds are isomeric. In other cases it is really necessary to compare four compounds, the possible combinations of two Lewis acids with two bases, as in eqn. (2). An example might be

$$2LiBu(l) + ZnO(s) \rightleftharpoons Li_2O(s) + ZnBu_2(l) \qquad (20)$$

The value of $\Delta H = -17$ kcal shows that Zn^{2+} is softer than Li^+, which is what we would conclude from their outer electronic structure. Notice also that it is likely that Zn^{2+} is a stronger acid than Li^+, and that O^{2-} is a stronger base than n-$C_4H_9^-$. However, the stable products do not contain the strongest acid combined with the strongest base.

The point has been made that the intrinsic strength of an acid or base is of comparable importance to its hardness or softness. Methods were described for estimating the strength of an acid or a base in terms of its size and charge, etc. It follows from what was said that the strongest acids are usually hard (not all hard acids are strong, however). Many, but not all, soft bases are quite weak (benzene, CO, etc.). One expects, in general, that the strongest bonding will be found between hard acids and hard bases. The strength of the coordinate bond in such cases may range up to hundreds of kilocalories.

Many combinations of soft acids with soft bases are held together by very weak bonds, perhaps only several kilocalories per bond. Examples would be some charge transfer complexes. With such weak overall bonding, one wonders why some soft-soft combinations are formed at all. A partial answer lies in considering eqn. (2) which, as mentioned before, represents the more common kind of chemical reaction actually occurring.

$$A:B' + A':B \rightleftharpoons A:B + A':B' \qquad (2)$$

The usual rule for a double exchange of the type above is that the strongest bonding will prevail. Thus if A and B are the strongest acid and base in the system, reaction will occur to form A:B. The product A':B' is necessarily formed as a by-product, even though its bonding may be quite weak.

It is in cases where the two acids or the two bases, or both, are of comparable strength that the effect of softness or hardness becomes most important. This can be seen from a consideration of eqn. (10). Applied to reaction (2), this leads to the predicted equilibrium constant

$$\log K = (S_A - S_{A'})(S_B - S_{B'}) + (\sigma_A - \sigma_{A'})(\sigma_B - \sigma_{B'}) \qquad (21)$$

Thus the I_3^- complex is formed in aqueous solution not

so much because of the strength of the binding between I^- and I_2, but because I_2 and H_2O are both weak acids and I^- and H_2O are both weak bases. Hence the first term on the right hand side of eqn. (21) must be small, and the second term must dominate. This is an alternative way of saying that the soft I^- and I_2 are weakly solvated by water, whereas water molecules solvate each other well by hydrogen bonding. Both A' and B' in eqn. (2) are water molecules, in this case.

Solubility may obviously be discussed in terms of hard-soft interactions. The rule is that hard solutes dissolve in hard solvents and soft solutes dissolve in soft solvents. This rule is actually a very old one when used in the form "like things dissolve each other." Hildebrand's rule for solubility is that substances of the same cohesive energy density ($\Delta E_{vap}/V$) are soluble in each other (25). Hard complexes, composed of hard acids and bases, have a high cohesive energy density, and soft complexes have a low cohesive energy density, as a rule.

Water is a very hard solvent, both with respect to its acidic and basic functions. It is the ideal solvent for hard acids, hard bases and hard complexes. Alkyl substituents, such as in the alcohols, reduce the hardness in proportion to the size of the alkyl group. Softer solutes then become soluble. For example oxalate salts are quite insoluble in methanol. Dithiooxalate salts are quite soluble. Benzene would be a very soft solvent, containing only a basic function, however. Aliphatic hydrocarbons are rather soft complexes, but have no residual acid or basic properties to help solvate solutes.

The solvation of cations by water is of paramount importance in determining the electromotive series of the metals. If one examines the series, one finds at the bottom of the list in reactivity the metals Pt, Hg, Au, Cu, Ag, Os, Ir, Rh, and Pd. All of these form soft metal ions in their normal oxidation states. Their softness is responsible for their lack of chemical reactivity in aqueous environment.

This can be seen by breaking up the process

$$M(s) \rightarrow M^+(aq) + e^- \qquad E° \qquad (22)$$

into three hypothetical parts:

$$M(s) \rightarrow M(g) \qquad\qquad \Delta H_{sub} \qquad (23)$$

$$M(g) \rightarrow M^+(g) + e^- \qquad IP \qquad (24)$$

$$M^+(g) \rightarrow M^+(aq) \qquad \Delta H_{hyd} \qquad (25)$$

the first two of these require energy: the heat of sublimation and the ionization energy, respectively. Only the third step gives energy back to drive the entire process. If the hydration energy is relatively weak, the metal will have a low $E°$ value and be unreactive. Soft metal ions will indeed have a low hydration energy compared to the energy requirements of the first two steps.

This suggests that these unreactive metals may be made reactive by using a different environment: a softer solvent or mixture of solvents. It is clear that in a mixed solvent, metal ions of different hardness or softness will sort out the mixture. For example, in very concentrated solutions of chloride ion in water, hard ions such as Mg^{2+} and Ca^{2+} will bind to H_2O, whereas softer ions such as Ni^{2+}, Cu^{2+}, Zn^{2+}, and Cd^{2+} will bind to Cl^- (26). Adding chloride ions to water should increase the reactivity of soft metals more than the reac-

tivity of hard metals.

It is of interest to note that the difference between the sum of the ionization potentials and the heat of hydration of an ion forms a series almost exactly like those of Table 5. The difference in energy must be divided by n, the number of electrons lost or gained by the ion to make ions of different charges comparable (Stan Ashland, private communication).

A useful rule is used by inorganic chemists when they wish to precipitate an ion as an insoluble salt. The rule is to use a precipitating ion of the same size, shape, and of opposite, but equal, charge. For example, Cr(NH_3)$_6^{3+}$ is used to precipitate Ni(CN)$_5^{3-}$ (27); PF$_6^-$ is used to precipitate Mo(CO)$_6^+$; but CO$_3^{2-}$ precipitates Ca^{2+}; S^{2-} precipitates Ni^{2+}; I^- precipitates Ag^+; etc. In the latter cases a good lattice energy results from the combination of small ions.

The insolubility of the large ions does not result so much from a good lattice energy, but from the poor solvation of the large ions, which may be regarded as soft, weak acids and bases. Even when precipitates are not formed, it is known that large cations form complexes, or ion-pairs, with large anions (28).

Consider the solid-state reaction

$$LiI(s) + CsF(s) \rightarrow LiF(s) + CsI(s) \qquad \Delta H° = -33 \text{ kcal} \quad (26)$$

The final combinations of hard Li^+ and hard F^- combined, as well as soft Cs^+ and soft I^-, is much more stable than the mismatched combination of hard and soft LiI and CsF. However, simple lattice energy considerations show that it is the high stability of LiF (solid) which drives the reaction. The weakly bound CsI is just along for the ride, so to speak.

In addition to solubility of salts, the tendency to form salt hydrates can be discussed from the HSAB viewpoint. To form a hydrate, we generally need a cation or an anion which is hard, so that it has an affinity for H_2O. However, if both the cation and anion are hard, the lattice energy will be too great and a hydrate will not form.

The alkali halides provide a nice example. We find the greatest tendency to form hydrates with LiI, and least with LiF, which is rather insoluble, in fact. At the other end, we find that CsF is one of the few simple cesium salts which does form a hydrate, whereas CsI does not. In the latter case, both ions are soft and, even though the lattice is weak, water has no tendency to enter.

The simple chemical reaction in eqn. (26) is an extremely informative one. Let us examine it in another way, by converting it to the gas phase.

$$LiI(g) + CsF(g) \rightarrow LiF(g) + CsI(g) \qquad (27)$$

In this case the heat of the reaction is -17 kcal, so it is still strongly favored to go to the right as shown. Again the strong bond between Li and F is decisive. This is of interest because Pauling (29) has a celebrated rule for predicting the heats of reactions such as in eqn. (27). According to this rule, a reaction is exothermic if the products contain the most electronegative element combined with the least electronegative element. Since Cs is more electronegative than Li, this rule predicts that reaction (27) will be endothermic! Pauling's rule is supposed to be a quantitative one.[5]

---

[5] However, it is not considered to be quite as reliable for bonds between two atoms of greatly different electronegativities.

	$\Delta H_{exp.}$	$\Delta H_{calc.}$[a]
Table 6. Heats of Gas Phase Reactions at 25°C		
$BeI_2 + SrF_2 = BeF_2 + SrI_2$	−48 kcal	+35 kcal
$AlI_3 + 3NaF = AlF_3 + 3NaI$	−94	+127
$HI + NaF = HF + NaI$	−32	+76
$HI + AgCl = HCl + AgI$	−25	+5
$NOI + CuF = CuI + NOF$	−10	+76
$LaF_3 + AlI_3 = AlF_3 + LaI_3$	−9	+84
$CaO + H_2S = CaS + H_2O$	−37	+25
$CS_2 + 2H_2O = CO_2 + 2H_2S$	−16	+37
$CS + PbO = PbS + CO$	−71	+64
$CH_3HgCl + CH_4 = (CH_3)_2Hg + HCl$	−40	+5
$CH_3HgCl + HI = CH_3HgI + HCl$	−11	+5
$COBr_2 + HgF_2 = COF_2 + HgBr_2$	−85	+66
$2CuF + CuI_2 = 2CuI + CuF_2$	−25	+14
$2TiF_2 + TiI_4 = TiF_4 + 2TiI_2$	−51	+56
$CH_3OH + CH_3OH = CH_2(OH)_2 + CH_4$	−20	+13
$CH_3F + CF_3I = CH_3I + CF_4$	−22	+69
$CH_3F + CF_3H = CH_4 + CF_4$	−19	+88

[a] Calculated from eqn. (29).

For a reaction (where A and C are the more metallic elements)

$$AB + CD \rightarrow AD + CB \qquad (28)$$

the heat of reaction in kcal/mole becomes[6]

$$\Delta H = 46(X_C - X_A)(X_B - X_D) \qquad (29)$$

where the X's are the electronegativities. This gives a value of ΔH equal to $46(1.0 - 0.7)(4.0 - 2.5) = +21$ kcal, for reaction (27).

Table 6 shows a number of heats of reaction calculated by Pauling's eqn. (29), compared to the experimental results. It can be seen that the equation is totally unreliable in that it gives the sign of the heat change incorrectly. Many other examples can be chosen, some of which will agree with eqn. (29) and some of which will not, as to the sign of ΔH. However, it is easy to tell in advance when the equation will fail (30).

Among the representative and early transition elements, X always decreases as one goes down a column in the periodic table. This leads to the Pauling prediction that for heavier elements in a column, the affinity for F will increase relative to that for I. The prediction is also made for preferred bonding to O compared to S, and N compared to P. The facts are always otherwise.

Similarly, if one goes across the periodic table, the electronegativity of the elements increases steadily. This leads to the Pauling prediction that in a sequence such as Na, Mg, Al, Si the affinity for I will increase relative to that for F. Similarly, bonding to S and P atoms will be preferred relative to O and N. However, as long as the elements have the positive group oxidation states, the facts are the opposite with very few exceptions.

Even more serious, eqn. (29) will almost always predict incorrectly the effect of systematic changes in A and C. For example, what happens to the heat of reaction in eqn. (28) if the oxidation state of the bonding atoms change, or if the other groups attached to these atoms are changed? Such changes affect the electronegativity in a predictable way. For example, the X's of Pb(II) and Pb(IV) are 1.87 and 2.33, respectively, (31). Similarly, the X value of carbon is 2.30 in CH_3, 2.47 in CH_2Cl and 3.29 in CF_3 (32). Increased positive oxida-

[6] This equation comes from the Pauling (29) bond energy equation

$$D_{AB} = \frac{1}{2}(D_{AA} + D_{BB}) + 23(X_A - X_B)^2$$

where D_{AB} is the bond energy of an AB bond, etc.

tion state and substitution of less electronegative atoms by more electronegative atoms always increases X of the central bonding atom. From eqn. (29), such changes again are predicted to decrease the relative affinity for F, O, and N, compared to I, S, and P. For all of the elements, except a few of the heavy post-transition elements (Hg, Tl, etc.), the reverse is true.

If organic chemistry is considered in terms of the HSAB concept, it becomes clear that a simple alkyl carbonium ion is a much softer Lewis acid than the proton (33). In an equilibrium such as

$$CH_3OH + HA \rightleftharpoons CH_3A + H_2O \qquad (30)$$

the equilibrium constant will be large when A^- is a base in which the donor atom is soft, such as C, P, I, S. Since carbon is more electronegative than hydrogen ($X = 2.1$), and since oxygen ($X = 3.5$) is more electronegative than any of the soft donor atoms, this could be explained by the use of eqn. (29), which works in this case (34).

However eqn. (29) predicts that if carbon becomes more electronegative than carbon in a methyl group, it will have an even greater affinity for soft donor atoms of low electronegativity. This is exactly the reverse of what is found. The more electronegative a carbon atom becomes, the less it wants to bind to soft atoms. Certainly the carbon of an acetyl cation is more electronegative than that of a methyl cation. Yet in the reactions

$$CH_3COOH + HA \rightleftharpoons CH_3COA + H_2O \qquad (31)$$

we now find that the equilibrium constant is small if A has C, P, I, S, etc., as a donor atom.

The poor results of Table 6 are not due to a poor choice of the X values of the elements. No reasonable adjustment of these values will improve the situation. If new parameters X_A, X_B, etc., are found for the elements to give the best fit to eqn. (28), they will no longer be identifiable as electronegativities. They would necessarily vary with position in the periodic table, with oxidation state, and with substitution effects in a way directly opposite from what one would expect of simple electronegativities.

The Principle of Hard and Soft Acids and Bases may be used to predict the sign of ΔH for reactions such as in eqn. (28). The Principle may be recast to state that, to be exothermic, the hardest Lewis acid, A or C, will coordinate to the hardest Lewis base, B or D. The softest acid will coordinate to the softest base. Softness of an acceptor increases on going down a column in the periodic table; hardness increases on going across the table, for the group oxidation state; hardness increases with increasing oxidation state (except Tl, Hg, etc.), and as electronegative substituents are put on the bonding atoms A or C. For donor atoms X may be taken as a measure of the hardness of the base, donors of low X being soft. Accordingly, the HSAB Principle will correctly predict heats of reaction where the electronegativity concept fails. Some exceptions will occur since it is unlikely that any single parameter assigned to A, B, C, and D will always suffice to estimate the heat of reaction.

It was not the purpose of this paper to discuss many applications of the HSAB principle. This has been done in previous papers (1, 33). A number of further

83

interesting applications to organic chemistry will appear shortly in papers by Saville (35). One could go on giving examples of the HSAB principle almost without limit, since they may be picked from any area of chemistry. It is to keep this generality of application that we have purposely avoided a commitment to any quantitative statement of the principle, or any special theoretical interpretation.

Whatever the explanations, it appears that the principle of Hard and Soft Acids and Bases does describe a wide range of chemical phenomena in a qualitative way, if not quantitative. It has usefulness in helping to correlate and remember large amounts of data, and it has useful predictive power. It is not infallible, since many apparent discrepancies and exceptions exist. These exceptions usually are an indication that some special factor exists in these examples. In such cases the principle can still be of value by calling attention to the need for further consideration.

Acknowledgment

The author wishes to thank the U. S. Atomic Energy Commission for generous support of the work described in this paper. Thanks are also due to Professor F. Basolo and to Dr. B. Saville for many helpful discussions.

Literature Cited

(18) JORGENSEN, C. K., *Structure and Bonding*, 1, 234 (1966).
(19) DRAGO, R. S., AND PURCELL, K. F., *Prog. Inorg. Chem.*, 6, 217 (1965).
(20) CHATT, J., *Nature*, 177, 852 (1956); CHATT, J., *J. Inorg. Nucl. Chem.*, 8, 515 (1958).
(21) PITZER, K. S., *J. Chem. Phys.*, 23, 1735 (1955).
(22) MULLIKEN, R. S., *J. Am. Chem. Soc.*, 77, 884 (1955).
(23) KLOPMAN, G., *J. Am. Chem. Soc.*, 90, 223 (1968).
(24) AHRLAND, S., *Helv. Chem. Acta*, 50, 306 (1967).
(25) HILDEBRAND, J. H., AND SCOTT, R. L., "Solubility of Non-Electrolytes," Dover Publications, Inc., New York, N. Y., 1964.
(26) ANGELL, C. M., AND GRUEN, D. M., *J. Am. Chem. Soc.*, 88, 5192 (1966).
(27) BASOLO, F., AND RAYMOND, K., *Inorg. Chem.*, 5, 949 (1966).
(28) PRUE, J., "Ionic Equilibria," Pergamon Press, Oxford, 1965, p. 97.
(29) PAULING, L., "The Nature of the Chemical Bond" (3rd ed.), Cornell University Press, Ithaca, N. Y., 1960, pp. 88–105.
(30) PEARSON, R. G., *Chem. Comm.*, 2, 65 (1968).
(31) ALLRED, A. L., *J. Inorg. Nucl. Chem.*, 16, 215 (1961).
(32) HINZE, H. J., WHITEHEAD, M. A., AND JAFFÉ, H. H., *J. Am. Chem. Soc.*, 85, 148 (1963).
(33) PEARSON, R. G., AND SONGSTAD, J., *J. Am. Chem. Soc.*, 89, 1827 (1967).
(34) HINE, J., AND WEIMAR, R. D., *J. Am. Chem. Soc.*, 87, 3387 (1965).
(35) SAVILLE, B., *Angew. Chem.* (International Edition), 6, 928 (1967).

Part III

Editor's Comments on Papers 10–18

In this section are a number of papers which are loosely woven together by a common theme: why are there two classes of acceptors (whether we call them (a) and (b) or hard and soft is immaterial), and what properties of the donors cause them to prefer one class of acceptor or the other?

The paper by Williams and Hale is a presentation of much interesting information and many reasonable ideas. The authors do not like the hard and soft concept because it is too qualitative, and because hard and soft have definite meanings which do not always fit the properties of the donors and acceptors. These are legitimate complaints.

Ahrland's 1966 paper discusses the key role of d electrons in producing soft, or class (b), behavior. Jørgensen's two papers are enjoyable and stimulating reading, as always. Ahrland's 1968 paper and the one by Schwarzenbach are concerned with conclusions to be drawn from a study of the thermodynamics of complex formation in aqueous solution. The important result is that the bonding between hard acids and bases must be largely electrostatic, and the bonding between soft acids and bases must be much more covalent.

The last three papers deal with the bond energy problem by calling attention to a clear failure of Pauling's well-known bond energy equation. From a practical point of view, it is the heats, or free energy changes, of reactions such as

$$AB + CD \rightleftharpoons AD + CB$$

which are important in chemistry. The idea of ionic resonance energy, more often than not, leads to a wrong estimate of the *sign* of ΔH, much less its magnitude. The HSAB principle, while it cannot predict the magnitude of ΔH, at least has the virtue of almost always getting the sign right. This is most helpful in preparative chemistry.

The papers by Huheey and Evans analyze the various factors going into bond energy changes, and explain the failure of Pauling's equation. They also make the useful point (actually a rather obvious one) that the HSAB principle can often be paraphrased as small donors like small acceptors, and large donors, willy nilly, are forced to be content with large acceptors.

Considering the enormous number of theoretical papers dealing with the chemical bond, the number that attempt to explain the experimental facts upon which HSAB is based is quite small. One would have thought that such a pervasive set of observations would have drawn more interest among theoreticians. One difficulty is that the heavy atoms, which show soft behavior, contain too many electrons for *ab initio* calculations. Among papers which have appeared, works by Gray (Rich and Davidson, 1968) and by Hollebone (Hollebone, 1971), while not *ab initio,* do give considerable insight by considering the molecular orbitals of ligands, metals, and complexes.

References

Hollebone, B. R., *J. Chem. Soc. A,* 3009 (1971).
Rich, A., and N. Davidson, Eds. "Structural Chemistry and Molecular Biology," W. H. Freeman, San Francisco, 1968, p. 783.

Reprinted from "Structure and Bonding," Volume 1, C. K. Jørgensen et al., eds., pp. 249–281 (1966)
Copyright 1966 by Springer-Verlag, New York, Inc.

10

The Classification of Acceptors and Donors in Inorganic Reactions

Dr. R. J. P. Williams and Dr. J. D. Hale

Wadham College, Inorganic Chemistry Laboratory, Oxford (Great Britain)

Table of Contents

This review is an attempt to clarify the classification of donors and acceptors, sometimes called *Lewis* bases and acids. Many divisions have been proposed such as (a)- and (b)-class and "soft" and "hard". It is our intention to show that these classifications have strictly limited applicability, being of value only within specified groups of donors or acceptors: Thus a series of *ionic* acceptors, such as metal cations in water, may owe their (a) or (b) character to different electronic properties from those which control the character of the *neutral* acceptors XR_n in organic solvents or in the gas phase, where X is an atom from Periodic Groups II, III or IV and R is an alkyl group. By attempting to generalise classes of donors and acceptors so as to include all chemical species in many different environments much confusion has arisen in recent chemical literature (*4, 11, 12*). When discussion is limited to ionic systems in water then the (a) and (b) classification may be shown to be closely linked with the balance between ionic and covalent bonding.

249

I. Introduction

In 1941 *Sidgwick* (*1*) summarized existing stability data in order to discuss the relative affinities of common donor ligands for various acceptor molecules and ions. He pointed out that the predominant order of bond strength occurring with most metal halides was $MF_n > MCl_n > MBr_n > MI_n$ but that in some cases the metal fluoride was *less* stable than the metal chloride. No attempt was made to generalize about the character of the acceptor which brought about this change of order nor was a clear statement made concerning the effect of solvent on these orders. In 1954 *Williams* (*2*) reviewed available thermochemical data for metal complex formation in aqueous solution. He discussed the effects of entropy change on the stability of metal complexes, and stressed the need to use ΔH_{aq} rather than ΔG_{aq} in the comparison of sequences of stabilities. He also discussed the effects of hydration on the stability of metal ion complexes and stated that the heat of formation of complexes in aqueous solution was the result of (1) the nature of the final complex, and (2) the nature of the hydration of ions and molecules involved in the reaction. He noted that cations such as Ag^I and Hg^{II}, which react strongly exothermically with halide ions, have the following order of stability: $I^- > Br^- > Cl^-$. Most other cations react with halide ions with large heat uptake and have the opposite order of stability. *Williams* also gave his reasons why the different orders occurred (see below).

Also in 1954 *Carleson* and *Irving* (*3*) reported the stability constants of indium halides and included in their study a review of the stability constants in water of metal halides located near indium in the Periodic Table. These authors stated that cations fell into two distinct groups: (i) those which formed metal halides whose stability order was $F^- > Cl^- > Br^- > I^-$, and (ii) those which formed metal halides whose stability order was $I^- > Br^- > Cl^- > F^-$. Cations showing the latter order were Cu(I), Ag(I), Cd(II), Hg(II) and probably Tl(III). No rigid classification scheme was proposed however until in 1958 *Ahrland, Chatt* and *Davies* (*4*) made an extensive review of the relative affinities of ligand atoms for acceptor molecules and ions. They stated that, since the amount of experimental data had increased enormously, it seemed profitable to attempt, revise, and extend correlations involving all the ligand atoms except hydrogen. These authors divided both acceptor *ions* and *molecules* into two general classifications. Class (a): Those which form their most stable complexes with the first ligand atom of each Periodic Group, i. e. N, O, F. Class (b): Those which form their most stable complexes with the second or a subsequent ligand atom of each Periodic Group. Stability was measured by the free energy of a reaction, usually quoted just as an equilibrium

250

constant, without regard to the solvent. It was noted that class (b) *cationic* acceptors (*in water*) in their normal valence states were found together in a more or less triangular shape in a particular part of the Periodic Table. The base of this triangle stretched, in the sixth Period from about tungsten to polonium, and had its apex in the first Period at copper. *Ahrland, Chatt* and *Davies* proposed that this order of affinity of class (b) metal ions depended on the availability of electrons from $(n-1)d$ sub-shell orbitals of the metal for dative π-bonding. *Craig, Maccoll, Nyholm, Orgel,* and *Sutton* (5) had earlier indicated from orbital overlap calculations that dative π-bonding could play a significant role in bond formation of B-group metals. The general idea that π-bonding to halides is of importance has found its way into text-books (6), though it has been much criticized.

In 1960 *Williams* (7) repeated his earlier alternative explanation (2) for the behaviour of class (b) metal ions in water. He suggested that the high polarizing power (ionization potentials) of these cations, relative to their size and charge, together with the increasing polarizability of the halide, accounted for the observed order of halide stability. *Williams* showed that a correlation existed between the stability constants of class (b) halides and a quantity, R, defined as the ratio of the ionization potential, IP, to the ionic function, z^2/r, where z is the ionic charge and r the ionic radius of the cation. R was taken to indicate the relative importance of covalent and ionic bonding in donor-acceptor systems (7). The effect of ion size was directly discussed and it was stated that there was no need to introduce the idea of dative π-bonding and, moreover, as the symmetries of the complexes were of many kinds, the symmetry of the wave functions of the possible orbitals concerned in the binding was much less important than their associated energies. He concluded that (b) — class behaviour could be explained by polarization of the ligand by the cation, in which the ligand donates electrons to the cation rather than the reverse, which is implied by dative π-bonding.

In 1961 *Poë* and *Vaidya* (9) discussed the relative stabilities of halogeno-complexes in water using new experimental material. They repeated an earlier statement by *Basolo* and *Pearson* (10) that absolute bond strengths of all metal halides in the gas phase had the order $F^- > Cl^- > Br^- > I^-$. The inversion of this order observed in solvents occurred as a result of hydration or other solvation effects (2, 8, 10). *Poë* and *Vaidya* then associated class (b) behaviour with a narrowing of bond strength differences in the series $F^- > Cl^- > Br^- > I^-$. They suggested that neither polarization, caused by increased effective charge on the metal ion, nor $d/p\pi$-bonding (back donation) could explain class (b) behaviour. By analysing the effects on thermochemical data in aqueous solution caused by varying selected properties, they concluded that ionic size was the major

251

90

factor involved in determining relative bond strengths and thus indirectly the degree of class (b) character. (That size is undoubtedly important cannot be denied, but it is clear that pairs of ions such as Ca^{2+} and Cd^{2+} or Ba^{2+} and Hg^{2+}, which have very comparable sizes, owe their differences to properties other than size.)

More recently still *Pearson (11)* has reviewed various explanations for class (a) and class (b) character referring to species of all charge-types[1] in all sorts of solvents. (He also related their behaviour directly to that of *Lewis* acids.) *Pearson* repeats his earlier statement that all *Lewis* acid *ions* would be class (a) if judged by their affinities for halide ions in the gas phase, but that such a classification is of little value since the reactions are purely hypothetical. Class (b) character in water arises then as a result of the differences in affinity along a sequence, F^- to I^-, which is being viewed against concomitant changes in the heats of hydration. This will be further illustrated below and now appears to be generally agreed *(2, 8, 9)*. By comparing ΔH values for the *gaseous* replacement reaction, i. e. $MX_{(g)} + Y^-_{(g)} = MY_{(g)} + X^-_{(g)}$ where X^- and Y^- are halides, one can measure the amount of stability which the fluoride, Y^-, has over the iodide, X^-, and this *Pearson (11)* considered to be a measure of class (b) character. He then lists a variety of metal ions and *Lewis* acids as class (a), class (b) or borderline. From polarization considerations he chooses to call class (a) "hard", and class (b) "soft". This generalized terminology has been extended to all types of acids and bases by *Pearson* and others. Unfortunately, it is doubtful if many authors using the terminology know *exactly* what they mean by "soft" and "hard", which *Pearson* himself relates to polarizability only in a general, vague, way. (In fact it cannot be convincingly shown that class (a) and class (b) character is related to polarizability (see below). Some of the possible meanings of "soft" and "hard" are also discussed later). Furthermore it is not clear that such definitions as "soft" and "hard" really refer to the same physical properties in neutral acceptors as in ions. For example in a recent symposium *(12)* much discussion was devoted to encourage the use of this qualitative language but the different authors at the meeting clearly used "soft" and "hard" in different senses. With such a long, polemical, and complicated history it is understandable that there should be some confusion as to the meaning and explanation of class (a) and class (b) behaviour. In an attempt to clarify the position we shall treat gas phase and aqueous ionic systems first and entirely separately from other systems for we believe that much of the confusion can be avoided by a separation of reactions of different charge-type.

---

[1] Charge-type refers to the formal charge on the donors and acceptors, contrast $F_3B \leftarrow NH_3$ and $Th^{4+} \leftarrow F^-$.

252

II. Ionic Donors and Acceptors

A. A working definition of class (a) and (b) for cations

For the following *limited* discussion of ions in water it is advisable to define class (a) and class (b) in the earliest historical sense. Class (a) metal ions form halides whose stability in water is of the order $MF_n > MCl_n > MBr_n > MI_n$. Class (b) metal ions form halides whose stability is in the reverse order. Table I classifies metal ions in this way. This definition clearly leaves out of consideration (on experimental grounds) acceptor properties, especially of neutral species, which could not be studied in this way and to which we return later. We now need some quantitative experimental assessment of the degree of class (a) or (b) character. For *simplicity* we shall use $-\Delta G^{\circ}_{aq}$/per ligand for the first group of more or

Table 1. *Classification of metal ions.*

Charge	(b)-class cations	Indeterminate	(a)-class cations
+1	Au > Ag > Cu > Tl	Cs	K < Li < H
+2	Hg (Pt Pd) > Cd > Pb		Mn < Sn, Ba, VO, UO_2 < Be Zn
+3	Au > Tl > (Rh)		In < La, Fe, Cr, Ga, Sc, Al
+4	Pt		U < Th < Zr

The classification is based upon the stability data given in *A. E. Martell* and *L. G. Sillen:* Stability Constants. Special Publication No. 17. London: Chemical Society 1964. The exact position of a cation relative to another in a series is not always clear as the variety of the data in the tables does not always lead to a well-defined position of a cation. The general classification is not affected by these vagaries.

less equally bound ligands. Table 1 lists the ions in the order of the class character as far as is known and can be reasonably guessed from general solution studies. This quantity is then to be compared with some theoretical parameter, which itself is just a property of the central atom acceptor. The quantity which is most readily treated theoretically is ΔH°_{aq}/per ligand, see below, and therefore we shall often refer to ΔH°_{aq} rather than ΔG°_{aq}. It should be observed that the entropy changes, ΔS°_{aq}, always favour (a)-class behaviour (*2*). Some authors have introduced other quantities in order to assess empirically the amount of class (a) or (b) character. Thus as mentioned above it has been suggested (*13*) that a measure of the amount of class (a) or (b) character is given by comparing the stability of a particular gaseous metal halide with the stability of a different gaseous halide of the same metal. For example

253

Pearson (11) used the absolute value of $\Delta H(g)$ for the following reaction as a means of measuring the amount of class (a) or (b) character: $MI_n(g) + nF^-(g) \rightarrow MF_n(g) + nI^-(g)$. In some cases this method classifies ions differently from the methods given above and as generally used by *Ahrland, Chatt, Davies, Irving, Williams*, and others. This is in part due to the historical use of ΔG_{aq} rather than ΔH_{aq} but also to the fact that the original classifications are based on the behaviour of *aquated* species. Thus if heat data are to be preferred in a classification, aqueous solution measurements refer to the reaction $MI_n(aq) + nF^-(aq) \rightarrow MF_n(aq) + nI^-(aq)$, which can give rise to different orders from those obtained from gas phase data. Unfortunately the use of ΔH_{aq} is not possible because of the lack of experimental data. We have also observed that a better though not perfect correlation exists between the (a)- and (b)-classification and the absolute heat of the gas reaction, as written above, when the latter is divided by the *mean* heat of formation of the fluoride and the iodide in question. We call this quantity, Q. Now despite the fact that neither Q nor $\Delta H(g)$ can be expected to give a very direct correlation with the (a)/(b) classification it remains undeniably true that the classification itself can best be understood by first analysing these gas phase quantities. This we shall do immediately below. Before doing so it is worth noting again that as we have defined them class (a) and class (b) behaviour are *strictly* related to aquated ions and that change of ligand bound to the cation, e. g. from $M(H_2O)_n$ to $M(CN)_n$, or even change of solvent can increase or decrease the number of cations in either of the two classes depending only on the nature of the ligands and the solvent. This is also true of neutral acceptor/donor systems.

B. Experimental evidence for gas-ion reactions

In the following discussion we shall define the stability of a metal halide in terms of its bond strength (bond energy) as being the enthalpy change for the following reaction: $M_{(g)}^{n+} + nX^-_{(g)} = MX_{n(g)}$. This enthalpy change is referred to as the heat of ionization or $\Delta H_{ion}(g)$.[2]

In the first instance we will examine the $\Delta H_{ion}(g)$ for metal dihalides. The metal dihalides have been chosen since they are an intermediate oxidation state for which stability data in the gas phase are well known.

---

[2] The bond strength or bond energy is usually defined as the heat of association of the neutral atoms rather than the ions; however, the two heats differ only by the ionization potential and electron affinity of the reactants. We have chosen this form of the reaction since it is in the same form as that which has been used to discuss equilibria in aqueous solution, i. e. $M_{(aq)}^{n+} + nX^-_{(aq)} = MX_{n(aq)}$.

These data appear in Table 2, in which Q is the ratio of column 6 to column 7.

Table 2. $-\Delta H_{ion}(g)$ *in kcal./mole for* $M^{++}{}_{(g)} + 2X^-{}_{(g)} = MX_{2(g)}$ *at* 298·15°

Values in parentheses are estimated

M	MF_2	MCl_2	MBr_2	MI_2	MI_2-MF_2	$(MI_2+MF_2)^{1/2}$	Q
Be	775	685	662	636	139	705	0.20
Mg	617	545	(523)	501	116	559	0.21
Ca	523	461	444	(423)	100	475	0.21
Sr	491	(438)	417	(398)	93	444	0.21
Ba	470	415	394	(374)	96	422	0.23
Zn	655	616	601	583	72	619	0.12
Cd	591	560	552	540	51	565	0.09
Hg	637	607	600	594	43	615	0.07
Sn	(565)	514	499	480	85	523	0.16
Pb	543	493	480	470	73	506	0.14

(Heats of ionization were taken from *L. Brewer, G. R. Somayajulu,* and *E. Brackett: Chem. Rev.* 63, III, (1963)).

Upon examination of the data it is apparent that all metal halides listed have the same order of stability, i. e., $MF_2 > MCl_2 > MBr_2 > MI_2$, in the gas phase.[3] It is also apparent, however, that there are interesting energy differences between metal ions which are classified in aqueous solution as class (a) and those classified as class (b), (compare Table 1). Firstly for metal ions of the same size (or for molecules of the same internuclear distance) class (b) metal halides have a larger $\Delta H_{ion}(g)$ than class (a). Secondly the differences in the $\Delta H_{ion}(g)$ of the halides of a particular class (b) metal are less than the differences in the $\Delta H_{ion}(g)$ of the halides of a corresponding class (a) metal (8, 9). As was mentioned previously, the differences listed in column 6 of Table 2 have been used by *Pearson* to assess in an alternative fashion the amount of class (a) and (b) character. By measuring, for example, the difference between the $\Delta H_{ion}(g)$ for MF_2 and MI_2 of Table 2 a metal ion order was established. The order for divalent metal ions would then appear as follows:
Class (a). Be > Mg > Ca, Sr, Ba > Sn > Pb > Zn > Cd > Hg. Class (b).[4]

---

[3] This is also the case for the heat of association of neutral atoms in the gas phase.

[4] This order would vary slightly if other halide differences were used rather than $\Delta(MF_2-MI_2)$. (In some other systems, MX_n, however, the order can be made to vary greatly depending on which halide differences are taken (see later).) Since the atomic replacement reaction $(MI_2+2F - MF_2+2I)$ differs only from the ionic replacement reaction $(MI_2+2F^- = MF_2+2I^-)$ by a constant which is the difference between the electron affinities of F and I, the order which we obtain here is the same order which would be obtained had we originally used ΔH of association of neutral *atoms*.

255

This order and that of Q in column (8), Table 2, can be compared with the order of class (a) character as defined above and given in Table 1.

It would be useful for our purposes to examine in similar fashion all other oxidation states of metals forming metal halides. However, the thermochemical data in the gas phase for these oxidation states are less well known. Where $\Delta H_{ion}(g)$ data for MF_n or MI_n are not available, other halide differences may be used to indicate approximate comparable orders. Also where no thermochemical data in the gas phase are known, lattice energies, i. e., $M^{n+}_{(g)} + nX^-_{(g)} = MX_{n(s)}$, may be used to determine an order. This is possible when the heats of sublimation for the various metal halides are either approximately constant or vary in such a fashion that the final metal order will not be affected.

Such examination gives the following metal orders which can again be compared with Table 1.

(1) Monovalent:

Class (a). B > Al > H > Li > Na, K, Rb, Cs, Tl > I > Cu, Ag, Au, Hg > Bi. Class (b).

(2) Trivalent:

Class (a). B > Al, Sc, Y, La > Ga(As, Sb, Bi) > In > Tl > Ir, Rh, Au. Class (b).

(3) Tetravalent:

Class (a). Si > Ge > Sn > C > Te > Pt. Class (b).
      Zr Ti

Although the order determined from $\Delta H_{ion}(g)(MF-MI)$ for monovalent ions does not vary greatly in general if other halide differences are used, the positions of Cu^+ and I^+ for example do vary considerably within the first series depending on which halide differences, i. e. $\Delta H_{ion}(g)(MF-MCl)$ or $\Delta H_{ion}(g)(MBr-MI)$, are chosen.

Now that the orders in the gas phase have been obtained we return to the aqueous solution data.

C. Experimental data in aqueous solution

The *heat* of ionization in aqueous solution, ΔH_{aq}, represents the enthalpy change for the following reaction: $M^{+n}_{(aq)} + nX^-_{(aq)} = MX_{n(aq)}$. Although much ΔH_{aq} data exist for class (b) metal chlorides, bromides and iodides, few data are available for class (b) fluorides and class (a) halides in general. This is because $MX_{n(aq)}$ in these cases is not a stable species. It is therefore difficult to compare class (a) and (b) halides in aqueous solution in a manner which is entirely consistent with $\Delta H_{ion}(g)$. It is easy to show, however, that in aqueous solution most metal ions, which are class (b) by

256

definition based on ΔG_{aq}, would be class (b) if the definition had been based on ΔH_{aq}. A second method for obtaining a rough measure of class (a) character in aqueous solution is to use the equilibrium data for solution of *anhydrous* solid halides. Since this subject has been discussed in earlier papers (2), only an example of a member of each class will be given. Table 3 compares data for two ions for both the defining solution equilibrium and the solid/solution equilibrium.

Table 3. $\Delta H^{o}_{298.15}$ *in kcal./mole for* $M^{++}_{(aq)} + 2X^{-}_{(aq)} = MX_{2(y)}$

M	MCl_2	MBr_2	MI_2	
Sr	+11	+16	+20	y = solid for SrX_2
Hg	−13	−21	−34	y = aqueous for HgX_2
Hg	−16	−25	−41	y = solid for HgX_2

From Table 3 it is apparent that Hg has the order: $MI_2 > MBr_2 > MCl_2$, and is therefore clearly a class (b) metal. Sr has the reverse order and is class (a). (It should be pointed out that any consideration of the entropy change for $SrX_{2(s)} \rightarrow SrX_{2(aq)}$ will only make the class (a) character of Sr more pronounced.)

The data given in Tables 2 and 3 are, of course, related to one another through a thermochemical cycle. $\Delta H_{ion}(g)$ and ΔH_{aq} differ only by the heats of solvation (hydration in aqueous solution) of the reactants and product, and therefore these heats of solvation must affect the absolute bond energies in the gas phases in such a manner as to cause an inversion in the order of stability in cases of class (b) behaviour (see below).

Summary

In the preceeding sections we have discussed the historical development of the concept of class character. It has been shown that if the earliest historical definition of class character is employed, then metal ions can be arranged in class groups and listed in order of varying class character. This metal ion order is established by comparing free energy differences between halides of specific metals. These and other energy differences in thermodynamic properties, such as $\Delta H_{ion}(g)$, reflect increased stability measured in Kcal of one halide (MF_n) over another halide (MI_n). The orders of cations are reasonably similar regardless of which thermodynamic values are employed. (It can be shown that $\Delta H_{f(s)}$, $\Delta H_{f(g)}$, ΔH_{diss}, U, ΔH_{ion}, $\Delta H_{(aq)}$, etc. will give approximately the same metal order of class character). Trends in these differences (class character) can be understood and explained by analysis of the ΔH values themselves. In the following sections we shall attempt to explain what energies go to make up ΔH values and how they can be explained in terms of atomic parameters. Having done this, differences in ΔH values (class character) reduce simply to the differences in the atomic parameters involved.

Qualitatively it can be seen almost immediately that these energy differences $\Delta H_{ion}(g)(MF_n - MI_n)$ are affected to a great extent by ionic size or internuclear distance. When we use these absolute energy differences to indicate a property (class character), we may be measuring mainly the differences in r_e (Z^2e^2/r_e [electrostatic energy]).

In order to eliminate this direct size dependence, another quantity, Q, was calculated. Q is not an absolute energy difference in kilocalories but is a percentage. By dividing absolute energy differences by their means we anticipate that a quantity more closely related to percentage ionic character will be obtained. In fact Q does follow ionic character quite closely as may be seen in the following sections. The Q order for divalent cations is $Ba > Sr, Ca, Mg, > Be > Sn > Pb > Zn > Cd > Hg$. Thus although Be (and other very small cations) would appear to be very high on any thermodynamic scale of energy differences $(MF_2 - MI_2)$ it does not show the biggest difference measured on a percent basis. It is known that these small cations are not the extreme cases of ionic behaviour. Both Cs and Ba may well be examples of greater ionic character but surprisingly they will be shown to be more (b) class than some smaller cations. Thus we can only conclude that it is the combination of many factors (amongst which covalence will usually be dominant) which gives rise to the (a)/(b) classification.

D. Properties of gaseous ions

Before turning to theories of the bonding in molecules some atomic properties can be compared with the orders given above. Fig. 1 (16). plots the Σ ionization potentials/z^2 where z is the charge, against ionic radius (*Pauling*) for various ions. The ordinate is then a measure of the lack of screening of low energy orbitals by inner electron shells which can become of obvious importance in covalence if the orbitals concerned extend into the overlap region. This will always be true of ns and np orbitals, is hardly ever true of $(n$-$2)f$ orbitals, and is true to an increasing extent for $3d < 4d < 5d$ amongst $(n$-$1)d$ orbitals. The abscissa of Fig. 1 is only a rough guide to ion size as we shall see later but no really gross alternations will be required in this property.

258

Interestingly on comparing Fig. 1 with Table 1 it is clear that ions of (b)-class congregate in the top righthand corner of the Figure. This is

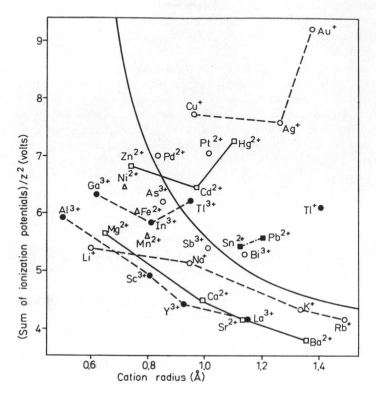

Fig. 1. A plot showing the way in which atomic properties distinguish metal cations in the gas phase. The full line is drawn in an arbitrary manner so as to separate (a) and (b) class cations

the basis of *Williams'* remarks that the relative magnitude of ionization potential to z^2/r i. e. Ir/z^2 decides (b) class character. In detail Fig. 1 does not, nor ever could be expected to, bring out all facets of (b) class character[5]. Other factors of importance are revealed by analysis of binding in molecules.

[5] Prof. *R. S. Nyholm* has pointed out to us that recently and independently he has used the same ratio to discuss (a) and (b) classifications. See *D. P. Craig* and *R. S. Nyholm*: 'Chelating Agents and Metal Chelates', ed. F. P. Dwyer and D. P. Mellor. Academic Press, New York, 1964, 51.

E. General approaches to bond energy

The first point to be stressed in any account of metal ion bonding is the great success of simple ionic calculations of bond energies for the IA and IIA metal halides. (At the same time we should note that no covalent model has ever achieved quantitative reliability.) Thus Fig. 2 shows that

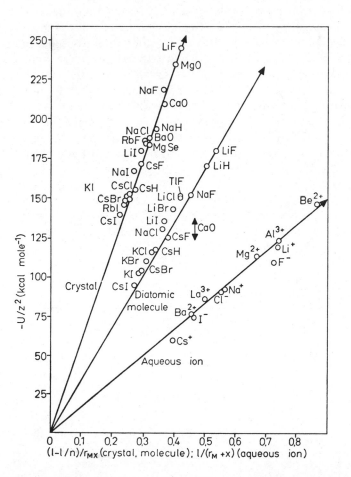

Fig. 2.* Plots testing simple ionic model equations for diatomic molecules in the gas phase, ionic crystals and the hydration of ions. The slopes of the lines coincide with those of the simple theory, see *Phillips* and *Williams*. U is the binding energy from free gas ions.

---

*Figs. 2, 5, 6, 7, and 8 have been taken from ref. *16* with permission.

260

for systems AB all the gaseous halides from LiF to CsI lie close to the predicted value given by the equation

$$\Delta H_{ion} = - \frac{A z_1 z_2}{r_e} \left(1 - \frac{1}{n}\right) \tag{1}$$

where r_e is the equilibrium distance and n is the exponent of the repulsive energy, $+B/r_e^n$, and A is a *Madelung*-type constant. The value of r_e is obtained experimentally. This agreement means that the ionic model must be the most reasonable starting ground for any discussion of cation bond energies. It works equally well for anhydrous IA solid halides or IA hydrates, Fig. 2, and therefore will be used first in examining both these systems too. The most marked exceptions to the ionic equation amongst these gaseous halides are all *high* e. g. LiI (and TlF), in Fig. 2. It is important to realise however that because empirical values of r_e are used *unexpected* changes in r_e which are caused by real deviations from the ionic model may not be seen in a plot such as Fig. 2. However such changes would contribute to unexpected *differences* between $\Delta H_{ion}(g)$ for any MF and MI respectively of the same cation M.

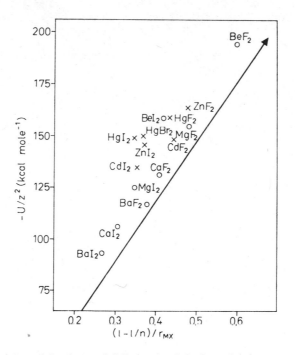

Fig. 3. A test of the theory, full line — simple ionic model, for gaseous MX_2 systems. U is the binding energy from free gas ions.

261

Although the above equation works well for IA cations it fails somewhat for IIA cations, Fig. 3, and much more significantly for IB and IIB cations. A series of modified equations have been proposed in which new terms are added to the ionic energy.

F. Ionic model plus London energy

We shall consider first a special case of the ionic model writing the energy of the molecules

$$E = -\frac{f}{r_e} - \frac{k}{r_e^6} + \frac{B}{r_e^n} \tag{2}$$

where f is a constant allowing for the number of atoms in the molecule and its geometry, k is a constant dependent on the product of polarizabilitis making the term $-k/r_e^6$ a London attraction. Without a number of assumptions all the constants in equation (2) can not be evaluated. The results of such calculations (e. g. *Pearson* and *Gray* (*13*)) make it clear, Table 4, that although the ionic model may be further improved in this way for (a)-class cations it is not much improved for (b)-class cations. Even amongst (a)-class cations this modified ionic model is slightly more exact for lithium than rubidium chloride and it is much worse for beryllium or magnesium chloride than for barium chloride. Again the observed deviations from the ionic model unlike the calculated ones do not change in a systematic way down all the Groups of the Periodic Table. The reason for this is clear enough.

Table 4. *Calculation of modified ionic energies in electron volts (Pearson and Gray (13))*

	+London Energy	+Polarisation	Experiment
LiCl	6.57	7.06	6.70
RbCl	4.23	4.57	4.51
$BeCl_2$	13.71	16.65	14.64
$BaCl_a$	8.39	8.70	8.94
$ZnCl_2$	10.51	12.33	13.00
$CdCl_2$	9.47	11.42	11.99
$HgCl_2$	9.33	11.19	13.05

The important quantity in the above equation which causes deviations from the ionic model is the constant, k, of the London energy which is proportional to the product of the polarizabilities of the interacting charge clouds. This energy increases down *all* Periodic Groups and is greatest in

262

such systems as CsI, see Tables 5 and 6. The deviation increases despite the dependence on distance ($\alpha\ 1/r_e^6$) for the changes in polarizability overcome the changes in $(1/r_e)^6$. Thus London energies will make *all* cations of greater (b) class than they would appear on a simple ionic model for polarizability increases $F^- < Cl^- < Br^- < I^-$. The enhancement of (b)-class character will be the greater the heavier the metal element in a Group, or elsewhere in the order of polarizabilities shown in Table 5. In this sense (b)-class character, polarizability (and "softness", which is not yet defined), move together down the metal Groups.

Table 5. *Ion Polarizabilities (10^{-24} cm^3)*

Ion	Polarizability	Ion	Polarizability
Li$^+$	0.03	Ag$^+$	1.9
K$^+$	1.00	Tl$^+$	3.9
Cs$^+$	2.40		
Be^{2+}	0.01	Zn^{2+}	0.5
Ca^{2+}	0.60	Cd^{2+}	1.15
Ba^{2+}	1.69	Hg^{2+}	2.45
La^{3+}	1.3	Pb^{2+}	3.6
F$^-$	Cl$^-$	Br$^-$	I$^-$
0.81	2.98	4.24	6.45

Table 6. *London energies (eV)*

	F	Cl	Br	I		
Li	0.29	0.44	0.48	0.57		
Na	0.34	0.40	0.44	0.47		
Rb	0.64	0.62	0.62	0.62		Solid Halides
Ag		1.70	1.60	1.74		
Tl		1.66	1.67	1.75		
	BeO	MgO	CaO	SrO	BaO	
	0.1	0.1	0.2	0.2	0.3	gas molecules

Data from *E. J. W. Vervey* and *J. H. de Boer:* Rec. Trav. Chim. Pays-Bas *55*, 431 and 443 (1936); and *J. E. Mayer:* J. Chem. Phys. *1*, 327 (1933).

For the sake of clarity it is necessary to digress here so as to eliminate some possible confusion about "softness". If coefficients of expansion or elastic constants are used to measure softness (this is a meaningful physical definition) then Table 7 shows that "softness" does increase down a Group of the Periodic Table much as would be expected from ionic model considerations. Expansion or elasticity will be assisted by large London energy terms which may help to *flatten* the potential energy minimum due to the opposing coulombic and core-core repulsion energies. Coefficients of expansion generally do not follow (b)-character however as can be

263

seen from Table 7 and the orders in Table 1. Thus if softness is given its simplest physical significance, "squashy-ness", it is a property unrelated to (b)-class character. (The term soft is used by metallurgists in this sense and see also reference (7)).

Table 7. *Physical properties of solids*

	Linear Coef. (0° C) Thermal Expansion x 10^{-6}	Elastic Constant dynes/cm^2 x 10^{11}
LiF	32	11.1
LiCl	43	4.9
LiBr	49	3.9
LiI	58	2.8
KCl	35.7	4.9
KI	38	2.8
CsI	47	—
TlBr	—	3.8
MgO	9.8	28.6
ZnS	6.3	10.5
PbS	18.0	12.7
CdS	4.0	8.1

G. Ionic model plus polarization

An alternative modification to the ionic model is to add a polarization energy, $-\dfrac{\Sigma \; \alpha(\mathrm{f})}{2r_e^4(1 + \dfrac{n\alpha}{r_e^3})}$ to the ionic energy, $\dfrac{Ae^2}{r_e}(1-\dfrac{1}{n})$ where α is the polarizability of the polarizable ion (usually assumed to be the halide), f is the same constant as used above, and n is a geometric factor. The sum in the equation allows for polarization of the cation also. There is no entirely satisfactory way of assessing the merits of this model relative to the previous one for it has dropped the London energy while introducing another form of polarization the permanent polarization of one electron cloud by a neighbouring ion. The calculations, Table 4, show that all the MX energies of Group IA halides can now be satisfactorily accounted for but that MX_2 energies of Group IIA halides are still in error by $\pm 10\%$ i. e. even for (a) class cations. The even larger deviations for other cations indicate the importance of some other energy.

Now this consideration of permanent polarization of a charge cloud of an anion or cation by a field of a gegenion gives rise to the constant α/r_e^4 where α is the polarizability. The maximum value of this term, unlike the maximum in London energies, arises for the smallest cation in

264

partnership with the largest anion and conversely for the largest anion in partnership with the smallest cation. Thus (b)-class character of a polarizing, but non-polarizable, cation is increased by this energy term. Conversely for a *large polarizable* cation the *strong* polarization by a *small* anion aids (a)-class character. The polarizable cation has an increased affinity $F^- > Cl^- > Br^- > I^-$ from this term. A good example of this effect is shown in Fig. 2 where both LiI and TlF are seen to deviate equally from the ionic model but more convincing evidence that the high polarizability of the low-valence cations of B-subgroups enhances (a) class character comes from the heats of formation of a wide range of compounds (see below).

Again we digress to discuss soft and hard (*11*). It has been said that polarizability measures softness and that the permanent charge-cloud distortion is responsible for (b)-class behaviour. Unfortunately polarization energies of this kind work in two ways — they generate (i) (b)-class behaviour due to anion polarizability but (ii) (a)-class behaviour due to the cation polarizability. Thus if softness and polarizability are related they are not necessarily related to (b)-class behaviour. Moreover on the basis of polarizability of this kind there can be no preferential soft-soft or hard-hard interaction.

Yet a third way in which qualitative thinking about softness and polarizability can be shown to be confusing is to examine both an angular crystal field and a radial polarization of *d*-orbitals. Here spectroscopic and magnetic parameters can be obtained directly from experiment in order to give a quantitative expression to polarization. The angular polarization of a cation is expressed by a ligand order — the spectrochemical series — which is $F^- > Cl^- > Br^- > I^-$. Thus on grounds of angular polarizability, transition metal cations prefer the hard anion, fluoride. Thus where direct measurement is apparently available the preferred soft-soft interaction is disproved. It is all the more disquieting that when the nephelauxetic series is used as a measure of radial polarization of the same cations by the same anions exactly the reverse order of polarization is apparently obtained — $I^- > Br^- > Cl^- > F^-$. Soft-soft interactions are dominant? The total effect due to these two opposing polarizations is not determinable, showing how confusing the loose use of polarization, as an all-embracing term, can be.

H. Covalence

The third general model (cf *Pearson* and *Gray* (*13*), *Klopman* (*14*), and *Ferreira* (*15*)) considers a partial ionic bond. In this model both the *Van der Waals* and the polarization energies are discarded in an attempt to see

265

if covalence can account for the gross failure of the earlier models with B-Group cations. The initial equation is that for the energy, W, of a *pair* of bonding electrons,

$$W = (1 + x)q_a + 2(1 - x^2)^{1/2}\beta + (1 - x)q_c$$

where x is the fractional ionic character of the bond, q_c is the cation coulomb integral and is very closely dependent on the ionization energy of the cation for $x \simeq O$, q_a is the anion coulomb integral and, for $x \simeq O$, depends on the ionization potential of the anion, its electron affinity, and upon the ionic energy $(\dfrac{xf}{r_e})$, which arises through the interaction of the anion with the cation; β is the exchange integral and is taken to be proportional to the geometric mean of the AA and BB bond energies:

$$2\beta = -1.2 \sqrt{E_{AA} \cdot E_{BB}}$$

The energy per bond of a molecule MX_2 for example now becomes

$$E = -(1 + x) \left\{ -(1 - x)IP_a - x\frac{(IP_a + EA)}{2} - \frac{x f}{2r_e} \right\} - 1.2(1 - x^2)^{1/2} \sqrt{E_{AA} \cdot E_{BB}}$$

$$+ (1 - x) \left\{ x(IP)_2 + (1 - x)\frac{\Sigma(IP)_c}{2} + \frac{VSPE}{2} \right\} - (IP_a + EA) - \frac{B}{r_e^n}$$

where IP_a is the ionization potential of the anion and EA is its electron affinity, xf/r_e is the *Madelung* energy, IP_2 and IP_c are the second and total ionization potentials of the divalent cation, and B/r_e^n is the repulsion energy used previously. In order to evaluate this expression dE/dx is put equal to zero, assuming such a differential is meaningful at constant r_e, and x found for this maximum E.

Crudely we may illustrate what is happening in the following manner. Take a diatomic system and remove one electron from the cation M and from the anion X. Returning the electrons at the equilibrium internuclear distance of MX gives us, as a first approximation and assuming that $(1 + x)$ of the electrons settle on the anion,

$$IP_a + x\, EA \quad \text{(i. e. anion energy)} \tag{1}$$

$$(IP_c)x \quad \text{(i. e. cation energy)} \tag{2}$$

$$\sqrt{E_{aa}E_{cc}}\, (1 - x^2)^{1/2} \quad \text{(i. e. covalence)} \tag{3}$$

$$\frac{xf}{r_e} \quad \text{(i. e. ionic energy)} \tag{4}$$

The sum of (1) + (2) + (4) is very like the attractive energy of an ionic bond formed from atoms A and B and involving a partial charge, x, while (3) is a covalence energy for a bond of fractional order $(1 - x)$. The

266

sum of Σ (1) + (2) + (4) and (3) is therefore the energy of a partial ionic/covalent bond. The separate functions are shown graphically in Fig. 4. Now we need to determine what decides the value of x for this gives the

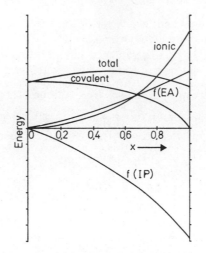

Fig. 4. Plots showing the change in three contributions to total binding energy (a) ionic contribution (b) covalent contribution due to reduction of charge, i. e. Σ ionisation and electron affinity, and (c) covalent contribution from overlap, as the charge varies

deviation from ionic behaviour. *Pearson* and *Gray* (*13*) give all the solutions required for MX_n systems. For example for a diatomic system, **MX**, the fractional ionic character is

$$x = \frac{(IP_{(a)} + EA_{(a)})/2 - IP_c + \frac{1}{2}/r_e}{IP_{(a)} - EA_{(a)} - 2\beta/(1 - x^2)^{1/2} - 1/r_e}$$

The condition which is most easily appraised is that for small x, when

$$x = -\frac{1}{2} + \frac{IP_a - IP_c}{IP_a - EA - 1/r_e}.$$

For a series of cations and a fixed anion this is given by

$$x = -\frac{1}{2} - \frac{constant - IP_c}{constant - 1/r_e}.$$

Thus it is the relationship of $\frac{IP_c/1}{r_e}$ which decides the ionic character.

This conclusion is in agreement with that of several other models for partial ionic bonds. So long as x is small the same result follows for MCl_n molecules.

267

Table 8 shows data for Σ IP(c), up to the charge n, divide by $\dfrac{f}{r_e}$ together with x of *Pearson* and *Gray* using the full formulae. The agreement is good even for very ionic systems. It is worth noting that (1) x is small for n is small, and that this arises because the ionization potentials of s electrons, the first to be ionized, can be disproportionately large (especially in B subgroups) compared with $\dfrac{1}{r_e}$. The s electrons are strongly penetrating. It is also worth noting that (2) x is large and does *not* change much down a Periodic Group in Groups IA to IIIA but that it decreases down Groups VII A, VIIIA, IB, IIB, IIIB, etc. Thus in the Periodic Table if we take elements in their normal oxidation states and use *observed* interionic distances in halides then IP/$\dfrac{1}{r_e}$ generates a triangle of compounds apex Cu(I) and base near Re(III) and Bi(III) for which x is small. [This triangle matches that of *Chatt's* (a) and (b) classification (4).]

Table 8. *Camparison of two procedures for classifying cations*

	I. P. (kcals)	r_e internuclear distance (Å)	(IP x r_e x 10^{-3})	$\dfrac{1}{(IP)r_e}$	$\dfrac{0.85}{(IP)r_e}$	x
$BeCl_2$	634	1.74	1.10	.90	.77	.79
$MgCl_2$	523	2.18	1.14	.87	.75	.75
$CaCl_2$	414	2.54	1.05	.95	.81	.80
$SrCl_2$	385	2.70	1.04	.96	.82	.80
$BaCl_2$	351	2.77	.97	1.03	.86	.84
$ZnCl_2$	631	2.12	1.34	.74	.64	.63
$CdCl_2$	597	2.24	1.34	.74	.64	.64
$HgCl_2$	673	2.28	1.53	.65	.55	.59

Note. The sixth column has been arbitrarily adjusted by the factor 0.85 to make a close comparison with *Pearson* and *Gray's*, x (13).

Now that a theoretical order of partial covalence has been established it is as well to examine physical methods to see if they confirm these covalence orders. Table 9 gives some evidence in support of the general theoretical analysis.

Thus the partial ionic model of *Pearson* and *Gray* (13), and very similar models of greater sophistication (14, 15) show that a very major consideration in causing deviations from the ionic model is the size of the ionization potential of the *exposed* orbitals of the cations, as compared with the ionic potential of a negative charge placed distance r_e from the

268

cation. There is here a rough and ready agreement with Williams' suggestion that covalent character and (b) class behaviour are related to the ratio $IP/\frac{z^2}{r}$ where r was the radius of the cation. The relationship was shown earlier in Fig. 1. On the other hand if softness and hardness are to retain anything of their simple physical connotation it is difficult to see how they can be related to these simple properties of gaseous cations.

Table 9. *Physical parameters and the (a)/(b) classification*
% covalency ("Experimental")

Metal			Halide		Method
	F$^-$	Cl$^-$	Br$^-$	I$^-$	
Sn(IV)	—	50—60	60—70	80—90	(a) (b)
Fe(III)					(b)
Pd(IV)		55	63		(a)
Pt(IV)		55	62	70	(a)
Se(IV)		45	55		(a)
Te(IV)		32	42	52	(a)
Ir(IV)		30	35		(c)
Pd(II)		40			(a)

Code of Method
(a) N.Q.R.   (c) E.S.R.
(b) *Mossbauer*

Now this discussion does not take into account the classical electrostatic polarization already invoked above, especially that of the cation. Thus we might suspect that the covalence argument would not hold equally well for such polarizable cations as Tl^+, Pb^{2+}, when compared with Cu^+ and Hg^{2+}. That this is the case can be seen from a study of heats of formation of a large variety of compounds.

I. Heats of formation of compounds

Phillips and *Williams* (16) have proposed a general method of noting thermodynamic evidence for deviation from the ionic model presuming the latter to be closely followed by all IA and IIA cations.

They compare heats of formation of A-metal and B-metal compounds by selecting pairs of ions (one A and one B) of the same charge and of very similar ionic radii. In such cases the lattice energies calculated on the simple ionic model will be very similar. B-metal ions cannot be assigned constant radii to the same extent that this is possible for A-metal ions,

269

but the consequences of this, although significant, are largely another illustration of these deviations. Suitable pairs of ions are Ca^{2+} and Cd^{2+} (*Pauling* ionic radii 0.99 and 0.97 Å), Sr^{2+} and Hg^{2+} (1.13 and 1.10 Å), and somewhat less satisfactorily Na^+ and Ag^+ (interatomic distance in NaCl, 2.81 Å, 2.77 Å in AgCl, but note the strong variability of the apparent radius of the Ag^+ ion). The first pair will illustrate the simple A-B relationship, the second pair the enhanced effects found in the Au-Hg-Tl-Pb-Bi series, while a comparison of the Ca/Cd case with the Na/Ag case will help to bring out the effects of charge. We begin with CaO and CdO both of which crystallize in the NaCl structure. The relevant data are given in Table 10.

Table 10. *Lattice energies of CaO and CdO (kcal mole^{-1})*

	Ca	Cd	Difference Cd — Ca
Sum of ionization potentials (I + II) (kcal mole^{-1})	+413	+596	
Metal sublimation energy (kcal mole^{-1})	+ 46	+ 27	
Thus $\Delta H_f M^{2+}$ (kcal mole^{-1})	+459	+623	+164
Experimental ΔH_f MO crystal (kcal mole^{-1})	−152	− 61	+ 91
Experimental lattice energy (kcal mole^{-1}) (ΔH_f O^{2-} = 217 kcal mole^{-1})	−828	−901	− 73
Interatomic distance in MO crystal (Å)	2.40	2.35	
Repulsive exponent, n	8	$8^1/_2$	
Calculated ionic lattice energy, MO (kcal mole^{-1})	−842	−866	− 24

The heat of formation of the gaseous Cd^{2+} ion is seen to be greater than that of the Ca^{2+} ion by 164 kcal mole^{-1}. In view of their similar radii, we should expect on an ionic model that the heats of formation of the crystalline oxides would have a very similar difference, but experimentally this is only 91 kcal. Part (24 kcal) of the 73 kcal discrepancy arises from the slightly different 'interionic' distances in CaO and CdO, but there still remains an extra 49 kcal. That is, although the heat of formation of CdO is less exothermic than that of CaO, we should on the ionic model have expected CdO to be even less exothermic. We may say that the Cd^{2+} ions, with their greater electron affinity, bond in part covalently to the O^{2-} ions. This tendency to covalence may be seen thus in terms of the difference between 164 kcal and 91 (more strictly 91 + 24) kcal. The effect is illustrated in Fig. 5, not only for CaO and CdO, but for a series of Ca and Cd compounds, and also for a series of compounds of the Sr-Hg pair, and the Na-Ag pair.

Now for Ca and Cd it will be seen that the deviation from the gas-ion prediction is least for the heats of formation of CaF_2 and CdF_2 and greatest for CaTe and CdTe. In fact the observed sequence of ligands

270

$$F^- \leqslant SO_4^{2-} \leqslant NO_3^- < H_2O < CO_3^{2-} < OH^- < Cl^- < O^{2-} < Br^- < I^- <$$

$$S^{2-} < Se^{2-} < Te^{2-}$$

forms a reasonable series of increasing polarizability. Relatively speaking Ca^{2+} ions will tend to combine with early members of the series, while Cd^{2+} ions will tend to combine with later members. Or we can say that there is a relative preference of Cd^{2+} for Cl^- rather than F^-, or for S^{2-} rather than O^{2-}, compare the (a)/(b) classification.

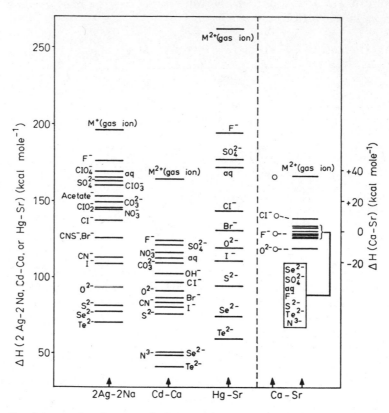

Fig. 5. A comparison between the heats of formation of pairs of compounds M_1X and M_2X, where M_1 and M_2 have very similar ionic radii, using the difference in ionization potential as a guide to the difference in the case of an extreme ionic system

The polarizability sequence for the Sr/Hg pair is almost exactly the same as for Ca and Cd, except that Br^- has moved up relative to oxide. This is shown in Fig. 6 where the Ca/Cd differences are plotted against

271

110

the Sr/Hg differences, all the points lying close to the smooth curve. However, the sequence is now much more spread out (i. e. the gradient of Fig. 6 is a good deal less than 1), and the divergence from the predicted gas-ion difference (262 kcal) even greater. This may be compared with the increased difference in ionization potential, i. e. the much greater covalence of the Hg^{2+} compared with Sr^{2+} compounds.

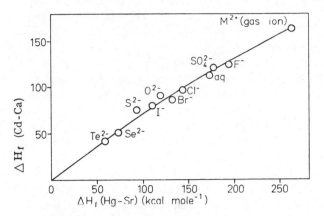

Fig. 6. A comparison between differences in heats of formation between Cd and Ca compounds on the one hand and Hg and Sr compounds on the other

The Na/Ag pair may similarly be compared with the Ca/Cd pair (the former values in each case corresponding to 2 metal atoms so as to compare the same number of anions or charges). Again a broadening of the

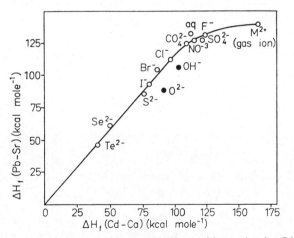

Fig. 7. A comparison between differences in heats of formation for Cd and Ca compounds on the one hand and Pb and Sr compounds on the other

272

sequence of values may be noted, and in general it would appear that divergence from the ionic model is *relatively* more important with ions of lower charge. This is a consequence of the product $(IP)r_e$, see above.

At the right-hand side of Fig. 5, a comparison is made between two A-metal ions, Sr^{2+} and Ca^{2+} (radii 1.13 and 0.99 Å). Here the difference in radii is much greater than in any of the A/B pairs with which we have been concerned. However, the differences between one anion and another are now a good deal smaller; moreover, these differences agree very well with values calculated on the basis of the differences in ionic size alone. The theoretical predictions given by ionic lattice-energy calculations are shown as circles to the left of the Sr/Ca points.

A similar comparison between the difference in the heats of formation of Sr^{2+} and Pb^{2+} compounds and the difference in the heats of formation of Ca^{2+} and Cd^{2+} compounds is made in Fig. 7. That is, the polarizability sequence shown by the (N-2) ions, where N is the Group valence, is compared with that shown by the 'normal' B-metal ions. The same type of result is achieved in Fig. 8 by comparing the differences in the heats of formation of K^+ and Tl^+ compounds with the difference between the heats of formation of Na^+ and Ag^+ compounds. (The Pb^{2+} ion is a little larger than the Sr^{2+} ion: the Sr^{2+}/Sn^{2+} comparison would have been better, but the data for Sn^{2+} are not so extensive as those for Pb^{2+}. Again the Tl^+ ion is a little larger than the K^+ ion and is better compared with an average of K^+ and Rb^+ (Tl^+ ionic radius is 1.40, K^+ 1.33, Rb^+ 1.48 Å), but the results obtained are essentially the same as for the Tl^+/K^+ comparison.)

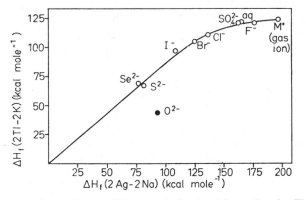

Fig. 8. A comparison between differences in heats of formation for Tl and K on the one hand and Ag and Na on the other

Two main points emerge from Figs. 7 and 8. The first is that, by comparison with Fig. 6, the curvatures are much more pronounced. Thus it would appear that with anions of low polarizability (N-2)-valence ions

273

are relatively inefficient at drawing back electrons. In other words, the ionic model would appear to be moderately satisfactory for the chemistry of Pb^{2+} and Tl^+ with sulphate, water, nitrate, etc. We may note, for example, the insolubility of PbF_2 and $PbSO_4$ (compare Sr^{2+} and contrast Cd^{2+}). On the other hand, Pb^{2+} and Tl^+ begin to show considerable deviations from the ionic model with the more polarizable ions such as sulphide and selenide, and in fact now behave very much more like B-metal ions.

The second point is that oxide falls well below the line in both cases (compare also OH^- in Fig. 7; there are no data for AgOH). In fact the comparison between Sr^{2+} and Sn^{2+} (ionic radii 1.13 and 1.12 Å) shows that the difference in the heats of formation of their oxides (73 kcal mole^{-1}) is actually less than that between their sulphides (83 kcal mole^{-1}), suggesting at first sight that Sr^{2+} is a more polarizing ion than Sn^{2+}. It would be more plausible to suspect that the (N-2)-valence ions have unusually strong bonding to oxide. There is some evidence for this in the structures of SnO, PbO, and Bi_2O_3, and partly hydrolysed Bi^{3+} and Sb^{3+} compounds somewhat reminiscent of the oxocations formed by the early transition metals. There would appear to be a similar, though less pronounced, effect in fluorides: thus we may note the unusual structures of TlF and $KSbF_4$. The effect is just that to be expected from the high

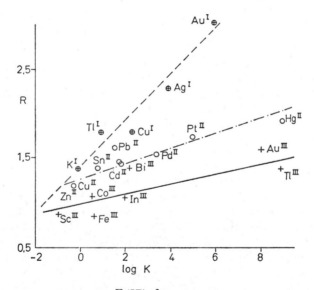

Fig. 9. A plot of R, the ratio of $\dfrac{\Sigma \, (IP) \, z^2}{r}$ (see Fig. 1), against the first stability constants for Br^- complexes in water

274

polarizability of the cations. This would lead to an increase in affinity for small anions due to a decrease in the distance from the negative charge.

A further way of showing that Tl^+, Sn^{2+}, Pb^{2+} and Bi^{3+} are abnormal is provided by Fig. 9, which also illustrates *Williams'* way of assessing (b) class strength. All the 'inert pair' cations form weaker bromide complexes than the value of $R = IP./100z^2/r$ would have led us to expect.

It should be pointed out again that the transition metal compounds provide an ideal opportunity for examining polarization energies as the *d–d* optical absorptions are directly connected with these energies on an electrostatic model. The observed values of $\Delta(10Dq)$ are $F^- > Cl^- > Br^- > I^-$ and $O^{2-} > S^{2-} > Se^{2-}$. It is the greater σ-bonding of the heavier atoms which causes the development of class (b) character, for example Pt(II), and not the electrostatic core polarization of the *d*-electrons. The ligand-field energies calculated in a conventional way are against (b) class character.

Now that a general agreement between covalence orders and class (a) and (b) character has been found it remains to be shown that solvents can in fact generate *two* stability sequences —

i. e. for class (a) $F^- > Cl^- > Br^- > I^-$ and
for class (b) $I^- > Br^- > Cl^- > F^-$.

J. Solvation energy

The simplest model for a solvent is one of a continuous dielectric which effectively reduces charge-charge interactions but does not influence energies due to covalence. No matter how naive this model may be it stresses a very simple point. As the dielectric constant increases and ionic energies diminish then the effect of covalence in bringing about association becomes dominant. Consider the equilibrium

$$A^+ + B^- \rightleftharpoons \overset{+\;-}{AB}$$

In a medium of high dielectric constant the coulombic A^+B^- interaction energy becomes very small. (Looking at the solvent in another way, the model of a solvent as a system of molecules with high dipole moments, then the solvation of A^+ and B^- is such that $^+AB^-$ is not stable.) This is the general reason why A-subgroup cations do not form strong Cl^-, Br^- or I^- complexes in water. That the simple, uniform dielectric solvation model is quite good for these cations is shown in Fig. 10. Now for B-metal cations, halide complexes still form in solvents of high dielectric constant but they are only notably stronger than A-metal cation complexes with such anions as Cl^-, Br^-, I^-. With F^- stability is similar in the two parts

275

114

of the Periodic Table. Assuming that fluorides are at least as ionic as aquo-complexes, this is easily understood by reference to polarizabilities, $H_2O \gg F^-$.

The slope of the ΔH_{ion} plot as a function of halide for the gaseous ions, plotting the halides with equal increment on the abscissae,

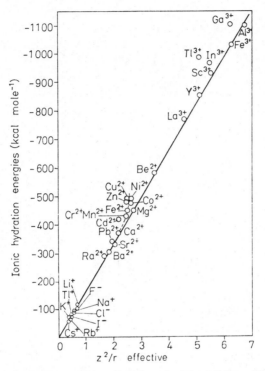

Fig. 10. Ionic hydration energies plotted against the full line of the *Born* equation

is dependent on r, the size of the cation. The slope of the hydration energy of the anions in this plot is independent of cation of course, Fig. 11. For Li, Na, (K) and all the Group IIA cations the slope of MX (gas) is greater than for X⁻ (hydrate). Thus these cations are different from very large cations such as Rb⁺ and Cs⁺ for which the reverse is true. The same difference between the two sets of cations is also found if MX (solid) lattice energy is compared with the X⁻ hydration energies. Thus it is clear that size alone can produce differences between cations which are very closely allied to those used in (a)/(b) classification, and this is so even for *purely ionic* systems. However only cations of unit charge and radius greater than 1.3 Å have a smaller change in ionic energy along the sequence F⁻ > Cl⁻ > Br⁻ > I⁻ than the fall in hydration energy of the

276

hydrates of the anions. In fact many other cations belong to class (b), Table 1. This comes about as the plot of $\Delta H_{(ion)}$ for these cations, Fig. 11, has a small slope through the influence of covalence. In this way the distinction between (a) and (b) classes is most commonly generated.

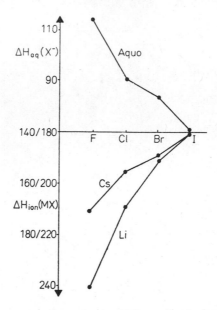

Fig. 11. Plot of energy of gas ion molecules, MX_n, against hydration energy of nX^-, showing how hydration brings about the distinction between (a)-and (b)-class behaviour

Reduction of the dielectric constant of the solvent decreases the slope of the hydration energy such that the solvation energy becomes less important as a background against which the equilibrium

$$A^+ + B^- \rightleftharpoons AB$$

is observed. As a consequence all cations revert to class (a) character as dielectric constant approaches unity — that is, for example, in the gas phase. (See also *D. C. Luehrs, R. T. Iuomato,* and *J. Kleinberg:* Inorganic Chemistry, *5,* 201 (1966)).

Of course, change of ligand around the cation does not affect the hydration energy of the anion in the equilibrium

$$MY_5H_2O + X^- \rightleftharpoons MY_5X + H_2O.$$

However, it affects the slope of MX in Fig. 10. From the above considerations any atoms Y which reduce the charge (or increase the size of M) without equally lowering the electron affinity of the orbital acting as

277

acceptor to X^- will increase (b) class character. Thus it is not surprising that $[Co(H_2O)_6]^{3+}$ is (a) class but $[Co(CN)_5H_2O]^{2-}$ and Vitamin B_{12a} are (b) class. In vitamin B_{12a} chloride it is known that the $Co^{3+}-Cl^-$ distance is much longer than in simpler Co^{3+} complexes, e. g. $[Co(NH_3)_5Cl]^{2+}$.

Another clear example is the effect of methyl groups on B-subgroup metals. Table 11 compares stability data for mercury, thallium, and lead systems. There is now an effective reduction of (b)-class character due to the extra ligands CH_3^-. This must be attributed to the lowered electron affinity of the metal orbitals if our discussion is correct.

Table 11. *Character of substituted cations*

	log (stability constant)						
	log β_1 (a)		log β_2 (b)			log β_2 (c)	
	Tl^+	$Hg(CH_3)^+$	Sn^{2+}	Cd^{2+}	$Sn(CH_3)_2^{2+}$	Pd^{2+}	$Pd(sin)_2^{2+}$
Cl^-	0.5	5.2	1.2	1.6	0.4	7.0	4.3
Br^-	1.0	6.6	0.7	1.8	−1.0	8.5	5.4
I^-	1.5	8.6				14.0	6.5

Data are from
(a) *M. Eigen, G. Geier,* and *W. Kruse:* Essays in Coordination Chemistry, 164. Basel: Birkhauser Verlag 1964.
(b) *H. N. Farrer, M. M. McGrady,* and *R. S. Tobias:* J. Amer. Chem. Soc. 87, 5019 (1965).
(c) *G. Gidden, F. J. C. Rossotti,* and *L. M. Venanzi:* to be published. sin is a sulphinic acid derivative.

K. Oxidation state and (a) and (b) class

As we have seen there is no easy way of assessing (a)-and (b)-class without experimental data. Table I suggests that as oxidation state is increased (a) class character increases. This is by no means a good generalisation. There seems to be a general increase in (a) class character for ions of the same Periodic Group IA < Group IIA < Group IIIA and this is also true in the B groups Ag^+ < Cd^{2+} < In^{3+}. However for any one element the series are quite unpredictable Cu(I) < Cu(II), Tl(I) > Tl(III), Sn(II) < Sn(IV), Pt(II) > Pt(IV) in (a) class character. All seems to depend on the relative magnitude not only of electron affinity of the cation and the exposure of the relevant orbitals (overlap), interatomic distance and charge, but also on cation polarizability.

L. Donors other than halides

Ahrland, Chatt and *Davies* suggested that those ions which gave complexes with a stability order F^- < Cl^- < Br^- < I^- would also give com-

278

117

plexes of a stability order $RO^- < RS^- < RSe^-$, $R_2O < R_2S$, and $R_3N < R_3P$. Good experimental evidence is lacking on most systems. Supposing that measurements can be made in water (we have already seen that a change to another solvent would invalidate the comparison) there is good ground for thinking that the order $RO^- < RS^- < RSe^-$ would parallel that of $F^- < Cl^- < Br^- < I^-$. The ionic/covalent contributions must be similar. However this is *not* true of $O^{2-} < S^{2-} < Se^{2-}$ or $R_2O < R_2S < R_2Se$. It is quite unclear how rapidly the ionic/covalent factors will change in these series as compared with solvation energies, see Fig. 10, and we are not in a position to draw such Figures for these anions and molecules. Until we are, there is no certainty that the (a)/(b) classification will apply in the same simple way to all donors.

Now quite apart from the ionic/covalence (single bond) acceptor-donor picture we have used, there is the problem of π-bonding. Although most simple anions must be largely donors, I^-, F^-, O^{2-}, RS^-, this is not true of R_2S and R_3P. It must not be thought too surprising if some cations will bind $R_3P > R_3N$ but $Cl^- < F^-$.

III. Non-ionic Systems

Finally, as *Ahrland, Chatt* and *Davies* (4) extended their observations to non-ionic systems, we must also look at such equilibria as

$$MR_n + DR_n \rightleftharpoons R_nD \rightarrow MR_n$$

$$\text{e. g. } BF_3 + NH_3 \rightleftharpoons H_3N \rightarrow BF_3$$

The question now is whether there is a simple possible classification. In this case the ionic contribution has been reduced to a dipole-dipole interaction and, obviously enough, the covalent contribution will become more important. (In all systems we are really looking at differences between series of equations for different ligands: E = ionic energy + London energy + classical polarization + covalence.) When the ionic term becomes insignificant then (b) class behaviour is favoured by the London term but not necessarily by the other two. Much hangs upon whether it is the acceptor or the donor which is the more highly polarizable and the values of E_{aa}. Without much more experimental evidence no firm conclusion can be drawn for calculations of all three energies are notoriously unreliable.

IV. Kinetics and (a)/(b) Classifications

The concept of "soft" and "hard" acids and bases

Pearson has also attacked the problem of the classification of acids and bases in a much more qualitative way. Realising that the pattern of rate

279

constants for nucleophilic substitution was not related to either "ionic charge density" or to the ability of the attacking group to form strong covalent bonds he proposed that it was mutual polarizability which controlled the rate constants. A polarizable group was then called "soft" and a non-polarizable group "hard". He extended the idea to thermodynamic data and attempted to embrace within the concept "soft" the (b)-class acceptors of *Chatt*. The (a)-class acceptors then would be called "hard". In so far as "soft" is synonymous with polarizable *Pearson* and *Gray* had already effectively disposed of this argument, see above. It is worth noting that some of the most polarizable atoms do *not* belong in the (b)-class, Table 1. However there are other objections to the whole idea that kinetic and thermodynamic orders can be one and the same. Consider a positive centre X which is approached by a group Y, the rate constants are

$$X + Y \; \underset{k_{-1}}{\overset{k_1}{\rightleftharpoons}} \; XY$$

Soft acids, X, and bases, Y, give rise to large k_1, by definition, but it is also true that good attacking groups are generally good leaving groups, (k_{-1} is large), whence k_1/k_{-1} can be large, small or utterly unrelated to k_1. Thus the equilibrium constant k_1/k_{-1} need not classify acids and bases in the same way as k_1 does. Another way of seeing this is by reference to Fig. 11.

V. Conclusions

This article has been very largely concerned with thermodynamic data on the classification of inorganic donors and acceptors. All the discussion has been semi-empirical using only the simplest approximate models. At the present time more sophisticated models are being developed, e. g. by Mr. *A. F. Orchard* (Oxford) and by *Jorgensen, Horner, Hatfield,* and *Tyree* (Cyanamid Laboratory Report). However even without these models the gross controlling factors of the classification into (a) and (b) classes is clearly one of the relative importance of ionic and covalent bonding. Undoubtedly too the percentage ionic character at which an acceptor of one type goes over to an acceptor of the other depends upon the charge type of the reactants and the solvent (or ligands) bound to them. To some degree London energies and classical polarizations are also important but as with π-bonding no quantitative demonstration of their effect has been demonstrated. It is only a hazard to clear-thinking if these effects are stressed before a proper appreciation of the consequences of the balance

280

between σ-bonding and ionic bonding has been obtained for heavier atoms. We can only deplore the use of even more qualitative notions, such as "soft" and "hard", which are without clear physical meaning as far as we can ascertain. It is just not good enough to confuse a *simple* classification into (a) and (b) types with a classification into "hard" and "soft" types for although (a) and (b) have no meaning "soft" and "hard" have clear descriptive significance in English.

References

1. *Sidgwick, N. V.:* The Chemical Elements and Their Compounds. Oxford: Oxford University Press 1950, 1170—4.
2. *Williams, R. J. P.:* J. Phys. Chem. *58,* 121 (1954).
3. *Carleson, S. L.,* and *H. Irving:* J. Chem. Soc. *1954,* 4390.
4. *Ahrland, S., J. Chatt,* and *N. R. Davies:* Quart. Rev., (London) *12,* 265 (1958).
5. *Craig, D. P., A. Maccoll, R. S. Nyholm, L. E. Orgel,* and *L. E. Sutton:* J. Chem. Soc. *1954,* 332 .
6. *Basolo, F.,* and *R. G. Pearson:* Mechanisms of Inorganic Reactions. New York: Wiley 1958, 179—182.
7. *Williams, R. J. P.:* Proc. Chem. Soc., 1960, 20; and *Williams, R. J. P.:* Ind. Chem. Belg. *4,* 389 (1963).
8. *Christensen, J. J.: R. M. Izatt, L. D. Hansen,* and *J. D. Hale:* Inorganic Chemistry *3,* 130 (1964). See also *Hale, J. D., Ph. D. Thesis,* Brigham Young University, July, 1963.
9. *Poë, A. J.,* and *M. S. Vaidya:* J. Chem. Soc. *1961,* 1023.
10. *Basolo, F.,* and *R. G. Pearson:* Mechanisms of Inorganic Reactions. New York: Wiley 1958, 180.
11. *Pearson, R. G.:* J. Amer. Chem. Soc., *85,* 3533 (1963).
12. Chemistry and Engineering News *1965* (May), 90.
13. *Pearson, R. G.,* and *H. Gray:* Inorganic Chemistry, *2,* 358 (1963).
14. *Klopman, G.:* J. Amer. Chem. Soc. *86,* 1463 and 4550 (1964).
15. *Ferreira, R.:* J. Phys. Chem. *68,* 2240 (1964).
16. *Phillips, C. S. G.,* and *R. J. P. Williams:* Inorganic Chemistry, Vol. II. Oxford: Oxford University Press 1966, 502—8.

(Received May 18, 1966)

Footnote

Discussions of donor/acceptor systems by *Fajans* earlier in this century have much in common with our reasonings. Unfortunately these have been confused by the "quanticule" hypothesis and we have therefore avoided direct reference to *Fajans'* views much though we wish to acknowledge them.

281

Reprinted from "Structure and Bonding," Volume 1, C. K. Jørgensen et al., eds., pp. 207–220 (1966)

11

Reports on Hard and Soft Acids and Bases

Factors Contributing to (b)-behaviour in Acceptors*

Dr. St. Ahrland

Institute of Inorganic and Physical Chemistry, University of Lund, Lund, Sweden

Table of Contents

Two Classes of Acceptors, (a) and (b)

Acceptors termed (a), or *hard*, behave just in the manner expected when the coordination to various donors is governed mainly by the electrostatic interaction existing between charges of opposite signs, *i. e.* the higher the charge and the smaller the radius of the acceptor, and of the donor atom to be coordinated, the stronger are generally the complexes formed. The most typical (a)-acceptors are therefore such ions as Be^{2+}, Al^{3+}, Ti^{4+}, and they prefer donor atoms like F (as F^-), O (in *e. g.* H_2O, OH^-, $R \cdot COO^-$, ethers and phenols), or, to a less marked degree, N (in *e. g.* NH_3, ammines and pyridine). Also H^+ is a rather typical (a)-acceptor in many respects, though, on the other hand, it shows quite a high affinity for some other types of donors as well. Acceptors termed (b), or *soft*, rather follow the opposite pattern. Their coordinating ability does not regularly increase with increasing charge or decreasing radius of the acceptor. Nor do they prefer the small donor atoms F, O and N, but rather their heavy congeners. This strongly indicates that the complexes formed contain bonds of a mainly covalent character. This paper deals with the important factors concerned in the formation and strength of these bonds.

---

    * Revised version of a talk, given at the Symposium on "Soft and Hard Acids and Bases", arranged by Cyanamid European Research Institute, Cologny (Geneva), May 10.–12., 1965.

207

There are certainly quite a number of causes which may contribute towards the formation of essentially covalent bonds between the donor and the acceptor, *i. e.* towards a display of (*b*)-behaviour. Though the matter is thus fairly complex, it is nevertheless possible to distinguish, from established experimental facts, some factors which obviously are of great significance. These will be discussed in the following, especially with regard to their relative importance, and the mutual influence they exert on each other.

It should be remembered, however, that an electrostatic interaction must always exist between acceptor and donor as soon as ions or dipoles are involved. This means that the (*b*)-type of bond will seldom be purely covalent. Its characteristic features prove beyond doubt, however, that the covalent contribution must be the dominating one.

By definition, the acceptors (*a*) and (*b*) are characterized by the following affinity sequences of the various groups of donor atoms (*1*):

Donor group	Oxidation state	(*a*)	(*b*)
7B	−I	$F \gg Cl > Br > I$	$F \ll Cl < Br < I$
6B	−II	$O \gg S > Se > Te$	$O \ll S \sim Se \sim Te$
5B	−III	$N \gg P > As > Sb$	$N \ll P > As > Sb$

The (*a*)-sequences are all alike. The (*b*)-sequences have the feature in common that the first, light donor atom is much more weakly bound than the second one, which is just contrary to what characterizes the (*a*)-sequences. The order between the heavier donor atoms is not as regular for the (*b*)- as for the (*a*)-sequences. The monotonous increase of stability with the atomic number, observed for the halide complexes, has disappeared in the next group of donors and even reversed for phosphorous and its congeners.

Fundamentally, acceptors should be termed more typical (*b*) (or (*a*)), the more strongly marked the differences of affinity are which are expressed by the sequences above. In addition, however, the absolute strength of the complex formation should also be taken into account, because a very high affinity to those heavy ligands which exert only a weak electrostatic attraction must involve a strong covalent interaction, *i. e.* be tantamount to a pronounced (*b*)-character. Besides these criteria which follow direct from the definition of (*b*)-character, there are further ones which use other properties of evidently the same ultimate origin as the affinity conditions described above. Thus the coordination of uncharged ligands of negligible proton affinity, such as carbon monoxide, olefins, acetylenes and aromatic hydrocarbons, is highly characteristic for (*b*)-acceptors. Also the rates of nucleophilic displacements can be used for classification. For (*b*)-acceptors, the rate will depend little upon the basicity of the donor, but much on its polarizability, while the

208

opposite behaviour is displayed by (*a*)-acceptors. Especially if the stabilities are difficult to obtain, such auxiliary criteria may be extremely useful, as has been demonstrated particularly by *Pearson* (2).

In the following treatment of the various factors contributing to (*b*)-behaviour, metal ions have as a rule been used in order to illustrate the various points. To a great extent, however, the arguments will also apply to other types of acceptors. On the other hand, factors additional to those of importance for metal ions undoubtly have to be considered for other acceptors, above all for the numerous organic acceptors containing multiple bonds.

All Metal Ions of Class (*b*) have a Large Number of d-electrons in their Outer Shell

In order to develop (*b*)-character, a metal ion must evidently possess a large number of d-electrons on high energy levels. Many of the most typical (*b*)-acceptors even have their outer d-shells completely filled, *i. e.* the configuration d^{10}, but also the configurations d^8 and d^6 often result in very strong (*b*)-properties. This is evident from the survey contained in Table 1, where all oxidation states containing many d-electrons, as well as a few containing none, have been entered, *viz.* acceptors of the configurations d^0, d^{10}, d^0s^2 and $d^{10}s^2$ in Table 1 A, and of the configurations d^9, d^8, d^6 and d^5 in Table 1 B. The acceptors exhibiting (*b*)-properties have been identified according to a code explained in the head of the Table. On account of experimental difficulties, the quantitative evidence is scarce for acceptors containing less than eight d-electrons and most of the assignments for the d^6 and d^5 groups have had to be founded on qualitative estimates of the stabilities of various complexes. Thus the (*b*)-character of the platinum metal acceptors possessing these electron configurations has primarily been inferred from their marked ability to form strong complexes with the heavy halides (for a survey see *e. g.* (*3, 4*)). At least for the d^6 acceptors Rh(III) and Ru(II) there are also strong indications (*5, 6*) that the halido complexes really exhibit the (*b*)-sequence of stabilities Cl $<$ Br $<$ I. Also carbonyl and/or olefin complexes are formed by several of these acceptors, *e. g.* by all members of the d^6 group Fe(II), Ru(II) and Os(II) (*3, 6, 7*).

Acceptors with less than six d-electrons never exhibit very strong (*b*)-properties, and less than five means (*a*)-properties in all but very few cases. The exceptions are found among the elements of the highest (6th) period considered here where the tendency to (*b*)-type bonding is especially marked, on account of the high polarizability, as will be more fully

St. Ahrland

Table 1. *Classification of acceptors of various electron configurations.*

Acceptors of class (*b*) are underlined, the most typical ones with a double line, doubtful or borderline cases with a dashed line. Unknown oxidation states, as well as known states of unknown properties, are indicated by a dotted line. The figures below the symbols give the polarizability of the respective acceptors, (in the unit Å^3) and are taken from the compilation of Jørgensen (*11*).

Table 1 A. *Acceptors of the configurations d^0, d^{10}, d^0s^2 and $d^{10}s^2$.*

Group →	1B	2B	3B	4B	5B	6B	7B
d^0			Al(III) 0.06	Si(IV)	P(V)	S(VI)	Cl(VII)
d^{10}	Cu(I) 1.6	Zn(II) 0.8	Ga(III)	Ge(IV)	As(V)	Se(VI)	....
	Ag(I) 2.4	Cd(II) 1.8	In(III) 1.2	Sn(IV)	Sb(V)	Te(VI)	I(VII)
	Au(I)	Hg(II) 2.9	Tl(III)	Pb(IV)	....	....	....
d^0s^2			....	....	P(III)	S(IV)	
$d^{10}s^2$			....	....	As(III)	Se(IV)	
			....	Sn(II)	Sb(III)	Te(IV)	
			Tl(I) 5.2	Pb(II) 4.8	Bi(III) 4.5(?)	....	

Table 1 B. *Acceptors of the configurations d^9, d^8, d^6 and d^5.*

Group →	7A	8			1B
d^9					Cu(II) ....
d^8				Ni(II) 1.0	....
				Pd(II)	....
				Pt(II)	Au(III)
d^6		Fe(II) 1.2	Co(III)	....	
		Ru(II)	Rh(III)	Pd(IV)	
		Os(II)	Ir(III)	Pt(IV)	
d^5	Mn(II) 1.2	Fe(III) 2.2	....		
	....	Ru(III)	Rh(IV)		
	....	Os(III)	Ir(IV)		

210

discussed below. A case in point is provided by Os(IV), of the configuration d^4, which shows a clear (b)-character. In the lower periods, however, an acceptor must seemingly possess at least a half-filled d-shell in order to form bonds of that kind which is embodied in the term (b)-behaviour.

On the whole the (b)-character thus increases with the number of d-electrons. It follows that, for elements situated within that area of the periodic system where the d-shells are still being built up, the (b)-properties of a certain element will be more marked, the lower its oxidation state. Moreover, if an element possesses only five d-electrons, only the zero state will be able to unfold (b)-properties, in accordance with the fact that any oxidation to a higher state would mean a decrease of the number of d-electrons below what is generally demanded for the formation of the fairly covalent (b)-type bond. The creation, or increase, of a positive charge on the acceptor as a consequence of the oxidation will also *per se* be of great importance for the bond formation, as will be further discussed later on.

Also electrons of other inner orbitals than d may of course give rise to covalent bonding, provided they are on suitable energy levels. It is thus very likely that the abnormal stability of certain actinide complexes is due to partially covalent bonds involving $5f$ electrons (*8*). On the other hand there are no signs that the $4f$ electrons of the lanthanides are able to participate in such bonds; evidently these are situated too deep within the atom.

From table 1 it is also clear, however, that the presence of many d-electrons does not necessarily constitute a (b)-acceptor. For one thing, further electrons in shells outside the d-shell cause a considerable weakening of the (b)-properties, as can be seen by comparing the acceptors of configuration d^{10} and $d^{10}s^2$ of table 1. For elements situated between the end of a transition series and the beginning of next period, an increase of the oxidation state will therefore often tend to enhance the (b)-character, as a result of the decrease in shielding of the d-shell by outer electrons. From this point of view it is quite clear why Tl(III) is a much more marked (b)-acceptor than Tl(I) (*9*). Also Sn(IV) and Pb(IV) seem to have more (b)-character than Sn(II) and Pb(II), respectively, to judge from the relative stabilities of their halido (*3*), and in the case of tin, also sulphido complexes. The quantitative evidence is admittedly still sparse on these points (*4*).

For elements at or beyond the end of the transition series, an increase of the oxidation state will thus lead to a strengthening of the (b)-properties, while exactly the opposite is true for elements containing fewer d-electrons.

It should be noted however, that this generalization is valid only if the d-electrons are essential for the formation of the covalent bond. For

14*   211 elements near the end of the long periods, in their low oxidation states, this is presumably not the case, and they will therefore not conform to the rule stated, as will be further discussed below.

However, not even a configuration containing many unshielded d-electrons, such as d^{10}, d^8 or d^6 can alone make certain that a (b)-acceptor really results. On the contrary, about half of e. g. the d^{10}-acceptors entered in Table 1 A are either of (a)-type, or are borderline cases. It is obvious, however, that the borderline is coincident with a sharp decrease in polarizability on passing from the (b)- into the (a)-area.

A Combination of many d-electrons and a high Polarizability is Apt to Produce a (b)-acceptor

This statement expresses the rather natural fact that the mere existence of d-electrons is not sufficient to create the covalent bonding involved in (b)-type coordination. In order to be available for bond formation, the d-electrons must further be on energy levels about as high as those of other bonding electrons. If they are, it will be possible to displace them considerably relative to the nucleus by an outer field of modest strength, which just means that the acceptor will be highly polarizable. For acceptors of a given net charge and outer electronic configuration, the polarizability will increase with the number of electron shells. The (b)-properties of an acceptor of a certain type will therefore be more marked, the higher the period it belongs to. This accounts for the well-known triangular shape of the area within the Periodic System comprising the most typical (b)-acceptors $(1, 10)$.

On the other hand it must be stressed that a high polarizability alone, without the presence of a well-filled d-shell, does not confer (b)-properties on a metal ion. This is obvious from a comparison between the (b)-acceptors of Table 1 and some metal ions of comparable polarizability which do not contain any d-electron, $viz.$ Cs^+, Ba^{2+} and La^{3+}. Their polarizabilities are quite high, 3.1, 2.5, and 1.6 Å^3, respectively (11), yet they do not exhibit any (b)-properties.

Influence of the Acceptor Charge

Another factor of great importance for the character of the acceptor is its charge. Now the influence of the charge is fairly complex, and several aspects have to be considered. First, varying the charge of a certain element, $i. e.$ its oxidation state, means changing its electron configuration,

212

126

which by itself will cause a very significant change of the bonding properties, as has already been discussed. A meaningful investigation of the influence of charge alone must therefore involve only comparisons of acceptors of identical outer electron configuration. It must further be remembered that with a variation of charge invariably goes a change of polarizability, and moreover generally in that direction that the acceptor will be less polarizable, the higher the charge. For this reason, an increase of charge should therefore tend to weaken the (b)-character. On the other hand, such an increase will also act in the opposite direction, by enhancing the polarization of the ligand to be coordinated. In order that a covalent bond should be formed, the bonding electrons of the ligand has to be available, and they will be more so, the stronger the polarizing influence of the acceptor. An increase of the charge on the acceptor will thus make its own bonding electrons less available, but those of the donor more available for bonding. Whether the net result will be a strengthening or a weakening of the bond, *i. e.* of the (b)-character, will evidently depend upon the relative magnitudes of these influences. The following rule can thus be formulated:

For acceptors of identical electron configurations, an increase of the charge will bring about a strengthening of the (b)-properties, by increasing the polarization of the donors, provided the accompanying decrease of their own polarizability is not too large.

A good example of such a strengthening is provided by the $d^{10}s^2$ series Tl(I), Pb(II), Bi(III), where the (b)-properties, as shown by the stabilities of the halide complexes (4), increase markedly with the charge. As demanded, the decrease of polarizability along this row is also quite modest (*cf.* Table 1 A, where the figure for Bi(III) is an educated guess). Still more striking cases of the increase of (b)-character with charge are found among the $d^{10}s^2$ and d^{10} acceptors of the lower periods. Thus Sb(III) and Te(IV) are (b)-acceptors, as indicated by their formation of seemingly rather strong halido and, in the case of Sb(III), also sulphido complexes, while Sn(II) certainly is an (a)-acceptor. Similarly, Sn(IV) and Sb(V) belong to class (b), while In(III) is typically class (a). Also As(V) is probably a (b)-acceptor, while the three previous d^{10} acceptors of that period, Zn(II), Ga(III) and Ge(IV), all are in class (a). Admittedly no values of the polarizability are known for most of these acceptors, but the interpretation seems nevertheless quite certain.

If the increase of charge leads to a sharp decrease of polarizability, a weakening of the (b)-character is invariably found, however, as is amply shown by comparing *e. g.* Cu(I) with Zn(II), or Ag(I) with Cd(II) or In(III). The set of electric polarizabilities entered in Table 1 have been compiled by *Jørgensen (11)*, on the assumption that the very small values previously reported for Zn(II) and In(III) aqua ions and negative

213

values obtained for Mg(II) and Al(III) actually are caused by the water molecules being slightly less polarizable in the presence of highly charged cations. This description allows much closer agreement with the polarizabilities of crystalline compounds.

Finally, an increase of charge will also enhance the electrostatic interaction between acceptor and donor, which may result in a strengthening of the electrostatic contribution to the bond relative to the covalent one. This may be the reason for the general weakening of the (b)-character observed for d^{10} and $d^{10}s^2$ acceptors of the groups 6 B and 7 B relative to those of group 5 B (Table 1 A).

The results of the various balancing influences, as found from the experimental facts and visualized in Table 1, are quite interesting. Besides the well established triangle (1, 10) of (b)-acceptors, with its apex at Cu(I), *a new triangular extension of the (b)-area emerges, with its apex at As.*

Influence of the Donor; Hard and Soft Donors

Whether a mainly covalent bond is formed or not will evidently depend not only on the acceptor, but also on the donor. It follows that the character of the acceptor will vary to a certain degree with the properties of the donor. As a first rule we can state:

The more polarizable the donor, the more pronounced the (b)-character of a certain acceptor.

It is therefore to be expected that acceptors, especially those on the borderline between the (a)- and the (b)-areas, should tend more to (b)-behaviour in conjunction with the heavy chalcogens than with the heavy halogens, the former being much more polarizable than the latter (11). This is confirmed experimentally in case of Co(II), Ni(II) and Cu(II) which are certainly of class (a) as far as halide ion donors are concerned, but decidedly, though mildly, class (b) when coordinated to chalcogen donors (1, 12, 13). It must be admitted, however, that Cd(II) seems to be an exception, displaying more (b)-behaviour towards the halide ions than towards donors of the chalcogen group (13). This unexpected conduct is presumably due to some additional factor which still remains obscure.

The polarizability of a certain donor is determined by the electronegativity of its coordinating donor atom, in conjunction with the properties of neighbouring atoms within the ligand. If the donor atom exists as an elementary ion, its own electronegativity is therefore the only factor of importance. A decrease of this quantity will imply an increase of the polarizability and, consequently, a stronger tendency to form

214

covalent bonds. In accordance with this, the affinity of the halide ions for covalently bonding (b)-acceptors follows the sequence Cl⁻ $<$ Br⁻ $<$ I⁻. Donors of the chalcogen and nitrogen groups cannot exist as elementary ions, however, but only as the donor atoms of more or less complex ligands. Under such circumstances, the decrease of the electronegativity with increasing number of shells will also cause an electron flow away from the donor atom to its neighbours. The result may well be that the electronic charge available for further covalent bonding does not increase with the number of shells, and hence not the affinity for (b)-acceptors. In fact no large variations are observed between the heavy chalcogens, and for the heavy members of the nitrogen group even the reverse sequence P $>$ As $>$ Sb results, as has been pointed out already in the introduction of this paper.

Besides a high polarizability, however, a ligand has to possess further qualities in order to coordinate (b)-acceptors by bonds of a strongly covalent character, *i. e.* in order to perform as a *soft donor*, according to the *Pearson* (2) classification. To be able to accomodate those d-electrons from the acceptor which are seemingly of decisive importance for the formation of an essentially covalent bond, the ligand must possess empty orbitals on suitable energy levels. This applies at least to all (b)-acceptors of the types listed in Table 1, characterized by their tendency to form bonds of an order higher than one, just by back-donating d-electrons to such ligands which are able to accomodate them.

It may further be presumed that in many instances the ligand will use vacant d-orbitals to accomodate the extra electrons, because a great jump in affinity to (b)-acceptors is observed between the first donor atom of each group (F, O, N) which has no d-orbitals of the same principal quantum number as the bonding electrons, and the following ones where such d-orbitals exist, and are vacant (*cf.* (*1, 2, 10*) and earlier references quoted therein). The jump is generally far greater than should be expected from the increase of polarizability alone.

On the other hand, spectroscopic data do not suggest any vacant d-orbitals on low energy levels for such an important group of soft ligands as the heavier halide ions (*14*). It is also evident that soft acceptors are able to coordinate, by bonds of a fairly covalent character, even ligands with light donor atoms such as O and N which certainly do not possess any low-lying d-orbitals. This is borne out *e. g.* by the affinities displayed for such ligands by various members of the zinc group of acceptors, according to the following reasoning.

Donors coordinating by F, O or N possess a general ability to coordinate both hard and soft acceptors, showing no special preference for soft ones. In the *Pearson* (2) classification such donors are termed *hard*. If their coordination are governed only, or at least mainly, by the electro-

215

static interaction, one would expect the tendency for hydrolysis, or formation of ammine complexes, always to decrease with increasing ionic radius of the acceptor as has been pointed out already in the introduction of this paper. At least as far as hydrolysis is concerned, this certainly also occurs when groups of hard acceptors are considered. Thus, within the titanium group, the hydrolysis becomes less prominent in the order (4) Ti(IV) $>$ Zr(IV) $>$ Hf(IV) ($>$ Th(IV)). For the zinc group, however, quite a different pattern is found, *viz.* Zn(II) $\sim$ Cd(II) $<<$ Hg(II). Especially the second dissociation constant of the hydrated mercury(II) ion is very large. The coordination of ammonia shows very similar features, including the formation of amido complexes by Hg(II) if the value of pH is not kept low enough. These circumstances certainly indicate a markedly covalent character of the metal to ligand bond, especially for Hg(II), but to a certain extent also for Cd(II). One may thus conclude:

A purely electrostatic bond hardly exists when soft acceptors are involved, not even in connexion with rather hard donors.

One group of typically soft donors, *viz.* the uncharged ligands of negligible proton affinity mentioned above (p. 208), certainly accommodate in a quite specific way the d-electrons donated by a soft acceptor. Besides the carbon ligands which have already been enumerated, this group also comprises nitrogen monoxide. Also the bond between soft acceptors and the cyanide ion seems to be of much the same kind, though this ligand in other respects differ strikingly, above all by its very strong proton affinity.

On coordination of these donors, a metal to ligand bond is formed of an order considerably higher than one (in carbonyls and nitrosyls approximately a double bond). This is unambiguosly shown by structure determinations which indicate a distance between metal and donor much shorter than expected for a single bond. Simultaneously the internal ligand bond is somewhat weakened, as is revealed by a decrease of its IR-stretching frequency (15). These facts, as well as the structural features of the complexes, are best described by postulating a metal to ligand bond made up partly of electrons from the ligand, forming a σ-bond, and partly of d-electrons from the metal, forming a π-bond. For the latter bond, empty p-orbitals on the donor are presumably utilized. Such views were first advocated by *Pauling* (16) for carbonyl and cyanide complexes, later on they have been successfully extended to olefin and acetylene complexes, by *Dewar* (17) and by *Chatt* (18).

It thus seems evident that those d-electrons which are responsible for the essentially covalent character of the bonds formed by soft acceptors are accomodated in different ways by different ligands. The easier they are fitted into the electron configuration of the ligand, the more prominent the covalent contribution to the bond, and by definition, the softer

216

the ligand. In practice, the degree of covalency is difficult to assess. It may be maintained, however, that the stronger the complexes formed by a ligand with a typically soft acceptor, relative to those formed with a typically hard acceptor, the softer the ligand. In this way, a rough measure of the softness of donors is obtained. As is the case also for acceptors, the concept is admittedly only a qualitative one, but (also as for acceptors) nevertheless quite useful.

The factors constituting a soft ligand can thus be summarized as follows:

A soft ligand combines high polarizability with vacant electron orbitals on suitable energy levels. The more available these orbitals, the softer the ligand. Extreme softness is presumably connected with the existence on the ligand of d-orbitals, or, for soft donor atoms in the second period, p-orbitals, of particularly favourable extension and energy.

Simultaneous Coordination of Donors of Various Softness

If a (b)-acceptor is offered various soft donors at the same time, it will as a rule coordinate them into a mixed complex, in proportions depending upon the relative affinities. This is the well-known phenomenon that soft ligands tend to come together in the mixed complexes formed by (b)-acceptors (as do, of course, hard ligands in the mixed complexes of (a)-acceptors). If an extremely soft ligand is coordinated, however, it may take care of all d-electrons available on the acceptor, leaving none to less soft ligands. The result may be that the residual coordination capacity of the acceptor can be used only by hard ligands, using mainly electrostatic forces for bonding. We thus state:

The coordination of very soft ligands to a (b)-acceptor will decrease its (b)-character, or even turn it into an (a)-acceptor.

A very nice example of this is provided by Burmeister and Basolo (*19*) who find that the very soft ligand triphenylphosphine causes Pt(II) and Pd(II) to coordinate the hard nitrogen end of the thiocyanate ion, forming an isothiocyanate complex, while the less soft triphenylstibine causes these acceptors to attach the thiocyanate ion by its soft sulphur end, in accordance with the more general rule that soft ligands flock together. As to triphenylarsine, whose softness lies between that of its congeners, both the thiocyanate and the isothiocyanate complex can be isolated, furnishing an interesting example of linkage isomerism.

The type of coordination prevailing in the various complexes has been found by IR-measurements. Especially the stretching frequency of

217

131

the C—S bond indicates very clearly whether the thiocyanate ion is coordinated by its sulphur or by its nitrogen donor atom (19, 20) as is evident from the results presented in table 2. As to the C—N bond, its stretching frequency is not very sensitive to this change, but the integrated absorption of the band is (21). An evaluation of this quantity would certainly have furnished further support for the proposed structures.

Table 2. *The coordination of very soft ligands to a (b)-acceptor will decrease its (b)-character, or even turn it into an (a)-acceptor (Burmeister and Basolo (19); Turco and Pecile (20)).*

Complex	C—N stretch, cm^{-1}	C—S stretch, cm^{-1}	Bond
K[SCN$^-$]	2066	749	
(NH$_4$)$_2$[Cr(NH$_3$)$_2$(NCS)$_4$]	2110, 2042	828	N
K$_2$[Pd(SCN)$_4$]	2118, 2086	703, 707, 696	S
Pd(SbPh$_3$)$_2$(SCN)$_2$	2119, 2115	no band in the region 780—860; thus no N— bonding *	S
Pd(AsPh$_3$)$_2$(SCN)$_2$	2119		S
Pd(AsPh$_3$)$_2$(NCS)$_2$	2089	854	N
Pd(PPh$_3$)$_2$(NCS)$_2$	2093	853	N

* Strong phenyl ring adsorption around 700 cm^{-1} precludes the detection of S-bonding band situated in that region.

From the facts just presented it is clear that the behaviour of an acceptor depends considerably both on the ligands to be coordinated, and on those already attached to it. Especially in borderline cases, ligands of different properties may rather easily tip the scale in favour of either (a) or (b)-behaviour, as has just been pointed out in the case of the coordination of halide ions *vs* chalcogen donors to Co(II), Ni(II) and Cu(II). If the donors possess very extreme properties, however, even the nature of otherwise very typical acceptors may be changed, as has been demonstrated in the case of the phosphine complexes of Pt(II) and Pd(II). Still further examples can be quoted to illustrate the influence of the ligand on the character of the acceptor. Thus *Schmidtke* (22) has recently found that Rh(III) and Ir(III), when present as pentammine complexes, coordinate the thiocyanate ion preferentially through its nitrogen atom. When coordinated to ammonia, these acceptors thus exhibit (a)-character, though they normally are quite typically class (b), as has been pointed out above.

The exceptional reactions of the pentammine complexes probably depend upon that the ammonia ligands are able to neutralize only a rather small fraction of the charge of the central ion. The high effective charge thus remaining there will greatly immobilize those d-electrons which would otherwise be available for the formation of (b)-type bonds.

218

In conclusion, the classification of acceptors given in Table 1 should be considered as a somewhat crude generalization, which does not account for exceptions caused by special ligand factors. Especially the assignments made for borderline cases are likely to be upset by such influences. It has been shown, however, that even in its present form the classification can serve as a useful guide for estimations of the probable stabilities of various types of complexes. In the first place it is valid for reactions involvning aquo ions, as most of the assignments are founded on data referring to such ions. In order to obtain a more precise forecast of the properties of various complexes, one has to find a quantitative measure for softness (and hardness), as has been pointed out by *Pearson* (*23*) and by *Schwarzenbach* (*24*). This presumes that a fairly accurate estimate can be made of the covalent and electrostatic contributions to the bonds between various acceptors and donors, a task which does not seem at all easy.

Acceptors Formed by Non-metallic Elements

Most of the (*b*)-acceptors so far discussed are more or less typical metal ions. For these, a large number of *d*-electrons seems to be indispensible in order to bring about (*b*)-behaviour, as has already been stated. This evidently also applies to the few non-metal (*b*)-acceptors entered in Table 1, which are iso-electronic with metal (*b*)-acceptors. For many non-metal acceptors, however, which have lately been classified as (*b*) by *Pearson* (*2*) the *d*-electrons are either non-existant (*e. g.* HO^+, RO^+ and various benzene derivatives), or too well shielded to be of any consequence for the bonding (*e. g.* RSe^+, RTe^+, Br^+, I^+). Evidently all these acceptors are able to form essentially covalent bonds using only the electrons available in the *s*- and *p*-shells. It is therefore not surprising that they often constitute exceptions from the rule that a lowering of the oxidation state means more of class (*a*)-character for elements beyond the transition series. This rule evidently ceases to be valid as soon no *d*-electrons participate in the formation of the covalent bond.

For (*b*)-acceptors bonding by *sp*-electrons alone, the experimental data still seems a little scarce to allow a very elaborate systematization. The charges and polarizabilities of acceptor and donor must obviously still be very important, however. It may further safely be presumed that for certain classes of non-metal (*b*)-acceptors, such as benzene derivatives and quinones, the existence of multiple bonds within the acceptor plays a vital part for the formation of a mainly covalent acceptor to donor bond.

219

References

1. *Ahrland, S., J. Chatt,* and *N. R. Davies:* Quart. Rev. *12,* 265 (1958).
2. *Pearson, R. G.:* J. Am. Chem. Soc. *85,* 3533 (1963).
3. *Remy, H.:* "Treatise on Inorganic Chemistry" (translated by J. S. Anderson) Elsevier Publishing Co, 1956.
4. *Sillén, L. G.* and *A. E. Martell:* (Ed.) Stability Constants of Metal-Ion Complexes, The Chemical Society, London, 1964.
5. *Chatt, J., N. P. Johnson,* and *B. L. Shaw:* J. Chem. Soc. 2508, *1964.*
6. *Chatt, J., B. L. Shaw,* and *A. E. Field:* J. Chem. Soc. 3466, *1964.*
7. *Abel, E. W., M. A. Bennet,* and *G. Wilkinson:* J. Chem. Soc. 3178, *1959.*
8. *Diamond, R. M., K. Street Jr,* and *G. T. Seaborg:* J. Am. Chem. Soc. *76,* 1461 (1954).
9. *Ahrland, S., I. Grenthe, L. Johansson,* and *B. Norén:* Acta Chem. Scand. *17,* 1567 (1963).
10. *Leden, I.* and *J. Chatt:* J. Chem. Soc. 2936, *1955.*
11. *Jörgensen, C. K.:* this volume, p. 234.
12. *Sandell, A.:* Acta Chem. Scand. *15,* 190 (1961).
13. *Yamasaki, K.* and *K. Suzuki:* Proc. 8ICCC Vienna 1964, **357.**
14. *Jörgensen, C. K.:* Inorganic Complexes, Academic Press, London 1963, p. 7.
15. *Wells, A. F.:* Structural Inorganic Chemistry, 3rd Ed., Clarendon Press, Oxford, 1962.
16. *Pauling, L.:* The Nature of the Chemical Bond, 3rd Ed., Cornell University Press, Ithaca, N. Y., 1960.
17. *Dewar, J. S.:* Bull. soc. chim. France, S 71, *1951.*
18. *Chatt, J.* and *L. A. Duncanson:* J. Chem. Soc. 2939, *1963.*
19. *Burmeister, J. L.* and *F. Basolo:* Inorg. Chem. *3,* 1587 (1963).
20. *Turco, A.* and *C. Pecile:* Nature *191,* 66 (1961).
21. *Fronaeus, S.* and *R. Larsson:* Acta Chem. Scand. *16,* 1447 (1962).
22. *Schmidtke, H.-H.:* J. Am. Chem. Soc. *87,* 2522 (1965).
23. *Pearson, R. G.:* Science *151,* 172 (1966).
24. *Schwarzenbach, G.:* Chem. Eng. News, 92, May 31, *1965.*

(Received January 26, 1966)

Reprinted from "Structure and Bonding," Volume 1, C. K. Jørgensen et al., eds., pp. 234–248 (1966)

12

Electric Polarizability, Innocent Ligands and Spectroscopic Oxidation States*

Dr. C. K. Jørgensen

Cyanamid European Research Institute, Cologny (Geneva), Switzerland

Table of Contents

Unexpected connections are found between inorganic chemistry, spectroscopy, applied group theory and formal logics allowing the definition of innocent ligands, spectroscopic and magnetochemical oxidation states (independant of the actual charge distribution) and preponderant one-electron configurations. It is discussed how "softness" has less to do with electronegativities and ionization energies than with electric polarizability and density of adjacent excited states; but chemical "deformability" is nevertheless somewhat different from the weak-field first-order polarizabilities, as a comparison between Ag(I) and Cs(I) clearly demonstrates.

I. Softness and Polarizability

As many other chemical concepts, *Pearson's* hard and soft Lewis acids and bases (*1*) have a long and complicated prehistory. *Fajans'* ideas of polarizability and polarizing ability (*e. g.* how AgCl forms a more stable lattice then NaCl in spite of equal *Madelung* potentials (*2*)), and *Ahrland* and *Chatt's* classification (*3*) of central atoms being type A(Mg(II), Al(III), Th(IV), *i. e.* hard; type B(Cu(I), Ag(I), Pt(II), Hg(II), Tl(III)), *i. e.* soft; and intermediate cases, including most of the transition group atoms with partly filled d shells. As criterion for the A or B character of a given central atom with a definite oxidation number, *Chatt* and *Ahrland* used comparison within the same group (columns in *Mendelejev's* Periodic Table, horizontal line in the *Thomsen-Bohr* representation), so that for A central atoms, the complex formation constants vary:

$$F \gg Cl > Br > I \tag{1}$$
$$O > S \sim Se \gtrsim Te$$

---

* Contribution to the Symposium on "Soft and Hard Acids and Bases" organized at the Cyanamid European Research Institute by Dr. *R. F. Hudson*, 10.–12. May 1965.

234

Many properties vary in the way indicated by the inequality signs. The most striking is perhaps the *electronegativity* which was originally defined by *Pauling* on the basis of bond energies but later reformulated by *Mulliken* as the average value of electron affinity and ionization energy of the atom promoted to a definite valence state. It is recently accepted (*4–8*) that the electronegativity is essentially the differential quotient of energy dE/dz with respect to the charge of the atom considered.

However, it would be a serious error to confuse electronegativity or ionization energy with "hardness". The clearest counter-example is Tl(III) which is a rather soft central atom but which must have a rather high ionization energy. There are other physical properties which accentuate the opposite inequality signs of (1) even more, for instance the electric *polarizability*. Table 1 contains many values for this quantity α in the unit 10^{-24} cm³. Gaseous atoms and positive ions have polarizabilities which can be calculated from wavefunctions (*9*) by evaluating the sum of matrix elements of induced dipole moment:

$$a = \frac{2}{3} \sum_k \frac{(E_k - E_0) \, | \, (\Psi_0 \, | \, R \, | \, \Psi_k) \, |^2}{(E_k - E_0)^2 - h^2 \, \nu^2} \tag{2}$$

The main contribution to α sometimes comes from discrete excited states Ψ_k (such as one ns-electron of the groundstate Ψ_0 being replaced by one np-electron) and in other cases from the continuum from states with energy E_k higher than the first ionization energy. In a one-electron description, the linear electric field has the symmetry $C_{\infty v}$ and produces a mixture of odd and even σ-states, usually of adjacent l-values (l with $l - 1$ and $l + 1$). It is seen from (2) that under equal circumstances, α is larger for smaller ionization energies (the denominator) and larger for larger average radius of the loosest bound orbital. Actually, α varies much more dramatically than the ionization energy, and is also dependent on intrinsic properties of the wavefunctions. Gaseous ions such as O^{--} and S^{--} (but not the halides X^-) have zero ionization energy and infinitely high polarizability. (It is only ions confined in a closed box (*10*) or restricted to a definite Hartree-Fock configuration, such as $1s^2 2s^2 2p^6$, which are capable of having negative ionization energies. This is an interesting restriction on the validity of the *Glockler-Lisitzin* parabolic relations (*5*)).

It is not easy to evaluate the utility of the concept of polarizability for chemistry. It is certainly true, as *Fajans* first pointed out, that a description involving, at the same time, polarization and partly covalent bonding, to a great extent takes the same thing into account twice. On the other hand, (2) is valid for weak electric fields, and the stronger interactions occurring in molecules certainly modify the polarizability

235

Table 1. *Atomic and molecular polarizabilities (mostly extrapolated from the visible region to* $\nu \to 0$ *in eq. 2) in the unit* 10^{-24} *cm³ (Å³). Frequently, the unit cm³/mole, which is* $\frac{4\pi}{3} N_L \cdot 10^{-24} = 2.55$ *times smaller, is used. It may be noted that these values always are smaller than the actual molar volumes.*

	calc.	obs.		obs.		obs.		obs.
H	0.664[a]	0.66[a]	H(−I)	1.5	In(III)	1.2[k]	CS_2	8.74[a]
H⁻	31.3[b,c]	34[b]	Li(I)	0.029[i]	Te(−II)	~9[i]	CCl_4	10.5[a]
He	0.205[c]	0.205[c]	O(−II)	1.3–2.0[i]	I(−I)	6.2[i]	SiF_4	3.3[a]
Li⁺	0.030[c]	0.023[c]	F(−I)	0.8[a,i]	Cs(I)	3.1[i]	SF_6	4.4[a]
Li	25[d,e]	21[e]	Na(I)	0.25[a,i]	Ba(II)	2.5[i]	$TiCl_4$	15.0[a]
Be⁺	3.6[d]	—	Mg(II)	0.1[a]	La(III)	1.6[k]	$GeCl_4$	12.5[a]
Be	5.5[b]	—	Al(III)	0.06[a]	Hg(II)	2.9[k]	$SnCl_4$	13.8[a]
F⁻	3.1[f]	—	S(−II)	5–6[i]	Tl(I)	5.2[i]	$SnBr_4$	19.5[a]
Ne	—	0.39[a]	Cl(−I)	3.0[i]	Pb(II)	4.8[i]	SnI_4	26.1[a]
Na⁺	0.15[f]	—	K(I)	1.2[i]	H_2	0.79[a]	OsO_4	6.3[a]
Mg⁺⁺	0.072[g]	—	Ca(II)	1.1[i]	N_2	1.76[a]	$HgCl_2$	9.0[a]
Na	23[e]	21[e]	Mn(II)	1.25[k]	O_2	1.60[a]	$HgBr_2$	11.5[a]
Mg	7[h]	—	Fe(II)	1.2[k]	Cl_2	4.61[a]	HgI_2	16.3[a]
Cl⁻	5.6[f]	—	Fe(III)	2.2[k]	HF	0.82[a]	CO_3^{--}	4[i]
Ar	—	1.63[a]	Co(II)	1.2[k]	HCl	2.63[a]	NO_3^-	3.7[i]
K⁺	1.26[f]	—	Ni(II)	1.0[k]	HBr	3.61[a]	SO_4^{--}	4.8[i]
Ca⁺⁺	0.8[f]	—	Cu(I)	1.6[i]	HI	5.45[a]	ClO_4^-	4.5[k]
K	45[e]	37[e]	Zn(II)	0.8[a,i]	BF_3	2.4[a]	IO_3^-	7[k]
Ca	20[h]	—	Se(−II)	6–7.5[i]	CH_4	2.60[a]		
Zn	—	~6[a]	Br(−I)	4.1[i]	CO	1.95[a]		
Kr	—	2.5[a]	Rb(I)	1.8[i]	CO_2	2.65[a]		
Rb	50[e]	~50[a]	Sr(II)	1.7[i]	CF_4	2.9[a]		
Sr	28[h]	—	Ag(I)	2.4[i]				
Cd	—	~8[a]	Cd(II)	1.8[i]				
Xe	—	4.0[a]						
Cs	68[e]	~60[a]						
Ba	38[h]	—						
Hg	—	5.1[a]						

Literature:

[a] *Landolt-Börnstein:* tables *1*, part 1, p. 399 and *1*, part 3, p. 509. Springer Verlag, Berlin, 1950 and 1951.

[b] *Kolker, H. J.,* and *H. H. Michels:* J. Chem. Phys. *43*, 1027 (1965).

[c] *Dalgarno, A.,* and *A. L. Stewart:* Proc. Roy. Soc. (London) *A 247*, 245 (1958).

[d] *Flannery, M. R.* and *A. L. Stewart:* Proc. Phys. Soc. *82*, 188 (1963).

[e] *Sternheimer, R. M.:* Phys. Rev. *127*, 1220 (1962).

[f] *Fischer-Hjalmars, I.,* and *M. Sundbom:* Acta Chem. Scand. *11*, 1068 (1957).

[g] *Bockasten, K.:* Phys. Rev. *102*, 729 (1956).

[h] *Altick, P. L.:* J. Chem. Phys. *40*, 238 (1964).

[i] *Tessman, J. R., A. H. Kahn* and *W. Shockley:* Phys. Rev. *92*, 890 (1953).

[k] *Fajans, K.* and *G. Joos:* Z. Physik *23*, 1 (1924) (the values for M⁺⁺ in aqueous solution have been increased 0.7 and for M⁺³ 1.1, obtaining agreement with the solids studied by i).

236

considerably by second-order effects with important coefficients. The experimental technique for determining α for a molecule involves the so-called electronic polarization term, different from the permanent dipole moment (if any) and the induced dipole moment due to modified vibrational motion. Actually, the values (2) calculated have been confirmed in that way for the noble gases and monatomic mercury vapour. However, in polyatomic molecules, the value of α obtained is usually very uncertain. Another, indirect but more accurate (or at least more reproducible) technique involves the refractive index for visible light. It is possible to define a fairly consistent set of molar refractivities R which are additive

$$R = \frac{n^2 - 1}{n^2 + 2} \cdot \frac{M}{\rho} = \frac{4\pi}{3} N_L (\alpha_1 + \alpha_2 + \ldots) \tag{3}$$

when n is the refractive index (usually extrapolated towards the limit $\nu \to 0$, assuming (2)), M the molar weight, ρ the density, and α the polarizability of the constituent atoms or ions, in the same way as the diamagnetic corrections to magnetic susceptibilities.

Table 1 includes such α values, derived from molar refractivities, with Roman numerals for the chemical species and Arabic numerals M^{+z} for isolated, monatomic ions. The α values for O(–II) and S(–II) are not infinite, as for O^{-2} and S^{-2}, but they depend on the crystalline environment (larger *Madelung* energy producing lower α values; *cf.* also the nephelauxetic effect in oxides (*11*), and different authors frequently give rather different results. In the case of experimental values, these discrepancies are connected with the problem of distributing α on cations and anions in salts, though it is easier than the distribution of ionic radii, because the contribution from most cations to the molar refractivities is rather negligible. Actually, a genuine variation is observed in the case of oxide (–II), α being 1.7 for MgO, 2.4 for CaO, 2.6 for SrO, 3.0 for BaO* and 1.35 for Al_2O_3. The latter value is comparable to the quantities 1.2 derived for SO_4^{--} and 1.1 for ClO_4^-. Quite generally, the gaseous molecules containing oxide, fluoride and chloride in Table 1 can be considered as if the α values for the anions were slightly smaller in molecules than ionic salts. In Fajans' opinion, one can describe these decreased values by a theory involving large *Madelung* constants and covalent bonding both subtracting from the values extrapolated for gaseous ions. However, this extrapolation from (Ne, Ar, Kr, Xe) to the isoelectronic (F^-, Cl^-, Br^-, I^-) probably does not have much physical significance, and Fajans was wrong with an infinite factor in the case of O^{--} and S^{--}.

---

* However, when the cation polarizabilities estimated by *Tessman et al.* (Table 1, ref. i) are used consistently, oxide (– II) varies from 1.65 in MgO to 2.05 in BaO.

237

The theoretical values are also uncertain by a factor 1.5 to 2 except cases such as H and He. It is particularly alarming that the *Hartree-Fock* wavefunction for H^- gives only 11.8 to 14 (even here, the authors disagree) whereas a much more elaborate wavefunction gives 31.34 in fair agreement with the experimental value 34.1 obtained from the cross-section for photodetachment. The molar refractivities of crystalline NaH and KH suggest $\alpha = 1.6$ for H(–I), whereas the stronger *Madelung* potential (and perhaps more pronounced covalent bonding) in LiH corresponds to $\alpha = 1.3$. With regard to *Ahrland's* contribution to this volume, it is worth emphasizing that H(–I) chemically is a very soft ligand, but is not expected to have low-lying, empty d orbitals except those of the continuum. We shall return below to the question of the relation between softness and polarizability.

The halogens or chalcogens exhibit variations such as (1) which every chemist will recognize. However, the inorganic chemists are familiar with elements of highly varying oxidation number, such as V, Cr, Mn, Ru, Re or Os. *Pearson* (*1*) suggested that the low oxidation numbers z correspond to the softest behaviour, and that an element gradually becomes harder when z increases, and electrovalent bonding becomes more important. On the other hand, there is much evidence (*12*) that manganese (see the qualitative Fig. 1) reaches a maximum of hardness at Mn(II), whereas, admittedly, Mn(–I) and Mn(I) are soft (*e. g.* $Mn(CO)_5I$ and $Mn(CO)_5H$ are more stable than $Mn(CO)_5Cl$, and $Mn(CO)_5F$ is not known) but there is, in my opinion, also a beginning tendency to softness in Mn(IV), Mn(V), Mn(VI) and Mn(VII). It is not easy to present such evidence in a clearcut way for formation of compounds with soft ligands. Whereas ReO_3Cl and ReO_3Br exist, only species such as MnO_4^- and MnO_3F are known of Mn(VII). However, the competition with redox reactions, preventing MnO_3Br and MnO_3I from being isolated, is no sufficient argumentation against a moderate softness of Mn(VII). It is quite definite that the partly covalent bonding is more pronounced, not only in MnO_4^-, but also in MnF_6^{--}, TcF_6^{--} and ReF_6^{--} when judged from the nephelauxetic effect (*13*), than it is in Mn(II) complexes of soft ligands, such as $MnBr_4^{--}$ and MnI_4^{--}. Now, it is not quite certain whether the extent of covalent bonding has so much to do with the softness discussed here. It would appear that the nephelauxetic series (*14*) and the variation of the optical electronegativities (*15*), though following (1), are rather connected with ionization energies or diagonal elements of one-electron operator energies. Fig. 2 represents an alternative explanation, the softness being produced by close approach of the energies of filled and empty orbitals (3d and 4s in Mn(I), ligand orbitals and 3d in Mn(VII)). Several authors (*16, 17*) have noted the correlation between polarizabilities and the nephelauxetic effect, which may be a somewhat artificial relation.

238

Fig. 1. Qualitative variation of the chemical hardness of manganese as function of the oxidation state, showing a maximum for Mn(II).

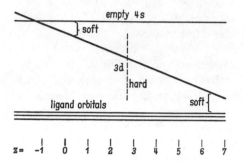

Fig. 2. The chemical softness may be connected with only small energy differences between empty and occupied orbitals. Hence, Mn(I) should be soft because of adjacent 4s and 3d orbitals of the central atom, and Mn(VII) should be soft because of adjacent 3d central atom and p ligand orbitals, whereas Mn(II) and Mn(III) with a fairly isolated, partly filled 3d shell are expected to be hard.

Already *Ahrland (18)* pointed out that Tl(III) is much softer than Tl(I). However, the s^2-family Sn(II), Sb(III), Te(IV), I(V), Xe(VI), Au(−I), Tl(I), Pb(II), Bi(III) and Po(IV) seems to belong to class B in a somewhat different way from that of the other B central atoms *(19)* and in particular seems to bind H_2O very weakly, increasing the complex formation constants of the halides. The relatively high polarizability α of Tl(I) seems to give thallium (I) salts somewhat different properties from Rb(I) salts, though the ionic radii are nearly identical. The differences are of the same kind as between Pb(II) and Ba(II), or between the soft d^{10}-system Ag(I) and Na(I) or K(I). *Orgel (20)* suggested that the rather unusual stereochemistry of the $6s^2$-family is caused by the relative closeness of the excited configuration 6s6p. The corresponding anisotropic polarizability tends to remove the center of inversion from such systems (producing the *Gillespie-Nyholm* phenomenon *(21)* of lone-pairs imitating ligands) and it is not even certain whether PbS, PbSe and PbTe, crystallizing in the NaCl-type, are not statistically distorted in the sense that the lead atoms tend to occupy, in equal numbers, the eight positions on the trigonal axes or the six positions on the tetragonal axes, somewhat removed from the center of each unit cell. Normally, such lattices distort

239

collectively by sufficient cooling (as when the high-temperature CsCl-modification of TlI is cooled; it has a *very* high refractive index) but $O°K$ may not always be sufficient. Unfortunately, the α-value derived from the molar refractivity of Tl(III) is not known; it may very well be smaller than for Tl(I), since already the value for Hg(II) seems to be smaller than for Tl(I). This is one example of a definite discrepancy between the physical concept, polarizability and the chemical softness, which is even more apparent in Cs(I) being more polarizable than Ag(I) though probably less polarizable than isolated Cs^+.

One way out of the dilemma of non-monotonic variation of the softness as function of the oxidation number z (Fig. 1) has implicitly been proposed by *Burmeister* and *Basolo* (22) *i. e.* that the hardness is a monotonic function of the *physical, fractional charge* $+ \delta$ on the central atom, which may not be a monotonic function of z. Actually, the nephelauxetic effect (13, 14) tends to suggest that ReF_6^{--} has a *higher* δ than the isoelectronic IrF_6; that $Mn(II)F_6$ in crystalline MnF_2 or $KMnF_3$ have δ comparable to the isoelectronic FeF_6^{-3} and probably higher than MnF_6^{--}; and that $MnCl_4^{--}$ has a higher δ than $FeCl_4^-$. On the other hand, we remember that the optical electronegativities of M in hexahalide complexes MX_6^{+z-6} are invariantly a weakly increasing function of z. (*cf.* Fig. 2).

This working hypothesis would suggest that δ has the maximum value, say 1.8, in typical Mn(II) complexes, whereas it might be $+0.2$ in $Mn(CO)_5^-$ and $+1.2$ in MnO_4^-. This would explain the inductive effect of softness, because the maximum value of δ certainly is lower in MnI_2 than in MnF_2, and it would, at least in part, explain the *symbiosis of ligands* (12), the tendency of soft ligands to flock together. For instance, δ of Mn in $Mn(CO)_5^+$ is already so low that this fragment appeals far more to the soft ligands H^- and I^- than to F^-. The behaviour of $Co(CN)_5X^{-3}$ (not to speak of vitamin B12, which probably is a Co(III) complex (23)) is definitely soft, whereas $Co(NH_3)_5X^{++}$ is hard, X = F being strongest bound. There may be some reluctance to accept Burmeister and Basolo's explanation (22) in detail in the special case of $Pd(As(C_6H_5)_3)_2(NCS)_2$ and Pd dip $(NCS)_2$ being most stable in the N-bound isomer, whereas only $Pd(NH_3)_2(SCN)_2$ and $Pd(SCN)_4^{--}$ S-bound isomers are known. The ambidentate ligand NCS^- is probably softer on the S-atom than on the N-end, though the nitrogen atom has more than one lone-pair and attains nearly as low an electronegativity. *Schmidtke* (24) has recently shown that $Rh(NH_3)_5SCN^{++}$ can be prepared, but is less stable than the N-bound isomer. It is of course true that low-lying, empty π^* orbitals of the ligands triphenylarsine and dipyridyl may accept some electronic density from Pd(II) and increase the value of δ and hence decrease the softness; but the explanation appears at the first view somewhat *ad hoc* because

240

the symbiotic carbon monoxide, hydride, phosphines, arsines and stibines usually cooperate nicely in conferring softness to their typical central atoms with low oxidation numbers. The truth may be that the balance is extremely delicate in the border-line cases, and that, as *Basolo* (*25*) emphasized, steric factors may become very important. A somewhat more typical ambidentate ligand is thiosulphate $S_2O_3^{--}$ which is expected to bind soft central atoms with S and hard central atoms with O. It is clear from spectroscopic studies (*26*) that the hard Eu(III) binds one or two oxygen atoms of hypophosphite $H_2PO_2^-$; it is an interesting question whether sufficiently soft central atoms would bind through hydrogen.

It has been customary to ascribe certain deviations from the principle of additivity of ionic colours to polarization effects (*27*). Thus, AgI, Ag_2CO_3, Ag_3PO_4 and Ag_3AsO_3 are yellow and Ag_3AsO_4 is red though the constituent ions are colourless in aqueous solution. In certain cases (*28*), such as the red Ag_2CrO_4, blue $AgMnO_4$, orange Ag_2ReCl_6 and blue Ag_2IrCl_6, it is possible to ascribe these anomalous colours to electron transfer bands from the filled $4d^{10}$-shell of Ag(I) to available, lowlying orbitals of the anions. This explanation cannot so easily be adapted to the cases mentioned above, because Ag_2C_2, AgCN, $AgNO_3$, Ag_2SO_3, AgSCN, $AgS_2P(OC_2H_5)_2$, $AgClO_3$ and $AgIO_3$ are white. Though this statement does not resolve the problem of the assignment of the optically excited states of the silver(I) salts, there is some physical reason in connecting:

hardness	softness
low polarizability	high polarizability
isolated groundstate	high density of low-lying states

Actually *Pearson* (*1*) was the first to discuss that "good" metals (such as Na, K, Al, Cu and Ag) are *ipso facto* soft. The metallic state introduces in chemistry the wavefunctions of the continuum, of the freely moving electron in empty space, with the concomitant crowded (over-all dense) distribution of states. Incidentally, it may be mentioned that chemists determining complex formation constants (*29*) already previously used the name "soft metals" for those elements rapidly coming in electrochemical equilibrium with their ions in aqueous solution, such as Cu, Ag, Cd, In, Hg, and "hard metals" for those not giving reproducible potentials, such as V, Cr, Ni, Mo or Ir. It may be remarked that the latter elements are hard in the normal sense of the word, usually are brittle, and have very high boiling points and heats of atomization. The solid state physicists have developed other criteria for typical metals, mainly based on the electric conductivity, and there is general agreement that whereas the number of conduction electrons liberated per atom has the order of magnitude of one in the alkali and coinage metals, it may be very small in "accidental metals" such as Bi, $CoSe_2$, $NiTe_2$, TiO (at room temperature) and superconducting semiconductors at very low temperature. The distinction between metals and non-metals is based on the approximate invariance (or minus-first order dependence) of the conductivity on the absolute temperature, whereas the intrinsic semiconductors have an Arrhenius activation energy for conduction, hence increasing strongly with the temperature. Experimentally, it is not easy to distinguish between accidental metals and non-stoichiometric, extrinsic semiconductors; thus, the energy gap of CrSb is very small and difficult to determine. For our purpose, the "good" metals with a great concentration of conduction electrons are typically soft materials.

II. Preponderant Configurations

Coming back to the monatomic species, it is remarkable to what extent the groundstate is only accompanied by a few discrete, lowlying states, whereas there is an infinite number of discrete states just before the ionization limit. It is possible to make a statistics of the number of discrete multiplet terms (including the groundlevel) in the interval between the groundlevel and half the first ionization energy for the 54 species consisting of M^0, M^+ and M^{++} containing from one to eighteen electrons (30). The results are:

1 + 0 :	15 cases	3 + 0 :	7 cases
1 + 1 :	12 cases	3 + 1 :	5 cases
1 + 2 :	1 case	3 + 2 :	2 cases
1 + 3 :	4 cases	3 + 3 :	1 case
1 + 5 :	2 cases	3 + 4 :	2 cases
1 + 6 :	1 case	3 + 5 :	1 case
		3 + 6 :	1 case

The numbers $a + b$ indicate that one normally ascribes a terms to the ground configuration ($a = 3$ for p^2, p^3 and p^4 plus closed shells) and b terms to other configurations. The statistics is entirely different for more than 18 electrons because of the near coincidence of the 3d and 4s orbital energies; thus, $a + b$ are for the 22-isoelectronic series:

Ti⁰   4 + 32   V⁺   15 + 50   Cr⁺⁺   10 + 34

After the end of the first transition group, smaller values for $a + b$ reappear:

28 electrons: Ni⁰ 4 + 9   Cu⁺ 1 + 9   Zn⁺⁺ 1 + 8
29 electrons: Cu⁰ 1 + 2   Zn⁺ 1 + 2   Ga⁺⁺ 1 + 1
30 electrons: Zn⁰ 1 + 1   Ga⁺ 1 + 2   Ge⁺⁺ 1 + 2

It is quite clear that this statistics is connected with *Scheibe's* observation (31) that most atoms and molecules have their first excited level

242

roughly a quarter of a *Rydberg* unit (*i. e.* 27400 cm^{-1}) below the first ionization energy, rather independently of the size of the latter quantity.

We have several reasons to want define *preponderant configurations*. In an atom or molecule containing several electrons, the wavefunction Ψ of a given state can be expanded on a series of mutually orthogonal configurations:

$$\Psi = a_0 \Psi_0 + a_1 \Psi_1 + a_2 \Psi_2 + \ldots; \qquad \Sigma a_k^2 = 1 \qquad (4)$$

and the basis should be chosen in such a way that a_0 is *as close as it can come to 1*. In molecules at normal internuclear distances, the preponderant M. O. configuration is a far better approximation than just one valence structure. It may be remembered that at large internuclear distance a hydrogen molecule is *better* described as two (1s) hydrogen atoms than by the M. O. configuration $(\sigma_g)^2$ whereas the latter is by far the better approximation close to the minimum of the potential curve. The preponderant configuration of a many-electron system has many surprising properties. One of them is that it *classifies* correctly the lowest-lying excited states. For instance, whereas a_0 in (4) is only some 0.996 for the preponderant configuration 1s^2 of the groundstate ^{1}S of helium, it is true that the first excited states have symmetry types ^{3}S, ^{1}S, ^{3}P, ^{1}P, . . . *as if* the configurations were 1s2s, 1s2p, . . . Another phenomenon is that auto-ionizing states exist inside the continuum, where one electron seems to be excited or ionized away from the inner shells. Such states can be studied by photoelectron spectroscopy (*32*) or by spectroscopy in the vacuo ultraviolet (*33*) or X-ray region (*34*). Though there is an infinity of other states having the same symmetry types below such inner-shell excitations, they seem quite well-defined in many cases and produce rather sharp absorption lines.

It may be useful to apply *Bertrand Russell's* theory of types (which I discuss critically in some detail elsewhere (*35*)) to propositions about preponderant configurations. He proposed that sentences about classes are of a higher type than sentences about the members of this class. Thus, certain properties belong to the class and not to the members, e. g. "to be numerous". It is clear that the proposition "Ψ_0 is the configuration of the state in (4)" is only approximately valid, whereas "Ψ_0 is the preponderant configuration" is absolutely valid even in the case where we cannot succeed in getting a_0 larger than 0.8. The point is that being the preponderant configuration, it determines the possible order of the lower excited states to quite a high extent, and hence it is of much higher intrinsic interest to us than Ψ_k with positive k. It is possible to define (*19*) *chromophores* MX$_N$ with definite oxidation state of the central atom M on the condition that the preponderant M. O. configuration only contains

16*

one partly filled shell. In other words, the spectroscopic characteristics of the chromophore $Cr(III)O_6$ in $Cr(H_2O)_6^{+3}$, $Cr(C_2O_4)_3^{-3}$ or in the ruby clearly demonstrate the presence of three electrons in the partly filled shell approximately being a d shell concentrated on the central chromium atom, but delocalized to some extent on the six nearest neighbour atoms, and *hence* we call it Cr(III). We do not suggest that it is a good approximation to say that the fractional, physical charge δ is close to $+3$, but we define spectroscopic oxidation states from the presence of an integral number of electrons in the partly filled shell, *e. g.* in chromium:

Cr(0)	Cr(I)	Cr(II)	Cr(III)	Cr(IV)	Cr(V)	Cr(VI)
d^6	d^5	d^4	d^3	d^2	d^1	d^0

(Dr. *Gwyneth Nord* was so kind as to point out to me that "oxidation state" can be used for the concept derived from the preponderant M. O. configuration, in contrast to the formal "oxidation number" used as a tool for writing reaction schemes). There are good reasons from the nephelauxetic effect (*14*) to expect that the higher limit of δ does not hold for transition group central atoms in Pauling's electroneutrality principle (*36*) formulated as $-1 < \delta < +1$ though the lower limit may seem quite plausible. However, we emphasize once more that the oxidation state has no direct connection with δ. It is a valuable by-product of the ligand-field theory which has not been sufficiently elaborated that when the preponderant M. O. configuration assigns an integral number of electrons to one partly filled shell, this implies a well-defined oxidation state.

III. Innocent Ligands

We define *innocent ligands* as those existing in a colourless* standard form with closed-shell preponderant configuration. This is not only true for species such as O^{--}, F^- and Cl^- isoelectronic with the noble gases, but also for instance CO, CN^-, NH_3, NCS^-, SO_3^{--}, $(C_2H_5O)_2PS_2^-$ *etc.* Innocent ligands permit the evaluation of the oxidation state of a single central atom. (Professor *R. Dannley* suggested kindly the alternative words "rigorous" for innocent and "ambiguous" for the contrary case). NO is not innocent in all situations because of the low-lying, partly filled π-orbital, and might occur as NO^+, NO or NO^- according to the circumstances. Many biological ligands containing conjugated heterocyclic systems may not always be innocent because of the comparable one-

---

* We use "colourless" in the slightly generalized sense of a well-isolated (say by 20000 cm^{-1}) groundstate. We remark that the conditions for temperature-independent paramagnetism in coloured species (say MnO_4^-) are the same as for softness in the representation Fig. 2.

244

electron energies of the delocalized orbitals in the ligands and of the partly filled shell. Recently, there has been much interest in the sulphur-containing ligands (37, 38).

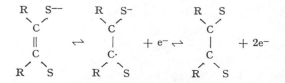

where R = CF$_3$, CN or C$_6$H$_5$. The complexes Ni(S$_2$C$_2$R$_2$)$_2^{--}$ and Pd (S$_2$C$_2$R$_2$)$_2^{--}$ probably contain quadratic chromophores Ni(II)S$_4$ and Pd(II)S$_4$ quite comparable to, say (39), Ni(S$_2$P(OC$_2$H$_5$)$_2$)$_2$. But the oxidized forms such as Ni(S$_2$C$_2$R$_2$)$_2$ or octahedral V(S$_2$C$_2$R$_2$)$_3$ and V(S$_2$C$_2$R$_2$)$_3^{--}$ pose perplexing problems, and it is probable that the ligands are not innocent in these cases.

In view of the desirability of approximately colourless (i. e. no low-lying excited states) closed-shell groundstates as a qualification for being innocent ligands, it is possible to make the generalization:

The softer a ligand is, the more probably the ligand is not innocent. However, this depends on the company of the ligand, i. e. the central atom. Thus, the conduction electrons, the non-innocent ligands par excellence, do not prevent most metallic alloys or black low-energy-gap semiconductors formed by the lanthanides from having magnetic properties indicating in nearly all cases that the number of 4f electrons is an integer, corresponding to M(II) or M(III). Professor *K. A. Jensen* has proposed to use sharp brackets M[II] or M[III] for such oxidation states, in order to avoid confusion with the Roman numerals of the Stock nomenclature.

For instance, the monosulphides (40) LaS, CeS, PrS, NdS, GdS, TbS, DyS, HoS, ErS and TmS have magnetic moments corresponding to M[III] and are metals with roughly one conduction electron per 4f group atom, whereas SmS, EuS and YbS are non-metallic and have the magnetic moments appropriate for M[II]. The distribution of the [II] and [III] states can be understood in terms of the large spin-pairing energy in the 4f shell (41). The metallic transition can be extremely sharp (42); NdSe and Nd$_{1-x}$Sm$_x$Se with x < 0.11 are metallic; with x > 0.13, they are semi-conducting. *Goodenough* (43) elaborated a theory of magnetic moments which are localized if the interatomic distances are much larger than the average radii of the partly filled shell, and otherwise do not correspond to a definite value for S of the transition group atoms. We would say that the preponderant configuration is atomic in the former case and delocalized in the latter case.

245

Hulliger and Mooser (*44, 45*) have described cases where sulphur and arsenic ligands in semi-conducting compounds such as FeS_2 and CoAsS still are innocent, the diamagnetic d^6-chromophores $Fe(II)S_6$, $Ru(II)S_6$, $Co(III)As_3S_3$ can be recognized, whereas the metallic compounds RhS_2 or NiAsS would have been d^7-systems, if they had not been metallic. CoS_2 can be recognized as having $S = \frac{1}{2}$, but is at the same time an accidental metal.

It is a very interesting question whether a given preponderant M. O. configuration is compatible with more than one distribution of oxidation states. For instance (*41*), diatomic hydrides MH would show the same order of M. O. energies when described as M(I)H(−I) or as M(−I)H(I). This is also true for the series BH_4^-, CH_4 and NH_4^+, the closed-shell groundstate of methane can be ascribed to $C(IV)H_4(−I)$ (slightly preferable for several reasons) or to $C(−IV)H_4(I)$. By the same token, the diatomic interhalogens allow a continuous translation from X(I)Y(−I) to X(−I)Y(I). This is not always true; we have discussed the reasons why IrF_6 must be written as a $5d^3$-system $Ir(VI)F_6(−I)$. It is instructive to remark that in such cases, the oxidation state of X in MX_N jumps two units at a time and of M, the jump is 2N units. Thus, the tetrahalides CF_4 and CCl_4 have the choice between $C(IV)F_4(−I)$ or the (unacceptable) $C(−IV)F_4(I)$. The M. O. symmetry types γ_n would be the same in the two cases, because the four F(I) atoms each have the preponderant configuration π^4. In the tetroxo complexes PO_4^{-3}, SO_4^{--} and ClO_4^-, the oxygen atoms have a similar choice between O(−II) and O(0) π^4. In the similarly colourless SF_6, only $S(VI)F_6(−I)$ and the absurd $S(−VI)F_6(I)$ would fulfill the group-theoretical requirements of symmetry types.

If oxidation states can be satisfactorily defined for transition group complexes of innocent ligands and for non-transition group complexes such as CCl_4, SO_4^{--} and SF_6, they cannot be reasonably used in all cases. The homonuclear molecules are perhaps the worst dilemma. It is no accident that the colourless hydrogen molecule with well isolated groundstate $^1\Sigma_g$ gets trapped as hydrogen(I) hydride(−I) in our oxidation state description of the preponderant M. O. configuration σ_g^2. Actually, hydrogen(O) hydrogen(O) clearly corresponds to two adjacent energy levels $^1\Sigma_g$ and $^3\Sigma_u$ as realized at large internuclear distances. It is possible to remove this difficulty by studying the symmetry types of the bonding σ-orbitals, each accommodating a pair of electrons. Then, H_2 would be formulated $H^+(\sigma^2)H^+$ and methane $C^{+4}(\sigma^2)_4H_4^+$ in close analogy to *Fajans'* quanticules (*46*). This description is satifactory with regard to the symmetry types $^1\Sigma_g$ and $^1\Gamma_1$ of the groundstates, but neglects the experimental fact (*32*) that the three 2p-like γ_5 orbitals in CH_4 has a much lower ionization energy than the 2s-like γ_1 orbital. If we describe diamond as $C^{+4}(\sigma^2)C^{+4}$ with a strongly bound electron pair between each pair of

246

adjacent C^{+4} cores, we cannot express the latter feature. Effectively, the preponderant configuration of diamond involves $2s2p^3$ per carbon atom, but because of the ostensive closed-shell structure of this crystal, it is permitted to re-arrange the orbitals in linear combinations *as soon* we are not interested in spectroscopic excitations or ionizations. This transformation is frequently used as an excuse for the valence-bond treatment. However, unfortunately, this is only valid for the closed-shell determinant Ψ_0 itself, and not for the actual wavefunction Ψ.

We may conclude that we are just at the beginning of an analysis of the concepts of soft and hard monatomic species and molecules. It is extremely fortunate that the classification of energy levels in preponderant one-electron configuration sometimes work, though admittedly this is not always the case. The occurrence of groundstates with positive S values is normally a consequence of necessarily (of group-theoretical reasons) degenerate orbitals in the preponderant configuration. The simplest wellknown molecule having this property is $O_2(S=1)$, but the phenomenon is extremely frequent in the transition groups. The ligand-field theory, in its recent amalgamation with M. O. theory, permits then usually the definition of axidation states.

In the present volume, contributions by *R. F. Hudson* and *S. Ahrland* are also included. A general survey of the symposium was published in the 31. May 1965 issue of "Chemical and Engineering News."

References

1. *Pearson, R. G.:* J. Am. Chem. Soc. *85*, 3533 (1963).
2. *Fajans, K.:* J. Chem. Phys. *9*, 281 (1941).
3. *Ahrland, S., J. Chatt,* and *N. R. Davies:* Quart. Rev. (London) *12*, 265 (1958).
4. *Iczkowski, R. P.,* and *J. L. Margrave:* J. Am. Chem. Soc. *83* 3547 (1961).
5. *Jørgensen, C. K.:* Orbitals in Atoms and Molecules, Academic Press, London (1962).
6. *Hinze, J., M. A. Whitehead,* and *H. H. Jaffe:* J. Am. Chem. Soc. *85*, 148 (1963).
7. *Ferreira, R.:* Trans. Faraday Soc. *59*, 1064 and 1075 (1963).
8. *Klopman, G.:* J. Am. Chem. Soc. *86*, 1463 and 4550 (1964).
9. *Sternheimer, R. M.:* Phys. Rev. *96*, 951 (1954).
10. *Baughan, E. G.:* Trans. Faraday Soc. *57*, 1863 and (1961) *59*, 635 and 1481 (1963).
11. *Jørgensen, C. K., R. Pappalardo,* and *E. Rittershaus:* Z. Naturforsch. *19a*, 424 (1964) and *20a*, 54 (1965).
12. — Inorg. Chem. *3*, 1201 (1964).
13. — and *K. Schwochau:* Z. Naturforsch. *20a*, 65 (1965).
14. —Progress Inorg. Chem. *4*, 73 (1962).
15. — Mol. Phys. *6*, 43 (1963).
16. *Zahner, J. C.* and *H. G. Drickamer:* J. Chem. Phys. *35*, 1483 (1961).
17. *Weakliem, H. A.:* J. Chem. Phys. *36*, 2117 (1962).

247

C. K. Jørgensen

18. *Ahrland, S., I. Grenthe, L. Johansson, and B. Norén:* Acta Chem. Scand. *17,* 1567 (1963).
19. *Jørgensen, C. K.:* Inorganic Complexes. Academic Press London 1963.
20. *Orgel, L. E.:* J. Chem. Soc. *1959,* 3815.
21. *Gillespie, R. J.* and *R. S. Nyholm:* Quart. Rev. (London) *11,* 339 (1957) and Progress Stereochem. *2,* 261 (1958).
22. *Burmeister, J. L.* and *F. Basolo:* Inorg. Chem. *3,* 1587 (1964).
23. *Pratt, J. M.* and *R. G. Thorp:* J. Chem. Soc. *1966A,* 187.
24. *Schmidtke, H.-H.:* J. Am. Chem. Soc. *87,* 2522 (1965).
25. *Basolo, F., W. H. Baddley* and *J. L. Burmeister:* Inorg. Chem. *3,* 1201 (1964).
26. *Barnes, J. C.:* J. Chem. Soc. *1964,* 3880.
27. *Palmer, W. G.:* Valency. Cambridge University Press, 1948 (2. Ed. 1959).
28. *Jørgensen, C. K.:* Acta Chem. Scand. *17,* 1034 (1963).
29. *Bjerrum, J.:* Chem. Rev. *46,* 381 (1950).
30. *Moore, C. E.:* Atomic Energy Levels. Nat. Bur. Stand. Circ. 467, Vol. I, II and III. Washington, 1949, 1952 and 1958.
31. *Scheibe, G., D. Brück,* and *F. Dörr:* Chem. Ber. *85,* 867 (1952).
32. *Al-Jouboury, M. I.* and *D. W. Turner:* J. Chem. Soc. *1963,* 5141 and *1964* 4434.
33. *Codling, K.* and *R. P. Madden:* Phys. Rev. Letters *12,* 106 (1964).
34. *Cauchois, Y.:* J. Phys. Radium *13,* 113 (1952) and *16,* 253 (1955).
35. *Jørgensen, C. K.:* Logique et Analyse (Louvain) *7,* 233 (1964).
36. *Pauling, L.:* J. Chem. Soc. *1948,* 1461.
37. *Maki, A. H., N. Edelstein, A. Davison,* and *R. H. Holm:* J. Am. Chem. Soc. *86,* 4580 (1964).
38. *Shupack, S. I., E. Billig, R. J. H. Clark, R. Williams,* and *H. B. Gray:* J. Am. Chem. Soc. *86,* 4594 (1964).
39. *Jørgensen, C. K.:* J. Inorg. Nucl. Chem. *24,* 1571 (1962).
40. *McClure, J. W.:* J. Phys. Chem. Solids *24,* 871 (1963).
41. *Jørgensen, C. K.:* Mol. Phys. *7,* 417 (1964).
42. *Reid, F. J., L. K. Matson, J. F. Miller,* and *R. C. Himes:* Phys. Chem. Solids *25,* 969 (1964).
43. *Goodenough, J. B.:* Magnetism and the Chemical Bond. Interscience, New York, 1963.
44. *Hulliger, F.:* Nature *204,* 644 (1964).
45. — and *E. Mooser:* Progress Solid State Chem. *2,* 330 (1965).
46. *Fajans, K.:* Chimia *13,* 439 (English translation available from Ulrich's Bookstore, Ann Arbor, Michigan) (1959).

(Received March 18, 1966)

248

Reprinted from "Structure and Bonding," Volume 3, C. K. Jørgensen et al., eds., pp. 106–115 (1967)

13

Relations between Softness, Covalent Bonding, Ionicity and Electric Polarizability

Dr. C. K. Jørgensen*

Cyanamid European Research Institute, Cologny (Geneva), Switzerland

In this paper, only static and no kinetic properties are discussed. In the writer's opinion, kinetics is like medicine or linguistics — it is fascinating, it is useful, but it is too early to hope to understand much of it.

Since *Pearson* (*1, 2*) introduced the words "soft" and "hard" in their new sense, there has been a persistent argument that it has something to do with electric dipole polarizabilities. Certainly, in most cases, the softness and high polarizability a run roughly parallel, in the same way as in many, but not all, cases, the soft ligands have low electronegativities. However, though a connection undoubtedly exists, a closer analysis of recent results (*3*) makes a a far more physical than chemical quantity.

In an isoelectronic series of gaseous, monatomic ions, a decreases dramatically as a function of increasing charge z. Thus, the best calculation (*4*) available for H⁻ gives 30.4 Å³, and for the groundstates of He 0.2045 Å³ (this has been confirmed experimentally to three significant figures) and of Li⁺ 0.0284 Å³. In the special case of H⁻, it is rather horrifying that the Hartree-Fock function gives the much lower value 14 Å³. However, this must be due to the extraordinarily low ionization energy of H⁻; in other cases, the discrepancy caused by neglecting correlation effects is far smaller. The chemical species H(-I) in crystalline hydrides (*3*) has a between 1.3 and 1.6 Å³.

Recently, much progress has been made studying a for excited states. Thus, the configuration 1s2s of He has the terms ³S (46.5 Å³) and ¹S (119 Å³). (*4*) The level $^3P_2$ of the configuration terminating np^5 $(n+1)$s has the a-values increasing mildly (*5*) from 27.6 Å³ for Ne to 64.4 Å³ for Xe in close analogy to the corresponding neutral alkali metal atoms (*6*). *Kelly* (*7*) analyzed the individual contributions of the one-electron transitions in the oxygen atom.

---

* The first symposium on soft and hard (*Lewis*) acids and bases was organized by Prof. *R. F. Hudson* (now at the University of Kent, Canterbury) at our institute in May 1965, and the proceedings partly published in *Structure and Bonding 1*, pp. 207—248. A second symposium was organized by Professor *Malcolm J. Frazer*, Northern Polytechnic, Holloway, London, 29.–31. March 1967. The proceedings of this discussion were not published, but the present paper is based on the invited lecture, partly modified after further helpful comments by Professors *J. Bjerrum*, *K. Fajans* and *G. Klopman*.

106

Whereas widely extended orbitals occur in the *Rydberg* states of gaseous molecules, and in many solids, they do not usually occur in solution, and the very large α values just discussed would not have so much chemical significance.

According to *Fajans* a molecule is either covalent, the atomic cores being bound by binuclear or multinuclear quanticules consisting of two or more electrons, or it is ionic and the constituent ions polarized according to a general physical scheme. This opinion is a very intelligible defence against some of the confusing consequencies (*8*) of "resonance between ionic and covalent structures". and in final analysis, it is contributing to the concept of preponderant electronic configurations (*3*) classifying correctly the symmetry types of the lowest levels of a system. However, in quantum mechanics, the distinction between covalent and electro-valent molecules *cannot* be made that sharp, and the main trouble for the valence-bond description is the exorbitant overlap integrals between the covalent and ionic structures. *Fajans* (*9*) has recently discussed why, among gaseous diatomic molecules, the polarity is higher for NaF than for any other MX and higher for SrO than for BaO, when defined from the reduction of the dipole moment μ from that expected for two point charges at the known internuclear distance. *Fajans* emphasizes that the high α for Cs(I) and Ba(II) contributes to the decreased polarity and low μ. We have entertained a long correspondence about the deviations from additivity of α values in salts and molecules. It is beyond any doubt that such an additivity cannot be obtained exactly (*10, 11*). The writer completely agrees with *Fajans'* qualitative statement that anions are tightened by the cations occurring in compounds, and cations are loosened by the anions (incidentally, this runs parallel to the ionization energies (*19*) of anions being increased and of cations being decreased by the *Madelung* potential). However, a closer analysis (*3*) shows that *Fajans* and *Joos'* rules produce larger deviations from additivity than needed; this is a problem somewhat analogous to the evaluation of ionic radii. Thus, α for O(-II) can be made (*3*) to vary only between 1.1 Å^3 for ClO_4^- to 2.05 Å^3 for solid BaO, neglecting these rules. *Fajans* (*9*) assumes a standard state for the oxide ion 2.63 Å^3 obtained by extrapolation from the neon atom (0.39 Å^3). When compared with the frenetically varying values for monatomic species mentioned at the beginning, this value for oxide cannot be considered to be entirely unique. Already gaseous F$^-$ may have a higher α than inferred from the fluorides, and because gaseous O^{--} spontaneously looses one electron, it would be expected to have extremely high α values if it was only weakly stabilized by the surrounding cations, *e.g.* in the hypothetical compound $[N(CH_3)_4]_2O$.

An interesting by-product of the new values (*3*) is that α does not decrease so rapidly with increasing oxidation state in chemical species

107

as it decreases with ionic charge in gaseous ions. Thus, the isoelectronic sequence Hg atom 5.1 Å³, Tl(I)5.2 and Pb(II) 4.8 Å³ is nearly invariant, and if the new rule for hydration effects is accepted (3), Fe(III) has a higher value, 2.2 Å³, than Mn(II) 1.25 and Fe(II) 1.2. This reversal is easy to understand; monatomic species have excited states at highly increasing energy and decreasing average radius as a function of increasing ionic charge, both contributing to decrease a, whereas this is not true for chemical species. The energy of the excited configuration 6s6p remains roughly constant (8) in Hg, Tl(I), Pb(II), Bi(III), and there is some evidence that the delocalization on the ligands increases in direction of Bi(III) and Po(IV). In particular, the electron transfer bands have much lower wavenumbers in Fe(III) than in Fe(II) or Mn(II) complexes of the same (non-conjugated) ligands, and it is quite conceivable that $Fe(H_2O)_6^{+3}$ of this reason has a higher a than $Mn(H_2O)_6^{++}$. However interesting these results are for the investigation of electric polarizabilities, and suggesting further experimental studies of molar refractivities, it must once more be emphasized (3): *Chemical softness is not exactly* a; the values 3.1 Å³ for Cs(I) and 2.5 Å³ for Ba(II), when compared with 2.4 Å³ for Ag(I) and 1.8 Å³ for Cd(II), do not correspond to the chemical classification. *Fajans* (33) describes this fact as Ag⁺ having a higher polarizing ability than Na⁺ and being penetrated more by the electrons of the adjacent atoms. In the writer's opinion, the experimental fact that the ionic radii are roughly equal, producing a much greater overlap between the filled 4d shell of silver with the ligand orbitals in AgCl than between the filled 2p shell of sodium and the 3p shell of chlorine in NaCl, is an essential characteristic of what *Pearson* calls soft behaviour. But *Fajans* and *Pearson*'s naming does not really include a physical explanation.

There have been several attempts to explain "chemical softness is nothing but...". *Williams* (12, 13) argues that electrovalent bonding (favouring "hard" tendencies) and covalent σ-bonds (favouring "soft" tendencies) explain most of the effects observed, and that one should not use the neologisms. It is certainly true that when comparing MX with MX', or MX with M'X, the variation may be rather weak relative to strong, general bonding effects. But it is also true that all animals are not equally equal, and there is increasing evidence (14) that the appropriate quantities to consider are differences of differences, such as

$$(M'X'—M'X) — (MX'—MX) = (M'X'—MX') — (M'X—MX)$$

and to deny that Au(I), Hg(II), Tl(III) and Ag(I) have a different chemistry (12, 15) from that of Cs(I), Ba(II), La(III) and Na(I) is impossible. But why?

108

Fundamentally, we admit *Mulliken's* ionicity in the molecular orbital (M.O.) description. This is not the same thing as the polarity derived from experimental dipole moments μ but has a close connection with the concept of preponderant electron configuration. It is wellknown that if one chooses a much more complete basis set of atomic orbitals for one of the two atoms in MX, the electrons tend to concentrate in this larger set and decrease the positive charge of the corresponding atom. However, in the case of non-transition group atoms, it is normally possible to choose s- and *p*-valence orbitals in such a way that the atomic population is fairly well defined. In complexes of the 3d, 4d and 5d groups, the parameters of interelectronic repulsion give a measure of a kind of effective charge for the partly filled shell, and the nephelauxetic or cloud-expanding effect (*16, 17*), *i.e.* the decrease of these parameters relative to the corresponding gaseous ion, is an observable change. It is an interesting question (*18*) whether the occupation of the 4s and 4p orbitals in a 3d-group complex should be counted in the central atom population; the average radii of these orbitals are comparable with the distance to the ligand nuclei, and there is a sense (*19*) in which the appropriate ionicity is connected with the 3d population only. By the same token, the empty 6s and 6p orbitals in a 4f-group compound are on the limit of belonging to the ligands anyhow. On the other hand, the empty 5d orbitals are more concentrated and must produce a lower ionicity than the very high value inferred from the weak nephelauxetic effect found for the partly filled 4f shell (*17*).

Neglecting this question of broadly extended basis orbitals, and neglecting the *Madelung* potential produced by interatomic *Coulomb* interactions (*19, 20*), the approximate M.O. calculations based on the *Wolfsberg-Helmholz* model (*21*) predict central atom charges between 0 and 1 in typical *chromophores* MX_N. This trend towards exaggerated electroneutrality is accompanied by an enhanced delocalization of all M.O., and it has been argued (*22*) that the calculated extent of π-antibonding effects on the partly filled shell was larger than can be brought in agreement with experimental evidence. If the ionicity ξ is defined (*19*) in MX_N^{-q} by the charge on each X atom being $-\xi$ and on the central atom (N ξ-q), it is possible to minimize the sum of the integrated differential ionization energies for the atoms (*20*) and the *Madelung* energy as a function of ξ and obtain the optimum ionicity. Since the *Madelung* energy rather accurately compensates for the energy needed for obtaining a large positive charge of M, the function produced is very shallow, and its minimum can be readily displaced towards somewhat smaller values for ξ. Actually, it can be shown (*19*) that the covalent bonding in most cases decreases ξ to some extent. This model adding the effects of covalent bonding as a second-order correction to the starting point of *Madelung*

109

potentials (favouring charge separation at small distances) and what is effectively electronegativities (20, 23) is sensible in the case, typical for inorganic chemistry, of fairly heteronuclear molecules. The opposite is true for the treatment of heteronuclear substitution in aromatic systems where the covalent bonding is the predominant effect. Quite generally, it can be said that the *Madelung* energy affects the potential energy of the molecule, whereas the covalent bonding mainly represents decreased kinetic energy in the bond region (24). In both cases, the virial theorem is satisfied by rather minor contractions of the radial functions in the atomic cores. The *angular overlap model* (25, 26) rationalizing most of the valid results of ligand field theory is connected with strongly increased kinetic energy of the anti-bonding orbitals close to the additional nodes necessitated by the orthogonalization on the filled orbitals of the ligands (27).

However, all of these major effects depend on the electronegativity and not on the polarizability of the atoms. A borderline case may be the π-back-bonding in certain complexes. From a spectroscopic point of view (15, 28) this type of stabilization of Au(I) and Hg(II) complexes of iodide is not very appealing; the lowest excited levels of ionic iodides occur above 44000 cm^{-1} or 5.5 eV. It must in all fairness be admitted that complexes of CO and, in the case of central atoms in very low oxidation states, CN$^-$ present such π-back-bonding according to the force constants observed, though CO and CN$^-$ have their first excited levels at even higher energy than iodides. However, the situation is rather different, because CO and N_2 are among the rare molecules where the hybridization of the kind needed in M.O. theory is of great importance; exceptionally, the chemical bonding has effects comparable to the energy difference between the 2p and 2s orbitals. It is also known from ligand field theory (8) that CN$^-$ has an extraordinarily large σ-anti-bonding effect on partly filled d-shells, and CO may very well have far larger nondiagonal elements between empty π-anti-bonding orbitals and central atom orbitals than has I$^-$. In phosphine and arsine complexes, the situation is not as clearcut because of low-lying excited levels of different origin, but the general consensus seems to be against extensive π-back-bonding (29).

Any molecule has an infinity of excited orbitals in the continuum above the first ionization energy. The electric dipole polarizability is connected partly with a few of these continuum orbitals and partly with the valence orbitals (7). If the simultaneous formation of σ-bonds in direction from X to M and of backbonding in the opposite direction is connected, not with definite, discrete, empty orbitals of X, but with the continuum, it is reasonable to think of M being polarized by X. The population of the continuum orbitals of X is expected to be the more

110

pronounced, the lower the ionization energy of X. The reason why the chemical softness of H(-I), R_3P, R_2S and I(-I) might have something to do with empty orbitals not present to the same degree in H_2O and F(-I) is not a question of π- back-bonding to 3d-orbitals of R_3P, R_2S and Cl(-I) but a question of the continuum starting at lower energy in these ligands compared to the analogous R_3N, R_2O and F(-I).

Quite recently, *Klopman (34)* elaborated his semi-empirical theory for heteronuclear molecules *(23)* to a second- order pertubation formula reproducing, with the exception of H^+, *Ahrland, Chatt* and *Pearson's* series of hard and soft central atoms:

$$Al(III) > La(III) > Be(II) > Mg(II) > Ca(II) > Fe(III) > Sr(II) >$$
$$Cr(III) > Ba(II) > Ga(III) > Cr(II) > Fe(II) > Li(I) > Ni(II) >$$
$$Na(I) > Cu(II) > H^+ > Tl(I) > Cd(II) > Cu(I) > Ag(I) >$$
$$Tl(III) > Au(I) > Hg(II)$$

This expression *(34)* involves solvation energy from a medium having the reciprocal dielectric constant close to zero and assuming *Latimer*-type ionic radii, being roughly the crystallographic radii for anions but increased 0.82 Å for cations (cf. also *(8)*, Table 24, p. 236). This surprising agreement is obtained in a model applying only data from atomic spectroscopy in the form of the ionization energy of the loosest bound orbital of the base and the electron affinity of the lowest empty orbital of the *Lewis* acid (to be called anti-base according to the proposal by *J.Bjerrum* discussed below) corrected for the solvation potential just mentioned. *Klopman's* description suggests that the soft-soft interactions are essentially covalent bonding between an empty and a filled orbital of comparable energy, where as the hard-hard interactions are essentially determined by the *Madelung* potential. The reason why this treatment is not trivial is that the solvation potential modifies very strongly the order of softness, and it is quite conceivable that it will be highly solvent-dependent.

Unfortunately, the M.O. description adequate for assignments of excited levels observed using absorption spectra *(8, 20)* is not sufficiently accurate for prediction of the very small energy differences of interest determining complex formation constants *(30)*. It is quite conceivable that small effects of the type characterizing charge-transfer complexes in organic chemistry are superposed the main effects which we agree are electrovalent and straight covalent bonding. As *Fajans (9)* and *Williams (12, 13)* correctly point out, the high polarizabilities of Cs(I) and of I(-I) are sufficient to decrease the polarity of diatomic CsF and LiI considering F^- and Li^+ only as sources of a Coulombic potential. However, it would appear that it is rather the exception that this physical polarizability in a definite atom is the most important contribution to what *Pearson*

111

calls "soft" behaviour. In my opinion, it is far more frequent that one atom has relatively low ionization energy and that another atom has continuum orbitals suitable for taking over a small amount of electron density from the first atom. This "chemical" effect only occurs because of the mutually concording properties of the two adjacent atoms. In a way, this very statement is a criticism of naïve L.C.A.O. with too few basis A.O. For instance, CsI and CsH are less typical cases for this co-operative effect than diatomic AgI and AgH. In both cases, the ligands have appropriate continuum orbitals, but the ionization energy of Cs(I) is far larger than of Ag(I) making the "chemical" effect more important for silver (I) compounds, though the local polarizability of Cs(I) is the largest. It may be added that a simultaneous invasion of silver continuum orbitals by electrons from iodide and hydride may contribute to the undoubted stabilization observed. *Ahrland* (*30*) pointed out that a complete or nearly filled d- shell is an additional condition besides high polarizability for a central atom to have "soft" characteristics.

In aqueous solution, the exchange of water ligands with halide anions gives raise to relatively small changes of free energy (proportional to the logarithm of the formation constant) and of enthalpy (corresponding to the heat evolved). *Ahrland* (*31*) pointed out a quite interesting difference between the formation of "hard" and "soft" complexes. In the former case involving *e.g.* the reaction between Al(III) or Fe(III) hexa-aqua ions and F^-, heat is absorbed, probably because of the breaking of hydrogen bonds between the ligands and the solvent water molecules, and only the increase of entropy due to decreased order of the system makes the reaction go nearly to completion(*cf.* the dissolution of solid NH_4NO_3 in water). On the other hand, the reaction between Hg(II) and ligands such as I^- and CN^- produces much heat, corresponding to actual covalent bonding, and much smaller entropy changes.

At present, there are not sufficient data available for the change of enthalpy by formation of hydroxo complexes which are further complicated by precipitation and polynuclear behaviour. However, some of *Julius Thomsen's* calorimetric measurements seem to indicate that heat frequently is evolved with hard central atoms in contrast to formation of the corresponding fluoro complexes. On the other hand, sulphate and carboxylates behave like F^- according to *Ahrland*.

J. Bjerrum (*35*) proposes to call *Lewis* acids for *anti-bases*, reserving the word acids for the species participating in *Brønsted* equilibria with solvated protons. Since a *Brønsted* acid hence contains hydrogen, many hydrogen-free anti-bases exist. Actually, *Berzelius* called species such as SO_3 acids; but for about 100 years, this word was reserved for compounds containing hydrogen which can be replaced by metallic elements. *J. Bjerrum* then argues that all *Lewis* bases are also *Brønsted* bases,

112

because they may add protons, at least in principle, and one needs only one word, base. This is consistent; and we know from the gaseous state (36) remarkably stable proton adducts such as CH_5^+ and KrH^+. However, CO is certainly much more of a *Lewis* base than a *Brønsted* base. It might be argued that the ligand PF_3 acts rather as a *Lewis* acid with the central atom as a base. An interesting case was mentioned by *Brønsted*, the free electron, which definitely is a *Brønsted* base but so to say half of a *Lewis* base. This is the origin of the border-line confusion between reducing and basic substances (*e.g.* sodium metal) and oxidizing and acidic substances (fluorine; or H_3O^+ oxidizing $Co(CO)_4^-$ to the Co(I) hydride $Co(CO)_4H$).

Some people argue that *Lewis* acids should be called electron-pair acceptors and the bases electron-pair donors. This may be concentrating the attention on electron pairs in chemistry more than they deserve; M.O. theory suggests a much more nuanced view, and there is no sense in which exactly six electron pairs are involved in the bonding of CrF_6^{-3} and IrF_6 according to spectroscopic observations (20). On the other hand, to call *Lewis* acids and bases simply acceptors and donors invite to confusion with *Mulliken's* typical charge-transfer complexes (37).

At the London meeting. Professor *R. P. Bell* asked whether soft interactions are not simply covalent bonding and hard interactions ionic bonding. This is not easy to answer in a clear-cut way; however, most chemists would feel that the strong covalent bond between two hydrogen atoms is not exactly the same situation as the reactions between iodine or iodide and sulphur-containing compounds. As Professor *G. Schwarzenbach* pointed out in 1965, the concept has its origin in the alchimistic time, the affinity between idealized sulphur and idealized mercury being a typical soft-soft interaction. *Goldschmidt's* geochemical concepts of lithophilic and chalkophilic elements also strongly resemble hard and soft behaviour, as well as the separation in qualitative analysis using sulphides. There is little doubt that whatever name will be given to the effect, it will occur in chemistry books (if any) for centuries to come. *Pearson's* names are short, and it may console him that some people once thought that "weak acid" also is a bad name, because the titrating capability is the same as of a strong acid.

Pearson (2) and *Drago* and *Weyland* (38) proposed a two-parameter equation for the equilibrium constant K for the reaction between A and B

$$\log K = S_A S_B + \sigma_A \sigma_B$$

(admittedly only a fair approximation) where S is related to the hard and σ to the soft behaviour. It is then a strong temptation to ascribe a complex number to each reactant and to write

C. K. Jørgensen

$$-\ln K = (a_A+ib_A)(a_B+ib_B) = (a_Aa_B-b_Ab_B) + i(a_Ab_B)+b_Aa_B)$$

the imaginary part $ic = i(a_Ab_B+b_Aa_B)$ representing that factor K_{osc}

$$K_{osc} = e^{ic} = \cos(c) + i\sin(c)$$

which "explains" the experimental deviations from the real part having $a_A=S_A\ln 10$ and $a_B=-S_B\ln 10$.

It is quite clear that electrovalent bonding is the easiest case to discuss corresponding to the behaviour of "hard" central atoms and ligands. The influence of covalent bonding is more variable from case to case. It has been argued by many authors that it is unfortunate to use the words "hard" and "soft" for this distinction. However, the many fine details of the fluctuating behaviour of covalent bonding may very well make these terms permissible in chemistry, though they cannot be directly measured, at least not in a unique way. A similar situation is found for chemical "side-group character". It is evident that As, Se and Br are closer analoga to P, S, Cl than are V, Cr and Mn. It is already less clear whether Ti or Ge is eka-silicon, or whether Sc or Ga resembles Al the most. One of the finer details is the tendency towards symbiosis of soft ligands (32) making the mixed complexes of definitely soft and hard ligands less stable than expected according to statistics, and hence, the mixed complexes tend to disproportionate, whereas the opposite is true for mixed complexes of similar ligands (39). Chemists cannot be prevented from inventing words for such a situation, even though theory is not yet sufficiently sophisticated to explain it.

References

1. *Pearson, R. G.:* J. Am. Chem. Soc. *85*, 3533 (1963).
2. — Science *151*, 172 (1966), and Chemistry in Britain *3*, 103 (1967).
3. *Jørgensen, C. K.:* Structure and Bonding (Springer-Verlag) *1*, 234 (1966).
4. *Chung, K. T.,* and *R. P. Hurst:* Phys. Rev. *152*, 35 (1966).
5. *Robinson, E. J., J. Levine,* and *B. Bederson:* Phys. Rev. *146*, 95 (1966).
6. *Dalgarno, A.,* and *R. M. Pengelly:* Proc. Phys. Soc. *89*, 503 (1966).
7. *Kelly, H. P.:* Phys. Rev. *152*, 62 (1966).
8. *Jørgensen, C. K.:* Absorption Spectra and chemical Bonding in Complexes. Oxford: Pergamon Press 1962.
9. *Fajans, K.:* Structure and Bonding *3*, 88 (1967).
10. —, and *G. Joos:* Z. Physik *23*, 1 (1924).
11. — Z. physik. Chem. *B 24*, 103 (1934).
12. *Phillips, C. S. G.,* and *R. J. P. Williams:* Inorganic Chemistry. Vol. I and II. Oxford: Clarendon Press 1965 and 1966.
13. *Williams, R. J. P.,* and *J. D. Hale:* Structure and Bonding *1*, 249 (1966).
14. *Hudson, R. F.:* Structure and Bonding *1*, 221 (1966).

114

15. *Jørgensen, C. K.:* Inorganic Complexes. London: Academic Press 1963.
16. — Progress Inorg. Chem. *4*, 73 (1962).
17. — Helv. Chim. Fasciculus extraordinarius Alfred Werner 131 (1967).
18. *Ros, P.,* and *G. C. A. Schuit:* Theoret. chim. Acta *4*, 1 (1966).
19. *Jørgensen, C. K., S. M. Horner, W. E. Hatfield.* and *S. Y. Tyree:* Int. J. Quantum Chem. *1*, 191 (1967).
20. — Orbitals in Atoms and Molecules. London: Academic Press 1962.
21. *Basch, H., A. Viste,* and *H. B. Gray:* J. Chem. Phys. *44*, 10 (1966).
22. *Fenske, R. F., K. G. Caulton, D .D. Radtke,* and *C. C. Sweeney:* Inorg. Chem. *5*, 951 and 960 (1966).
23. *Klopman, G.:* J. Chem. Phys. *43*, S 124 (1965).
24. *Ruedenberg, K.:* Rev. Mod. Phys. *34*, 326 (1962).
25. *Jørgensen, C. K., R. Pappalardo,* and *H.-H. Schmidtke:* J. Chem. Phys. *39*, 1422 (1963).
26. *Schäffer, C. E.,* and *C. K. Jørgensen:* Mol. Phys. *9*, 401 (1965).
27. *Jørgensen, C. K.:* Chem. Phys. Letters *1*, 11 (1967).
28. — Halogen Chem. (Academic Press) *1*, 265 (1967).
29. *Canadine, R. M.:* J. Chem. Soc., in press.
30. *Ahrland, S.:* Structure and Bonding *1*, 207 (1966).
31. — Helv. Chim. Acta *50*, 306 (1967).
32. *Jørgensen, C. K.:* Inorg. Chem. *3*, 1201 (1964).
33. *Fajans, K.:* J. Chem. Phys. *9*, 281 (1941).
34. *Klopman, G.:* J. Am. Chem. Soc., submitted.
35. *Bjerrum, J.:* Angew. Chem. *63*, 527 (1951); Naturwiss. *38*, 461 (1951).
36. *Field, F. H.,* and *J. L. Franklin:* J. Am. Chem. Soc. *83*, 4509 (1961).
37. *Briegleb, G.:* Elektronen-Donator-Acceptor-Komplexe. Berlin: Springer-Verlag 1961.
38. *Drago, R. S.,* and *B. B. Wayland:* J. Am. Chem. Soc. *87*, 3571 (1965).
39. *Bjerrum, J.:* Metal Ammine Formation in Aqueous Solution. 2. Ed. Copenhagen: Haase and Son 1957.

Received April 24, 1967

Satz und Druck: Druck- und Verlagshaus Hans Meister KG, Kassel

Reprinted from "Structure and Bonding," Volume 5, C. K. Jørgensen et al., eds., pp. 118–149 (1968)

14

Thermodynamics of Complex Formation between Hard and Soft Acceptors and Donors

Dr. S. Ahrland

Division of Inorganic Chemistry, Chemical Center, University of Lund, Lund, Sweden

Table of Contents

I. Nature and Scope of the Classification of Acceptors and Donors as Hard and Soft

The primary purpose of classifying acceptors in (*a*), or hard, and (*b*), or soft, is to correlate a large mass of experimental facts. All the criteria used for the classification are thus purely empirical; they simply express the very different chemical behaviour of various acceptors (*1,2*).

The fundamental criteria are the affinity sequences found for the complexes formed in aqueous solution by various groups of ligand atoms. These sequences turn out to have very important features in common for the halide, chalcogen and nitrogen groups. For acceptors termed (*a*) and (*b*), respectively, they are as follows:

Donor group	Oxidation state	(*a*)	(*b*)
7 B	−I	$F \ggg Cl > Br > I$	$F \lll Cl < Br < I$
6 B	−II	$O \ggg S > Se > Te$	$O \lll S \simeq Se \simeq Te$
5 B	−III	$N \ggg P > As > Sb$	$N \lll P > As > Sb$

The (*a*)-sequences are thus all alike, with the first donor atom within each group forming much stronger complexes than the second, or any of

118

the following ones. The (*b*)-sequences are not alike, but have the common trait that, contrary to the (*a*)-sequences, the first donor atom within each group forms much weaker complexes than the second one.

It is further striking that acceptors which are classified as strongly (*b*) according to those criteria, *e. g.* the platinum and noble metals in their normal oxidation states, also form strong complexes with uncharged ligands of negligible proton affinity, such as carbon monoxide, olefins, acetylene and aromatic hydrocarbons (*2,3*). Surely enough, such complexes are not formed by all (*b*)-acceptors, but the important point is that they are not formed at all by (*a*)-acceptors (for a recent survey of these complexes, see (*4*)). Acceptors forming complexes with such ligands can therefore certainly be classified as (*b*).

Also the rate of nucleophilic displacements, *i.e.* reactions where a ligand is substituted by another, can be naturally correlated with the fundamental criteria, as pointed out by *Pearson* (*2*). For (*b*)-acceptors, the rate will depend much upon the polarizability, but little upon the basicity of the ligands concerned, for (*a*)-acceptors the opposite will be the case.

Auxiliary criteria like these are of course especially valuable in cases where, for some reason, the fundamental criteria cannot be applied.

When metal ions are classified according to these criteria, all the (*b*)-acceptors turn out to be situated within two roughly triangular areas of the periodic system (*1,5*). The bases of these triangles stretch along the highest period, and their apices are at copper and arsenic (or possibly antimony), respectively. For the noble metals, and the elements to the left of these, the (*b*)-character increases as the oxidation state decreases. For the elements to the right of the noble metals, the opposite trend is observed, though the data available within this region are still too few, too qualitative, and partly too uncertain to allow any final conclusion.

Consequently, the elements to the left of the noble metals show strongest (*b*)-character in their zero-valent oxidation state. Thus iron(0), cobalt(0) and nickel(0) are typically (*b*), forming *inter alia* strong carbonyl complexes, while the higher oxidation states of these elements have no marked (*b*)-character at all. Elements in zero-valent state in fact display (*b*)-character as far left in the periodic system as chromium, or even vanadium, which in higher oxidation states behave as very typical (*a*)-acceptors. To the right of the noble metals, on the other hand, the metals in their zero-valent states do not show any marked (*b*)-character; they do not form *e.g.* carbonyl or olefin complexes.

The very regular distribution of (*a*) and (*b*) acceptors within the periodic system affords in itself a good criterion whether a certain acceptor will behave as (*a*) or (*b*). This may in fact be fairly accurately predicted from its position, and oxidation state.

119

As has been pointed out before, all (b)-acceptors formed by metallic elements, possess a large number of d-electrons in their outer shell which must moreover be easily dislocated, i.e. a (b)-acceptor has always a high polarizability (5). Also the ligands they prefer are generally characterized by a higher polarizability than that displayed by the ligands rejected. This striking influence of the polarizability upon the complex formation has prompted *Pearson* (2) to advance the words "hard" and "soft" (suggested to him by *D. H. Busch*) to characterize the behaviour of the various classes of acceptors and donors. Acceptors of class (a) are hard, and so are the donor atoms they prefer, i.e. F, O, and, to a lower degree, N; acceptors of class (b) are soft, and so are the donor atoms they prefer, i.e. the heavier congeners of each group, as well as carbon.

It should be stressed, however, that soft and polarizable are not synonymous; a soft acceptor is certainly always highly polarizable, but a highly polarizable acceptor need not necessarily be soft, i.e. have (b)-properties (5). For metal ion acceptors, the outer d-electrons are as essential as the polarizability.

The differences in chemical behaviour characterized above as hard and soft are of course due to differences in the nature of the bonds formed between acceptor and donor. So far, nothing has been postulated about these differences. This is natural in view of the fact that the classification primarily serves to systematize and correlate a vast mass of experimental observations. The classification is *a priori* empirical.

Nevertheless, the chemical behaviour actually observed strongly suggests that the bonds formed are more electrostatic in nature, the harder the acceptor and donor, and more covalent, the softer the acceptor and donor. This generalization fits both to the fundamental and to the auxiliary criteria given above for hard and soft behaviour, and is further strengthened by the fact that typically hard acceptors are always characterized by high charge and/or small radius. Generally, the charge densities on acceptor and donor are the prime factors, as far as hard-hard interactions are concerned, while the availability of easily dislocated electrons of suitable energy is of paramount importance for soft-soft interactions (5).

Once it is realized that the acceptor to donor bond must be of very different nature in complexes of different types, it is also clear that its character very much depends upon the medium used, as the electrostatic interaction between any two particles carrying charges is stronger, the lower the dielectric constant, D, in the environment of these charges. As water has about the highest D of all ordinary liquid, or gaseous, media, aqueous solutions will represent an extreme of low electrostatic interaction. In all solvents of lower D, and especially in gaseous phases of $D \simeq 1$, the electrostatic interaction will be stronger than in water. The bond formed by a certain acceptor-donor pair will thus tend to become

120

more electrostatic in character, the lower the value of D of the medium, and it will moreover tend to become stronger.

The bonding forces are of appreciable strength only within rather short ranges, a few tens of Ångströms. For particles within such distances, an interposed medium cannot exert its full influence, especially not when the distance becomes very short. The actual value of the dielectric constant will be smaller, even much smaller, than the ordinary macroscopic value. How small is not so easy to determine exactly, but a good estimate can be obtained by a rather ingenious method originally due to *Schwarzenbach* (6), and later on further developed by *Schwarzenbach* and *Schneider* (7). According to this method, the standard change of free energy, ΔG^0, is measured for the reactions of a certain acceptor (the proton, or a metal ion) with two ligands which essentially differ only by one positive charge. The difference in ΔG^0 between the two reactions is approximately equal to the energy spent in order to overcome the repulsion due to this extra charge, and as the distance between the interacting charges is known, the actual value of D can be calculated. As expected, the values becomes smaller with decreasing distance for nearly all acceptors measured, though not at all at the same rate for different acceptors. These differences are interpreted as most probably due to variations between the acceptors in the freedom of orientation of the water molecules within their hydration shells (7).

The macroscopic value of D should thus not be used for the calculation of the electrostatic interactions. On the other hand it is clear, most convincingly just from the work of *Schwarzenbach* and *Schneider* (7) that a medium of a high macroscopic value of D will nevertheless lower the strength of interaction between acceptor and donor very considerably. Thus, from the constants measured in aqueous solution for complexes between various acceptors and ligands with a straight carbon chain, where the distance between the charges is $\simeq 5$ Å, the actual values of D obtained vary from a lowest value of $\simeq 12$, for H^+, to $\simeq 25$ for Cu^{2+}, $\simeq 35$ for Ni^{2+}, Zn^{2+}, Cd^{2+}, Pb^{2+}, and > 50 for Ca^{2+} and Sr^{2+}. For Ba^{2+}, no significant difference is even found from the macroscopic value of 80.

As has already been stressed, the classification of acceptors as (*a*) and (*b*) has been founded on their behaviour in aqueous solution. For media of lower D, and consequently stronger electrostatic interactions, any given acceptor will display more (*a*)-character than in water, judged by the fundamental criteria given above. In the gas phase, almost all metal ion acceptors seem in fact to show (*a*)-sequences. This was pointed out by *Pearson* already in his first paper (2), and has later on been very convincingly elaborated by *Pearson* and *Mawby* (8), as will be more discussed below.

Very important is moreover that the action of a solvent is not limited to its influence upon the strength of the electrostatic interaction between

121

acceptor and donor. A high value of D just indicates that the molecules of the solvent are strongly polar. They will therefore interact with ions, and also with other dipoles present.

This interaction is especially strong with metal cation acceptors as these are generally small and moreover often of a high charge. Also the hydrogen ion interacts strongly with the solvent. These ions are therefore more or less extensively solvated in all solvents of high D, *i.e.* the solvent molecules act as ligands, in competition with other ligands present in the solution.

The interactions between the solvent dipoles and other ligands are on the other hand fairly weak in most cases, as the typical ligand is either an anion with a rather large radius, and often of low charge, or a neutral molecule, with no net charge at all. Some anions, however, characterized by an unusually high ratio of charge to radius, *e.g.* F^- and, to a lower degree, SO_4^{2-}, do interact as strongly with the solvent molecules as many cations, which is of paramount importance for their complex formation, as will be discussed below.

These various influences of the solvent will make the classification of acceptors dependent upon the medium used, as far as the formal definition goes, *i.e.* according to the criteria given above. But independent of this, that tendency for covalent bond formation which is ultimately responsible for (b)-behaviour will remain, as it is a property linked to the electronic configuration of the acceptor. It should therefore perhaps be possible to order the acceptors in a softness sequence, *i.e.* in a sequence of ability to form covalent bonds.

To avoid arbitrariness, such an order must be founded on some property that can be measured quantitatively so that the degree of softness can be expressed by a figure. There has been several such attempts, notably by *Pearson* and *Mawby* (*8*), by *Klopman* (*9*), and by *Drago* and *Wayland* (*10*), the last one referring to nonmetal, mostly organic, acceptors. These constitute a vast field of great theoretical and practical interest; in the main, however, they have been considered as outside the scope of the present treatise. The sequences proposed are all founded on energy relations. Their discussion will therefore be postponed until the relation between the thermodynamics of the complex formation reaction and the nature of the bond formed has been examined.

In a similar way, it should be possible to arrange donors in an order of softness. Different approaches to establish such an order has been suggested by *Klopman* (*9*), for inorganic, and by *Drago* and *Wayland* (*10*), for organic donors.

At this point it must be realized that all attempts to establish an order of ability for covalent bond formation for acceptors and donors are beset by a serious principal difficulty, namely that it cannot *a priori* be presu-

122

med that such an order for, say acceptors, will not to a certain degree depend upon the donor offered. Various soft donors may conceivably not fit the acceptors in the same order; some acceptors may form covalent bonds preferentially with, say, the heavier halides; other with the heavier chalcogens. Various orders will then be obtained, depending upon the choice of reference donor.

From the consistent trend of affinities between donors of various groups expressed by the fundamental criteria quoted above, it is on the other hand obvious that donors of various groups do have much the same preferences. There is, therefore, a fair chance that they will arrange the acceptors in about the same order of softness. Any really drastic reversals are anyway not likely to occur. Still it remains an open question how large an influence upon the bond that will in fact be exerted by very specific interactions between acceptors and donors.

II. The Nature of the Coordinate Bond and the Thermodynamics of Its Formation

Reactions involving bonds of very different nature are likely to have different thermodynamic characteristics. This must especially apply if reactions take place in markedly polar solvents, where the solvent molecules are likely to interact more strongly with hard acceptors and donors than with soft ones. Such differences must be reflected in the overall energy changes observed.

As far as halide and cyanide complexes in aqueous solution are concerned, this conclusion has been shown to be amply verified by experiment (11). Complexes of the extremely hard fluoride ion are formed in endothermic reactions, at least as far as the first step is concerned[1]. The enthalpy change thus counteracts the complex formation which takes place only because of a strongly positive entropy term. This term is especially large when very hard acceptores are involved, i.e. cations of high charge and/or small radius. On the other hand, complexes of the very soft cyanide ion are formed in strongly exothermic reactions. The enthalpy change thus strongly favours the complex formation, whereas in this case the entropy term is as a rule negative; i.e. counteracting[2].

---

[1] According to measurements now in progress (12), this also applies to the first complex of UO_2^{2+}, contrary to what has been found before (11). A value of $\Delta H_1^0 = 0.49$ kcal/mole has now been measured (at $I = 1$ (NaClO$_4$)), as compared with the previous value of ~ -2 kcal/mole. This further stresses the unreliability of many values of ΔH found from the temperature coefficient of the stability constant.

[2] Since these rules were formulated from the data at hand in 1966, they have been further substantiated by recent measurements of the copper(I) and silver(I) cyanide systems (13).

123

The other halides form a transitional sequence between these extremes, with the properties fairly smoothly changing in the order of increasing softness: $Cl^- < Br^- < I^-$.

These experimental facts have been interpreted as follows (*11*). For the formation of a complex between a hard donor and a hard acceptor, strong bonds between these and the water molecules of their hydration shells have first to be broken. This takes much energy which is not completely regained by formation of the predominantly electrostatic acceptor to donor bond. The net reaction therefore tends to be endothermic. On the other hand, the resulting liberation of several water molecules from the hydration shells implies a large gain of entropy which in fact constitutes the driving force of the reaction between hard particles.

For soft donors and acceptors, the conditions are about contrary. Most of these interact only weakly with the water dipoles; they are as a rule little, if at all, hydrated. The complex formation will consequently not imply any extensive liberation of water molecules and hence no large entropy gain. In most cases a net loss of entropy is in fact found, most probably due to several factors acting in concert. The formation of the complex will mean a decrease in the number of particles, and presumably also in the rotational and vibrational freedom of the ligands. Possibly the water structure may also become more ordered as a result of the coordination of strongly structure-breaking ligands into a complex. On the other hand, the formation of a markedly covalent acceptor to donor bond is evidently accompanied by a large evolution of heat which constitutes the driving force of the reaction between soft particles.

If this interpretation is valid, the formation of complexes in aqueous solution by other hard donors should occur according to a pattern very similar to that found for the fluoride complexes, while, conversely, complexes of other soft donors should be formed according to a pattern similar to that found for cyanide, or iodide, complexes. Moreover, donors of intermediate softness, or hardness, should show a transitional behaviour between the extreme types also in the matter of thermodynamics of reaction. It has been a chief aim of this treatise to investigate whether this is in fact true.

An important difference between halogens and other donor atoms is that the former present themselves only as monoatomic, and monovalent, ions, while the latter are present in a great variety of ligands. Among these are only in few cases, and under exceptional conditions, the simple monoatomic ions (such as S^{2-} at pH $\gtrsim 13$, (*14, 15*). The environment of a certain donor atom is very different in different types of ligands which may result in considerable modifications of its capacity for covalent bonding. To assess such influences, it is necessary to compare complex

124

formation reactions of the same donor atom, present in ligands of distinctly different structures.

The intended comparison should include ligands coordinating by donor atoms of the chalcogen, nitrogen and carbon groups. As compared with the previous investigation (11), the present one is hampered by the added difficulty that the number of reliable free energy and, especially, enthalpy data is very limited indeed, as far as the heavier donor atoms, and also carbon, are concerned. Only for oxygen and nitrogen donors, the material available is as rich as for the halides.

From the strong preferences shown by hard acceptors for oxygen donors it is evident that these are generally very hard, next to the fluoride ion seemingly the hardest ones that exist. Nitrogen donors form complexes in aqueous solution with most soft acceptors, and also with acceptors of medium hardness, such as cobalt(III) and zinc(II), but not with very hard acceptors. Nitrogen should therefore be classified as a considerably softer donor atom than oxygen, though not yet typically soft. Very soft are on the other hand the heavy chalcogens, and the heavy congeners of nitrogen which are all strongly preferred by soft acceptors. This also applies, as already mentioned, to all carbon donors.

In the following, complex formation by oxygen and nitrogen donors have been treated first (including the mixed oxygen-nitrogen ligand EDTA) and then their heavy congeners. Finally data are presented for some carbon (olefinic) ligands.

Like in the halide-cyanide compilation, the values of ΔG^0, ΔH^0 and ΔS^0 presented mostly refer to the first step of the complex formation, as data for this step are generally most numerous, most reliable, and least objectionable to compare (11). In some cases, however, quantities referring to the sum of several steps have been calculated, in order to make possible a comparison with results for chelating ligands. In some other cases, data referring to later steps have been presented, either because they have been considered to be of special interest, or because they have been the only ones available for the system of complexes in question.

III. Complex Formation with Oxygen and Nitrogen Donors

As oxygen donors, sulphate, acetate, glycolate and malonate ions have been chosen. The three latter ones represent three important types of carboxylate ligands, viz. a simple carboxylate, an α-hydroxycarboxylate where the hydroxy group presumably gives rise to a weak chelate bond resulting in a five-membered ring, and a dicarboxylate, certainly forming a six-membered ring with two equally strong oxygen to metal bonds.

125

The oxygen donors are surely in a rather different environment in those four ligands.

For one of the most important oxygen donors, *viz.* the hydroxide ion, the determination of reliable values of ΔH^0, and also of ΔG^0, have been much hampered by the complicated nature of most hydrolytic equilibria. Polynuclear complexes are as a rule formed and even the formulas of these have often been disputed. Recently, however, quite a few enthalphy data have been determined so that a comparison between metal ions of fairly different types is now possible. Such a comparison, comprising hydroxide complexes of lead(II), indium(III), iron(III), uranyl(VI), thorium(IV) and beryllium(II) has been undertaken by *Mesmer* and *Baes* (16). The value of ΔH^0 per hydroxide group attached is then found to be strongly negative, —6 to —9 kcal, *i.e.* contrary to what is generally to be expected for hard donors. The values of ΔH^0 and ΔS^0 are moreover not very dependent upon the nature of the metal ion, being about the same for *e.g.* the various complexes formed by lead(II) and uranyl(VI). Evidently the thermodynamics of hydroxide complex formation is quite peculiar. This is rather to be expected, however, considering the peculiar character of hydrolytic reactions. At least as far as inner complexes are concerned, and most complexes of any appreciable strength are of this type, formation of a complex ordinarily implies the substitution of water of hydration with a ligand. This means breaking of bonds between the metal ion and the oxygen of the water and formation of bonds between the metal ion and the ligand. The formation of hydroxide complexes in acid solution is, on the other hand, presumably best described as a transfer of protons from water molecules within the hydration shell to water molecules outside this shell. In such a process, metal to oxygen bonds are neither broken, nor formed. The dissociation is often followed by a formation of oxo or hydroxo bridges, joining two or more metal ions into the polynuclear aggregates mentioned above. The character of hydroxide complex formation thus differs appreciably from that of other oxygen donors. It is therefore not surprising that the thermodynamic characteristics differ from those expected for oxygen donors in general.

The data found for the formation of sulphate complexes are presented in Table 1. Only a few of these have been determined by calorimetry, the method generally to be preferred. Most of them have been found from measurements of the temperature coefficient of the equilibrium constant which is an inherently less reliable method[3]). The main conclusion of the

---

[3]) In the following Tables, *"cal"* in the column "Method ΔH°" signifies a calorimetric determination and *"T"* a determination from the variation of the equilibrium constant, between the temperatures stated (°C).

126

sulphate compilation is nevertheless not in doubt, *viz.* that sulphate complexes are formed according to very much the same pattern as fluoride complexes. The reactions are thus characterized by strongly positive values of ΔH_1^0 counteracting the complex formation, and strongly positive values of ΔS_1^0 promoting it. Strong complexes are formed in so far as the latter term predominates.

Table 1. *Thermodynamic functions for the formation of sulphate complexes at 25° C*

Acceptor	Method ΔH^0	I^b)	ΔG_1^0	ΔH_1^0	ΔS_1^0	Ref.
		0	−2.71	5.74	28.4	(17, 18)
H+	cal	1	−1.65	5.15	22.8	(19, 12)
		2	−1.47	5.54	23.5	(20, 21)
Mg2+	T 0—45	0	−3.07	4.8	26	(22)
Ca2+	T 0—40	0	−3.14	1.7	16	(23)
La3+	T 20—35	0	−5.2	2.5	26	(24)
	T 19—40	1	−1.90	2.5	15	(25)
Ce3+	T 15—45	0	−4.58	4.7	31	(27)
		1	−1.68	3.6	18	
Th4+	cal	2	−4.49	5.00	31.9	(20, 21)
U4+	T 10—40	2	−4.90	2.3	24	(20, 21, 28)
Np4+	T 10—35	2	−4.79	4.4	31	(20, 21, 29)
	T 25—200	0	−3.70	~4	~26	(30)
UO2 2+	cal	1	−2.42	4.30	22.6	(31, 19, 12)
	T 10—40	2	−2.56	2.3	16	(20, 32)
Cr3+	inner a) T 48—84	1	−2.58	7.2	29	(33)
	outer a)	1	−1.64	~0	~5	(33)
Co(NH3)6 3+	outer T 4—49	0	−4.53	0.4	17	
	outer	0	−4.47	~0	~15	(34)
		1	−1.43	0.3	4	
Co(NH3)5H2O 3+	T 25—44	0	−4.5	4	28	
	inner	1	−1.48	3.7	17	
Mn2+	T 0—45	0	−3.07	3.4	23	
Co2+	T 0—45	0	−3.21	1.7	17	(35)
Ni2+	T 0—45	0	−3.16	3.3	22	
Zn2+	T 0—45	0	−3.25	4.0	24	(22)

a) at 60° C.   b) values of the ionic strength $I \neq 0$ refer to perchlorate media.

In some cases, involving inert complexes of chromium(III) and cobalt(III), it has been possible to distinguish unambigously between inner and outer sphere complexes (33, 34). In the latter type of complexes, no water of hydration is displaced by the ligand on complex formation. No bonds directly involving the metal are thus broken or formed, nor are any large number of water molecules set free from the hydration shells. If those interpretations are true which have been given above for the

127

high positive values of ΔH_1^0 and ΔS_1^0 found for the formation of inner sphere complexes of hard ligands like fluoride and sulphate, than the formation of outer sphere complexes of these ligands must be characterized by much lower values of both ΔH_1^0 and ΔS_1^0. This is evidently the case, Table 1, which of course confirms the views expressed about the influence of solvation. The value of ΔH_1^0 turns out to be $\simeq 0$ in all the four instances of outer complex formation tabulated. The formation of the corresponding inner complexes has been investigated in three instances with values of ΔH_1^0 around 4 and 7 kcals, *i.e.* of the same order of magnitude as for other sulphate complexes presumed to be inner sphere[4]).

In the case of labile complexes, an equilibrium between inner and outer sphere complexes may exist, as has been concluded by *Eigen* and *Tamm* (*36*) from measurements of ultrasonic absorption. They have studied sulphate complexes of Be^{2+}, Mg^{2+} and the divalent ions of the first series of transition elements from Mn^{2+} on. In all those cases, the outer sphere complexes have been found to predominate, the inner sphere complexes being only $10-20\%$ of the total. The values of ΔH_1^0 reported for these systems (*22, 23, 35*) would certainly suggest a higher proportion of inner sphere complexes, being closer to the values of ΔH_1^0 found for complexes which are more or less firmly established as inner sphere than to those found for complexes established as outer sphere. This question merits further investigation.

For the formation of acetate complexes, Table 2, the values of ΔH_1^0 are still quite positive not only for the hard lanthanoid ions, where a markedly electrostatic bond is certainly expected, but also for Cu^{2+}, Zn^{2+} and Cd^{2+} which are all in the border region between hard and soft acceptors. Only for the soft Pb^{2+}, and also for H^+, ΔH_1^0 is $\simeq 0$. All acetate complexes so far investigated are thus entropy stabilized, *i.e.* oxygen behaves as a very hard donor atom also in the environment presented by the acetate ion.

For glycolates, data are available only for lanthanoid ions, and for H^+, Table 3. In this case ΔH_1^0 is just below 0 even for the hard lanthanoid acceptors, thus slightly favouring the complex formation. The main contribution to the stability is still provided by gain of entropy, however, though also this term has a significantly lower value than for the acetate complexes.

If the glycolate ion acts as a chelate which is very plausible (*49, 50*), then the values of ΔH_1^0 and ΔS_1^0 should rather be compared with the

---

[4]) At least HSO_4^- and UO_2SO_4 are certainly inner sphere complexes, and so are most probably also the strong sulphate complexes formed by the tetravalent cations Th^{4+}, U^{4+} and Np^{4+}.

123

Table 2. *Thermodynamic functions for the formation of acetate complexes, in perchlorate media at 25° C. All values of ΔH^0 determined calorimetrically except for acetic acid at $I = 0$ (temperature coefficient 0—60° C)*

Acceptor	I	ΔG_1^0	ΔH_1^0	ΔS_1^0	$\Delta G_{\beta 2}^0$	$\Delta H_{\beta 2}^0$	$\Delta S_{\beta 2}^0$	Ref.
H^+	3	−6.84	−0.75	20.3	−6.09	−0.06	20.1	(37, 38)
	1		−0.36					(39)
	0.5		−0.22					(39)
	0	−6.487	0.112	22.1				(40)
Y^{3+}	2	−2.14	3.26	18.1	−3.73	5.39	30.6	
La^{3+}	2	−2.16	2.18	14.6	−3.44	3.79	24.2	
Ce^{3+}	2	−2.33	2.09	14.8	−3.73	3.66	24.8	
Pr^{3+}	2	−2.50	1.72	14.2	−3.91	4.15	27.0	
Nd^{3+}	2	−2.63	1.71	14.6	−4.17	3.49	25.7	
Sm^{3+}	2	−2.77	1.45	14.2	−4.49	2.88	24.7	(41, 42)
Gd^{3+}	2	−2.54	1.87	14.8	−4.31	3.25	25.3	
Dy^{3+}	2	−2.33	2.93	17.6	−4.13	4.44	28.7	
Ho^{3+}	2	−2.28	3.17	18.3	−3.99	5.01	30.2	
Er^{3+}	2	−2.24	3.28	18.5	−3.96	5.51	31.8	
Yb^{3+}	2	−2.30	3.51	19.5	−3.97	6.07	33.7	
UO_2^{2+}	1	−3.40	3.05	21.7	−6.10	4.20	34.6	(12)
Cu^{2+}	3	−2.56	1.04	12.1	−4.25	1.45	19.2	(43, 44)
Zn^{2+}	3	−1.29	2.04	11.2	−1.83	5.26	23.7	(37, 44)
Cd^{2+}	3	−1.78	1.46	10.9	−3.11	1.85	16.7	(45, 46)
	2	−1.69	1.50	10.7	−2.70	1.58	17.7	
	1	−1.60	1.62	10.8	−2.48	2.85	17.9	(47, 46)
	0.5	−1.62	1.76	11.3	−2.59	3.23	19.5	
	0.25	−1.73	1.56	11.0	−2.73			
Pb^{2+}	3	−3.18	−0.06	10.5	−4.91	−0.15	16.0	(48, 44)

sum of these quantities for the first two steps of the acetate systems, denoted by the subscript $\beta 2$ in Table 2. For both ΔH_1^0 and ΔS_1^0, the glycolate values are very much lower. This indicates that if a second bond is formed by the glycolate ion via its hydroxy group, it is certainly very different from the bond formed via the carboxylate group, as may also be expected. In the case of equivalent bonds, one would expect that ΔH_1^0 for glycolate and $\Delta H_{\beta 2}^0$ for acetate would have about the same value while ΔS_1^0 should be higher (by $\simeq 8$ e.u.[5]) than $\Delta S_{\beta 2}^0$, the increase being due to the cratic entropy term (see *e.g.* (51)). As already stated, this is very far from being the case. It may be that the hydroxy group is bonded to the metal ion via a water molecule (42) which would certainly mean a decrease of both ΔH_1^0 and ΔS_1^0.

Malonate complexes of metal ions should, on the other hand, be chelates with two equivalent bonds, *i.e.* one would expect values of ΔH_1^0

---

[5]) e. u. = cal/degree.

Table 3. *Thermodynamic functions for the formation of glycolate complexes at $I = 2\ M$ ($NaClO_4$) and 25° C. All values of ΔH_1^0 calorimetrically determined*

Acceptor	ΔG_1^0	ΔH_1^0	ΔS_1^0	Ref.
H^+	−5.08	−0.56	15.2	(*49, 42*)
Y^{3+}	−3.37	−0.07	11.1	(*41, 42*)
La^{3+}	−2.98	−0.63	7.9	
Ce^{3+}	−3.20	−0.81	8.0	
Nd^{3+}	−3.40	−1.19	7.4	
Sm^{3+}	−3.47	−1.04	8.1	(*49, 42*)
Gd^{3+}	−3.37	−0.61	9.3	
Dy^{3+}	−3.45	−0.16	11.0	
Er^{3+}	−3.55	−0.19	11.3	
Yb^{3+}	−3.70	−0.29	11.4	

of about the same magnitude as $\Delta H_{\beta 2}^0$ for acetate, and values of ΔS_1^0 about 8 e. u. higher than $\Delta S_{\beta 2}^0$ for acetate, provided no extra stabilization takes place. For H^+, forming no chelate, ΔH_1^0 and ΔS_1^0 should not be very different for the two ligands.

Data are available for malonate complexes of the divalent ions of the first transition series from Mn^{2+} on, and also for H^+, Table 4. For Cu^{2+}, Zn^{2+} and H^+, a comparison with the corresponding acetate systems is thus possible. Unfortunately, the comparison is made somewhat uncertain by the fact that the acetate and malonate data, except for H^+, refer to media of very different ionic strength, which presumably has a consi-

Table 4. *Thermodynamic functions for the formation of malonate complexes, at $I = 0$ and 25° C. All values of ΔH^0 calorimetrically determined, except for H^+ (temperature coefficient 0—60° C)*

Acceptor	ΔG_1^0	ΔH_1^0	ΔS_1^0	Ref.
H^+	−7.77	1.16	30.0	(*52*)
Mn^{2+}	−4.48	3.68	27.4	
Co^{2+}	−5.13	2.90	27.0	(*53, 54*)
Ni^{2+}	−5.60	1.88	25.0	(*53, 55*)
Cu^{2+}	−7.69	2.85	35.6	(*56ª*), *55*)
Zn^{2+}	−5.20	3.13	28.0	(*57, 55*)
	ΔG_{2H}^0	ΔH_{2H}^0	ΔS_{2H}^0	
II^+	−11.63	1	42	(*58*)

ª) Value of ΔG_1^0 evidently a mean of several determinations (Ref. *56*, and earlier Refs. cited there).

130

derable influence especially upon the entropy. It is nevertheless clear that both ΔH^0 and ΔS^0 of the malonate systems are of that order of magnitude which should be expected according to the reasoning above.

The general chemistry of nitrogen donors classify them as donors of intermediate properties, considerably softer than the oxygen donors, as has already been stated. The connexion between chemical softness and thermodynamics so far found to be valid thus demands that both ΔH^0 and ΔS^0 should be markedly lower for the formation of complexes by nitrogen than by oxygen donors. The data for complexes formed by ammonia and ethylenediamine presented in Tables 5 and 6, will show that this is true.

Table 5. *Thermodynamic functions for the formation of ammonia complexes at 25° C*

Ac-ceptor	Method ΔH^0	$I^{a)}$	ΔG_1^0	ΔH_1^0	ΔS_1^0	$\Delta G_{\beta 2}^0$	$\Delta H_{\beta 2}^0$	$\Delta S_{\beta 2}^0$	Ref.
H⁺	T 10—40	2	−12.95	−12.95	0				(59)
	cal	0	−12.61	−12.64	−0.1				(60—63)
	T 0—50	0	−12.61	−12.48	0.4				(60)
Ni²⁺	cal	2	− 3.86	− 4.01	−0.5	− 6.99	− 7.59	− 2.1	(64, 65)
Cu²⁺	cal	2	− 5.76	− 5.43	1.1	−10.60	−11.04	− 1.5	(64, 66)
	T 10—40	2.15	− 5.82	− 5.9	0	−10.72	−11.5	− 3	(59)
Zn²⁺	cal	2	− 3.29	− 2.6	2	− 6.66	− 5.7	3	(64, 67)
	T 10—40	2.15				− 6.84	− 6.7	0	(68)
Cd²⁺	cal	2	− 3.67	− 3.5	1	− 6.58	− 7.0	− 1	(64, 67)
	T 10—40	2.15	− 3.76	− 3.7	0	− 6.71	− 7.5	− 2	(59)
Hg²⁺	cal	2				−23.6	−24.7	− 4	(64, 67)
Ag⁺	cal	0				− 9.86	−13.40	−11.9	(69)

a) Values of $I \neq 0$ refer to ammonium nitrate media.

For ammonia complexes, $\Delta S_1^0 \simeq 0$, and ΔG_1^0 will thus be determined almost exclusively by the value of ΔH_1^0 which is fairly negative for all acceptors of Table 5. Especially low is the value for H⁺, as should be expected on account of the strongly covalent character of the bonds within the ammonium ion. The magnitude of $\Delta H_{\beta 2}^0$ is approximately twice that of ΔH_1^0 which indicates that the second ligand is bonded with about the same strength as the first. This is rather to be expected when an uncharged ligand is being coordinated by a bond of a fairly covalent character. Like ΔS_1^0, $\Delta S_{\beta 2}^0 \simeq 0$.

Ethylenediamine (en) is coordinated to metal ions as a chelate. The values of ΔH_1^0 should therefore be of about the same order of magnitude as $\Delta H_{\beta 2}^0$ for ammonia, while ΔS_1^0 should be some 8 e. u. higher than $\Delta S_{\beta 2}^0$

9*

131

173

S. Ahrland

Table 6. *Thermodynamic functions for the formation of ethylenediamine complexes, at 25° C*

Acceptor	Method	I	ΔG_1^0	ΔH_1^0	ΔS_1^0	Ref.
H^+	T 10—40	2 c)	−14.01	−12.77	4.1	(59)
		0	−13.55	−11.5	7	(70)
Mn^{2+}	cal	1 b)	− 3.75	− 2.80	3.0	} (64, 71)
Fe^{2+}	cal	1 b)	− 5.90	− 5.05	3.0	
Co^{2+}	cal	1 b)	− 8.10	− 6.90	4.0	(72, 71)
Ni^{2+}	cal	1 a)	−10.23	− 9.01	4.1	(73)
	cal	1 a)	−10.25	− 9.05	4.0	(73, 71)
	cal	1 b)	−10.50	− 8.90	5.5	(72, 71)
	T 10—40	0	−10.10	− 9.5	2	(70)
Cu^{2+}	T 10—40	2.15 c)	−15.04	−14.6	1	(68)
	cal	1 a)	−14.60	−13.0	5	(73)
	T 10—40	0	−14.34	−12.8	5	(70)
Zn^{2+}	T 10—40	2.15 c)	− 8.39	− 6.6	6	(68)
	cal	1 a)	− 8.05	− 7.00	3.5	(74, 71)
	cal	1 b)	− 7.90	− 6.65	4.0	(75, 71)
	T 10—40	0	− 7.71	− 5.0	9	(70)
Cd^{2+}	T 10—40	2.15 c)	− 7.96	− 7.0	3	(68)
		0	− 7.38	− 6.2	4	(76)

			ΔG_{2H}^0	ΔH_{2H}^0	ΔS_{2H}^0	
H^+	T 10—40	2 c)	−24.4	−24.5	0	(59)
		0	−22.98	−21.8	4	(70)

The following salts have been used for the medium:
a) KNO_3, b) KCl, c) NH_4NO_3.

for ammonia. This is evidently not far from what really happens, though especially the $Cu\,en^{2+}$ is markedly stronger than would be expected from these simple rules. The extra strength comes from a value of ΔH_1^0 lower than expected (68). The coordination of H^+ to en and to ammonia releases almost the same amount of energy. As H^+ must be in very much the same situation in both cases, this seems natural enough. About the same decrease of enthalpy is also found for the coordination of the second H^+ to en. In view of the fairly long distance to the proton already coordinated, this is certainly not unexpected either.

A mixed oxygen nitrogen donor like ethylenediaminetetraacetate (EDTA) should combine the affinities of oxygen and nitrogen donors and thus be able to form strong complexes with a great variety of acceptors. As the ligand acts as a multidentate, the strength of the complexes will be further enhanced by a large cratic entropy term. In most cases a hexadentate is formed, corresponding to a cratic entropy gain of $\simeq$ 40 e.u. It is also obvious from the data compiled in Table 7

132

that, for most acceptors, by far the largest contribution to the stability of the complex comes from the entropy term. Only very soft acceptors, such as Hg^{2+}, form bonds that are so covalent that the enthalpy term becomes negative enough to contribute the main part. For border line acceptors like Ni^{2+}, Cu^{2+}, Cd^{2+} and Pb^{2+}, the contributions from enthalpy and entropy terms are of the same order of magnitude. As is obvious from the behaviour of pure oxygen and nitrogen donors, the covalent bonding should be presumed to take place preferably to the nitrogen atom.

Table 7. *Thermodynamic functions for the formation of ethylenediaminetetraacetate complexes at $20°$ C and $I = 0.1$ (KNO$_3$). All values of ΔH^0 calorimetrically determined (77)*

Acceptor	ΔG^0	ΔH^0	ΔS^0
Mg^{2+}	−11.65	3.49	51.0
Ca^{2+}	−14.35	− 6.55	26.6
Sr^{2+}	−11.57	− 4.08	25.6
Ba^{2+}	−10.41	− 4.93	18.7
La^{3+}	−20.79	− 2.80	61.4
Mn^{2+}	−18.51	− 4.56	46.6
Co^{2+}	−21.87	− 4.20	60.3
Ni^{2+}	−24.97	− 7.55	59.4
Cu^{2+}	−25.21	− 8.15	58.2
Zn^{2+}	−22.13	− 4.85	59.0
Cd^{2+}	−22.07	− 9.05	44.4
Hg^{2+}	−29.23	−18.90	35.5
Pb^{2+}	−24.19	−13.20	37.5

IV. Complex Formation with Sulphur and Phosphorus Donors

As to the heavy congeners of oxygen and nitrogen, enthalpy data are only available for a few complexes of one sulphide and one phosphine. The ligands are of much the same type *viz.* aliphatic (ethyl) compounds, substituted with an alcoholic hydroxy group in order to be sufficiently water soluble. The complexes should therefore be well comparable. The sulphide is the monovalent anion of a mercaptan while the phosphine is a neutral molecule. Both are strongly basic, the phosphine almost as much as the sulphide, Table 8. In this respect, the aliphatic phosphine differs strikingly from the previously investigated aromatic one, of very weakly basic properties (*80*). As to their basicity, soft ligands may evidently be divided in two rather sharply divided categories: those

133

Table 8. *Thermodynamic functions for the formation of complexes of two very soft ligands, viz. a sulphide (78) and a phosphine (79). All values of ΔH^0 calorimetrically determined*

Ligand	$HOCH_2CH_2S-$			$HOCH_2CH_2P(C_2H_5)_2$		
Medium	20° C; $I=0.1$ (KNO$_3$)			22° C; $I=1.0$ (KNO$_3$)		
Acceptor	ΔG_1^0	ΔH_1^0	ΔS_1^0	ΔG_1^0	ΔH_1^0	ΔS_1^0
H^+	−12.8	− 6.5	21.3	−10.95	− 8.3	9.0
Ag^+				−15.97	−19.3	−11.3
CH_3Hg^+	−21.6	−19.8	6.2	−19.72	−22.6	− 9.8
				$\Delta G_{\beta 2}^0$	$\Delta H_{\beta 2}^0$	$\Delta S_{\beta 2}^0$
Ag^+				−28.15	−35.8	−25.9
Hg^{2+}				−50.37	−52.8	− 8.2
				$\Delta G_{\beta 3}^0$	$\Delta H_{\beta 3}^0$	$\Delta S_{\beta 3}^0$
Ag^+				−34.71	−44.9	−34.6

which are not basic at all, such as the heavy halides, aromatic phosphine, carbon monoxide, olefins and benzene, and those which are strongly basic, such as sulphide ions, aromatic phosphine and cyanide ion. There are indications that the latter category is generally less exclusive than the former in their choice of acceptors for covalent bond formation, compare *e.g.* the bridged (and hence slightly soluble) sulphide complexes formed by several acceptors otherwise classified as (*a*), such as Fe^{2+}, Co^{2+}, Ni^{2+}, Cu^{2+} and Zn^{2+}, or the cyanide complexes formed not only by these but also by Fe^{3+} and Co^{3+}.

The proton complex of the sulphide is stabilized by about equal contributions from the enthalpy and entropy terms, that of the phosphine mainly by the enthalpy loss.

The high stability of the sulphide and phosphine complexes formed by soft acceptors is entirely due to strongly negative values of ΔH^0, exactly as expected. For all phosphine complexes, ΔS^0 is in fact < 0, *i.e.* counteracting the reaction. For the only sulphide complex of a soft acceptor so far investigated, that of CH_3Hg^+, ΔS^0 is only slightly > 0, *i.e.* of little importance for the formation of the complex.

Especially for the phosphine, the thermodynamics of complex formation follows very much the same pattern as has been found before for the likewise very soft cyanide ligand. Evidently the bonds formed by these ligands must be of a rather similar nature, all markedly covalent according to the arguments advanced above.

134

V. Complex Formation of Trihalogenoplatinate(II) Acceptors with Various Soft Ligands

By a combination of recent measurements, it is possible to compute the thermodynamic functions for the coordination of some olefin ligands (all of them allyl compounds, containing the group $CH_2=CH-CH_2-$, abbreviated "all") to the acceptors trichlorido- and tribromidoplatinate (II). The result is to be found in Table 9, where the corresponding quantities for the halide ligands Cl^- and Br^- have also been entered for comparison. In the Table, the way of calculation has also been indicated.

Table 9. *Thermodynamic functions for the formation of complexes between trihalogenidoplatinate(II) acceptors and various soft ligands at 25° C*

Acceptor	PtCl$_3^-$			PtBr$_3^-$		
Ligand	ΔG_4^0	ΔH_4^0	ΔS_4^0	ΔG_4^0	ΔH_4^0	ΔS_4^0
Cl$^-$	−2.54[a]	− 4.4[a]	− 6	−3.47[b]	− 5.7[b]	− 7
Br$^-$	−4.12[b]	− 6.1[b]	− 7	−3.77[a]	− 7.8[a]	−14
allNH$_3^+$	−7.37[c]	−11.5[c]	−14			
all(Et)NH$_2^+$	−7.14[c]	−10.3[c]	−11	−7.01[c]	−12.6[c]	−19
all(Et)$_3$N$^+$	−5.83[c]	− 9.3[c]	−12			
allOH	−8.28[c]	−12.5[c]	−14			

[a]) From measurements of K_4 from 15 to 60° C (chloride, (81, 82)) or from 15 to 35° C (bromide, (83)); $I = 0.5$ (HClO$_4$).

[b]) Data of (81, 83), combined with measurements of the substitution equilibria $[PtCl_{4-n}Br_n]^{2-} + Br^- \leftrightharpoons [PtCl_{3-n}Br_{n+1}]^{2-} + Cl^-$ at $I = 0.318$ (NaClO$_4$); ΔH^0 estimated, (84).

[c]) Data of (81—83), combined with measurements of the substitution equilibria $PtL_4 + ol \leftrightharpoons PtL_3ol + L$, where $L = Cl^-$ or Br^- and ol = charged or uncharged olefin ligand containing the allyl group $CH_2=CHCH_2-$ (all); ΔH^0 from temperature coefficients (between 25 (or 30) and 60° C for $L = Cl^-$, between 0 and 35° C for $L = Br^-$); $I = 2$ (NaCl + HCl and KBr + HBr, respectively), Refs. (85—87) for $L = Cl^-$, (87) for $L = Br^-$.

These platinum (II) acceptors are typically soft, and so are the ligands, with the softness increasing in the order $Cl^- < Br^- <$ olefins, to judge from their general chemical behaviour. The complexes should therefore throughout be formed in exothermic reactions, ΔH_4^0 becoming more negative in the sequence stated. At least for the olefins, ΔS_4^0 should moreover be < 0, to judge from the experience with the phosphines and cyanides. As seen from Table 9, the thermodynamics of all these reactions is really as anticipated. Also olefins thus conform nicely to the general rules found above to be valid for other ligands.

135

The values of ΔH_4^0 are throughout more negative for $PtBr_3^-$ than for $PtCl_3^-$, in keeping with the rule that the coordination of soft ligands tends to make a soft acceptor still softer. This is the symbiotic effect of *Jørgensen*, also expressed by the slogan "soft (hard) ligands flock together" (*88*). This effect is most likely due to a decrease of the effective charge on the central atom, more efficiently brought about the softer the ligand. For most elements, such a decrease of charge implies an increase of softness (p. 119). This presupposes, however, that the ligands first coordinated do not engage the total covalent bonding capacity of the central ion. If they do, the acceptor will instead become harder by their coordination, as has been very strikingly shown by the studies of *Burmeister* and *Basolo* (*89*) of platinum(II) and palladium(II) complexes of the general formula ML_2T_2 where L means a phosphine, arsine or stibine and T the thiocyanate ion. In the stibine complexes, the soft sulphur end of the thiocyanate ion is always coordinated to the metal, according to the general rule that soft ligands flock together. In the complexes containing the still softer phosphine ligand, however, the thiocyanate is always coordinated by its harder nitrogen end. The phosphine evidently uses up so much of the covalent bonding capacity of the metal that not enough is left for the coordination of the sulphur. Also in the arsine complexes, nitrogen bonding of the thiocyanate is preferred, but at least in the case of palladium, the unstable sulphur bonded isomer can be isolated, providing an interesting example of linkage isomerism.

The values of ΔS_4^0 in Table 9 are < 0 not only for the olefin but also for the halide complexes. The reason is certainly that the present acceptors are very little hydrated. The coordination of a ligand will thus not imply any extensive liberation of water. The entropy change will therefore rather tend to be negative.

VI. Complex Formation in Media of Low Dielectric Constant, and especially in the Gas Phase

As pointed out above (p. 120), the bond between an acceptor and a donor should generally be stronger, and more electrostatic in nature, the lower the dielectric constant D of the medium. The variation of the bond strength with D will be more marked the higher the charges and the smaller the radii of the acceptor and donor involved, *i.e.* the stronger the electrostatic attraction they can potentially exert upon each other. The properties mentioned tend to result in hardness, because a strong electrostatic attraction will often be the predominating bonding factor even in media of such a high dielectric constant as water. On the other

136

hand it must be realized that the resultant character of the bond depends upon the *relative* strength of the electrostatic and covalent interactions. Even a very strong electrostatic attraction may not become predominating if the covalent bonding capacity is exceptionally large. Consequently, soft acceptors, or donors, may well exert stronger electrostatic attraction than soft ones. So does *e.g.* the soft acceptor Tl^{3+}, as compared with the hard Tl^+, and the soft donors Cl^-, Br^- and CN^-, as compared with the hard NO_3^- and ClO_4^-. This can be concluded *inter alia* from their hydration enthalpies ΔH_h^0 (90—92) which should give a fairly good measure of the electrostatic interaction between the various ions mentioned and a common reference species, *viz.* the polar water molecule[6]).

The increase of the bond strength with decreasing D of the medium will thus as a rule occur at a faster rate for interactions between hard acceptors and hard donors, but there will also be quite a few cases of rapid change involving acceptors and donors classified as soft according to the conventional definition based on the conditions in aqueous solution.

As far as solvents are concerned, it must also be remembered that a change in the dielectric properties is always accompanied by a change in the solvation of the species present. Though it may be postulated (p. 122) that a lower value of D generally implies a lower degree of solvation, the variation is certainly not a monotonic one. Solvent molecules are certainly prone to show special preferences for certain species, solvating them to a higher degree than would be expected from the purely electrostatic interaction arising from the dipole properties of the solvent. Thus, if the solvent molecules can act as soft donors, a strong extra solvation of soft acceptors will occur. Also, solvents able to form hydrogen bonds are prone to solvate species which can participate in such bonds, *i.e.* those containing strongly electronegative atoms like F, O and possibly Cl[7]).

Systematic investigations of complex equilibria in a variety of solvents of different dielectric and solvating properties are still scarce. The most informative study so far seems to be that performed by *Luehrs, Iwamoto,* and *Kleinberg* (93) on the solubility of silver halides, and the

---

6) It should then be observed that the value of $\Delta H_h^0(NO_3^-)$ given in (92) is presumably 8—10 kcal too positive, *cf.* (91). The more probable value of —70 kcal is nevertheless markedly lower than the values of $\Delta H_h^0(L^-) = -87$, —80 and —82 kcal found for $L^- = Cl^-$, Br^-, and CN^- respectively. Under the same conditions, $\Delta H_h^0(ClO_4^-) = -57$ kcal (91). — The value of $\Delta H_h^0(Tl^{3+}) = -989$ kcal is of quite a different order of magnitude than $\Delta H_h^0(Tl^+) = -74$ kcal (90).

7) These arguments do not invalidate the conclusion drawn above that the electrostatic interaction of Cl^-, Br^- and CN^- with water is stronger than that of NO_3^- or ClO_4^-. A possible hydrogen bond formation would evidently rather favour the latter ones. Still the energy gained on their hydration is lower.

137

S. Ahrland

stability of silver dihalogenido complexes, in five organic solvents, *viz.* dimethylsulfoxide (DMSO), acetonitrile, methanol, acetone and nitroethane. The authors also recorded the formal reduction potentials e_c of the Ag^+/Ag couple in the various solvents, measured *vs* the aqueous saturated calomel electrode ($e = 0.2415$ V). This potential serves as a measure of the solvation of the silver ion. The lower its value, *i.e.* the more stable the silver ion, the higher the degree of solvation to be postulated.

The results of *Luehrs et al.* are presented in Table 10, together with corresponding data for aqueous solutions (*94—97*). The Table also contains some results found by *Parker et al.* (*98*) for other organic solvents, *viz.* formamide and dimethylformamide. Though no values of e_c are available in these cases, they are nevertheless of considerable value for a comparison. For all solvents, the value of D has also been entered (*93, 99*).

Unfortunately, no quantity related to the solvation of the halide ions in the various media has been measured, but these ions are certainly by far most solvated in hydrogen-bonding, *i.e.* protic, solvents. Among these, water forms the strongest hydrogen bonds, but the effect should also be noticeable for methanol (*93*) and formamide (*98*). The solvation due to hydrogen-bonding should further decrease fairly rapidly in the

Table 10. *Solubility products of silver halides, and formation constants of dihalogenido complexes of the silver ion in various solvents*

Solvent	D	e_c	I	pK_{so}			$\log \beta_2$		
				Cl^-	Br^-	I^-	Cl^-	Br^-	I^-
H_2O	79	0.54[a]	[b]	9.42	12.10	16.50	4.72	7.11	11.74
$HCONH_2$	106	...	[c]	9.4	11.4	14.5	...	...	...
$HCON(CH_3)_2$	37	...	[c]	14.5	15.0	15.8	16.3	16.6	17.8
$(CH_3)_2SO$	47	0.03	0.1[d]	10.4	10.6	12.0	11.9	11.7	13.1
CH_3CN	38	0.13	0.1[d]	12.4	13.2	14.2	12.6	13.4	14.6
CH_3OH	33	0.34	1[e]	13.0	15.2	18.2	8.0	10.9	14.8
$(CH_3)_2CO$	21	0.40	0.1[e]	16.4	18.7	20.9	16.7	19.7	22.2
$C_2H_5NO_2$	28	0.61	0.1[d]	21.1	21.8	22.6	22.2	22.5	22.5

a) At $I = 0.1$ M and $25°$ C (*97*), *i.e.* at conditions fairly representative for the measurements of K_{so} and β_2 in water.

b) The values of K_{so} and β_2 in water are valid for $NaClO_4$ media of $I = 0.2$ M (chloride, (*94*)) and 0.1 M (bromide, (*95*)), both at $25°$ C, and for $I = 0$ at $18°$ C (iodide, (*96*)).

c) $I = 0.005 - 0.01$ M, $25°$ C (*98*).

d) $(C_2H_5)_4NClO_4$ and e) $LiClO_4$, all measurements at $23°$ C. Of the two sets of values of β_2 reported by *Luehrs et al* (*93*), those determined potentiometrically have been entered here.

138

order Cl⁻ > Br⁻ > I⁻. This has been experimentally confirmed *inter alia* by the decrease of $|\Delta H_h^0|$ in that order (*11, 91*).

As expected, the strength of the complexes formed between silver and halide ions generally increases with decreasing value of D and decreasing solvation of the acceptor and donor, Table 10. The variation is moreover most marked for Cl⁻ and decreases in the order Cl⁻ > Br⁻ > I⁻. This is to be expected, as both the increase of the electrostatic attraction with decreasing value of D, and the decrease of the solvation should be most marked for Cl⁻, and then less marked in the order mentioned. These two factors are thus acting in concert to form that pattern of complex stability which is just observed.

Due to the lack of information on the solvation of the silver ion in two of the solvents (where no value of e_c is available) and, above all, due to the general lack of quantitative data on the solvation of the halides, it is difficult to assess the relative importance of the dielectric and solvating properties of the solvent. It is rather clear, however, that the low values of β_2 found especially for AgCl₂⁻ in methanol and, even more, in water are primarily due to the strong solvation of the halide ions in those media. As far as β_2 is concerned, this factor is evidently strong enough to offset the influence of the low solvation of the silver ion which should tend to tip the equilibrium in favour of the halide complexes. As to K_{so}, the solvation of the halide ions is of less importance (*93*). In the aqueous system, only Cl⁻ is enough hydrated to make $pK_{so} = 9.42$ lower than for most other solvents, while for I⁻ $pK_{so} = 16.50$ is in the middle of the field, in spite of the high value of D. For K_{so}, the solvation of the silver ion evidently counts very much.

In the aprotic nitroethane, of low D and low solvation, the values of K_{so} and β_2 are about the same for all the halides. The values of β_2 correspond to a value of $\Delta G_{\beta 2}^0 \simeq -30$ kcal. As the solvation is low, the entropy term should be of minor importance, and $\Delta G_{\beta 2}^0$ should therefore also give the order of magnitude of $\Delta H_{\beta 2}^0$. This would mean an energy gain of $\simeq 15$ kcal per bond formed, indicating bonds of a rather respectable strength.

A comparison between the protic formamide and the aprotic dimethylformamide also provides a good example of that smoothing out of the differences between the values of K_{so} and β_2 for different halides which occurs for aprotic solvents, Table 10.

In the gas phase where $D \simeq 1$, the bonds formed between ions, or between ions and marked dipoles, should be very strong and predominantly electrostatic. All metal acceptors would then be expected to show (*a*)-sequences. Further, as there is no solvation, the large amount of energy gained by the formation of the strong coordinate bond is to no extent used up for breaking other bonds. The complex formation will

139

therefore be very strongly exothermic, in contrast to the endothermic reactions met in highly polar solvents like water when complexes containing predominantly electrostatic bonds are formed. Also in contrast to the conditions in polar solvents, the lack of solvation will moreover make the entropy terms small for reactions in the gaseous state, especially relative to the huge values of ΔH^0. For all practical purposes, one may thus presume that $\Delta G^0 \simeq \Delta H^0$.

Though the electrostatic interactions certainly predominate the bonds between charged particles in the gaseous state, the covalent contribution may nevertheless not be negligible, especially not in bonds formed between acceptors and donors of otherwise markedly soft behaviour, i.e. of especially high covalent bonding capacity.

As positive and negative ions in the gas phase are not stable relative to the neutral atoms, ionic reactions equivalent to those occurring in solutions cannot be studied directly. The energy of formation of a gaseous complex from its component ions, or its equivalent of opposite sign, the coordinate bond energy (CBE) of *Basolo* and *Pearson* (*100*), must therefore be calculated by a suitable combination of available energy data. Extensive calculations of CBE for metal halides, comprising both neutral and anionic complexes, have been performed by *Pearson* and *Mawby* (*8*) whose work should also be consulted about details and conditions of the method of calculation. It should only be remarked here that calculations of CBE are certainly more practicable, or at least more reliable, for halides than for complexes of any other ionic ligands. The reason is primarily that the electron affinities which have to be introduced in the calculations are better known for the halogens than for any other atoms, or groups of atoms, but also as far as other energy data are concerned, the situation is presumably most favourable for the halides.

A selection of the CBE values calculated by *Pearson* and *Mawby* are presented in Table 11[8]). In addition values of CBE for the hydrogen halides have been entered, calculated by the present author[9]). The values entered are the enthalpy change for the total reaction of n steps. Division of these values by n will give an average CBE per bond which should be possible to use for a fairly just comparison of the strength of the acceptor to donor bond in halides of different types.

As expected, the formation of halide complexes in the gaseous state is always a strongly exothermic reaction. The net evolution of heat per

---

[8]) The values have been recalculated in kcal, in order to make them immediately comparable with the data of the previous Tables.

[9]) By combining the standard enthalpies of formation of the gaseous hydrogen halide, and of the gaseous hydrogen and halogen atoms, with the ionization potential of hydrogen and the electron affinity of the halogen. The data compiled in (*101*) have been used.

140

Table 11. *Coordinate bond energies (CBE)*[a]*), i.e. ΔH^0 of reactions $MX_n(g) \to M^{n+}(g)$* *$+ nX^-(g)$, for halides at 25° C (kcal). — Total ionization potentials*[b]*) ΣI_n, and* *hydration energies*[c]*), ΔH_h^0, for the acceptors tabulated (eV)*

M	r[d])	CBE F	Cl	Br	I	ΣI_n	$-\Delta H_h^0$
			M+				
Li	0.68	183.8	152.9	146.9	138.4	5.39	5.40
K	1.33	138.1	117.8	113.2	106.1	4.34	3.34
Cs	1.67	129.8	112.5	108.6	101.5	3.89	2.87
H	...	367.5	331.7	321.9	312.7	13.60	11.32
Cu	0.95	(198.5)	181.0	179.4	(176.4)	7.72	...
Ag	1.13	178.7	166.2	166.0	165.6	7.57	3.5
Au	...	205.9	194.2	194.9	196.9	9.22	...
			M2+				
Be	0.30	777	689	669	643	27.53	25.2
Ca	0.94	523	464	450	(429)	17.98	16.3
Ba	1.29	466	413	401	(380)	15.21	14.1
Cr	...	(602)	553	537	517	23.25	19.2
Mn	0.80	592	553	535	519	23.07	19.1
Fe	0.76	(627)	581	563	547	24.08	19.9
Co	0.70	(639)	588	574	(556)	24.91	21.3
Ni	0.68	657	604	588	574	25.78	21.8
Cu	...	(666)	620	613	(597)	27.99	21.8
Zn	0.69	664	618	604	588	27.35	21.2
Cd	0.92	595	565	558	547	25.89	18.8
Hg	0.93	(643)	613	606	602	29.18	19.9
			M3+				
Al	0.45	1411	1282	1252	1220	53.24	48.4
Sc	0.68	(1204)	1098	1065	(1035)	44.09	40.5
Y	0.89	(1114)	(1008)	(980)	950	39.12	37.2
La	1.02	(1065)	952	(918)	883	36.21	34.6
Fe	0.53	1348	1256	1238	1217	54.7	45.3
			M4+				
Ti	0.60	2342	2179	2142	2103		
Zr	0.77	2082	1916	1877	1831		

[a]) Values in paranthesis are based on estimated heats of vaporization of the halides and therefore somewhat more uncertain.

[b]) According to *Moore* (as cited in Ref. (90)). Strictly, these values refer ro 0° K, but it has been judged unnecessary to introduce the small temperature correction.

[c]) In the main according to *Halliwell* and *Nyburg* (91), or from Ref. (90), with the values recalculated with the value of $\Delta H_h^0 = -260.7$ kcal, recommended by *Halliwell* and *Nyburg*. For a few ions, however, values have been compiled from other sources (for a comprehensive list of references, see Ref. (91)). For

141

S. Ahrland

Erläuterungen zu Tabelle 11 (Fortsetzung)

Be^{2+}, a somewhat lower value than that found from Refs. (*90, 91*) seems more probable, *viz.* —580 kcal. For Ag$^+$ and Hg^{2+}, not listed in Refs. (*90, 91*), $\Delta H_h^0 =$ —80 and —460 kcal, respectively, have been adopted as the most reasonable estimates. All values have been recalculated to eV, in order to facilitate the comparison with *Klopman's* energy data (Table 12.)

d) According to *Goldschmidt* (see *e.g.* Ref. (*4*)); for the rare earth ions values of *Spedding* and *Gschneider* (see Ref. (*101*)).

coordinate bond is in fact enormous compared with all reactions considered previously. For those, 25 kcal per bond was a very high value while here all values exceed 100 kcal, the largest ones even 500 kcal. Also as expected, virtually all sequences are of type (*a*), indicating the strongly electrostatic character of the bonds. The predominance of the electrostatic interaction is further confirmed by the strong increase of the CBE per bond with increasing charge n of the acceptor. Within a certain group of acceptors (the alkali ions, the rare earth ions etc.), the value of CBE also generally increases with decreasing radius r of the acceptor which also fits into the same picture. The latter rule is not without exceptions, however. Thus Hg^{2+}, in spite of its larger radius, shows a markedly stronger interaction with all the halide ions than does Cd^{2+}. The same relation presumably exists between Au$^+$ and Ag$^+$, though in this case the radius of the heavier ion is not certainly known. It is also striking that the relation of the values of CBE between different groups of acceptors is not the one expected from the relation between the radii. Thus, the bonds formed by monovalent ions of the noble metals are much stronger than those formed by alkali ions of the same size, and the same applies to the divalent zinc group ions in relation to the alkaline earth ions.

There are presumably at least two reasons for these reversals. First, the transition metal ions have very plausibly a higher effective charge than the ions of noble gas configuration, because of the poorer shielding exerted by the d-shell on the nuclear charge. This shielding should moreover be poorer, the more extensive the d-shell, and may thus at least partly account also for the reversals within the transition metal groups which have no counterpart within the groups of ions with noble gas shells. Second, the capacity for covalent bonding displayed by transition metal ions should increase their CBE relative to ions with noble gas shells, and more so the softer the acceptor. This influence should therefore be most marked for Au$^+$ and Hg^{2+}, *i.e.* just in those cases where abnormally high values of CBE are in fact found. It is thus throughout probable that the reversals are caused by the concerted action of increasing effective charge on the acceptor, and increasing covalent bond formation.

Whether the latter factor is really of importance will be unambiguously shown by the relation between the values of CBE found for complexes

142

of different halides. A covalent contribution to the bond will increase in the order known as (b), i.e. $F^- < Cl^- < Br^- < I^-$. If the contribution is perceptible, the decrease of the (still predominantly electrostatic) CBE in the order mentioned will thus generally be slower, the softer the acceptor. It may even be possible to observe a partial reversal of the (a)-sequence towards the end of the series, where the change of radius between two consecutive halide ions is relatively small (and consequently their electrostatic interaction with a certain acceptor not very different) while the increase of softness (i.e. of covalent bonding capacity) is considerable. Such partial reversals should preferably be observed when the relative importance of the electrostatic interaction is at its lowest, i.e. for very soft acceptors of low charge. If, on the other hand, no significant covalent bonding occurs which would mean that the effective charges would be the only factors of importance, then the rate of decrease along the series of halides would be about the same for all acceptors.

As is immediately evident from Table 11, the rate of decrease is in fact generally slower, the softer the acceptor. Even a partial reversal is observed. viz. for Au^+ which is just the acceptor most likely to display such a behaviour, according to the criteria given. For Ag^+, the trend towards a reversal is very evident; the CBE stays almost constant for the last three halides. Also for Hg^{2+}, the rate of decrease of CBE is very slow.

Thus, even in the gaseous state, covalent bonding still contributes significantly to the overall strength of the bond, at least as far as soft acceptors and donors are concerned.

This conclusion is further strengthened considerably by the theoretical calculation of CBE originally performed by *Pearson* and *Gray* (102) and later on somewhat modified by *Pearson* and *Mawby* (8). Values of CBE are calculated according to three models, viz. the hard sphere model, the polarizable ion model and the localized molecular orbital model. Only the last one, treating the bonds as covalent, is able to account in a satisfactory way for the values found experimentally for such halides as $HgCl_2$ and $CdCl_2$. For LiCl and NaCl, on the other hand, an acceptable fit with the experimental values is obtained already by the hard sphere model, which certainly indicates a predominantly electrostatic interaction.

To express the variation in the rate of decrease of CBE along the series of halides, *Pearson* and *Mawby* (8) has defined a parameter [CBE(F) −CBE(I)]/CBE(F). An acceptor will evidently have a lower value of this parameter (here called σ_P), the slower the rate of the decrease. The value of σ_P is thus closely connected with the softness of the acceptor and will therefore provide a measure of this quantity. It is obvious from its definition, however, that the strength of the electrostatic interactions is also important for its value which will therefore not be a function

143

of the softness only. This question will be further discussed in the next section where two other softness parameters will also be introduced.

VII. Scales of Softness

The softness parameter σ_P of *Pearson* and *Mawby* has been entered in Table 12 for a large number of acceptors of varying charge and character. Within each charge group, its variation rather faithfully reflects the order of softness arrived at from the chemical behaviour, as embodied in the introductory criteria. Thus, for monovalent ions, high values of σ_P are found for alkali ions, and much lower values for noble metal ions.

Table 12. *Comparison of the softness parameters σ_P, σ_K and σ_A for ion acceptors of various charge*

	σ_P	σ_K	σ_A		σ_P	σ_K	σ_A
		M+				M2+	
Li	0.247	−0.49	−0.01	Be	0.172	−3.75	1.2
Na	0.211	0.000	0.93	Mg	0.167	−2.42	1.4
K	0.232	...	1.00	Ca	0.180	−2.33	0.9
Rb	0.229	...	1.06	Sr	0.172	−2.21	0.6
Cs	0.218	...	1.02	Ba	0.184	−1.89	0.5
H	0.149	−0.42	2.28	Cr	0.142	−0.91	2.0
				Mn	0.124	...	2.0
Cu	0.112	2.30	...	Fe	0.127	−0.69	2.1
Ag	0.073	2.82	4.1	Co	0.130	...	1.8
Au	0.044	4.35	...	Ni	0.126	−0.29	2.0
				Cu	0.104	0.55	3.1
Ga	0.267	...	...				
In	0.213	...	...	Zn	0.115	...	3.1
Tl	0.215	1.88	...	Cd	0.081	2.04	3.5
				Hg	0.064	4.64	4.6
		M3+		Sn	0.148	...	...
				Pb	0.131	...	4.1
Al	0.136	−6.01	1.6				
Sc	0.140	...	1.2			M4+	
Y	0.147	...	0.6				
La	0.171	−4.51	0.5	Ti	0.102	−4.35	...
				Zr	0.121	...	...
Cr	0.107	−2.06	...				
Fe	0.097	−2.22	3.1				
Ga	0.099	−1.45	2.9				
In	0.100	...	3.3				
Tl	...	3.37	4.3				

144

These also decrease rapidly in the order $Cu^+ > Ag^+ > Au^+$, reflecting the slower and slower rate of decrease of CBE along the halide series with increasing softness of the acceptor. For H^+, a very reasonable intermediate value is found. The pattern is very similar for the divalent ions. High values are found for the alkaline earths, and low values for the ions of the zinc group, the latter ones decreasing in the order $Zn^{2+} > > Cd^{2+} > Hg^{2+}$, while the transition metal ions are in an intermediate position, as expected.

On the other hand, the values of σ_p for acceptors of a certain softness systematically come out lower, the higher the charge. As a result, the value of found for the very hard Ti^{4+} is in fact lower than that found for the very soft Cu^+. The reason is evidently that the value of CBE increases very rapidly with the charge, as has been pointed out above. This will cause a systematic decrease of the ratio σ_p with the charge, by unduly increasing its denominator.

The values of σ_p will thus provide a quantitative scale of softness only for ions of the same charge.

Another parameter related to softness has been introduced by *Klopman* (9) from theoretical considerations involving a polyelectronic perturbation treatment. In agreement with the views expressed above, he concludes that hard-hard interactions are essentially electrostatic ("charge controlled") while soft-soft interactions are essentially covalent ("frontier controlled"). He further finds that the harder the acceptor, the lower the energy of its empty frontier orbital relative to the energy wanted for the de-solvation of the acceptor, and vice versa. Provided these energies can be calculated, their difference would be a quantitative measure of softness. By introducing certain empirical, or semi-empirical, assumptions, *Klopman* has in fact been able to calculate them, with reference to water. The difference[10]), here denoted σ_K, has been entered in Table 12 (in eV). In most cases, σ_K reflects very well the general chemical behaviour. Large negative values are found for very hard acceptors, large positive values for very soft ones. The order within the groups is as expected, and moreover the values of σ_K seem to be unbiased by the charge.

The order found has one rather questionable feature, however, *viz.* the position of the alkali ions. These occupy here an intermediate position, with $\sigma_K \simeq 0$. It could be argued that this position reflects their very indifferent acceptor properties. On the other hand this argument

---

[10]) The values of σ_K of Table 12 has been formed by subtracting the de-solvation from the orbital energy, while *Klopman* has it the other way round. The present procedure has been adopted primarily in order to facilitate a comparison between the softness scales obtained by the *Klopman* parameter and by the new parameter σ_A defined below. It seems moreover natural that the value of a softness parameter should rather increase with the softness.

should apply with almost the same force to the alkaline earth ions, but these have values of $\sigma_K \simeq -2.5$, as would rather be expected from a chemical point of view. If softness should be a concept closely linked to the general chemical behaviour of acceptors which seems natural, then it is also rather misleading to characterize the alkali ions with a number close to that found for Ni^{2+} or Cu^{2+}.

It is interesting to explore the following very simple empirical approach to find that quantitative measure of the extent of covalent bonding which should be expressed by a softness parameter. It seems reasonable to postulate that the more completely the energy spent on the formation of a ion is regained by introducing the ion in a hard solvent like water, the harder is the ion. This means that the difference between the total ionization potential for the formation of M^{n+}, ΣI_n, and the dehydration energy $-\Delta H_h^0$ should be larger, the softer the ion. For the comparison of ions of different charges, the differences should further be divided by n, in order to express the difference per interacting charge.

The total difference $n\sigma_A$ is evidently the enthalpy change for the reaction $M(g) \rightarrow M^{n+} (aq) + ne^-$, i.e. for the formation of an aqueous ion out of a gaseous atom of the element[11]). The more endothermic this reaction per electron split of, the softer the acceptor M^{n+}.

A practical difficulty is that values of ΔH_h^0 are missing for quite a few ions of interest, and are fairly unreliable for others. This is unfortunate, as ΣI_n and $-\Delta H_h^0$ are not very different, as is evident from the values entered in Table 11. Especially for the high-valent ions, the differences are therefore not as certain as one could have wished.

The result (Table 12, σ_A in eV) is nevertheless rather illuminating. The values of σ_A vary generally according to expected pattern, and are seemingly not systematically influenced by the charge of the acceptor. For hard ions, including the alkali and alkaline earth ions, σ_A is as a rule between 0 and 1 eV[12]).

For very soft donors, e.g. Hg^{2+}, a value close to 5 eV is reached. The transition metal ions, and H^+, are intermediate with $\sigma_A \simeq 2$ eV.

If the alkali ions are excepted, the parameters σ_K and σ_A arrange the acceptors in very much the same order of softness. It may be argued that this is not so very unexpected, as the ionization potential is an important term in *Klopman's* orbital energy, and ΔH_h^0 has evidently

---

[11]) The quantity $n\sigma_A$ has before been calculated by *Jørgensen* (Ref. *103*, p. 236; a different set of ΔH_h^0-values has been used). *Jørgensen* notes that $n\sigma_A$ is sometimes strongly positive for cations, and asks the question why then metal atoms are not separated. The general answer given is that only electrons of too low energies are available.

[12]) The somewhat higher values of Al^{3+} and Mg^{2+} may very likely be due to an error in ΔH_h^0.

146

much to do with his solvation energy. On the other hand it should be remembered that σ_K and σ_A have been derived by very different approaches, that their calculation involve different, and partly rather crude, assumptions, and that they both emerge as a fairly small difference between large numbers. The extensive agreement between the results is therefore both astonishing and gratifying.

This work was initiated during my stay in 1966 at the Laboratorium für Anorganische Chemie, Eidgenössische Technische Hochschule, Zürich. The results have partly been presented before at the International Symposium on Soft and Hard Acids and Bases, Northern Polytechnic, London, March 1967. — My sincere thanks are due my friends *Gerold Schwarzenbach* and *Chr. Klixbüll Jørgensen* for many interesting and fruitful discussions on the topics of this treatise, and for their great hospitality during my stay in Switzerland. The financial support of Statens naturvetenskapliga forskningsråd (The Swedish Natural Science Research Coucil) is also gratefully acknowledged.

VIII. References

1. *Ahrland, S., J. Chatt,* and *N. R. Davies:* Quart. Rev. (London) *12*, 265 (1958).
2. *Pearson, R. G.:* J. Am. Chem. Soc. *85*, 3533 (1963).
3. *Leden, I.,* and *J. Chatt:* Chem. Soc. *1955*, 2936.
4. *Cotton, F. A.,* and *G. Wilkinson:* Advanced Inorganic Chemistry, 2nd Edit. New York, London, Sydney: Interscience Publishers 1966.
5. *Ahrland, S.:* Struct. Bonding *1*, 207 (1966).
6. *Schwarzenbach, G.:* Z. Physik. Chem *A 176*, 133 (1936).
7. —, and *W. Schneider:* Helv. Chim. Acta *38*, 1931 (1955).
8. *Pearson, R. G.,* and *R. J. Mawby:* Halogen Chemistry, Vol. 3, p. 55 (Ed. *V. Gutmann*). London and New York: Academic Press 1967.
9. *Klopman, G.* : CERI—TIC—P 142, Cyanamid European Research Institute, Cologny—Genève, 1967, and J. Am. Chem. Soc. *90*, 223 (1968).
10. *Drago, R. S.,* and *B. B. Wayland:* J. Am. Chem. Soc. *87*, 3571 (1965).
11. *Ahrland, S.:* Helv. Chim. Acta *50*, 306 (1967).
12. —, and *L. Kullberg:* Work in progress.
13. *Izatt, R. M., H. D. Johnston, G. D. Watt,* and *J. J. Christensen:* Inorg. Chem. *6*, 132 (1967).
14. *Widmer, M.,* and *G. Schwarzenbach:* Helv. Chim. Acta *47*, 266 (1964).
15. *Schwarzenbach, G.,* and *M. Widmer:* Helv. Chim. Acta *49*, 111 (1966).
16. *Mesmer, R. E.,* and *C. F. Baes, Jr.:* Inorg. Chem. *6*, 1951 (1967).
17. *Dunsmore, H. S.,* and *G. H. Nancollas:* J. Phys. Chem. *68*, 1579 (1964).
18. *Austin, J. M.,* and *A. D. Mair:* J. Phys. Chem. *66*, 519 (1962).
19. *Ahrland, S.,* and *L. Brandt:* Acta Chem. Scand. In press.
20. *Zebroski, E. L., H. W. Alter,* and *F. K. Heumann:* J. Am. Chem. Soc. *73*, 5646 (1951).
21. *Zielen, A. J.:* J. Am. Chem. Soc. *81*, 5022 (1959).
22. *Nair, V. S. K.,* and *G. H. Nancollas:* J. Chem. Soc. *1958*, 3706.
23. *Bell, R. P.,* and *J. H. B. George:* Trans. Faraday Soc. *49*, 619 (1953).
24. *Jones, H. W.,* and *C. B. Monk:* Trans. Faraday Soc. *48*, 929 (1952).

S. Ahrland

25. *Mattern, K. L.:* (Diss.) Univ. Calif. Berkeley UCRL — 1407, 1951 (as cited in (*26*)).
26. *Sillén, L. G.,* and *A. E. Martell* (Ed.): Stability Constants, 2nd Edit. London: Chemical Society 1964.
27. *Newton, T. W.,* and *G. M. Arcand:* J. Am. Chem. Soc. *75*, 2449 (1953).
28. *Day, R. A., R. N. Wilhite,* and *F. D. Hamilton:* J. Am. Chem. Soc. *77*, 3180 (1955).
29. *Sullivan, J. C.,* and *J. C. Hindman:* J. Am. Chem. Soc. *76*, 5931 (1954).
30. *Lietzke, M. H.,* and *R. W. Stoughton:* J. Phys. Chem. *64*, 816 (1960).
31. *Ahrland, S.:* Acta Chem. Scand. *5*, 1151 (1951).
32. *Day, R. A.,* and *R. M. Powers:* J. Am. Chem. Soc. *76*, 3895 (1954).
33. *Fogel, N., J. M. J. Tai,* and *J. Yarborough:* J. Am. Chem. Soc. *84*, 1145 (1962).
34. *Taube, H.,* and *F. A. Posey:* J. Am. Chem. Soc. *75*, 1463 (1953); *78*, 15 (1956).
35. *Nair, V. S. K.,* and *G. H. Nancollas:* J. Chem. Soc. 1959, 3934.
36. *Eigen, M.,* u. *K. Tamm:* Z. Electrochem. *66*, 107 (1962).
37. *Perssson, H.:* Private communication (cited in (*38, 44*)).
38. *Gerding, P.:* Acta Chem. Scand. *21*, 2007 (1967).
39. *Hansson, E.:* Private communication.
40. *Harned, H. S.,* and *R. W. Ehlers:* J. Am. Chem. Soc. *55*, 652 (1933).
41. *Sonesson, A.:* Acta Chem. Scand. *12*, 165, 1937 (1958); *14*, 1495 (1960).
42. *Grenthe, I.:* Acta Chem. Scand. *18*, 283 (1964).
43. *Gerding, P.:* Acta Chem. Scand. *20*, 2624 (1966).
44. — Acta Chem. Scand. *21*, 2015 (1967).
45. *Leden, I.:* Potentiometrisk undersökning av några kadmiumsalters komplexitet (Diss.) Univ. Lund 1943.
46. *Gerding, P.,* and *B. Johansson:* Private communication.
47. — Acta Chem. Scand. In press.
48. *Gobom, S.:* Acta Chem. Scand. *17*, 2181 (1963).
49. *Sonesson, A.:* Acta Chem. Scand. *13*, 998 (1959).
50. *Fronæus, S.:* Komplexsystem hos koppar (Diss.)Univ. Lund 1948.
51. *Rossotti, F. J. C.:* Modern Coordination Chemistry (Ed. *J. Lewis* and *R. G. Wilkins*). New York and London: Interscience Publishers, 1960.
52. *Harmer, W. J., J. O. Burton,* and *S. F. Acree:* J. Res. Bur. Stand. *24*, 269 (1940).
53. *Nair, V. S. K.,* and *G. H. Nancollas:* J. Chem. Soc. *1961*, 4367.
54. *McAuley, A.,* and *G. H. Nancollas:* J. Chem. Soc. *1963*, 989.
55. — —, and *K. Torrance:* J. Nucl. Chem. *28*, 917 (1966).
56. *Gelles, E.,* and *G. H. Nancollas:* J. Chem. Soc. *1956*, 4847.
57. *Nair, V. S. K.:* J. Chem. Soc. *1965*, 1450.
58. *Gelles, E.,* and *G. H. Nancollas:* Trans. Faraday Soc. *52*, 680 (1956).
59. *van Panthaleon van Eck, C. L.:* Hydration and Complex Formation of Ions in Solution (Diss.). Univ. Leiden 1958, from data in *Spike, C. G.* Diss. Univ. Michigan 1953.
60. *Bates R. G.,* and *G. D. Pinching:* J. Res. Bur. Stand. *42*, 419 (1949)
61. *Pitzer, K. S.:* J. Am. Soc. *59*, 2365 (1937).
62. *Papee, H. M., W. J. Canady,* and *K. J. Laidler:* Can. J. Chem. *34*, 1677 (1956).
63. *Ackermann, T.:* Z. Electrochem. *62*, 411 (1958).
64. *Bjerrum, J.:* Metal Ammine Formation in Aqueous Solution (Diss.). Univ. Copenhagen 1941.
65. *Schultz, J. L.:* Diss. Univ. Minnesota 1959 (as cited in Ref. *26*).
66. *Scott, P. C.:* Diss. Univ. Minnesota 1959 (as cited in Ref. *26*).
67. *Yatsimirskii, K. B.,* and *P. U. Milyukov:* Zh. Neorgan. Khim. *2*, 1046 (1957).

148

68. *Spike, C. G.*, and *R. W. Parry:* J. Am. Chem. Soc. *75*, 3770 (1953).
69. *Smith, W. V., O. L. I. Brown*, and *K. S. Pitzer:* J. Am. Chem. Soc. *59*, 1213 (1937).
70. *McIntyre, G. H., Jr., B. P. Block*, and *W. C. Fernelius:* J. Am. Chem. Soc. *81*, 529 (1959).
71. *Ciampolini, M., P. Paoletti*, and *L. Sacconi:* J. Chem. Soc. *1960*, 4553.
72. *Edwards, L. J.:* Diss. Univ. Michigan 1950 (as cited in Ref. *71*).
73. *Poulsen, I.*, and *J. Bjerrum:* Acta Chem. Scand. *9*, 1407 (1959).
74. *Bjerrum, J.*, and *P. Andersen:* Kgl. Danske Videnskab Selskab, Mat.-Fys. Medd. *22*, No. 7 (1945).
75. *Carlson, G. A., J. P. McReynolds*, and *F. H. Verhoek:* J. Am. Chem. Soc. *67*, 1334 (1945).
76. *Droll, H.:* Diss. Pennsylvania State Univ. as cited in *C. R. Bertsch, W. C. Fernelius*, and *B. P. Block:* J. Phys. Chem. *62*, 444 (1958).
77. *Anderegg, G.:* Helv. Chim. Acta *46*, 1831 (1963).
78. *Schwarzenbach, G.*, and *M. Schellenberg:* Helv. Chim. Acta *48*, 28 (1965).
79. *Meier, M.:* Phosphinokomplexe von Metallen (Diss. Nr. 3988) Eidgenössische Technische Hochschule Zürich 1967.
80. *Wright, G.*, and *J. Bjerrum:* Acta Chem. Scand. *16*, 1262 (1962).
81. *Elding, L. I.*, and *I. Leden:* Acta Chem. Scand. *20*, 706 (1966).
82. *Drougge, L., L. I. Elding*, and *L. Gustafson:* Acta Chem. Scand. *21*, 1647 (1967).
83. *Elding, L. I.:* Private communication.
84. *Dunning, W. W.*, and *D. S. Martin, Jr.:* J. Am. Chem. Soc. *81*, 5566 (1959).
85. *Denning, R. G., F. R. Hartley*, and *L. M. Venanzi:* J. Chem. Soc. (A) *1967*, 324.
86. *Hartley, F. R.*, and *L. M. Venanzi:* J. Chem. Soc. (A) *1967*, 330.
87. *Denning, R. G.*, and *L. M. Venanzi:* J. Chem. Soc. (A) *1967*, 336.
88. *Jørgensen, C. K.:* Inorg. Chem. *3*, 1201 (1964).
89. *Burmeister, J. L.*, and *F. Basolo:* Inorg. Chem. *3*, 1587 (1964).
90. *Phillips, C. S. G.*, and *R. J. P. Williams:* Inorganic Chemistry, Vol. I. Oxford: Clarendon Press 1965.
91. *Halliwell, H. F.*, and *S. C. Nyburg:* Trans. Faraday Soc. *59*, 1126 (1963).
92. *Ladd, M. F. C.*, and *W. H. Lee:* J. Inorg. Nucl. Chem. *13*, 218 (1960).
93. *Luehrs, D. C., R. T. Iwamoto*, and *J. Kleinberg:* Inorg. Chem. *5*, 201 (1966).
94. *Berne, E.*, and *I. Leden:* Svensk Kem. Tidskr. *65*, 88 (1953).
95. — — Z. Naturforsch. *8a*, 719 (1953).
96. *Lieser, K. H.:* Z. Anorg. Allgem. Chem. *292*, 97 (1957).
97. *Owen, B. B.*, and *S. B. Brinkley:* J. Am. Chem. Soc. *60*, 2233 (1938).
98. *Alexander, R., E. C. F. Ko, Y. C. Mac*, and *A. J. Parker:* J. Am. Chem. Soc. *89*, 3703 (1967).
99. Landolt-Börnstein, 6th Ed., Vol. II: 6, (Table 2745). Berlin, Göttingen, Heidelberg: Springer 1959.
100. *Basolo, F.*, and *R. G. Pearson:* Mechanisms of Inorganic Reactions. New York: John Wiley 1958.
101. Handbook of Chemistry and Physics, 48th Ed. Cleveland, Ohio: Chemical Rubber Co. 1967—1968.
102. *Pearson, R. G.*, and *H. B. Gray:* Inorg. Chem. *2*, 358 (1963).
103. *Jørgensen, C. K.:* Absorption Spectra and Chemical Bonding in Complexes. Oxford: Pergamon Press 1962.

Received February 23, 1968

149

Reprinted from *Pure and Applied Chemistry*, **24**, 307–334 (1970)

15

©

ELECTROSTATIC AND NON-ELECTROSTATIC CONTRIBUTIONS TO ION ASSOCIATION IN SOLUTION

G. Schwarzenbach

Swiss Federal Institute of Technology, Zürich, Switzerland

ABSTRACT

The coordination behaviour of d^0-cations on the one hand and that of d^{10}-cations of low charge on the other hand is designated as A- and B-character respectively. For A-character interactions the charges and radii and for B-character interactions the ionization potentials and electronegativities of the combining atoms are decisive (electrovalent and covalent behaviour). Generally A- and B-character may be developed to any extent and seem to be mixed in a complicated manner. The thermodynamic functions of the association processes are often a good criterion, the formation constant of the adduct being large because of a dominant positive $T \Delta S$ in the case of A-character interactions and because of a dominant negative ΔH in the case of B-character interactions.

The increase in entropy which is generally observed when the association process is due to electrostatic forces, is caused by the negative temperature coefficient of the dielectric constant (DK) of the solvent. However, the calculation leads to reasonable results only, if a smaller effective DK ε_e is used in place of the DK of the bulk of the solvent ε. Furthermore, $- \delta\varepsilon_e/\delta T$ is considerably smaller than $- \delta\varepsilon/\delta T$ and $\delta(\ln\varepsilon_e)/\delta(\ln T)$ becomes less negative the more the electrostriction increases. This statement is substantiated experimentally with the aid of proton transfer reactions, which are controlled entirely by simple electrostatics. The greater the electrostriction around the species associating on account of purely electrostatic forces, the more exothermic the reaction will be. The structural changes exerted by a given metal cation on its solvation shell depend not only on its charge and radius, but also on its individuality.

---

1. COORDINATION SELECTIVITY

An examination of the large body of facts on the solution stability of metal complexes which has accumulated during the past 30 years[1] reveals markedly different preferences of the various metal ions for the ligands offered to them in aqueous solution.

The d^0-cations react only with fluoride and with oxygen donors to an appreciable extent. Insoluble fluorides are precipitated with alkali fluoride and often can be dissolved in an excess of the reagent. The hydroxides, carbonates and phosphates of almost all of the multivalent cations are

307

insoluble, and again mononuclear hydroxo- and carbonato-complexes are sometimes formed if the ligand is added in high enough concentration. Acetate is quite a general, although weak, complexing agent. The deprotonated carboxylic acid group reveals its potentialities more clearly as a ligand group of a multidentate agent such as oxalate, tartrate, citrate and the anions of aminopolycarboxylic acids such as EDTA, which are quite universal sequestering agents. All the d^0-cations with a charge of more than one form of either insoluble oxalates or oxalato chelates and can be masked with tartrate, citrate and EDTA. However, d^0-cations do not react with the heavy halide anions even when present in large excess. The basic ligands ammonia, cyanide and sulphide do not coordinate to d^0-cations but act as deprotonating agents of the aquo complexes, so that again the metal hydroxides are precipitated.

The degree of condensation with fluoride and oxygen donors depends characteristically on the charges and radii of the interacting species, as is to be expected if the formation of the adduct is the result of simple electrostatic forces. The higher the charge of the cation, the more stable is usually the complex formed with a given ligand or the smaller is the solubility product of a precipitate. The stability of the complex increases also with decreasing radius in a series of metal ions of a given charge. Exceptions are encountered only with multidentate ligands when the smaller metal ion is unable to accommodate sterically the many ligand atoms offered by the chelating agent and is thereby at a disadvantage compared to the larger metal ion. It is in line with this electrovalent behaviour that the ammonia molecule with the smaller dipole moment cannot compete successfully with the larger dipole H_2O which is present in such an overwhelming excess. Furthermore it is understood readily that the smaller fluoride is preferred to the larger halide ions and that oxygen donors which bring up their charged ligand atom within a smaller distance to the metal cation are preferred to comparable sulphur donors. It is surprising, however, that even the anions of the heavy halides have difficulty in competing with the dipole molecules of the solvent. The d^0-cations hardly form halogeno complexes even in concentrated solutions of HCl, HBr and HI (e.g. hexaaquoaluminium chloride $[Al(H_2O)_6]Cl_3$ crystallizes from fuming hydrochloric acid which hardly contains any unprotonated H_2O but an overwhelming excess of Cl^-).

The complex formation of the low charged d^{10}-cations Cu(I), Ag(I), Au(I) reveals a totally different behaviour. The very soluble AgF (CuF and AuF are unstable to disproportionation) is highly dissociated in solution, but the heavy halides have small solubility products and dissolve when an excess of the corresponding alkali halide is added with the production of mononuclear chloro-, bromo- and iodo-complexes which are of an appreciable stability. The heavy halides therefore are strongly preferred as ligands to fluoride, and analogously sulphur donors to oxygen donors. The less polar molecule NH_3 is more strongly coordinated than the more polar H_2O and the phosphine adducts are even more stable than the ammine complexes. The cyano complexes of these noble metals have especially large stability constants.

It is obvious that this type of interaction is not governed by electrostatic forces. The larger Au(I) forms more stable associates than the smaller Cu(I), and of the halide ligands the one with the greatest radius is preferred to the smaller anions. It is instructive to consider an isoelectronic series of d^{10}-cations, such as Ag(I), Cd(II), In(III), which demonstrates that the chloro-, bromo-, iodo-, ammine- and phosphine-complexes become less stable with increasing charge (*Table 1*), whereas the stability of the fluoro complexes as well as the stability of the adducts with oxygen donors increases in the series. The quantities which are decisive for the type of interaction of these low charged d^{10}-cations, seem to be the ionization potential of the metal (or the tendency of the metal cation to take up electrons) and the electronegativity of the element furnishing the ligand (or the tendency of the ligand atom to donate electrons). The production of these noble metal complexes must be due to the formation of new and more stable covalent bonds in the course of the reaction.

Instead of using expressions such as electrovalent and covalent behaviour, it commits one less to speak of *A-character* and *B-character* if one wishes to state that a given metal cation is behaving like a d^0-cation or like a low charged d^{10}-cation. Anyhow, even the metal–oxygen and the metal–fluorine bonds may be covalent to a considerable extent, but nevertheless in aqueous solution the formation of the complex species involved is governed almost exclusively by electrostatic forces because of the replacement of one metal–oxygen bond ($M-OH_2$) by another metal–oxygen bond ($M-O-Donor$) or by the similar metal–fluorine bond. Furthermore, the formation of the adduct may benefit from ligand field stabilization effects which are non-electrostatic contributions to the stability although no covalency necessarily has to be involved.

The letters A and B used for characterizing the coordination selectivity have been borrowed from the assignments of the columns of the periodic table where the metals forming d^0- and d^{10}-cations respectively have their places, just as the designation of the classes a and b by Ahrland, Chatt and Davies[2]. However, the more vague expressions A- and B-character which had already been suggested[3] before the appearance of the paper of Ahrland *et al.* seem more appropriate because it is impossible to classify the experimental facts into two categories only. There are many cases where a certain metal ion falls into class 'a' according to its behaviour with a first and into class 'b' with a second series of ligands. More recently the adjectives 'hard' and 'soft' have been suggested for A- and B-behaviour respectively[4]. It is certainly true that the A-character metal ions are generally less polarizable than the B-character metal ions and it is often also serviceable to be able to have a shorthand expression for gradations like hard, harder, hardest and soft, softer, softest. However, one should never forget that there is no simple correlation between coordination selectivity and polarizability[5].

It goes without saying that the association energies due to simple electrostatic forces as well as those due to covalency both may be large or both may be small or that the one kind or the other may be dominant to any degree. It seems that in aqueous solution only F^- and oxygen donors are able to compete successfully with the solvent molecules on account of electrostatics. To an extent depending on the charge, these ligands coordinate to every metal cation. All the other ligands are selective and a certain degree of B-character is needed to obtain an adduct. B-character builds up in the series of transition metal cations nd^q with increasing q and increasing n and is the more pronounced the smaller the charge of the cation. The extreme B-charac-

309

194

ter of the univalent d^{10}-cations, on the other hand, is progressively replaced by A-character when the charge is increased (*Table 1*). However, there are many exceptions to these rules and somewhat different results may be obtained depending on the charge-type of the selective ligand used for the investigation of the degree of B-character[9]. For a more satisfactory description of coordination selectivity it would be highly desirable to find a method which would allow a separation of the electrostatic contribution to the free energy from the non-electrostatic part.

Table 1. Approximate stabilities of 1:1-complexes (log K_1) of isoelectronic d^{10}-metal cations

Ligand	Ag^+	Cd^{2+}	In^{3+}	*Ref.*
Cl^-	3·4	1·9	~2	1
Br^-	4·2	2	1·8	1
I^-	7	2·3	1·7	1
HO—CH_2—CH_2—S^-	13·2	?	9·0	6·7
H_3N	3·2	2·5	ppt.	1
HO—CH_2—CH_2—$P\begin{smallmatrix}Et\\\\Et\end{smallmatrix}$	12	4	ppt.	8
F^-	−0·2	0·5	3·8	1
OH^-	2·9	4	10	1
EDTA-anion	7	16	25	1

ppt.: precipitation of metal hydroxide.

2. THE ENTHALPY–ENTROPY CRITERION

Our discussion so far has been based entirely on free energy data ($\Delta G = - RT \ln K$). The enthalpy changes for the associations were difficult to obtain before the development of thermistors, which allowed accurate direct calorimetric determinations of ΔH at low concentrations of the reactants in media of constant ionic strength. Reliable ΔH-values for metal complex formation reactions were still very scarce about ten years ago[10], but the situation is much more favourable today. A survey of these data reveals that the enthalpy change taking place during the association is an excellent criterion to decide whether we are dealing with a mainly electrovalent or mainly non-electrovalent interaction[11, 12]. Associations with fluoride or oxygen donors are usually somewhat endothermic or only slightly exothermic and the stability of the adduct is almost entirely due to a large and positive entropy change. The addition of a selective ligand on the other hand is always of considerable exothermicity and ΔH is the dominant factor in the Gibbs–Helmholtz equation for making ΔG negative[13].

$$\Delta G = \Delta H - T \, \Delta S \qquad (1)$$

A-character associations: $T \, \Delta S$ dominant and positive
B-character associations: ΔH dominant and negative

310

195

In *Table 2* the thermodynamic quantities for the formation of some halogeno complexes are listed. The monofluoroaquo cations have stability constants as high as 10^5 to 10^6 which originate entirely in the large entropy term. The enthalpy change is even unfavourable for the association in two of the three examples. The additions of heavy halide anions to extreme B-character cations, on the other hand, are all exothermic and ΔH is now the decisive factor for the stability of the complex, although $T \Delta S$ usually is again positive and makes a minor contribution to the negative ΔG.

Table 2. Formation of halide complexes, 25°C, kcal mole^{-1}

Reaction	μ	ΔG	ΔH	$T \Delta S$	Ref.
$Be^{2+} + F^- \rightarrow BeF^+$	1·0	− 6·7	− 0·4	+6·3	14
$Al^{3+} + F^- \rightarrow AlF^{2+}$	0·5	− 8·4	+ 1·1	+9·5	15
$Fe^{3+} + F^- \rightarrow FeF^{2+}$	0·5	− 7·1	+ 2·3	+9·4	16
$Ag^+ + Cl^- \rightarrow AgCl$	0	− 4·5	− 2·7	+1·8	17
$Hg^{2+} + Cl^- \rightarrow HgCl^+$	0·5	− 9·2	− 5·5	+3·7	18
$Hg^{2+} + Br^- \rightarrow HgBr^+$	0·5	−12·3	−10·2	+2·1	18
$Hg^{2+} + I^- \rightarrow HgI^+$	0·5	−17·5	−18·0	−0·5	18

Table 3. Complex formation with oxygen- and sulphur-donors, 25°C, kcal mole^{-1}

Reaction	μ	ΔG	ΔH	$T \Delta S$	Ref.
$Fe^{3+} + OH^- \rightarrow FeOH^{2+}$	1	−16·1	− 3·0	+13·1	19
$Cr^{3+} + OH^- \rightarrow CrOH^{2+}$	0·1	−10·6	− 3·0	+ 7·6	20
$CH_3Hg^+ + SR^- \rightarrow CH_3HgSR$	0·1	−21·6	−19·8	+ 1·8	21
$Cd^{2+} + AcO^- \rightarrow CdOAc^+$	2	− 1·7	+ 1·5	+ 3·2	22
$Y^{3+} + AcO^- \rightarrow YOAc^{2+}$	2	− 2·1	+ 3·3	+ 5·4	23
$La^{3+} + AcO^- \rightarrow LaOAc^{2+}$	2	− 2·1	+ 2·2	+ 4·3	23
$UO_2^{2+} + SO_4^{2-} \rightarrow UO_2SO_4$	1	− 2·4	+ 4·3	+ 6·7	24, 27
$La^{3+} + SO_4^{2-} \rightarrow LaSO_4^+$	1	− 1·9	+ 2·5	+ 4·4	25, 27
$Ce^{3+} + SO_4^{2-} \rightarrow CeSO_4^+$	1	− 1·7	+ 3·6	+ 5·3	25, 27
$Th^{4+} + SO_4^{2-} \rightarrow ThSO_4^{2+}$	2	− 4·5	+ 5·0	+ 9·5	26, 27

The formation of $FeOH^{2+}$ and $CrOH^{2+}$ (*Table 3*) is slightly exothermic, but $-\Delta H$ contributes only 20 to 30 per cent to $-\Delta G$. In the addition of a mercaptide anion to the B-character cation CH_3Hg^+ on the other hand, $-\Delta H$ contributes more than 90 per cent to the negative free energy change. The comparison of the tervalent Fe^{3+} and Cr^{3+} with the univalent CH_3Hg^+ certainly is not satisfactory, but there are no other data available which would be more suitable for the comparison of a sulphur analogue with OH^-. The study of the formation of simple hydroxy- and thio-complexes is made very difficult by the overwhelming tendency to form polynuclear species.

The central part of *Table 3* deals with 1:1-acetato and the lower part with 1:1-sulphato complexes. It is especially notable that the formation of these adducts, with only small stability constants, are reactions of considerable endothermicity. However, $+\Delta H$ is overcompensated by the still larger $-T \Delta S$, so that ΔG becomes negative.

311

The replacement of H_2O from the aquo shell of a metal cation by an uncharged ligand (*Table 4*) always seems to be an exothermic process. Because of the circumstance that ammonia, amines, phosphines and thioethers all have considerably smaller dipole moments than H_2O, it would be hard to imagine such a substitution to take place on the basis of Coulomb-forces. Obviously the bonds from the metal to N, P and S must be more covalent and more stable than the metal–oxygen bond which is reflected in the negative value of ΔH. The association becomes the more exothermic the more pronounced the B-character of the metal and the lower the electronegativity of the ligand atom. The large amount of heat produced in the addition of an aliphatic phosphine to Ag(I) and Hg(II) is remarkable[8].

Table 4. Complexes of uncharged ligands, kcal mole^{-1}

Reaction	μ	ΔG	ΔH	$T\Delta S$	$t°C$	Ref.
$Ni^{2+} + 2NH_3 \rightarrow Ni(NH_3)_2^{2+}$	2	$-7\cdot0$	$-7\cdot6$	$-0\cdot6$		1
$Cu^{2+} + 2NH_3 \rightarrow Cu(NH_3)_2^{2+}$	2	$-10\cdot6$	$-11\cdot0$	$-0\cdot4$		1
$Zn^{2+} + 2NH_3 \rightarrow Zn(NH_3)_2^{2+}$	2	$-6\cdot7$	$-5\cdot7$	$+1\cdot0$	$25°$	1
$Hg^{2+} + 2NH_3 \rightarrow Hg(NH_3)_2^{2+}$	2	$-23\cdot6$	$-24\cdot7$	$-1\cdot1$		1
$Ag^+ + 2NH_3 \rightarrow Ag(NH_3)_2^+$	0	$-9\cdot9$	$-13\cdot4$	$-3\cdot5$		28
$Hg^{2+} + 2PR_3 \rightarrow Hg(PR_3)_2^{2+}$	1	$-50\cdot4$	$-52\cdot8$	$-2\cdot4$		8
$Ag^+ + 2PR_3 \rightarrow Ag(PR_3)_2^+$	1	$-28\cdot2$	$-35\cdot8$	$-7\cdot6$	$20°$	8
$Ag^+ + 2SR_2 \rightarrow Ag(SR_2)_2^+$	1	$-8\cdot2$	$-14\cdot8$	$-6\cdot6$		29

$$PR_3: \quad \overset{C_2H_5}{\underset{C_2H_5}{>}} P-CH_2-CH_2-OH \qquad SR_2: HO-CH_2-CH_2-S-CH_2-CH_2-OH$$

The entropy change in the reactions of *Table 4*, on the other hand, opposes the formation of the product. After all, it is normal that ΔS is negative in a condensation process and a comparison with the data of *Tables 2* and *3* reveals that a positive entropy change must be caused by a compensation of charges. Indeed, the lowering of the electrostatic fields around the solutes which takes place in the course of a charge compensation will increase the mobility of the molecules of the solvent exposed to these fields and must cause an entropy increase. Therefore it is to be expected that the entropy change in A-character associations, which are due to electrostatic forces mainly, is more positive than in B-character associations caused mainly by covalency or ligand field stabilization effects.

3. THE MOST SIMPLE HARD AND THE MOST SIMPLE SOFT ACID

In the following sections the cations H^+ and CH_3Hg^+ play a prominent role because of the stoichiometric simplicity of their reactions. Both cations add practically only one single ligand. Coordination number two certainly has been observed, but the equilibria constants K_2 for the processes $HL^{(1-\lambda)} + L^{\lambda-} \rightarrow HL_2^{(1-2\lambda)}$ as well as $CH_3HgL^{(1-\lambda)} + L^{\lambda-} \rightarrow CH_3HgL_2^{(1-2\lambda)}$

312

are exceedingly small. In the case of H^+, the formation of hydrogen bridged HF_2^- takes place even in aqueous solution, but to a negligible extent only[30]. Also the formation of $CH_3HgX_2^-$, where X is a heavy halogen or SCN, must be taken into account only at very high concentrations[31] of X^-. While K_2 is very small, K_1 for the reactions $H^+ + L^{\lambda-} \rightarrow HL^{(1-\lambda)}$ and $CH_3Hg^+ + L^{\lambda-} \rightarrow CH_3HgL^{(1-\lambda)}$ is often extremely large[21]. The very exceptional position of the hydrogen ion in the history of chemistry as the real carrier of acidity is due to this peculiarity[32] and the methylmercury cation is its counterpart.

Table 5. The most simple hard and the most simple soft acid, kcal mole^{-1}, 20°C

$B^{\lambda-} + H^+ \longrightarrow HB^{1-\lambda}$				$B^{\lambda-} + CH_3Hg^+ \longrightarrow CH_3HgB^{1-\lambda}$				
$HB^{1-\lambda}$	ΔG	ΔH	$T\Delta S$	$CH_3HgB^{1-\lambda}$	ΔG	ΔH	$T\Delta S$	Ref.
HF	$-4\cdot0$	$+2\cdot6$	$+6\cdot6$	CH_3HgF	$-2\cdot0$	?	?	21
HCl	$(+9)$	$(+14)$	$(+5)$	CH_3HgCl	$-7\cdot0$	$-6\cdot0$	$+1\cdot0$	21
HOH	$-21\cdot3$	$-13\cdot7$	$+7\cdot6$	CH_3HgOH	$-12\cdot6$	$-8\cdot5$	$+4\cdot1$	21
HSR	$-12\cdot8$	$-6\cdot5$	$+6\cdot3$	CH_3HgSR	$-21\cdot6$	$-19\cdot8$	$-1\cdot8$	
HNH_3^+	$-12\cdot9$	$-12\cdot8$	$+0\cdot1$	$CH_3HgNH_3^+$	$-10\cdot3$	?	?	21
HPR_3^+	$-11\cdot0$	$-8\cdot3$	$+2\cdot7$	$CH_3HgPR_3^+$	$-19\cdot7$	$-22\cdot6$	$-2\cdot9$	8
HCN	$-12\cdot3$	$-10\cdot9$	$+1\cdot4$	CH_3HgCN	$-18\cdot8$	$-22\cdot1$	$-3\cdot3$	21

$$PR_3: \quad \begin{matrix} C_2H_5 \\ \diagdown \\ \diagup \\ C_2H_5 \end{matrix} P{-}CH_2{-}CH_2{-}OH \qquad HSR: HS{-}CH_2{-}CH_2{-}OH$$

From *Table 5* it is apparent that the proton has dominant A-character whereas CH_3Hg^+ has pronounced B-character. This follows from the coordination selectivity, that is, from the fact that H^+ prefers F^- to Cl^-, OH^- to SH^- and an amine to a phosphine, whereas CH_3Hg^+ forms the more stable adduct with the ligand deriving from the second row element Cl, S or P. However, the A-character of H^+ is not pronounced. A cation of extreme A-character such as Al^{3+} does not coordinate at all to sulphur donors in aqueous solution and does not form any amine-, phosphine- or cyano-complexes.

The fact that H^+ has only a moderate although dominant A-character, whereas CH_3Hg^+ has extreme B-character, follows also from an inspection of the values ΔH and ΔS listed in *Table 5*. The formation of the proton complexes usually is a reaction of appreciable exothermicity, but $T\Delta S$ is also considerable if the ligand is an anion. For the formation of the methylmercury complexes, on the other hand, the term $T\Delta S$ makes a small contribution to $(-\Delta G)$ only and sometimes is even opposed to the production of the adduct.

The position of H^+ as somewhere between the cations of extreme A- and extreme B-character fits in with the high charge density on the surface of the small proton (creating electrovalent behaviour) on the one hand and the appreciable covalency of even the most polar bonds between the ligand atom and H of the proton complexes (creating covalent behaviour) on the other hand.

313

4. ION ASSOCIATION IN A STRUCTURELESS DIELECTRIC MEDIUM

The electrostatic energy to be gained by bringing two gaseous hard spherical ions of opposite charge ev^+ and $e\lambda^-$ from infinity up to a distance 'a' is per mole:

$$A_{el} = Ne^2 v\lambda/a \qquad (2)$$

where N is Avogadro's number, and the thermodynamic functions of the process are given by:

$$\Delta G = -T\Delta S_t - A_{el} \qquad (3)$$

$$\Delta S = \Delta S_t \qquad (4)$$

$$\Delta H = -A_{el} \qquad (5)$$

ΔS_t is the difference of the translational entropies of product and reactants of about -30 e.u. A_{el} is a large quantity in comparison to the entropy term $T\Delta S$ (~ 10 kcal/mole at $T = 298°$K) and amounts even in the case of singly charged ions ($v = \lambda = 1$) of usual size ($a \approx 2$ Å) to about 200 kcal/mole. In the electrostatic ion association in vacuum ΔG and ΔH therefore are both strongly negative and of the same order of magnitude. For the reaction 6,

$$Al^{3+}(g) + F^-(g) \rightarrow AlF^{2+}(g) \qquad (6)$$

taking for 'a' the sum of the atomic radii ($= 1·8$ Å), the result is:
$\Delta G = -550$ kcal/mole and $\Delta H = -560$ kcal/mole in comparison to an 'experimental' value of $\Delta H \approx -600$ kcal/mole.

Let us now approach the spherical ions in a medium of dielectric constant ε which we will consider to be structureless and homogeneous[33]. The electric work is now much smaller and given by 7, and for the thermodynamic quantities we obtain equations 8 to 10:

$$A_{el} = Ne^2 v\lambda/a\varepsilon \qquad (7)$$

$$\Delta G = -T\Delta S_t - A_{el} \qquad (8)$$

$$\Delta S = \Delta S_t - A_{el}\{(1/\varepsilon)(\delta\varepsilon/\delta T)\} \qquad (7)$$

$$\Delta H = -A_{el}\{1 + (T/\varepsilon)(\delta\varepsilon/\delta T)\} \qquad (10)$$

It is important to recognize that for the solvent reaction, not only is ΔG very much less negative than in the corresponding vacuum process, but also ΔS is no longer given simply by the loss in translational entropy, but depends on the temperature coefficient of the dielectric constant. For water at 25°C ($\varepsilon = 78·5$) the temperature coefficient is negative and numerically quite large ($\delta\varepsilon/\delta T = -0·36$) so that the quantity within the brackets of 10 becomes negative ($= -0·36$) and ΔH positive. As a result of this simple consideration, therefore, we expect that the ion association in water indeed should be an endothermic process as is often found to be the case for A-character interactions (Tables 2 and 3).

With the equations 7 to 10, we have formulated what has been concluded already: through the neutralization of charges the solvent molecules surrounding the ions become more mobile causing an increase in ΔS. The large dielectric

314

constant ε of polar solvents (which are polarized in an electric field mainly by orientation of the molecules) as well as their temperature coefficients, are expressions of this mobility. The dielectric constant of water becomes smaller when the temperature is raised, with the consequence that the forces between the ions become larger, resulting in a displacement of the equilibrium towards the adduct, so that the association is expected to be an endothermic process.

However, the equations 7 to 10 fail to give quantitative results. Applied to process 6, now considered to take place in aqueous solution,

$$Al(aq)^{3+} + F(aq)^- \rightarrow AlF(aq)^{2-} \tag{11}$$

we find: $\Delta G \approx 0$ and $\Delta H = 2\cdot6$ kcal/mole, which is to be compared with the experimental data of *Table 2*: $\Delta G = -8\cdot4$ and $\Delta H = +1\cdot1$ kcal/mole. Although we are somewhat in doubt about the value to be introduced in 8 for ΔS_t (reasonable values ranging from 8 to 40 e.u.), there is no doubt that 7 and 8 furnish a value for ΔG which is much too positive, corresponding to an association constant far too small. The electrostatic work A_{el} therefore must be very much larger than that given by 7 and the discrepancies are so large that we cannot obtain a considerable improvement by changing the distance 'a' between the ions in the adduct. It is quite evident that the failure is due to the use of too large a value for the dielectric constant. Because of the electrostriction and dielectric saturation effects an effective constant ε_e should be used in equation 7 and must be considerably smaller than the dielectric constant of the bulk of the solvent ($\varepsilon_e < \varepsilon$).

Furthermore, this effective dielectric constant must be less dependent on temperature than the macro constant: $-(\delta\varepsilon_e/\delta T) < -(\delta\varepsilon/\delta T)$. It is not only reasonable to assume this to be so because of the expected reduced mobility of the electrostricted water molecules; it follows also from equation 10 which would furnish a much too positive ΔH if the replacement of ε by the smaller ε_e were not supplemented by substitution of $\delta\varepsilon/\delta T$ by a less negative $\delta\varepsilon_e/\delta T$. For reaction 11 for instance the experimental values for ΔG and ΔH are obtained from 7 to 10 with $\varepsilon_e \approx 30$ and $\delta\varepsilon_e/\delta T = -0\cdot11$, using again 30 e.u. for ΔS_t.

The model suggested therefore is to treat the solvent as a *structureless*, but *non-homogeneous* medium in which the dielectric constant to be used and its temperature coefficient vary with the distance of the interacting ions. In the following sections we shall see whether we can obtain some information on this local dielectric constant.

5. THE INFLUENCE OF CHARGE ON BASICITY

The role of solvent electrostriction and the dielectric saturation effects in the enormous electric fields occurring in the vicinity of ions are difficult to assess. According to calculations on models[34, 35, 36], the local dielectric constant ε_e' is believed to rise abruptly from low values in the immediate neighbourhood of the ion and to reach the usual macroscopic constant within a distance of a few Ångstrøm units. The quantity ε_e to be introduced in

315

200

equation 7 is a sort of mean value of the local constant ε'_e, as becomes evident from 12

$$\int_{\infty}^{a} (e^2 v\lambda/a^2 \varepsilon'_e)\, da = e^2 v\lambda/a\varepsilon_e \qquad (12)$$

Because it is impossible to obtain any reliable values for ε_e from theory, it is interesting to calculate the effective dielectric constant and its temperature coefficient from the thermodynamic quantities of reactions, which are controlled entirely by Coulomb forces. Imagine two basic molecules B and B★ with identical basic groups X, which differ from one another solely by a charge situated somewhere on the molecule of B★ at a distance 'a' from X[37]:

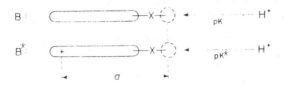

On protonation of X, the free energy changes of the reactions with B and B★ differ from each other only because of the energy needed to bring up the hydrogen ion in the additional field of the charge situated on B★. This electrostatic energy therefore can be obtained from $\Delta pK = (pK - pK★)$ and we can calculate an experimental value of ε_e when the distance 'a' is known. Correspondingly the temperature coefficient of ε_e is accessible from the enthalpy change of the proton transfer from the base B and B★. Three different types of such pairs have been investigated.

Type I: The uncharged B is a symmetrically built diamine molecule and B★ its first protonation product. In the case of an unbranched, primary polymethylenediamine, the formulae are:

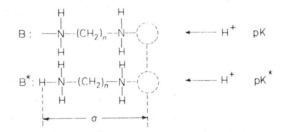

It will be readily recognized that the second hydrogen ion during the protonation of B★ has to be brought up in the field of the charge of the first proton attached during the protonation of B and the length 'a' therefore is the intramolecular distance between the two acidic protons in the diammonium ion H_2B^{2+} (or $HB^{★2+}$), which almost certainly will be a fully stretched zigzag chain because of the repulsion of the two charges, so that 'a' is easily obtained from models.

Type II: The uncharged B here is a monoamine and the formula of B★ is

316

obtained by replacing a carbon atom of B by nitrogen with its higher nuclear charge. The base B for instance may be a primary non-branched aliphatic amine and B★ the monoprotonated form of the polymethylenediamine with one carbon atom less than the molecule of B:

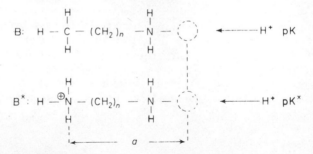

Obviously, the length 'a' is now somewhat smaller than the intramolecular distance of the two acidic protons in the diammonium ion $[HB★]^{2+}$, namely the distance between one of the acidic protons of the first ammonium group and the nitrogen nucleus of the second ammonium group.

Type III: The pair is analogous to the type I pair, but the basic atoms are sulphur of thioether groups, which are very selective ligands, coordinating only to metal cations of extreme B-character. Instead of a proton, the equally singly charged CH_3Hg^+ is used as the electrophilic cation[38].

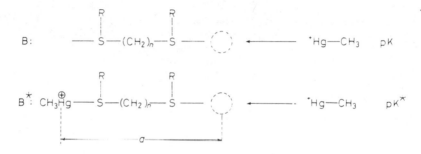

Corresponding to type I, the length 'a' is the intramolecular distance between the two metal atoms in the cation of dimercurated dithioether. As substituent R, the hydroxyethyl group $HO—CH_2—CH_2—$ has been chosen which makes the dithioethers water soluble. The pKs are analogously defined as with the proton acceptors type I and II and are the logarithms of the stability constants of the methylmercury complexes according to 13:

$$pK = \log\frac{[CH_3HgB]}{[CH_3Hg][B]} \qquad pK★ = \log\frac{[(CH_3Hg)_2B]}{[CH_3Hg][CH_3HgB]} \qquad (13)$$

The free energy of the transfer of H^+ (or CH_3Hg^+) from the base B★ to the base B:

$$[HB★]^{2+} + B \rightarrow HB^+ + [B★]^+$$
$$(\text{or } CH_3HgB★^{2+} + B \rightarrow CH_3HgB^+ + [B★]^+ \qquad (14)$$

317

202

is calculated from $\Delta pK = (pK - pK^\star)$ and is given by 15:

$$\Delta G = - 2\cdot305RT\,\Delta pK = - A_{el} - TS_{st} \tag{15}$$

A_{el} now is the electrostatic energy gained by removing to infinity the charge of the proton (or the methylmercury cation) from the equally positive charge 'e' situated at the distance 'a' on the molecule of base $B^\star$

$$A_{el} = Ne^2/a\varepsilon_e \tag{16}$$

Equation 15 is valid if the basic groups X on B and $B^\star$ are exactly alike. The additional charge on $B^\star$ should not change the group X electronically (it should not change its internal basicity). Furthermore, in the case of the pairs types I and III, the electron densities around N and S should not change appreciably on protonation or mercuration respectively. The latter condition would be poorly fulfilled if carboxylate groups[39] were used as basic groups

X because of the transition from the symmetrically built $-C\begin{smallmatrix} O \\ O \end{smallmatrix} \ominus$ to the

asymmetric $-C\begin{smallmatrix} O \\ OH \end{smallmatrix}$ on protonation. We therefore did not calculate the

electrostatic energy A_{el} from the pK-difference of dicarboxylic acids.

There is no change in the number of solute particles during process 14 and no 'cratic' or translational entropy terms have to be taken into consideration. The last term in equation 15 merely takes care of the circumstance that H^+ or CH_3Hg^+ can add to either of the two basic sites of B (pairs I and III) and that either of the two acidic protons of $HB^\star$ (pairs I and II) or either of the two methylmercury cations of $CH_3HgB^\star$ (pair III) may be given off. Therefore: $S_{st} = R \ln 4$ in the case of pairs types I and III and $S_{st} = R \ln 2$ in the case of pair type II. For the entropy and enthalpy change of process 14 we obtain:

$$\Delta S = - A_{el}[(1/\varepsilon_e)\,(\delta\varepsilon_e/\delta T)] + S_{st} \tag{17}$$

$$\Delta H = - A_{el}[1 + (T/\varepsilon_e)\,(\delta\varepsilon_e/\delta T)] \tag{18}$$

Table 6 contains the results of equilibrium and calorimetric measurements which were all carried out in the same solvent of ionic strength $1\cdot0$. The pKs of the unbranched primary mono- and di-amines have been obtained with good precision. The stability constants of the methylmercury complexes of the three S,S'-bis (hydroxyethyl)-dithioethers which were investigated, however, could not be obtained so accurately[38]. An indirect pH-method was applied[21] by studying the displacement of OH^- from CH_3HgOH by thioethers, the methylmercuryhydroxide being in equilibrium with CH_3Hg^+ and $(CH_2Hg)_2OH^+$. Also the results of the calorimetric measurements are not as accurate as desirable, especially the ΔH-values of the CH_3Hg^+-additions.

A combination of the data of Table 6 furnishes ΔG and ΔH of the transfer reaction 14 given in Table 7, which contains furthermore the quantities obtained with the equations 15 to 18.

318

Table 6. Logarithms of formation constants and enthalpy changes (kcal/mole) at $\mu = 1\cdot0$ (KNO$_3$) and 20°C

n	Protonation of H_2N-(CH$_2$)$_n$-NH$_2$				Protonation of CH_3-(CH$_2$)$_n$-NH$_2$		Methylmercuration of R–S–(CH$_2$)$_n$–S–R			
	pK$_1$	pK$_2$	$-\Delta H_1$	$-\Delta H_2$	pK	$-\Delta H$	pK$_1$	pK$_2$	$-\Delta H_1$	$-\Delta H_2$
2	10·226 ± 0·01	7·485 ± 0·01	12·4 ± 0·3	10·8 ± 0·3	10·960 ± 0·01	10·8 ± 0·3	4·26 ± 0·005	2·83 ± 0·1	6·85 ± 0·5	4·5 ± 0·5
3	10·832 ± 0·01	9·125 ± 0·01	13·15 ± 0·3	12·85 ± 0·3	11·005 ± 0·01	13·05 ± 0·3	3·81 ± 0·005	2·70 ± 0·015	6·78 ± 0·5	6·0 ± 0·5
4	11·122 ± 0·01	9·858 ± 0·01	13·1 ± 0·3	13·05 ± 0·3	11·041 ± 0·01	13·35 ± 0·3	—	—	—	—
5	11·258 ± 0·01	10·173 ± 0·01	13·2 ± 0·3	13·2 ± 0·3	11·046 ± 0·01	13·3 ± 0·3	4·09 ± 0·05	3·28 ± 0·1	6·82 ± 0·5	7·8 ± 0·5

Table 7. Thermodynamic quantities for reaction 13(kcal/mole) $\mu = 1\cdot0$, (KNO$_3$), 20°C. ε_e is to be compared with the bulk dielectric constant: $\varepsilon = 80\cdot36$ and $\delta\varepsilon_e/\delta T$ with $\delta\varepsilon/\delta T = -0\cdot368$; $\delta \ln \varepsilon/\delta \ln T = -1\cdot34$

n	ΔG	ΔH	$T S_{ss}$	A_{el}	a(Å)	ε_e	$-\delta\varepsilon_e/\delta T$	$-\dfrac{\delta \ln \varepsilon_e}{\delta \ln T}$
Pairs of type I								
2	−3·69 ± 0·03	−1·6 ± 0·6	0·81	2·88 ± 0·03	5·35	22	0·034	0·45
3	−2·30 ± 0·03	+0·2 ± 0·6	0·81	1·49 ± 0·03	6·60	34	0·13	1·13
4	−1·71 ± 0·03	+0·3 ± 0·6	0·81	0·90 ± 0·03	7·90	47	0·21	1·33
5	−1·46 ± 0·03	+0·1 ± 0·6	0·81	0·65 ± 0·03	9·20	55	0·22	1·16
Pairs of type II								
2	−4·68 ± 0·03	−1·6 ± 0·6	0·40	4·28 ± 0·03	4·50	17	0·036	0·63
3	−2·54 ± 0·03	−0·1 ± 0·6	0·40	2·14 ± 0·03	5·75	27	0·09	0·96
4	−1·59 ± 0·03	+0·2 ± 0·6	0·40	1·19 ± 0·03	7·00	40	0·16	1·17
5	−1·18 ± 0·03	+0·1 ± 0·6	0·40	0·78 ± 0·03	8·30	51	0·19	1·13
Pairs of type III								
2	−1·93 ± 0·2	−2·3 ± 1·0	0·81	1·12 ± 0·2	8·65	~35		
3	−1·50 ± 0·3	−0·8 ± 1·0	0·81	0·69 ± 0·3	9·95	~50		
5	−1·09 ± 0·2	+1·0 ± 1·0	0·81	0·28 ± 0·2	12·35	~90		

319

In *Figure 1*, the electrostatic work A_{el} has been plotted as a function of the distance 'a'. It is very satisfactory that the data from base pairs of types I and II apparently form a single smooth curve as it should be, if the assumptions made are correct. However, the data obtained with the methylmercury complexes of the dithioethers (pairs type III) do not fit in as nicely. The small deviations are almost certainly caused by chelation. The methylmercury cation coordinates only one single ligand strongly. However, the formation of weak 1:2-complexes is noticeable in many cases[31]. Therefore it is to be suspected that the dithioethers form chelates with CH_3Hg^+ which makes pK_1 larger than it would be otherwise and pK_2 smaller. The deviation from the expected values should be largest if the ring formed is five-membered ($n = 2$) and should become smaller with increasing ring size, and the experimental results (*Figure 1*) are in harmony with this expectation.

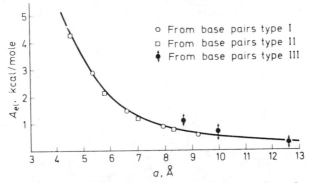

Figure 1. Electric work to be gained if a positive electronic charge situated on one end of an unbranched chain of CH_2-groups is removed from another such charge situated on the other end of the chain at a distance of aÅ.

In *Figure 2* the effective dielectric constant ε_e obtained from A_{el} by means of equation 16 has been plotted as a function of the distance 'a'. It is interesting to compare the experimental values with the local dielectric constant ε_e' obtained theoretically from models[35, 36]. According to equation 12 ε_e' should be smaller than ε_e which is in conspicuous contrast to our results. However, this deviation is not unexpected. In order to investigate the dielectric shielding experimentally, a carrier for the charges influencing one another was required. A_{el} is the electrostatic work needed to bring the charge $+e$ from infinity up to a distance 'a' from another charge $+e$, but the two charges are sitting at the ends of a zigzag chain of n CH_2-groups after this has been done. We can imagine that a part of the solvent between the two charges is replaced by a hydrocarbon medium of much lower polarizability. That this actually is one of the causes of the small values found for ε_e will be made clear in the next section by demonstrating that the effective dielectric constant ε_e decreases with increasing bulkiness of the organic part of the molecule used to carry the charges whose electrostatic influence is investigated.

Another comparison is also of interest. Born's equation 19 for the free energy of hydration of gaseous ions of charge v and radius r produces much

320

205

too negative values[40] if crystal radii and the macro dielectric constant ε of the solvent are used.

$$\Delta G_{\mathrm{hyd}} = (e^2 v/2r)(1 - 1/\varepsilon)N \tag{19}$$

In order to get better agreement with experimental values, it has become customary to use effective radii r_e in 19 which are considerably larger than the crystal radii[41]. Noyes[42], on the other hand, proposed to correct 19 by introducing an effective dielectric constant which is much smaller than the macro constant. These data, ε_e^N, are shown on *Figure 2* at the lower left hand corner, plotted as a function of the crystal radius of the ion. Most people consider that the Noyes method produces too low values for the effective dielectric constant.

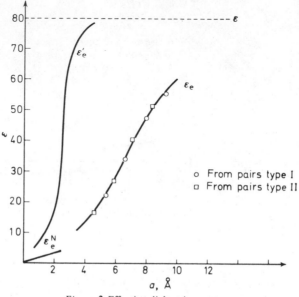

Figure 2. Effective dielectric constant ε_e.

Because of the organic molecule needed to carry the charges the interaction of which is to be studied by means of the proton transfer 14, the results hardly allow us to draw any definite conclusions concerning the effective dielectric constant to be used in association processes of ions of opposite charges (see section 7 of this paper 327). However, the thermodynamic functions of reaction 14 not only furnish values for ε_e (equations 15 and 16), but also for the temperature coefficient of the effective dielectric constant (equation 18). Up to the present, no information whatsoever seems to be available on that quantity.

Again, the proton transfer 14 will not furnish numbers for $\delta\varepsilon_e/\delta T$ which can be used at once for an evaluation of ΔH of ion associations (equation 10). But the relation of the quantities ε_e and $\delta\varepsilon_e/\delta T$ is most interesting and in this respect the experience gained by means of process 14 is remarkable. Both

321

quantities will depend on the strength of the electric field in the neighbourhood of the charges and it is to be expected that a diminution of ε_e will be accompanied by a lowering of $-\delta\varepsilon_e/\delta T$. That this actually is so can be seen from *Figure 3* in spite of the relatively large uncertainties (lengths of the vertical lines) of the experimental values. Furthermore, there is no doubt that the quantity $-\delta\varepsilon_e/\delta T$ decreases relatively more than ε_e, with the consequence that $\delta \ln \varepsilon_e/\delta \ln T$ becomes less negative the stronger the electrostriction (*Table 7*). Therefore, an ion association will be much less endothermic than predicted by equation 10, using the bulk dielectric constant and its temperature coefficient.

This phenomenon actually is observed (see page 315) and we now have an explanation for it.

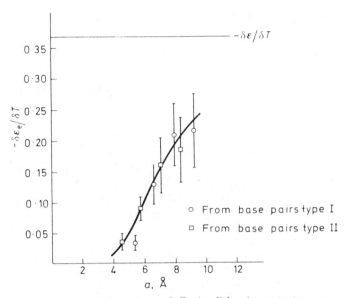

Figure 3. Temperature dependence of effective dielectric constant ε_e.

6. DIELECTRIC SHIELDING IN STRONG ELECTRIC FIELDS

Figures 2 and *3* demonstrate that the effective dielectric constant and its temperature coefficient diminish to an extraordinary extent in an increasing electrostatic field. Unfortunately no data are available for distances below 4 Å of the two interacting positive ammonium groups, because of the impossibility of carrying out measurements with $H_2N–CH_2–NH_2$, which is unstable to hydrolysis. However, it is possible to increase the strength of the electric field to which one of the charges exposes the other charge by making the organic part of the charge carrier more bulky. In the mean the medium between the two charges will become less polarizable by doing so.

322

In the diprotonated diethylenediamine I (piperazine) and triethylenedia-mine II the charges not only are somewhat closer than in H_2en^{2+}, they are also dielectrically shielded from one another to a smaller extent.

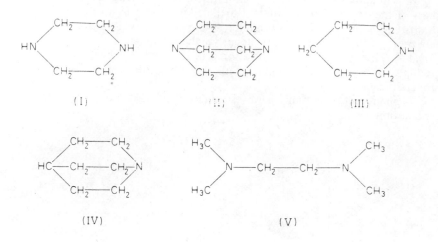

By comparing the first step of protonation of I[43] or II[44] with the second step, we obtain data for the interaction of two $+e$-charges at the intramolecular distance of the two acidic protons (base pair type I). Two base pairs of type II are obtained by comparing III[45] or IV[38] with the respective first protonation products of I or II; the two interacting charges being now at a distance proton–nitrogen. The results listed in *Table 8* are most interesting. The increase in bulkiness of the hydrocarbon part between the two charges causes a strong increase of the electrostatic energy A_{el} which is caused by a decrease of the effective dielectric constant ε_e. Again $-\delta\varepsilon_e/\delta T$ decreases relatively more than ε_e, so that $\delta \ln \varepsilon_e/\delta \ln T$ becomes less negative and process 14 more exothermic.

In reaction 14 two charges of equal sign are separated from one another, which is to be compared with the approach of two opposite charges (ion association), which also causes an intensity decrease of the electric field to which the solvent is exposed. These processes are endothermic, if the electro-striction is small in magnitude (*Table 7*, pair type I, $n = 3, 4, 5$ and pair type II, $n = 4, 5$), but become exothermic with increasing electrostriction when the quantity $-(\delta \ln \varepsilon_e/\delta \ln T)$ drops below 1 (equation 18). The electrostriction increases (ε_e and $-\delta\varepsilon_e/\delta T$ decrease) when the number of carbon atoms n in $H_3N–(CH_2)_n–NH_3$ becomes smaller (*Table 7*) and when the bulkiness of the organic part of the charge carrier gets larger (*Table 8*). If the protonated form of II is compared with the protonated form of V (*Table 8*, lines 3 and 5), it is recognized that the increase of the organic part between the charges is much more effective in lowering ε_e and $-\delta\varepsilon_e/\delta T$ than an increase outside the charges, as would be expected.

The electrostriction may be increased also by placing more than two charges on the charge carrier. Their influence on ε_e and $\delta\varepsilon_e/\delta T$ can be estimated

323

208

Table 8. Free energy and enthalpy changes in the proton transfer from base B★ to base B (kcal/mole), $\mu = 0.1$, 25°C. ε_e, $\delta\varepsilon_e/\delta T$ and $\delta \ln \varepsilon_e/\delta \ln T$ are to be compared with these quantities of the bulk solvent $\varepsilon = 78.54$; $\delta\varepsilon/\delta T = -0.361$; $\delta \ln \varepsilon/\delta \ln T = -1.37$

B	B★	Ref.	Pair type	$-\Delta G$	$-\Delta H$	TS_{st}	A_{el}	a (Å)	ε_e	$-\dfrac{\delta\varepsilon_e}{\delta T}$	$\dfrac{\delta \ln \varepsilon_e}{\delta \ln T}$
$HN=(CH_2-CH_2)_2=N$	$H_2^\oplus N=(CH_2-CH_2)_2=NH$	43	I	5.62	3.05	0.82	4.80	4.75	14	0.017	0.36
$H_2C=(CH_2-CH_2)_2=N$	$H_2^\oplus N=(CH_2-CH_2)_2=NH$	43, 45	II	7.54	5.67	0.41	7.13	3.8	12	0.008	0.21
$N\equiv(CH_2-CH_2)_3\equiv N$	$NH^\oplus\equiv(CH_2-CH_2)_3\equiv N$	44	I	7.94	4.29	0.82	7.15	4.5	10	0.014	0.40
$HC\equiv(CH_2-CH_2)_3\equiv N$	$HN^\oplus\equiv(CH_2-CH_2)_3\equiv N$	44, 38	II	10.69	8.29	0.41	10.28	3.5	9	0.006	0.19
$(CH_3)_2N-CH_2-CH_2-N(CH_3)_2$	$(CH_3)_2^\oplus NH-CH_2-CH_2-N(CH_3)_2$	50	I	4.30	0.76	0.82	3.48	5.35	18	0.047	0.78
$(H_2N-CH_2-CH_2-)_2NH$	$(H^\oplus N-CH_2-CH_2)_2NH$	47	I	7.55	4.00	0.66	6.89	5.35	18	0.025	0.42
$(H_2N-CH_2-CH_2-CH_2-)_2NH$	$(H_3^\oplus N-CH_2-CH_2-)_2NH$	47	I	4.00	1.82	0	4.00	6.6	25	0.046	0.55
$(H_2N-CH_2-CH_2-CH_2-)_3N$	$(H_3^\oplus N-CH_2-CH_2-CH_2-)_3N$	48	I	6.70	5.02	0	6.70	6.6	22	0.019	0.25
$[Ac_2=N-C_2H_4-N=Ac_2]^{4-}$	$[Ac_2=NH-C_2H_4-N=Ac_2]^{3-}$	49	I	5.55	1.33	0.81	4.74	5.35	13	0.031	0.72
$[Ac_2=N-C_6H_{10}-N=Ac_2]^{4-}$	$[Ac_2=NH-C_6H_{10}-N=Ac_2]^{3-}$	49	I	8.34	4.59	0.81	7.52	5.0	9	0.012	0.39
PO_4^{3-}	HPO_4^{2-}	51	I	6.47	2.70	0.58	5.89	3.8	15	0.027	0.54
HPO_4^{2-}	$H_2PO_4^-$	51	I	6.93	2.68	0.48	6.45	3.8	14	0.026	0.56
PO_4^{3-}	SO_4^{2-}	51	II	13.6	8.9	0	13.6	2.2	11	0.013	0.35

Upper: $\mu = 0.1$, 20°C: lower: $\mu = 0.0$, 25°C.

324

by comparing the basicity of the members of pairs VI, VII and VIII, wherein the central nitrogen of B is to be protonated as well as that of B*:

VI $\Big\{$

B:
$$H_2N-CH_2-CH_2$$
$$\searrow$$
$$NH \leftarrow H^+ \ pK$$
$$\nearrow$$
$$H_2N-CH_2-CH_2$$

$$\varepsilon_e = 18, \ -\delta\varepsilon_e/\delta T = 0\cdot025$$

B*:
$$H_3^\oplus N-CH_2-CH_2$$
$$\searrow$$
$$NH \leftarrow H^+ \ pK\star$$
$$\nearrow$$
$$H_3^\oplus N-CH_2-CH_2$$

VII $\Big\{$

B:
$$H_2N-CH_2-CH_2-CH_2$$
$$\searrow$$
$$NH \leftarrow H^+ \ pK$$
$$\nearrow$$
$$H_2N-CH_2-CH_2-CH_2$$

$$\varepsilon_e = 25, \ -\delta\varepsilon_e/\delta T = 0\cdot046$$

B*:
$$H_3^\oplus N-CH_2-CH_2-CH_2$$
$$\searrow$$
$$NH \leftarrow H^+ \ pK\star$$
$$\nearrow$$
$$H_3^\oplus N-CH_2-CH_2-CH_2$$

VIII $\Big\{$

B:
$$H_2N-CH_2-CH_2-CH_2$$
$$\searrow$$
$$H_2N-CH_2-CH_2-CH_2-N \leftarrow H^+ \ pK$$
$$\nearrow$$
$$H_2N-CH_2-CH_2-CH_2$$

$$\varepsilon_e = 22, \ -\delta\varepsilon_e/\delta T = 0\cdot019$$

B*:
$$H_3^\oplus N-CH_2-CH_2-CH_2$$
$$\searrow$$
$$H_3^\oplus N-CH_2-CH_2-CH_2-N \leftarrow H^+ \ pK\star$$
$$\nearrow$$
$$H_3^\oplus N-CH_2-CH_2-CH_2$$

The central nitrogen atoms of VII and of VIII are the most basic of the non-protonated triamine and tetramine because they are secondary and tertiary amino groups respectively[46]. The proton again is bonded by the central N in the third protonation step of VII and the fourth protonation step of VIII. The central nitrogen becomes available again in a tautomeric change, induced by the repulsion of charges, which takes place when the second proton is added:

VII: $(H_2N-CH_2-CH_2-CH_2-)_2 \overset{\oplus}{N}H_2 \xrightarrow{\ H^+\ }$

$$(H_3^\oplus N-CH_2-CH_2-CH_2-)_2 \, NH$$

VIII: $(H_2N-CH_2-CH_2-CH_2-)_3 \overset{\oplus}{N}H \xrightarrow{\ 2H^+\ }$

$$(H_3^\oplus N-CH_2-CH_2-CH_2-)_3 \, N$$

325

In order to obtain the thermodynamic functions for the proton transfer 14 from B★ to B, we simply have to subtract ΔG (ΔH) of the third protonation step (pair VII)[47] or the fourth protonation step (pair VIII)[48] from ΔG (ΔH) of the first protonation step. Again we obtain A_{el} from 15, S_{st} being zero, and ε_e from 16 after introducing a factor of two (two e-charges on B★ in pair VII) or three (three charges on B★ in pair VIII). Equation 18 is applicable without any change. From the results (*Table 8*) we learn that three charges at the ends of chains of three CH_2-groups (pair VIII) have approximately the same influence on ε_e and $\delta\varepsilon_e/\delta T$ as a single charge at the end of the shorter ethylene chain (*Table 7*, line 1).

For bis-(β-aminoethyl)-amine VI[47], it has been assumed that the three nitrogens all have the same 'internal basicity', which leads to a statistical entropy term in 15 of $S_{st} = R \ln 3$. The result (line 6, *Table 8*) is as expected: two ammonium charges separated by an ethylene bridge reduce ε_e and $-\delta\varepsilon_e/\delta T$ considerably more than a single charge at the same distance (line 1, *Table 7*).

The anions of EDTA (IX) and DCTA (X) are also symmetrical amine bases which form pairs B and B★ of type I.

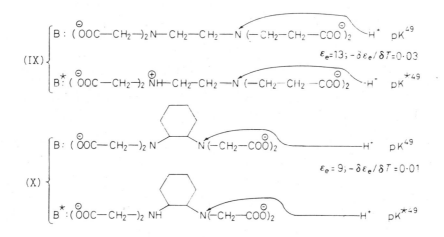

The positive charges of the protons whose electrostatic interaction gives rise to the basicity difference (pK − pK★) approach one another here in the additional field of the four negative carboxylate groups. This must cause a more intense electrostriction of the solvent and a comparison of IX (line 9, *Table 8*) with V[50], which is also a N,N'-tetrasubstituted ethylene diamine (line 5, *Table 8*), reveals that ε_e as well as $-\delta\varepsilon_e/\delta T$ have indeed become considerably smaller. Also the results obtained with pair X are as expected: the two nitrogens are somewhat closer than in IX, they are separated by a more bulky organic carrier and the four $COO^\ominus$- groups are restricted to a smaller area, with the effect that ε_e drops to a very low value and also the temperature coefficient of ε_e is extraordinarily small.

The reduced basicity of HPO_4^{2-} with respect to PO_4^{3-} and of $H_2PO_4^-$ with respect to HPO_4^{2-} may be considered as being caused by electrostatic forces only: the second proton has to be approached in the additional field of the

326

211

first and the third proton in the additional field of the second. This constitutes the two second last pairs of type I in *Table 8* which have to be evaluated with $S_{st} = R \ln 8/3$ and $R \ln 9/4$ respectively in equation 15. The assumption that the 'internal basicity' of the oxygens remaining unprotonated does not change from PO_4^{3-} to HPO_4^{2-} and $H_2PO_4^-$ is of course difficult to justify. However, the results obtained for ε_e and its temperature coefficient are very reasonable indeed.

The bases PO_4^{3-} and SO_4^{2-} differ from each other by the nuclear charge of the central atom and constitute a pair of type II (bottom line, *Table 8*). On protonation of SO_4^{2-}, the hydrogen ion has to be brought up within a distance of only 2·2 Å of this additional charge and it is reasonable again to find very low values of ε_e as well as of its temperature coefficient.

7. THE THERMODYNAMIC FUNCTIONS OF COMPLEX FORMATION

Before finishing we will consider a general association process taking place in aqueous solution with reactants which may be ions or uncharged dipoles or multipoles and atomic or molecular species of any kind. Charges of opposite or like signs will approach one another during such a reaction and an electrostatic energy A_{el} is to be gained, which has to be formulated as a sum according to 20:

$$A_{el} = N \sum_i \frac{e^2 z_+ z_-}{a_i \varepsilon_{ei}} = (1/\varepsilon_e) \, \text{constant} \tag{20}$$

z_+ and z_- are the charges on the individual atoms of the associating ions and molecules which are brought up to distances a_i from one another. The solvent again will be treated as a structureless but non-homogeneous medium in using local dielectric constants. For each of the pairs of charges in the sum of 20 we need an individual effective dielectric constant ε_{ei} to account for their interaction and these are replaced by a mean value ε_e in the expression at the RHS of 20.

Equations 21 to 24 are obtained for the changes in free energy, entropy and enthalpy of the association process:

$$\Delta G = -T(\Delta S_t + \Delta S_c + \Delta S_r) - A_{el} - E_n \tag{21}$$

$$\Delta S = (\Delta S_t + \Delta S_c + \Delta S_r) - A_{el}(1/\varepsilon_e)(\delta \varepsilon_e / \delta T) \tag{22}$$

$$\Delta H = -A_{el}(1 + \delta \ln \varepsilon_e / \delta \ln T) - E_n \tag{23}$$

The entropy change is composed of several parts. The changes of translational entropy ΔS_t will be negative for an association reaction. On coordination flexible molecular reactants, especially chelating ligands, will lose their flexibility to a large extent which is accounted for by the conformational entropy change ΔS_c which again will be a negative quantity. The reactants furthermore will no longer be able to rotate independently in the adduct and therefore the change in rotational entropy ΔS_r is also negative. Changes in vibrational entropy are probably negligible and a fourth quantity ΔS_v, taking care of these, has been omitted from the first bracket on the right of 21 and 22.

327

The term E_n accounts for non-electrostatic interactions between the reactants, such as covalency and ligand field stabilization effects, and it is assumed that this quantity is independent of temperature.

Unfortunately, all the quantities in 21 to 23 are difficult to assess. We are mainly interested in the electrostatic contribution A_{el} and the non-electrostatic contribution E_n. However, neither ΔG nor ΔH is a simple function of these quantities.

Let us have a look at the entropy terms in 21. The translational entropy of a solute species, which will be solvated to an unknown extent, certainly cannot be obtained simply with the Sackur–Tetrode equation. It has become customary[52] to use instead of ΔS_t the cratic entropy[53] ($nR \ln 55 = n \times 8$ e.u., n being the difference of the number of product and reactant molecules in the chemical equation). However, the cratic entropy is a mixing and not a translational entropy.

From the magnitude of the chelate effect[54] I would like to conclude that ΔS_t for a 1:1 association must be considerably larger than eight entropy units. The chelate effect for a bidentate ligand Z has been defined as 24

$$\text{Chel} = \log K_{MZ} - \log \beta_{MA_2} \qquad (24)$$

where K_{MZ} is the stability of the chelate MZ and β_{MA} the product $K_1 K_2$ of the individual stability constants of MA and MA_2 with a unidentate ligand A; Z and A having like ligand atoms. An equivalent equation is

$$2 \cdot 3 \, RT \, \text{Chel} = \Delta G_{MA_2} - \Delta G_{MZ} \qquad (25)$$

which can be combined with 21. In doing so, we may neglect A_{el}, if the ligands Z and A are uncharged

$$(A_{el})_{MZ} = (A_{el})_{MA_2} \approx 0 \qquad (26)$$

Furthermore, the same bonds are formed between the metal atom and the ligand atoms of MZ and MA_2 respectively, which leads to

$$(E_n)_{MZ} = (E_n)_{MA_2} \qquad (27)$$

Equations 26 and 27 are equivalent to the statement that Chel is an entropy effect

$$2 \cdot 3 \, R \, \text{Chel} = \{(\Delta S_t)_{MZ} - (\Delta S_t)_{MA_2}\} + \{(\Delta S_c)_{MZ} - (\Delta S_c)_{MA_2}\}$$
$$+ \{(\Delta S_r)_{MZ} - (\Delta S_r)_{MA_2}\} \qquad (28)$$

Considering that $(\Delta S_t)_{MZ} = (S_t)_{MZ} - (S_t)_M - (S_t)_Z$ (the difference of the translational entropies of MZ, M, Z) and $(\Delta S_t)_{MA_2} = (S_t)_{MA_2} - (S_t)_M - 2(S_t)_A$, we recognize that the quantity within the first brace of 28 is positive and constitutes a mean value of translational entropy S_t of a solute particle.

The differences within the second and third braces on the other hand are negative quantities. The chelating agent Z will lose more flexibility on coordination than the unidentate ligand A, and the more the longer the carbon chain connecting the two ligand atoms of Z. The chelate effect therefore decreases with the size of the chelate ring formed. There are hundreds of examples which demonstrate that the largest stability increase by chelation is achieved with a five-membered chelate ring; a four-membered ring being

328

213

probably strained, so that 27 is no longer valid. The stability of the complexes with uncharged chelating polyamines furthermore shows unambiguously that the stability increase gained by chelation is essentially due to an increased entropy term of the Gibbs–Helmholtz equation.

Equation 28, therefore, has a reasonably good experimental basis for uncharged ligands, and can be used for an estimation of the translational entropy S_t, which is identical with the difference $(\Delta S_t)_{MZ} - (\Delta S_t)_{MA_2}$. According to 28, the chelate effect can never be larger than $S_t/2.3\,R$. With ethylenediamine for Z, in comparison with ammonia or an aliphatic amine for A, the chelate effect amounts to two or three units. However, Chel can become as large as four or five units with rigid chelating agents such as phenanthroline in comparison with pyridine. Rigidity of the chelating ligand reduces the negative value of the difference within the second brace in 28. From these experimental chelate effects we conclude that the translational entropy S_t of a solute particle in aqueous solution must be of the order of magnitude of 30 entropy units and is therefore about as large as the translational entropy of a gas molecule.

Also the complexes of negatively charged ligands can be stabilized enormously by chelation. However, the magnitude of the stability increase is now very much more difficult to interpret because of the importance of the electrostatic energy terms. Certainly, equation 26 is now no longer valid, which influences not only ΔG of the exchange of two A by Z but also ΔH of this process and indeed, it is found experimentally that Chel (equation 24) is now no longer only an entropy effect[55].

A statement that the stability increase achieved by chelation was to a large extent a cratic effect[56] and would almost disappear by using unitary quantities, has caused considerable confusion[57]. Using unitary quantities is equivalent to expressing the concentrations in mole fractions. Now, the chelate effect (equation 24) has the dimension of the logarithm of a concentration and its numerical value depends on the definition of the concentration unit. The value is very much smaller (by $\log 55 = 1.7$) when mole fractions are used instead of mole per litre, just as it is very much larger (by $\log 10^3$) when the concentrations are expressed in millimoles per litre. By no means does the chelate effect disappear through using unitary quantities; it is just measured by a very much larger unit and becomes numerically smaller. The translational entropy depends on the volume which the molecules have at their disposal (Sackur–Tetrode equation) and the numerical value of the difference ΔS_t in equation 28 depends therefore on the reference state.

Finally, what is to be said concerning the contributions of the electrostatic energy A_{el} and the non-electrostatic term E_n to ΔG on the one hand and to ΔH on the other? From 21 we learn that A_{el} and E_n are of equal importance for the free energy change, whereas for the enthalpy change 23, A_{el} has to be multiplied with a factor which may be positive or negative. This factor contains the effective dielectric constant as well as its temperature coefficient, both of which are very difficult to assess.

From experience (section 2) we know that associations are either slightly exothermic or slightly endothermic when the reaction is caused mainly by Coulombic forces (which was decided by the selectivity criterion, section 1). As a mean we can assume ΔH to be around zero when no covalency is

329

involved ($E_n = 0$), and therefore—according to 23: $\delta \ln \varepsilon_e/\delta \ln T \approx -1$. With reasonable entropy terms in 21 an approximate value for A_{el} is found, and with 20 we obtain the result that an effective dielectric constant of about 30 does account for many observed ΔGs and a value around -0.1 for $\delta\varepsilon_e/\delta T$ is needed to account for the ΔH-values of common A-character associations (Reaction 11 may serve as an example, page 315). In the adduct formed, the reactants of opposite charge are about 2 to 3 Å apart and glancing now at *Figures 2* and *3* we recognize a conspicuous discrepancy.

The method outlined in section 5 apparently furnishes values for ε_e and its negative temperature coefficient which are substantially too small to be useful in common ion association processes. To a large extent this is due to the organic molecular carrier needed to investigate the interaction of charges by means of the proton transfer 14. There is of course no organic molecule between the associating ions forming an adduct and it is understandable that ε_e as well as $-\delta\varepsilon_e/\delta T$ will be larger than the data presented in *Figures 2* and *3*.

But, in addition to the influence of the organic carrier, there must be still another reason for the discrepancy between the values for ε_e and $\delta\varepsilon_e/\delta T$ found with base pairs and those needed to account for the thermodynamic functions of A-character associations. The electrostriction exerted on the solvent (characterized by ε_e and $\delta\varepsilon_e/\delta T$) apparently depends not only on the magnitude of the charge of the ion influencing the solvent and its radius but also on the structure of the solvation shell as well. The various ions seem to be shielded dielectrically in a very individual manner. This is demonstrated clearly by a study of the influence of a charge situated on a chelating ligand on the stability of its metal complexes. The anions of the following two substituted iminodiacetic acids Z and Z★ have been compared in determining the stability of the adducts with some cations[58].

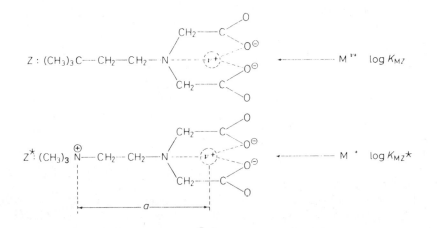

The ligands Z and Z★ are a pair of bases as described in section 5. However, they can bind not only the proton but also metal cations M^{v+} with completely identical donor groups. The difference ($\log K_{MZ} - \log K_{MZ\star}$) corresponds to the quantity ΔpK of base pairs type II and is proportional to the free energy

330

change of the transfer of M^{v+} from the complex with $Z^{\star-1}$ to the ligand Z^{2-}

$$MZ^{\star v-1} + Z^{2-} \rightarrow MZ^{v-2} + Z^{\star-1} \tag{29}$$

and in analogy with 15 we get

$$\Delta G_{29} = -2.305\, RT\, \Delta \log K = -A_{el} \tag{30}$$

There is no statistical term to be added on the RHS side; and no translational entropy, which is such a questionable quantity, brings in any uncertainty. For the electrostatic energy we write

$$A_{el} = Ne^2 v/a\varepsilon_e \tag{31}$$

The intramolecular distance 'a' between the site of the metal cation within the complex and the nitrogen nucleus of the quaternary ammonium group is obtained from models and so the quantity ε_e can be calculated. The results are collected in *Table 9*.

Table 9. Free energy and some enthalpy changes in the metal transfer process between ligands Z and Z* (reaction 29): ionic strength $\mu = 0.1$; 20°C; kcal/mole

M^{v+}	$\Delta \log K$	ΔG	ΔH	a (Å)	ε_e	$-\delta\varepsilon_e/\delta T$	$\dfrac{\delta \ln \varepsilon_e}{\delta \ln T}$
H$^+$	4.8	−6.4	−4.8	4.5	12	0.01	−0.255
Mg^{2+}	2.2	−3.0	+0.3	5.1	44	0.17	−1.11
Ca^{2+}	1.8	−2.4	n.d.	5.4	51	—	—
Mn^{2+}	2.7	−3.6	n.d.	5.2	35	—	—
Ni^{2+}	2.7	−3.6	n.d.	5.1	36	—	—
Cu^{2+}	3.8	−5.1	−2.2	5.1	25	0.05	−0.57
Zn^{2+}	2.6	−3.5	n.d.	5.1	37	—	—
Cd^{2+}	2.5	−3.4	n.d.	5.3	37	—	—
Pb^{2+}	2.8	−3.8	n.d.	5.5	32	—	—

n.d. = not determined

The ΔGs are not very accurate (± 0.2 kcal/mole) and there are uncertainties also in the distance 'a'. But even if it is admitted that the ε_e-values obtained are uncertain to about ten per cent, it is quite obvious that the various cations behave very individually with respect to the effective dielectric constant. Different cations apparently have a very different 'order producing capacity' with respect to the surrounding solvent, which is reflected also in their free energy of hydration. Most surprising is the result that considerably more energy is needed to bring up the singly charged hydrogen ion within a certain distance of the positive quaternary ammonium group than to approach any of the doubly charged metal cations.

The charge of A-character metal cations seems to be especially well shielded dielectrically. The charge of the hydrogen ion on the other hand is shielded to an extraordinarily small extent, perhaps because the proton fits exactly into the water structure thus causing formation of an iceberg around it, which may be of special rigidity. This makes it understandable that the experimental values for ε_e obtained from the proton transfer process 14 cannot *per se* be used for complex formation of metal ions.

331

The enthalpy change of 29 has only been determined[59] for the transfer of the proton as well as the transfer of the metal ions Mg^{2+} and Cu^{2+}. From ΔH again the temperature coefficient of the effective dielectric constant has been obtained (equation 18) and the results (*Table 9*) demonstrate anew that $\delta \varepsilon_e / \delta T$ deviates relatively more from the normal value $\delta \varepsilon / \delta T$ ($= -0.36$) the more ε_e deviates from the bulk dielectric constant ε, with the consequence that with increasing electrostriction the quantity ($-\delta \ln \varepsilon_e / \delta \ln T$) decreases and the reaction becomes more exothermic. This seems to be quite generally so. The transfer of Mg^{2+} in 29 is slightly endothermic, but the transfer of Cu^{2+} is somewhat and the transfer of H^+ is considerably exothermic. It is remarkable that this parallels generally the complex formation of these three cations with anionic ligands although the covalency contributions cancel out in process 29.

A very intimate knowledge of the structure of the solvent in the vicinity of the ions seems to be needed for a full understanding of the thermodynamic functions of complex formation in aqueous solution.

Table 10. A-character associations which are ΔH stabilized, kcal/mole

	Reaction		μ	ΔG	ΔH	$T \Delta S$	$t°C$	Ref.
Lu^{3+}	$+ \; digl^{2-}$	$\rightarrow Lu(digl)^+$	1	$- 7.71$	$+1.23$	$+ 8.97$		
	$+ \; digl^{2-}$	$\rightarrow Lu(digl)_2^-$	1	$- 6.70$	-0.45	$+ 6.23$		
	$+ \; digl^{2-}$	$\rightarrow Lu(digl)_3^{3-}$	1	$- 3.48$	-4.60	$- 1.05$	25	60
Lu^{3+}	$+ \; dipic^{2-}$	$\rightarrow Lu(dipic)^+$	0.5	-12.27	-2.21	$+10.05$		
	$+ \; dipic^{2-}$	$\rightarrow Lu(dipic)_2^-$	0.5	-10.54	-3.82	$+ 6.71$		
	$+ \; dipic^{2-}$	$\rightarrow Lu(dipic)_3^{3-}$	0.5	$- 6.28$	-6.65	$- 0.37$		
GdY^-	$+ \; ida^{2-}$	$\rightarrow GdY(ida)^{3-}$	0.1	$- 5.77$	-6.61	$- 0.84$		
TbY^-	$+ \; ida^{2-}$	$\rightarrow TbY(ida)^{3-}$	0.1	$- 5.34$	-8.10	$- 2.76$		
GdY^-	$+ \; X^{3-}$	$\rightarrow GdY(X)^{4-}$	0.1	$- 6.52$	-7.92	$- 1.40$	20	62
TbY^-	$+ \; X^{3-}$	$\rightarrow TbY(X)^{4-}$	0.1	$- 6.24$	-8.49	$- 2.25$		

One more word should be said concerning the rule stating that electrostatic adducts are mainly entropy stabilized, whereas covalent complexes are enthalpy stabilized. Equation 23 raises some doubt concerning the general validity of this rule. The factor with which A_{el} has to be multiplied in order to obtain ΔH is bound to become positive in strong electric fields because of the stronger relative decrease of $-\delta \varepsilon_e / \delta T$ in comparison to ε_e. We expect therefore exothermic associations with highly charged reactants. This actually proves to be so. Examples are the complexes of many A-character metal ions with the anion of EDTA. Even the calcium–EDTA-complex is produced in a reaction which is exothermic by 6.5 kcal/mole and it certainly

332

would be wrong to assume that this is due to the formation of covalent bonds.

Further examples are given in *Table 10*. The complexes of the rare earth cations—which are d^0 and of pronounced A-character—with the doubly negative anions of diglycolic acid and dipicolinic acid[60], are very informative. The 1:1-adduct is almost entirely ΔS-stabilized; the addition of the second ligand, however, is already appreciably exothermic and the formation of ML_3^{3-} from ML_2^- is caused by a large negative ΔH, while ΔS is unfavourable. Certainly, the rare earth cation is not acquiring more and more B-character from step to step in the complexes ML^+, ML_2^- and ML_3^{3-}. The 1:2-complex ML_2^- does not add any chloride, cyanide or ammonia and does not react with sulphur donors. The selectivity criterion and the ΔS–ΔH-criterion, therefore, do not give the same answer with respect to the third step.

The explanation is obvious. In the third step the two reactants ML_2^- and L^{3-} are both anions and repel one another at longer distances. At these distances A_{el} is negative and $-(\delta \ln \varepsilon_e/\delta \ln T)$ will probably be greater than one (little electrostriction), so that the approach of the two anions is exothermic as long as they are repelling one another. At shorter distances the repelling forces give way to attraction and A_{el} becomes positive, but $-(\delta \ln \varepsilon_e/\delta \ln T)$ will now become smaller than one, so that ΔH again is negative. The large electrostriction in the range of attraction between the two anions is caused, of course, by an almost complete orientation of the water molecules between the many close ionic charges to a rigid solvent structure, making ε_e small and its temperature coefficient almost zero. The electrostatic association of two anions resembles to some extent the formation of an adduct in the gas phase, as in both cases the electrostatic attraction forces depend little on temperature. For the lanthanide complexes (for which E_n is negligible) it is generally observed[61] that the more exothermic the complex formation, the less positive is the entropy change, which is explained by equations 21 and 22.

The lower part of *Table 10* contains further examples of associations between an anionic complex of a 'hard' metal and a 'hard' anionic ligand[62]. Again it is found that these reactions are strongly exothermic and proceed only because of a dominant negative ΔH, whereas the entropy change is unfavourable. All of the reactions of *Table 10* are A-character associations which do not follow the general rule, the adducts formed being enthalpy and not entropy stabilized.

REFERENCES

[1] L. G. Sillén and A. E. Martell (Eds), *Stability Constants of Metal Ion Complexes*, 2nd ed. The Chemical Society: London (1964).
[2] S. Ahrland, J. Chatt and N. R. Davies, *Quart. Revs. (London)*, **12**, 265 (1958).
[3] G. Schwarzenbach, *Experientia Suppl.* **5**, 162 (1956).
[4] R. G. Pearson, *J. Am. Chem. Soc.* **85**, 3533 (1963).
[5] C. K. Jörgensen, *Rev. Chim. min.* **6**, 183 (1969).
[6] G. Schwarzenbach, O. Gübeli and H. Züst, *Chimia*, **12**, 84 (1958).
[7] G. Schwarzenbach and K. Tunaboylu, unpublished.
[8] M. Meier, *Dissertation Nr. 3988.* ETH: Zürich (1967).
[9] S. Ahrland, *Chem. Phys. Letters*, **2**, 303 (1968).
[10] F. J. C. Rossotti in *Modern Coordination Chemistry*, J. Lewis and R. G. Wilkins (Eds). Interscience: New York (1960).
[11] E. L. King, *J. Chem. Educ.* **30**, 72 (1953).
[12] R. J. P. Williams, *J. phys. Chem.* **58**, 121 (1954).
[13] S. Ahrland, *Helv. Chim. Acta*, **50**, 306 (1966).

333

[14] R. E. Messmer and C. F. Baes, *Inorg. Chem.* **8**, 618 (1969).

[15] C. Brosset, *Dissertation.* Stockholm (1942).

[16] R. E. Connick, W. M. Latimer *et al. J. Am. Chem. Soc.* **78**, 1827 (1956).

[17] J. H. Jonte and D. S. Martin, *J. Am. Chem. Soc.* **74**, 2052 (1952).

[18] L. G. Sillén, *Acta Chem. Scand.* **3**, 359 (1949).
R. M. Izatt *et al. Inorg. Chem.* **2**, 1243 (1963); **3**, 130 (1964).

[19] K. Schlyter, *Trans. Roy. Inst. Tech. Stockholm* (1962).

[20] G. Schwarzenbach, M. Waibel and M. Zobrist, unpublished.

[21] G. Schwarzenbach and M. Schellenberg, *Helv. Chim. Acta*, **48**, 28 (1963).

[22] S. Ahrland in Vol. V of *Structure and Bonding*, C. K. Jörgensen, J. B. Neilands, R. S. Nyholm, D. Reinen and R. J. P. Williams (Eds), Springer: Berlin (1968).

[23] A. Sonesson, *Acta Chem. Scand.* **12**, 165 and 1937 (1958); **14**, 1495 (1960).
I. Grenthe, *Acta Chem. Scand.* **18**, 283 (1964).

[24] S. Ahrland, *Acta Chem. Scand.* **5**, 1151 (1951).

[25] T. W. Newton and G. M. Arcand, *J. Am. Chem. Soc.* **75**, 2449 (1953).

[26] E. L. Zebroski *et al. J. Am. Chem. Soc.* **73**, 5646 (1951).
A. J. Ziden, *J. Am. Chem. Soc.* **81**, 5022 (1959).

[27] R. M. Izatt *et al. J. Chem. Soc. A*, 47 (1969).

[28] K. S. Pitzer *et al. J. Am. Chem. Soc.* **59**, 1213 (1937).

[29] G. Schwarzenbach and M. Widmer, unpublished.

[30] F. J. C. Rossotti *et al. Proceedings of the Seventh International Conference on Coordination Chemistry*, Stockholm (1962).

[31] R. Barbieri and J. Bjerrum, *Acta Chem. Scand.* **19**, 469 (1965).

[32] G. Schwarzenbach, *Chimia*, **3**, 1 (1949); *Svensk Kem. Tidskr.* **79**, 290 (1967).

[33] J. E. Prue, *J. Chem. Educ.* **46**, 12 (1969).

[34] H. Sack, *Phys. Z.* **27**, 206 (1926); **28**, 199 (1927).

[35] J. B. Hasted, D. M. Ritson and C. H. Collie, *J. Chem. Phys.* **16**, 1 and 11 (1948).

[36] F. Booth, *J. Chem. Phys.* **19**, 391, 1327 and 1615 (1951).

[37] G. Schwarzenbach, *Z. Phys. Chen.. A.* **176**, 133 (1936).

[38] Th. Landis, *Dissertation*, ETH: Zürich (1969).

[39] N. Bjerrum, *Z. Phys. Chem.* **106**, 219 (1923).

[40] G. H. Nancollas, *Interactions in Electrolyte Solutions*, Elsevier: Amsterdam (1966).

[41] W. M. Latimer, K. S. Pitzer and C. M. Stansky, *J. Chem. Phys.* **7**, 108 (1939).

[42] R. M. Noyes, *J. Am. Chem. Soc.* **84**, 513 (1962).

[43] P. Paoletti *et al. J. phys. Chem.* **67**, 1067 (1963).

[44] P. Paoletti *et al. J. phys. Chem.* **69**, 3759 (1965).

[45] R. G. Bates *et al. J. Res. Nat. Bur. Stds*, **57** (No. 3) (1956).

[46] J. Clark and D. D. Perrin, *Quart. Revs. (London)*, **18**, 295 (1964).

[47] P. Paoletti *et al. J. Chem. Soc. A*, 1385 (1966).
P. Paoletti *et al. Inorg. Chem.* 1384 (1966).

[48] A. Vacca and P. Paoletti, *Progress in Coordination Chemistry*, p. 588. Edited by M. Cais. Proceedings of the Eleventh International Conference on Coordination Chemistry: Haifa (1968).

[49] G. Anderegg, *Helv. Chim. Acta*, **46**, 1831 (1963).

[50] P. Paoletti, private communication.

[51] K. S. Pitzer, *J. Am. Chem. Soc.* **59**, 2365 (1937).

[52] H. A. Bent, *J. Phys. Chem.* **60**, 123 (1956).

[53] R. W. Gurney, *Ionic Processes in Solution*, McGraw-Hill: New York (1953).

[54] G. Schwarzenbach, *Helv. Chim. Acta*, **35**, 2344 (1952).

[55] G. Anderegg, *Helv. Chim. Acta*, **47**, 1801 (1964); **48**, 1718 and 1722 (1965).

[56] A. W. Adamson, *J. Am. Chem. Soc.* **76**, 1578 (1954).

[57] A. E. Martell, *Essays in Coordination Chemistry*, p 52. W. Schneider, G. Anderegg, R. Gut (Eds) Birkhauser: Basel (1964).

[58] G. Schwarzenbach and W. Schneider, *Helv. Chim. Acta*, **38**, 1931 (1955).

[59] J. Zobrist, *Diplomarbeit*, ETH: Zürich (1965) (directed by W. Schneider and G. Anderegg).

[60] I. Grenthe, *Acta Chem. Scand.* **17**, 2487 (1963).

[61] I. Grenthe, *Acta Chem. Scand.* **18**, 293 (1964).

[62] G. Geier and U. Karlen, *Progress in Coordination Chemistry*, p 159. M. Cais (Ed.) (Proceedings of the Eleventh International Conference on Coordination Chemistry: Haifa) (1968).

334

Reprinted from CHEMICAL COMMUNICATIONS, 1968, page 65.

16

Failure of Pauling's Bond Energy Equation

By RALPH G. PEARSON

(*Department of Chemistry, Northwestern University, Evanston, Illinois* 60201)

ONE of the best known empirical relations in chemistry is Pauling's bond energy equation

$$D_{AB} = \tfrac{1}{2}(D_{AA} + D_{BB}) + 23(X_A - X_B)^2 \quad (1)$$

where D_{AB}, the bond energy of an AB bond, is related to the mean of the AA and BB bond energies and the difference in electronegativities, $X_A - X_B$, of the two bonded atoms or groups.[1] The most important application of equation (1) has been to establish values of the average electronegativities (EN's) of the elements. These useful numbers can be correlated with many chemical and physical properties of the elements.[2]

If equation (1) is applied to a chemical reaction (A and C are the more metallic elements)

$$AB + CD = AD + CB \quad (2)$$

we can immediately write the heat of the reaction in kcal. as

$$\Delta H = 46(X_C - X_A)(X_B - X_D) \quad (3)$$

This equation makes a clear prediction which is widely taught and accepted: reaction (2) will be exothermic if the products contain the least electronegative element, A, combined with the most electronegative element, D. Pauling used equation (3) to predict the heats of various isomerizations,[3] and Hine and Weimar[4] used it to explain why carbon, compared to hydrogen, prefers to bond to the less electronegative elements such as C, S, P, and I.

The Table shows a number of experimental heats of reaction of type (2) for polar molecules, and the heats calculated from Pauling's equation (3). The equation is totally unreliable in that it gives the sign of the heat change incorrectly.

The examples chosen are gas-phase reactions. For this reason most of the data refer to halides, for which the gas-phase heats of formation are very complete.[5] It is clear from the heats of formation of solids, and from the heats of reaction in solution, that the same kind of results will be found for Group V and Group VI compounds and complexes.

Of other examples, some will agree with equation (3) and some will disagree as to the sign of ΔH. The important point is that it is possible to predict in advance when the equation will fail. Among the representative and early transition elements, electronegativity always decreases down a column in the Periodic Table. This leads to the Pauling prediction that for heavier elements in a column, the affinity for F will increase relative to that for I. The prediction is also made for preferred bonding to O compared to S, and N compared to P. The facts are always otherwise.

Consider the simple example

$$LiF(g) + CsI(g) = LiI(g) + CsF(g) \quad (4)$$

The Pauling equation predicts that ΔH is 46 $(0\cdot7 - 1\cdot0)(4\cdot0 - 2\cdot5) = -21$ kcal., "because the least electronegative element, Cs, wants to bond to the most electronegative element, F." The

Heats of gas-phase reactions at 25°

				ΔH (exp.)[a]	ΔH (calc.)[b]
$BeI_2 + SrF_2 = BeF_2 + SrI_2$	..	..	..	-48 kcal.	$+35$ kcal.
$AlI_3 + 3NaF = AlF_3 + 3NaI$				-94	$+127$
$HI + NaF = HF + NaI$	..	..	..	-32	$+76$
$HI + AgCl = HCl + AgI$	..	..	..	-25	$+5$
$NOI + CuF = CuI + NOF$	..	..	..	-10	$+76$
$LaF_3 + AlI_3 = AlF_3 + LaI_3$	..	..	..	-9	$+84$
$CaO + H_2S = CaS + H_2O$	..	..	..	-37	$+25$
$CS_2 + 2H_2O = CO_2 + 2H_2S$	..	..	..	-16	$+37$
$CS + PbO = PbS + CO$	..		..	-71	$+64$
$MeHgCl + CH_4 = Me_2Hg + HCl$			..	-40	$+5$
$MeHgCl + HI = MeHgI + HCl$			..	-11	$+5$
$COBr_2 + HgF_2 = COF_2 + HgBr_2$			..	-85	$+66$
$2CuF + CuI_2 = 2CuI + CuF_2$			..	-25	$+14$
$2TiF_2 + TiI_4 = TiF_4 + 2TiI_2$	..		..	-51	$+56$
$MeOH + MeOH = CH_2(OH)_2 + CH_4$				-20	$+13$
$CH_3F + CF_3I = CH_3I + CF_4$	..		..	-22	$+69$
$CH_3F + CF_3H = CH_4 + CF_4$	..		..	-19	$+88$

[a] Experimental heats of formation from reference 5 for halides; others from NBS Circular 500 and its revision Technical Notes 270–1 and 270–2, D. D. Wagman, *et al.*, National Bureau of Standards, Washington, D.C., Oct., 1965, May 1966. CF₃ derivatives from A. Lord, C. A. Gray and H. O. Pritchard, *J. Phys. Chem.*, 1967, **71**, 1086.
[b] Calculated from equation (3).

experimental value is $+17$ kcal. A rough electrostatic calculation shows that the dominant energy term is the attraction of the small Li and F atoms for each other.

Similarly, across the Periodic Table, the electronegativity of the elements increases steadily. This leads to the Pauling prediction that in a sequence such as Na, Mg, Al, Si the affinity for I will increase relative to that for F. Similarly, bonding to S and P atoms will be preferred relative to O and N. However as long as the elements have the positive group oxidation states, the facts are the opposite with very few exceptions.

Even more serious, equation (3) will almost always predict incorrectly the effect of systematic changes in A and C. For example, what happens to the heat of reaction (2) if the oxidation state of the bonding atoms change, or if the other groups attached to these atoms are changed? Such changes affect the electronegativity in a predictable way. For example, the electronegatives of Pb^{II} and Pb^{IV} are 1·87 and 2·33, respectively.[6] Similarly, the electronegativity of carbon is 2·30 in CH_3, 2·47 in CH_2Cl, and 3·29 in CF_3.[7] Increased positive oxidation state and substitution of less electronegative atoms by more electronegativity atoms always increases the electronegativity of the central bonding atom.

From equation (3), such changes again are predicted to decrease the relative affinity for F, O, and N, compared to I, S, and P. For all of the elements, except a few of the heavy post-transition elements (Hg, Tl, *etc.*[8]), the reverse is

true. The impression that bonding tendencies of carbon in organic chemistry can be understood in terms of electronegativities becomes quite incorrect as soon as a range of organic radicals is considered. The more electronegative a carbon atom becomes, the *more* it prefers to bond to other electronegative atoms.

The poor results in the Table are not due to a poor choice of the electronegativity values of the elements. No reasonable adjustment of these values will improve the situation. If new parameters X_A, X_B, etc., are found for the elements to give the best fit to equation (3), they will no longer be identifiable as electronegativities. They would necessarily vary with position in the Periodic Table, with oxidation state, and with substitution effects in a way directly opposite from what one would expect of simple electronegativities.

The Principle of Hard and Soft Acids and Bases (HSAB)[9] may be used to predict the sign of ΔH for reactions such as (2). The Principle states that, to be exothermic, the hardest Lewis acid, A or C, will co-ordinate to the hardest Lewis base, B or D. The softest acid will co-ordinate to the softest base. Softness of an acceptor increases on descending a column in the Periodic Table; hardness increases on traversing the Table, for the group oxidation state; hardness increases with increasing oxidation state (except Tl, Hg, *etc.*), and as electronegative substituents are placed on the bonding atoms A or C. For donor atoms the electronegativity may be taken as a measure of the hardness of the base, donors of low electronegativity being soft. Accordingly, the HSAB Principle will correctly predict heats of reaction where the electronegativity concept fails. Some exceptions will occur since it is unlikely that any single parameter assigned to A, B, C, and D will always suffice.

(*Received, November 27th*, 1967; *Com.* 1267.)

[1] L. Pauling, "The Nature of the Chemical Bond," Cornell University Press, Ithaca, N.Y., 1960, 3rd edn., pp. 88–105.

[2] A. L. Allred and E. G. Rochow, *J. Inorg. Nuclear Chem.*, 1958, 5, 264, 269.

[3] Ref. 1, p. 103–105.

[4] J. Hine and R. D. Weimar, jun., *J. Amer. Chem. Soc.*, 1965, 87, 3387.

[5] R. C. Feber, Los Alamos Report LA-3164, 1965.

[6] A. L. Allred, *J. Inorg. Nuclear Chem.*, 1961, 17, 215.

[7] H. J. Hinze, M. A. Whitehead, and H. H. Jaffe, *J. Amer. Chem. Soc.*, 1963, 85, 148.

[8] S. Ahrland, "Structure and Bonding," 1967, vol. 1, p. 207.

[9] R. G. Pearson, *J. Amer. Chem. Soc.*, 1963, 85, 3533; *Science*, 1966, 151, 172; R. G. Pearson, *Chem. in Brit.*, 1967, 3, 103; R. G. Pearson and J. Songstad, *J. Amer. Chem. Soc.*, 1967, 89, 1827.

J. inorg. nucl. Chem., 1970, Vol. 32, pp. 373 to 381. Pergamon Press. Printed in Great Britain

ELECTRONEGATIVITY, ACIDS, AND BASES—I

HARD AND SOFT ACIDS AND BASES AND PAULING'S ELECTRONEGATIVITY EQUATION

17

ROBERT S. EVANS and JAMES E. HUHEEY

Department of Chemistry, University of Maryland, College Park Md. 20742

(*Received* 28 *April* 1969)

Abstract—The apparent conflict between the empirical HSAB rule and Pauling's Electronegativity Equation is examined and discussed in terms of more recent (Mulliken-Jaffé) electronegativity theory. It is concluded that no basic conflict exists but that Pauling's concept of "ionic resonance energy" neglects the variation in bond length from molecule to molecule, an error that is not serious in calculating electronegativities from bond energies but which is emphasized when applied to the prediction of enthalpies of HSAB reactions. ·

INTRODUCTION

THE DEVELOPMENT of acid-base concepts has seen an increasing emphasis on electronic aspects beginning with the Lewis[1] definition of acids and bases in terms of electron pair donation and acceptance. Increasingly, basicity has been viewed in terms of available electron density and acidity in terms of ability to accept electron density[2–4]. Examples of the importance of inductive effects, e.g., the basicity of amines[5] or the acidity of substituted organic and inorganic acids[6] are numerous and need not be discussed further here. The importance of electronegativity in determining the direction and magnitude of inductive effects has long been recognized. In this and subsequent papers we shall apply recent developments in electronegativity theory[7–10] to the problem of correlating acidity-basicity with electron density. These methods have proven useful in correlating inductive effects of groups[11], nmr data[12], and various reaction rates and molecular properties[13], and should be expected to provide information on acid-base systems as well.

Pearson[14] has suggested that at least two factors are involved in acid-base interactions. One, the inherent strength, can perhaps be most closely related

1. G. N. Lewis, *Valence and the Structure of Molecules*. The Chemical Catalogue Co., New York (1923).
2. M. Usanovich, *Zh. obshch. Khim.* **9**, 182 (1939).
3. R. S. Mulliken, *J. phys. Chem.* **56**, 801 (1952).
4. J. E. Huheey, *J. inorg. nucl. Chem.* **24**, 1011 (1962).
5. F. E. Condon, *J. Am. chem. Soc.* **87**, 4481, 4485, 4491, 4494 (1965).
6. G. E. K. Branch and M. Calvin, *The Theory of Organic Chemistry*, pp. 183–220. Prentice-Hall, New York (1944).
7. R. T. Sanderson, *J. chem. Educ.* **31**, 2 (1945); *Chemical Periodicity*, Reinhold, New York (1960).
8. J. Hinze and H. H. Jaffé, *J. Am. chem. Soc.* **84**, 540 (1962).
9. J. Hinze, M. A. Whitehead and H. H. Jaffé, *J. Am. chem. Soc.* **85**, 148 (1963).
10. J. E. Huheey, *J. phys. Chem.* **69**, 3284 (1965).
11. J. E. Huheey, *J. org. Chem.* **31**, 2365 (1966).
12. J. E. Huheey, *J. chem. Phys.* **45**, 405 (1966).
13. W. J. Considine, *J. chem. Phys.* **42**, 1130 (1965).
14. R. G. Pearson, *J. Am. chem. Soc.* **85**, 3533 (1963); *J. chem. Educ.* **45**, 581, 643 (1968).

to electron density. The second factor suggested by Pearson is the hardness or softness of the species involved. Pearson has pointed out that the principle of hard and soft acids and bases (*hard acids prefer to bond to hard bases and soft acids prefer soft bases*) is a pragmatic one and he has stressed its application as a predictive and correlative principle. Nevertheless, various factors have been suggested to account for hard-hard interactions (electrostatic forces) and soft-soft interactions (polarizability, pi-bonding, London forces). The factors for a given type of interaction are not mutually exclusive and are related to size. In general, hard-hard and soft-soft interactions might just as well be termed small-small and large-large interactions, respectively. Recently, in an interesting note[15] Pearson has called attention to a fundamental conflict between the principle of hard and soft acids and Pauling's equation relating bond energy and electronegativity[16]. Since the latter predicts that ionic resonance energy will be proportional to the *square* of electronegativity differences, the stabilization of the bond between the *most electropositive* element (A) and the *most electronegative* element (D) should dominate the energetics and one might therefore expect all reactions of the type:

$$AB + CD = AD + CB \qquad (1)$$

to favor the formation of the compound containing the bond between A and D. Pearson has shown that many equilibria such as $LiF + CsI = CsF + LiI$ violate this expectation, but that the principle of hard and soft acids and bases (HSAB) correctly predicts the results. Admittedly, Pauling never intended his equation to be applied quantitatively to bonds such as Cs–F, etc., but still the failure of the rearranged Pauling equation:

$$\Delta H = 46(\chi_C - \chi_A)(\chi_B - \chi_D) \qquad (2)$$

is puzzling. In the present paper we shall apply the methods of electronegativity equalization[7–10] to the estimation of heats of reaction of the type shown in Equation (1) in an attempt to eliminate the apparent disparity between these two widely applicable principles of chemistry.

METHODS

The reactions studied are those discussed by Pearson[15] who has given values for the experimental heats of reaction as well as those calculated from Equation (2). Electronegativity data for our calculations are those of Jaffé *et al.* for the elements H–Cl[8], representative elements K–I[17], and the first transition series, Sc–Co[18], plus those of Mulliken[19] for Cu and Zn. For those reactions cited by Pearson that we have not been able to study because of lack of accurate electronegativity data of the Mulliken-Jaffé type (unavailable for heavy elements such as Cs, Ag, Hg, etc.) we have substituted similar compounds and reactions, e.g.,

15. R. G. Pearson, *Chem. Comm.* 65 (1968).
16. L. Pauling, *The Nature of the Chemical Bond*, 3rd Edn. Cornell University Press, New York (1960).
17. J. Hinze and H. H. Jaffé, *J. phys. Chem.* **67**, 1501 (1963).
18. J. Hinze and H. H. Jaffé, *Can. J. Chem.* **41**, 1315 (1963).
19. R. S. Mulliken, *J. chem. Phys.* **2**, 782 (1934); *J. chim. Phys.* **46**, 497 (1935).

Table 1. Experimental and calculated enthalpies of reactions of hard and soft bases

A. Reaction	B. Experimental enthalpy*	C. Calc Equation (2) χ_P†	D. Calc Equation (2) χ_{MJ}‡	E. Method I§	F. Method II§	G. Method III
1 $CsF + LiI = LiF + CsI$	-17	+21	+21	+1 (p); -3 (te)	-4 (p); -6 (te)	-22 (p); ¶
2 $(RbF + LiI = LiF + RbI)$	-14	(+14)		0 (p); -1 (te)	-2 (p); -2 (te)	-11 (p); ¶
3 $(NaF + LiI = LiF + NaI)$	-12	(+7)	+6	+10 (p); -1 (te)	+14 (p); -9 (te)	-79 (p); ¶
4 $BeI_2 + SrF_2 = BeF_2 + SrI_2$	-48	(+69)	+70	-15 (p); -2 (te)	-22 (p); +2 (te)	-22 (p); ¶
5 $(CaI_2 + ZnF_2 = CaF_2 + ZnI_2)$	-28	(-83)	-99	-14 (p); -2 (te)	-28 (p); -7 (te)	-148 (p); ¶
6 $(BeI_2 + CrF_2 = BeF_2 + CrI_2)$	-43	(-14)	-104	+10 (p); +1 (te)	+18 (p); +3 (te)	+49 (p); ¶
7 $(CrI_2 + ZnF_2 = CrF_2 + ZnI_2)$	-23	(0)	+64	+10 (p); +3 (te)	+20 (p); +6 (te)	+50 (p); ¶
8 $(MnI_2 + ZnF_2 = MnF_2 + ZnI_2)$	-8	(-14)	+64	-7 (p)**; -1 (te)**	-15 (p)**; -3 (te)**	-58 (p)**; ¶
9 $(MgI_2 + SnF_2 = MgF_2 + SnI_2)$	-32	(-83)	-66	+13 (p); -3 (te)	+16 (p); -12 (te)	-22 (p); ¶
10 $HI + NaF = HF + NaI$	-32	(+83)	+93	+36 (p); +12 (te)	+50 (p); +15 (te)	+60 (p); ¶
11 $NOI + CuF = NOF + CuI$	-10	+76	+189	+15	+7	-91
12 $CaO + H_2S = H_2O + CaS$	-37	(+9)	+13	+22	+42	+147
13 $CS_2 + 2H_2O = CO_2 + 2H_2S$	-16	(+120)††	+268			‡‡
14 $CF_3I + CH_3F = CF_4 + CH_3I$	-22	+69††	+75	+14 (p); +2 (te)	+19 (p); +4 (te)	‡‡
15 $CF_3H + CH_3F = CF_4 + CH_4$	-19	+88††	+92	+17	+23	‡‡

Reactions are listed in increasing order of approximate covalency. Enthalpies are in kcal/mole.
*Enthalpies are from Ref. [15] for reactions given therein; remainder from Refs. [20] and [21].
†Values are from Ref. [15] except for values in parentheses which are new calculations or recalculations by us, using Pauling electronegativities.
‡Values are calculated using Mulliken-Jaffé electronegativities.
§Hybridizations in parentheses refer to iodine.
¶Calculations using tetrahedral hybridization of iodine are inappropriate for "completely ionic" compounds.
**Hybridization of Sn = 8·75% s.
††Although this value is listed here since it comes from Ref. [15], it is obtained by combining Pauling (F. 1) and Mulliken-Jaffé (CH₃, CF₃) electronegativities.
‡‡Choice of appropriate formal charges in these predominantly covalent compounds is ambiguous and without physical meaning.

RbF and RbI for CsF and CsI in the reaction discussed above. These substitutions are enclosed in parentheses in Table 1. In order to enlarge the scope of the discussion, we have also included some reactions not mentioned by Pearson although they serve to illustrate the HSAB principle. Values of experimental reaction enthalpies (all gas phase) are those given by Pearson where applicable and from Brewer et al.[20, 21] for other reactions. Values of enthalpies predicted by Equation (2) are those of Pearson where applicable[22]. As Pearson has pointed out[15], the qualitative results are independent of electronegativity scale employed, but for comparison, we have computed the remaining values using both Pauling[16] and Mulliken-Jaffé[8, 17–19] electronegativities.

We have calculated ionic resonance stabilization energies (those energies resulting from differences in electronegativity) by three methods: one based on electronegativity energy alone, one based on electronegativity and Madelung energy, and a third, compromise method intended to avoid difficulties inherent in the other two procedures.

Since electronegativity depends upon hybridization[8], some assumptions must be made concerning the appropriate hybridization. For nonlinear molecules, hybridizations of the central atoms were estimated from bond angles by the method of Coulson[23]. For linear, triatomic molecules such as BeF_2, digonal hybridization of the metal was assumed, and for molecules such as CO_2, a bent bond approach was employed using tetrahedral orbitals since this is computationally simpler and mathematically equivalent to a sigma-pi approach[24]. Diatomic molecules were treated using s orbitals for the alkali metals and a tetrahedral bent bond approach for the presumed double bond in molecules such as CaO. This last is the most uncertain of the assignments but since it appeared in only one calculation that did not appear to differ appreciably in results from the others, it was accepted. The halogens were assumed to be using pure p orbitals, except for iodine for which both p and tetrahedral hybridizations were employed (see Discusssion).

(I) *Electronegativity energy resulting from electronegativity equalization.* When electron density is transferred from one atom to another, one atom gains in energy while the other loses. If a unit charge is involved, the energy of the transfer is given simply by the sum of the ionization energy of atom A and the electron affinity of atom B. Since the ionization energy of the most electropositive element (Cs, 3·89 ev) is greater than the electron affinity of the most electronegative elements (F, 3·44 eV; Cl 3·60 eV [25]) complete transfer of a unit charge is endothermic unless compensated by other effects[26]. However, transfer of electronic

20. L. Brewer and E. Brackett, *Chem. Rev.* **61**, 425 (1961).
21. L. Brewer, G. R. Somayajulu and E. Brackett, *Chem. Rev.* **63**, 111 (1963).
22. The rearranged equation (Equation (2)) is strictly applicable only to reactions involving diatomic species. For reactions involving polyatomic species, the original Pauling equation for a bond A—B, $\chi_A - \chi_B = (\Delta/23·06)^{1/2}$, or a generalized form of Equation (2) must be used.
23. C. A. Coulson, *Valence*, pp. 193–195. Oxford University Press, Oxford (1952).
24. J. E. Huheey, *J. phys. Chem.* **70**, 2086 (1966).
25. The *apparently* anomalous electron affinity of fluorine is well known and will not be discussed here.
26. A complete discussion of the other factors involved will be reserved for the "Discussion" sections of this and subsequent papers.

density in an appropriate amount *less* than that of a total electronic charge is *always* exothermic when the charge is transferred from the less electronegative element to the more electronegative element. This can be shown simply as follows: The ionization energy-electron affiinity curve for all elements thus far studied has been shown to be very close to perfectly quadratic over a range of several oxidation states as long as all electrons being removed are in the same energy level and same type of orbital[27]. Jaffé *et al.*[8, 9] have determined valence state ionization energies for many elements and have assumed a quadratic relation between energy and charge or occupancy[28]:

$$E_A = a_A \delta_A + \tfrac{1}{2} b_A \delta_A{}^2. \tag{3}$$

For any two atoms, A and B, that total ionization energy-electron affinity energy $(E_A + E_B)$ will have a minimum. The partial charge (δ) corresponding to the minimum energy may be found:

$$E_A + E_B = a_A \delta_A + \tfrac{1}{2} b_A \delta_A{}^2 + a_B \delta_B + \tfrac{1}{2} b_B \delta_B{}^2 \tag{4}$$

$$\delta_A = - \delta_B \tag{5}$$

$$dE/d\delta = a_A + b_A \delta_A - a_B + b_B \delta_A = 0 \tag{6}$$

$$\delta_A = \frac{a_B - a_A}{b_A + b_B}. \tag{7}$$

The charge on atom A (δ_A) will be *positive* and $E_A + E_B$ *negative* as long as the electronegativity of $B(a_B)$ is greater than the electronegativity of $A(a_A)$. If the electronegativities are reversed, the charges are reversed, but *there is always a minimum* in the curve described by Equation (4) as can be shown by substituting the charge found in Equation (7) into Equation (4):

$$(E_A + E_B)_{\min} = \frac{-(a_B - a_A)^2}{2(b_A + b_B)} = E_\chi. \tag{8}$$

This gain in energy obtained by partial transfer of electronic charge we have termed the *electronegativity energy* (E_χ).

Equation (7) may be obtained alternatively by applying the principle of electronegativity equalization[7]. Electronegativity may be defined[9, 10, 27] as the derivative of energy with respect to charge:

$$E_A = a_A \delta_A + \tfrac{1}{2} b_A \delta_A{}^2 \tag{9}$$

$$\chi_A = dE/d\delta = a_A + b_A \delta_A. \tag{10}$$

Equalization of electronegativity is thus identical to minimizing the electronegativity energy and also yields Equation (7).

27. R. P. Iczkowski and J. L. Margrave, *J. Am. chem. Soc.* **83**, 3547 (1961).
28. Jaffé *et al.* have used the occupancy number, n, whereas in this and previous papers[10–12, 24], we have used the partial charge, δ. The relation between the two is $\delta = 1 - n$.

Using electronegativity parameters (*a* and *b*) obtained from the work of Jaffé *et al.*[29] it is possible to estimate the *electronegativity* energies of the various compounds assuming complete equalization of electronegativity. We have computed electronegativity energies by this method and obtained the *net* electronegativity energies for reactions of the type shown in Equation (1), assuming no other energy terms affect the overall enthalpy of reaction. The results are listed in Table 1 together with the values given by Pearson for Pauling's equation (Equations 2) and experimental results. It will be noted that although the wrong sign is still obtained in a majority of cases,[30] the results are considerably better than those predicted by Equation (2). For predominantly ionic compounds, the results are reasonably good, but for essentially covalent compounds the results are rather poor.

(II) *Electronegativity energy plus Madelung energy*. Although electronegativity equalization accounts completely for *atomic* energies, molecules contain other energies. One of these is the Madelung energy, E_M[31]. Using the partial charges (δ_A, δ_B) calculated by the method of electronegativity equalization (I) and experimental values for bond lengths (r_{AB}) we have obtained estimates for ionic resonance energies ($E_X + E_M$) by adding the Madelung energy:

$$E_M = \frac{\delta_A \delta_B A}{r_{AB}}\left(1 - \frac{1}{n}\right) \tag{11}$$

where *n* is the Born exponent and A is the Madelung constant, based solely on geometry. The results are listed in Table 1 and will be discussed below.

(III) *Minimization of total electronegativity and Madelung energy*. The calculation given previously in (II) is artificial to the extent that it assumes a charge obtained by mimimization of the *electronegativity energy*, E_X, alone, and then uses this to calculate an additional Madelung energy, E_M. To be rigorous, one must demand that the *sum* of both energies be minimized ($d(E_X + E_M)/d\delta = 0$). We have performed calculations of this type as well and the results are listed in Table 1. It is found that with exception of the most covalent molecules, the driving force of the large Madelung energy causes the minimum energy to be that corresponding to full transfer of electrons, i.e. the compounds are completely ionic[33]. This phenomenon has been observed and commented upon previously[32].

DISCUSSION

From a simplistic point of view, bonding energy may be thought to be the sum of three interactions: a Madelung term, a covalent term, and an electronegativity

29. Although all of the electronegativity data used are exactly those given by Jaffé *et al.*[8, 17, 18], the symbolism is somewhat different. See Ref.[10] for a comparison of the symbolism.
30. The exact percentage depends upon whether *te* or *p* hybridization is used for iodine.
31. We follow Jørgensen[32] in preferring the terms *Madelung energy* and *Madelung interaction* to the more common usage of "electrostatic energy" and interaction inasmuch as all chemical forces are basically electrostatic in nature.
32. C. K. Jørgensen, *Orbitals in Atoms and Molecules*. Academic Press, New York (1962).
33. The *mathematical* minimum in energy actually occurs beyond that corresponding to complete transfer of electrons, for reasons discussed below. Such a minimum has no physical meaning since a discontinuity occurs when a noble gas configuration is reached[27].

term. The principle of electronegativity equalization, neglecting as it does the Madelung term and the covalent term, is as successful as it is only because the latter terms are antagonistic and tend to cancel each other to a certain degree [10]. A completely covalent molecule (no net charge on any atom) will have zero Madelung energy, and a completely ionic "molecule" (ion pair, etc., with charges on the atoms equal to their formal oxidation state) will have no covalent energy.

In compounds that are very polar, neglect of the Madelung energy (I) will always underestimate the stability of small, principally ionic species. We might expect this method to fail to illustrate the stability of hard-hard interactions, often attributed to strong electrostatic stabilization. On the other hand, we might expect that using the sum of electronegativity energy and Madelung energy while neglecting the *loss* of covalent energy attendant to increases in ionicity (III) in a very polar compound will undoubtedly *overestimate* the stabilization of small, ionic species. We can therefore assume that, for strongly polar compounds at least, these two methods should provide reasonable estimates that *bracket* the true values. For predominantly ionic compounds the totally artificial method (II) of assuming a charge based upon electronegativity equalization and then using it to calculate Madelung energy is probably the most accurate, surprising as it may seem. As stated previously, the relative success of electronegativity equalization methods relies on the antagonistic effects of ionic and covalent energies. To minimize the energy expression containing Madelung terms but not covalent terms (III) clearly biases the results towards increased ionicity. This, of course, is the crux of the method of treating "ionic" lattices developed by Born, Mayer, and others [34]. Such lattices have long been considered to be essentially 100 per cent ionic whereas there is probably considerable covalent bonding involved. Although "completely ionic" models have been used with considerable success in treating gaseous alkali halide molecules [35, 36], it must be pointed out that these calculations involve arbitrary repulsion parameters and polarization energy terms [20]. *Ab initio* calculations on molecules such as LiH and LiF clearly show an appreciable "covalent" contribution to the total bonding energy [37].

Of our calculations listed in Table 1, the greatest success has come from reactions containing predominantly ionic species. Those reactions containing predominantly covalent compounds (H_2S, CH_4, etc.) give the poorest results, again indicating the necessity of taking the variation in covalent bonding into account. We explore this point further in the following paper. Nevertheless, *in almost every case studied by (I) and the majority of those by (II), the results of our calculations are closer to experiment than those based on Equation (2).*

Calculations involving tetrahedrally hybridized iodine *usually* give better results than those using pure *p* hybridization, although the differences are not clear cut. Although there is evidence [9, 11] that iodine uses some *s* character in covalent bonding, it need not necessarily be as much as 25 per cent, nor constant from compound to compound.

34. For a review of the Born-Mayer model of ionic lattices see T. C. Waddington, *Adv. Inorg. Chem. Radiochem.* (Edited by H. J. Emeléus and A. G. Sharpe) p. 158 (1959).
35. E. S. Rittner, *J. chem. Phys.* **19**, 1030 (1951).
36. J. Berkowitz, *J. chem. Phys.* **29**, 1386 (1958).
37. See B. J. Ransil and J. J. Sinai, *J. chem. Phys.* **46**, 4050 (1967); G. Doggett, *J. chem. Soc.* 229 (1969) and references cited therein.

The "failure" of Pauling's equation when applied to reactions of the type given in Equation (1) results from the fact that the "ionic resonance energy" is the sum of several energy terms. Increased bonding energy will result from electronegativity energy and Madelung energy as the ionicity increases. However, this will result in a smaller covalent contribution[38].

The Madelung energy and the electronegativity energy do not depend upon electronegativity differences in the same way. The electronegativity energy is given by Equation (8). Since the values for b for various elements do not differ too widely, the denominator of Equation (8) may be considered roughly constant and hence E_χ varies approximately as the square of the differences in inherent electronegativities of the elements (a_A, a_B).

The Madelung energy depends on the square of the charges and the distance, and using the approximation of electronegativity equalization, is given by:

$$E_M = \frac{-(a_B - a_A)^2}{(b_A + b_B)^2 r_{AB}}. \tag{12}$$

Again the energy goes as the *square* of the electronegativity difference but in this case the distance, r_{AB}, is also involved in the denominator. Most examples given by Pearson[15] are those in which the difference between r_{AD} and r_{CB}, etc., are maximized; in other words small-small and large-large interactions compared with two small-large interactions. Pauling's relation has long been known to work best when comparing atoms of similar size and electronegativity, such as O–F, Cl–F, etc. The extent of the failure of Pauling's equation will be the extent to which the summation of Equations (8) and (12) into a single equation:

$$E_\chi + E_M = -\frac{(a_B - a_A)^2}{2(b_A + b_B)} - \frac{(a_B - a_A)^2}{(b_A + b_B)^2 r_{AB}} \tag{13}$$

represented by the approximation:

$$\Delta = k(\chi_B - \chi_A)^2 \tag{14}$$

(the original Pauling electronegativity equation) ignores the variation in bond length and to a lesser degree, the small variation that occurs in the charge coefficient, b, from element to element. It appears that Pauling's relation (Equation (14)) is an amazingly accurate one considering the information available when it was first formulated, although the advantages of a Mulliken-Jaffé type approach based on valence state ionization energies are apparent both with respect to increased accuracy and the ability to cope with variable hybridization and charge.

Ionic resonance energies calculated by Equation (13) are greater than those calculated from Equation (14) to the extent that hard-hard interactions involve a short bond length. Pearson[15] has suggested that electrostatic forces might be expected to favor hard-hard (or "small-small") interactions such as that in

38. The covalent contribution will be greatest in a homopolar bond, A–A. It will be smaller in a heteropolar molecule $A^{\delta+}$–$B^{\delta-}$, and for a hypothetical completely ionic molecule, A^+B^-, the contribution will be zero.

LiF. A comparison of the values in columns B and F of Table 1 indicate that the small-small interaction is favored by more than a simple Madelung energy. Even in reactions containing only "ionic" molecules such as LiF, the electronegativity energy and Madelung energy are *not* sufficient to account completely for the hard-hard stabilization.

In summary, it may be said that Pauling's Equation can be shown to have a firm basis in terms of energies derived from current electronegativity theory. It is an approximation that ignores differences in bond lengths and differences in the rate at which different atoms change electronegativity with charge (the *b* parameter). When used to calculate inherent electronegativities of elements from the stabilization of a bond between two atoms, AB, this neglect is not serious. However, the rearranged equation (Equation (2)) as applied by Pearson to reactions of the type $AB + CD = AD + CB$ emphasizes the errors accompanying the neglected terms. If they are taken into account, a much closer approximation to experimentally found reaction energies can be made. Nevertheless, it is apparent that a size effect involving more than the obvious Madelung interaction is involved in hard-hard and soft-soft bonding. This non-Madelung size effect must involve covalent bonding.

J. inorg. nucl. Chem., 1970, Vol. 32, pp. 383 to 390.   Pergamon Press.   Printed in Great Britain

18

ELECTRONEGATIVITY, ACIDS, AND BASES—II

SIZE EFFECTS AND THE PRINCIPLE OF HARD AND SOFT ACIDS AND BASES

JAMES E. HUHEEY and ROBERT S. EVANS

Department of Chemistry, University of Maryland, College Park, Md. 20742

(*Received* 28 *April* 1969)

Abstract—The effect of size upon the strength of hard and soft acid–base interactions is examined and it is concluded that hard–hard interactions are especially stabilized by small size that favors enhanced Madelung *and* covalent bonding energies. For the ionic species studied, the soft–soft interaction appears to be a negative one, i.e. it is imposed by the tendency of hard species to associate with each other. The homopolar bond energy may be related quite accurately ($r = 0.975$) to the bond length as $E_H = 132.5 - 37.1\,r$ (E_H = energy in kcal/mole; r = bond length in Ångstroms). Consideration of lone pair repulsions allows even better prediction of homopolar bond energy as a function of bond length.

PEARSON [1] has divided acids and bases into two broad categories: *hard*, those with small, non-polarizable centers, and *soft*, those with larger, polarizable centers. Pearson has suggested an empirical rule for predicting relative stabilities: *hard acids prefer hard bases and soft acids prefer soft bases*. Electrostatic ("ionic" or Madelung) energies have often been offered as the principle stabilization of hard–hard interactions. Various factors such as covalency and London forces resulting from polarizability have been suggested to explain the apparent stability of soft–soft interactions. π-Bonding, currently one of the most controversial topics in inorganic chemistry [2–6], may also be important in soft–soft bonding.

In the previous paper [7] we have discussed hard and soft interactions with special reference to Pauling's electronegativity equation. One of the conclusions drawn there was that size effects were important and that hard–hard interactions might equally well be called small–small interactions. Pearson [8] has suggested that the driving force for many hard–hard interactions may be the enhanced electrostatic (i.e., Madelung) energies favored by the small species. In this paper we would like to explore the importance of size on both Madelung (ionic) and covalent energies.

One interesting aspect of hard–soft interactions that has not been stressed

1. R. G. Pearson, *J. Am. chem. Soc.* **85**, 3533 (1963); *J. chem. Educ.* **45**, 581, 643 (1968).
2. F. A. Cotton and G. Wilkinson, *Advanced Inorganic Chemistry*, 2nd Edn., p. 745. Interscience, New York (1966).
3. W. D. Horrocks, Jr. and R. C. Taylor, *Inorg. Chem.* **2**, 723 (1963).
4. S. O. Grim, D. A. Wheatland and W. McFarlane, *J. Am. chem. Soc.* **89**, 5573 (1967); S. O. Grim, R. L. Keiter and W. McFarlane, *Inorg. Chem.* **6**, 1133 (1967).
5. D. J. Darensbourg and T. L. Brown, *Inorg. Chem.* **7**, 959 (1968).
6. S. S. Zumdahl and R. S. Drago, *J. Am. chem. Soc.* **90**, 6669 (1968).
7. R. S. Evans and J. E. Huheey, *J. inorg. nucl. Chem.* **32**, 373 (1970).
8. R. G. Pearson, *Chem. Comm.* 65 (1968).

sufficiently is that the hard–hard (A–D) interaction is the major driving force in a reaction of the type:

$$AB + CD = AD + CB. \qquad (1)$$

In Table 1 we have listed several reactions containing predominantly ionic species as given in Pearson's note[8] together with others we have compiled from the

Table 1. Enthalpies of atomization and enthalpies of reaction
associated with some hard–soft equilibria

Hard–soft	+	soft–hard	⟶	hard–hard	+	soft–soft	ΔH_{exp}
LiI	+	CsF	⟶	LiF	+	CsI	
84·6		119·6		137·5		82·4	−15·7
BeI$_2$	+	SrF$_2$	⟶	BeF$_2$	+	SrI$_2$	
150		265		300		161	−46
AlI$_3$	+	3 NaF	⟶	AlF$_3$	+	3 NaI	
182		3(114)		423		3(73)	−118
HI	+	NaF	⟶	HF	+	NaI	
71		114		135		73	−23
BeI$_2$	+	ZnF$_2$	⟶	BeF$_2$	+	ZnI$_2$	
150		184		300		100	−66
2 LiI	+	HgF$_2$	⟶	2 LiF	+	HgI$_2$	
2(84·5)		125		2(137·5)		69	−50
2 CsI	+	HgF$_2$	⟶	2 CsF	+	HgI$_2$	
2(82·5)		125		2(119·5)		69	−18
CrI$_2$	+	SnF$_2$	⟶	CrF$_2$	+	SnI$_2$	
122		218		229		121	−10

thermodynamic data of Brewer *et al.*[9] for alkali halides and metal dihalides. The hard–soft acid–base (HSAB) rule correctly predicts the qualitative enthalpy of the reaction (positive or negative) and in each case the hard–hard interaction is the strongest. In almost every case the soft–soft interaction is the *weakest* of the four. These reactions proceed as they do because the sum of the strongest bonding (hard–hard) and the weakest bonding (soft–soft) is greater than the sum of two moderate (hard–soft) situations. We are forced to conclude that in these predominantly ionic species, at least, the driving force for the HSAB rule is the extremely stable hard–hard interaction. The soft–soft interaction takes place not because "soft acids prefer soft bases" but because the soft species are "left over" after the hard species have combined with each other.

Although hard–hard interactions such as Li^+F^- and $Be^{2+}2F^-$ may appear to be readily explainable in terms of Madelung energy[10], it has been pointed out[9] that such models involve arbitrary repulsion parameters, and *ab initio* calculations [11] indicate considerable "covalent" character even in molecules such as LiF.

9. L. Brewer and E. Brackett, *Chem. Rev.* **61**, 425 (1961); L. Brewer, G. R. Somayajulu and E. Brackett, *Chem. Rev.* **63**, 111 (1963).
10. E. S. Rittner, *J. chem. Phys.* **19**, 1030 (1951); J. Berkowitz, *J. chem. Phys.* **29**, 1386 (1958).
11. See B. J. Ransil and J. J. Sinai, *J. chem. Phys.* **46**, 4050 (1967) and G. Doggett, *J. chem. Soc.* 229 (1969) and references cited therein.

For other hard–hard interactions an "ionic" model is clearly insufficient and, for reasons given in the previous paper[7] and discussed further below, we believe Madelung energy is insufficient to explain *all* of the extra stability of hard–hard interactions. For this reason we should like to investigate the role of covalent bonding in these molecules.

One aspect of covalent bonding that has been widely known but too little emphasized is the loss in overlap that occurs as one progresses from the lighter elements to their heavier congeners, e.g., $C-C$, $Si-Si$, $Ge-Ge$, $Sn-Sn$, $Pb-Pb$. This factor has been suggested as the cause of the "inert pair effect" found in heavier elements[12].

Covalent bond energies[13] for single, homopolar bonds are plotted as a function of bond length[14] in Fig. 1. The correlation is surprisingly close except for the larger alkali metals using diffuse s orbitals and for $N-N$, $O-O$, and $F-F$. The relation between bond energy (kcal/mole) and bond length (Å) for homopolar molecules may be expressed as:

$$E_H = 132 \cdot 5 - 37 \cdot 1 \; r \tag{2}$$

with an overall correlation coefficient for the 15 elements excepting those just mentioned of $0 \cdot 975$, i.e., 95 per cent of the variance in bond energy is accounted for on the basis of size along, leaving only 5 per cent to be accounted for by all other causes, including experimental error in the bond energies. For the four congeners in Group IV, $C-C$, $Si-Si$, $Ge-Ge$, and $Sn-Sn$, the correlation coefficient is $0 \cdot 994$. For the seven larger elements containing one or more non-bonding electron pairs ($P-P$, $S-S$, etc.), $r = 0 \cdot 929$. The displacement of this line from that of the Group IV elements results from lone pair–lone pair repulsions similar to the larger repulsions responsible for the weakness of $N-N$, $O-O$, and $F-F$ bonds[15].

The good correlation between single, homopolar bond energy and bond length for both types of elements (Group IV and lone-pair elements) suggests that the attractive forces for all of these homopolar bonds are inversely proportional to bond length if the lone pair repulsions are considered. Presumably overlap decreases with increasing bond distance. As a result, small–small interactions gain from increased covalent energy as well as Madelung energy.

Figures 2, 3, and 4 illustrate heteropolar bond energies as a function of distance for metal halides. There are several interesting points to be seen on these graphs. First, all of the bond energies except one for these heteropolar bonds lie above those expected on the basis of homopolar energies. This increased energy is related to Pauling's ionic resonance stabilization energy. The one exception appears to be the unusually weak bonds in mercuric iodide, a "typical" soft–soft

12. R. S. Drago, *J. phys. Chem.* **62**, 353 (1958).
13. T. L. Cottrell, *The Strengths of the Chemical Bond*, 2nd Edn. Butterworths, London (1958).
14. *Tables of Interatomic Distances and Configuration in Molecules and Ions*, (Edited by L. E. Sutton). Spec. Publ. No. 11, The Chemical Society, London (1958); idem, Supplement 1956–59, Spec. Publ. No. 18 (1965).
15. K. S. Pitzer, *J. Am. chem. Soc.* **70**, 2140 (1947); R. S. Mulliken, *ibid.* **72**, 4493 (1950), **77**, 884 (1955); L. Pauling, *The Nature of the Chemical Bond*, 3rd Edn, pp. 142–144. Cornell University Press, Ithaca (1960).

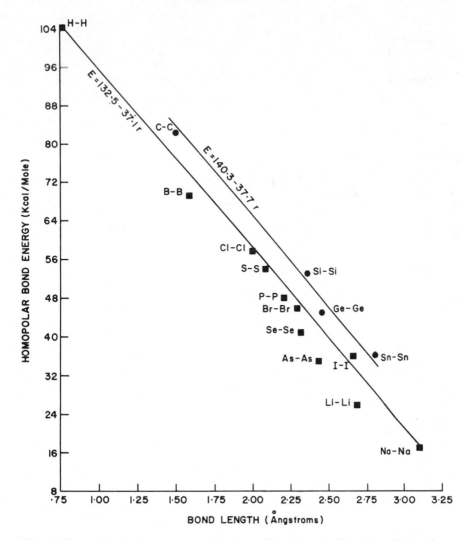

Fig. 1. Homopolar bond energy as a function of bond length. Experimental data from Cottrell[13] and Sutton[14].

interaction that might have been supposed to have an "extra" soft–soft stabilization energy. Rather it appears that the Hg–I bond is an essentially purely covalent bond with little or no ionic resonance energy stabilization ($\chi_{Hg} = 2\cdot00$; $\chi_I = 2\cdot66$)[16].

A second interesting feature is that the four halides of a given metal fall on a straight line; the only significant deviations from linearity come from the fluorides which are sometimes slightly less stable than might have been expected.

The slopes of the lines for the various elements are rather similar, ranging from that of the mercuric halides which does not differ significantly from the average

16. Pauling thermochemical electronegativities calculated by A. L. Allred, *J. inorg. nucl. Chem.* **17**, 215 (1961).

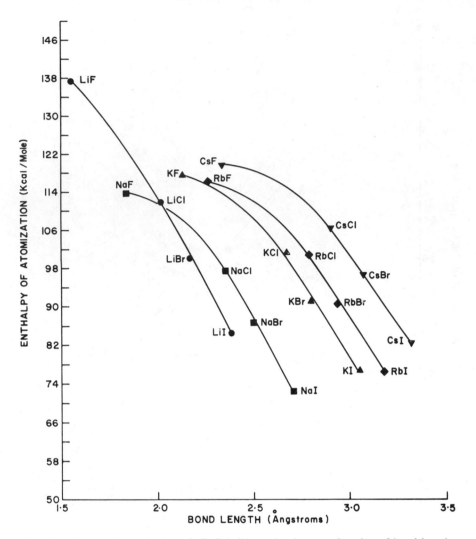

Fig. 2. Enthalpy of atomization of alkali halide molecules as a function of bond length.
Experimental data from Brewer and Brackett[9].

covalent line, to that of Be^{2+}, the "hardest" acid listed. This change in slope is
that expected on the basis of the HSAB rule. Soft acids will have lesser slopes
since they are not particularly stabilized as fluorides (hard base), but hard species
(Be^{2+}, etc.) will have greater slopes since bonding with hard fluoride will be
proportionately stronger than with soft iodide, hence the slope of the line is
changed considerably from that of the "normal" covalent bond. Nevertheless,
"hardness" *per se* is not sufficient to explain all of the stability trends since the
compounds of Na^+ and Mg^{2+}, reasonably hard species, are consistently less
stable than those of their congeners.

The slopes of the transition metal dihalides are of particular interest. As is
well known, crystal field effects alter both the bond energy and the bond length

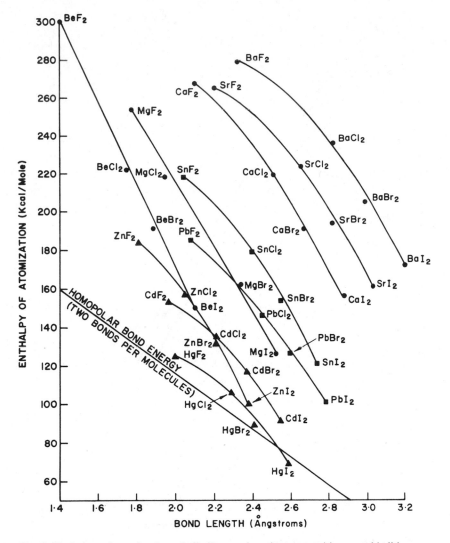

Fig. 3. Enthalpy of atomization of alkaline earth and post-transition metal halides as a function of bond length. Experimental data from Brewer, Somayajulu and Brackett[9].

from an otherwise monotonic progression across the series[17]. Nevertheless, the lines are remarkably parallel, the only significant deviation in slope being that of Cu^{2+}. This lack of significant differences in the bonding energies of the dihalides of the elements Ti to Ni is of special interest since Ahrland[18] has suggested that softness should be proportional to the number of d electrons, with d^6 or greater necessary for truly "soft" behavior. From the set of lines in Fig. 3, Zn^{2+} (d^{10}) does not appear to be significantly softer (bonding relatively more strongly to iodide than fluoride) than Ti^{2+} (d^2) or Cr^{2+} (d^4). Instead the slope of

17. L. Brewer, G. R. Somayajulu and E. Brackett, *Chem. Rev.* **63**, 111 (1963); R. A. Berg and O. Sinanoglu, *J. chem. Phys.* **32**, 1082 (1960).
18. S. Ahrland, *Structure and Bonding*, **1**, 207 (1966); **5**, 118 (1968).

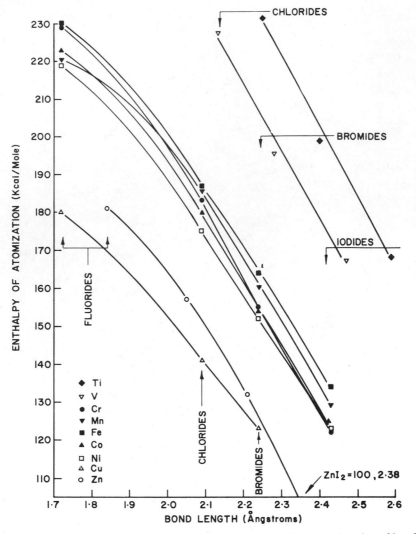

Fig. 4. Enthalpy of atomization of first transition metal dihalides as a function of bond length. Experimental data from Brewer, Somayajulu and Brackett[9].

the line is approximately the same with a *weaker* bond for a particular internuclear distance. On the other hand, Cu^{2+} may be a significant "soft-bonder" to the extent of perhaps 15–20 kcal/mole in the bromide.

Finally, not only do all of the points (except HgI_2) lie above and to the right of the covalent energy line, but the lines for the various elements are displaced upwards and to the right, even when having comparable slopes, e.g., $Be^{2+} < Mg^{2+} < Ca^{2+} < Sr^{2+} < Ba^{2+}$. This seems to be a direct effect of electronegativity and indicates that Pauling's ionic resonance energy is responsible for the displacement. For example, although the small and hard beryllium fluoride has a greater stability than any of the other alkaline earth halides, all of the remaining members of Group IIA with the possible exception of strontium *reverse* this trend with the *larger* elements forming the stronger bonds. Specifically, the heats of atomization decrease in order: $BaF_2 > SrF_2 \approx CaF_2 > MgF_2$; $BaI_2 > SrI_2 > CaI_2 > MgI_2$, etc. Any adequate theory of bonding will have to explain the unusual behavior of these compounds. Certainly a simple "hard–soft" classification will not satisfactorily account for the relative lack of stability of Mg^{2+} compounds. The factors responsible are more subtle: a decrease in covalent and Madelung energy with respect to Be^{2+}; a more unfavorable electronegativity situation with respect to Ca^{2+}, Sr^{2+}, and Ba^{2+}. We will show in the following paper how these various energies may be treated with a resultant analysis of the energetics of hard–soft bonding.

In summary, there appear to be a number of conclusions that can be made concerning HSAB interactions that are primarily ionic. First, soft–soft interactions tend to be weak and the tendency of soft species to combine with each other appears to be a negative one, i.e., it is imposed by the tendency of hard species to associate with each other. The latter interaction is stabilized by two factors: the inherent stability of small–small interactions, i.e., the inherent strength of short bonds, *whether ionic or covalent*, plus the extra ionic resonance energy stabilizing bonds of metals of low electronegativity. The latter consists of enhanced Madelung energy and the electronegativity energy (ionization energy-electron affinity energy). These energies may be readily calculated and the succeeding paper will present an attempt to account for all of the energies in HSAB interactions in terms of a simple model.

The apparent absence of a stabilizing "soft–soft" effect in the compounds studied here removes the necessity for finding an additional bonding factor (π-bonding, London forces, etc.) to account for it. This casts grave doubt on the possibility that there is significant stabilization of the heavier halides through π-bonding. This is of particular interest with respect to the π-bonding controversy. It should be pointed out that halides have been considered weak π-bonders in the past and it may well be that in soft bases such as CO and R_3P π-bonding is significant.

Part IV

Editor's Comments on Papers 19–23

These papers in this section are listed together because they all proposed scales whereby a numerical value might be assigned to hardness or softness. It is by no means a complete list of such scales. Several others are referenced in the various papers. I have a strong doubt in my own mind about the usefulness of making the HSAB concept quantitative. What we gain in precision, we will surely lose in generality. However I would like to make some comments on the scales which have been suggested.

My first comment is an obvious one. No one scale is going to be applicable for all solvents and all temperatures. My second comment is that there should be only one such parameter for each Lewis acid or base. This is a number which represents hardness at one end of the scale and softness at the other. Two independent numbers, one for hardness and one for softness, are logically absurd.

Of all the scales which have been proposed, the most intriguing is that due to Ahrland. This is based on the difference between the ionization energy or electron affinity in the gas phase and the heat of hydration.

$$M_{(g)} \rightarrow M_{(aq)}^{n+} + ne \qquad\qquad \sigma_A = \Delta H/n$$

$$L_{(g)} + ne \rightarrow L_{(aq)}^{n-} \qquad\qquad \sigma_B = \Delta H/n$$

These numbers are closely related to redox potentials, e.g.,

$$M_{(s)} \rightarrow M_{(aq)}^{n+} + ne \qquad\qquad E_0$$

Examination of the possible four-parameter equation

$$\log K = S_A S_B + \sigma_A \sigma_B$$

with the Edwards equation

$$\log K = \alpha E_n + \beta H$$

242

suggests that $S_A = \beta$, $S_B = H$, $\sigma_A = \alpha$, and $\sigma_B = E_n$ (it would be necessary to rescale Ahrland's σ_B values to make his value for F$^-$ be small, or zero). As Table I in paper 21 by Yingst and McDaniel shows, these analogies are not far out of line.

Ahrland's values also bear a great similarity to those derived by Klopman in article 22. The latter author has carried out a very ambitious quantum mechanical treatment of chemical reactivity in solution. Considering the difficulty of the problem, the success in interpretation of diverse chemical phenomena must be considered excellent. Klopman's work represents the most detailed theoretical explanation of the HSAB principle available. Softness is defined as the "energy of the lowest empty molecular orbital for cations, and the energy of the highest occupied orbital for anions." Corrections must be made for solvation and for orbital occupancy.

The last paper is concerned with the kind of four-parameter equation first proposed by Drago and Wayland (Drago and Wayland, 1965):

$$- \Delta H = E_A E_B + C_A C_B$$

Instead of the original paper, I include a later one by Drago, Vogel, and Needham. This has a large number of E and C values, which are useful. It also has a discussion of the HSAB principle and its relationship to the E–C equation, which is far from useful and for which a rebuttal is in order.

Basically what the authors have done is to claim that the idea of intrinsically strong acids and bases is missing from the HSAB concept. Only one parameter, hardness–softness, is allowed to describe each acid and base. This, of course, is just the opposite of what I have argued from my first paper onward.

The equation I have suggested,

$$\log K = S_A S_B + \sigma_A \sigma_B$$

assigns at least equal importance to strength and to hardness–softness. If we consider the two equations as essentially the same (which they are not), then it is clear that $S_A = E_A$, $S_B = E_B$, $\sigma_A = C_A$, and $\sigma_B = C_B$. This is quite different from Drago, Vogel, and Needham's mysterious decision to call C softness and E hardness.[1]

An examination of Tables I and II now shows that the C values do *not* form a very reasonable set of softness parameters. For example $C_B = 6.55$ for $(CH_3)_3P$ and $C_B = 11.54$ for $(CH_3)_3N$. Since the E_B values are virtually the same (0.838 and 0.808), we see that, according to these numbers, it will be impossible for the phosphine ever to form a more stable complex than the amine. While this prediction is presumably correct for the limited data on which the constants are based, it must surely fail when new reference acids are included.

It must also be appreciated that empirical four-parameter equations do not possess unique solutions for the parameters. At least four values must be set beforehand. Two of these simply fix the scale. The other two are selected with some model in mind, and automatically bias all other values to fit that model.

---

[1]In a footnote to one of my early papers I incautiously remarked that the ratio of C_B/E_B values given by Drago and Wayland seemed to fit an order of softness. Apparently this offhand comment has made a greater impression on Drago than all the statements on HSAB which I have written or uttered, before or after.

A paper by Jolly, Illige, and Mendelsohn (Jolly *et al.,* 1972) shows how a four-parameter equation can fit gas phase dissociation energies, even for very polar molecules. The reader may be interested in also reading a similar paper by McMillin and Drago (McMillin and Drago, 1972) in which essentially the same data are analyzed, but with a different model for setting the arbitrary parameters. As expected, the E and C parameters that result are quite different from the E and C parameters of Jolly *et al.*

A much more stringent test of any equation for bond energies, empirical or otherwise, will be its ability to calculate the heat of a reaction which involves differences in bond energies, e.g.,

$$AB + CD \rightleftharpoons AD + CB$$

As mentioned previously, it is such differences which are important in real chemistry.

References

Drago, R. S., and B. B. Wayland, *J. Amer. Chem. Soc.* **87**, 3571 (1965).
Jolly, W. L., J. D. Illige, and M. H. Mendelsohn, *Inorg. Chem.* **11**, 869 (1972).
McMillin, D. R., and R. S. Drago, *Inorg. Chem.* **11**, 872 (1972).

⌈ Reprinted from the Bulletin of the Chemical ⌉
⌊ Society of Japan, Vol. 43, No. 12, pp. 3680–3684 (1970) ⌋

Evaluation of Softness from the Stability Constants of Metal-ion Complexes[*1]

19

Makoto MISONO and Yasukazu SAITO

Department of Synthetic Chemistry, Faculty of Engineering, The University of Tokyo, Hongo, Bunkyo-ku, Tokyo

(Received May 23, 1970)

Utilizing the dual parameter equation we derived previously, we have evaluated quantitatively softness or class (b) character of metal ions on the basis of the original criteria of soft-hard classification, *viz.*, the stability of metal-ion complexes.   Various softness parameters proposed so forth have been proved insufficient to describe softness quantitatively, although they are useful for a rough classification of metal ions into soft and hard acids.   Discussion is also given concerning origin of softness, especially from the viewpoint of its close correlation with the solvation effect.

The general idea of classifying acids and bases into class (a) and (b) (hard and soft) acids and bases has been proposed,[2] according to which hard acids prefer to coordinate with hard bases and soft acids with soft bases. This concept is based on the general trend in inorganic stability constants, that is, some acids (hard) form stable complexes in the sequence $F>Cl>Br>I$, while others in $I>Br>Cl>F$. This classification has been shown to be useful even in organic chemistry, although prediction remains qualitative, owing to the lack of quantitative scales for the acid or base strength.

Attempts were made to calculate the softness quantitatively from the fundamental properties of elements such as ionization potentials and ionic radii.[3-7] It has been claimed that a certain dual parameter scale is inevitably required in order to express the acid or base strength completely.[8,9]

The purpose of the present work is to determine the softness of metal ion on the basis of the original criteria for the soft-hard classification, utilizing the

dual scale equation proposed before.[4] The most commonly used criterion for the soft-hard classification is the stability constants of halogeno-complexes. The concept of softness recently seems to be more or less subjective. The present attempt would make objective evaluation possible.

The dual parameter scale proposed by the authors is as follows.[4] One parameter X is closely related to "hardness" or the electronegativity, and another parameter Y to "softness." With these parameters, the log of stability constants of metal-ion complexes, $\log K$,[*2] is expressed well for hard-hard complexes and, to a lesser extent, for soft-soft complexes by the equation

$$\log K = \alpha X + \beta Y + \gamma \qquad (1)$$

where α and β are the dual basicity parameters of a ligand corresponding to X and Y, respectively, and γ is a constant determined for each ligand. The instability constants of hard-hard complexes are mainly determined by αX term and there is a good linear correlation between X and $\log K$.[4] However, the correlation between Y and $\log K$ of soft-soft complexes was less satisfactory. It was, therefore, suggested that if one can adjust Y values, Eq. (1) will be much improved for stability constants.[4] Thus, we attempted to improve Y values using the dual parameter equation of the form of Eq. (1) and the regularities found among $\log K$ of halogeno-complexes of metal ions.

Results

Procedure for the Derivation of Softness Parameters, Y'. The stability constants of metal-

*1 A part of this work has been referred elsewhere.[1]

1) M. Misono, *Shokubai (Catalyst)*, **9**, 252 (1967).

2) a) R. G. Pearson, *J. Amer. Chem. Soc.*, **85**, 3533 (1963). b) S. Ahrland, J. Chatt and N. R. Davies, *Quart. Rev.*, **12**, 265 (1958).

3) R. S. Nyholm, Proceeding of the Third International Congress on Catalysis, Amsterdam, 1964, p. 25, North-Holland, Amsterdam (1965).

4) M. Misono, E. Ochiai, Y. Saito and Y. Yoneda, *J. Inorg. Nucl. Chem.*, **29**, 2685 (1967).

5) R. J. P. Williams and J.D. Hale, *Structure and Bonding*, **1**, 249 (1966).

6) G. Klopman, *J. Amer. Chem. Soc.*, **90**, 223 (1968).

7) A. Yingst and D. H. McDaniel, *Inorg. Chem.*, **6**, 1067 (1967).

8) J. O. Edwards, *J. Amer. Chem. Soc.*, **76**, 1540 (1954); **78**, 1819 (1956).

9) R. S. Drago and B. B. Wayland, *ibid.*, **87**, 3571 (1965).

*2 The stability constant, K, is the equilibrium constant of $M^{n+}+mL \rightarrow ML_m^{n+}$. In the previous paper,[4] K was the equilibrium constant of the reverse reaction, that is, the instability constant. Therefore, $\log K$ in the present paper is equal to pK in the previous paper.

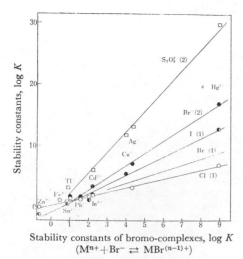

Stability constants of bromo-complexes, log K
$$(M^{n+} + Br^- \rightleftharpoons MBr^{(n-1)+})$$

Fig. 1. Regularity among the stabilities of soft-soft complexes.

  Log of stability constants of several soft ligand complexes are plotted against those of bromo-complexes. Numbers in parentheses indicate coordination numbers.

ion complexes with various soft ligands, log K, are plotted against those of bromo-complexes for several metal ions in Fig. 1. A fairly good regularity is found among these stability constants. In this figure two apparent trends should be noted: (i) soft metal ions have large log K values ($Hg^{2+} > Cd^{2+} > Zn^{2+}$), and (ii) soft ligands have steep slopes ($I^- > Br^- > Cl^-$). The trends are well interpreted with the softness sequences and are also expected from the βY term of Eq. (1), suggesting that this type of equation holds among these complexes.

  Usefulness of a dual scale equation like Eq. (1) and the regularity in Fig. 1 demonstrated for these complexes prompted us to estimate the softness upon these bases. X values have already been proved to be satisfactory as the hardness parameter in Eq. (1). Thus, if one can choose such Y values that the equation gives satisfactorily the observed log K values, they may be considered to be good softness parameters. In order to avoid confusion, Y', α', β' and γ' will be used hereafter for the new parameters. Then the equation is written as

$$\log K = \alpha' X + \beta' Y' + \gamma' \qquad (2)$$

where X values are common to Eq. (1). From $\beta' Y'$ term which is obtained by the substraction of $\alpha' X + \gamma'$ from log K value, the softness of each metal ion can be estimated with the following procedure.

  1) The log K of chloro-complexes is taken, to a first approximation, as Y' value respectively as follows: $Hg^{2+}(7.2)$, $Tl^{3+}(6.5)$, $Ag^+(3.8)$, $Cu^+(3.5)$, $Cd^{2+}(2.1)$, $In^{3+}(1.9)$, $Fe^{3+}(1.6)$, $Pb^{2+}(1.6)$, $Sn^{2+}(1.6)$, $Tl^+(1.2)$ and $Zn^{2+}(1.0)$. Since there are linear correlations among the stability constants of halogeno-

complexes (Fig. 1), the stability constants of any halogeno-complexes may be used as the basis for Y' values. Here, chloro-complexes are chosen, simply because a good number of reliable stability constants are available and they give Y' values whose magnitude is better to handle. When there is no experimental log K value of chloro-complex, estimation is made from log K of other halogeno-complexes, considering the regularity in Fig. 1.

  2) By the method of least squares, α', β', and γ' are then calculated by Eq. (2) for several soft ligands, including different coordination numbers.

  3) Improved Y' values are tentatively obtained from the slopes in log $K - \alpha' X - \gamma'$ vs. β' plots (Fig. 2).

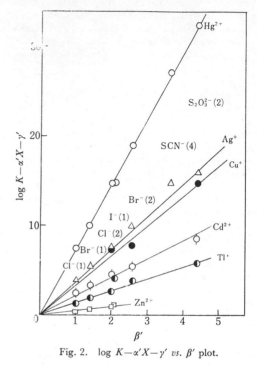

Fig. 2. log $K - \alpha' X - \gamma'$ vs. β' plot.

  Repeating the above procedure one can get the most reasonable value of Y' for each metal ion. Iteration was not actually continued, since the agreement of the second Y' with the first ones was good enough. The final Y' values are Hg^{2+} (7.3), Tl^{3+} (6.0), Ag^+ (3.8), Cu^+ (3.3), Cd^{2+} (2.0), In^{3+} (1.8), Sn^{2+} (1.8), Pb^{2+} (1.6), Fe^{3+} (1.4), Tl^+ (1.3), and Zn^{2+} (0.7). Linearity of the plots in Fig. 2 is significantly improved compared to that in Fig. 1, which provides a good support for Y' as a good softness parameter which is based on the original criteria for the soft-hard classification. Additional Y' values are determined in a similar manner, using only log K of halogeno-complexes, as follows: Au^+ (10.9), Au^{3+} (9.4), Pt^{2+} (6.9), Pd^{2+} (5.6), Co^{2+} (1.2) Fe^{2+} (0.3) and Mn^{2+} (0.0). Although these values might be less reliable because of the limited number

246

TABLE 1. CALCULATED AND OBSERVED STABILITY
CONSTANTS OF METAL-ION COMPLEXES

	Y'[b]	OH⁻ (calcd)	OH⁻ (obsd)	NH₃ (calcd)	NH₃ (obsd)	pyridine (calcd)	pyridine (obsd)	Br⁻ (calcd)	Br⁻ (obsd)	I⁻ (calcd)	I⁻ (obsd)	S₂O₃²⁻ (calcd)	S₂O₃²⁻ (obsd)
Na⁺	0	−1.3	−0.5										
Ag⁺	3.8	2.7	2.3	2.9	3.2	1.9	2.0	4.3	4.3			13.7	13.0
Cu⁺	2.3											11.5	11.7
Tl⁺	1.3							0.8	0.9	0.9	0.7	2.6	3.1
Mg²⁺	0	2.5	2.6	0.6	0.2								
Ca²⁺	0	1.4	1.3	0.0	−0.2								
Fe²⁺	0.3	4.1	3.9	1.8	1.4	0.7	0.7						
Ni²⁺	1.7	5.4	4.6	3.3	2.8	1.7	1.8					3.7	1.0
Cu²⁺	1.9	6.0	6.5	3.8	4.2	1.9	2.5	1.5	∼0				
Zn²⁺	0.7	4.5	4.4	2.2	2.4	1.0	1.0	−0.1	−0.6	−0.7	−1.3	2.6	3.1
Cd²⁺	2.0	5.3	2.3?	3.4	2.7	1.8	1.3	1.8	2.2	2.0	2.3	6.1	5.8
Hg²⁺	7.3	9.4	10.3	8.8	8.8	5.3	5.1	9.1	9.0	12.9	12.8	29.5	29.7
Fe³⁺	1.4	11.1	11.8					0.9	0.5				
In³⁺	1.8	9.8	10.3					1.4	2.0	1.2	1.2		
Number of log K data used for calculation		17		13		8		9		6		7	
α'		1.50		0.89		0.41		−0.01		−0.12		0.12	
β'		0.59		0.90		0.60		1.40		2.08		4.40	
γ'		−2.79		−2.50		−1.29		−0.97		−1.63		−3.31	

a) The values on the left part of each row are those calculated by $\log K = \alpha'X + \beta'Y' + \gamma'$ and those on the right are the observed ones[4] which are the same as in the previous paper[4] except for some revision and supplements using data in Ref. 14. The values for $S_2O_3^{2-}$ are the overall instability constants of $M(S_2O_3)_2^{(n-4)+}$. Other values are of mono-coordinated complexes.

b) Since very few $\log K$ values have been reported for the complexes of less soft metal ions, such as Fe^{2+}, Co^{2+} and Cu^{2+}, Y' was estimated from $Y' = 2.8Y - 6.2$ which is found to hold approximately between Y and Y'. Y' was put to zero when $2.8Y - 6.2$ became negative. This estimation seems sufficient for $\log K$ calculation, becuase $\beta'Y'$ term in Eq. (2) contributes little in these cases.

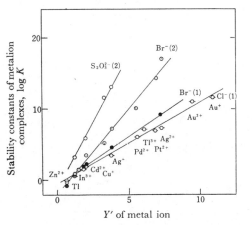

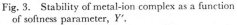

Fig. 3. Stability of metal-ion complex as a function of softness parameter, Y'.

Numbers in parentheses are the same as in Fig. 1.

of reported $\log K$, satisfactory linear correlation of new parameters, Y', with $\log K$ values of soft-soft complexes is obtained as demonstrated in Fig. 3. This good correlation is comparable with that of X with $\log K$ of hard-hard complexes.[4] A part of the deviation from the straight line found in Fig. 3 may be due to the minor contribution from $\alpha'X$ term in Eq. (2). Some of the calculated $\log K$ values by Eq. (2) are given in Table 1 for several metal-ion complexes. The results for the complexes of ligands with considerable softness are given in this table. For hard ligand complexes, the agreement between the calculated and the observed $\log K$ values[4] was as good as the present one. Agreements between the calculated and the measured $\log K$ in this table are excellent as a whole. Remarkable improvements in agreement are observed especially for I^- and $S_2O_3^{2-}(\beta'>1)$. As for the magnitude of β there is little change; both β and β' follow the softness sequence and are linearly correlated with n_{Pt}, the reactivity index of a nucleophile for the displacement reaction of Pt^{2+}-complex.[10]

10) U. Belluco, L. Cattalini, F. Basolo, R. G. Pearson and A. Turco, *J. Amer. Chem. Soc.*, **87**, 241 (1965).

Discussion

Comparison of Y' with Other Softness Parameters. The proposed parameter Y' thus obtained semi-empirically on the basis of the original criteria for the soft-hard classification, may reasonably be considered as a quantitative expression of softness (see Figs. 1 and 2). It seems of interest to compare such various parameters proposed for the quantitative expression of softness of metal ion, as $R(N)$,[3] $R(W)$,[5] Y,[4] α/β[7] and E_n^+,[6] with the present parameter Y' and look into the physical meaning of softness. All of these parameters are useful as rough criteria for the soft-hard classification of metal ions, although a few exceptions are always inevitable.

However, it is apparent that the correlations are only qualitative. If one plots these parameters against Y', only fair correlations are obtained for E_n^+ and Y, and rather poor ones for $R(N)$, $R(W)$ and α/β. This means that the approximation used for the derivation of the parameters previously proposed are too simple for quantitative use.

Origin of Softness. The soft-hard classification is based on the fact that, reviewing inorganic stability constants in *aqueous* solution, metal ions fall into two distinct groups: those which form halogeno-complexes whose stability order is (i) $F>Cl>Br>I$, and (ii) $I>Br>Cl>F$. This situation, however, contrasts with that observed in the gas phase, where the sequence of molecular stability is always found to be $MF_2>MCl_2>MBr_2>MI_2$. This obviously demonstrates the importance of solvent effect, that is, the change in solvation state during the complex formation. Klopman[6] has recently derived his softness parameter as the difference between the bond energy and the desolvation energy accompanying complex formation. The bond energy consists of electrostatic and charge-transfer interaction energies. In his treatment, the extent of softness of metal ion depends on how much the bond energy overcomes the energy loss due to desolvation. We have previously discussed the bond energy of coordination and the soft-hard properties, dividing the total bond energy into electrostatic, σ-bond and π-bond energies.[4] In this treatment the heat of solvation was assumed to be proportional to the electronegativity X, since the bond energy of metal ion to aquo ligand is mainly correlated to the αX term.

The heat of complex formation, $M^{n+}+mI^-\rightarrow ML_m^{(n-m)+}$ ($L=F$, Cl, Br and I) is equal to the difference between the heat of M-L bond formation and the desolvation energy. Since the latter is roughly equal, with opposite sign, to the solvation energies of M^{n+} and L^-, the stability sequence for a given metal ion, either $F>Cl>Br>I$ or $F<Cl<Br<I$, may be determined by the change in M-L bond energy relative to that in solvation energy of

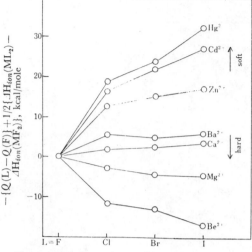

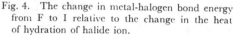

Fig. 4. The change in metal-halogen bond energy from F to I relative to the change in the heat of hydration of halide ion.

$Q(L)$: Heat of hydration of halide ion (L^-).
$Q(F)$: Heat of hydration of fluoride ion.
$\Delta H_{ion}(ML_2)$: Heat of reaction of M^{2+}(gas)+ $2L^-$(gas)$\rightarrow ML_2$(gas).

L^- from F to I.

In Fig. 4, this situation is demonstrated for some metal ions. The coordinate represents the difference between the changes among halogens from F to I in M-L bond energies and those in solvation energies of L^- by refering to fluoride ion in both cases. In contrast to the M-L bond energy which is approximated by half of the heat of the reaction of $M^{2+}+2L^-\rightarrow ML_2$ in the gas phase, $F>Cl>Br>I$, the desolvation energy of L^- suggests a favorable order for the stable complex formation ($F<Cl<Br<I$). It may be stated, therefore, that the main factor to determine the stability sequence of metal-halogeno complexes is how much the decrease in M-L bond energy in F, Cl, Br, I series is compensated by the decrease in desolvation energy in the same series. Thus Hg^{2+} and Cd^{2+} are soft, because their M-L bond energies decrease more moderately than those of metal ions such as Be^{2+} and Ca^{2+}.

Roughly speaking, both the solvation energy and

TABLE 2. CALCULATED STABILITY CONSTANTS OF THE COMPLEXES OF METHYLMERCURIC ION, CH_3Hg^+ BY Eq. (2) AND THE OBSERVED VALUES

	OH^-	Ac^-	NH_3	pyridine	Cl^-	Br^-	I^-
calcd[a]	12.9	4.0	10.6	5.9	5.9	7.4	10.4
obsd[b]	12.8	4.4	10.5	6.1	5.9	7.6	10.3

a) Calculated using $X=6.52$ and $Y'=5.35$ together with α', β' and γ' in Table 1.
b) Taken from Ref. 15.

TABLE 3. CALCULATED REACTIVITY INDICES BY Eqs. (3), (4) AND (5) AND REPORTED VALUES

		Ac⁻	pyridine	Cl⁻	Br⁻	OH⁻	NH₃	I⁻	S₂O₃²⁻ ᵃ⁾
n	calcd	1.48	2.51	2.79	3.63	5.00	4.40	5.15	6.67
	reported[b]	2.72	3.6	3.04	3.89	4.20	—	5.04	6.36
E_n	calcd	0.57	0.98	1.12	1.45	1.92	1.71	2.06	2.65
	reported[c]	0.95	1.20	1.24	1.51	1.65	1.84	2.06	2.52
H	calcd	6.86	4.56	−0.5	−2.1	20.0	10.5	−4.6	−2.8
	reported[c]	6.46	7.04	−3.0	−6.0	17.5	11.2	−9.0	3.6

a) The values of α', β' of $S_2O_3^{2-}$ are estimated as $\alpha'=0.06$ and $\beta'=2.50$ for mono-coordinated complexes, from those of dicoordinated ones considering the dependency of α' and β' of other ligands on the changes in the coordination number.

b) Taken from Ref. 11.

c) Taken from Ref. 8.

the σ-bond energy linearly depend on X,[4] probably the latter, to a lesser degree. As shown in Fig. 4, the more ionic the M-L bond, the more strongly the bond energy depends on X, resulting in a more negative slope in Fig. 4. In other words, an increase in covalency increases softness. The ability of π-bond formation (back donation) strongly favors the sequence I>Br>Cl>F. The reason why E_n^* of Pd²⁺, Pt²⁺ and Au⁺ are lower than those expected from Y' may be explained by the fact that π-bonding was neglected in E_n^* calculation. Thus, if the decrease in M-L bond energy from F to I of a certain metal ion becomes moderate, by the enhancement of covalency, especially of π-bonding in M-L bond, class (b) character or softness, increases.

Changes in softness may also occur by the solvent effect in the way pointed out by Williams and Hale.[5] Reduction of dielectric constant of solvent decreases the change in the solvation energy from F to I, so that the solvation becomes less important. As a consequence all cations revert to class (a) character as dielectric constants approach unity, that is, for example, in the gas phase.

Some Examples of Applications. The values of softness and hardness, (X, Y'), of CH_3Hg^+, CF_3-Hg^+ and $C_3F_7Hg^+$ are calculated by Eq. (2) as $(X=6.52, Y'=5.35)$, $(6.79, 5.80)$ and $(6.54, 5.89)$, respectively. These calculated log K values agree well with the reported ones (Table 2). Y' are large as expected, but it is to be noted that X values are also large. Belluco et al.[10] reported poor correlation of the log K of CH_3Hg^+ complexes with n_{Pt}. However, n_{Pt} correlates well with β[4] or β'. This discrepancy is understandable, if one takes into account the considerable hardness of CH_3Hg^+; n_{Pt} and β are determined by softness[4] while log K of CH_3Hg^+ complexes depends on both softness and hardness. The ability of mercuric ion to form strong σ-bond may partly account for the hardness, as σ-bond energy depends on X.[4]

The reactivity indices of a nucleophile in various displacement reactions on carbon, such as n,[11] E_n and H,[8] have been related to the present α' and β' by the formulas

$$n = 2.3\alpha' + 2.6\beta' \qquad (3)$$

$$E_n = 0.9\alpha' + 1.0\beta' \qquad (4)$$

$$H = 13.2\alpha' - 1.4\beta' \qquad (5)$$

n and E_n are the nucleophilic constants (approximately, $n\propto E_n$) and H the basicity of a nucleophile (oxibase scale[12]). The values calculated by the above equations are in fair agreement with the reported values of n and E_n, but not with H (Table 3). Poor argeement for H may not be serious, because H is usually of little importance in the reactivity which depends mostly on n (or E_n). Equations (3) and (4) demonstrate that the reactivity of a nucleophile depends almost equally on hardness α' and softness β'. The ratio of these two coefficients, which gives the relative dependency of the reactivity on hardness to that on softness, is expected to vary from one type of reaction to another. In fact, in the case of the ligand displacement of Pt²⁺ complexes, the reactivity n_{Pt} is mainly controlled by softness.[4] The reactivities for the displacement reactions of CH_3I and CH_3Br by a nucleophile, n_{CH_3I} and n_{CH_3Br}, are determined by both hardness and softness, since they are proportional to E_n.[13]

The authors are indebted to Professor Y. Yoneda for his helpful discussions.

11) C. G. Swain and C. B. Scott, *J. Amer. Chem. Soc.*, **75**, 141 (1953).

12) R. E. Davis, *ibid.*, **87**, 3010 (1965).

13) R. G. Pearson, H. Sobel and J. Songstad, *ibid.*, **90**, 319 (1968).

14) Chem. Soc. Special Publication, No. 17, Stability Constants of Metal-ion Complexes, London (1964).

15) R. S. Tobias, *Organometal. Chem. Rev.*, **1**, 93 (1966).

Volume 2, number 5   CHEMICAL PHYSICS LETTERS   September 1968

20

SCALES OF SOFTNESS FOR ACCEPTORS AND DONORS

S. AHRLAND

Department of Inorganic Chemistry, Chemical Center.
University of Lund, Sweden

Received 23 July 1968

Various parameters designed to express quantitatively the softness of chemical species, i.e. their tendency for covalent bond formation, are discussed. Two new parameters are defined which give a good measure of the softness of acceptors and donors, respectively.

The classification of acceptors and donors as soft or hard has been found to be a useful means of correlating a large mass of chemical information [1-5]. The empirical criteria leading to this classification further indicate that the relative softness, or hardness, of a species is related to the nature of the bonds it tends to form. The softer an acceptor, and donor, the more covalent the bond formed between them; the harder they are, the more electrostatic the bond. The new concepts thus describe very fundamental properties of the reacting species.

Initially, this description was largely a qualitative one. The lack of a quantitative measure of softness or hardness has been strongly felt [2], however, and it cannot be denied that as long as no such measure exists, any proposed order of softness must be arbitrary to a certain extent.

As a consequence, several attempts have lately been made to establish a quantitative scale of softness. Nearest at hand would be to use equilibrium data, i.e. free energy relations [3]. For all reactions taking place in markedly polar solvents, however, the interactions of the acceptors, and donors, with the solvent molecules are so strong that the change of free energy will depend very much upon these. The terms referring to the acceptor-donor interaction will thus be difficult to discern [4,5].

It turns out, however, that in a strongly polar solvent like water the enthalpy change of the total reaction will generally be more negative (or less positive), the softer the acceptor and the donor involved [4,5]. Thus, for the complexes formed by various acceptors with one and the same donor, the enthalpy change measured for e.g. the first step, ΔH_1^O, will order the acceptors in a fairly reasonable softness sequence. Conversely,

the enthalpy change for the formation of the first complex between various donors and one and the same acceptor will provide a scale of softness for the donors. To be allowed, such comparisons have of course to involve only complexes with the same number of bonds between the acceptor and the ligands; preferably the complexes should be monodentates. In practice, extensive comparisons are difficult, as an enthalpy change determination demands the formation of complexes of suitable strength and solubility. A particular ligand does not generally form such complexes with a wide variety of acceptors, especially if chelate ligands should be avoided. Coherent scales, characterizing each acceptor and donor with a certain softness number, are therefore practically impossible to obtain in this way. In table 1, enthalpy changes for a few selected ligands (taken from refs. [4,5]; all given in kcal/mole) have been listed. These will be further discussed below.

The empirical fact that the enthalpy changes for reactions taking place in aqueous solution provide a direct measure of the softness of the acceptors and donors involved evidently depends upon a balance of influences peculiar to a strongly polar solvent like water. In less polar solvents, where the electrostatic interaction between acceptor and donor is stronger, and the interactions between these and the solvent molecules are weaker, this balance is completely upset, and even more so in the gasphase [5]. At least for all reactions involving ions or marked dipoles, the enthalpy changes in such media are very much dominated by the electrostatic interactions, and will therefore not provide any direct measure of the softness, i.e. of the covalent contribution to the bond.

Drago and Wayland [6] have suggested a proce-

303

Table 1

Comparison between the softness parameters σ_A, σ_K and σ_P, and the values of ΔH_1^0 for the formation of complexes with donors of varying softness. The acceptors are listed in the order of increasing value of σ_A, the donors in the order of increasing softness from left to right. (Taken from refs. [4,5], all given in kcal/mole.)

	σ_A	σ_K	σ_P	F^-	O [a)	O,N [b)	Cl^-	I^-	P [c)
						ΔH_1^0			
Ba^{2+}	0.5	−1.89	0.183			− 4.93			
La^{3+}	0.5	−4.51	0.171		2.2	− 2.80			
Sr^{2+}	0.6	−2.21	0.172			− 4.08			
Y^{3+}	0.6		0.147	2.2	3.3				
Ca^{2+}	0.9	−2.33	0.180			− 6.55			
Na^+	0.93	0.00	0.211						
Mg^{2+}	1.4	−2.42	0.167			3.49			
Co^{2+}	1.8		0.130			− 4.20			
Mn^{2+}	2.0		0.124			− 4.56			
Ni^{2+}	2.0	−0.29	0.126			− 7.55			
H^+	2.28	−0.42	0.149	2.8	−0.8				− 8.3
Fe^{3+}	3.1	−2.2	0.097	2.4			4.2		
Zn^{2+}	3.1		0.115	1.5	2.0	− 4.85			
Cu^{2+}	3.1	0.55	0.104		1.0	− 8.15			
Cd^{2+}	3.5	2.04	0.081	1.2	1.5	− 9.05	−0.1	− 2.3	
Pb^{2+}	4.1		0.131		−0.1	−13.20			
Ag^+	4.2	2.82	0.073				−2.7		−19.3
Tl^{3+}	4.3	3.37					−6.0		
Hg^{2+}	4.6	4.64	0.064	0.9		−18.90	−5.8	−18.0	∼ −30

a) O = acetate.
b) O, N = ethylendiaminetetraacetate (EDTA).
c) P = diethyl-β-hydroxyethyl-phosphine, $(C_2H_5)_2PCH_2CH_2OH$.

dure for distinguishing between the electrostatic and covalent contributions to the enthalpy change of acceptor-donor reactions taking place in non-polar solvents like carbon tetrachloride. The parameters found yield a good and coherent picture of the facts observed, but at least as applied by the authors, the method is restricted to species soluble in non-polar solvents. This means that ionic species are virtually excluded. In view of the paramount theoretical and practical importance of such species, it seems necessary to look for an approach including them too.

Pearson and Mawby [7] have calculated the coordinate bond energies (CBE) for a large number of metal halides, i.e. $-\Delta H^0$ for the gas phase reactions $M^{n+}(g) + nL^-(g) \rightarrow ML_n(g)$. Such reactions do not occur in nature, but ΔH may be computed by a combination of various thermodynamic quantities. The most negative values of ΔH^0 are then found for the formation of complexes with the hardest among the ligands, i.e. F^-, while the reactions of the softest ligand, I^-, are generally accompanied by the least negative ΔH^0. The trend is thus just contrary to that observed for reactions in aqueous solution. This evidently depends upon the predominating electrostatic contribution to ΔH^0 in the gas phase reactions [5]. Still, the covalent contribution to the bonding will make it itself more felt, the softer the acceptor and the donor involved. Consequently, the relative difference in CBE between fluoride and iodide complexes will be smaller, the softer the acceptor. This has been used by Pearson and Mawby to form a softness parameter $[CBE(F^-) - CBE(I^-)]/CBE(F^-)$, here called σ_P. This parameter reflects faithfully the degree of softness, as indicated by the chemical behaviour, but only if the comparison is restricted to acceptors of the same charge, table 1. The value of σ_P systematically decreases with the charge, evidently of the accompanying rapid increase of $CBE(F^-)$ [5].

304

Another softness parameter has been introduced by Klopman [8] from a theoretical treatment based of polyelectronic perturbation theory. In agreement with the conclusions already drawn from chemical evidence, he finds that hard-hard interactions are essentially electrostatic ("charge controlled") while the soft-soft interactions are essentially covalent ("frontier controlled"). He further concludes that the lower the energy of the empty frontier orbital is relative to the energy wanted for the desolvation of the acceptor, the harder the acceptor. The difference (in eV) between these two quantities will therefore be a suitable softness parameter. It is introduced in table 1 as σ_K. To facilitate a comparison with the parameter σ_A defined below, the difference has been formed so that σ_K increases with increasing softness (which is opposite to the procedure adopted by Klopman). The softness sequence found is on the whole very reasonable, though the intermediate position occupied by the alkali ions seems questionable. If the softness number should express the general chemical behaviour of an acceptor, then Na^+ should evidently not be placed between Ni^{2+} and Cu^{2+}.

Also for donors, Klopman defines a softness parameter, obtained as the negative sum of the frontier orbital energy of the donor and its desolvation energy. The values of this parameter (in eV), here denoted σ_L, are listed in table 2.

It is now interesting to explore the following simple empirical approach to find a softness parameter for acceptors. As a reasonable starting point it is postulated that the more completely the energy spent on the formation of a positive ion in the gasphase is regained by the introduction of the ion in a hard solvent like water, the harder the ion. Thus, the larger the difference between the total ionization potential for the formation of $M^{n+}(g)$ and the dehydration energy $-\Delta H^O$ (both > 0), the softer the ion. For the comparison of ions of different charges, the difference should further be divided by n, in order to express the difference per interacting charge, here denoted σ_A. The total difference $n\sigma_A$ is evidently equal to ΔH^O of the reaction $M(g) \rightarrow M^{n+}(aq) + ne^-$. For a given value of n, the formation of an aqueous ion out of a gaseous atom will thus be more endothermic, the softer the ion.

A practical difficulty is that values of ΔH_h^O are missing, or unreliable, for quite a few ions of interest. The values selected for the present calculation are presented in ref. [5]. The resulting set of σ_A is listed in table 1, where the acceptors have been arranged in the order of increasing softness, as indicated by increasing values of σ_A.

Table 2
Comparison between the softness parameters σ_B and σ_L. The donors are listed in the order of increasing values of σ_B. (All is given in eV.)

L^-	EA for L	$-\Delta H_h^O$	σ_B	σ_L
F^-	3.45 a)	5.25 e)	−8.70	−12.18
ClO_4^-	5.8 b)	2.4 e)	−8.2	
OH^-	2.8 c)	4.8 e)	−7.6	−10.45
Cl^-	3.62 a)	3.76 e)	−7.38	− 9.94
Br^-	3.36 a)	3.48 e)	−6.84	− 9.22
I^-	2.6 a)	3.5 e)	−6.13	− 8.31
CN^-	3.07 a)	3.06 e)	−6.1	− 8.78
SH^-	2.3 a)	3.5 e)	−5.8	− 8.59
H^-	0.72 d)	3.41 f)	−4.13	− 7.37

a) Handbook of Chemistry and Physics, 48th ed. (Chemical Rubber Co., Cleveland, Ohio, 1967-68) p. E68 (compilation by H. O. Pritchard).
b) H. O. Pritchard, Chem. Rev. 52 (1953) 529.
c) From ref. [8].
d) Hylleraas, as quoted in C. S. G. Phillips and R. J. P. Williams, Inorganic Chemistry, Vol. 1 (Clarendon Press, Oxford, 1965) p. 27.
e) H. F. Halliwell and S. C. Nyburg, Trans. Faraday Soc. 59 (1963) 1126.
f) Put equal to the desolvation energy of Klopman, which seem reasonable for ions of low ΔH_h^O.

The sequence certainly reflects the variation in chemical behaviour along the series quite well. It also agrees with the sequences indicated by σ_K or σ_P, except on these points which have already been mentioned. The parameter σ_A thus places the alkali ions among the hard acceptors, and the acceptor charge does not seem to exert any undue influence upon its value.

Also the values of ΔH_1^O listed in table 1 show trends which are on the whole very compatible with the σ_A-sequence. For each ligand, independent of its softness, the values of ΔH_1^O generally become less positive, or more negative with increasing value of σ_A. Considering the complex nature of the net enthalpy change ΔH_1^O, the trends are rather more consistent than would be expected. This applies especially to the chelate ligand EDTA where special influences on ΔH_1^O should be expected, due to changes in the number and/or arrangement of the coordinated donor atoms.

Pearson has pointed out to me that a softness parameter analogous to σ_A can be defined also for donors. In this case the enthalpy change of the reaction $L(g) + ne^- \rightarrow L^{n-}(aq)$, divided by n, should provide a measure of the softness of the donor L^{n-}. This enthalpy change would be equal

305

to $-EA + \Delta H_h^O$, where EA is the electron affinity of L, defined according to the usual convention so that $EA > 0$ when $L(g) + ne^- \rightarrow L^{n-}(g)$ is exothermic.

Unfortunately, reliable values of the softness parameter $\sigma_B = (-EA + \Delta H_h^O)/n$ thus defined cannot be calculated for very many donors, on account of the lack of good values of ΔH_h^O, and, especially, EA. In fact it seems possible to obtain reasonably certain values of σ_B only for a few donors, all monovalent anions. These have been listed in table 2, together with the values of EA and ΔH_h^O used for their calculation (all given in eV). The donors are arranged in the order of σ_B becoming less negative, which should also be the order of increasing softness. The sequence found seems on the whole quite reasonable. The only questionable feature is the almost similar values found for I^- and CN^-. From the values of ΔH^O found for their complex formation with the same acceptors [4], and also from their general chemical behaviour, one would rather expect CN^- to be markedly softer than I^-. The σ_B sequence also so coincides well with that indicated by σ_L except for I^- which occupies an even more unexpected

position in Klopman's sequence. The general agreement between the sequences, and their compatibility with the general chemical behaviour of the donors involved are strong indications, however, that these softness parameters really express fairly faithfully the relative ability of the various donors to form covalent bonds.

REFERENCES

[1] R. G. Pearson, J. Am. Chem. Soc. 85 (1963) 3533.
[2] R. G. Pearson, Science 151 (1966) 172.
[3] R. G. Pearson, Chem. Brit. 3 (1967) 103.
[4] S. Ahrland, Helv. Chim. Acta 50 (1967) 306.
[5] S. Ahrland, Structure and Bonding, Vol. 5, Eds. C. K. Jørgensen et al. (Springer-Verlag, Heidelberg, 1968) in press.
[6] R. S. Drago and B. B. Wayland, J. Am. Chem. Soc. 87 (1965) 3571.
[7] R. G. Pearson and R. J. Mawby, Halogen Chemistry, Vol. 3, Ed. V. Gutmann (Academic Press, London and New York, 1967) p. 55.
[8] G. Klopman, J. Am. Chem. Soc. 90 (1968) 223.

306

[Reprinted from Inorganic Chemistry, **6**, 1067 (1967).]

21

Use of the Edwards Equation to Determine Hardness of Acids

Sir:

The utility of the concept of soft and hard acids and bases[1] was recently emphasized by an international symposium held on this subject.[2] One limitation to the application of this concept would appear to be the difficulty of unambiguously classifying acids and bases as hard or soft. In the case of bases, a numerical ordering is possible based on the logarithm of the rate constant for the reaction of various nucleophiles with *trans*-dichlorodipyridineplatinum(II).[3] It is also possible to get an ordering of softness[4] from the C_B/E_B ratios of the Drago–Wayland equation,[5] where C_B and E_B are constants assigned to Lewis bases in a fashion designed to make C_B reflect the covalent contribution and E_B reflect the electrostatic contribution of the base to the enthalpy of formation of a molecular addition compound. Neither of these approaches is applicable to the determination of the hardness of metal ions. We wish to report here the use of α/β ratios of the Edwards equation[6] as a measure of the hardness of metal ions. Recent stability constant data,[7] β_n, were used along with what we felt to be the best assignments[8] of H (ligand $pK_a + 1.74$) and E_n (the ligand nucleophilicity parameter) to obtain α and β values of the Edwards equation. A computer program was used to fit a rearranged form of the Edwards equation

$$\log [\beta_n + 1.74n]/H = \alpha E_n/H + \beta$$

Although this method of fitting the Edwards equation has been questioned by Edwards,[9] its simplicity has led us and others[10] to use it. Our results are shown in Table I along with the Pearson classification. Hard

(1) R. G. Pearson, *J. Am. Chem. Soc.*, **85**, 3533 (1963).
(2) Condensation of the symposium papers appeared in *Chem. Eng. News*, **43** 90 (May 31, 1965).
(3) R. G. Pearson, *Science*, **151**, 172 (1966).
(4) See ref 3, particularly footnote 29.
(5) R. S. Drago and B. B. Wayland, *J. Am. Chem. Soc.*, **87**, 3571 (1965).
(6) J. O. Edwards, *ibid.*, **76**, 1541 (1954).
(7) L. G. Sillén and A. E. Martell, "Stability Constants of Metal Ion Complexes," The Chemical Society, London, 1954.
(8) Sources the same as those cited for calculations in D. H. McDaniel and A. Yingst, *J. Am. Chem. Soc.*, **86**, 1334 (1964).
(9) J. O. Edwards, *ibid.*, **78**, 1819 (1956).
(10) R. E. Davis, R. Nehring, S. P. Molnar, and L. A. Suba, *Tetrahedron Letters*, 885 (1966).

TABLE I
A COMPARISON OF α/β WITH HARDNESS

Metal ion	α	β	α/β^a	Hardness[b]
Hg^{2+}	5.786	−0.031	187	S
Cu^+	4.060	0.143	28.4	S
Ag^+	2.812	0.171	16.5	S
Pb^{2+}	1.771	0.110	16.1	B
Sr^{2+}	1.382	0.094	13.0	H
Cd^{2+}	2.132	0.171	12.5	S
Cu^{2+}	2.259	0.233	9.7	B
Mn^{2+}	1.438	0.166	8.7	H
In^{3+}	2.442	0.353	6.9	H
Mg^{2+}	1.402	0.243	5.8	H
Zn^{2+}	1.367	0.252	5.4	B
Ga^{3+}	3.795	0.767	5.0	H
Ba^{2+}	1.786	0.411	4.4	H
Fe^{3+}	1.939	0.523	3.7	H
Ca^{2+}	1.073	0.327	3.5	H
Al^{3+}	−0.749	1.339	0.6	H
H^+	0.000	1.000	0.0	H

[a] Ratio from monoligated complex; sign neglected. [b] Pearson classification;[3] S = soft, H = hard, B = borderline. [c] Not classified by Pearson; predicted from α/β ratio.

acids in general have a low α (sensitivity to nucleophilic ligand character, or polarizability) and a high β (sensitivity to basicity of ligand toward protons). The α/β ratio allows simultaneous consideration of both factors and should be of value in assigning hardness to more metal ions as further stability constant data become available. Negative α or β values are interpretable as due to an overlap of the factors making up H and E_n.[8] Since these negative α or β values will always be small, the absolute value of α/β will unambiguously put metals with negative α in the hard classification and metals with negative β in the soft classification.

Acknowledgment.—We wish to thank The Ethyl Corp. for a fellowship supporting A. Y. for the academic year 1964–1965, Myron J. DeLong for assistance in the computer work, and the National Science Foundation for a grant to the University of Cincinnati Computing Center (Grant No. 19281).

DEPARTMENT OF CHEMISTRY   AUSTIN YINGST
UNIVERSITY OF CINCINNATI   DARL H. MCDANIEL
CINCINNATI, OHIO 45221

RECEIVED JULY 29, 1966

[Reprinted from the Journal of the American Chemical Society, **90**, 223 (1968).]

22

Chemical Reactivity and the Concept of Charge- and Frontier-Controlled Reactions

G. Klopman[1]

Contribution from the Cyanamid European Research Institute, Cologny–Geneva, Switzerland. Received July 17, 1967

Abstract: A general treatment of chemical reactivity is described. It is based on a polyelectronic perturbation theory involving both reactants and the solvent. The perturbation equation reproduces the known qualitative features of the concept of hard and soft Lewis acids and bases, of nucleophilic order, and of other reactivity indices. It emphasizes the importance of charge- and frontier-controlled effects, connected with charge transfer or partly covalent bonding in the transition state. Several specific examples are examined such as nucleophilic addition and electrophilic substitution on heterocyclic molecules. Satisfactory agreement with unexpected experimental evidence is obtained. The treatment seems to be particularly appropriate for the study of generalized ambient reactivity.

It is the usual practice to relate the chemical reactivity of organic compounds to a particular MO index such as free valence, charge density, Z value, and localization energies.[2] This procedure, however, does not account for the nature of the reagent and fails to reproduce the changes in relative reactivity of various positions of attack. Thus, changes in orientation observed in the electrophilic attack on aromatic derivatives by several acceptors cannot be explained satisfactorily[3] (Table I).

Table I. Isomer Distribution in Electrophilic Substitution on Toluene

Reaction	o	m	p
Chlorination[a]	75	2	23
Nitration[a]	59	4	37
Bromination[b]	33	1	66
Mercuration[c]	19	7	74

[a] Reference 2. [b] V. Gold and M. Whittaker, *J. Chem. Soc.*, 1184 (1951). [c] W. J. Klapproth and F. H. Westheimer, *J. Am. Chem. Soc.*, **72**, 4461 (1950).

(1) Chemistry Department, Case-Western Reserve University, Cleveland, Ohio 44106.
(2) J. Koutecky, R. Zahradnik, and J. Cizek, *Trans. Faraday Soc.*, **57**, 169 (1961).

Steric effects are clearly responsible for the anomalous behavior of some reagents but can certainly not account for all observed differences. The situation is rather similar for all ambient reactions, where no fundamental explanation could be found for why a given reagent attacks a particular position and another reagent a different one,[4] *e.g.*

$$MeNO_2 \xleftarrow{MeI} O{=}N{-}O^- \xrightarrow{*BuCl} *BuONO$$

$$MeSCN \xleftarrow{MeI} N{\equiv}C{-}S^- \xrightarrow{RCOX} RCONCS$$

However, experience has shown that some atoms have a specific affinity for some other atoms, and an order of nucleophilicity could be determined to fit this requirement.[5,6] Nevertheless, it appeared that the nucleophilic order itself is different for various reaction centers. Thus, it also varies with the type of reaction and is, for example, quite different for alkylation than for acyla-

(3) See, however, R. O. C. Norman and G. K. Radda, *J. Chem. Soc.*, 3610 (1961).
(4) N. Kornblum, R. A. Smiley, R. K. Blackwood, and D. J. Iffland, *J. Am. Chem. Soc.*, **77**, 6269 (1955).
(5) C. G. Swain and C. B. Scott, *ibid.*, **75**, 146 (1953).
(6) J. O. Edwards, *ibid.*, **76**, 1540 (1954).

223

tion or phosphonylation.[7] In the latter case, it is found to be related to the pK_a of the nucleophile.

The interpretation of inorganic reactivity and of stabilities of complexes is in an even worse situation since no direct calculations could yet be made to help understanding the chemical behavior of inorganic species. For example, no theoretical explanation can be given for the fact that, in solution, some metals M form their stable complexes with halogens as in the following sequence, F > Cl > Br > I, while others as I > Br > Cl > F. This is illustrated in Table II for the stability constants (log K) of several complexes in water.[8]

Table II. Stability Constants (Log K) of Complex Formation in Water

	F⁻	Cl⁻	Br⁻	I⁻
Fe⁺³	6.04	1.41	0.49	...
Zn⁺²	0.77	−0.19	−0.6	−1.3
Cd⁺²	0.57	1.59	1.76	2.09
Hg⁺²	1.03	6.74	8.94	12.87

This situation is encountered quite commonly in solution but contrasts with that observed in the gas phase where the sequence of stability is usually found to be $MF_n > MCl_n > MBr_n > MI_n$. This undoubtedly demonstrates the fundamental importance of solvent effects. The main problem lies in the difficulty of handling quantitatively the simultaneous effects of the solvent and of the ligand interaction with M. To our knowledge, such an approach has not yet been described, but qualitative attempts have been made to rationalize the behavior observed. Here also, specific affinities have been found between particular atoms in definite oxidation states. Ahrland, Chatt, and Davies thus compiled available data and suggested dividing the acceptors into (a) and (b) categories.[8] The (a) acceptors are those which form their most stable complex with the first ligand of each group: F > Cl > Br > I; N > P > As > Sb > Bi; O > S > Se > Te. The (b) acceptors are those which show approximately the reverse order.

This general idea of classifying reagents with respect to their chemical behavior stimulated further research on the physical properties of complexes and has more recently been extended by Pearson to the general acid–base reaction.[9] He has suggested the name soft bases for those bases (donors or nucleophiles) whose valence electrons are easily polarizable, and hard bases for those whose valence electrons are not. Hard acids (acceptors or electrophiles) are recognized as small sized, highly positively charged, and not easily polarizable; soft acids are defined as those possessing the reverse properties. Pearson then formulates a general principle based on experimental observations according to which hard acids prefer to coordinate with hard bases and soft acids prefer to coordinate with soft bases.

This concept, which found its first applications in rationalizing inorganic stability constants, quickly developed and was shown to be useful even in organic

chemistry. Thus, Hudson showed that by considering hard and soft reagents, some ambident reactions can be rationalized.[10] Saville, using this concept, interpreted electrophilic catalysis and described a number of new reactions.[11]

However, predictions remain qualitative, as the general concept suffers from the lack of physical basis, and the hardness or softness of an acid or a base to some extent remains a matter of personal appreciation.[12] This weakness of the concept has been noted by several authors; quantitative physical definition have subsequently been proposed, and analogies found between the idea one has of hardness and softness and several properties such as polarizability, low-lying d orbitals, and oxidizing properties. None of these, however, correlates very well with the experimental facts nor stands on a well-defined physical basis, and none accounts for all properties attributed to hardness and softness.

A more interesting idea relates the hard–hard and soft–soft character respectively to ionic and covalent interaction.[13] An empirical equation was even suggested by Drago[14] to correlate heats of formation of acid–base complexes, such as

$$-\Delta H = E_A E_B + C_A C_B \tag{1}$$

where the E terms represent the susceptibility of the acid or the base to undergo electrostatic interaction, and the C terms represent their ability to participate in covalent bonding.

This equation seems to give excellent agreement with experiment and compares very favorably with that suggested by Pearson,[9] where the pK_a and a softness index are used for the correlation. However, its empirical nature and the number of independent parameters involved in the calculations make it very impractical to use. Also, no physical reason or explanation for hard and soft behavior is provided by such an approach. Nevertheless, in spite of all the possible criticisms, the HSAB concept has undoubtedly proved to be useful, and therefore we have tried to look more deeply into its physical implications. In this paper, we show how the polyelectronic perturbation treatment of chemical reactivity, published elsewhere,[15] can be used in this context and accounts for most of the phenomena described above. Such a treatment leads to a reasonable definition of hardness and softness; it implies Pearson's principle and provides a general interpretation of ambident reactivity.

Theory

When two reactants approach each other, a mutual perturbation of the molecular orbitals of both reactants occurs. The resulting change in energy can be estimated from SCFMO calculations. When the bonds are completely formed and when the systems are simple enough, then good accuracy can be obtained for the calculations of the heats of formation. We have previ-

(7) R. F. Hudson, *Chimia*, (Aaraw), **16**, 173 (1962).

(8) S. Ahrland, J. Chatt, and N. R. Davies, *Quart. Rev.* (London), **12**, 265 (1958).

(9) R. G. Pearson, *J. Am. Chem. Soc.*, **85**, 3533 (1963); *Chem. Eng. News*, **43**, 90 (May 31, 1965); *Chem. Brit.*, **3**, 103 (1966).

(10) R. F. Hudson, *Chem. Eng. News*, **43**, 102 (May 31, 1965); *Struct. Bonding* (Berlin), **1**, 221 (1966).

(11) B. Saville, *Chem. Eng. News*, **43**, 100 (May 31, 1965); International Conference on Hard and Soft Acids and Bases, London, 1967.

(12) C. K. Jørgensen, *Struct. Bonding* (Berlin), **1**, 234 (1966).

(13) R. J. P. Williams and J. D. Hale, *ibid.*, **1**, 249 (1966).

(14) R. S. Drago and B. B. Wayland, *J. Am. Chem. Soc.*, **87**, 3571 (1965).

(15) G. Klopman and R. F. Hudson, *Theoret. Chim. Acta*, **8**, 165 (1967).

Journal of the American Chemical Society / *90:2* / *January 17, 1968*

ously used such a method to calculate heats of formation of diatomic[16] and polyatomic molecules.[17]

A polyelectronic perturbation treatment consistent with the previous calculation has also been described.[15] It has two advantages. The first one is that it can be handled very easily even for complicated species and thus allows predictions to be made for systems which are too complicated to be treated in a complete SCF calculation. The second advantage is that it allows one to visualize the various phenomena which occur during the process of bonding and to gain more insight into the factors governing the reaction rate. These factors are masked by the numerical complexity when a full SCF treatment is performed.

Thus, let two systems R and S interact through their atoms r and s. The total perturbation energy is produced by two distinct effects (a) the neighboring effect which accounts for the interaction due to the formation of an ion pair without any charge or electron transfer, and (b) the partial charge transfer usually accompanied by covalent bonding

$$R^-_{solv} + S^+_{solv} \xrightarrow{a} R^-_{solv} S^+_{solv} \xrightarrow{b} (R^{\delta-} - S^{\delta+})_{solv}$$

Neglecting all intermolecular ion-pair interactions involving atoms other than r and s, these effects can be evaluated with the same approximations and symbolism as used in previous calculations[16] as follows.

A. Neighboring Effects. These are essentially produced by the perturbation operator H_1 on the unperturbed molecular orbitals ψ_R and ψ_S of R and S.

$$\Delta E_{(1)} = \int \psi_R \psi_S H_1 \psi_R \psi_S \, d\tau$$

The resulting energy change accounts for the Coulomb interaction (Madelung energy) between charged species and possibly also for some partial desolvation $\Delta solv_{(1)}$ which follows the union of R and S

$$\Delta E_{(1)} = -q_r q_s \frac{\Gamma}{\epsilon} + \Delta solv_{(1)} \tag{2}$$

where q_r and q_s are the total initial charges respectively of atoms r and s, Γ is the Coulomb repulsion term[16] between atoms r and s, and ϵ is related to the local dielectric constant of the solvent (see the section on solvation).

B. Electron-Transfer Effects. These effects are produced by the direct interaction between overlapping molecular orbitals. They lead to covalent bonding and decrease the ionicity of the reactants. They are thus also responsible for the desolvation of initially charged species. The procedure consists here in calculating the perturbation produced on each orbital ψ_m of one species (R) by each orbital ψ_n of the other (S). Such a perturbation involves new partly perturbed molecular orbitals ψ_{mn} such that for only two orbitals

$$\psi_{mn} = a\psi_m + b\psi_n$$

where a and b are variational parameters such that $a^2 + b^2 = 1$ (the overlap being neglected in this treatment). Only those pairs of orbitals which contain together at least one but at most three electrons contribute to the change in energy. When ψ_m was orig-

(16) G. Klopman, *J. Am. Chem. Soc.*, **86**, 4550 (1964).
(17) G. Klopman, *ibid.*, **87**, 3300 (1965); M. J. S. Dewar and G. Klopman, *ibid.*, **89**, 3089 (1967).

inally doubly occupied, with ψ_n empty, the change in energy ΔE_{mn} produced by this partial perturbation is equal to the difference between the energy of the two electrons in the new molecular orbitals minus their energy in the absence of any mixing between the two orbitals ψ_m and ψ_n ($a = 1$, $b = 0$ or $b = 1$, $a = 0$), that is, their energy in the isolated molecule plus that produced by the neighboring effect

$$\Delta E_{mn} = \int \psi_{mn}(1)\psi_{mn}(2)H\psi_{mn}(1)\psi_{mn}(2) \, d\tau_1/d\tau_2 -$$
$$\int \psi_m(1)\psi_m(2)H_0\psi_m(1)\psi_m(2) \, d\tau_1 d\tau_2 + \Delta solv_{mn}$$

where H is the total Hamiltonian operator of the joined RS system, and $\Delta solv_{mn}$ is the desolvation produced by the partial transfer of electrons from ψ_m to ψ_n.

The first two terms of the right-hand side of this equation can easily be evaluated (see Appendix I for the procedure and symbolism), and the resulting energy change becomes that shown in eq 3. The last step consists now

$$\Delta E_{mn} = 2b^2\left\{IP_n - EA_m + (c_s{}^n)^2[q_r + 2(c_r{}^m)^2]\frac{\Gamma}{\epsilon} - \right.$$
$$(c_r{}^m)^2 q_s \frac{\Gamma}{\epsilon} - (c_r{}^m)^2(c_s{}^n)^2\Gamma\left(\frac{2}{\epsilon} - 1\right)\left.\right\} +$$
$$4ab[c_r{}^m c_s{}^n \beta] - b^4\left[IP_n - EA_n + IP_m - \right.$$
$$EA_m - 2(c_r{}^m)^2(c_s{}^n)^2\Gamma\left(\frac{2}{\epsilon} - 1\right)\left.\right] + \Delta solv_{mn} \tag{3}$$

in the minimization of the energy with respect to the variational parameters a and b.

Before proceeding, one needs to find a reasonable estimate of the desolvation, $\Delta solv_{(1)}$ and $\Delta solv_{mn}$.

Influence of Solvent. There have been only a few attempts in the past to include systematically solvation energy into the quantum mechanical treatments of molecules or ions. However, as demonstrated in our introductory section, this term is essential for a correlation of reactivity and stability constants. The previous attempts usually consisted in simply adding the experimental solvation energy to the calculated heats of formation in the gas phase. It has been recognized that some chemical properties such as the basicity of heteronuclear aromatic derivatives[18] and the properties of charge-transfer complexes can only be correlated when the solvation energy is being included in the treatment.

A more reasonable approach for our treatment can be provided by the recently suggested solvaton theory[19] which allows the minimization of the energy to be made in the presence of the interaction forces of the solvent. This theory makes use of the Born[20] equation.

$$E_{solv} = \frac{q^2}{2R_{eff}}\left(1 - \frac{1}{\epsilon}\right) \tag{4}$$

where q is the total charge, R_{eff} the effective radius[21] of the ions, and ϵ the dielectric constant of the medium.[22] Although eq 4 does not seem to give a good account

(18) R. Daudel, *Tetrahedron Suppl.*, **2**, 351 (1963).
(19) G. Klopman, *Chem. Phys. Letters*, **1**, 5 (1967).
(20) M. Born, *Z. Physik*, **1**, 45 (1920).
(21) W. M. Latimer, K. S. Pitzer, and C. M. Slansky, *J. Chem. Phys.*, **7**, 108 (1939).
(22) The same equation remains valid for enthalpies of solvation but ϵ then is related to the dielectric constant D of the medium by[21] $\epsilon = D/[1 + (\partial \ln D/\partial \ln T)]$.

Klopman | Concept of Charge- and Frontier-Controlled Reactions

for changes in solvent, it nevertheless reproduces fairly well the experimental heats of solvation of several ions in a given solvent (Table III).

Table III. Experimental and Calculated (Born Equation) Heats of Hydration of Ions

Ion	Radius[a] + 0.82 (A)	Heat of solvation, kcal/mole	
		Exptl[b]	Calcd
Be^{+2}	1.17	560	573
Al^{+3}	1.33	1109	1082
Ga^{+3}	1.44	1024	1088
Mg^{+2}	1.48	443	438
Li^+	1.50	109	112
Fe^{+2}	1.56	420	438
Sr^{+2}	1.94	338	324
Ag^+	2.08	79	103
Ba^{+2}	2.16	303	290
Tl^+	2.22	74	67

[a] L. Pauling, "Nature of the Chemical Bond," 3rd ed, Cornell University Press, Ithaca, N. Y., 1960. [b] F. D. Rossini, et al., "Selected Values of Chemical Thermodynamic Properties," National Bureau of Standards Circular 500, U. S. Government Printing Office, Washington, D. C.

In order to be able to deal with the desolvation which must be estimated, the following hypotheses are made. (a) The desolvation, if any, produced by the union of R and S without any charge transfer can possibly be attributed to a steric inhibition of solvation and might be accounted for either by increasing the effective radius of the ion by a small constant or alternatively by decreasing the dielectric constant (ϵ) around the reacting species. (b) The desolvation produced by the partial electron transfer from ψ_m to ψ_n results from a decrease of the total charge of the reagents.[23] It will tentatively be given the form[19]

$$\Delta\text{solv}_{mn} = \frac{[q_r + 2b^2(c_r{}^m)^2 x]^2 - q_r{}^2}{2R_r}\left(1 - \frac{1}{\epsilon}\right) +$$
$$\frac{[q_s - 2b^2(c_s{}^n)^2 x]^2 - q_s{}^2}{2R_s}\left(1 - \frac{1}{\epsilon}\right) \quad (5)$$

where b is the variational parameter defined before, and x is empirically set equal to $q - (q - 1)\sqrt{\kappa}$ for $q > 0$, κ being a universal constant. This formalism is introduced so that when complete charge transfer occurs then the solvation energy of the remaining ions is

$$E_{\text{solv}} = \frac{(q \pm 1)^2\kappa}{2R}\left(1 - \frac{1}{\epsilon}\right) \quad (6)$$

and accounts for the experimental observation that the solvation energy decreases faster with decreasing charge than would have been calculated by the Born equation. This is probably due to the fact that charge variations alter the ionic radius and are accounted for by setting $\kappa < 1$.

We can now go back to the variational treatment and calculate the perturbation energy (see eq A3, Appendix II) which becomes

$$\Delta E_{mn} = E_n{}^* - E_m{}^* + b^2 [(E_m{}^* - E_n{}^*)_{b^2} -$$
$$(E_m{}^* - E_n{}^*)_{b^2 = 0}] + \sqrt{(E_m{}^* - E_n{}^*) + 4(c_r{}^m)^2 (c_s{}^n)^2\beta^2}$$
$$(7)$$

where $E_m{}^*$ and $E_n{}^*$ can be associated with the energy of the molecular orbitals ψ_m of molecule R and ψ_n of molecule S under their mutual influence.

Results and Discussion

A. Chemical Reactivity. Our treatment of chemical reactivity is based, as already discussed briefly in a previous paper,[15] on the limits of eq 8 for small perturbations, i.e., small β (eq A4, Appendix II)

$$\Delta E_{\text{total}} = \Delta E_{(1)} + \frac{1}{2}\sum_{mn}\Delta E_{mn}$$

$$\Delta E_{\text{total}} = -q_r q_s \frac{\Gamma}{\epsilon} + \Delta\text{solv}(1) +$$
$$\sum_{m}^{\text{occ}}\sum_{n}^{\text{unocc}}\left[\frac{2(c_r{}^m)^2(c_s{}^n)^2\beta^2}{E_m{}^* - E_n{}^*}\right] \quad (8)$$

It follows essentially from the consideration of the difference in energy, $E_m{}^* - E_n{}^*$, between the highest occupied orbital of the nucleophile (donor), ψ_m, and the lowest empty orbital of the electrophile (acceptor), ψ_n, i.e., the frontier orbitals.[24] In this respect, the treatment is a generalization of Fukui's F.O.D.[24] and of Brown's[25] Z values, but the analogy is only superficial since the frontier orbitals are used here only as a criterion for determining the type of controlling effect and do not necessarily determine the reactivity. Superdelocalizability,[26] on the other hand, would have been conceptionally closer to this treatment if it had accounted also for the nature of the reactant.

B. Hard and Soft Behavior.[27] When the difference between $E_m{}^*$ and $E_n{}^*$ for the frontier orbitals is large, $|E_m{}^* - E_n{}^*| \gg 4\beta^2$, then obviously $E_m{}^* - E_n{}^*$ for all pairs of orbitals will be large, all b^2's tend to zero, and very little charge transfer occurs. The small energy differences between the various molecular orbitals of each molecule can be neglected (Figure 1). The total perturbation energy then becomes

$$\Delta E_{\text{total}} = -q_r q_s\frac{\Gamma}{\epsilon} + \Delta\text{solv}(1) + 2\sum_{m}^{\text{occ}}(c_r{}^m)^2\sum_{n}^{\text{unocc}}(c_s{}^n)^2\gamma \quad (9)$$

where $\gamma = \beta^2/(E_m{}^* - E_n{}^*)_{\text{average}}$. It is apparent that in such a case the perturbation energy is primarily determined by the total charges on the two reagents. Very little electron transfer occurs, and the reaction will thus be called a *charge-controlled reaction*. Such an effect reflects an ionic type of interaction; it is predominant between highly charged species, when $E_m{}^*$ is very low, that is, when the donor is difficult to ionize or polarize, and when $E_n{}^*$ is very high, that is, when the acceptor has a low tendency to accept electrons and when both reactants are strongly solvated, i.e., are of small size. It is also enhanced by small values of β, corresponding to low tendency to form covalent bonds and high Γ, again favored by small radius and low polarizability of the two reactants.

(23) R. M. Noyes, J. Am. Chem. Soc., 84, 513 (1962).
(24) K. Fukui, T. Yonezawa, and H. Shingu, J. Chem. Phys., 20, 722 (1952); 22, 1433 (1954).
(25) R. D. Brown, J. Chem. Soc., 2232 (1959).
(26) K. Fukui, T. Yonezawa, and C. Nagata, J. Chem. Phys., 27, 1247 (1957).
(27) Part of the material presented in this section has been the subject of a communication at the 2nd Symposium on Hard and Soft Acids and Bases, London, 1967.

Table IV. Type and Rate of Reaction between Hard and Soft Reagents

Case	Donor E_m^*	Acceptor E_n^*	$E_m^* - E_n^*$	Γ	Conclusion β		Reactivity
1	High (soft) large orbital	High (hard) small orbital	Medium	Small	Very small	Undef	Low
2		Low (soft) large orbital	Small	Very small	Large	Frontier controlled	High
3	Low (hard) small orbital	High (hard) small orbital	Large	Large	Small	Charge controlled	High
4		Low (soft) large orbital	Medium	Small	Very small	Undef	Low

All these properties correspond perfectly to those associated with hard–hard interaction, and the charge-controlled effect can thus be directly identified with it.

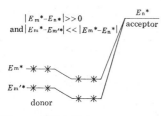

Figure 1. Charge-controlled effect.

On the other hand, when the two frontier orbitals are nearly degenerate, i.e., $|E_m^* - E_n^*| \approx 0$, then, their interaction becomes predominant (Figure 2 and eq 8), and strong electron transfer occurs between them. When such a case happens, we will call the reaction a *frontier-controlled reaction*, and the total perturbation energy can be approximated by

$$\Delta E = 2c_r^m c_s^n \beta \qquad (10)$$

It occurs only in reactions between nucleophiles of low electronegativity and electrophiles of high electronegativity with a good overlap of the interacting orbitals. The reactivity in this case is essentially determined by the frontier electron density ($c_r^{m^2}$ and $c_s^{n^2}$), when the reaction occurs between uncharged or weakly charged species. It is enhanced by high polarizability of the reagents, low solvation energies, and, in fact, by all properties reverse to those producing the previous case. It leads to covalent bonding ($b^2 \to 0.5$) and can be associated with soft–soft interaction. In fact, it is easy to predict qualitatively from eq 8 which properties of the reagents would favor one or another controlling effect. In Table IV several possible cases have been summarized.

Only two combinations, cases 2 and 3, lead to high reactivity, namely, that occurring between hard reagents and that between soft reagents. Table IV thus shows that hard acceptors will tend to complex hard donors and soft acceptors will prefer soft donors. In other words, Pearson's principle[9] is a direct result of such a treatment.

Another interesting conclusion follows from the previous considerations. Hard–hard interactions are charge controlled and depend mainly on the ionic interaction of the reagents. In water such an interaction leads to the most favorable decrease of free energy, but

to an increase of the enthalpy as $\partial \ln D / \partial \ln T$ is equal to[23] -1.357.

$$\Delta H_{hh} = -\frac{q_x q_y \Gamma}{D}\left(1 + \frac{\partial \ln D}{\partial \ln T}\right) = \frac{e^2}{r}\left(\frac{0.357}{80}\right) > 0$$

In addition, the steric inhibition of solvation which may occur during the union of the ions further accentuates the endothermicity of the reaction characteristic of hard–hard interaction.

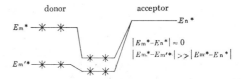

Figure 2. Frontier-controlled effect.

On the other hand, in the case of soft–soft interaction (eq 10), these phenomena are counterbalanced by the stabilization brought about by covalent bonding, and the reaction thus becomes exothermic.

$$\Delta H_{ss} = \Delta H_{hh} + 2c_x c_y \beta < 0$$

This is exactly what has been observed by Ahrland[28] and is illustrated in Table V for several inorganic complexes in water. For each halogen, the enthalpy decreases regularly with the softness of the metal, and for each metal, with the softness of the halogen.

Table V. Enthalpy (kcal/mole) of Formation of Mono Complexes in Water

	Cl^-	Br^-	I^-
Cr^{+3}	6.6	5.1	...
Fe^{+3}	4.2	...	...
Sn^{+2}	2.6	1.4	...
Cd^{+2}	−0.10	−1.0	−2.26
Tl^+	−1.1	−2.5	...
Ag^+	−2.7	...	...
Hg^{+2}	−5.79	−9.57	−18
Tl^{+3}	−6.04	−8.96	...

The question which arises now is whether this treatment allows a theoretical scale to be established for the hard or soft character of a reagent. Until now, we have based all our qualitative conclusions on the relative values of E_m^* and E_n^*. If these are correct, then these values should also provide a quantitative basis for es-

(28) S. Ahrland, *Helv. Chim. Acta*, **50**, 306 (1967).

Klopman / Concept of Charge- and Frontier-Controlled Reactions

Table VI. Calculated Softness Character (Empty Frontier Orbital Energy) of Cations

X[a]	IP,[a] ev	EA,[a] ev	Orbital energy, ev	r^b + 0.82, A	Desolvation	$E_n{}^{\ddagger}$, ev	
Al+3	28.44	18.82	26.04	1.33	32.05	6.01	↑
La+3	19.17	11.43	17.24	1.96	21.75	4.51	
Ti+4	43.24	28.14	39.46	1.50	43.81	4.35	Hard
Be+2	18.21	9.32	15.98	1.17	19.73	3.75	
Mg+2	15.03	7.64	13.18	1.48	15.60	2.42	
Ca+2	11.87	6.11	10.43	1.81	12.76	2.33	
Fe+3	30.64	15.95 (16.18)	26.97	1.46	29.19	2.22	
Sr+2	11.03	5.69	9.69	1.94	11.90	2.21	
Cr+3	30.95	16.49	27.33	1.45	29.39	2.06	
Ba+2	10.00	5.21	8.80	2.16	10.69	1.89	
Ga+3	30.70	20.51	28.15	1.44	29.60	1.45	
Cr+2	15.01[d] (16.49)	7.28[d] (6.76)	13.08	1.65	13.99	0.91	Borderline
Fe+2	16.18	7.90	14.11	1.56	14.80	0.69	
Li+	5.39	0.82	4.25	1.50	4.74	0.49	
H+	13.60	0.75	10.38	...	10.8[c]	0.42	
Ni+2	17.11[d] (18.15)	8.67[d] (7.63)	15.00	1.51	15.29	0.29	
Na+	5.14	0.47	3.97	1.79	3.97	0	
Cu+2	17.57[d] (20.29)	9.05[d] (7.72)	15.44	1.54	14.99	−0.55	
Tl+	6.10	(2.0)	5.08	2.22	3.20	−1.88	
Cd+2	16.9	8.99	14.93	1.79	12.89	−2.04	
Cu+	7.72	2.0	6.29	1.78	3.99	−2.30	
Ag+	7.57	2.2	6.23	2.08	3.41	−2.82	Soft
Tl+3	29.30	20.42	27.45	1.77	24.08	−3.37	
Au+	9.22	2.7	7.59	2.19	3.24	−4.35	
Hg+2	18.75	10.43	16.67	1.92	12.03	−4.64	↓

[a] C. E. Moore, "Atomic Energy Levels," National Bureau of Standards Circular 467, U. S. Government Printing Office, Washington, D. C., 1949. [b] See footnote a, Table III. [c] Experimental value; see footnote b, Table III. [d] The values refer to ionization of s orbital.

tablishing a theoretical scale for the hard and soft character of reactants. The quantities $E_m{}^*$ and $E_n{}^*$ have been given an algebraic form (see eq A2, Appendix II) and can thus be evaluated directly. However, their evaluation requires the knowledge of the molecular Coulomb interaction Γ which will vary slightly from molecule to molecule. We can, however, define an intrinsic quantity $E^{\ddagger}$ (softness) independent of the other reagent by setting $\Gamma = 0$ for all acids or bases. The softness character of the reagents will thus be established within a constant factor. This procedure does not affect to a first approximation their relative order.

$$E_m{}^{\ddagger} = IP_m - a^2(IP_m - EA_m) - \frac{x_r(c_r{}^m)^2}{R_r}\left(1 - \frac{1}{\epsilon}\right)[q_r + 2b^2 x_r(c_r{}^m)^2]$$

$$E_n{}^{\ddagger} = IP_n - b^2(IP_n - EA_n) - \frac{x_s(c_s{}^n)^2}{R_s}\left(1 - \frac{1}{\epsilon}\right)[q_s - 2b^2 x_s(c_s{}^n)^2]$$

(11)

A hard base (nucleophile) is characterized by a low value for the energy of the occupied frontier orbital, $E_m{}^*$, a soft base by a higher value of $E_m{}^*$. Accordingly, the hardness of a base increases with the decrease of $E_m{}^{\ddagger}$. A hard acid on the contrary is characterized by a high value for the energy of the empty frontier orbital $E_n{}^*$, and its hardness will decrease with the decrease of $E_n{}^{\ddagger}$.

When a complete charge transfer occurs (frontier-controlled reaction), then $b^2 = \frac{1}{2}$ and $a^2 = \frac{1}{2}$. When no charge transfer occurs (charge-controlled reaction), then $b^2 = 0$ and $a^2 = 1$. The softness $E^{\ddagger}$ which measures the tendency for a given reagent to form either covalent or ionic interactions is calculated, by analogy with gas phase, for the intermediate situation such as $b^2 = \frac{1}{4}$, $a^2 = \frac{3}{4}$. This is illustrated below for Ba+2 and

done in Table VI for several acids or acceptors and in Table VII for bases or donors. The calculations refer to water ($\epsilon = 80$), and the value of κ (see eq 6) which led to these results was empirically set equal to 0.75. For example, for Ba+2

$$Ba^{+2} \xrightarrow{IP_n = -10.0 \text{ ev}} Ba^{+} \xrightarrow{EA_n = -5.21 \text{ ev}} Ba$$

ionic radius of Ba+2 = 1.34 A

$$R_s = 1.34 + 0.82 = 2.16 \text{ A}$$

$$b^2 = \frac{1}{4} \qquad q_s = 2 \qquad (c_s{}^n)^2 = 1$$

$$x_s = q_s - (q_s - 1)\sqrt{x} = 2 - \sqrt{0.75}$$

(a) orbital energy = $IP_n - b^2(IP_n - EA_n) = 8.80$ ev

(b) desolvation = $\frac{x_s(c_s{}^n)^2}{R_s}\left(1 - \frac{1}{\epsilon}\right) \times [q_s - 2b^2 x_s(c_s{}^n)^2] = 10.69$ ev

softness $E_n{}^{\ddagger} = -$ (a) + (b)

= −8.80 + 10.69 = 1.89 ev

The result agrees extremely well with the qualitative description of softness and hardness given by Pearson[9] and established on the consideration of the chemical properties of these acids and bases, and can be used as a direct criterion of such properties for other species. The treatment does not require any preestablished knowledge of the *chemical* properties of the ions, contrary to most of the previous attempts to correlate hard and soft behavior.[9,13,27,29] Hence the definition of softness becomes the tendency for an acid to accept electrons (or a base to give electrons) in chemical bonds *in solution*. The analogy with electronegativity is obvious not only qualitatively but also quantitatively.[30]

(29) A. Yingst and D. H. McDaniel, *Inorg. Chem.*, 6, 1067 (1967).
(30) G. Klopman, *J. Am. Chem. Soc.*, 86, 1463 (1964).

Table VII. Calculated Softness Character (Occupied Frontier Orbital Energy) of Donors

X^z	$IP,^a$ ev	$EA,^a$ ev	Orbital energy, ev	r, A^b	Desolvation, ev	$E^{\ne}{}_m,$ ev
F^-	17.42	3.48	6.96	1.36	5.22	−12.18
H_2O	25.4	12.6	15.8	(1.40)	(−5.07)^c	−(10.73)
OH^-	13.10	2.8	5.38	1.40	5.07	−10.45
Cl^-	13.01	3.69	6.02	1.81	3.92	−9.94
Br^-	11.84	3.49	5.58	1.95	3.64	−9.22
CN^-	14.6	3.2	6.05	2.60	2.73	−8.78
SH^-	11.1	2.6	4.73	1.84	9.86	−8.59
I^-	10.45	3.21	5.02	2.16	3.29	−8.31
H^-	13.6	0.75	3.96	2.08	3.41	−7.37

(right side of table: an arrow pointing up labeled "Hard" and down labeled "Soft")

^{a} See footnote a, Table VI. ^{b} See footnote a, Table III. ^{c} This value is negative, as it would be in general for neutral ligands, because the solvation increases rather than decreases during the removal of the first electron. The numerical value has been put equal to the value for OH^- in absence of more reliable data.

Equation 12 was shown to be the mathematical definition of electronegativity for a *neutral* atom in the gas phase. Equation 13 represents the softness which

$$\chi = \left(\frac{\partial E_{at}}{\partial a^z}\right)_{a^z = 1/2} \quad (12)$$

$$E^{\ne} = \left(\frac{\partial E_{at}}{\partial a^z}\right)_{a^z = 1/4} \quad (13)$$

hence can be defined as the electronegativity of the *ion* in solution. As for electronegativity, however, care should be taken when using the values derived from this treatment in dealing with chemical reactions. For example, the alkali metals are found to be on the borderline between hard and soft on the $E^{\ne}$ criterion. However, in these cases the value of β is so small[16] that, although $E^{\ne}$ is relatively high, they still behave as hard acceptors.

An interesting application is also provided by the study of the variation of $E^{\ne}$ with the local or microscopic dielectric constant ϵ. Although no precise correlation should be made between the value ϵ and the nature of the solvent,[31] it is seen that when ϵ decreases the softness parameter is shifted toward higher values.

This does not necessarily mean that the softness of all acids increases by decreasing the dielectric constant since other factors like those associated with Γ are varying at the same time. But the relative order of $E^{\ne}$ may remain an important qualitative indication of the character of the reagent. The magnitude of the shift is approximately proportional to the oxidation number and therefore, when ϵ decreases, a drastic modification of the softness scale occurs (Figure 3).

Acids with high oxidation numbers thus become relatively softer in nonpolar solvents as is illustrated, for example, by the fact that Ti^{IV} coordinates diarsines (soft bases) in dioxane to a higher extent than Fe^{III} or Cr^{III}.[32]

A last application is provided by the study of the stability of $B(CH_3)_3$ adducts with various amines. The softness parameter for amines cannot be calculated accurately as their second ionization potential is not known. A reasonable guess can, however, be made by assuming that it is equal to the first ionization poten-

(31) E. Grunwald, G. Baughan, and G. Kohnstam, *J. Am. Chem. Soc.*, **82**, 5801 (1960).
(32) R. J. H. Clark, personal communication.

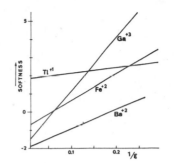

Figure 3. Variation of the softness ($E^{\ne}$) of acids with changes of dielectric constant (ϵ).

tial plus the atomic electronic interaction. Since desolvation is to the first approximation constant for all amines, it appears that the hardness increases proportionally to the ionization potentials of the amine. This conclusion can be checked by considering the data of Drago[14] based on his equation (eq 1) which correlates the enthalpy of formation of adducts with two parameters; C is related to the susceptibility of amines to form covalent bonds and E to undergo electrostatic interaction.

Table VIII. Correlation between the Ionization Potential of Amines and the Ratio of Ionic to Covalent Bonding

Amines	$EA_m,$ ev	$(E/C) \times 10^{-2}$
NH_3	10.15	39
CH_3NH_2	8.97	19
$(CH_3)_2NH_2$	8.24	13
$(CH_3)_3N$	7.82	5

As shown before the softness parameter can be considered as a measure of the ratio of these two tendencies. This is actually found as shown in Table VIII where the first ionization potentials (EA_M) are compared with the E/C ratios.

C. Nucleophilic Order. Providing that the necessary interaction terms β and Γ are known, the treatment allows an estimation of the reactivity order toward a particular reagent. This order is determined by both reactants, and it is intrinsically incorrect to relate reactivity to a particular reactivity index. It is well known for example that electrophilic centers possessing a high positive charge, such as $RC(=O)-$, $RS(=O)_2-$, $(RO)_2P(=O)-$, and H^+, react rapidly with F^- or OH^- whereas electronically saturated centers, such as RCH_2-, R_2P-, $RS-$, and $Br-$, react preferentially with I^- or R_3P. The treatment is likely to reproduce these phenomena, and quantitative calculations based on eq 2 and 7 for specific reactions are being investigated. However, it appeared that even a crude model may be used to provide interesting conclusions.

Neglecting the minor contribution of $\Delta solv(1)$, the total perturbation as represented by eq 8 can be calculated for the reaction of several nucleophiles with electrophiles defined by their value of E_n^*.

The calculations have been made with the following values: $q_r = -1$, $q_s = +1$, $c_r^{m^2} = c_s^{n^2} = 1$, $\epsilon = 80$, Γ

Klöpman / *Concept of Charge- and Frontier-Controlled Reactions*

Table IX. Reactivity Scales

	I⁻	Br⁻	Cl⁻	F⁻	SH⁻	CN⁻	OH⁻
R_{XH}, A	1.60	1.41	1.27	0.92	1.34	1.06	0.96
$-2\beta_X$, ev	3.16	3.60	4.10	4.48	4.0	3.8	4.50
$-E_m^*$, ev	8.31	9.22	9.94	12.18	8.59	8.78	10.45
$-\Delta E$ ($E_n^* = -7$ ev)	2.52	1.75	1.54	1.06	2.64	2.30	1.49
($E_n^* = -5$ ev)	1.07	0.98	0.97	0.82	1.25	1.17	1.01
($E_n^* = +1$ ev)	0.451	0.479	0.516	0.535	0.551	0.557	0.58

Table X. Nucleophilic Order

E_n^*		Nucleophilic order
−7	Calcd	HS⁻ > I⁻ > NC⁻ > Br⁻ > Cl⁻ > HO⁻ > F⁻
	$k \times 10^4$, reaction with	Too
	peroxide oxygen	fast 6900 10 0.23 0.0011 ≈ 0
−5	Calcd	HS⁻ > NC⁻ > I⁻ > HO⁻ > Br⁻ > Cl⁻ > F⁻
	$k \times 10^6$, attack on	
	saturated carbon	(25) 10 12 1.2 0.5 0.11
	E (Edwards)	2.79 2.06 1.65 1.51 1.24 1.0
+1	Calcd	HO⁻ > NC⁻ > HS⁻ > F⁻ > Cl⁻ > Br⁻ > I⁻
	k for attack on carbonyl	
	carbon (acylation)	890 10.8 ... 0.001 ... Unreactive ...
	pK_a	15.7 9.1 7.1 3.2 ... (−4.3) (−7.3)

$= e^2/R_{XH}$, where R_{XH} is the bond distance in the XH molecule, and the β_X's are the values obtained in previous calculations[16] and such that $\beta_{XR} = \sqrt{\beta_X\beta_R}$. β_R was set equal to 1. The results are summarized in Table IX. Although these results involve crude approximations and remain questionable until the full treatment confirms them, it is interesting to find that the various reactivity scales agree closely with experiments. Thus, when $E_n^* = -7$ ev, the order of reactivity which is found refers to a reaction toward an extremely soft electrophile. This is the case, for example, of the reaction with peroxides,[33] and it is seen from Table X that the predicted order agrees very well with the results. Similarly $E_n^* = -5$ ev refers to a moderately soft center such as a saturated carbon atom. Here again the correlation with reactivity is satisfactory as shown in Table X. Also, the data fit very well with Edward's nucleophilic order[6] for the various basis.

Finally, for a hard center characterized by a value of E_n^* of $+1$ ev, a good correlation is found both with pK_a[6] and with the nucleophilic reactivity order toward carbonyl carbon atoms.[34] The treatment thus reproduces the experimental reactivity numbers and provides a theoretical basis for the four-term free-energy equations.[6]

Another important analysis which can be made would be based on the variation of the nucleophilic order with the nature of the solvent. It is clear, for example, that the smaller ϵ is, the bigger the charge-controlled effect will be. In other words, solvents with low dielectric constants are bound to enhance charge-controlled interactions. Very polar solvents, on the other hand, favor frontier-controlled reactions.

D. Ambient Reactivity. Reasonable nucleophilic orders are thus provided by the above treatment. Various scales of nucleophilicity are found, for example, for alkylation and for acylation. One may therefore

(33) J. O. Edwards, *J. Am. Chem. Soc.*, **84**, 22 (1962).
(34) W. P. Jencks and J. Carriuolo, *ibid.*, **83**, 1743 (1961).

expect that the field of ambient reactivity in its most general sense might be particularly fruitfully studied by such a treatment. It is of interest to consider in detail the case of reactions between uncharged or weakly charged species ($q_r q_s \approx 0$) which are often encountered in organic chemistry.

For large molecules, accurate values of E_m^* and E_n^* are difficult to obtain although semiempirical SCF methods might be promising for approaching this problem. But even such results have not yet been sufficiently established for heterocyclic molecules. However, reasonable Hückel one-electron energies are readily available and might possibly be used to approximate the necessary values of E_m^* and E_n^*. In such a first approximation the reaction will be primarily controlled by any of the two limits of the last term of eq 8

$$\Delta E = \sum_{\substack{m \\ occ}} \sum_{\substack{n \\ unocc}} \frac{2(c_r^m)^2(c_s^n)^2\beta^2}{E_m^* - E_n^*} \tag{14}$$

This equation is identical with the chemical perturbation equation, but the point we are interested in studying here and which has not attracted much attention so far is the variation of this perturbation energy with that of $E_m^* - E_n^*$. As before, when $|E_m^* - E_n^*|$ for the frontier orbitals of the reactants is large, then the reaction is charge controlled (see eq 9) (in this case, it is the total electronic charge density which determines the reactivity)

$$\Delta E = 2\sum_{\substack{m \\ occ}}(c_r^m)^2\sum_{\substack{n \\ unocc}}(c_s^n)^2\gamma$$

otherwise it is frontier controlled (see eq 10) and leads to strong covalent bonding between the centers possessing the highest charge in the frontier orbitals, *i.e.*, $\Delta E = 2c_r^m c_s^n\beta$. The fact that the two alternatives still exist for weakly charged organic substances is particularly important to realize.[35] However, their origin differs from that producing charge- or frontier-controlled effects for ionic reagents. They result from the fact that in organic compounds the total electronic charge of an atom is often distributed among various delocalized molecular orbitals of different energy. Therefore, in contrast to ions, the frontier electron density is usually not proportional to the total electronic charge density. Equation 14 automatically includes this duality as will be shown for a few examples at the end of this section. However, approximations are conceivable such as linear combinations of the two limiting cases. This may give rise to an equation such as

$$\Delta E = \left(\sum_{\substack{m \\ occ}} (c_r{}^m)^2 \sum_{\substack{n \\ unocc}} (c_s{}^n)^2\right)\alpha_{RS} + (c_r{}^m)^2(c_s{}^n)^2\beta_{RS} \quad (15)$$

where α_{RS} and β_{RS} are now variable parameters, characterizing the reaction of R and S in such a way that, when α_{RS} is large and β_{RS} small, the reaction is charge controlled, whereas, when α_{RS} is small and β_{RS} large, the reaction is frontier controlled.

Such equations as already mentioned have been empirically suggested and in fact used successfully by Drago,[14] whose equation (eq 1) can be recognized by making the following transformations

$$\sum_{\substack{m \\ occ}} (c_r{}^m)^2\alpha_R = E_R$$

$$\sum_{\substack{n \\ unocc}} (c_s{}^n)^2\alpha_s = E_s$$

$$(c_r{}^m)^2\beta_R = C_R$$

$$(c_s{}^n)^2\beta_S = C_S$$

and

$$\alpha_R\alpha_S = \alpha_{RS}$$

$$\beta_R\beta_S = \beta_{RS}$$

More recently, Rogers[36] has studied several reactions of aromatic species using eq 15 and found that the use of both indexes, total electron density and frontier electron density, largely improves the correlation.

As would have been expected from the previous discussion he found that the best correlation for deuterium exchange (hard) with conjugated vinyl ethers is obtained when $\alpha/\beta \sim 2.8$, but the free-radical phenylation of 2-methylnaphthalene is best correlated for $\alpha/\beta = 0.31$.[36] Free radicals show very soft properties as would have been expected from the fact that they interact mainly through their nonbonding orbital. This behavior has been extensively discussed by Dewar,[37] who showed that radical recombination largely involves this particular orbital.

Finally it may be noted that the fact that Fukui's frontier orbital density is working at all for aromatic hydrocarbons is easily understood by noticing that the absence of total charge differences in such compounds leaves as the only controlling factor the frontier electron density. On the other hand, the lack of correlation for reactions of heterocyclic molecules using Fukui's treatment can

(35) A preliminary discussion of these effects has already been published: R. F. Hudson and G. Klopman, *Tetrahedron Letters*, 1103 (1967).
(36) N. A. J. Rogers, personal communication.
(37) M. J. S. Dewar, *J. Am. Chem. Soc.*, 74, 3345 (1952).

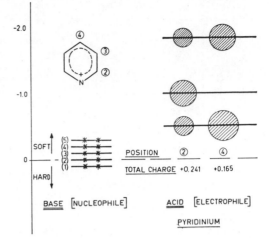

Figure 4. Schematic representation of the interaction between several nucleophiles and the antibonding orbitals of pyridinium salts.

certainly be attributed to the fact that in such cases a definite variation in total charge density occurs. This will be illustrated in the following examples where eq 14 is used to show how several similar centers are behaving toward a particular reagent and how the nature of this reagent modifies their relative reactivity through an intimate mixing of charge- and frontier-controlled effects.

The procedure consists in calculating in the Hückel approximation the various molecular orbital energies of conjugated organic reagents. The perturbation which is produced by the other reactant R at each possible center of the conjugated species is then calculated from eq 14. Various degrees of softness of the attacking reagent are taken into account by varying E_m^*. The parameters used for these calculations are those suggested by Streitwieser[38] except for O^- (in pyridine oxide) where the high negative charge suggests a lower value for the Coulomb integral (Table XI).

Table XI

$\alpha_z = \alpha_0 + k_z\beta_0$	$-\beta_{xy} = h_{xy}\beta_0$		
$k(-\overset{+}{\text{O}}-) = 2.0$	$k(\text{O}=) = 1.0$	$k(\text{O}-) = 0.1$	
$k(-\text{N}=) = 2.0$	$k(-\text{N}-) = 1.5$	$k(-\text{N}=) = 0.5$	
$h(\text{C}=\text{N}) = 1.0$	$h(\text{N}-\text{O}) = 0.75$	$h(\text{C}=\text{O}) = 1.0$	

(1) Nucleophilic Addition to the Pyridinium Ion. Figure 4 shows on the right the antibonding molecular orbitals of pyridinium ions. The circles represent the $(c_s{}^n)^2$, i.e., the square of the coefficients of the atomic orbitals for positions 2 and 4 which are the two possible reacting centers. On the left side, five nucleophiles of increasing hardness ($k_m = -0.3, -0.2, -0.1, 0, 0.1$) are represented through their highest filled orbital. The perturbation produced by each of these nucleophiles both on positions 2 and 4 is calculated by means of eq

(38) A. Streitwieser, Jr., "Molecular Orbital Theory," John Wiley and Sons, Inc., New York, N. Y., 1961, p 120.

Klopman | Concept of Charge- and Frontier-Controlled Reactions

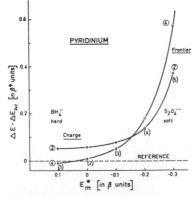

Figure 5. Nucleophilic addition on pyridinium ions.

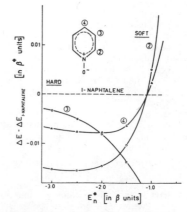

Figure 6. Electrophilic substitution on pyridine oxide.

14, and the result is plotted in Figure 5. In order to bring all the results into a single representative graph, the values of ΔE for the two possible reacting centers are compared to that obtained for the reactivity of a reference compound whose orbital E_m^* would be at $\alpha - 0.65\beta$. Since we are interested only in comparing the possible reacting centers of pyridine, this procedure does not modify the conclusions. This plot shows that for hard nucleophiles ($k_m = 0.1$), the most reactive center is position 2 ($\Delta E_{(2)} > \Delta E_{(4)}$), but as the softness of the nucleophile increases (k_m decreases), a progressive change to position 4 occurs; $\Delta E_{(4)} > \Delta E_{(2)}$. This is exactly what has been observed as BH_4^-, aniline, and hydroxide ions, which are hard, attack position 2 whereas the soft CN^- and $S_2O_4^{-2}$ react at position 4.[39] Kosower had suggested that reactions occurring at position 4 may go through an activated state consisting of a charge-transfer complex,[40] but this explanation could not account for the fact that CN^- attacked position 4 as no evidence for the complex formation was found in this case. Our explanation thus provides a reasonable alternative to the charge-transfer mechanism but does not oppose it as it also suggests that the soft nucleophiles are the most likely to produce charge-transfer complexes with the pyridinium ion.

(2) Electrophilic Substitution on Pyridine Oxide. The procedure is similar to that described in the previous example. The perturbation is calculated here for the possible interaction of positions 2, 3, and 4

with electrophiles of increasing hardness ($k_n = -1$, -1.5, -2, -2.5, -3). The results, compared to that for attack on position 1 of naphthalene as reference, are plotted in Figure 6 and lead to the unexpected conclusion that all three positions may become the most reactive depending on the softness of the electrophile. Very soft reagents attack position 2, rather hard reagents, position 4, and very hard ones, position 3. This conclusion is entirely verified experimentally as mercura-

tion was shown to occur at position 2, nitration at position 4, and sulfonation at position 3.[41]

These predictions are valid for the free base; the conjugated acid which is preponderant in acidic conditions (sulfonation?) is found to react at position 3 irrespective of the softness of the reagent.

Other applications of such a treatment are provided by the study of most reactions of conjugated centers but can also be applied to classical ambient ions such as thiocyanate

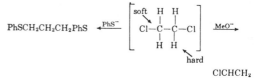

(39) E. M. Kosower, *J. Am. Chem. Soc.*, **78**, 3497 (1956).
(40) E. M. Kosower, *Progr. Phys. Org. Chem.*, **3**, 106 (1965).

(41) A. R. Katritzky and C. D. Johnson, Chemical Society Anniversary Meeting of the Exeter, 1967.

Journal of the American Chemical Society / *90:2* / *January 17, 1968*

Finally, the Woodward–Hofmann rule and the orientation in Diels–Alder additions between asymmetric dienes and dienophiles might possibly also be explained satisfactorily by such an approach.

In the latter case a four-center attack should be considered and eq 14 should be replaced by

$$\Delta E = \sum_m \sum_n \frac{2[(c_r^m)^2(c_s^n)^2 + (c_{r'}^m)^2(c_{s'}^n)^2]\beta^2}{E_m^* - E_n^*}$$
$$\text{occ unocc}$$

Conclusion

The various applications discussed in this paper are by no means restrictive but are rather intended to provide models for the wide variety of cases where the information from general eq 8 can be used. Even if a part of the above agreement is fortuitous, as might well be the case, since one-electron treatments are after all very crude and depend so much on parameters, it remains that the general trend found in these examples shows the potential applicability of the method. The treatment defines hardness and softness of reactants and explains the particular behavior associated with these properties.

The classification of reagents into hard and soft classes is a useful one, but as demonstrated above it does not necessarily lead to a universal order of reactivity. The more general treatment based on eq 8 has a far greater potential applicability to the study of the factors influencing reactivity such as attacking reagent, leaving group, substituent, and solvent. Further examples need, however, to be investigated, and a more quantitative examination of the data provided by semiempirical SCF procedures might lead to a better understanding of chemical reactivity. This work is now in progress for accounting quantitatively for such properties as equilibrium constants, pK_a, and reaction rates.

Acknowledgments. The author wishes to thank Dr. C. K. Jørgenson and Professor R. F. Hudson for continuous help in the preparation of the manuscript and for numerous helpful suggestions.

Appendix I

The calculation of the change in energy produced during the interaction of two systems, R and S, by the partial transfer of electrons from an initially doubly occupied ψ_m of R to an initially empty orbital ψ_n of S is given here. The interaction occurs through atoms r (atomic orbital φ_r) of R and s (atomic orbital φ_s) of S.

$$\psi_m = \sum_\rho c_\rho^m \varphi_\rho \quad \psi_n = \sum_\sigma c_\sigma^n \varphi_\sigma \quad \psi_{mn} = a_m + b\psi_n$$

$$\Delta E_{mn}' = \int \psi_{mn}(i)\psi_{mn}(j)H\psi_{mn}(i)_{mn}(j)\,d\tau_i d\tau_j -$$
$$\psi \int \psi_m(i)\psi_m(j)H\psi_m(i)\psi_m(j)\,d\tau_i d\tau_j$$

where φ_ρ and φ_σ are atomic orbitals, respectively, of systems R and S. ψ_{mn} is the new molecular orbital; i and j are the two electrons initially occupying orbital ψ_m. a and b are variational parameters, and H is the total Hamiltonian operator of the joined RS system.

$$\Delta E_{mn}' = \int [a\psi_m(i) + b\psi_n(i)] \times$$
$$\left[-\tfrac{1}{2}\nabla^2 + \frac{q_r + 2(c_r^m)^2}{R_{ri}} + \frac{q_s}{R_{si}} \right] \times$$
$$[a\psi_m(i) + b\psi_n(i)]\,d\tau_i + \int [a\psi_m(j) + b\psi_n(j)] \times$$
$$\left[-\tfrac{1}{2}\nabla^2 + \frac{q_r + 2(c_r^m)^2}{R_{rj}} + \frac{q_s}{R_{sj}} \right][a\psi_m(j) + b\psi_n(j)\,d\tau_j -$$
$$\int [a^2\psi_m^2(i) + b^2\psi_n^2(i)]\frac{1}{R_{ij}}[a^2\psi_m^2(j) + b^2\psi_n^2(j)\,d\tau_i d\tau_j -$$
$$\int \psi_m(i)\left[-\tfrac{1}{2}\nabla^2 + \frac{q_r + 2(c_r^m)^2}{R_{ri}} + \frac{q_s}{R_{si}} \right]\psi_m(i)\,d\tau_i -$$
$$\int \psi_m(j)\left[-\tfrac{1}{2}\nabla^2 + \frac{q_r + 2(c_r^m)^2}{R_{ri}} + \frac{q_s}{R_{si}} \right]\psi_m(j)\,d\tau_j +$$
$$\int \psi_m^2(i)\frac{1}{R_{ij}}\psi_m^2(j)\,d\tau_i d\tau_j$$

where c_r^m and c_s^n are the coefficients of atomic orbitals φ_r and φ_s, respectively, in the unperturbed orbitals ψ_m and ψ_n, and q_r and q_s are the total initial charges of atoms r and s and include the core and all electrons: $q_r = Z_r - 2\Sigma_{occ}(c_r^m)^2$; $q_s = Z_s - 2\Sigma_{occ}(c_s^n)^2$. In establishing this equation, the neglect of differential overlap between ψ_m and ψ_n was assumed, and the only transmolecular interactions taken into account are those involving atoms r and s. We may now make the following substitutions

$$\int \psi_m(i)\left[-\tfrac{1}{2}\nabla^2 + \frac{q_r + 2(c_r^m)^2}{R_{ri}} + \frac{q_s}{R_{si}} \right]\psi_m(i)\,d\tau_i =$$
$$IP_m + (c_r^m)^2 q_s\Gamma$$

$$\int \psi_n(i)\left[-\tfrac{1}{2}\nabla^2 + \frac{q_r + 2(c_r^m)^2}{R_{ri}} + \frac{q_s}{R_{si}} \right]\psi_n(i)\,d\tau_i =$$
$$IP_n + (c_s^n)^2 [q_r + 2(c_r^m)^2]\Gamma$$

$$\int \psi_m(i)\left[\frac{q_r + 2(c_r^m)^2}{R_{ri}} + \frac{q_s}{R_{si}} \right]\psi_n(i)\,d\tau_1 = c_r^m c_s^n \beta$$

$$\int \psi_m^2(i)\frac{1}{R_{ij}}\psi_m^2(j)\,d\tau_i d\tau_j = IP_m - EA_m$$

$$\int \psi_n^2(i)\frac{1}{R_{ij}}\psi_n^2(j)\,d\tau_i d\tau_j = IP_n - EA_n$$

$$\int \psi_m^2(i)\frac{1}{R_{ij}}\psi_n^2(j)\,d\tau_i d\tau_j = (c_r^m)^2(c_s^n)^2\Gamma$$

where IP_m, IP_n, EA_m, and EA_n are the energies of ionizations of electrons for the following processes

$$R \xrightarrow[-EA_m]{-e} R^+ \xrightarrow[-IP_m]{-e} R^{2+} \text{ (for removal of electrons from } \psi_m)$$

$$S \xrightarrow[IP_n]{+e} S^- \xrightarrow[EA_n]{+e} S^{2-} \text{ (for adding electrons on } \psi_n)$$

and Γ is the Coulomb interaction between r and s in a central field approximation.[16] In solution, all these Coulomb interactions Γ become Γ/ϵ except that for the interaction between electrons initially belonging to the same orbital. In this case Γ must be replaced by $\Gamma \cdot [(2/\epsilon) - 1]$ as required by solvaton theory.[19]

Klopman / Concept of Charge- and Frontier-Controlled Reactions

Finally, by replacing a^2 by $1 - b^2$, one obtains

$$\Delta E_{mn}' = 2b^2 \left\{ IP_n - EA_m + (c_s^n)^2[q_r + 2(c_r^m)^2]\frac{\Gamma}{\epsilon} - (c_r^m)^2 q_s \frac{\Gamma}{\epsilon} - (c_r^m)^2(c_s^n)^2\Gamma\left(\frac{2}{\epsilon} - 1\right) \right\} + 4ab(c_r^m c_s^n \beta) - b^4\left[IP_n - EA_n + IP_m - EA_m - 2(c_r^m)^2(c_s^n)^2\Gamma\left(\frac{2}{\epsilon} - 1\right) \right]$$

$$\tag{A1}$$

The interaction between all other combinations of orbitals can be found easily by a similar treatment.

Appendix II

Minimization of the electron transfer energy with respect to the variational parameters a and b is included here.

$$\Delta E_{mn} = \Delta E_{mn}' + \Delta solv_{mn}$$

where $\Delta E_{mn}'$ has the value derived in Appendix I and $\Delta solv_{mn}$ that of eq 5. The variational treatment applied to this expression leads to the following equation for the perturbational energy

$$\Delta E_{mn} = M(m,m) + M(n,n) + \sqrt{[M(m,m) - M(n,n)]^2 + 4M(m,n)^2}$$

where the matrix elements M are

$$M(m,m) = 0$$

$$M(n,m) = c_r^m c_s^n \beta$$

$$M(n,n) = IP_n - EA_m + (c_s^n)^2(q_r + 2(c_r^m)^2)\frac{\Gamma}{\epsilon} - (c_r^m)^2 q_s \frac{\Gamma}{\epsilon} - (c_r^m)^2(c_s^n)^2\Gamma\left(\frac{2}{\epsilon} - 1\right) - b^2\Big[IP_n - EA_n + $$

$$IP_m - EA_m - 2(c_r^m)^2(c_s^n)^2\Gamma\left(\frac{2}{\epsilon} - 1\right) \Big] - \sum_s \frac{x_s(c_s^n)^2}{R_s}\left(1 - \frac{1}{\epsilon}\right)[q_s - 2b^2 x_s(c_s^n)^2] + \sum_r \frac{x_r(c_r^m)^2}{R_r}\left(1 - \frac{1}{\epsilon}\right)[q_r + 2b^2 x_r(c_r^m)^2]$$

and with the following transformations

$$E_m^* = IP_m + q_s(c_r^m)^2\frac{\Gamma}{\epsilon} - a^2\left[IP_m - EA_m - (c_r^m)^2(c_s^n)^2\Gamma\left(\frac{2}{\epsilon} - 1\right) \right] - \sum_r \frac{x_r(c_r^m)^2}{R_r}\left(1 - \frac{1}{\epsilon}\right)[q_r + 2b^2 x_r(c_r^m)^2]$$

$$E_n^* = IP_n + [q_r + 2(c_r^m)^2](c_s^n)^2\frac{\Gamma}{\epsilon} - b^2\left[IP_n - EA_n - (c_r^m)^2(c_s^n)^2\Gamma\left(\frac{2}{\epsilon} - 1\right) \right] - \sum_s \frac{x_s(c_s^n)^2}{R_s}\left(1 - \frac{1}{\epsilon}\right)[q_s - 2b^2 x_s(c_s^n)^2] \tag{A2}$$

The final perturbation energy becomes (after correction for double account of electron–electron interaction)

$$\Delta E_{mn} = E_n^* - E_m^* + b^2[(E_m^* - E_n^*)_{b^2} - (E_m^* - E_n^*)_{b^2=0}] + \sqrt{(E_m^* - E_n^*)^2 + 4(c_r^m)^2(c_s^n)^2\beta^2} \tag{A3}$$

When the perturbation is small, i.e., $4\beta^2 << (E_m - E_n)^2$ and therefore $b^2 \to 0$, then this expression can be further reduced and approximated by

$$\Delta E_{mn} = \frac{2(c_r^m)^2(c_s^n)^2\beta^2}{E_m^* - E_n^*} \tag{A4}$$

A Four-Parameter Equation for Predicting Enthalpies of Adduct Formation

Russell S. Drago,* Glenn C. Vogel,[1] and Terence E. Needham

*Contribution from the William A. Noyes Laboratory,
University of Illinois, Urbana, Illinois 61801. Received September 24, 1970*

23

Abstract: A computer-fitted set of parameters for acids and bases is presented which accurately correlates over 280 enthalpies of adduct formation. The parameters can be used to predict over 1200 enthalpies of interaction. A matrix formulation of the problem is presented to illustrate the relationship among the various solutions to this problem. Our selection of a particular solution is justified on the basis of our intuitive understanding of the nature of molecular interactions. Means of transforming this solution to any model are presented. The conditions which lead to the Hammett and other two-parameter equations, which are subsets of the four-parameter equation, are derived from the matrix formulation. The fact that the donor number approach proposed by Gutman is not a generalized approach for estimating solvent donor strengths is established mathematically. The relationship of our parameters to the soft–hard acid–base model is discussed, and it can be demonstrated that our approach is not simply a quantitative manifestation of this concept, but the soft–hard concept as it is generally applied is incomplete. Furthermore, it is shown that this incompleteness can often lead to incorrect qualitative predictions of the magnitude of interaction.

A double-scale enthalpy equation was originally proposed to correlate (and predict) the enthalpy of adduct formation in gas-phase or poorly solvating media for several Lewis acid–base systems.[2] This empirical correlation is represented by eq 1. Two

$$-\Delta H = E_A E_B + C_A C_B \qquad (1)$$

empirically determined parameters, E_A and C_A, are assigned to each acid and two, E_B and C_B, are assigned to each base such that when substituted into eq 1, they give the enthalpy of adduct formation for the acid–base pair. E_A and E_B were originally interpreted as the susceptibility of the acid and base, respectively, to undergo electrostatic interaction and C_A and C_B as the susceptibility of the acid and base, respectively, to form covalent bonds. Equation 1 was found to correlate the enthalpies of interaction of donor–acceptor systems where reversals in donor strength are observed[3] (*e.g.*, oxygen donors interact more strongly with acids like phenol than with I_2, whereas analogous sulfur donors prefer I_2 more than acids like phenol). For several of the systems employed in the qualitative classification of acids and bases as type B or "soft" and type A or "hard,"[4,5] the E and C equation was in accord with the observation that combination of "like" acids and bases gives the most effective interaction.

At the time eq 1 was proposed, the amount of reliable enthalpy data was very limited. Since then, much enthalpy data have become available, both through direct calorimetric measurement and enthalpy-spectral parameter correlations. A linear relationship for certain types of donors between the shift in the frequency of the O–H stretching vibration of phenols[6–9] and

aliphatic alcohols,[10,11] $\Delta \bar{\nu}_{OH}$, and the enthalpy of adduct formation has been reported. A similar correlation has been shown to exist for pyrrole.[12] In addition a linear relationship between the tin–proton nmr coupling constant, $J_{Sn^{119}-CH_3}$, and the enthalpy of adduct formation of trimethyltin chloride has been cited.[13] With this large amount of reliable enthalpy data, the E and C correlation can be extended to many different acids and bases. In this paper, a refined set of E and C parameters, based upon the newly available enthalpy data, is reported. A computer fit of the data has been employed instead of the hand solution previously utilized. Furthermore, a matrix formulation of eq 1 is discussed which provides considerable insight into the quantitative correlation.

Experimental Section

Infrared Measurements.[11,15] The hydroxyl group frequency shifts, $\Delta \bar{\nu}_{OH}$, were obtained from spectra run on a Perkin-Elmer 521. Sodium chloride liquid cells of various path lengths were used with solvents carbon tetrachloride and tetrachloroethylene (used with amines[15]).

The concentration of the phenol was always kept less than 0.02 M to ensure against self-association. The base concentrations, which were always in excess of the acid concentration, were varied over a wide range. Thus, all frequency shifts were checked for concentration dependence and, when necessary, the reported value was obtained by extrapolation to infinite dilution. The experimentally determined shift is converted to an enthalpy using the equation

$$-\Delta H = 0.0103 \Delta \nu_{OH} \ (cm^{-1}) + 3.08$$

and reported as such in Table III (indicated by footnote *n*). This procedure has been demonstrated to be valid for oxygen and some nitrogen donors.[8–11]

(1) Abstracted in part from the Ph.D. thesis of G. C. Vogel, University of Illinois, Urbana, Ill., 1970.
(2) R. S. Drago and B. B. Wayland, *J. Amer. Chem. Soc.*, **87**, 3571 (1965).
(3) R. S. Niedzielski, R. S. Drago, and R. L. Middaugh, *ibid.*, **86**, (1964).
(4) R. G. Pearson, *Chem. Brit.*, **3**, 103 (1967); *J. Chem. Educ.*, **45**, 581, 643 (1968), and references cited therein.
(5) S. Ahrland, *Chem. Phys. Lett.*, **2**, 303 (1968), and references cited therein.
(6) M. D. Joesten and R. S. Drago, *J. Amer. Chem. Soc.*, **84**, 3817 (1962).

(7) T. D. Epley and R. S. Drago, *ibid.*, **89**, 5770 (1967).
(8) T. D. Epley and R. S. Drago, *ibid.*, **91**, 2883 (1969).
(9) G. C. Vogel and R. S. Drago, *ibid.*, **92**, 5347 (1970).
(10) R. S. Drago, N. O'Bryan, and G. C. Vogel, *ibid.*, **92**, 3924 (1970).
(11) K. F. Purcell, J. A. Stikeleather, and S. D. Brunk, *ibid.*, **91**, 4019 (1969).
(12) M. Nozari and R. S. Drago, *ibid.*, **92**, 7086 (1970).
(13) T. F. Bolles and R. S. Drago, *J. Amer. Chem. Soc.*, **88**, 5730 (1966).
(14) T. D. Epley and R. S. Drago, *J. Paint Technol.*, **41**, 503 (1969).
(15) A. Allerhand and P. v. R. Schleyer, *J. Amer. Chem. Soc.*, **85**, 371 (1963).

Method of Calculation. The original set of E and C parameters was determined mainly from the enthalpy data of iodine and phenol. Using experimental enthalpies for iodine interacting with a series of alkylamines, eq 1 was solved and empirical values were obtained for the amine E and C parameters by setting $C_A = E_A = 1.00$ for iodine and $C_B = aR_D$ and $E_B = b\mu$ for the amines. Here R_D is the total distortion polarization and μ is the ground-state dipole moment. The four amine parameters were employed with four enthalpies of adduct formation toward phenol to calculate E_A and C_A parameters for phenol, in effect using two enthalpies as checks on the procedure. The parameters for other donors were obtained by using their enthalpy of adduct formation with iodine and phenol, yielding two equations in the two unknowns C_B and E_B for each base. Finally, by using the E_B and C_B parameter for these donors, the correlation was extended to other Lewis acids with several experimental enthalpies remaining to provide checks on the entire procedure.

In an attempt to get the best set of E and C parameters from the large number of enthalpy data presently available, a computer program was written employing a least-squares analysis to find the values of the parameters which give the best fit between the measured enthalpies and those calculated from eq 1. The previously reported values of E and C parameters are used as initial guesses, and corrections to the parameters are calculated from the matrix equation[17]

$$\Delta = (A'PA)^{-1}PF \qquad (2)$$

where Δ is the vector of corrections to the parameters, A is the matrix of partial derivatives $\partial(-\Delta H_{i_{calcd}})/\partial(\text{parameter } j)$, P is the weight matrix, and F is the vector of residuals, $-\Delta H_{i_{calcd}} + \Delta H_{i_{obsd}}$. P is taken to be a diagonal matrix. Since standard deviations for experimental enthalpies are rarely available, the diagonal elements of the weight matrix have been chosen to be $1/(4 - \Delta H_{i_{calcd}})^2$, a function which seems to give appropriate weights to both large and small enthalpies. Occasionally, when a measured enthalpy has seemed particularly good or bad, a different weight was given to that heat. All enthalpies estimated from constant acid frequency shift relations were weighted as $1/(10 - \Delta H_{i_{calcd}})^2$.

The corrected parameters are used to calculate a new A matrix and F vector, new corrections are calculated, and the process is repeated until the calculated corrections are essentially zero. As will be shown later, a total of four parameters must be specified in order to determine a unique solution for the E and C numbers; therefore, the following parameters were held fixed and not allowed to vary: iodine $E_A = 1.00$, iodine $C_A = 1.00$, DMA $E_B = 1.32$, diethyl sulfide $C_B = 7.40$.

Results

The parameters calculated for the acids are given in Table I and those for the bases in Table II. The weighted root-mean-square deviation between the experimental enthalpies and those calculated from the parameters in Tables I and II using eq 1 is about 0.016, corresponding to a deviation of about 0.2 kcal/mol for a heat of 8 kcal/mol. The excellent agreement between the experimental enthalpies of adduct formation and the calculated enthalpies for all of the interactions is shown in Table III. Newly determined frequency shifts used to estimate $-\Delta H$ are reported in Table IV.

Accuracy of the Parameters. Included in Tables I and II are marginal standard deviations and conditional standard deviations for each parameter. The method for calculating these numbers is outlined briefly in the Appendix. For a more complete treatment, the reader is referred to ref 16 and 17. The marginal standard deviations are large and reflect the fact that a significant change in one parameter can be compensated for by changes in many of the other parameters so that the heats calculated from all the parameters change very little. Thus, the errors in the parameters are

(16) W. C. Hamilton, "Statistics in Physical Science," Ronald Press, New York, N. Y., 1964.
(17) P. J. Lingane and Z. Z. Hugus, Jr., *Inorg. Chem.*, **9**, 757 (1970).

highly correlated and, for example, the C number given for phenol can be too large by 0.213 only if the E number for phenol is too small by nearly 0.51. The marginal standard deviation of a parameter is quite dependent on the extent that the parameter is connected to the standards through enthalpies. Most uses of E and C numbers do not require an exact knowledge of the absolute magnitude of the numbers themselves, but only an accurate knowledge of trends in the numbers, which are, in fact, much more accurately known. The conditional standard deviations given here are standard deviations for each parameter assuming that all the other parameters have their true values. They may be regarded as lower limits to the inaccuracies in the parameters for any use of the parameters. Because of the high correlation in the errors, the appropriate error limits for examining trends in a series of similar numbers are much closer to the conditional standard deviations than to the marginal standard deviations.

One very important use of E and C numbers is the calculation of heats of interaction for systems which have not been examined experimentally. From our knowledge of the standard deviations of the parameters and their correlation coefficients (see the Appendix), we have calculated the expected standard deviations for calculated heats for all possible combinations of all but a few of the acids and bases listed in Tables I and II. For the hydrogen-bonding acids and sulfur dioxide, these predicted standard deviations nearly all lie between 0.1 and 0.3 kcal/mol. For most other systems, the errors are somewhat worse than this, averaging around 0.7 kcal. Of course, the heats for many of these interactions are generally larger, too.

It should be noted that the above treatment of errors does not take into account the effects of any systematic errors in the experimental heats, which might arise, for example, from solvent effects or from the fact that many of the heats were taken from spectroscopic correlations. However, these effects should be smaller than the effects of the larger random errors.

Discussion

Uniqueness of the E and C Parameters. Equation 1 can be rewritten in matrix notation by assigning the acid parameters to a vector X_A and the base parameters to a vector Y_B, *i.e.*

$$X_A = \begin{bmatrix} E_A \\ C_A \end{bmatrix} \text{ and } Y_B = \begin{bmatrix} E_B \\ C_B \end{bmatrix}$$

and allowing the enthalpy of adduct formation, ΔH, to be a scalar function of the vector

$$-\Delta H = Y_B{}^T X_A = [E_B C_B]\begin{bmatrix} E_A \\ C_A \end{bmatrix} = $$
$$E_B E_A + C_B C_A \qquad (3)$$

The series of equations (3) has an infinite number of "best-fit" solutions for E_A, C_A, E_B, C_B, each of which predicts exactly the same ΔH for every interaction. This can be shown by defining a 2×2 transformation matrix A which leads to the new vectors X_A' and Y_B' in the following way

$$-\Delta H = Y_B{}^T X_A$$
$$= Y_B{}^T[A^{-1}A]X_A$$
$$= [A^{-1T}Y_B]^T[AX_A]$$

Drago, Vogel, Needham / Predicting Enthalpies of Adduct Formation

Table I. Acid Parameters

No.	Acid	No.a of enthalpies	C_A (Marginal)	C_A (Conditional)h	E_A (Marginal)	E_A (Conditional)
1.	Iodine	39	1.00 (d)	(d)	1.00 (d)	5.10 (d)
2.	Iodine monochlorideb	8	0.830 (0.334)	(0.05)	(0.62)	(0.10)
3.	Iodine monobromideb	3	1.56 (0.31)	(0.09)	2.41 (0.41)	(0.12)
4.	Thiophenolb,g	3	0.198 (0.059)	(0.022)	0.987 (0.193)	(0.082)
5.	p-tert-Butylphenol	5	0.387 (0.161)	(0.014)	4.06 (0.37)	(0.13)
6.	p-Methylphenolc	6	0.404 (0.166)	(0.021)	4.18 (0.39)	(0.17)
7.	Phenol	34	0.442 (0.170)	(0.010)	4.33 (0.35)	(0.06)
8.	p-Fluorophenol	9	0.446 (0.164)	(0.014)	4.17 (0.35)	(0.08)
9.	p-Chlorophenol	10	0.478 (0.170)	(0.015)	4.34 (0.37)	(0.11)
10.	m-Fluorophenol	10	0.506 (0.173)	(0.015)	4.42 (0.37)	(0.09)
11.	m-Trifluoromethylphenol	22	0.530 (0.174)	(0.014)	4.48 (0.37)	(0.08)
12.	tert-Butyl alcoholb	2 (11)e	0.300 (0.102)	(0.028)	2.04 (0.30)	(0.11)
13.	Trifluoroethanol	12	0.434 (0.159)	(0.018)	4.00 (0.33)	(0.06)
14.	Hexafluoroisopropyl alcohol	12	0.509 (0.221)	(0.018)	5.56 (0.46)	(0.08)
15.	Pyrroleb (C_4H_4NH)	4 (6)e	0.295 (0.100)	(0.013)	2.54 (0.26)	(0.10)
16.	Isocyanic acidb (HNCO)	4	0.258 (0.147)	(0.025)	3.22 (0.39)	(0.10)
17.	Isothiocyanic acid (HNCS)	9	0.227 (0.217)	(0.011)	5.30 (0.47)	(0.11)
18.	Boron trifluorideb,f,i	5	3.08 (0.36)	(0.07)	7.96 (1.13)	(0.24)
19.	Boron trifluoride (g)b,f	4	1.62 (0.44)	(0.06)	9.88 (1.48)	(0.31)
20.	Boron trimethylf	6	1.70 (0.26)	(0.03)	6.14 (0.86)	(0.18)
21.	Trimethylaluminum	18	1.43 (0.67)	(0.02)	16.9 (1.4)	(0.2)
22.	Triethylaluminumc	4	2.04 (0.53)	(0.08)	12.5 (1.4)	(0.3)
23.	Trimethylgallium	4	0.881 (0.54)	(0.030)	13.3 (1.5)	(0.4)
24.	Triethylgallium	5	0.593 (0.53)	(0.030)	12.6 (1.2)	(0.2)
25.	Trimethylindium	2	0.654 (0.66)	(0.043)	15.3 (2.5)	(0.5)
26.	Trimethyltin chloridef	10	0.0296 (0.312)	(0.0273)	5.76 (0.64)	(0.07)
27.	Sulfur dioxideb	6	0.808 (0.043)	(0.020)	0.920 (0.164)	(0.079)
28.	Bis(hexafluoroacetylacetonate)copper(II)b	7	1.40 (0.13)	(0.03)	3.39 (0.36)	(0.11)
29	Antimony pentachloridef	4	5.13 (0.42)	(0.14)	7.38 (0.85)	(0.24)
30.	Chloroform	10	0.150 (0.138)	(0.009)	3.31 (0.30)	(0.15)
31.	1-Hydroperfluoroheptane [$CF_3(CF_2)_6H$]b	2	0.226 (0.101)	(0.014)	2.45 (0.35)	(0.15)

a Number of heats used to determine the parameters for the specified acid. b Tentative value calculated from limited data or data limited to bases with similar C/E ratios. In latter case, can be confidently used only with bases with C/E ratios less than 4.0. c Tentative value calculated from estimated enthalpies. d Parameter is a standard. e This number of enthalpies estimated from infrared frequency shifts agree with these parameters. f Steric effects commonly encountered. g Accuracy of input data estimated to be at best 10%. h Marginal and conditional standard derivations. i Data from 1,2-dichloroethane displacement reactions.

where

$$A = \begin{bmatrix} a_{11} & a_{12} \\ a_{21} & a_{22} \end{bmatrix}$$

The enthalpy in terms of the new vectors X_A' and Y_B' is given by

$$-\Delta H = Y_B'^T X_A'$$

where

$$X_A' = AX_A; \quad \begin{bmatrix} E_A' \\ C_A' \end{bmatrix} = \begin{bmatrix} a_{11} & a_{12} \\ a_{21} & a_{22} \end{bmatrix} \begin{bmatrix} E_A \\ C_A \end{bmatrix} \quad (3a)$$

$$Y_B' = A^{-1}{}^T Y_B; \quad \begin{bmatrix} E_B' \\ C_B' \end{bmatrix} =$$

$$\frac{1}{(a_{11}a_{22} - a_{21}a_{12})} \begin{bmatrix} a_{22} & -a_{21} \\ -a_{12} & a_{11} \end{bmatrix} \begin{bmatrix} E_B \\ C_B \end{bmatrix} \quad (3b)$$

with the conditions that

$$\det A = a_{11}a_{22} - a_{21}a_{12} \neq 0$$

and the a_{ij}'s have finite values which can be determined. Hence, if one transforms all of the E and C parameters by eq 3a and 3b using an arbitrary matrix A (whose inverse exists), one arrives at a new set of parameters which predict the same enthalpies of adduct formation, ΔH.

A is a linear-transformation matrix which allows one to transform from one best-fit solution to another (from the unprimed set to the primed set mentioned above). In order to specify a particular solution (and a unique transformation matrix A to get to that solution from some arbitrary solution), four E and C parameters may[18] be chosen and assigned specific values, as long as they are chosen in such a way that the transformation from our arbitrary solution is completely defined and finite; i.e., the four parameters must be chosen so that the elements of A, a_{ij}, are completely defined and the determinant of A must be *nonzero* in order that A^{-1T} exist, $[a_{11}a_{22} - a_{12}a_{21} \neq 0]$. These requirements are not always met for an arbitrary choice of parameters to be fixed. Once a set of parameters is procured which gives a best fit between experimental enthalpies of adduct formation and the ones calculated from eq 3, one can attempt to impose any model on the parameters by finding the transformation matrix A which allows one to map the vectors (X_A, Y_B) of one solution respectively onto those (X_A', Y_B') of another. Imposing a model corresponds to finding a solution for which the E or C parameters, or some function of them, correspond to some physical property; for example, one may want the $E_A E_B$ product to correspond to the electrostatic interaction or one may want the C_B's to be proportional to the polarizability of the bases. If a transformation matrix which maps one solution onto another does not exist, then the model cannot be imposed on the set of parameters which gives the best fit between experimental enthalpies and those calculated by eq 3.

As mentioned above, the initial model[2] was chosen so as to break the enthalpy of adduct formation into electrostatic and covalent contributions, i.e., $-\Delta H = E_A E_B + C_A C_B$. The fact that more than one solution can exist enables us to *attempt* to break up the enthalpy of adduct formation into a physically meaningful model other than the electrostatic and covalent model initially employed. One apparently obvious break-up of the enthalpy of adduct formation which is of chemical significance would be that of σ and π contributions.

$$-\Delta H = \sigma_A \sigma_B + \pi_A \pi_B \quad (4)$$

(18) This is not the only way of specifying a particular solution. For example, one might instead specify three E and C numbers and then require that two amine E numbers be proportional to their dipole moments.

The idea here is that reversals occur when there is extensive π back-bonding from the acid to the base. The enthalpy data of phenol and iodine interacting with several alkylamines can be used to test the feasibility of breaking up the enthalpy of adduct formation into σ and π contributions. Since it has presumably no π-bonding capabilities, phenol can be assigned the following parameters and used as the standard: $\sigma_{phenol} = 1.00$ and $\pi_{phenol} = 0.0$. Using the phenol–amine enthalpy data, the following equations result in which it is immediately obvious that the amine σ parameter is equal to its enthalpy of interaction with phenol.

$$NH_3 \qquad -\Delta H = 7.8 = \sigma_B(1.00)$$
$$CH_3NH_2 \qquad -\Delta H = 8.6 = \sigma_B'(1.00)$$
$$(CH_3)_2NH \qquad -\Delta H = 8.6 = \sigma_B''(1.00)$$
$$(CH_3)_3N \qquad -\Delta H = 8.8 = \sigma_B'''(1.00)$$

Since alkylamines have no low-energy orbitals for π back-bonding interactions, the iodine–amine enthalpies can be represented by the following equations.

$$NH_3 \qquad -\Delta H = 4.8 = 7.8\sigma_{I_2}$$
$$CH_3NH_2 \qquad -\Delta H = 7.1 = 8.6\sigma_{I_2}$$
$$(CH_3)_2NH \qquad -\Delta H = 9.8 - 8.6\sigma_{I_2}$$
$$(CH_3)_3N \qquad -\Delta H = 12.1 = 8.8\sigma_{I_2}$$

Since there is no solution for σ_{I_2}, the model described by eq 4 is too simple to accurately describe the systems in Table III. In matrix notation, this means that no transformation matrix exists which can transform our present best-fit set of E and C parameters into a set of σ and π parameters.

Two-Parameter Equations. One may question the need of a four-parameter enthalpy equation, i.e., whether describing an acid or base by two parameters is redundant. The following simple matrix algebra shows the conditions whereby a four-parameter model reverts to a less redundant two-parameter equation. Letting A be the transformation matrix, E and C represent the parameters for the four-parameter model, and α represent the acid parameters for the two-parameter model, the following equation results.

$$\begin{bmatrix} \alpha \\ 0 \end{bmatrix} = \begin{bmatrix} a_{11} & a_{12} \\ a_{21} & a_{22} \end{bmatrix} \begin{bmatrix} E \\ C \end{bmatrix}$$

After multiplication, eq 5a and 5b are obtained.

$$Ea_{11} + Ca_{12} = \alpha \quad (5a)$$
$$Ea_{21} + Ca_{22} = 0 \quad (5b)$$

For any two acids i and j, eq 5b yields

$$E_i a_{21} + C_i a_{22} = 0$$
$$E_j a_{21} + C_j a_{22} = 0$$

Rearranging and eliminating a_{22} and a_{21} gives

$$C_i/E_i = C_j/E_j = k \quad (6)$$

Thus, for a transformation from a model describing an acid and base by two parameters each to one describing an acid and base by one parameter each, it is found that the C/E ratio for all the acids (or alternatively all the bases) in the two-parameter set must be the

Table II. Base Parameters

Base	No.[a] of enthalpies	C_B (Marginal)	C_B (Conditional)	E_E (Marginal)	E_E (Conditional)
1. Pyridine	21 (H)	6.40 (0.29)	(0.11)	1.17 (0.18)	(0.02)
2. Ammonia	5 (H)	3.46 (0.24)	(0.16)	1.36 (0.07)	(0.03)
3. Methylamine	5 (H)	5.88 (0.30)	(0.19)	1.30 (0.15)	(0.03)
4. Dimethylamine	4 (H)	8.73 (0.46)	(0.27)	1.09 (0.29)	(0.03)
5. Trimethylamine[d]	7 (H)	11.54 (0.58)	(0.22)	0.808 (0.433)	(0.021)
6. Ethylamine[b]	4 (H)	6.02 (0.36)	(0.21)	1.37 (0.18)	(0.07)
7. Diethylamine[b]	2 (H)	8.83 (0.48)	(0.28)	0.866 (0.314)	(0.045)
8. Triethylamine[d]	9 (H)	11.09 (0.57)	(0.23)	0.991 (0.402)	(0.026)
9. Acetonitrile	11 (C)	1.34 (0.10)	(0.06)	0.886 (0.033)	(0.016)
10. Chloroacetonitrile	6 (C)	0.53((0.16)	(0.051)	0.940 (0.089)	(0.025)
11. Dimethylcyanamide	7 (C)	1.81 (0.16)	(0.10)	1.10 (0.04)	(0.03)
12. Dimethylformamide	4 (C)	2.48 (0.22)	(0.17)	1.23 (0.07)	(0.05)
13. Dimethylacetamide	13 (C)	2.58 (0.14)	(0.10)	1.32 (c)	(c)
14. Ethyl acetate	14 (C)	1.74 (0.12)	(0.07)	0.975 (0.033)	(0.017)
15. Methyl acetate	4 (C)	1.61 (0.17)	(0.15)	0.903 (0.041)	(0.032)
16. Acetone	9 (C)	2.33 (0.18)	(0.14)	0.987 (0.036)	(0.018)
17. Diethyl ether[d]	11 (C)	3.25 (0.22)	(0.13)	0.963 (0.069)	(0.017)
18. Isopropyl ether[b,d]	3 (C)	3.19 (0.25)	(0.19)	1.11 (0.09)	(0.06)
19. n-Butyl ether[d]	8 (C)	3.30 (0.23)	(0.17)	1.06 (0.07)	(0.03)
20. p-Dioxane [(CH₂)₄O₂]	6 (C)	2.38 (0.20)	(0.13)	1.09 (0.04)	(0.02)
21. Tetrahydrofuran [(CH₂)₄O]	10 (C)	4.27 (0.23)	(0.12)	0.978 (0.110)	(0.021)
22. Tetrahydropyran	4 (C)	3.91 (0.27)	(0.17)	0.949 (0.112)	(0.038)
23. Dimethyl sulfoxide	14 (C)	2.85 (0.18)	(0.10)	1.34 (0.04)	(0.02)
24. Tetramethylene sulfoxide [(CH₂)₄SO]	5 (C)	3.16 (0.20)	(0.15)	1.38 (0.06)	(0.04)
25. Dimethyl sulfide	5 (H)	7.46 (0.41)	(0.20)	0.343 (0.293)	(0.021)
26. Diethyl sulfide	10 (H)	7.40 (c)	(c)	0.339 (0.265)	(0.015)
27. Trimethylene sulfide [(CH₂)₃S]	5 (H)	6.84 (0.38)	(0.19)	0.352 (0.266)	(0.021)
28. Tetramethylene sulfide	12 (H)	7.90 (0.42)	(0.13)	0.341 (0.313)	(0.014)
29. Pentamethylene sulfide	5 (H)	7.40 (0.41)	(0.20)	0.375 (0.288)	(0.022)
30. Pyridine N-oxide	4 (C)	4.52 (0.23)	(0.16)	1.34 (0.11)	(0.04)
31. 4-Methylpyridine N-oxide	5 (C)	4.99 (0.25)	(0.13)	1.36 (0.13)	(0.04)
32. 4-Methoxypyridine N-oxide[b]	3 (C)	5.77 (0.30)	(0.19)	1.37 (0.18)	(0.05)
33. Tetramethylurea[d]	3 (C)	3.10 (0.24)	(0.19)	1.20 (0.09)	(0.06)
34. Trimethylphosphine[b]	6 (H)	6.55 (0.61)	(0.18)	0.838 (0.219)	(0.017)
35. Benzene	5 (CorH)	0.707 (0.12)	(0.087)	0.486 (0.038)	(0.025)
36. p-Xylene[b]	2 (CorH)	1.78 (0.20)	(0.13)	0.416 (0.097)	(0.036)
37. Mesitylene[b]	3 (CorH)	2.19 (0.19)	(0.11)	0.574 (0.109)	(0.040)

Table II (*Continued*)

Base	No.[a] of enthalpies	C_B (Marginal)	C_B	C_B (Conditional)	E_B (Marginal)	E_B	E_B (Conditional)
38. 2,2,6,6-Tetramethylpyridine N-oxyl[d] ($C_9H_{18}NO$)	7 (H)	(0.50)	6.21	(0.15)	(0.202)	0.915	(0.025)
39. 1-Azabicyclo[2.2.1]octane [$HC(C_2H_4)_3N$]	5 (H)	(0.70)	13.2	(0.31)	(0.518)	0.704	(0.038)
40. 7-Oxabicyclo[2.2.1]heptane ($C_6H_{10}O$)	3 (C)	(0.24)	3.76	(0.15)	(0.10)	1.08	(0.048)
41. Dimethyl selenide	3 (H)	(0.48)	8.33	(0.20)	(0.344)	0.217	(0.019)
42. 1-Phospha-4-ethyl-1,5,7-trioxabicyclo[2.2.1]octane [$C_2H_5C(CH_2O)_3P$]	3 (H)	(0.73)	6.60	(0.23)	(0.253)	0.515	(0.033)
43. Hexamethylphosphoramide[b]	5 (C)	(0.11)	1.33	(0.37)	(0.13)	1.73	(0.030)

[a] Number of heats used to determine the parameters for the specified base. The solvent recommended for getting enthalpies for comparison is indicated in parentheses; H stands for cyclohexane or gas phase and C represents carbon tetrachloride or gas phase: M. S. Nozari and R. S. Drago, submitted for publication. [b] Tentative value calculated from data limited to acids with similar C/E ratios. [c] Parameters considered a standard for the purpose of calculating standard deviations. [d] Steric effects often expected.

Table III

Acid	Base	Measd	Calcd	Ref	Acid	Base	Measd	Calcd	Ref
I_2	C_5H_5N	7.8	7.6	a	C_6H_5SH	C_5H_5N	2.4	2.4	k
	NH_3	4.8	4.8	a		$HC(O)N(CH_3)_2$	1.8	1.7	k
	CH_3NH_2	7.1	7.2	a		C_6H_6	0.5	0.6	k
	$(CH_3)_2NH$	9.8	9.8	a	$p\text{-}tert\text{-}C_4H_9C_6H_4OH$	C_5H_5N	7.2	7.2	l
	$(CH_3)_3N$	12.1	12.3	a		$(C_2H_5)_3N$	8.3	8.3	l
	$C_2H_5NH_2$	7.4	7.4	a		$CH_3C(O)N(CH_3)_2$	6.4	6.4	l
	$(C_2H_5)_2NH$	9.7	9.7	a		$(C_2H_5)_2S$	4.2	4.2	m
	$(C_2H_5)_3N$	12.0	12.1	a		$(CH_2)_4S$	4.5	4.4	m
	$HC(C_2H_4)_3N$	13.9	13.9	rr	$p\text{-}CH_3C_6H_4OH$	C_5H_5N	7.8	7.5	l, n
	CH_3CN	1.9	2.2	a		$CH_3C(O)N(CH_3)_2$	6.4	6.6	l, n
	$ClCH_2CN$	1.5	1.5	b		$CH_3C(O)OC_2H_5$	4.6	4.7	l, n
	$(CH_3)_2NCN$	2.8	2.9	c		$(C_4H_9)_2O$	5.9	5.8	m
	$HC(O)N(CH_3)_2$	3.7	3.7	a		$(C_2H_5)_2S$	4.3	4.4	m, n
	$CH_3C(O)N(CH_3)_2$	4.0	3.9	a		$(CH_2)_4S$	4.6	4.6	m
	$CH_3C(O)OC_2H_5$	2.8	2.7	a	C_6H_5OH	C_5H_5N	8.0	7.9	l
	$CH_3C(O)OCH_3$	2.5	2.5	a		NH_3	7.8	7.4	o
	$CH_3C(O)CH_3$	3.3	3.3	a		CH_3NH_2	8.6	8.2	o
	$(C_2H_5)_2O$	4.2	4.2	a		$(CH_3)_2NH$	8.6	8.6	o
	$(C_4H_9)_2O$	4.4	4.4	ll		$(CH_3)_3N$	8.8	8.6	o
	$(CH_2)_4O$	5.3	5.2	a		$C_2H_5NH_2$	8.6	8.6	o
	$[(CH_3)_2CH]_2O$	4.3	4.3	ll		$(C_2H_5)_3N$	9.1	9.2	p
	$(CH_2)_5O$	4.9	4.9	d		$HC(C_2H_4)_3N$	9.0	8.9	pp
	$(CH_2)_4O_2$	3.5	3.5	a		CH_3CN	4.6	4.4	p
	$C_6H_{10}O$	4.9	4.8	ww		$ClCH_2CN$	4.2	4.3	l, n
	$(CH_3)_2SO$	4.4	4.2	a		$(CH_3)_2NCN$	5.4	5.6	n, u
	$(CH_2)_4SO$	4.4	4.5	a		$CH_3C(O)N(CH_3)_2$	6.8	6.8	p
	$(C_2H_5)_2S$	7.8	7.7	a		$HC(O)N(CH_3)_2$	6.1	6.4	q
	$(CH_2)_4S$	7.1	7.2	e		$CH_3O(O)OC_2H_5$	4.8	5.0	p
	$(CH_2)_4S$	8.3	8.2	e		$CH_3C(O)OCH_3$	4.8	4.6	l, n
	$(CH_3)_2S$	7.8	7.8	ll		$CH_3C(O)CH_3$	5.1	5.3	p
	$(CH_2)_5S$	7.8	7.8	e		$C_9H_{18}NO$	6.9	6.7	oo
	$(CH_3)_2Se$	8.5	8.6	ss		$(C_2H_5)_2O$	6.0	5.6	q
	$[(CH_3)_2N]_2CO$	4.3	4.3	a		$(C_4H_9)_2O$	6.0	6.0	m
	C_5H_5NO	5.9	5.9	f		$[(CH_3)_2CH]_2O$	6.2	6.2	q
	$4\text{-}CH_3C_5H_4NO$	6.3	6.3	qq		$(CH_2)_4O$	6.0	6.1	l
	$4\text{-}CH_3OC_5H_4NO$	7.2	7.1	qq		$(CH_2)_5O$	6.1	5.8	l, n
	C_6H_6	1.3	1.2	a		$(CH_2)_4O_2$	5.6	5.8	l, n
	$p\text{-}(CH_3)_2C_6H_4$	2.2	2.2	h		C_9H_{10}	6.4	6.3	vv
	$s\text{-}(CH_3)_3C_6H_3$	2.9	2.8	h		$(CH_3)_2SO$	6.9	7.1	r
	$C_6(CH_3)_6$	3.7	3.7	h		$(CH_2)_4SO$	7.2	7.4	a
ICl	$CH_3C(O)N(CH_3)_2$	9.2	8.9	a		$(C_2H_5)_2S$	4.6	4.7	m
	$CH_3C(O)OC_2H_5$	6.1	6.4	i		$(CH_3)_2S$	4.6	4.8	s
	$(CH_2)_4O_2$	7.5	7.5	a		$(CH_3)_2S$	4.5	4.5	s
	$ClCH_2CN$	5.3	5.3	b		$(CH_2)_4S$	4.9	5.0	m
	$(CH_3)_2NCN$	7.3	7.1	c		$(CH_2)_5S$	4.7	4.9	s
	C_6H_6	2.8	3.1	a		$[(CH_3)_2N]_2CO$	6.6	6.6	l, n
	$p\text{-}(CH_3)_2C_6H_4$	3.6	3.6	j		C_5H_5NO	7.9	7.8	t
	$s\text{-}(CH_3)_3C_6H_3$	4.7	4.7	j		$4\text{-}CH_3C_5H_4NO$	8.4	8.1	r
IBr	CH_3CN	4.1	4.2	b	$p\text{-}FC_6H_4OH$	C_5H_5N	7.9	7.7	l
	$ClCH_2CN$	3.1	3.1	b		$(C_2H_5)_3N$	9.0	9.1	l
	$(CH_3)_2NCN$	5.6	5.5	c					

Table III (*Continued*)

Acid	Base	Measd	Calcd	Ref	Acid	Base	Measd	Calcd	Ref
	CH₃C(O)OC₂H₅	4.9	4.8	*l*	C₄H₄NH	C₂H₅N	5.0	4.8	*z*
	(CH₃)₂SO	6.6	6.9	*tt*		(C₂H₅)₃N	5.9	5.8	*z*
	(CH₂)₄O	5.6	6.0	*tt*		(CH₃)₂SO	4.2	4.2	*z*
	(C₂H₅)₂O	5.6	5.5	*tt*		HC(C₂H₄)₃N	5.6	5.7	*pp*
	[(CH₃)₂N]₃PO	8.0	7.8	*tt*	(CH₃)₃SnCl	CH₃CN	4.8	5.1	*aa*
	(C₂H₅)₂S	4.7	4.7	*m*		CH₃C(O)N(CH₃)₂	7.9	7.7	*aa*
	(CH₂)₄S	5.0	4.9	*m*		CH₃C(O)CH₃	5.7	5.8	*aa*
p-ClC₆H₄OH	C₅H₅N	8.1	8.1	*l*		[(CH₃)₂N]₃PO	10.1	10.0	*aa*
	(C₂H₅)₃N	9.5	9.6	*l*		CH₃C(O)OCH₃	5.2	5.2	*g*
	CH₃C(O)N(CH₃)₂	7.0	7.0	*n*		(CH₃)₂SO	8.2	7.8	*aa*
	CH₃C(O)OC₂H₅	5.0	5.1	*l*		(CH₃)₂NCN	6.4	6.4	*u*
	CH₃C(O)CH₃	5.4	5.4	*n*		C₅H₅NO	7.8	7.0	*g*
	(C₄H₉)₂O	6.3	6.2	*n*		4-CH₃C₅H₄NO	7.9	8.0	*g*
	(CH₃)₂SO	7.2	7.2	*n*		4-CH₃OC₅H₄NO	8.1	8.1	*g*
	(CH₃)₂NCN	5.6	5.7	*u*	BF₃ (DCE)	(CH₃)₂SO	19.5	19.5	*g*
	(C₂H₅)₂S	5.0	5.0	*m*		C₅H₅NO	24.5	24.5	*g*
	(CH₂)₄S	5.3	5.2	*m*		4-CH₃C₅H₄NO	26.2	26.1	*g*
m-FC₆H₄OH	C₅H₅N	8.4	8.4	*v*		4-CH₃OC₅H₄NO	28.5	28.7	*g*
	CH₃CN	4.9	4.6	*v*		C₉H₁₈NO	26.5	26.4	*oo*
	CH₃C(O)OC₂H₅	5.2	5.2	*v*	B(CH₃)₃	C₅H₅N	17.6	18.0	*bb*
	CH₃C(O)N(CH₃)₂	7.0	7.1	*v*		NH₃	14.3	14.2	*bb*
	(CH₃)₂SO	7.3	7.4	*v*		CH₃NH₂	18.2	18.0	*cc*
	(C₂H₅)₂S	5.2	5.2	*m*		C₂H₅NH₂	18.6	18.6	*bb*
	(CH₂)₄S	5.5	5.5	*m*		(CH₃)₃P	16.5	16.3	*ss*
	C₉H₁₈NO	7.5	7.2	*oo*		C₂H₅C(CH₂O)₃P	14.4	14.4	*pp*
	(C₄H₉)₂O	6.0	6.4	*mm*	Al(CH₃)₃	C₅H₅N	26.7	28.9	*dd*
	C₂H₅C(CH₂O)₃P	5.6	5.6	*pp*		NH₃	27.6	28.0	*dd*
m-CF₃C₆H₄OH	C₅H₅N	8.5	8.6	*l*		CH₃NH₂	30.0	30.4	*dd*
	CH₃CN	4.9	4.7	*n*		(CH₃)₂NH	30.8	30.8	*dd*
	ClCH₂CN	4.4	4.5	*n*		(CH₃)₃N	30.0	30.2	*dd*
	(CH₃)₂NCN	5.8	5.9	*n*		(C₂H₅)₂NH	27.3	27.3	*dd*
	CH₃C(O)N(CH₃)₂	7.3	7.3	*l*		CH₃C(O)CH₃	20.3	20.0	*dd*
	CH₃C(O)C₂H₅	5.2	5.3	*n*		(C₂H₅)₂O	20.2	20.9	*dd*
	CH₃C(O)OCH₃	5.0	4.9	*n*		(CH₂)₄O₂	22.9	21.9	*dd*
	CH₃C(O)CH₃	5.9	5.7	*n*		(CH₃)₂SO	28.6	26.8	*ee*
	(C₂H₅)₂O	6.5	6.0	*n*		(CH₂)₄SO	28.2	27.9	*ee*
	[(CH₃)₂CH]₂O	6.7	6.7	*n*		(C₂H₅)₂S	16.7	16.3	*ff*
	C₉H₁₈NO	7.5	7.4	*oo*		(CH₃)₂S	16.7	16.5	*ff*
	(CH₂)₄O	6.5	6.6	*n*		(CH₂)₃S	16.0	15.8	*ff*
	(CH₂)₅O	6.5	6.3	*n*		(CH₂)₄S	17.0	17.1	*ff*
	(CH₂)₄O₂	6.0	6.2	*n*		(CH₂)₅S	17.0	16.9	*ff*
	(CH₃)₂SO	7.4	7.5	*n*		(CH₃)₂Se	16.0	15.6	*bb*
	(CH₂)₄SO	7.6	7.9	*n*		(CH₃)₃P	22.1	23.6	*dd*
	(C₂H₅)₂S	5.4	5.4	*m*	Al(C₂H₅)₃	C₅H₅N	27.0	27.6	*gg*
	(CH₃)₂S	5.4	5.5	*s*		(C₂H₅)₂O	18.8	18.7	*gg*
	(CH₂)₃S	5.4	5.2	*s*		(CH₂)₄O	21.6	20.9	*gg*
	(CH₂)₄S	5.7	5.7	*m*		(CH₂)₄O₂	18.3	18.5	*gg*
	(CH₂)₅S	5.6	5.6	*s*	Ga(C₂H₅)₃	NH₃	19.2	19.2	*hh*
	[(CH₃)₂N]₂CO	7.0	7.0	*n*		CH₃NH₂	19.8	19.8	*hh*
(CH₃)₃COH	C₅H₅N	4.3	4.3	*w*		(CH₃)₂NH	18.8	18.8	*hh*
	(CH₃)₂SO	3.6	3.6	*w*		(CH₃)₃N	17.0	17.0	*ii*
CF₃CH₂OH	C₅H₅N	7.8	7.4	*x*		(CH₃)₃P	14.5	14.4	*a*
	(C₂H₅)₃N	8.8	8.8	*x*	SO₂	C₅H₅N	6.0	6.2	*a*
	CH₃CN	4.4	4.1	*x*		(CH₃)₃N	10.3	10.1	*jj*
	(C₂H₅)₂O	5.1	5.3	*x*		CH₃C(O)N(CH₃)₂	3.3	3.3	*a*
	HC(O)N(CH₃)₂	6.1	6.0	*x*		(CH₂)₄SO	4.0	3.8	*a*
	CH₃C(O)N(CH₃)₂	6.4	6.4	*x*		C₆H₆	1.0	1.0	*a*
	CH₃C(O)OC₂H₅	4.4	4.6	*x*		*s*-(CH₃)₃C₆H₃	2.2	2.3	*a*
	CH₃C(O)CH₃	5.0	5.0	*x*	Cu(hfacac)₂	C₅H₅N	13.4	12.9	*kk*
	[(CH₃)₂N]₃PO	7.7	7.5	*x*		CH₃C(O)N(CH₃)₂	8.0	8.1	*kk*
	(C₄H₉)₂O	5.8	5.7	*n*		CH₃C(O)OC₂H₅	5.9	5.7	*kk*
	(CH₃)₂SO	6.3	6.6	*x*		(CH₃)₂SO	8.5	8.6	*kk*
	C₉H₁₈NO	6.2	6.4	*oo*		(CH₂)₄O	9.1	9.3	*nn*
(CF₃)₂CHOH	C₅H₅N	9.8	9.7	*y*		C₉H₁₈NO	11.7	11.8	*nn*
	(C₂H₅)₃N	11.5	11.0	*y*		C₆H₁₀O	8.8	8.9	*nn*
	CH₃CN	5.9	5.6	*y*	SbCl₅	ClCH₂CN	9.6	9.6	*oo*
	CH₃C(O)N(CH₃)₂	8.5	8.6	*y*		CH₃CN	13.9	13.4	*oo*
	CH₃C(O)OC₂H₅	6.5	6.3	*y*		4-CH₃C₅H₄NO	35.4	35.6	*oo*
	[(CH₃)₂N]₃PO	9.9	10.3	*y*		CH₃C(O)OC₂H₅	15.5	16.1	*oo*
	C₉H₁₈NO	7.9	8.2	*oo*	HCCl₃	(C₂H₅)₃N	4.8	4.9	*yy*
	(CH₂)₄S	5.8	5.9	*m*		(CH₂)₄S	2.4	2.3	*yy*
	CH₃C(O)CH₃	6.7	6.7	*y*		C₅H₅N	4.9	4.9	*yy*
	(C₂H₅)₂O	7.2	7.0	*y*		(CH₂)₄O	3.9	3.9	*yy*
	(C₄H₉)₂O	7.9	7.6	*n, y*		CH₃C(O)CH₃	3.6	3.6	*yy*
	(CH₃)₂SO	8.7	8.9	*y*					

Table III (*Continued*)

Acid	Base	Heat, kcal/mol Measd	Calcd	Ref	Acid	Base	Heat, kcal/mol Measd	Calcd	Ref
	HC(C2H4)3N	4.3	4.3	yy		(CH2)3S	3.3	3.4	uu
	C2H5C(CH2O)3P	2.7	2.7	pp		CH3CN	4.6	5.0	uu
	[(CH3)2N]3PO	5.9	5.9	yy		(C2H5)2O	6.3	5.8	uu
	CH3C(O)OC2H5	3.8	3.5	yy		(n-C4H9)2O	6.5	6.4	uu
	C6H6	2.0	1.7	yy		C4H8O	6.3	6.2	uu
CF3(CF2)6H	C5H5N	4.3	4.3	pp	Ga(CH3)3	(CH3)3N	21	20.9	bb
	HC(C2H4)3N	4.7	4.7	pp		(CH3)3P	18	16.9	bb
HNCO	(C2H5)2O	3.9	3.9	xx		(CH3)2Se	10	10.2	bb
	C4H8O	4.2	4.2	xx	In(CH3)3	(CH3)3N	19.9	19.9	bb
	CH3CN	3.2	3.2	xx			17.1	17.1	bb
	C5H5N	5.4	5.4	xx	BF3(g)	(CH3)3P	18.9	18.9	ss
HNCS	(CH3)2S	3.5	3.5	uu		(CH2)4O	16.8	16.6	ss
	(C2H5)2S	3.5	3.5	uu		(CH2)5O	15.4	15.7	ss
	(CH2)5S	3.7	3.7	uu		CH3COOC2H5	12.8	12.4	ss
	(CH2)4S	3.6	3.6	uu					

a See ref 2. *b* W. B. Person, *et al.*, *J. Amer. Chem. Soc.*, **85**, 891 (1963). *c* E. Augdahl and P. Klaeboe, *Acta Chem. Scand.*, **19**, 807 (1965). *d* M. Tamres and Sr. M. Brandon, *J. Amer. Chem. Soc.*, **82**, 2134 (1960). *e* M. Tamres and S. Searles, *J. Phys. Chem.*, **66**, 1099 (1962). *f* T. Kubota, *J. Amer. Chem. Soc.*, **87**, 458 (1965). *g* J. C. Hill, Ph.D. Dissertation, University of Illinois, 1968. *h* R. M. Keefer and L. J. Andrews, *J. Amer. Chem. Soc.*, **77**, 2164 (1955). *i* D. G. Brown, R. S. Drago, and T. F. Bolles, *ibid.*, **90**, 5706 (1968). *j* N. Ogimachi, *et al.*, *ibid.*, **77**, 4202 (1955). *k* R. Mathur, *et al.*, *J. Phys. Chem.*, **67**, 2190 (1963). *l* R. S. Drago and T. D. Epley, *J. Amer. Chem. Soc.*, **91**, 2883 (1969). *m* See ref 9. *n* Heats obtained from the appropriate linear ΔH–$\Delta \nu$ relationship and the frequency shifts in Table IV. *o* Frequency shifts from R. S. Drago, *et al.*, *Inorg. Chem.*, **2**, 1056 (1963). *p* T. D. Epley and R. S. Drago, *J. Amer. Chem. Soc.*, **89**, 5570 (1967). *q* M. D. Joesten and R. S. Drago, *ibid.*, **84**, 2037, 2096, 3817 (1962). *r* Frequency shift from T. S. S. R. Murty, Ph.D. Dissertation, University of Pittsburgh, 1967. *s* Estimated from results in ref 9 and $\Delta \nu_{OH}$. *t* Frequency shift from D. Herlocker, *et al.*, *Inorg. Chem.*, **5**, 2009 (1966). *u* Frequency shift from H. F. Henneike and R. S. Drago, *ibid.*, **7**, 1908 (1968). *v* M. Nozari, G. C. Vogel, and R. S. Drago, in preparation. *w* See ref 10. *x* A. D. Sherry and K. F. Purcell, *J. Phys. Chem.*, **74**, 3535 (1970). *y* See ref 11. *z* See ref 12. *aa* See ref 13. *bb* F. G. A. Stone, *Chem. Rev.*, **58**, 101 (1958). *cc* See ref 27. *dd* C. H. Henrickson, *et al.*, *Inorg. Chem.*, **7**, 1047 (1968). *ee* C. H. Henrickson, *et al.*, *ibid.*, **7**, 1028 (1968). *ff* C. H. Henrickson and D. P. Eyman, *ibid.*, **6**, 1461 (1967). *gg* E. Bonitz, *Chem. Ber.*, **88**, 742 (1955). The data taken from this reference were corrected for dimerization using the heat of dimerization given in ref *dd* and assuming 90% dimerization of the triethylaluminum. *hh* G. E. Coates, *J. Chem. Soc.*, 2003 (1951). *ii* L. G. Stevens, *et al.*, *J. Inorg. Nucl. Chem.*, **26**, 97 (1964). *jj* J. Grundes and S. D. Christian, *J. Amer. Chem. Soc.*, **90**, 2239 (1968); **93**, 20 (1971); ΔE converted to ΔH. *kk* W. Partenheimer and R. S. Drago, *Inorg. Chem.*, **9**, 47 (1970). *ll* Estimated from very similar alkyl-substituted donor; *e.g.*, butyl estimated 0.1 kcal greater than ethyl. *mm* M. S. Nozari and R. S. Drago, to be published. *nn* R. L. Chiang and R. S. Drago, to be published. *oo* Y. Y. Lim and R. S. Drago, to be published. Donor is a free-radical base, 2,2,6,6-tetramethylpiperidine *N*-oxyl. *pp* F. Slejko and R. S. Drago, to be published. *qq* R. C. Gardner and R. O. Ragsdale, *Inorg. Chim. Acta*, **2**, 139 (1968). *rr* A. M. Halpern and K. Weiss, *J. Amer. Chem. Soc.*, **90**, 6297 (1968). *ss* E. M. Arnett, *Progr. Phys. Org. Chem.*, **1**, 223 (1963). *tt* E. M. Arnett, *et al.*, *J. Amer. Chem. Soc.*, **92**, 2365 (1970). *uu* T. M. Barakat, *et al.*, *Trans. Faraday Soc.*, **62**, 2674 (1966); **65**, 41 (1969). *vv* E. M. Arnett and C. Y. Wu, *J. Amer. Chem. Soc.*, **84**, 1684 (1962). *ww* M. Tamres, *et al.*, *ibid.*, **86**, 3934 (1964). *xx* J. Nelson, *Spectrochim. Acta, Part A*, **26**, 109 (1970). *yy* F. Slejko, R. S. Drago, and D. Brown, submitted for publication.

same. Inspection of Tables I and II shows that this condition obviously does not exist in general.

In addition, rearranging eq 6 and combining with eq 5a illustrate that $\alpha_i = k'' \alpha_j$. Therefore, a two-parameter model would require that a plot of the enthalpies of adduct formation of one acid *vs.* the enthalpies of adduct formation of another acid for the same series of bases be linear with a zero intercept. The enthalpies of adduct formation for I_2 and phenol with the same series of bases is plotted in Figure 1. These acids have very different C/E ratios, and their enthalpies of adduct formation cannot be correlated by a two-parameter model. The lack of linearity and a nonzero intercept in Figure 1 support this. Furthermore, a two-parameter, one-term model could not incorporate systems in which reversals in donor–acceptor strength are observed.[3,9] However, it is possible to correlate enthalpies of adduct formation for acids with very similar C/E ratios, such as hydrogen-bonding acids, using a two-parameter equation. Correlations restricted to one particular type of acid are, of course, only a subset of the overall E and C correlation.

This brings the discussion to the concept of donor numbers proposed by Gutman[19] to order solvents with regard to their donor strength toward acidic solutes.

(19) V. Gutman, *Coord. Chem. Rev.*, **2**, 239 (1967).

The conditions under which eq 1 is valid have been clearly emphasized.[2,20] In spite of warnings in the literature about these conditions, the inability of our approach to correlate data in a solvating solvent (1,2-dichloroethane)[21] for an acid in which steric effects are potentially operative (antimony pentachloride) was used to reject the E and C approach for estimating enthalpies. In its place, the author proposed what amounts to a two-parameter equation. The idea is that when the enthalpy of adduct formation for SbCl3, SbBr3, C6H5OH, or I2 for a series of donors is plotted *vs.* the enthalpies for SbCl5 for the same series of donors, linear plots are obtained. A line for a new acid can be determined by measuring enthalpies for the new acid interacting with two or three donors, and the enthalpies for other donors can be interpolated from this line using the enthalpies of SbCl5 with the donors. Hence, enthalpies for SbCl5 are called donor numbers, and it is claimed that this is the essential property needed to characterize a solvent as a base.

A quick glance at Figure 1 illustrates the fallacy of this approach. Both I_2 and phenol enthalpy plots cannot give linear plots with SbCl5 unless they give linear plots with each other. The apparent success in the donor-

(20) R. S. Drago, *Chem. Brit.*, **3**, 516 (1967).
(21) M. Nozari, G. C. Vogel, and R. S. Drago, *J. Amer. Chem. Soc.*, in press; Y. Y. Lim and R. S. Drago, *Inorg. Chem.*, in press.

Drago, Vogel, Needham / *Predicting Enthalpies of Adduct Formation*

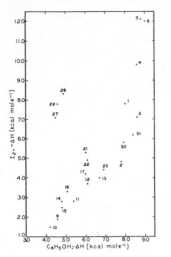

Figure 1. Enthalpies of adduct formation of donors with I_2 plotted *vs.* corresponding enthalpies of adduct formation for phenol with the same series of bases. The numbering of the bases refers to that in Table II.

number lies in utilizing a limited number of donors with similar C/E ratios and using several incorrect enthalpies of interaction with phenol. Clearly, the systems employed as the bases for the donor-number concept do not satisfy the requirements outlined above for a two-parameter equation to be operative. Consequently, the idea upon which the donor-number approach is based is incomplete, and it will work only with acids and bases that satisfy the requirements outlined above for a two-parameter equation.

Relationship of the E and C and Hammett Equations. The Hammett equation was originally proposed to describe the influence of polar meta or para substituents on the reactivity of the functional groups of many benzene derivatives.[22] It has been very successful in correlating and predicting relative reaction rates and equilibrium constants. In addition to correlating Gibbs free energy values, ΔG, the Hammett substituent constants, σ, have also been correlated with the infrared stretching frequency shifts of phenols, $\Delta\nu_{OH}$,[9,23,24] and with the enthalpy of adduct formation of phenols with oxygen, nitrogen, and sulfur donors.[8,9] The Hammett substituent constant relationship correlates and predicts enthalpies of adduct formation as well as eq 1 for meta- and para-substituted phenols. Consequently, it is of interest to ascertain how the Hammett equation is related to eq 1 and to determine the conditions whereby the parameters in Tables I and II can be transformed to parameters which obey an equation of the form of the Hammett equation.

The Hammett equation is a two-parameter equation with a constant of the general form $-\Delta H = \sigma\rho - \Delta H^0$ or $-\Delta H + \Delta H^0 = \sigma\rho$, where σ is the Hammett substituent constant and ρ is a parameter assigned to a

(22) J. Shorter, *Chem. Brit.*, **5**, 2969 (1969), and references cited therein.
(23) G. C. Pimentel and A. L. McClellan, "The Hydrogen Bond," W. H. Freeman, San Francisco, Calif., 1960.
(24) C. Laurence and B. Bruno, *C. R. Acad. Sci.*, **264**, 1216 (1967).

Table IV. Frequency Shift Data

Acid	Base	$\Delta\bar\nu_{OH}$,[a] cm^{-1}
p-CH$_3$C$_6$H$_4$OH	C$_5$H$_5$N	458 ± 10[b]
	CH$_3$C(O)N(CH$_3$)$_2$	336
	CH$_3$C(O)OC$_2$H$_5$	147[b]
	(C$_4$H$_9$)$_2$	277
	(C$_2$H$_5$)$_2$S	245
C$_6$H$_5$OH	ClCH$_2$CN	111
	(CH$_3$)$_2$NCN	222[c]
	CH$_3$C(O)OCH$_3$	171, 164[c]
	CH$_3$C(S)N(CH$_3$)$_2$	308[c]
	(CH$_2$)$_5$O	294
	(CH$_2$)$_4$O$_2$	240
	(C$_4$H$_9$)$_2$O	286
	(CH$_3$)$_2$SO	366
	(CH$_3$)$_2$S	253[c]
	(CH$_2$)$_3$S	246
	(CH$_2$)$_5$S	264[c]
	[(CH$_3$)$_2$N]$_2$CO	338[c]
p-ClC$_6$H$_4$OH	CH$_3$C(O)N(CH$_3$)$_2$	378 ± 10
	CH$_2$C(O)C$_2$H$_5$	190 ± 10
	CH$_3$C(O)CH$_3$	220[b]
	(C$_4$H$_9$)$_2$O	310
	(CH$_3$)$_2$SO	400
m-FC$_6$H$_4$OH	C$_5$H$_5$N	520 ± 10
	CH$_3$CN	175
	CH$_3$C(O)N(CH$_3$)$_2$	384 ± 10
	CH$_3$(O)OC$_2$H$_5$	194
	(CH$_3$)$_2$SO	402
m-CF$_3$C$_6$H$_4$OH	CH$_3$CN	181
	ClCH$_2$CN	131[c]
	(CH$_3$)$_2$NCN	260[c]
	CH$_3$C(O)OC$_2$H$_5$	207 ± 10
	CH$_3$C(O)OCH$_3$	182[c]
	CH$_3$C(O)CH$_3$	280 ± 10
	CH$_3$C(S)N(CH$_3$)$_2$	354[c]
	(C$_2$H$_5$)$_2$O	328[c]
	[(CH$_3$)$_2$CH]$_2$O	348[c]
	(CH$_2$)$_4$O	331[c]
	(CH$_2$)$_5$O	334
	(CH$_2$)$_4$O$_2$	286
	(CH$_3$)$_2$SO	416
	(CH$_2$)$_3$SO	425
	(CH$_3$)$_2$S	294
	(CH$_2$)$_3$S	298
	(CH$_2$)$_5$S	309
	[(CH$_3$)$_2$N]$_2$CO	382[c] (398)
(CF$_3$)$_2$CHOH	(C$_4$H$_9$)$_2$O	372 ± 10
	C$_6$H$_6$	70
	p-(CH$_3$)$_2$C$_6$H$_4$	78
	s-(CH$_3$)$_3$C$_6$H$_3$	96
	C$_6$(CH$_3$)$_6$	135
CF$_3$CH$_2$OH	(C$_4$H$_9$)$_2$O	254 ± 10

[a] These frequency shifts were measured by the procedure given in the Experimental Section unless otherwise noted. The error limits are ± 5 cm^{-1} unless otherwise stated. [b] Measured by Dr. H. F. Henneike in very dilute CCl$_4$ solutions. [c] Measured by Dr. W. Partenheimer in CCl$_4$ as solvent. Shift obtained from most dilute solutions.

constant reaction type. In terms of a primed set of E and C parameters, one may write the enthalpy as a function of two vectors

$$-\Delta H = \mathbf{Y}_{B}'^{T}\mathbf{X}_{A}' = E_{A}'E_{B}' + C_{A}'C_{B}'$$

To put this equation in the form of the Hammett equation for a series of substituted phenols, the restriction $-\Delta H_{B}^0 = C_{A}'C_{B}'$ is required, where ΔH_{B}^0 is the enthalpy of adduct formation for a given base with unsubstituted phenol. Now E_{A}' is comparable to σ and E_{B}' is comparable to ρ. Using the transformation equations (3a and 3b), this restriction in terms of the unprimed E and C parameters is given by

$$-\Delta H_B^0 = C_A' C_B' =$$
$$\frac{(a_{21}E_A + a_{22}C_A)(-a_{12}E_B + a_{11}C_B)}{a_{11}a_{22} - a_{12}a_{21}} \quad (7)$$

In addition, for the standard acid, it should be noted that $-\Delta H + \Delta H^0 = \Delta\Delta H = E_A^{0\prime} E_B'$, where $E_A^{0\prime}$ is the new E_A value for the reference acid phenol. Since $\Delta\Delta H$ for phenol is zero, $E_A^{0\prime} = 0 = a_{11}E_A^0 + a_{12}C_A^0$ and

$$a_{11}/a_{12} = -C_A^0/E_A^0 \quad (8)$$

where C_A^0 is the old C_A value for phenol. For eq 7 to be valid for different acids, the term C_A' ($= a_{21}E_A + a_{22}C_A$) must be constant, since all other terms are independent of the acid. Thus, for acids 1 and 2, it follows that

$$a_{21}E_{A1} + a_{22}C_{A1} = a_{21}E_{A2} + a_{22}C_{A2}$$
$$a_{21}(E_{A1} - E_{A2}) = -a_{22}(C_{A1} - C_{A2})$$
$$-a_{21}/a_{22} = (C_{A1} - C_{A2})/(E_{A1} - E_{A2})$$

or for a series of acids with the unsubstituted phenol as reference

$$-a_{21}/a_{22} = (C_A - C_A^0)/(E_A - E_A^0) \quad (9)$$

Thus, for a series of acids to conform to an equation of the form of the Hammett equation, the ratio given in eq 9 must be a constant for all acids. In addition, for eq 7 to be valid when the base is changed, the expression $(-a_{12}E_{B_i} + a_{11}C_{B_i})/-\Delta H_{B_i}^0$ must be constant. Next, let us consider the base restrictions, if any, on a Hammett-type correlation. Considering any two bases, 1 and 2, the following results are obtained.

$$\frac{-\Delta H_{B1}^0}{-\Delta H_{B2}^0} = \frac{-a_{12}E_{B1} + a_{11}C_{B1}}{-a_{12}E_{B2} + a_{11}C_{B2}} =$$

$$\frac{-E_{B1} + (a_{11}/a_{12})C_{B1}}{-E_{B2} + (a_{11}/a_{12})C_{B2}} = \frac{-E_{B1} - (C_A^0/E_A^0)C_{B1}}{-E_{B2} - (C_A^0/E_A^0)C_{B2}} =$$

$$\frac{-E_A^0 E_{B1} - C_A^0 C_{B1}}{-E_A^0 E_{B2} - C_A^0 C_{B2}} = \frac{-\Delta H_{B1}^0}{-\Delta H_{B2}^0}$$

Hence, this condition is always met, regardless of the base E and C parameters.

It would be of interest to see if our E and C parameters for the substituted phenol–base interactions can be transformed into Hammett σ and ρ parameters, i.e., to see how well the constraints of a Hammett treatment are adhered to by our E and C parameters derived from all types of donor–acceptor interactions. Unfortunately, the form of the ratios $(C_A - C_A^0)/(E_A - E_A^0)$, which will be designated R, is such that the uncertainty in these numbers is so large as to make comparisons meaningless. For the sake of completeness, the algebra will be continued for the transformation with the assumption that R is constant for the substituted phenols. From the transformation equation (3a)

$$E_A' = a_{11}E_A + a_{12}C_A$$

Combining this with eq 8, the following is obtained.

$$E_A' = a_{12}(C_A - (C_A^0/E_A^0)E_A) \quad (10)$$

The ρ values are a function of the base and the standard acid selected. From the transformation equation (3b)

$$E_B' = \frac{1}{(a_{11}a_{22} - a_{12}a_{21})}(a_{22}E_B - a_{21}C_B)$$

Dividing by a_{22} and substituting $-R = a_{21}/a_{22}$ yields

$$E_B' = \frac{1}{(a_{11} + a_{12}R)}(E_B + RC_B)$$

Dividing the numerator and denominator by a_{12}, substituting $a_{11}/a_{12} = -C_A^0/E_A^0$, and rearranging yields

$$E_B' = \frac{E_A^0(E_B + RC_B)}{a_{12}(RE_A^0 - C_A^0)} \quad (11)$$

Clearly, given R, a_{12} could be adjusted so that E_A' and E_B' correspond as closely as possible to the known values of σ and ρ.

The experimental values[8] of ρ for $(C_2H_5)_3N$, C_5H_5N, $CH_3CON(CH_3)_2$, $(CH_2)_4S$, and $(C_2H_5)_2S$ are 2.73, 2.01, 1.58, 1.84, and 1.84, respectively. One might expect that ρ is related to basicity, the larger ρ corresponding to the stronger base. Since the enthalpies of adduct formation of these bases with phenol are 9.1, 8.0, 6.8, 4.9, and 4.6 kcal/mol, respectively, the ρ values for this series do not parallel basicity. By considering our four-parameter equation instead of a two-parameter one, the cause of this apparent discrepancy becomes clear. The E_A values for the phenols are very large and, consequently, the large E_B values for the oxygen and nitrogen donors and the small values for sulfur donors determine the order for $-\Delta H$. However, the differences in the E_A values for the substituted phenols are very slight, $\sim 7\%$. The differences in the C_A values are larger, $\sim 30\%$. Consequently, a donor with a large C_B value will emphasize the differences in substituent even though the total $-\Delta H$ may be less. The ρ value measures the sensitivity of the interaction to substituent change and not the basicity, i.e., $\rho = k(E_B + RC_B)$ from eq 11 once a standard acid is selected.

$$(\partial\rho/\partial E_B)_{C_B} = k \text{ while } \partial\rho/\partial C_B = kR$$

Finally, we recall that $\Delta\Delta H = \sigma\rho = E_A' E_B'$. Substituting (10) and (11) for E_A' and E_B' into this expression for $\Delta\Delta H$ yields

$$\Delta\Delta H = \frac{(C_A E_A^0 - C_A^0 E_A)(E_B + RC_B)}{RE_A^0 - C_A^0}$$

In view of the large uncertainty in R, it seems likely that we could recompute all of the E and C parameters with the additional requirement that R be constant for all of the substituted phenols without greatly affecting the parameters or the calculated heats.

If a series of bases obeys a Hammett type of treatment toward a constant acid, the above discussion is applicable to this problem also. Now the transformed base parameter E_B' is related to the substituent constant σ and E_A' is related to ρ. The appropriate transformation equations result by simply interchanging the subscripts B and A in all of the above equations. It should also be mentioned that the constancy of $-R$, the requirement for a Hammett-type equation, is a different requirement than the constancy of the C/E ratio for a one-parameter equation. A limited set of data can obey a one-parameter equation and not be amenable to a Hammett type of approach. For example, the parameters of all of the alcohols (aliphatic and aromatic) undergoing a hydrogen-bonding interaction have a fairly constant C/E ratio and give fair agreement with a one-parameter treatment.

Interpretation of the E and C Parameters. As mentioned above, iodine was chosen as the reference acid and assigned parameters $E_A = 1.00$ and $C_A = 1.00$. Consequently, the absolute values of the E and C parameters in Tables I and II are meaningless, but the relative values and trends in a set of E parameters or in the set of C parameters are significant. It should be remembered that the parameters in Tables I and II are not known with equal confidence. The ability of acid parameters to predict accurate enthalpies of interaction when used with accurate base parameters depends upon the number of enthalpies for that acid which were included in the correlation and upon the range of C/E ratios for the bases involved in those interactions. This, of course, also applies to base parameters used with accurate acid parameters. For example, iodine, trimethylaluminum, and phenol have been studied with a large number of bases including bases with very different C/E ratios. Enthalpies predicted using the parameters for these acids should then be very accurate, as long as the base parameters are also well known. Di-*n*-butyl ether is an example of a base for which only enthalpies with hydrogen-bonding acids (with similar C/E ratios) have been measured. Hence, it should be possible to predict good di-*n*-butyl ether enthalpies with other hydrogen-bonding acids. In an attempt to improve our estimate of results with "soft" acids, we have assumed that the di-*n*-butyl ether and diethyl ether enthalpies with iodine are the same within experimental error. The dimethyl and diethyl sulfide enthalpies with iodine were also assumed to be the same.

Our interpretation of these parameters in terms of covalent and electrostatic bond forming tendencies is not proven and can be justified only in terms of our intuitive feelings for these quantities. For example, it is known from classical organic chemistry that substitution of an alkyl group for a hydrogen on a donor atom increases the nucleophilicity of the donor atom. This behavior is manifested in the C parameter for the alkylamines upon successive substitution of alkyl groups on ammonia. As pointed out above and graphically shown in Figure 1, one major success of eq 1 is its ability to correlate donor–acceptor systems where reversals in donor strength are observed, *e.g.*, sulfur and oxygen donors toward phenol and iodine. The magnitudes of the E and C parameters are in accord with our qualitative ideas about these systems.

At present, the correlation contains one transition metal complex, Cu(HFAcAc)$_2$. The results on this complex are very interesting and somewhat unusual for a transition metal in that enthalpies have been obtained in a poorly solvating solvent with nonionic donors,[25] in contrast to the typical stability constant study on a metal cation in some highly polar solvent. It is of considerable interest that a transition metal ion complex can be incorporated into the E and C scheme using the same base parameters that are used to correlate the enthalpies of formation of hydrogen-bonding and charge-transfer adducts.

(25) W. Partenheimer and R. S. Drago, *Inorg. Chem.*, **9**, 47 (1970).

The E and C correlation contains molecules which undergo drastic changes in their geometry upon adduct formation. Acids like (CH$_3$)$_3$SnCl, BF$_3$, and Al-(CH$_3$)$_3$ undergo extensive rearrangement from their structure as free acids when they form adducts. The fact that acids such as these fit into the E and C correlation illustrates the complexity of our E and C numbers. Many effects including the recently reported concept of unit acidity[26] must be evaluated before these numbers are completely understood.

Finally, it should be noted that the E and C parameters for a particular acid or base do not contain information concerning intermolecular steric effects, since these steric effects are a property of the geometry of the adduct and not a property of the individual acid or base. Molecules such as (CH$_3$)$_3$SnCl, B(CH$_3$)$_3$, Al-(CH$_3$)$_3$, and Cu(HFAcAc)$_2$ may encounter steric repulsive interaction with certain bases, and the discrepancy between calculated and observed enthalpies gives a quantitative estimate of the magnitude of this effect. Excellent agreement between experimental enthalpies of adduct formation and the calculated ones in Table III for these Lewis acids indicates that steric effects, as manifested through the enthalpy, are minimized with the donors selected for incorporation into this table. However, with donors such as (C$_2$H$_5$)$_3$N and (C$_2$H$_5$)$_2$O, which were omitted from the correlation for these acids, it has been suggested that steric effects are present.[2,13,25,27] Table V contains enthalpies calculated

Table V

Acid	Base	$-\Delta H_{\text{calcd}}$	$-\Delta H_{\text{measd}}$
B(CH$_3$)$_3$	(CH$_3$)$_3$N	24.6	17.6
(CH$_3$)$_3$SnCl	(C$_2$H$_5$)$_2$O	5.6	2.2
(CH$_3$)$_3$SnCl	(CH$_2$)$_4$O	5.8	5.1
Al(CH$_3$)$_3$	(C$_2$H$_5$)$_3$N	32.7	26.5
B(CH$_3$)	N(CH$_2$CH$_2$)$_3$CH	26.7	19.9
BF$_3$(g)	(C$_2$H$_5$)$_2$O	14.8	12.4

using eq 1 which are larger than the experimentally measured enthalpies; the difference is attributed to steric hindrance. In the case of (CH$_3$)$_3$SnCl adducts, one would expect the steric interaction to be greater for (C$_2$H$_5$)$_2$O than for (CH$_2$)$_4$O from examination of Shulman molecular models.[13] Accordingly, the (CH$_2$)$_4$O adduct gives closer agreement between ΔH_{calcd} and ΔH_{measd}.

The Hard–Soft Acid–Base (HSAB) Model. In our original work, we used an ionic–covalent model to interpret the E and C parameters. It has been suggested that our studies can be correlated with the hard–soft model.[4] "Softness" (or "hardness") can be considered[28] as a measure of the *ratio* of the tendency of a species to undergo covalent interaction to the tendency of a species to undergo electrostatic interaction. The relative softness or hardness is depicted in the C/E ratio.[28] It should be emphasized that if our ratio has this meaning, it is because of the model we originally imposed on the solution to the E and C equation.[2] The ratios for the acids and bases can be cal-

(26) D. G. Brown, R. S. Drago, and T. F. Bolles, *J. Amer. Chem. Soc.*, **90**, 5706 (1968).

(27) H. C. Brown, *J. Chem. Soc.*, 1248 (1956).

(28) G. Klopman, *J. Amer. Chem. Soc.*, **90**, 223 (1968); R. G. Pearson, *Science*, **151**, 172 (1966).

Journal of the American Chemical Society / *93:23* / *November 17, 1971*

culated from the data in Tables I and II. If the ratio C/E is comparatively large, the acid or base would be classified as type B, or soft. Conversely, if the ratio is comparatively small, the species is classified as type A, or hard.[28] Inasmuch as the relative ratios of C/E tell the relative importance of the two effects for various donors and acceptors, we agree that the hardness or softness discussed in the HSAB model is given by this ratio.

A qualitative classification of hardness and softness has been presented.[4] This classification divides the Lewis acids in Table I as follows: soft I_2, IBr, etc.; borderline $B(CH_3)_3$, SO_2, $Cu(HFAcAc)_2$; hard H-bonding acids, $Al(CH_3)_3$, $Ga(CH_3)_3$, BF_3, $(CH_3)_3SnCl$. The ratio C/E gives a quantitative order of relative hardness or softness for the various Lewis acids and agrees fairly well with the classification of Pearson.[4] The acids which do not follow the qualitative classification are BF_3 and SO_2. As mentioned above, the parameters for BF_3 were determined from data limited to oxygen donors. The qualitative placing of SO_2 is incorrect and, as will be shown shortly, when strong interactions are compared with weak ones, *the procedures used by Pearson to determine hardness and softness do not give the same result as the C/E ratio because the magnitudes of the C and E numbers which are important in determining the magnitude of an enthalpy are lost in the ratio; i.e.*, a large C divided by a large E can give the same ratio as a small C divided by small E.

The C/E ratios for I_2, IBr, and ICl are 1.0, 0.65, and 0.16, respectively, which is the ordering one would expect from considering ground-state dipole moments and electronegativities. According to the ratio, ICl is as hard as the alcohols which have ratios around 0.1, ranging from the softest, thiophenol, with $C/E = 0.20$, to the hardest, $(CF_3)_2CHOH$, with $C/E = 0.09$.

Table II contains the E and C constants for the donors presently in the correlation. Calculation of the C/E ratio from data in Table II indicates, as it does for the acids in Table I, that a large variety of *different* types of species are present. According to the soft and hard classification, the donors are categorized as follows:[4] soft R_2S, R_3P, benzene; borderline C_5H_5N; hard R_2O, RNH_2, NH_3. This is, in effect, a one-dimensional ranking of bases that goes from soft to borderline to hard. Recalling that the larger the C/E ratio, the softer the species, these qualitative observations agree with the ratios. However, the benzene ratio does not fit into the type B or soft classification, but rather into the type A, or hard, classification. Again, comparing this weak interaction with stronger ones, as in the case of SO_2, leads to an error when the procedures employed by Pearson are employed for a hard–soft ranking. Recently, it has been concluded that classical electrostatic forces make significant contributions to the stability of donor–acceptor complexes and perhaps are of predominant importance for the weak complexes.[29] Since benzene and its methyl derivatives are sacrificial donors,[31] they usually form weak complexes. The E and C parameters of benzene reflect this behavior, but the qualitative HSAB ranking of benzene does not.

In conclusion, the C/E ratios for donors (acids) indi-

cate whether hardness or softness is most important in interactions of a particular donor (acid), but softness or hardness so defined does not enable one to predict even the relative strength of interaction toward a soft or hard acid (base) because, as will be seen below, the magnitudes of the C and E numbers are lost.

Next, we shall describe why the magnitudes of the E and C numbers are not just quantitative manifestations of the HSAB concept, but give insight into intermolecular interactions which are absent in the qualitative soft–soft and hard–hard labeling of interactions. As can be seen from the data in Tables I and II, each acid and base has both a C and an E number which could be thought to correspond to possessing properties of softness and hardness. If this were the case, ammonia, which Pearson labels hard, has a larger C_B value than benzene, which is labeled soft, while dimethylamine which is soft (or borderline) has a larger E_B number than acetonitrile, which is labeled hard. Therefore, toward any hard acid, the soft base $(CH_3)_2NH$ will appear harder than one of the hardest bases in the correlation, CH_3CN. In other words, $(CH_3)_2NH$ is both harder and softer than CH_3CN, leading to the prediction that $(CH_3)_2NH$ will form stronger adducts with both hard and soft acids. Similarly, because of the magnitude of C, the borderline acid, $Cu(HFAcAc)_2$, will always be softer than the soft acid, I_2, when ordered by a soft base. In the original HSAB article, the importance of strength of interaction was recognized as the cause of not getting a complete reversal in donor order when a hard and a soft acid are compared. However, in subsequent HSAB applications in both this and other articles, most authors invariably ignore strength. In the HSAB treatment, one does not find the same substance being ranked as both soft and hard, but instead an ordering that goes from hard to intermediate to soft. Clearly, even though $Al(CH_3)_3$ is hard, its C number is greater than that for I_2, and it should interact more strongly with soft bases than I_2. Whenever weak bases are compared with strong bases (or acids) toward a soft acid (or base), the weak base (or acid) may appear to be not as soft as the strong base even if the weak base is actually softer as manifested in the C/E ratio. Furthermore, since no attempt is made in the HSAB model to factor out the hardness contribtion toward a soft reference acid (*e.g.*, CH_3Hg^+), orders of softness will vary when the reference soft acid is changed if the hard contributions are not identical for both acids. For example, if we consider the interaction of $(C_2H_5)_3N$, $(C_2H_5)_2O$, and $(C_2H_5)_2S$ toward the soft acid, I_2, the softness order would be $(C_2H_5)_3N > (C_2H_5)_2S > (C_2H_5)_2O$. Considering all the interactions and factoring the hard and soft contributions, the C/E ratio gives the order $(C_2H_5)_2S > (C_2H_5)_3N > (C_2H_5)_2O$. The hardness order deduced from interactions with the hard acid phenol is $(C_2H_5)_3N > (C_2H_5)_2O > (C_2H_5)_2S$, which is not the reverse of either soft order. Thus, a perfect order of softness (*i.e.*, one related to the intensity of softness) cannot be obtained from a qualitative examination of the enthalpies unless the E term for the reference soft acid is zero. It should be emphasized that our approach and the HSAB approach are very different because we attempt to factor the total interaction into the electrostatic (hardness) and covalent (softness) components which must accompany nearly all intermolecular

(29) R. S. Mulliken and W. B. Person, "Molecular Complexes," Wiley, New Yok, N. Y., 1969.

Drago, Vogel, Needham | Predicting Enthalpies of Adduct Formation

interactions. The HSAB approach makes no attempt to factor a given interaction into the two components and assign a magnitude to them. Consequently, the hard–soft description is found to be conceptually incomplete because the magnitude of hardness and softness as manifested by the magnitude of E and C is not taken into account, giving rise to all of the above-described difficulties even in a qualitative application. A substance whose interaction energy comes mainly from the soft–soft term would be labeled soft in the HSAB concept. However, if the intensity of the softness were low, a basically hard material with a significant C term would interact more strongly, in violation of the HSAB rule that "hard prefers hard and soft prefers soft." The absence of listings of the same substances as being both very hard and very soft indicates that this idea is missing from the HSAB concept.[30] The names themselves imply that this is missing because, by definition, if something gets less hard, it gets more soft. If hardness were given by H_A, then softness might be given by $1/H_A$. Consequently, we recommend abandoning the the hard–soft nomenclature because the words imply a two-parameter, two-term approach to acid–base chemistry, i.e.

$$-\Delta H = H_A H_B + k[(1H_A)(1H_B)]$$

Subsequent to our publication of the E and C equation, Pearson reported[4] the equation

$$\log K = S_A S_B + \sigma_A \sigma_B .$$

which he applied to equilibrium constant data in polar solvents. If solvation effects are constant and the entropy is proportional to the enthalpy, $\log K \sim -\Delta H$, and this would constitute an extension of our model[2] to a new body of data. If we replace $\log K$ by $-\Delta H$ and relabel S and σ, eq 1 results. It should be emphasized that this equation is basically our model and not the equation for the HSAB description. The equation in terms of $\log K$ was not generally applicable (i.e., constants could not be determined) as we suspected when our program began, because solvation and entropy contributions in polar solvents introduce too many variables for the few parameters. Contrary to claims in the literature,[4] there is no straightforward general connection between $\log K$ and ΔH for wide variations in donors and acceptors, as a plot of enthalpy data in Table III and equilibrium constants in the literature will show. Linear relations do exist when the systems treated are limited to a particular set of either acids or bases.

Acknowledgment. The authors appreciate stimulating discussion of the matrix formulation of this problem with Professor V. Schomaker. The support of this research by the National Science Foundation through Grant No. GP5498 and by the Paint Research Institute is also gratefully acknowledged.

(30) In mathematical terms, the C/E ratio gives the angle the acid or base vector makes with the E axis, but carries no information about the length of the vector which is needed to give $-\Delta H$.

Appendix

The conventional approach to a nonlinear least-squares problem is to "linearize" the equations by assuming that the current values of the parameters are approximately correct and that the calculated function (ΔH_{icalcd} changes linearly with changes in the parameters. Thus, we have

$$\Delta(-\Delta H_{icalcd}) = E_{Ai}\Delta E_{Bi} + E_{Bi}\Delta E_{Ai} + \\ C_{Ai}\Delta C_{Bi} + C_{Bi}\Delta C_{Ai} \quad (12)$$

(Note that this is valid only if the Δ's are small.) There is one such equation for each experimental heat and there are two parameters for each acid and base, except that four parameters are considered fixed (iodine E and C, DMA E_B, and diethyl sulfide C_B). Linear least-squares procedures lead to eq 3 for the calculation of the Δ's. When the process has been repeated until the Δ's are zero, the errors in the parameters are given by $\mathbf{Mx}$, the variance–covariance matrix for the parameters, which is computed[16] as $\sigma^2 \cdot (\mathbf{A^T P A})^{-1}$. $\mathbf{A}$ and $\mathbf{P}$ are the same as in eq 2 and σ^2 can be estimated as $(\mathbf{F^T P F})/(n - m)$, where $\mathbf{F}$ is again the vector of residuals, n is the number of experimental enthalpies, and m is the number of parameters (twice the number of acids and bases minus four). The variance–covariance matrix has as its diagonal elements the variances of the parameters, $(\sigma_i)^2$, that is, the squares of the (marginal) standard deviations. The off-diagonal elements are the covariances of pairs of parameters and can be written $\sigma_i \sigma_j \rho_{ij}$, where σ_i and σ_j are standard deviations of parameters i and j and ρ_{ij} is called the correlation coefficient and can be calculated since the σ's are known from the diagonal elements. For the E and C parameters, many of the ρ's are close to -1 or 1, indicating very high correlations among the parameter errors. To compute the conditional standard deviations, we assume that all of the parameters except one have exactly their true values, so that we have only one variable and the matrix $(\mathbf{A^T P A})$ is reduced to a single element. The variance of the parameter is then simply σ^2 divided by that element and the standard deviation is just the square root of the variance.

To calculate the variances for calculated ΔH's, eq 12 is squared and the Δ products are replaced by appropriate variances and covariances. Thus, $\sigma^2_{\Delta H_{icalcd}} = E^2_{Ai}\sigma^2_{EBi} + E^2_{Bi}\sigma^2_{EAi} + C^2_{Ai}\sigma^2_{CBi} + C^2_{Bi}\sigma^2_{CAi} + E_{Ai}E_{Bi}\sigma_{EBi}\sigma_{Ai}\rho_{EAEBi} + E_{Ai}C_{Ai}\sigma_{EBi}\sigma_{CBi}\rho_{EBCBi} + E_{Ai} \cdot C_{Bi}\sigma_{EBi}\sigma_{CBi}\rho_{EBCBi} + E_{Bi}C_{Ai}\sigma_{EAi}\sigma_{CBi}\rho_{EACBi} + E_{Bi}C_{Bi} \cdot \sigma_{EAi}\sigma_{CAi}\rho_{EACAi} + C_{Ai}C_{Bi}\sigma_{CBi}\sigma_{CAi}\rho_{CBCAi}$. The standard deviation is, as usual, just the square root of the variance. The ρ's used here are usually close to -1, so the estimated standard deviations for the calculated heats are very much smaller than one would expect from the marginal standard deviations of the parameters.

It should be remembered that this treatment assumes that eq 12 is valid. This is true only for small deviations from the true parameters, so a large standard deviation must also be a very inaccurate standard deviation; although we know, in that case, that the parameter in question is bad, we don't know exactly how bad.

Erratum: Page 268, equation 2 should read: $\Delta = (A'PA)^{-1}A'PF$

Part V

Editor's Comments on Papers 24–36

This section contains a number of papers which illustrate the use of HSAB in inorganic chemistry. The first paper, by Basolo, discusses a most useful rule for selecting a counter-ion to precipitate a given ion of any size, charge, or shape. It again emphasizes the point that small–small interactions are favored, and that large–large interactions are favored, often for negative reasons.

A paper by Lewis, Long, and Oldham (Lewis *et al.*, 1965) was quite astonishing when it appeared, since it showed that metal ions could combine with the acetylacetonate ligand at carbon, as well as at the familiar oxygen atom. I have selected the second paper of this series for inclusion. We can now recognize the results as an expected consequence when an ambidentate ligand reacts with a soft Lewis acid.

Hard metal ions react with the oxygen atoms of the ligands. Soft metal ions, as well as electrophiles such as the halogens, sulfur, oxygen, selenium, and tellurium, will react at the carbon atom. Examples in addition to Pt(II) include Ag(I), Au(I), Hg(II), Pd(II), Rh(I), and Pt(IV) (Gibson, 1969). The analog in organic chemistry is the reaction of (soft) CH_3I at C, and the reaction of (hard) CH_3COCl at O.

The next two papers discuss the peculiar behavior of the ambident thiocyanate ion. The first linkage isomers were prepared by Basolo and Burmeister, and many others have been made [for two review articles on linkage isomers, see Burmeister (1966; 1968)]. A rather delicate balance of factors, including steric ones, determines which isomer will be found.

While the HSAB principle usually works, there are notable exceptions which deserve comment. In particular an antisymbiotic effect is often found. Thus *cis*-$Pt(NH_3)_2(SCN)_2$ is S-bonded as expected, but $Pt[P(C_2H_5)_3]_2(NCS)_2$ is surprisingly N-bonded. Following Turco and Pecile (Turco and Pecile, 1961), this kind of result has been explained by invoking strong π-bonding, in which negative charge is removed from the central metal atom by phosphine ligands. It seems quite unlikely that alkylphosphines remove more charge by π-bonding than they donate by σ-bonding. Hence metal atoms must be made softer, rather than harder, by such ligands.

I have already pointed out that an antisymbiotic *trans* effect exists in planar and linear complexes (p. 50). Thus two soft ligands attempting to use the same orbital for covalent bonding can destabilize each other. That compounds such as *cis*-Pt(PR$_3$)$_2$(NCS)$_2$ are N-bonded is no surprise and accords with many other observations. On the other hand, *trans*-Pt(PR$_3$)$_2$(SCN)$_2$ is much more likely to be S-bonded. It may be mentioned that full structures are generally not known for the linkage isomers which have been prepared. A fascinating example is PdL(NCS)(SCN), where L is 1-diphenylphosphino-3-dimethylaminopropane. The structure is as predicted:the N-bonded thiocyanate is *trans* to P and the S-bonded thiocyanate is *trans* to N (Clark and Palenik, 1970).

The paper by Gutterman and Gray is a nice combination of the counter-ion stabilization effect and linkage isomerism. Similar results can be achieved by changing solvents (Burmeister, 1966; 1968).

The paper by Forster is another example of an antisymbiotic effect. It is found that *cis*-Rh(CO)$_2$Cl$_2^-$ is more stable than Rh(CO)$_2$I$_2^-$, even though Rh(I) is a typical soft acceptor. Also the stable form of *cis*-Ir(CO)$_2$X$_4^-$ is found to be Ir(CO)$_2$Cl$_2$I$_2^-$, with the two chloride ions *trans* to CO.

The two papers by Belluco and others on nucleophilic reactivity towards Pt(II) complexes illustrates several important points. One is simply the great advantage that soft nucleophiles have over hard nucleophiles, even when the latter is a strong base like methoxide ion. Another point relates to the symbiotic effect again. An examination of Table III and Figure 1 in paper 31 by Belluco *et al.* shows a clear working of the symbiotic effect. Thus *trans*-Pt[P(C$_2$H$_5$)$_3$]$_2$Cl$_2$ reacts very slowly with methanol, and very rapidly with SCN$^-$. But *trans*-Pt(py)$_2$Cl$_2$ reacts much more rapidly with methanol and less rapidly with SCN$^-$, in comparison.

Naively, one might expect a complex such as *trans*-Pt(py)$_2$CH$_3$Cl to show a very high reactivity towards soft nucleophiles, since CH$_3^-$ is one of the softest ligands available. Instead it is found that, while the complex is reactive, it does not react particularly rapidly with soft nucleophiles (Belluco *et al.*, 1966). This is, of course, another example of the antisymbiotic effect in planar complexes.

The second paper by Belluco *et al.* serves to show that nucleophilic reactivity sequences are not completely dominated by solvent effects. It is generally realized that in the gas phase all Lewis acids, probably, prefer fluoride ion to iodide ion. Strong solvation inverts the order in protic solvents for soft acids. A change to a dipolar aprotic solvent can cause a return to the gas phase order in some cases. However, this will only happen for a borderline soft acid, such as alkyl carbonium ions in alkyl halides. It does *not* happen for platinum(II).

The two papers by Jones and Clark represent a very useful application of HSAB ideas to catalysis of ligand substitution reactions. The examples are the inorganic analogs of the organic cases discussed by Saville (p. 365).

Paper 30 is also very closely related to catalytic processes. Cross points out how the HSAB principle provides a driving force for ligand migration reactions of organometallic compounds. This is particularly important, since it is now realized that the β-hydrogen migration is the main route for decomposition of transition metal organometallic compounds (Whitesides *et al.*, 1972). The stabilizing effect of fluorine substitutents on an alkyl group can be rationalized in terms of the instability of M——F bonds when M is a class (b) metal ion.

Paper 34 on metal ion–nitrene complexes is an example of how the HSAB principle was used to predict the behavior of an unknown species. We had guessed that coordinated nitrene, NH, was formed in certain reactions of azide complexes; the problem was to prove it. Reasoning by analogy with carbene, CH_2, we predicted that coordinated nitrene would be a soft electrophile, and would react preferentially with soft bases.

The last topic dealt with concerns redox reactions. The reduction of halide complexes of metal ions can show either a normal (F < Cl < Br < I) or inverse order of reactivity. Attempts to rationalize the orders in terms of mechanisms had proved contradictory. Haim's paper shows how certain puzzling results can be understood in terms of the expected hard or soft behavior of halo bridged activated complexes.

The paper by Jenkins and Kochi shows how one can predict whether a free radical will attack the metal atom or the ligand of a metal complex. Presumably the ideas expressed can be greatly extended.

References

Belluco, U., M. Graziani, and P. Rigo, *Inorgan. Chem.* **5**, 1123 (1966).
Burmeister, J. L., *Coord. Chem. Rev.* **1**, 205 (1966); **3**, 225 (1968).
Clark, G. R., and G. J. Palenik, *Inorg. Chem.* **9**, 2754 (1970).
Gibson, D., *Coord. Chem. Rev.* **4**, 225 (1969).
Lewis, J., R. F. Long, and C. Oldham, *J. Chem. Soc.,* 6740.
Turco, A., and C. Pecile, *Nature* **191**, 66 (1961).
Whitesides, G. M., J. F. Gaasch, and E. R. Stedronsky, *J. Amer. Chem. Soc.* **94**, 5258 (1972).

Coordination Chemistry Reviews
Elsevier Publishing Company, Amsterdam
Printed in The Netherlands

Paper presented at the Meeting on the Synthetic and Stereochemical Aspects of Coordination Chemistry, Nara, (Japan),
September 20-22, 1967.

24

STABILIZATION OF METAL COMPLEXES BY LARGE COUNTER-IONS

FRED BASOLO

Department of Chemistry, Northwestern University, Evanston, Illinois 60201 (U.S.A.)

INTRODUCTION

Science lends itself to various degrees of sophistication, and chemistry is no exception. Often it is the less sophisticated qualitative generalization that is of greater practical value. One such generalization in chemistry is that similar substances interact with each other to a greater extent giving a more stable system than do dissimilar substances. This is a well known phenomenon and many different types of examples can be cited to support it. For instance, polar compounds dissolve in polar solvent, whereas non-polar compounds dissolve in non-polar solvents. Similarly, like solvents are completely miscible, but unlike solvents may be immiscible. This type of behavior is of common occurrence in chemistry as in life and is adaquately described by the proverb "birds of a feather flock together."

Such a truism, amply supported by experimental fact was referred to as *symbiosis* by Jørgensen[1]. This is the mutual stabilizing effect on metal complexes observed for systems containing either a flock of soft ligands or a flock of hard ligands[2]. Thus, $[Co(NH_3)_5X]^{2+}$ is far better bound for X = F than I, whereas $[Co(CN)_5X]^{3-}$ is most stable with X = I and not even known for X = F. This is in accord with NH_3 being a hard base and symbiosis favoring the hard F^- over soft I^-, but CN^- is soft so for the cyanide system the preference is for I^-. Similarly CO substitution of $Cr(CO)_6$ is very slow compared with that for $Cr(CO)_4bipy$[3], presumably because the hexacarbonyl is more stable due to symbiosis.

Operationally, Berzelius was aware of the fact that certain metals occur on the earth's crust as oxides and others as sulfides. Approximately one hundred fifty years have elapsed and we now feel very certain that metals can be classified as class (a) and class (b) metals[4] or hard and soft Lewis acids, respectively. The class (a) or hard metals show a preference for halide ions in the order $F^- > Cl^- > Br^- > I^-$, whereas the reverse order is found for the class (b) or soft metals. Pearson[5] has cleverly pointed out that such an approach can also be extended to organic systems, and he stated the general principle: *Hard acids prefer to associate with hard bases and soft acids prefer to associate with soft bases.*

The theme of this review is the importance of flocking together of like substances in chemical systems, and of particular interest here is the useful general

Coordin. Chem. Rev., 3 (1968) 213–223

principle: *Solid salts separate from aqueous solution easiest for combinations of either small cation—small anion or large cation—large anion, preferably with systems having the same but opposite charges on the counterions.* This is a well known fact to coordination chemists and others, but it needs to be emphasized because it is sometimes overlooked (see last section on errors).

Considering the general case represented by reaction (1),

$$C^+_{(aq)} + A^-_{(aq)} \rightleftharpoons CA_{(s)} \tag{1}$$

it is seen that what determines whether or not the formation of solid compounds CA is favored is the difference between the hydration energies of the ions and the lattice energy of the solid. Clearly the driving force for the formation of the small cation–small anion system is the large lattice energy of CA, whereas for the large–large combination the driving force must be the small hydration energies of the large ions. This is illustrated by the values of the hydration energies of the ions and lattice energies of the solids in kcal/mole provided for the species in equations (2)–(5).

$$\underset{-124\qquad -121\qquad -245}{Li^+_{(aq)} + F^-_{(aq)} \rightleftharpoons LiF_{(s)}} \qquad \frac{\Delta H^\circ,\ \text{kcal/mole}}{0} \tag{2}$$

$$\underset{-124\qquad -70\qquad -178}{Li^+_{(aq)} + I^-_{(aq)} \rightleftharpoons LiI_{(s)}} \qquad +16 \tag{3}$$

$$\underset{-66\qquad -121\qquad -173}{Cs^+_{(aq)} + F^-_{(aq)} \rightleftharpoons CsF_{(s)}} \qquad +14 \tag{4}$$

$$\underset{-66\qquad -70\qquad -141}{Cs^+_{(aq)} + I^-_{(aq)} \rightleftharpoons CsI_{(s)}} \qquad -5 \tag{5}$$

The results are in accord with the solids containing counter-ions of the small–small (LiF) and large–large (CsI) combinations being favored over the less similar systems LiI and CsF, respectively. In practice the facts support this, LiF is less soluble than LiI and CsI is less soluble than CsF. The importance of the same but opposite charges on the ions is shown by comparing the solubility of salts such as *AgCl*, Ag_2SO_4, Ag_3PO_4; $BaCl_2$, *BaSO*$_4$, $Ba_3(PO_4)_2$; $LaCl_3$, $La_2(SO_4)_3$, *LaPO*$_4$. For each metal ion its least soluble salt is in italics.

The importance of large ions stabilizing large counter-ions in the solid state is not restricted to reactions in aqueous solution, but is fairly commonplace also in other systems. For example, beginning students learn that although the alkali metals have very similar chemical properties, their reactions with oxygen are dramatically different. Lithium gives the expected oxide Li_2O, whereas cesium forms the superoxide CsO_2. This is said to result from the large Cs^+ having a greater stabilization effect on the larger O_2^- compared with the smaller O^{2-} relative to the reverse for the small Li^+. Beginning textbooks also often include the fact that the thermal stabilities of the alkaline earth carbonates increase in the order Mg_2CO_3

$< CaCO_3 < SrCO_3 < BaCO_3$, and attribute this to the stabilizing influence on the larger CO_3^{2-} relative to O^{2-} as the size of the cation increases from Mg^{2+} to Ba^{2+}.

Other textbook examples would include $M[ICl_4]$, where the stability of the solid increases for changes in M in the order Li < Na < K < Rb < Cs. Similarly for $R_4N[ICl_4]$, the stability of the salt increases with the increasing size of R_4N^+. More recent examples may include reactions (6) and (7)[6].

$$NaCl + BCl_3 \rightarrow \text{No reaction} \tag{6}$$
$$CsCl + BCl_3 \rightarrow Cs[BCl_4] \tag{7}$$

The isolation of Cd(I) as Cd_2^{2+} was achieved[7] by the stabilizing influence of $AlCl_4^-$ to give $Cd_2[AlCl_4]_2$.

Students (and also some faculty) are surprised to find that reaction (8) is exothermic.

$$LiI_{(s)} + CsF_{(s)} \rightarrow LiF_{(s)} + CsI_{(s)} \tag{8}$$
$$-65 \quad -127 \quad -146 \quad -80 = \Delta H_f \text{ kcal mole}$$

The heats of formation of the individual compounds are given and it is seen that the reaction is exothermic to the extent of 34 kcal/mole. We are conditioned to thinking that the combination of the most electropositive and most electronegative elements form the most stable compounds and, therefore, it is surprising to find that reaction (8) is not in the direction favoring the formation of CsF. Nature has its way and the small Li^+ small F^- plus large Cs^+ large I^- combinations are preferred over the reverse arrangement. This is due to the large lattice energy of LiF.

STABILIZATION OF METAL COMPLEXES

This section will refer to a few examples in which the metal complex in question is not stable in aqueous solution but is stable in the solid state. Keeping in mind that metal complexes are large in size, the previously stated principle can be restated as follows: *Solid metal complexes are stabilized by large counter-ions, preferably ions of the same but opposite charge.* Some of the large counter-ions that have been used for this purpose are shown in Table 1.

In the mid-fifties when crystal field theory appeared to be the answer to chemical bonding in metal complexes, it became of interest to isolate and characterize some tetrahedral nickel(II) complexes. This was achieved for $[NiCl_4]^{2-}$ and $[NiBr_4]^{2-}$ using the large cations $(C_2H_5)_4N^+$ and $(C_6H_5)_2CH_3As^+$ and ethanol as a solvent. The method of synthesis was readily applied to other analogous systems of the type $R_2[MX_4]$, where M^{II} = Mn, Fe, Co, or Zn and X = Cl, Br or I. In all of these cases the complexes $[MX_4]^{2-}$ are stable in ethanol solution at the conditions of these experiments.
are readily established. Thus $CuCl_5^{3-}$ is not stable in solution but it is produced in

Coordin. Chem. Rev., 3 (1968) 213–223

TABLE 1

SOME LARGE COUNTER-IONS USED TO STABILIZE AND ISOLATE METAL COMPLEXES

Cations	Anions
Cs^+	ClO_4^-, I^-
R_4N^+	BF_4^-
$(C_6H_5)_4As^+$	PF_6^-
$(C_6H_5)_3CH_3As^+$	$AlCl_4^-$
$(C_6H_5)_4P^+$	$B(C_6H_5)_4^-$
$(C_6H_5)_3CH_3P^+$	$[Cr(NH_3)_2(NCS)_4]^-$
$[Co(NH_3)_4(NO_2)_2]^+$	$[Co(NH_3)_2(NO_2)_4]^-$
Ba^{2+}, $[Pt(NH_3)_4]^{2+}$	SiF_6^{2-}
$[Ni(phen)_3]^{2+}$	MCl_4^{2-} M = Co, Ni, Zn, Cd, Hg, Pt
$[Ni(bipy)_3]^{2+}$	$PtCl_6^{2-}$
$[Co(NH_3)_5NO_2]^{2+}$	$Fe(CN)_6^{2-}$
La^{3+}	$M(CN)_6^{3-}$ M = Fe, Co, Cr
$[M(NH_3)_6]^{3+}$ M = Co, Cr, Rh	$[M(C_2O_4)_3]^{3-}$ M = Co, Cr
$[M(en)_3]^{3+}$ M = Co, Cr, Rh	MF_6^{3-} M = Al, Fe, Co
$[M(pn)_3]^{3+}$ M = Co, Cr, Rh	
Th^{4+}	$[M(CN)_8]^{4-}$ M = Mo, W
$[Pt(NH_3)_6]^{4+}$	
$[Pt(en)_3]^{4+}$	

In contrast to this, the species $[CuCl_5]^{3-}$ can be isolated[9] from an aqueous solution that contains "none" of it by the addition of the cation $[Cr(NH_3)_6]^{3+}$. Since Cu^{II} complexes are substitution labile, it follows that the equilibria represented by (9)

$$CuCl_3^- \rightleftharpoons CuCl_4^{2-} \rightleftharpoons CuCl_5^{3-} \qquad\qquad (9)$$
$$\downarrow [Cr(NH_3)_6]^{3+}$$
$$[Cr(NH_3)_6] [CuCl_5]$$

aqueous solutions containing excess Cl^- because of the driving force for the formation of the stable solid $[Cr(NH_3)_6] [CuCl_5]$. This is an excellent illustration of the importance that the charge of the counter-ion be the same but opposite to that of the complex one wishes to stabilize. Note that the concentrations of $CuCl^{3-}$ and $CuCl_4^{2-}$ in solution far exceed that of $CuCl_5^{3-}$, but it is the 3- anion that is preferentially stabilized and brought out of solution.

It had also been assumed that $CuCl_5^{3-}$ is crystallized from solution by $[dienH_3]^{3+}$ (dien = $NH_2C_2H_4NHC_2H_4NH_2$), but a very recent x-ray study[10] shows that the substance obtained is a double salt and should be formulated as $[dienH_3]Cl[CuCl_4]$. This suggests that better results are expected if the counter-ion has a compact spherical shape similar to that of the metal complex to be stabilized, vis à vis $[Cr(NH_3)_6]^{3+}$ and not $[dienH_3]^{3+}$ relative to $CuCl_5^{3-}$. The structure of $CuCl_5^{3-}$ was determined by x-rays to be trigonal bipyramidal[11] and the

Cu–Cl distances[12] are shorter for the axial (2.296 Å) than for the equatorial (2.391 Å) positions.

Much the same discussion as that above applies to the six-coordinated complexes $[MCl_6]^{3-}$, where M = Cr, Mn or Fe, which are not stable in aqueous solution but can be stabilized in a crystal lattice[13] with $[Co(pn)_3]^{3+}$. The addition of a concentrated HCl solution containing these metal ions readily yields the salts $[Co(pn)_3][MCl_6]$. Because the reaction mixtures are strongly acidic, the choice of a 3+ counter-ion requires that it be a substitution inter metal ammine.

It has long been known that pale yellow solutions of $Ni(CN)_4^{2-}$ become an intense orange-red color upon the addition of excess CN^-. Investigations of these solutions have shown[14] that this is due to the formation of the species $Ni(CN)_5^{3-}$. However, it could not be isolated as the potassium salt, because evaporation of the orange-red solution yielded a mixture of $K_2Ni(CN)_4 + KCN$ in the solid phase. The complex $Ni(CN)_5^{3-}$ was stabilized and isolated[15] as the salts of $[Cr(NH_3)_6]^{3+}$ and $[Cr(en)_3]^{3+}$. It has also been found possible[16] to isolate $K_3Ni(CN)_5$ at $-15°C$, but it is not stable at room temperature. These results are summarized by scheme (10).

$[Cr(en)_3][Ni(CN)_5]$ stable solid

$\uparrow [Cr(en)_3]^{3+}$

$$Ni(CN)_5^{3-} \xrightarrow{\text{K}^+,\text{Conc.}} K_2Ni(CN)_4 + KCN \tag{10}$$

$-15° \downarrow K^+$ R.T.

$K_3Ni(CN)_5$

The corresponding Co^{III} ammines were not suitable due to their undergoing substitution reactions at the experimental conditions of high CN^- concentration. This may be caused by the presence of catalytic amounts of Co^{II} and/or base catalysis in the strongly alkaline solution[17]. Neither of these effects are as pronounced here for the Cr^{III} ammines. Attempts to isolate $[Cr(en)_3][Co(CN)_5]$ have not yet been successful[18]. The addition of $[Cr(en)_3]^{3+}$ to a concentrated solution of $[Co(CN)_5]^{3-}$ results in an immediate color change from intense green to yellow-orange. This suggests that the water reaction[19] (11) has been catalyzed.

$$2Co(CN)_5^{3-} + H_2O \rightleftharpoons [HCo(CN)_5]^{3-} + [Co(CN)_5OH]^{3-} \tag{11}$$

It was recently reported[20] that further reaction takes place to generate $[Co(CN)_6]^{3-}$, and in support of this the salt $[Cr(en)_3][Co(CN)_6]$ was isolated[18]. This is being repeated in other solvents and also attempts[21] are being made to stabilize $Pd(CN)_5^{3-}$, $Pt(CN)_5^{3-}$ and $Au(CN)_5^{2-}$.

The structure of $[Cr(en)_3][Ni(CN)_5]1.5H_2O$ has been determined by x-rays[22] and of special interest is the structure of $Ni(CN)_5^{3-}$. Theoretical chemists were reluctant to predict its structure, but at the author's insistence some did, and

Coordin. Chem. Rev., 3 (1968) 213–223

the predictions were 50% trigonal bipyramidal and 50% tetragonal pyramidal. Nature is on their side, for it is found that the symmetry cell contains two cations and two anions, one $Ni(CN)_5{}^{3-}$ is a tetragonal pyramid and the other is a distorted trigonal bipyramid. This means that the energy differences between the two structures must indeed be small[23]. Again the Ni-C bond distances in the axial positions are shorter than in the equatorial positions of the trigonal bipyramid. It begins to appear that this may be true for d^n systems where n is sufficiently large, whereas the reverse holds for d° compounds[24].

Substitution reactions of tetrahedral Si^{IV} are generally bimolecular displacements which are believed to proceed through a five-coordinated active intermediate[25]. Silicon chemists were very pleased a few years ago when the stable five-coordinated Si^{IV} cation $[(C_6H_5)_3Si(bipy)]^+$ was isolated[26] as the I^- salt. It now appears that this previously illusive coordination number for Si^{IV} is readily obtained if one makes the proper choice of system and counter-ion. Thus $SiF_5{}^-$ is stabilized by $(C_6H_5)_4As^+$ and the salt separates from an aqueous-methanol solution of a reaction mixture represented by (12)[27].

$$SiO_2 + 5HF + (C_6H_5)_4AsCl \rightarrow (C_6H_5)_4AsSiF_5 + HCl + 2H_2O \qquad (12)$$

This synthesis was prompted by the x-ray studies which suggested the presence of $SiF_5{}^-$ in a substance previously incorrectly formulated (see last section).

In addition to qualitative experimental observations of the type mentioned above, quantitative thermodynamic data are also available which support the principle that a large counter-ion stabilized a metal complex. Some such data are shown in Table 2. The results clearly show that the stabilities of $AlH_4{}^-$ and of $VF_6{}^-$ increase for the respective salts in the order Li < Na < K < Rb < Cs.

TABLE 2

THERMODYNAMIC DATA SHOWING THE INCREASE IN STABILIZATION OF METAL COMPLEXES WITH INCREASE IN SIZE OF COUNTER-ION

$MAlH_4{}^a{}_{(s)} \rightarrow MH_{(s)} + Al_{(s)} + 3/2H_{2(g)}$				
	Li	Na	K	Cs
ΔG°, kcal/mole	-3.9	$+3.0$	$+14.9$	$+16.5$

$MVF_6{}^b{}_{(s)} \rightarrow MF_{(s)} + VF_{5(s)}$					
	Li	Na	K	Rb	Cs
ΔG°, kcal/mole	-55	-25	$+1$	$+7$	$+14$

^{a}Ref. 28. ^{b}Ref. 29.

ISOLATION OF METAL COMPLEXES

Not as spectacular as some of the examples mentioned in the previous section, but perhaps even of greater importance to the coordination chemist is the proper choice of counter ion to permit the isolation of a metal complex that is stable in solution (so much so that most of its salts are extremely soluble and difficult to isolate). A past master at being able to isolate systems difficult to obtain was the late Professor Dwyer[30]. It has been said that if he could not resolve a metal complex then it must not be optically active. His approach was simply that of using with his gifted "green thumb", appropriate counter-ions of large size and the same but opposite charge. For example, he was able to start with $(+)-$ $[Co(en)_2(NO_2)_2]^+$ and resolve several other Co^{III} complexes as represented by scheme (13).

$$\begin{array}{c} \nearrow (+)-[Co(EDTA)]^- \rightarrow (+)-[Co(en)_2C_2O_4]^+ \\ (+)-[Co(en)_2(NO_2)_2]^+ \rightarrow (+)-[Co(en)(C_2O_4)_2]^- \rightarrow (+)-[Co(en)_2CO_3]^+ \\ \searrow (-)-[Co(en)(mal)_2]^- \qquad\qquad\qquad\qquad (13) \end{array}$$

Some of these complexes such as $[Co(EDTA)]^-$ had been previously resolved[31] but were not obtained optically pure until the proper choice of resolving agent was made.[32]

Aquo complexes often form very easily in water solution, but their isolation may not be easy because of their extreme solubility. For example this is true of cis-$[Cr(en)_2(H_2O)Cl]^{2+}$. However, it is possible to isolate this complex as the salt cis-$[Cr(en)_2(H_2O)Cl]ZnCl_4$ (or $ZnBr_4$) from an aqueous ethanol solution[33]. Similarly, the salt $[Co(tetraen)Cl]Cl_2$ is very soluble and when it was first prepared no attempt was made to separate its various geometrical isomers[34]. The α and β forms of this complex have now been obtained[35] as the salts of $ZnCl_4^{2-}$.

Many cationic and anionic metal carbonyls and their derivatives have been prepared and often their isolation is accomplished with a large counter-ion of the same but opposite charge. This is best illustrated by reaction (14).

$$Mn(CO)_5Cl + CO + AlCl_3 \rightarrow [Mn(CO)_6]AlCl_4 \qquad\qquad (14)$$

In this instance the investigators[36] were doubly fortunate since the Lewis acid required to remove the chloro group from $Mn(CO)_5Cl$ was also able to provide the necessary anion for the formation of a stable salt with the cation $[Mn(CO)_6]^+$. A few other examples with references but no comment are $R_4N[Mn(CO)_4X_2]^{37}$, $R_4N[Cr(CO)_5X]^{38}$, $R_4N[Rh(CO)_2X_2]^{39}$, $[Cr(Triars)(CO)_2X]B(C_6H_5)_4{}^{40}$, $[C_5H_5Cr(CO)_4]Br_4{}^{41}$, $[C_5H_5Fe(CO)_3]PF_6{}^{42}$.

ERRORS

Mistakes have been made because investigators have failed to give enough credence to the phenomenon that large counterions stabilize metal complexes. This author[43] is guilty of such an error and can use his naivete at the time as the first example. At the early stages of our investigations on the mechanisms of substitution reactions of metal complexes, it was of interest to design an experiment to determine if water is involved in a nucleophilic displacement process for aquation reactions of Co^{III} complexes (incidently, it is still of interest to design such an experiment.)[17]. For this purpose a series of complexes of the type *trans*-$[Co(AA)_2Cl_2]^+$ were prepared, where AA = C- and N-alkyl substituted ethylenediamines. The compounds were prepared by the method used to prepare the well known *trans*-$[Co(en)_2Cl_2]Cl$ and the results obtained were similar to those described for this compound. All of the compounds were green and had visible absorption spectra similar to that of *trans*-$[Co(en)_2Cl_2]^+$. This complex separates from its reaction mixture as *trans*-$[Co(en)_2Cl_2]Cl.2H_2O.HCl^{44}$, but the H_2O-HCl can be removed by heating it in an oven at 110° overnight. It was assumed that HCl was more difficult to remove from the substituted- en complexes, because even after heating for longer periods they still gave high chlorine analyses. The complexes were incorrectly formulated as *trans*-$[Co(AA)_2Cl_2]Cl.nHCl$. Recent studies show that the compounds were instead *trans*-$[CoI(AA)_2Cl_2]_2CoCl_4^{45}$. Lest someone gets overly concerned as to the validity of the aquation studies, these were performed with the pure ClO_4^- salts.

A very recent example of the mistaken identity of a compound due to failure to heed the essential theme presented here, is provided by the reaction of *trans*-$[Pt(PEt_3)_2HCl]$ with C_2F_4 in cyclohexane solution in a glass vessel at 120° for 50 hr. One of the minor products, obtained in about 10 % yield, was formulated[46] as $[Pt(PEt_3)_2(H)(Cl)(C_2F_4)]$. It was claimed that this formula was supported by its analysis (Pt, P, F – not C, H although C and H analyses were reported for the other products), its ir spectrum and its ^{19}F nmr spectrum. Although conductivity measurements in nitrobenzene showed the compound to be a 1:1 electrolyte, this was suggested to involve a dissociation of a H^+ or a Cl^-. The formula proposed for this substance seemed plausible on the basis of the available evidence, but even more so because such a compound is believed to be the active intermediate leading to the overall insertion (or ligand migration)[17] reaction to form the sigma bonded alkyl product.

A sample of the compound was examined by means of x-rays and it was quickly determined that it contained two aggregates, presumably a cation and an anion in keeping with its being a 1:1 electrolyte. It was also impossible to find C_2F_4 in the compound. The research group at Northwestern University[47] began to speculate that the ir band at 2100 cm.$^{-1}$ which had been assigned to Pt-H stretch was in fact due to C–O stretch and that the cation may be $[Pt(PEt_3)_2COCl]^+$ which is isoelectronic with the Vaska Ir^I compound. The anion had a regular tetrahedral structure, and it appeared this could be BF_4^- coming from the reaction of the Pyrex glass tube with HF. Also contained in this preparation was the anion SiF_5^-, which is the only anion when the preparation is carried out in silica tubes.

The correct formula for this compound, *trans*-$[Pt(PEt_3)_2COCl]BF_4$ (or SiF_5) is in accord with the large anion stabilization of the complex and agrees with its properties as previously reported. The final confirmation of this structure is the rational synthesis of the BF_4^- salt by the research group at the University of

Western Ontario. The compound *trans*-[Pt(PEt$_3$)$_2$HCl] reacts with CO and aqueous HBF$_4$ under pressure to yield *trans*-[Pt(PEt$_3$)$_2$COCl]BF$_4$. In addition the rational synthesis of (C$_6$H$_5$)$_4$AsSiF$_5$ was accomplished as shown by reaction (12). Further studies are in progress on the syntheses and reactions of these systems[48].

Often the (+)-tartrate ion is not a suitable resolving agent for metal complexes, and some years ago it was found that better results may be obtained using antimonyl (+)-tartrate ion[49]. It appeared that this worked best with 2+ cations such as [Ni(bipy)$_3$]$^{2+}$ and [Fe(phen)$_3$]$^{2+}$, which would be contrary to the same but opposite sign on the anion if it is correctly formulated as (+)-[SbO(C$_4$H$_4$O$_6$)]$^-$. The recent important synthesis and resolution[50] of *cis*-[Pt(en)$_2$Cl$_2$]$^{2+}$ made use of the diastereoisomeric salt (+)^{450}-*cis*-[Pt(en)$_2$Cl$_2$]-(+)-[SbO(C$_4$H$_4$O$_6$)]$_2$.

The x-ray structure determination of K[SbC$_4$H$_4$O$_7$].1/2H$_2$O shows the anion to be

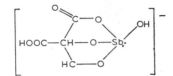

with a distorted tetrahedral structure around antimony[51]. Recent x-ray studies[52] on (+)-[Fe(phen)$_3$] [(+)-SbC$_4$H$_4$O$_7$]$_2$ affords the most "expected" result that the anion indeed has a 2− charge.

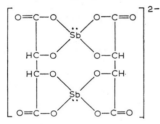

The structure at antimony is that of a distorted tetrahedron, and the absolute configuration of [Fe(phen)$_3$]$^{2+}$ was determined.

Finally, the reaction of K$_2$PtCl$_4$ with dien.3HCl yielded a pale yellow crystalline product which was assigned[53] the empirical formula PtC$_8$H$_{30}$ON$_6$Cl$_8$, perhaps as a typographical error because the analytical data reported require Pt$_3$C$_8$H$_{30}$ON$_6$Cl$_8$. In any case no attempt was made to designate the structure of this species. Recent detailed studies[54] of this system show it to be [Pt$_2$C$_8$H$_{28}$N$_6$Cl$_2$]PtCl$_4$.H$_2$O and that the cation has the structure

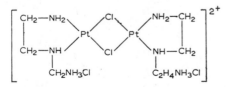

This provides yet another example of a large cation being stabilized by a large anion of the same but opposite charge.

In conclusion it need only be said that if the reader still remains skeptical of the importance and practical value of the general principle discussed here, then he is referred to the chemical literature which abounds with many more examples, some of which may better suit his fancy.

ACKNOWLEDGEMENT

I should like to thank Professor D. F. Shriver for the B and Al examples, Professor R. G. Pearson for keeping me straight on HSAB and Mr. K. N. Raymond for interesting discussions. Our research of the type described here is supported in part by a National Institutes of Health Grant.

REFERENCES

1   C. K. Jørgensen, *Inorg. Chem.*, 3 (1964) 1201.
2   R. G. Pearson, *J. Am. Chem. Soc.*, 85 (1963) 3533; *Science*, 151 (1966) 172.
3   R. J. Angelici and J. R. Graham, *J. Am. Chem. Soc.*, 87 (1965) 5586.
4   S. Ahrland, J. Chatt and N. R. Davies, *Quart. Rev. (London)*, 12 (1958) 265.
5   R. G. Pearson and J. Songstad, *J. Am. Chem. Soc.*, 89 (1967) 1827.
6   E. L. Muetterties, *J. Inorg. Nucl. Chem.*, 12 (1960) 355.
7   J. D. Corbett, W. J. Burkhard and L. F. Druding, *J. Am. Chem. Soc.*, 83 (1961) 76.
8   N. S. Gill and R. S. Nyholm, *J. Chem. Soc.*, (1959) 3997.
9   M. Mori, *Bull. Chem. Soc. Japan*, 33 (1960) 985.
10 B. Zaslow and G. L. Ferguson, *Chem. Comm.*, (1967) 822.
11 M. Mori, Y. Saito and T. Watanable, *Bull. Chem. Soc. Japan*, 34 (1961) 295.
12 K. N. Raymond, D. W. Meek and J. A. Ibers, *Inorg. Chem.*, in press.
13 W. E. Hatfield, R. C. Fat, C. E. Pfluger and T. S. Piper, *J. Am. Chem. Soc.*, 85, (1963), 265.
14 J. S. Coleman, H. Peterson, Jr. and R. A. Penneman, *Inorg. Chem.*, 4 (1965) 135, and references therein.
15 K. N. Raymond and F. Basolo, *Inorg. Chem.*, 5 (1966) 949.
16 W. C. Andersen and R. H. Harris, *Inorg. Nucl. Chem. Let.*, 2 (1966) 315.
17 F. Basolo and R. G. Pearson, *Mechanisms of Inorganic Reactions*, John Wiley and Sons, New York, 2nd. Ed., 1967.
18 K. N. Raymond, private communication.
19 B. DeVries, *J. Catalysis*, 1 (1962) 489.
20 M. G. Burnett, P. J. Connolly and C. Kemball, *J. Chem. Soc. (A)*, (1967) 800.
21 V. Myers, private communication.
22 K. N. Raymond, P. W. R. Corfield and J. A. Ibers, *Inorg. Chem.*, in press.
23 C. Furlani, *Coordin. Chem. Revs.*, 3 (1967) 25.
24 E. L. Muetterties and R. A. Schunn, *Quart. Revs. (London)*, 20, (1966) 245.
25 L. H. Sommer, *Stereochemistry, Mechanism and Silicon*, McGraw-Hill Book Company, New York, 1965.
26 J. Y. Corey and R. West, *J. Am. Chem. Soc.*, 85 (1963) 4034.
27 H. C. Clark and K. R. Dixon, *Chem. Comm.*, (1967) 717.
28 M. B. Smith and G. Bass, Jr., *J. Chem. Eng. Data*, 8 (1963) 342.

29 W. KLEMM, *J. Inorg. Nucl. Chem.*, 8 (1958) 532.
30 F. P. DWYER, *Advances in the Chemistry of the Coordination Compounds*, S. KIRSCHNER, Ed., MacMillan Co., New York, 1961, pp. 21–33.
31 D. H. BUSCH AND J. C. BAILAR, JR., *J. Am. Chem. Soc.*, 75 (1953) 4574.
32 F. P. DWYER AND F. L. GARVAN, *Inorg. Syn.*, 6 (1960) 192.
33 D. A. HOUSE AND C. S. GARNER, *J. Inorg. Nucl. Chem.*, 28 (1966) 904.
34 R. G. PEARSON, C. R. BOSTON AND F. BASOLO, *J. Phys. Chem.*, 59 (1955) 304. R. T. M. FRASER, *Proc. Chem. Soc.*, (1963) 262.
35 D. A. HOUSE AND C. S. GARNER, *Inorg. Chem.*, 5 (1966) 2097.
36 E. O. FISCHER AND K. ÖFELE, *Angew. Chem.*, 73 (1961) 581.
37 E. W. ABEL AND I. S. BUTLER, *J. Chem. Soc.*, (1964) 434. R. J. ANGELICI, *Inorg. Chem.*, 3 (1964) 1099.
38 P. BARRETT, Ph. D. Thesis, University of Toronto, Toronto, Canada, (1965).
39 L. M. VALLARINO, *Inorg. Chem.*, 4 (1965) 161.
40 R. S. NYHOLM, M. R. SNOW AND M. H. B. STIDDARD, *J. Chem. Soc.*, (1965), 6570.
41 E. O. FISCHER AND K. ULM, *Z. Naturforsch*, 16B (1961) 757.
42 E. O. FISCHER AND K. FICHTEL, *Chem. Ber.*, 94 (1961) 1200. R. B. KING, *Inorg. Chem.*, 1 (1962) 964.
43 F. BASOLO, *J. Am. Chem. Soc.*, 75 (1953) 227.
44 S. OOI, Y. KOMIYAMA, Y. SAITO, AND H. KUROYA, *Bull Chem. Soc. Japan*, 32, (1959), 263. R. D. GILLARD AND G. WILKINSON, *J. Chem. Soc.*, (1964) 1640.
45 E. W. GILLOW, PH. D. Thesis, State University of New York at Buffalo, Buffalo, New York, (1966)
46 H. C. CLARK AND W. S. TANG, *J. Am. Chem. Soc.*, 89 (1967) 529.
47 H. C. CLARK, P. W. R. CORFIELD, K. R. DIXON, AND J. A. IBERS, *J. Am. Chem. Soc.*, 89 (1967), 3360.
48 H. C. CLARK AND K. R. DIXON, *Abstracts of 154th ACS Meeting, September* 1967, Chicago, Illinois.
49 F. P. DWYER AND E. C. GYARFAS, *J. Proc. Roy. Soc.*, N. S. Wales 83 (1950) 263; 85 (1951) 135.
50 C. F. LIU AND J. DOYLE, *Chem. Comm.*, (1967) 412.
51 D. GRDENIC AND B. KAMENAR, *Acta Cryst.*, 19 (1965) 197.
52 D. H. TEMPLETON, A. ZOLKIN AND UEKI, *Acta Cryst.*, 21 (1966) A154.
53 F. G. MANN, *J. Chem. Soc.*, (1934) 466.
54 G. W. WATT AND W. A. CUDE, *Inorg. Chem.*, in press.

Reprinted from the *Journal of the Chemical Society, Section A—Inorganic, Physical, and Theoretical Chemistry*, 1453–1456 (1966)
Copyright 1966 by the Chemical Society, London

25

Metal–β-Diketone Complexes. Part II.* Carbon-bonded Platinum(II) Complexes of Trifluoroacetylacetone and Benzoylacetone

By D. Gibson, J. Lewis, and C. Oldham

The complexes K[Pt(diketone)$_2$X] (diketone = trifluoroacetylacetone and benzoylacetone, X = Cl or Br) have been prepared. The n.m.r. spectra of the trifluoroacetylacetone complexes and the infrared spectra of all the compounds indicate the presence of both oxygen- and carbon-bonded β-diketone groups. The factors which influence carbon- as opposed to oxygen-bonding of β-diketones to metals are discussed.

X-RAY structural determinations of platinum(II) and platinum(IV) complexes have established [1,2] that in addition to bonding to the oxygen atom, platinum is also capable of bonding to the 3-carbon atom of acetylacetone. In Part I [3] the combined application of infrared and proton magnetic resonance spectroscopy to platinum(II) acetylacetonates was discussed. It was shown that the presence of 3-carbon-bonded acetylacetone groupings were most readily detected in the n.m.r. by a large splitting of the 3-CH proton by platinum-195. In the infrared region a number of characteristic changes in the CH modes were observed and in particular the appearance of a band in the 500—600 cm.$^{-1}$ region was associated with the platinum–carbon stretching mode, the platinum–oxygen vibrations falling in the range 400—500 cm.$^{-1}$.[4] We now report spectroscopic information which establishes a similar structure for some other β-diketone complexes of platinum(II).

RESULTS AND DISCUSSION

Fay and Piper reported the preparation of a wide range of complexes between transition metals and the diketones trifluoroacetylacetone (3Facac)[5] and benzoylacetone (Bzac).[6] Spectroscopy indicated that all the complexes formed by these ligands were conventionally oxygen-chelated. In contrast, we prepared compounds analogous to the bisacetylacetonatoplatinum(II) anion which involve bonding of the β-diketone by the carbon rather than the oxygen atom.

* Part I, J. Lewis, R. F. Long, and C. Oldham, *J. Chem. Soc.*, 1965, 6740.
[1] B. N. Figgis, J. Lewis, R. F. Long, R. Mason, R. S. Nyholm, P. J. Pauling, and G. B. Robertson, *Nature*, 1962, 195, 1278.
[2] A. G. Swallow and M. R. Truter, *Proc. Roy. Soc.*, 1960, A, 254, 205.

[3] J. Lewis, R. F. Long, and C. Oldham, *J. Chem. Soc.*, 1965, 6740.
[4] R. D. Gillard, H. G. Wilver, and J. L. Wood, *Spectrochim. Acta*, 1964, 20, 63.
[5] R. C. Fay and T. S. Piper, *J. Amer. Chem. Soc.*, 1963, 85, 500.
[6] R. C. Fay and T. S. Piper, *J. Amer. Chem. Soc.*, 1962, 84, 2303.

Potassium chloro- and bromo-bis(trifluoroacetylacetonato)platinum(II) dihydrates can be prepared by reaction of potassium chloro- and bromo-platinite with trifluoroacetylacetone in alkaline solution. The corresponding chloro- and bromo-benzoylacetone complexes can be prepared similarly. Molecular weights in aqueous solution of 329 for the trifluorochloro-complex and of 308 for the trifluorobromo-complex indicate a monomer in solution. The molar conductances in aqueous solution are in the expected range for a uni-univalent electrolyte.

The benzoylacetone compounds were too insoluble to obtain adequate n.m.r. results. The proton magnetic resonance spectra of the bromo- and chloro-trifluoroacetone complexes are essentially the same in water and acetone solutions. Table 1 gives the results for the

TABLE 1

N.m.r. spectra of KPt(3Facac)₂Cl,2H₂O in deuterioacetone

Shift (c./sec.) [*]	Intensity	Multiplicity	Splitting (c./sec.)	Assignment
112	3	1	—	CH_3
133	3	3	8	CH_3
172	4	1	—	H_2O
339	1	3	120	3-CH
348	1	3	6	3-CH

[*] Relative to tetramethylsilane.

chloro-complex in aqueous solution. Each spectrum has a distinct pair of equally intense methyl proton signals together with two 3-CH proton resonances. Comparison of chemical shifts, intensity ratios, and coupling constants, in particular of the 3-CH proton at 339 c./sec., with those of the complex potassium chlorobis(acetylacetonato)platinum(II)[3] strongly suggests two dissimilar trifluoroacetylacetone groups as would be expected by a structure of the type (I).

With an asymmetric ligand such as trifluoroacetylacetone or benzoylacetone, two isomers are expected. From the n.m.r. spectra of the trifluoroacetylacetone complex, we have apparently obtained only one, although the multiplicity of bands in the infrared regions associated with carbonyl absorption and metal–carbon stretching modes, is consistent with the presence of both isomers. However, it is difficult here to differentiate between structural isomerism and solid-state effects, which are not uncommon in infrared spectral measurements.

Trifluoroacetylacetone exists as a mixture of keto and enol tautomers, although the main component is the enol (97% at 30°). The carbonyl frequencies of the keto-tautomer occur at 1775 and 1745 cm.⁻¹, whilst in agreement with the lower bond order of the carbon–oxygen bond in the enol form, the corresponding frequency is observed near 1600 cm.⁻¹. Benzoylacetone is

entirely enolic, as is reflected in the carbonyl region, the two highest bands being at 1600 and 1560 cm.⁻¹. In oxygen-bonded chelating complexes of these ligands the frequencies fall in the range observed for the enol form of the free ligand. For example, in the iron(III) complexes the highest frequencies found are at 1610 cm.⁻¹ for [Fe(3Facac)₃] and 1590 cm.⁻¹ for [Fe(Bzac)₃].

The infrared spectra of the two sets of compounds are in Table 2. The chloride and bromide complexes are closely similar and there are many overall similarities between the groups of complexes. In the trifluoro-compounds a strong band at 1721 cm.⁻¹ is 100 cm.⁻¹ higher than any other frequency reported by other workers for chelating trifluoroacetylacetonates. A strong band at 1688 cm.⁻¹ in the benzoylacetone compounds is 90 cm.⁻¹ higher than any band in the free

TABLE 2

Infrared spectra [*]

KPt(2Facac)₂Cl 2H₂O	KPt(3Facac)₂Br 2H₂O	KPt(Bzac)₂Cl	KPt(Bzac)₂Br
1721s	1721s	1688s	1688s
1658m	1656s	1636m	1642m
1655sh	1650sh	1590s	1590s
1590s	1585s	1550s	1555
1530m	1528m	1524s	1535
1422w	1400w	1480w	1390s
1360w	1355w	1450w	1310w
1310s	1310s	1375s	1280w
1285s	1280s	1310w	1220s
1230m	1231m	1300w	1180s
1200s	1202s	1213s	1155sh
1178s	1172s	1178s	1110m
1148s	1143s	1170m	1100sh
1075s	1068s	1152sh	1070m
1050sh	1045sh	1110m	1050s
1015w	1018w	1071m	1030w
945w	995w	1052s	1022s
875m	940w	1032w	998m
828m	870m	1020m	980m
812sh	821m	996m	890
802m	805m	950w	870
770m	762m	885w	825
745m	739m	860m	802
726s	720s	820s	770s
658sh	640m	798w	750sh
650s	610w	770s	720s
608s	595sh	720s	700s
545s	545s	710m	640m
513sh	513m	700s	564m
437m	437m	640m	547s
333m		565m	505m
		548s	482m
		505m	469m
		481w	439w
		470m	372w
		444w	350w
		374w	
		355m	
		342s	

[*] s = strong, sh = shoulder, m = medium, w = weak.

ligand or than any band observed in metal complexes. In every case, however, strong bands are also observed in the region (1520—1600 cm.⁻¹) associated with the presence of chelating β-diketone groupings. Thus this region suggests in all cases the presence of an oxygen-chelating ligand group and one bonding *via* the 3-carbon atom; the latter is stabilised in the keto-form. Consistent with this view, for the trifluoro-compounds a 2,4-dinitrophenylhydrazine adduct has been prepared in which the 1721 cm.$^{-1}$ band is considerably reduced in intensity. The 3-CH-in-plane and out-of-plane bending vibrations at, respectively, 1200 and 802 cm.$^{-1}$ in the trifluoro-compounds and 1213 and 820 cm.$^{-1}$ in the benzoylacetone compounds can be considered as characteristic of chelation *via* oxygen. The similarity of the remaining infrared absorptions above 600 cm.$^{-1}$ and those of potassium chlorobis(acetylacetonato)-platinum(II) suggest structural similarities between these compounds.[3]

The low-frequency region of the infrared spectrum is most diagnostic for a carbon-bonded species, with the appearance of bands at 500—600 cm.$^{-1}$, whose pattern is consistent with a structure of type I. For the tri-fluoro-complexes bands at 545 and 513 cm.$^{-1}$ are in the range associated with platinum–carbon stretching modes, and a broad band at 437 cm.$^{-1}$ corresponds to the platinum–oxygen vibration in agreement with the assignments of Gillard *et al.*[4] A frequency at 333 cm.$^{-1}$ in the complex KPt(3Facac)$_2$Cl,2H$_2$O, absent from the bromo-analogue, is in the range expected for platinum–chlorine vibrations.

The benzoylacetone complexes have similar frequencies in the 500—600 cm.$^{-1}$ region which can be assigned to platinum–carbon stretching modes and a band at 444 cm.$^{-1}$ to the platinum–oxygen vibration. A band at 342 cm.$^{-1}$ in the chloro-complex is absent from the bromo-analogue and can be identified as the platinum–chlorine stretching frequency.

The complexes K[Pt(acac)$_2$X] (X = Cl or Br) react with mineral acids forming water-insoluble light-sensitive compounds, soluble in organic solvents in which the previously unidentate carbon-bonded acetylacetone group rearranges to give an unsaturated ligand grouping.[7] The platinum–benzoylacetone complexes also reacts with acids to give complexes which may be similar to HPt(acac)$_2$X but are even less stable. It has not been possible to isolate a similar complex from the reaction of the trifluoroacetylacetone complex with acid.

As has been observed for all previous platinum acetyl-acetonates containing unidentate ligands the benzoyl-acetone complexes also yield coloured precipitates with transition-metal chlorides. The reactivity of these compounds will be discussed elsewhere.

It is pertinent to consider what influences carbon- as opposed to oxygen-bonding of the β-diketones to metals. With platinum, since both forms of bonding have been recognised, it may be difficult to decide on the specific factors which determine these different bonding patterns. In general there will be a balance between many factors which may depend upon a small energy difference between large energy terms. In any consideration of the energetics of these systems four bonding forms must be considered, as for each carbon and oxygen system the β-diketone may exist in either an enolic or ketonic form.

It is significant that in oxygen-bonded systems co-ordination occurs *via* the enolic form and utilises both oxygen atoms, whilst for the metal–carbon bonded systems considered so far the β-diketone is stabilised in the keto-form. Reaction of the β-diketone with non-metals appears to occur *via* attack on the 3-carbon atom. Thus the halogens, sulphur, selenium, and tellurium lead to carbon-substituted products, whilst metal co-ordin-ation mainly utilises oxygen co-ordination with formation of a six-membered chelate ring structure. One possible factor is, therefore, the electronegativity of the bonding group, high electronegativity appearing to favour carbon substitution. It has been recognised that the electro-negativity of the elements of the third row of the transition series is much higher than initially expected and carbon–metal bond formation may be partly associated with this.[8]

Swallow and Truter[2] suggested that in the metal–carbon bonded species the β-diketone can be considered to be behaving as a substituted methyl group. Chatt and Shaw[9] have shown that the stability of metal–methyl complexes may be related to the separation between the various *d*-orbitals of the metal which in turn may be associated with the presence of π-accepting groups as ligands. As the molecules containing metal–carbon bonded groups prepared so far have no strong π-bonding ligands present, the stability may be associated with the electronegativity of the platinum. Increasing the electronegativity of the metal will increase the interaction of the metal *d*-orbitals and ligand orbitals and may ensure a sufficient energy separation to stabilise the metal–carbon bond. It has also been shown that replacement of hydrogen by fluorine stabilises the metal–carbon bonds in the metal–alkyl systems; the acyl group will behave electronically in a similar manner to fluorine. The stability of the fluorine complexes has been associated in part with the enhanced ionic nature of the σ-bond of metal-carbon system and with π-bonding from the metal to the antibonding orbitals on the peralkyl group.[10] A similar mechanism can be envisaged for these systems.

It is significant that the n.m.r. results show in all the complexes studied with carbon-bond groups that the β-diketone is stabilised in the keto-form. The keto–enol equilibria of 3-carbon substitution systems has been related to either the electron-donor capacity of the substituent or steric effects.[11] The equilibrium has also been shown to depend upon the dielectric content of the solvent. However since all the results obtained in the platinum system have been obtained in deuterioacetone or deuterium oxide we can ignore this effect. Replace-ment of hydrogen by chlorine favours stabilisation of the enolic form whilst alkyl groups favour the ketonic form. As stabilisation of the keto-form occurs on co-ordination by the carbon to metal, this implies that the

[7] C. Allen, J. Lewis, R. Long, and C. Oldham, *Nature*, 1964, **202**, 589.
[8] R. S. Nyholm, *Proc. Chem. Soc.*, 1961, 273.
[9] J. Chatt and B. L. Shaw, *J. Chem. Soc.*, 1959, 705.

[10] F. G. A. Stone, *Pure Appl. Chem.*, 1965, **10**, 37.
[11] G. Allen and R. A. Dwek, *J. Chem. Soc.* (B), 1966, 161; J. A. Pople, W. G. Schneider, and H. J. Bernstein, " High Resolution Nuclear Magnetic Resonance," McGraw-Hill, New York, 1959.

metal is acting as an electron donor as would be expected if π-bonding is significant in these systems. Replacement of terminal methyl groups of acetylacetone by perfluoromethyl or benzoyl groups leads to complete stabilisation of the enol form, yet on co-ordination to platinum in both these molecules the equilibrium is shifted completely to the keto-form.

Stabilisation of the keto-form is also considered to be associated with steric effects. The change from keto to enolic forms involves a bond-angle change of the 3-carbon atom of 109 to 120° associated with the change in hybridisation from sp^3 to sp^2. This is considered to increase the interaction of substitutions on the 3-carbon atom with the terminal groups. However, structural determination of the β-diketone–platinum compounds reported all indicate that the angle in the keto-form is approximately 120°.[1,2] This poses the question of the nature of the 3-carbon–hydrogen bond. It may be significant that all our attempts to prepare carbon-bonded derivatives in which replacement of the hydrogen by other groups, such as allyl, phenyl, vinyl, alkyl, or chlorine, have failed and this may reflect the specific properties associated with the bond of a hydrogen atom to this position.

EXPERIMENTAL

Potassium Chlorobis(trifluoroacetylacetonato)platinum(II) *Dihydrate.*—Potassium chloroplatinite (1 mole) in water, was treated with an aqueous solution of potassium hydroxide (5 moles) together with trifluoroacetylacetone (7 moles).

After being shaken for 1 hr. at 55—60° and cooling, a yellow solid separated. This was washed with benzene and recrystallised from water to give the *complex* in 65% yield [Found: C, 19·5; H, 2·2; Cl, 6·1; Pt, 31·6%; M (H$_2$O), 329. $C_{10}H_{12}ClF_6KO_6Pt$ requires C, 19·6; H, 2·0; Cl, 5·8; Pt, 31·9%; M, 305]; conductivity (H$_2$O), 97 mho mole^{-1} cm.^{2}. The bromo-analogue is prepared similarly [Found: C, 18·7; H, 1·5; Br, 12·3%; M (H$_2$O), 308. $C_{10}H_{12}BrF_6KO_6Pt$ requires C, 18·3; H, 1·8; Br, 12·2%; M, 328].

Potassium Chlorobis(benzoylacetonato)platinum(II).— Potassium chloroplatinite (1 mole) in water (150 moles altogether) was treated with aqueous potassium hydroxide (8 moles) and benzoylacetone (7 moles). After being shaken for 2 hr. at 60° a yellow solid separated, which was filtered off, was washed with chloroform and ether, and recrystallised from ethanol–water to give the *complex* (Found: C, 41·0; H, 2·9; Cl, 6·1. $C_{20}H_{18}ClKO_4Pt$ requires C, 40·6; H, 3·0; Cl, 6·0%); conductivity (H$_2$O) 130 mho mole^{-1} cm.^{2}. The *bromo-analogue* was prepared similarly [Found: C, 37·7; H, 2·7; Br, 12·2; O, 10·02%; M (H$_2$O), 295. $C_{20}H_{18}BrKO_4Pt$ requires C, 37·8; H, 2·83; O, 10·01%; M (H$_2$O), 318]; conductivity (H$_2$O) 135 mho mole^{-1} cm.^{2}.

Spectroscopy.—Infrared spectra were measured on a Perkin-Elmer model 221 spectrometer. Proton magnetic resonance measurements were made on a Varian A60 spectrometer at 60 Mc./sec.

We thank the S.R.C. for a Research Scholarship for D. G. and C. O., and also Johnson, Matthey and Co. Limited for potassium chloroplatinite.

THE CHEMISTRY DEPARTMENT, THE UNIVERSITY,
    MANCHESTER 13.   [6/390 *Received, March 28th,* 1966]

[Reprinted from the Inorganic Chemistry, **3**, 1587 (1964).]

CONTRIBUTION FROM THE DEPARTMENT OF CHEMISTRY,
NORTHWESTERN UNIVERSITY, EVANSTON, ILLINOIS

Inorganic Linkage Isomerism of the Thiocyanate Ion

BY JOHN L. BURMEISTER[1] AND FRED BASOLO

Received June 17, 1964

Linkage isomers of metal complexes of the type M–SCN (thiocyanato) and M–NCS (isothiocyanato) are reported for the compounds $[Pd(As(C_6H_5)_3)_2X_2]$ and $[Pd(bipy)X_2]$. The method of preparation depends on the rapid isolation of the kinetic reaction product prior to its rearrangement to the stable form. It is suggested that both electronic and steric factors of the coordinated ligands contribute to the nature of SCN^- bonding in complexes of Pd(II) and Pt(II).

Introduction

This study was concerned with what is the least familiar of the types of isomerism found in coordination chemistry, namely, linkage isomerism. Linkage isomerism occurs whenever a given ligand can attach itself to the same central metal atom by bonding through either one of two different atoms within the ligand. For true linkage isomerism, all other factors, such as the geometrical configuration, just remain the same.

Prior to this work, the only ligand known to exhibit this type of isomerism in metal complexes was the nitrile ion, NO_2^-. The first example was reported in 1893 by Jørgensen,[2] who isolated the nitro (Co–NO_2) and nitrito (Co–ONO) pentaammines of cobalt-(III). Recently, Basolo and Hammaker[3] succeeded in extending nitro–nitrito linkage isomerism to include the pentaammines of rhodium(III), iridium(III), and platinum(IV).

The thiocyanate ion, $:N{\equiv}C{-}S:^-$, presents itself as the most logical candidate for a second example of linkage isomerism. It is known to form both thiocyanato (M–SCN) and isothiocyanato (M–NCS) complexes, depending on the central metal atom employed,[4] as well as bridged (M–SCN–M) species.[5] Several authors[6,7] have pointed out that the change from M–NCS to M–SCN bonding coincides approximately in the periodic table with the change in the relative bonding strengths of the halide ions from F^- > Cl^- > Br^- > I^- to I^- > Br^- > Cl^- > F^-. The first order is followed by halogeno complexes of metals of the first transition series, the second order by halogeno complexes of metals to the right of group VII in the second and third transition series, the class a and class b acceptors, respectively, proposed by Ahrland, Chatt, and Davies[8] for the coordination of metal ions with an extensive series of ligands. Several explanations have

been proposed for this change.[8,9] The most recent, by Pearson,[9c] suggests that S in SCN^- is soft and will prefer to coordinate with soft acids (class b metals) whereas N in SCN^- is hard and coordinates with hard acids (class a metals). The terms soft and hard are used to designate substances which are polarizable and nonpolarizable, respectively. The generalization of thiocyanate bonding given above applies only to complexes wherein thiocyanate is the sole ligand.

The only previously reported attempt to prepare inorganic linkage isomers of the thiocyanate ion was for chromium(III) complexes[10] and was not successful. The initial attempts to prepare thiocyanate linkage isomers in our study were formulated along the same lines and were also unsuccessful.

Success was finally realized, due largely to an interesting and significant observation made by Turco and Pecile.[11] They found that, for palladium(II) and platinum(II) complexes, coordinated thiocyanate ion is either S- or N-bonded, depending upon the nature of the other ligands present. Thus, the systems $[M(SCN)_4]^{2-}$ and $[M(NH_3)_2(SCN)_2]$ are S-bonded, whereas $[M(PR_3)_2(NCS)_2]$ is N-bonded. It follows that, in these systems, there should be some borderline ligands for which the energy difference between the M–SCN and M–NCS isomers is relatively small, permitting isolation of both. This paper reports the effects of a series of ligands on thiocyanate bonding in palladium(II) and platinum(II) complexes, and the synthesis,[12] characterization, and preliminary kinetics of the isomerization of thiocyanato and isothiocyanato isomers of some palladium(II) complexes.

Experimental

Preparation of Complexes.—Most of the complexes were prepared by essentially the same method. This involved mixing an aqueous solution of $K_2[MX_4]$, where M is Pd(II) or Pt(II) and X^- is SCN^- or Cl^-, with an alcoholic solution of a stoichiometric amount of a monodentate ligand, L, or a bidentate ligand, L', to give the corresponding products $[ML_2X_2]$ or $[ML'X_2]$. Aqueous solutions of L were used in the preparations of the pyridine and

(1) Abstracted in part from the Ph.D. Thesis of J. L. B., Northwestern University, 1964.

(2) S. M. Jørgensen, *Z. anorg. allgem. Chem.*, **5**, 169 (1893).

(3) F. Basolo and G. S. Hammaker, *J. Am. Chem. Soc.*, **82**, 1001 (1960); *Inorg. Chem.*, **1**, 1 (1962).

(4) F. Basolo and R. G. Pearson, "Mechanisms of Inorganic Reactions," John Wiley and Sons, Inc., New York, N. Y., 1958, p. 14.

(5) J. Chatt and L. A. Duncanson, *Nature*, **178**, 997 (1956).

(6) I. Lindqvist and B. Strandberg, *Acta Cryst.*, **10**, 176 (1957).

(7) P. C. H. Mitchell and R. J. P. Williams, *J. Chem. Soc.*, 1912 (1960).

(8) S. Ahrland, J. Chatt, and N. R. Davies, *Quart. Rev.* (London), **12**, 265 (1958).

(9) (a) K. B. Yatsimirski and V. P. Vasilev, "Instability Constants of Complex Compounds," Pergamon Press, Oxford, 1960; (b) C. K. Jørgensen, "Inorganic Complexes," Academic Press, New York, N. Y., 1963, p. 7; (c) R. G. Pearson, *J. Am. Chem. Soc.*, **85**, 3533 (1963).

(10) R. L. Carlin and J. O. Edwards, *J. Inorg. Nucl. Chem.*, **6**, 217 (1958).

(11) A. Turco and C. Pecile, *Nature*, **191**, 66 (1961).

(12) F. Basolo, J. L. Burmeister, and A. J. Poë, *J. Am. Chem. Soc.*, **85**, 1700 (1963).

TABLE I

STIRRING PERIODS, PER CENT YIELDS, AND ANALYTICAL DATA FOR THE $[ML_2X_2]$ AND $[ML'X_2]$ COMPLEXES

Complex[a]	Stirring period, hr.	Yield, %	Theory C	H	N	Found C	H	N
$[Pd(P(n\text{-}C_4H_9)_3)_2(NCS)_2]$	1.25[b]	31.2	49.79	8.68	4.47	49.50	8.23	4.74
$[Pd(P(C_6H_5)_3)_2Cl_2]$	0.05[c]	82.4	61.60	4.31		62.72	4.41	
$[Pd(P(C_6H_5)_3)_2(NCS)_2]$	3.0[c]	94.3	61.07	4.05	3.75	61.29	4.20	3.59
$[Pd(As(n\text{-}C_4H_9)_3)_2Cl_2]$	0.02[h]	Not det.	43.03	8.13		42.87	8.04	
$[Pd(As(n\text{-}C_4H_9)_3)_2(NCS)_2]$	0.12[b]	Not det.	43.67	7.61	3.92	45.10	7.53	2.92
$[Pd(As(C_6H_5)_3)_2Cl_2]$	0.08[c]	85.1	54.75	3.83		54.79	3.91	
$[Pd(As(C_6H_5)_3)_2(NCS)_2]$	3.0[c]	85.0	54.66	3.62	3.36	54.63	3.65	3.44
$[Pd(Sb(C_6H_5)_3)_2Cl_2]$	0.08[c]	58.9	48.95	3.42		48.89	3.26	
$[Pd(Sb(C_6H_5)_3)_2(SCN)_2]$	3.0[c]	12.8	49.15	3.26	3.02	49.54	3.30	2.99
$[Pd(py)_2Cl_2]$	0.08[b]	70.4	35.80	3.00		35.83	2.93	
$[Pd(py)_2(NCS)_2]$	0.17[b]	66.8	37.84	2.65	14.71	37.31	2.73	14.59
$[Pd(\gamma\text{-pic})_2Cl_2]$	0.08[b]	72.8	39.65	3.88		39.74	4.03	
$[Pd(\gamma\text{-pic})_2(SCN)_2]$	0.33[b]	81.2	41.15	3.45	13.71	41.39	3.57	13.43
$[Pd(4\text{-}n\text{-ampy})_2Cl_2]$	0.08[b]	Not det.	50.49	6.36		50.36	6.34	
$[Pd(4\text{-}n\text{-ampy})_2(SCN)_2]$	0.07[b]	Not det.	50.71	5.80	10.75	51.96	5.89	10.62
$[Pd(bipy)Cl_2]$	0.50[b]	96.5	36.02	2.42		35.96	2.63	
$[Pd(bipy)(NCS)_2]$	0.17[b]	94.0	38.04	2.13	14.79	38.01	2.34	14.86
$[Pd(phen)Cl_2]$	0.05[b]	Not det.	40.31	2.26		40.49	2.32	
$[Pd(phen)(SCN)_2]$	0.05[b]	Not det.	41.76	2.00	13.91	41.79	2.17	13.75
$[Pd(tu)_2(SCN)_2]$	24.0[c]	Not det.	12.82	2.15	22.43	12.96	2.07	22.58
$[Pd(etu)_2Cl_2]$	34.5[c]	Not det.	18.87	3.17		18.44	3.27	
$[Pd(etu)_2(SCN)_2]$	1.0[c]	Not det.	22.51	2.83	19.69	23.23	2.95	19.33
$[Pt(P(C_6H_5)_3)_2Cl_2]$	3.5[c]	98.0	54.69	3.83		55.07	3.36	
$[Pt(P(C_6H_5)_3)_2(NCS)_2]$	2.5[c]	93.0	54.60	3.62	3.35	53.16	3.66	3.27
$[Pt(As(C_6H_5)_3)_2(NCS)_2]$	1.0[c]	67.8	49.41	3.27	3.03	49.34	3.31	3.12
$[Pt(Sb(C_6H_5)_3)_2Cl_2]$	4.0[c]	69.1	44.48	3.11		44.56	3.10	
$[Pt(Sb(C_6H_5)_3)_2(SCN)_2]$	2.0[c]	49.0	44.86	2.97	2.75	44.69	2.99	2.79
$[Pt(py)_2Cl_2]$	6.5[b]	81.8	28.31	2.38		28.60	2.36	
$[Pt(py)_2(NCS)_2]$	0.5[b]	75.4	30.70	2.15	11.94	30.58	2.30	11.76
$[Pt(bipy)(NCS)_2]$	24.0[b]	<1	30.83	1.73	11.99	30.36	1.95	11.52

[a] Some of the compounds reported here had previously been prepared and are described in the literature. [b] Both reactant solutions were at room temperature before being mixed. [c] A room temperature solution of $K_2[MX_4]$ was mixed with an $\sim75°$ ethanolic solution of the stoichiometric amount of ligand. Abbreviations: py, pyridine; bipy, 2,2′-bipyridine; γ-pic, γ-picoline (4-methylpyridine); 4-n-ampy, 4-n-amylpyridine; phen, 1,10-phenanthroline; tu, thiourea; etu, ethylenethiourea. Whenever the symbol NCS⁻ is written in a compound this designates M–NCS bonding, whereas SCN⁻ designates M–SCN bonding.

4-methylpyridine complexes. The amounts used were of the order of 1 mmole of $K_2[MX_4]$, dissolved in 5 ml. of H_2O, and 2 mmoles of L or 1 mmole of L′, dissolved in 5 ml. of absolute C_2H_5OH. After stirring the reaction mixture with a magnetic stirrer for a period of time to allow for the completion of the reaction, the product was isolated by vacuum filtration, washed with H_2O, absolute C_2H_5OH, and ethyl ether, and dried *in vacuo* over P_2O_5. The products precipitated from solution immediately upon their formation. The rates of the reactions varied, those involving Pd(II) being essentially instantaneous, those involving Pt(II) being much slower. Usually, both reactant solutions were mixed at room temperature. In some cases, the relatively low solubility of the ligand in absolute C_2H_5OH necessitated the use of an ethanolic ligand solution whose temperature was close to the boiling point of C_2H_5OH. The stirring periods employed, per cent yields obtained, and analyses of the complexes prepared by the foregoing method are shown in Table I.

No attempt was made to ascertain the structures of these compounds beyond the determination of the bonding involved (M–SCN or M–NCS) in the thiocyanate-containing complexes. It is reasonable to assume that they are square-planar complexes. Furthermore, the Pd(II) complexes all probably have the *trans* configuration, except, of course, the complexes containing 2,2′-bipyridine and 1,10-phenanthroline. It was noted by Mann and Purdie,[13] who prepared a series of compounds of the general formula $[Pd(LR_3)_2X_2]$, where L is either P or As, R is an alkyl group, and X⁻ is Cl⁻, Br⁻, or I⁻, that with the exception of the

As$(CH_3)_3$ derivative, only the *trans* isomer could be made. The Pt(II) complexes probably have the *cis* configuration.[14]

Compounds which were not prepared by the above general method and which are not listed in Table I are discussed here separately.

cis- and *trans*-$[Pt(As(C_6H_5)_3)_2Cl_2]$.—These geometric isomers were prepared according to the method of Jensen.[14]

Anal. Calcd. for $PtAs_2C_{36}H_{30}Cl_2$: Pt, 22.21. Found, for the *cis* isomer: Pt, 22.1; for the *trans* isomer; Pt, 22.5.

$[Pt(bipy)Cl_2]$ and $[Pt(bipy)_2][PtCl_4]$.—These compounds were prepared according to the methods of Morgan and Burstall.[15]

Anal. Calcd. for $PtCl_{10}H_8N_2Cl_2$ and $Pt_2C_{20}H_{16}N_4Cl_4$: C, 28.45; H, 1.91; Pt, 46.30. Found, for $[Pt(bipy)Cl_2]$: Pt, 46.2; for $[Pt(bipy)_2][PtCl_4]$: C, 28.08; H, 2.59.

$[Pt(bipy)_2][Pt(SCN)_4]$ was prepared in a completely analogous manner. The analytical results, however, were poor for the brick-red solid obtained.

Anal. Calcd. for $Pt_2C_{24}H_{16}N_8S_4$: C, 30.83; H, 1.73; N, 11.99; Pt, 41.74. Found: C, 29.56; H, 2.01; N, 10.20; Pt, 41.9.

cis-$[Pt(NH_3)_2(SCN)_2]$.—To a solution of 0.89 g. (5.2 mmoles) of $AgNO_3$ in 5 ml. of H_2O was added 0.78 g. (2.6 mmoles) of *cis*-$[Pt(NH_3)_2Cl_2]$. The mixture was heated on a steam bath until the only solid remaining was a pure white precipitate of AgCl. The AgCl was removed by filtration and the filtrate placed in an ice bath. To the filtrate was added 0.89 g. (5.2 mmoles) of KSCN, dissolved in 2 ml. of H_2O, resulting in the precipitation of a bright yellow solid, which was collected, washed with H_2O, C_2H_5OH, and $(C_2H_5)_2O$, and dried *in vacuo* over P_2O_5.

(13) F. G. Mann and D. Purdie, *J. Chem. Soc.*, 1554 (1935); see also G. E. Coates and J. Parkin, *ibid.*, 421 (1963).

(14) K. A. Jensen, *Z. anorg. allgem. Chem.*, **229**, 242 (1936).
(15) G. T. Morgan and F. H. Burstall, *J. Chem. Soc.*, 965 (1934).

Anal. Calcd. for PtC₂H₆N₄S₂: C, 6.96; H, 1.75; N, 16.23. Found: C, 6.89; H, 1.51; N, 15.97.

trans-[Pt(NH₃)₂(SCN)₂], pale yellow in color, was prepared in the same manner, starting with *trans*-[Pt(NH₃)₂Cl₂].

Anal. Found: C, 7.14; H, 1.73; N, 16.18.

[Pd(bipy)₂](B(C₆H₅)₄)₂.—A mixture of 0.22 g. (0.7 mmole) of [Pd(bipy)Cl₂] and 0.22 g. (1.4 mmoles) of 2,2'-bipyridine in H₂O was heated on a steam bath until the [Pd(bipy)Cl₂] dissolved, due to the formation of [Pd(bipy)₂]²⁺. The solution was cooled in an ice–salt bath and passed through a filter. To the filtrate was added a solution of 0.51 g. (1.5 mmoles) of Na[B(C₆H₅)₄] in 25 ml. of H₂O. A light yellow precipitate formed immediately. This was collected on a filter and dried *in vacuo* over Mg(ClO₄)₂.

Anal. Calcd. for PdC₄₄H₃₆N₄B: C, 77.25; H, 5.34. Found: C, 78.53; H, 5.56.

[Pd(bipy)₂](ClO₄)₂ was prepared in a similar fashion, using Ba(ClO₄)₂·2H₂O as the precipitating agent.[16] Analysis by combustion was not attempted because such perchlorates explode when heated.

[Pd(bipy)₂][PdCl₄].—[Pd(bipy)₂](ClO₄)₂ (0.18 g., 0.3 mmole) and K₂[PdCl₄] (0.10 g., 0.3 mmole) were stirred in 20 ml. of H₂O for 45 min. Filtration yielded a pink solid which was dried *in vacuo* over Mg(ClO₄)₂. The product contained some entrapped [Pd(bipy)₂](ClO₄)₂, as determined from the infrared spectrum of the product.

[Pd(As(C₆H₅)₃)₂(SCN)₂].—K₂Pd(SCN)₄ (1.03 g., 2.47 mmoles), dissolved in 25 ml. of absolute C₂H₅OH and 5 drops of H₂O, and As(C₆H₅)₃ (1.52 g., 4.95 mmoles), dissolved in 25 ml. of absolute C₂H₅OH and 5 drops of (C₂H₅)₂O, were cooled to 0°, then mixed in a vessel surrounded by an ice bath. After the solution was stirred for 1 min., the yellow-orange S-bonded product was precipitated by the addition of 50 ml. of ice water, isolated by filtration, washed with ice-cold C₂H₅OH and (C₂H₅)₂O, and dried *in vacuo* over P₂O₅ (yield 1.87 g., or 91%).

Anal. Calcd. for PdAs₂C₃₈H₃₀N₂S₂: C, 54.66; H, 3.62; N, 3.36. Found: C, 54.50; H, 3.80; N, 3.27. Found, after heating at 156° for 30 min., resulting in complete isomerization to the bright yellow N-bonded isomer: C, 55.01; H, 3.86; N, 3.63.

[Pd(bipy)(SCN)₂].—K₂[Pd(SCN)₄] (0.42 g., 1 mmole), dissolved in 10 ml. of absolute C₂H₅OH and 5 drops of H₂O, and 2,2'-bipyridine (0.16 g., 1 mmole), dissolved in 10 ml. of absolute C₂H₅OH and 5 drops of (C₂H₅)₂O, were cooled to −78° in a Dry Ice–acetone bath, then mixed in a vessel surrounded by a Dry Ice–acetone bath. No formation of solid took place. The solution was removed from the bath and after about 10 min. had become opaque. An unidentified orange solid was removed by filtration and the desired light orange-yellow S-bonded isomer immediately separated in the filtrate. This was isolated, washed with −78° C₂H₅OH and (C₂H₅)₂O, and dried *in vacuo* over Mg(ClO₄)₂.

Anal. Calcd. for PdC₁₂H₈N₄S₂: C, 38.04; H, 2.13; N, 14.79. Found: C, 38.17; H, 2.48; N, 14.95. Found, after heating at 156° for 30 min., resulting in complete isomerization to the light yellow N-bonded isomer: C, 38.56; H, 2.43; N, 14.81.

[Pd(bipy)₂][Pd(SCN)₄].—An acetone solution of 0.11 g. (0.25 mmole) of K₂[Pd(SCN)₄] was cooled to −78° in a Dry Ice–acetone bath. To this solution was added 0.32 g. (0.3 mmole) of [Pd(bipy)₂](B(C₆H₅)₄)₂. Filtration yielded a pink solid. The product, dried *in vacuo* over Mg(ClO₄)₂, contained some entrapped K[B(C₆H₅)₄], as determined from the infrared spectrum of the product.

Sources of Additional Compounds.—The following compounds were obtained from the indicated individuals, Chemistry Department, Northwestern University: R. C. Johnson: *cis*- and *trans*-[Pt(NH₃)₂Cl₂], [Pt(en)X₂] (en = ethylenediamine; X⁻ = Cl⁻, SCN⁻), [M(tripy)X]X (M = Pd(II), Pt(II); tripy = 2,2',2''-tripyridine; X⁻ = NCS⁻, Cl⁻, Br⁻, I⁻); T. Schenach: [*trans*-Pd(P(*n*-C₄H₉)₃)₂Cl₂]; K. Stephen: [Pt(NH₃)₄][PtCl₄].

Analyses.—Platinum contents were determined gravimetrically by igniting a weighed sample of the compound in a porcelain crucible over a Meker burner. Carbon, hydrogen, and nitrogen analyses were performed by Miss H. Beck, Chemistry Department, Northwestern University.

Visible and Ultraviolet Spectra.—Visible and ultraviolet absorption spectra were measured on either a Beckman DK-2 or Cary Model 11 recording spectrophotometer. Solutions of the compounds in question in 0.1 or 1.0 cm. quartz cells were usually used, although some visible transmission spectra[17] were taken of Nujol mulls of solid samples, using quartz plates. Pieces of Whatman No. 1 filter paper, impregnated with Nujol, served as dispersing media in the sample and reference beams.

Infrared Spectra.—Infrared spectra of samples in KBr disks were measured on either a Baird-Associates Model AB-2 or Beckman IR-5 recording spectrophotometer. Both are double-beam instruments with NaCl optics. Infrared spectra of Nujol mulls of solid samples were obtained on either of the above-named instruments or on a Beckman IR-9 recording spectrophotometer, employing a KBr prism coupled with a diffraction grating. NaCl plates were used in preparing the mulls in all cases, except when the spectrum was taken on the Beckman IR-9 instrument, in which case KBr plates were used. A polystyrene standard was used for calibration in all cases. High resolution infrared spectra of Nujol mulls of solid samples in the thiocyanate C–N stretching frequency range were measured on a Perkin-Elmer Model 112 single beam recording spectrophotometer, using a CaF₂ prism and NaCl plates. The instrument was calibrated in this range on CO vapor by Dr. D. F. Shriver. Infrared solution spectra, using matched NaCl cells, were measured on a Baird-Associates Model 4-55 recording spectrophotometer.

Conductance Measurements.—Conductances of compounds dissolved in N,N-dimethylformamide (Eastman Spectro grade) at 25° were measured with an Industrial Instruments, Inc., Model RC-16 conductivity bridge and a cell with platinum electrodes.

Dipole Moment Measurements.—The dipole moments of [Pd(As(C₆H₅)₃)₂(SCN)₂] and [Pd(As(C₆H₅)₃)₂(NCS)₂], dissolved in benzene, were determined according to the approximate method of Jensen and Nygaard,[18] using only dielectric constant measurements. The apparatus employed was a General Radio Co. Capacitance Measuring Assembly, Type 1610-A, set at 1000 c.p.s. An oscilloscope was used to balance the bridge. The capacitance measurements of the benzene solutions were made in a 40-ml. specially designed cell. Neglecting electron and atom polarizations is said to introduce an error of only a few per cent.

Rates of Isomerization.—The rate of isomerization of [Pd(As(C₆H₅)₃)₂(SCN)₂] to [Pd(As(C₆H₅)₃)₂(NCS)₂] was studied in KBr disks by observing the decrease in intensity of the S-bonded thiocyanate C–N stretching band with time. A Perkin-Elmer Model 112 single-beam recording spectrophotometer was used. One KBr disk was used throughout an entire kinetic run. Its spectrum was measured over the 5000–650 cm.⁻¹ range on a Baird AB-2 double-beam recording spectrophotometer before and after the kinetic run to determine whether any decomposition of the complex had occurred. No Br⁻ substitution was observed. Throughout the course of the run, the disk, in a metal holder, was heated at the desired temperature in either a thermostated oven or electric coil heater. The spectra were measured *vs.* air.

For this study, the usual first-order rate expression becomes

$$k = -(2.303/t) \log (A - A_\infty) \qquad (1)$$

where k is the first-order rate constant in sec.⁻¹, t is the time in sec., A is the absorption at time t, and A_∞ is the absorption at infinite time. Although spectrally pure S⁻ isomer was used in preparing the KBr disks, the high pressure and heat encountered in their preparation caused partial isomerization to the N- isomer to occur and, as a result, both C–N stretching bands were always present in the initial spectra of the disks.

The concentration of the complex is proportional to log (A −

---

(16) S. Livingstone, *J. Proc. Roy. Soc. N.S. Wales*, **86**, 32 (1952).

(17) F. A. Cotton, D. M. L. Goodgame, M. Goodgame, and A. Sacco, *J. Am. Chem. Soc.*, **83**, 4157 (1961).

(18) K. A. Jensen and B. Nygaard, *Acta Chem. Scand.*, **3**, 479 (1949).

TABLE II

EFFECT OF OTHER LIGANDS ON THIOCYANATE BONDING.

C–S AND C–N STRETCHING FREQUENCIES OF Pd(II) AND Pt(II) THIOCYANATE COMPLEXES

Complex	C–N stretch, cm.$^{-1}$ (ν_3 SCN)	C–S stretch, cm.$^{-1}$ (ν_1 SCN)
	Thiocyanates	
$K_2[Pd(SCN)_4]$[a]	2118, 2086	703, 707 sh, 696 sh
$K_2[Pt(SCN)_4]$[a]	2120, 2089	697, 700 sh, 690 sh
$[Pd(Sb(C_6H_5)_3)_2(SCN)_2]$	(M) 2119 sh, 2115 s,sp	(D)[b]
$[Pt(Sb(C_6H_5)_3)_2(SCN)_2]$	(M) 2123 sh, 2120 s,sp	(D)[b]
$[Pd(\gamma\text{-pic})_2(SCN)_2]$	(M) 2109 s,sp	(D) 702 w
$[Pd(4\text{-}n\text{-ampy})_2(SCN)_2]$	(M) 2111 s,sp	(D) 707 w
$[Pd(phen)(SCN)_2]$	(M) 2114 s,sp	(D) 696 w
$[Pd(tu)_2(SCN)_2]$	(M) 2107 s,sp	(D) 703 w
$[Pd(etu)_2(SCN)_2]$	(M) 2101 s,sp	(D) 701 w
$[Pt(en)(SCN)_2]$	(M) 2114 s,sp	(M) 696 w
$cis\text{-}[Pt(NH_3)_2(SCN)_2]$	(M) 2116 s,sp	(M) 698 w
$trans\text{-}[Pt(NH_3)_2(SCN)_2]$	Not det.	(M) 706 w
	Isothiocyanates	
$[Pd(P(n\text{-}C_4H_9)_3)_2(NCS)_2]$	(M) 2102 s,br	(D) 847 m
$[Pd(P(C_6H_5)_3)_2(NCS)_2]$	(M) 2093 s,br	(D) 853 m
$[Pt(P(C_6H_5)_3)_2(NCS)_2]$	(M) 2097 s,br	(D) 859 m
$[Pd(As(n\text{-}C_4H_9)_3)_2(NCS)_2]$	(M) 2111 s,br	(M) 844 w
$[Pt(As(C_6H_5)_3)_2(NCS)_2]$	(M) 2090 s,br	(D) 861 m
$[Pd(py)_2(NCS)_2]$	(M) 2115 s,sp	(D) 865 w
$[Pt(py)_2(NCS)_2]$	(M) 2123 s,sp	(D) 847 w
$[Pt(bipy)(NCS)_2]$	(M) 2095 s,br	(D) 844 m
$[Pd(tripy)(NCS)_2]$[c]	(M) 2088 s,br	(D) 848 w
$[Pt(tripy)NCS]SCN$[d]	(M) 2089 s,br, 2040 s,br	(D) 863 w

[a] Frequencies taken from ref. 11.   [b] No C–S stretching peak in N-bonded range; strong phenyl ring absorption in S-bonded range.   [c] tripy coordinated as a bidentate ligand—no C–N stretching band found for ionic thiocyanate.   [d] tripy coordinated as a tridentate ligand—ionic thiocyanate C–N stretching band at 2040 cm.$^{-1}$. Abbreviations: M, Nujol mull; D, KBr disk; s, strong; m, medium; w, weak; sp, sharp; br, broad; sh, shoulder.

TABLE III

C–N AND C–S STRETCHING FREQUENCIES, COLORS, MELTING POINTS, MOLAR CONDUCTANCES, AND DIPOLE MOMENTS OF THIOCYANATE LINKAGE ISOMERS

Complex	Color	M.p. (dec.), °C.	C–N stretch, cm.$^{-1}$	C–S stretch, cm.$^{-1}$	Λ_m,[d] ohm^{-1} cm.$^{-1}$ mole^{-1}	Dipole moment, D.
$[Pd(As(C_6H_5)_3)_2(SCN)_2]$	Yellow-orange	195[a]	(M) 2119 s,sp	(M)[b]	13.3	3.8
$[Pd(As(C_6H_5)_3)_2(NCS)_2]$	Bright yellow	195	(M) 2089 s,br	(M) 854 m	13.8	3.6
$[Pd(bipy)(SCN)_2]$	Light orange-yellow	270[c]	(M) 2117 m,sp, 2108 s,sp	(M) 700 w	20.3	Not det.
$[Pd(bipy)(NCS)_2]$	Light yellow	270	(M) 2100 s,br	(M) 842 m, 849 sh	20.8	Not det.

[a] Complex, upon heating, became bright yellow in color at ~130°.   [b] No C–S stretching band in N-bonded range, strong phenyl ring absorption in S-bonded range.   [c] Complex, upon heating, became light yellow in color at ~120°.   [d] Concentration 1 mM. Molar conductances of approximately 70 correspond to 1:1 electrolytes, the values here are due to partial solvation (ionization) of the non-ionic complexes in DMF.

A_∞), since the cell length remains constant throughout a kinetic run. Straight lines, the slopes of which are equal to $-k/2.303$, were obtained from plots of t vs. log $(A - A_\infty)$. The time, t, refers to the time the disks were in the oven or heater. It was felt that, since the rates of isomerization were so slow at the temperatures employed, the isomerization reaction was effectively quenched when the disk was removed from the heat source. No serious scattering of points resulted from making this correction. It would have been desirable to study the isomerization reaction in Nujol mulls, but experimental difficulties (flowing of the Nujol layer at the temperatures employed and consequent scattering of points) prohibited this.

Results

The type of thiocyanate bonding in the thiocyanate-containing complexes was determined on the basis of the position of the thiocyanate C–S stretching fre-

quency for the complex in question. As has been pointed out by several authors,[11,19] this frequency is shifted to higher wave numbers in the spectra of iso-thiocyanates, and to lower wave numbers in the spectra of thiocyanates, both relative to the C–S stretching frequency of "ionic" thiocyanate, as in KSCN (749 cm.$^{-1}$).[20] Turco and Pecile[11] give the following ranges: M–SCN, 690–720 cm.$^{-1}$; M–NCS, 780–860 cm.$^{-1}$. The position of the C–S stretching frequency in each case was determined by comparing

(19) (a) M. M. Chamberlain and J. C. Bailar, Jr., *J. Am. Chem. Soc.*, **81**, 6412 (1959); (b) J. Lewis, R. S. Nyholm, and P. W. Smith, *J. Chem. Soc.*, 4590 (1961); (c) K. Nakamoto, "Infrared Spectra of Inorganic and Co-ordination Compounds," John Wiley and Sons, Inc., New York, N. Y., 1963, p. 175.
(20) L. H. Jones, *J. Chem. Phys.*, **25**, 1069 (1956).

the spectrum of the thiocyanate-containing complex with that of the analogous chloro complex. The C–N and C–S stretching frequencies found for the complexes are listed in Table II.

The C–N and C–S stretching frequencies of the two linkage isomeric pairs and other pertinent data concerning them are shown in Table III.

The absorption maxima found in the visible and near-ultraviolet spectra of [Pd(As(C₆H₅)₃)₂(SCN)₂] and [Pd(As(C₆H₅)₃)₂(NCS)₂], as well as those found in the spectra of other complexes germane to the discussion which follows, are shown in Table IV.

TABLE IV

VISIBLE AND NEAR-ULTRAVIOLET SPECTRA

Complex	Solvent	Absorption, maxima, 700–350 mμ ($\epsilon \times 10^{-4}$)
[Pd(As(C₆H₅)₃)₂(SCN)₂]	HCCl₃	351 (2.09)^a
[Pd(As(C₆H₅)₃)₂(NCS)₂]	HCCl₃	351 (2.09)
[Pd(As(C₆H₅)₃)₂(SCN)₂]	Nujol mull	395, 475 sh
[Pd(As(C₆H₅)₃)₂(NCS)₂]	Nujol mull	350
K₂[PdCl₄]	Nujol mull	475
[Pd(bipy)₂][PdCl₄]	Nujol mull	475
K₂[Pd(SCN)₄]	Nujol mull	500 (sh), 410 (sh)
[Pd(bipy)₂][Pd(SCN)₄]	Nujol mull	500 (sh), 410 (sh)

^{a} [Pd(As(C₆H₅)₃)₂(SCN)₂] evidently isomerizes to the stable N- isomer upon dissolution in HCCl₃ at room temperature.

The rate constants determined for the isomerization of [Pd(As(C₆H₅)₃)₂(SCN)₂] in KBr disks at 126 and 147° were, respectively, 3.47×10^{-5} and 7.81×10^{-5} sec.$^{-1}$. Qualitatively, it was observed that, in the pure solid state, complete isomerization took place in less than 2 hr. at 118°.

Discussion

Turco and Pecile[11] suggest that π-bonding ligands, such as phosphines, coordinated to Pd(II) or Pt(II) enhance M–NCS bonding in these four-coordinated systems. They point out that there are two sets of antibonding π-orbitals localized on the sulfur atom, which, along with the sulfur atom's vacant d-orbitals, can accept electron density from the metal's filled non-bonding d-orbitals, resulting in additional stability of the M–S bond. The importance of this additional stability will depend on the availability of the electrons of the metal and their relative energy as compared with that of the orbitals of the thiocyanate. Strong π-electron acceptors, as trialkylphosphines are believed to be,[21] can make the metal d-orbital electrons less available for donation to the thiocyanate, removing the source of additional stability for the M–S bond.

Another way of saying about the same thing is that π-bonding ligands in these systems tend to reduce the electron density on the metal and thereby change class b, or soft metals, to class a, or hard metals. This is accompanied by a change in M–SCN bonding to M–NCS, respectively. Such a π-bonding hypothesis finds support in the observation (Table II) that the

bonding is Pd–NCS in the pyridine complex, but Pd–SCN in the analogous γ-picoline and 4-*n*-amylpyridine complexes. Alkyl substitution donates electron density into the pyridine ring, opposing the withdrawal of d-electron density by the ring from the Pd(II), relative to pyridine. The greater σ-bonding of the more basic alkylpyridines may also increase the electron density on Pd(II), relative to pyridine, and further promote Pd–SCN bonding. It is important to note that, in these systems, the steric factors at the metal are kept constant. This then shows that the electronic factors of coordinated ligands can alter the nature of SCN⁻ bonding in metal complexes. It also suggests that small differences can sometimes change the type of bonding.

It has been shown recently[22] that steric factors can alter the nature of SCN⁻ bonding in these systems. In most comparisons, such as [Pt(NH₃)₂(SCN)₂] and [Pt(PR₃)₂(NCS)₂], both electronic and steric factors change with changes in the ligands. It is then often impossible to assess the importance of each factor to the type of SCN⁻ bonding. The point is that M–SCN bonding, because of the angular structure of M–S–C, has a larger steric requirement than does the linear structure of M–NCS.[23] Thus, because of the larger size of PR₃, it would tend to generate some strain in M–SCN and promote M–NCS bonding. The steric factor of PR₃ operates in the same direction as does its electronic effect, and both may well contribute. Similarly, the difference found between P(C₆H₅)₃ and Sb(C₆H₅)₃ may be the result of electronic and/or steric factors. This is because P(C₆H₅)₃ is the better π-bonder[24] and, because of the smaller size of phosphorus, it places the phenyl groups nearer to the metal and offers a greater steric hindrance at the metal than does Sb(C₆H₅)₃. That 1,10-phenanthroline and 2,2'-bipyridine cause different types of SCN⁻ bonding is not understood, except to repeat that small differences are sufficient to cause a change in bond type.

In all of these systems the type of bonding, M–NCS or M–SCN, was assigned on the basis of the infrared spectra in the C–S stretching region.[11,19] In most cases, the thiocyanato complexes exhibited very sharp well-formed C–N stretching peaks above 2100 cm.$^{-1}$, whereas the isothiocyanato complexes exhibited relatively broad, more intense peaks around or below 2100 cm.$^{-1}$. The apparent exceptions, where the positions of the C–S stretching frequencies support N-bonding, are [Pt(py)₂(NCS)₂], [Pd(py)₂(NCS)₂], and [Pd(As(*n*-C₄H₉)₃)₂(NCS)₂]. Most of the thiocyanato complexes which were thought to have the *cis* configuration exhibited an expected splitting of the C–N stretching band, the exceptions being [Pt(en)(SCN)₂] and [Pd(phen)(SCN)₂]. None of the *cis*-isothiocyanato complexes exhibited this splitting.

(21) E. W. Abel, M. A. Bennett, and G. Wilkinson, *J. Chem. Soc.*, 2323 (1959); W. D. Horrocks and R. C. Taylor, *Inorg. Chem.*, **2**, 123 (1963).

(22) F. Basolo, W. H. Baddley, and J. L. Burmeister, *ibid.*, **3**, 1202 (1964).

(23) See discussion and references in ref. 20b. It should be noted that in the solid state, some M–NCS systems are angular, with large angles of about 160°.

(24) J. Chatt, L. A. Duncanson, and L. M. Venanzi, *J. Chem. Soc.*, 4461 (1955).

Having observed that $As(C_6H_5)_3$ is a borderline ligand between $P(C_6H_5)_3$, which gives Pd–NCS, and $Sb(C_6H_5)_3$, which gives Pd–SCN, it was decided to use $As(C_6H_5)_3$ to prepare the first example of linkage isomers in these systems. This was achieved by the reaction shown in eq. 2.

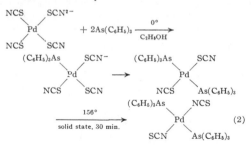

$$(2)$$

The experiment was designed such that the S-bonded kinetic product could be isolated prior to its rearrangement to the more stable N-bonded isomer. This required that the reaction be carried out at some low temperature under conditions such that the product could be separated rapidly from the reaction mixture. Thus, a nonionic complex of slight solubility was prepared. Pd-(II), rather than Pt(II), was used because the substitution reactions of Pd(II) are much more rapid than those of Pt(II). The infrared spectrum of the yellow-orange product obtained under these conditions showed the complete absence of the N-bonded C–S stretching peak at 854 cm.$^{-1}$, and, as predicted by Mitchell and Williams,[7] the C–N stretching peak was found at a higher frequency, 2119 cm.$^{-1}$, than that (2089 cm.$^{-1}$) of the N-isomer. Heating the S- isomer at 156° for 30 min. resulted in complete isomerization to the bright yellow N- isomer. It was then determined, by the use of high resolution infrared spectroscopy in the C–N stretching frequency range, that what results from the reaction at room temperature is actually a mixture of the two isomers.

In *ca.* 10^{-4} *M* chloroform solution, the visible–near-ultraviolet spectra of the two isomers are identical (see Table IV), indicating rapid isomerization to the same equilibrium mixture upon dissolution. The visible transmission spectra of Nujol mulls of the isomers are quite different, however (Table IV). The band corresponding to the d–d transition of the N- isomer is apparently hidden by its charge-transfer band at 351 mμ; that of the S- isomer is seen as a shoulder at 475 mμ.

In an attempt to obtain some information concerning the structure (*cis* or *trans*) of these linkage isomers, the dipole moments of both were determined in benzene solution. The results were inconclusive. The dipole moment of the S- isomer was found to be 3.8 D., that of the N- isomer 3.6 D. That both are the same, within the limits of experimental error, is without meaning because the infrared spectra of the two solutions are identical. This suggests again that there is rapid equilibration upon dissolution at room tempera-

ture. The solutions exhibit three C–N stretching peaks: ~2090, ~2120, and ~2160 cm.$^{-1}$. The high frequency of the peak at 2160 cm.$^{-1}$ indicates that thiocyanate bridging[5] takes place in solution. Thus, there is no evidence concerning the geometric structures of the linkage isomers, and the arbitrary choice of *trans* for both is made on the basis that, for steric reasons, the bulky $As(C_6H_5)_3$ groups would prefer to be at opposite positions in the complex. Support for this choice is found in the previously cited work of Mann and Purdie.[13]

Consequently, to obviate the possibility of concurrent geometric isomerism, a bidentate ligand, 2,2′-bipyridine, was employed in a manner similar to that used in preparing the triphenylarsine derivatives. Isolation of the light orange-yellow S- isomer, [Pd-(bipy)(SCN)$_2$], was accomplished by lowering the temperature of the reaction to −78° (the temperature of a Dry Ice–acetone bath). Complete isomerization to the light yellow N-bonded isomer, obtained from the room temperature reaction, was accomplished by heating the S- isomer at 156° for 30 min. The only other reasonable compound which would give the same analytical results as the S-bonded isomer is [Pd(bipy)$_2$]-[Pd(SCN)$_4$]. Although the conductivity measurements (Table III) seem to rule out this possibility, it was felt that the most convincing evidence would be gained by actually preparing this Magnus type complex. Its visible transmission spectrum (Nujol mull) exhibits shoulders at 500 and 410 mμ which are also found in the spectrum of $K_2[Pd(SCN)_4]$ (Table IV), but are absent in the spectrum of [Pd(bipy)-(SCN)$_2$] and [Pd(bipy)(NCS)$_2$], neither of which shows any absorption maxima in the visible range.

A comparison of the rates of isomerization of solid $[Pd(As(C_6H_5)_3)_2(SCN)_2]$ in KBr disks with qualitative observations of the rate of isomerization in the pure solid state shows that a large decrease in rate results from carrying out the isomerization in a KBr disk. It is of interest to note that the isomerization, nitrito → nitro, shows no difference in rate when carried out in a KBr disk.[3] It would, of course, have been desirable to investigate the rates and mechanism of isomerization in solution, but the isomerization in the solvents used was too fast to permit such a study by conventional techniques.

In conclusion, it should be mentioned that other examples of M–SCN and M–NCS linkage isomers have recently been reported. Zinc(II) forms Zn–NCS and mercury(II) forms Hg–SCN bonded complexes; cadmium(II) is intermediate in character in that both Cd–SCN and Cd–NCS bonded species have been observed in solution. This was first reported by Tramer[25] on the basis of infrared and Raman spectral studies. Plane[26] has made similar observations using Raman spectra. Similar results were recently re-

(25) A. Tramer, *J. chim. Phys.*, **59**, 232 (1962); "Theory and Structure of Complex Compounds," B. Jezowska-Trzebiatowska, Ed., Pergamon Press, New York, N. Y., 1964, p. 225.
(26) R. A. Plane, private communication.

ported[27] on the basis of n.m.r. investigations. Finally, Wojcicki and Farona[28] have just reported the isolation of the linkage isomers [Mn(CO)$_5$SCN] and [Mn(CO)$_5$NCS].

(27) O. W. Howarth, R. E. Richards, and L. M. Venanzi, in press.
(28) A. Wojcicki and M. F. Farona, 147th National Meeting of the American Chemical Society, Philadelphia, Pa., April, 1964; Proceedings of the Eighth International Conference on Coordination Chemistry, Springer-Verlag, Vienna and New York, 1964, pp. 262–264.

Acknowledgments.—This research was supported by the U. S. Atomic Energy Commission, COO-1087-80. We also wish to thank Professors R. G. Pearson and D. F. Shriver for stimulating and helpful discussions. Special expressions of gratitude are due the latter, for advice and assistance with some of the spectral measurements, and Dr. A. J. Poë, who supervised a year of this research.

[Reprinted from the Journal of the American Chemical Society, **91**, 3105 (1969).]
Copyright 1969 by the American Chemical Society and reprinted by permission of the copyright owner.

27

Cation-Induced Linkage Isomerism of the Thiocyanatopentacyanocobaltate(III) Complex

Sir:

The mode of bonding of the ambidentate ligand thiocyanate to transition metals has been shown to be subject to a variety of influences; these include the nature of the metal,[1-3] the steric and electronic characteristics of other ligands in the coordination sphere,[1,4-6] and the nature of the solvent.[7] In a recent report a noncoordinated anion was shown to influence the mode of bonding in [Pd(Et$_4$dien)NCS]$^+$.[8] We now wish to report that for the complex Co(CN)$_5$SCN^{3-} the nature of the countercation determines which of the two linkage isomers is the more stable in solid materials.

The complex K$_3$[Co(CN)$_5$SCN] was first prepared several years ago and shown to be stable to isomerization in the solid state.[9] More recently, the linkage isomer K$_3$[Co(CN)$_5$NCS] has been reported.[10] We have converted each of these compounds to a tetra-*n*-butylammonium salt by extracting an aqueous solution of complex with a solution of [(*n*-C$_4$H$_9$)$_4$N]Cl in methylene chloride, evaporating the organic layer, and recrystallizing the residue from methylene chloride–ether. We have also converted K$_3$[Co(CN)$_5$SCN] to a tetra-*n*-butylammonium salt by preparing the acid H$_3$[Co(CN)$_5$SCN], titrating it to neutral pH with [(*n*-C$_4$H$_9$)$_4$N]OH, and removing the water under reduced pressure.

The crystalline (*n*-C$_4$H$_9$)$_4$N$^+$ salts prepared by the three pathways have identical electronic absorption spectra in aqueous solution (Table I) and identical

Table I. Electronic Spectra of Co(CN)$_5$NCS^{3-} and Co(CN)$_5$SCN^{3-} in Aqueous Solution

Compound	λ_{max}, nm	ϵ_{max}
[(*n*-C$_4$H$_9$)$_4$N]$_3$[Co(CN)$_5$NCS]	363	500
	265	2,350
	202	28,000
K$_3$[Co(CN)$_5$NCS]	363	447
	265	2,360
	203	24,000
K$_3$[Co(CN)$_5$SCN]	378	191
	265	17,100
	227	4,300
	200	16,700
K$_3$[Co(CN)$_5$CNS]^a	363	260
	263	13,800

^{a} Spectrum of the compound reported in ref 10. This may be interpreted as a spectrum of a mixture of about 70% K$_3$[Co(CN)$_5$SCN] and about 30% K$_3$[Co(CN)$_5$NCS].

infrared spectra in the C≡N stretching region (2140 cm^{-1}, s, br; 2113 cm^{-1}, vs) and may therefore be presumed to contain the same anionic linkage isomer. An unambiguous assignment of the mode of bonding in this compound is complicated by the presence of absorptions of (*n*-C$_4$H$_9$)$_4$N$^+$ in the region of the infrared spectrum where the C–S stretching band is expected and absorptions of the [Co^{III}(CN)$_5$] unit in the regions of the other thiocyanate fundamentals. However, a potassium salt, prepared by metathesis of the compound with KNCS in ethanol, shows an infrared absorption at 812 cm^{-1} assignable[1,2] to the C–S stretching mode of a nitrogen-bound thiocyanate; this potassium salt must then be the K$_3$[Co(CN)$_5$NCS] isomer. The electronic spectra of K$_3$[Co(CN)$_5$NCS] and the tetra-*n*-butylammonium salt are essentially the same and both are considerably different from the spectrum of K$_3$[Co(CN)$_5$SCN]. Thus the crystalline tetra-*n*-butylammonium salt must contain exclusively the isothiocyanatopentacyanocobaltate(III) anion, Co(CN)$_5$-NCS^{3-}.[11]

In aqueous solution the linkage isomerization of either Co(CN)$_5$NCS^{3-} or Co(CN)$_5$SCN^{3-} is a relatively

(1) A. Turco and C. Pecile, *Nature*, **191**, 66 (1961).
(2) J. Lewis, R. S. Nyholm, and P. W. Smith, *J. Chem. Soc.*, 4590 (1961).
(3) M. K. Chamberlain and J. C. Bailar, Jr., *J. Am. Chem. Soc.*, **81**, 6412 (1959).
(4) A. Sabatini and I. Bertini, *Inorg. Chem.*, **4**, 1665 (1965); **5**, 1025 (1966).
(5) J. L. Burmeister and F. Basolo, *ibid.*, **3**, 1587 (1964).
(6) M. F. Farona and A. Wojcicki, *ibid.*, **4**, 1402 (1965).
(7) M. F. Farona and A. Wojcicki, *ibid.*, **4**, 857 (1965).
(8) J. L. Burmeister and J. C. Lim, *Chem. Commun.*, 1346 (1968); J. L. Burmeister, H. J. Gysling, and J. C. Lim, *J. Am. Chem. Soc.*, **91**, 44 (1969).

(9) J. L. Burmeister, *Inorg. Chem.*, **3**, 919 (1964).
(10) I. Stotz, W. K. Wilmarth, and A. Haim, *ibid.*, **7**, 1250 (1968).
(11) Satisfactory elemental analytical data have been obtained for all compounds reported here.

Communications to the Editor

3106

slow process. Electronic absorption spectra of aqueous solutions of the two isomers show only minor changes over a 24-hr period at 25°. It is also interesting that $Co(CN)_5^{3-}$ does not appear to affect significantly the isomerization rates; in an aqueous solution 0.001 M in complex and *ca.* 0.0005 M in $Co(CN)_5^{3-}$, the only spectral changes noted after 2 hr could be attributed to the expected[12] Co(II)-catalyzed formation of $Co(CN)_6^{3-}$. A comparison of the molar extinction coefficients reported here for the electronic absorption spectrum of $K_3[Co(CN)_5NCS]$ with those reported by Stotz, Wilmarth, and Haim[10] indicates that their compound, formed by the direct substitution of NCS^- into $Co(CN)_5H_2O^{2-}$, is a mixture of about 70% $Co(CN)_5SCN^{3-}$ and about 30% $Co(CN)_5NCS^{3-}$. This conclusion is supported by cyclic voltammetric studies which show distinctly different electrochemical behavior for $Co(CN)_5SCN^{3-}$ and $Co(CN)_5NCS^{3-}$, whereas the product prepared by the substitution procedure[10] gives overlapping waves which indicate that both isomers are present.[13,14]

It has been reported that $K_3[Co(CN)_5NCS]$ isomerizes to $K_3[Co(CN)_5SCN]$ on heating.[10] The thermogram of a sample of $K_3[Co(CN)_5NCS]$ prepared by our procedure shows two irreversible endotherms between 25 and 200°. The infrared spectra of KBr pellets of the original compound and of a sample which had been heated to 100° (before the onset of the first endotherm) are identical. Infrared spectra of samples heated to 140° (before the onset of the second endotherm) and to

200° (after the second endotherm) are identical with each other and with the spectrum of $K_3[Co(CN)_5SCN]$. In addition, the electronic spectrum of a sample heated to 200° is identical with that of $K_3[Co(CN)_5SCN]$. The peak at 128° thus represents a phase transition associated with the isomerization of $K_3[Co(CN)_5NCS]$ to $K_3[Co(CN)_5SCN]$. This is not a reversible transition, and we conclude that $K_3[Co(CN)_5SCN]$ is more stable than $K_3[Co(CN)_5NCS]$ in solid samples.[15]

In contrast to the situation with K^+ as counterion, it is apparently not possible at 25° to prepare a pure sample of $[(n-C_4H_9)_4N]_3[Co(CN)_5SCN]$. A precipitate that contains mainly $[(n-C_4H_9)_4N]_3[Co(CN)_5SCN]$ can be obtained by concentrating an aqueous solution at 0°; this material completely isomerizes to $[(n-C_4H_9)_4N]_3[Co(CN)_5NCS]$ at room temperature within 3 days. It is also possible by rapid evaporation of a fresh $(n-C_4H_9)_4N^+$–CH_2Cl_2 extract of $Co(CN)_5SCN^{3-}$ to obtain an oily residue that contains mainly the S-bonded isomer; however, no S-bonded isomer is left after the oily material is crystallized.

The role the countercation plays in the isomerization is of considerable interest. We suggest that the stabilization of the N-bonded isomer is due to an electronic effect in which the polarizable end of $-NCS^-$ is better accommodated by the nonpolar, hydrocarbon environment of the $(n-C_4H_9)_4N^+$ counterion. We further suggest that the stabilization of the S-bonded form in the compound containing K^+ as counterion is due to favorable interaction with the hard end of coordinated $-SCN^-$.[16] We are hopeful that the counterion stabilization feature, which allowed us to prepare materials containing exclusively the $Co(CN)_5NCS^{3-}$ isomer, will have further synthetic utility in this field.

Acknowledgment. We thank the National Science Foundation for support of this research.

(12) J. P. Birk and J. Halpern, *J. Am. Chem. Soc.*, **90**, 305 (1968).

(13) H. S. Lim and F. C. Anson, to be submitted for publication.

(14) We have allowed aqueous solutions containing pure samples of each of the linkage isomers to come to equilibrium. At 40°, the position of equilibrium as measured from electronic absorption spectra corresponds to about two times as much $Co(CN)_5SCN^{3-}$ as $Co(CN)_5NCS^{3-}$. Thus our results indicate that the product formed[10] by the NCS^- substitution into $Co(CN)_5H_2O^{2-}$ at 40° is essentially the appropriate equilibrium mixture for the given conditions. It is also of interest that in CH_2Cl_2 at 40° the equilibrium position is far to the side of the $Co(CN)_5NCS^{3-}$ isomer and is reached starting from $Co(CN)_5SCN^{3-}$ in about 4 hr.

(15) In the thermogram of $K_3[Co(CN)_5SCN]$ there is an irreversible endotherm at about 150°. Infrared spectra of the original compound and the heated sample are identical. This peak may represent the loss of the small amount of solvent which the elemental analysis indicates is present.

(16) A similar interpretation could be put forward to explain the observed[14] solvent effect on the relative stabilities of the two isomers.

Diane F. Gutterman, Harry B. Gray
Contribution No. 3813, The Arthur Amos Noyes Laboratory of Chemical Physics, California Institute of Technology Pasadena, California 91109
Received March 3, 1969

[Reprinted from Inorganic Chemistry, 11, 1686 (1972).]

CONTRIBUTION FROM THE CENTRAL RESEARCH DEPARTMENT,
MONSANTO COMPANY, ST. LOUIS, MISSOURI 63166

Relative Stabilities of Some Halide Complexes of Rhodium and Iridium

28

BY DENIS FORSTER

Received December 29, 1971

Halide-exchange equilibria in the systems $Rh(CO)_2I_2^--Cl^-$, $Ir(CO)_2I_2^--Cl^-$, $Ir(CO)_2I_2^--Br^-$, $Rh(Ph_3P)_2(CO)I-Cl^-$, and $Ir(Ph_3P)_2(CO)I-Cl^-$ have been investigated spectrophotometrically in both aqueous and nonaqueous media. Overall replacement constants have been determined for the systems and show that the lighter halides are preferred in nonaqueous solvents in all cases. In the aqueous media this preference is inverted for the anionic complexes but is retained for the phosphine complexes. Qualitative equilibrium data in nonaqueous solvents have also been obtained for the systems $Ir(CO)_2Cl_4^--I^-$, $Rh(CO)_2I_5^2--Cl^-$, and $Rh(CO)I_5-Br^-$ by infrared spectroscopy. The rhodium(III) systems show a preference for the heavier halide. The iridium(III) anion displays novel ambivalent acceptor behavior, displaying soft acceptor halide preference at two halide positions and hard acceptor behavior at the other halide positions (probably trans to the carbonyls).

Rhodium and iridium are regarded as soft acceptors and thus rhodium has been observed to prefer to bind heavier halides over lighter halides in its complexes in aqueous media.[1,2] However, the equilibrium constants observed for the rhodium systems were sufficiently small that it was apparent[2] that the halide solvation energy was most likely the dominant factor in the equilibria. No studies of relative halide complex stability in nonaqueous solvents have been reported and solvation effects are frequently ignored in consideration of the halide preference exhibited by a particular metal. We, therefore, undertook a limited study of halide exchange and preference in rhodium and iridium systems, in both protic and aprotic media.

Experimental Section

Electronic spectra were recorded with a Coleman-Hitachi 124 spectrophotometer. Infrared spectra were recorded with a Beckman IR-12 spectrophotometer.

Preparation of Complexes.—The following compounds were prepared according to the references indicated: $[(C_6H_5)_4As]$-$[Ir(CO)_2Cl_2]$, $[(C_6H_5)_4As][Ir(CO)_2Br_2]$, and $[(C_6H_5)_4As][Ir(CO)_2$-$I_2]$;[3] $[(C_6H_5)_4As][Rh(CO)_2Cl_2]$, $[(C_6H_5)_4As][Rh(CO)_2I_2]$, and $[(C_4H_9)_4N][Rh(CO)I_4]$;[4] $[Ir([C_6H_5]_3P)_2(CO)Cl]$;[5] $[Ir([C_6H_5]_3P)_2$-$(CO)I]$;[6] $[(C_2H_5)_4N][Ir(CO)_2I_4]$ and $[(C_2H_5)_4N][Ir(CO)_2Cl_4]$;[7] $[(C_2H_5)_4N]_2[Rh(CO)_2Cl_5]$;[8] $Rh[(C_6H_5)_3P]_2(CO)Cl$;[9] and Rh-$[(C_6H_5)_3P]_2(CO)I$.[10]

$[(C_2H_5)_4N][Ir(CO)_2Cl_2I_2]$.—$[(C_6H_5)_4As][Ir(CO)_2Cl_2]$ (0.5 g) and $(C_2H_5)_4NCl$ (0.15 g) were dissolved in chloroform (3 ml). Iodine (0.20 g) was dissolved in the minimum volume of cold chloroform and added to the above solution. Dark red crystals rapidly separated and were filtered off, washed with ethanol and diethyl ether, and air-dried; yield 0.25 g. *Anal.* Calcd for $C_{10}H_{20}Cl_2$-I_2IrN: C, 17.08; H, 2.87; Cl, 10.08; I, 36.08. Found: C, 17.53; H, 2.74; Cl, 9.71; I, 35.77.

Determination of Equilibrium Constants.—The preferred halide in a particular rhodium(I) or iridium(I) system was first established qualitatively and then the complex containing the preferred halide was studied in solutions containing known excess amounts of the less preferred halide (added as the tetrabutylammonium salt). Equilibrium data are expressed as overall replacement constants (γ_i), *i.e.*

$$\gamma_i = \frac{[MA_i][B]^i}{[MB_i][A]^i}$$

for $MB_i + iA \rightleftharpoons MA_i + iB$.

(1) H. L. Bott and A. J. Poë, *J. Chem. Soc.*, 5931 (1965).
(2) E. J. Bounsall and A. J. Poë, *J. Chem. Soc. A*, 286 (1966).
(3) D. Forster, *Inorg. Nucl. Chem. Lett.*, **5**, 433 (1969).
(4) L. M. Vallarino, *Inorg. Chem.*, **4**, 161 (1965).
(5) M. Kubota, *Inorg. Syn.*, **11**, 101 (1968).
(6) P. B. Chock and J. Halpern, *J. Amer. Chem. Soc.*, **88**, 3510 (1966).
(7) D. Forster, *Syn. Inorg. Metal-Org. Chem.*, **1**, 221 (1971).
(8) D. Forster, *Inorg. Chem.*, **8**, 2556 (1969).
(9) J. A. McCleverty and G. Wilkinson, *Inorg. Syn.*, **8**, 214 (1966).
(10) L. M. Vallarino, *J. Chem. Soc.*, 2287 (1957).

The following absorption peaks were used to determine the concentration of the various species studied: $Ir(CO)_2Cl_2^-$, λ_{max} 352 mμ (ϵ_{max} 2.8 × 10³), 317 mμ (ϵ_{max} 2.7 × 10³); $Ir(CO)_2$-Br_2^-, λ_{max} 355 mμ (ϵ_{max} 2.7 × 10³), 312 mμ (ϵ_{max} 3.0 × 10³); $Ir(CO)_2I_2^-$, λ_{max} 342 mμ (ϵ_{max} 3.7 × 10³); $Rh(CO)_2Cl_2^-$, λ_{max} 334 mμ (ϵ_{max} 3.2 × 10³); $Rh(CO)_2I_2^-$, λ_{max} 290 mμ (ϵ_{max} 2.0 × 10³); $Rh(Ph_3P)_2(CO)Cl$, λ_{max} 365 mμ (ϵ_{max} 3.2 × 10³), ϵ at 335 mμ is 3.0 × 10³; $Rh(Ph_3P)_2COI$, λ_{max} 360 mμ (ϵ_{max} 3.5 × 10³), ϵ at 335 mμ is 4.5 × 10³; $Ir(Ph_3P)_2COCl$, λ_{max} 385 mμ (ϵ_{max} 2.5 × 10³), 336 mμ (ϵ_{max} 2.9 × 10³); $Ir(Ph_3P)_2COI$, λ_{max} 395 mμ (ϵ_{max} 2.1 × 10³), 350 mμ (ϵ_{max} 2.4 × 10³). No attempt was made to determine concentrations of intermediate species. Equilibration was essentially instantaneous with all of the above systems.

Equilibrium positions in the iridium(III) and rhodium(III) systems studied were determined, qualitatively, by observation of the infrared spectra in the carbonyl region of the complexes dissolved in nonaqueous solvents containing equal amounts of the competing halides.

Results and Discussion

The results obtained for the iridium(I) and rhodium-(I) systems are summarized in Table I. It can be seen that the lighter halides are strongly preferred in all cases in nonaqueous solvents. These results presumably reflect the higher bond strength of metal-chloride and metal-bromide bonds when compared with metal-iodide bonds.[11] The importance of the relative halide ion solvation energies in these equilibria is illustrated by the inversion of halide preference which occurs with the anionic complexes in CH_3CN-H_2O.[12] In the case of the neutral phosphine complexes, the preference for the lighter halide is much reduced in the protic medium as compared to the aprotic medium but an inversion is not observed. This observation is noteworthy with respect to the preparative techniques used for $Rh(Ph_3P)_2(CO)I$[10] and $Ir(Ph_3P)_2(CO)I$[6] which both involve halide exchange with sodium iodide. In both cases it appears that the insolubility of sodium chloride in the reaction medium is actually the principal driving force rather than the halide preference of the transition metal.

The results obtained by infrared measurements with the systems $RhX_5(CO)^{2-} + Y^-$ [X and Y are either (a) iodide or chloride or (b) iodide or bromide] and IrX_4-$(CO)_2^- + Y^-$ (X and Y are iodide or chloride) are summarized in Scheme I. It is apparent that, in both rhodium systems studied, iodide is preferred in all

(11) R. G. Pearson and R. J. Mawby, "Halogen Chemistry," Vol. 3, V. Gutmann, Ed., Academic Press, New York, N. Y., 1967, p 55.
(12) The hardness of the position trans to the carbonyl in $[M(Ph_3P)_2$-$(CO)X]$ is also reflected in the bonding mode adopted by pseudohalides: see J. L. Burmeister and N. J. DeStefano, *Chem. Commun.*, 1698 (1970).

TABLE I
HALIDE PREFERENCES IN SOME RHODIUM(I)
AND IRIDIUM(I) SYSTEMS

System	Solvent	Favored halide	Replacement constant $(\pm 20\%)^a$
$Rh(CO)_2Cl_2^--I^-$	1,2-Dichloroethane	Cl^-	80
$Rh(CO)_2Cl_2^--I^-$	CH_3CN	Cl^-	200
$Rh(CO)_2Cl_2-I^-$	CH_3CN-H_2O (10% H_2O)	I^-	30
$Ir(CO)_2Cl_2^--I^-$	1,2-Dichloroethane	Cl^-	5000
$Ir(CO)_2Cl_2^--I^-$	CH_3CN	Cl^-	400
$Ir(CO)_2Cl_2^--I^-$	CH_3CN-H_2O (10% H_2O)	I^-	20
$Ir(CO)_2Br_2^--I^-$	1,2-Dichloroethane	Br^-	11
$Ir(CO)_2Br_2^--I^-$	CH_3CN	Br^-	50
$Ir(CO)_2Br_2^--I^-$	CH_3CN-H_2O (10% H_2O)	I^-	60
$[Rh(Ph_3P)_2(CO)Cl]-I^-$	DMF	Cl^-	250
$[Rh(Ph_3P)_2(CO)Cl]-I^-$	DMF-H_2O (10% H_2O)	Cl^-	20
$[Ir(Ph_3P)_2(CO)Cl]-I^-$	1,2-Dichloroethane	Cl^-	$>10^3$
$[Ir(Ph_3P)_2(CO)Cl]-I^-$	DMF-H_2O (10% H_2O)	Cl^-	60

^{a} Expressed with the preferred halide ion on the left-hand side of the equilibrium.

SCHEME I
INFRARED STUDIES OF HALIDE EXCHANGE IN THE
SYSTEMS $RhX_5(CO)^{2-}$ + Y^- AND $IrX_4(CO)_2^-$ + Y^-
IN NONAQUEOUS SOLVENTS (WHERE X AND Y
ARE DIFFERENT HALIDES)^a

Iridium

$IrI_4(CO)_2^-$ + $2Cl^-$ $\xrightarrow[CH_3NO_2]{\text{room temp}}$
2115, 2070^b

$IrI_2Cl_2(CO)_2^-$ $\underset{CH_3NO_2}{\overset{75°, 2 \text{ hr}}{\rightleftharpoons}}$ $IrCl_4(CO)_2^-$ + $2I^-$
2129, 2081   2143, 2094

$\uparrow CH_3NO_2$

$IrCl_2(CO)_2^-$ + I_2

Rhodium

1. $RhI_4(CO)^-$ + $10I^-$ $\xrightarrow[1 \text{ min}]{\text{room temp}}$ $RhI_5(CO)^{2-}$ $\underset{1.5 \text{ hr, } 80°}{\overset{CH_3NO_2}{\rightleftharpoons}}$
2076   2047

$\underset{80°}{\overset{1.5 \text{ hr}}{\uparrow}}$ $RhCl_5(CO)^{2-}$ + 2087 $5Cl^-$ + $10I^-$

$RhI_4(CO)^-$ + $14Cl^-$ + $10I^-$

$RhI_4(CO)^-$ + $10Cl^-$ $\xrightarrow[1 \text{ min}]{\text{room temp}}$ $RhI_4Cl(CO)^{2-}?$
2076   2057

(Similar results are observed in CH_3CN)

2. $RhBr_5(CO)^{2-}$ + $15Br^-$ + $20I^-$ $\xrightarrow[2 \text{ hr, } 75°]{CH_3NO_2}$
2072

$RhI_5(CO)^{2-}$ $\underset{2 \text{ hr, } 75°}{\overset{CH_3NO_2}{\rightleftharpoons}}$ $RhI_4(CO)^-$ + $16I^-$ + $20Br$
2047   2076

^{a} Stoichiometries in the equations refer to molar ratios used in the experiments. ^{b} Carbonyl frequencies (cm^{-1}) throughout the scheme refer to solution measurements.

halide coordination sites in the two aprotic solvents studied. These results contrast markedly with the results described above for rhodium(I) systems in nonaqueous solvents. This finding is somewhat unexpected since it has been predicted that higher valence states of metals would behave "harder" than lower

valence states,[13] although the differing charges and stereochemistry of the species involved make comparisons of doubtful value in this case.

The results obtained with the iridium(III) tetrahalodicarbonyl anions are particularly intriguing. The following conclusions may be reached from the experiments described in Scheme I: (1) two of the halide acceptor sites on iridium in $IrX_4(CO)_2^-$ prefer a lighter halide (Cl^-) over a heavier halide (I^-); (2) the other two halide acceptor sites prefer the heavier halide (I^-) over the lighter halide (Cl^-);[14] (3) two of the halide positions exchange rapidly at room temperature whereas the other two exchange very slowly at room temperature and require elevated temperatures to effect facile exchange. The two halide positions which exchange rapidly at room temperature are most probably those trans to the carbonyl groups in view of the well known trans-labilizing influence of carbonyls. These considerations strongly suggest that the species observed with the carbonyl bands at 2129 and 2081 cm^{-1} is $IrCl_2I_2(CO)_2^-$ with both chlorides trans to the carbonyl groups. Thus the equilibrium position in the system studied with equal amounts of chloride and iodide (in excess) available to the iridium is such that two chlorides and two iodides are coordinated. It is thus apparent that the iridium(III) is displaying "ambivalent"[15] acceptor behavior in the solvent used, with the positions trans to the carbonyls being "hardened" relative to the positions cis to the carbonyls.

In view of the above results obtained with the iridium(III) system it is worth reconsidering the rhodium(III) systems where only one type of acceptor behavior was found. It seems probable that there is a difference in relative "hardness" or "softness" of the two sites in $RhX_5(CO)^{2-}$ also but that a different solvent might be required to allow the rhodium to manifest "ambivalent" acceptor behavior. However, the greater electron density on the metal in the dinegative rhodium anion may serve to lessen differences between the acceptor sites since it is likely that the marked difference in the iridium system is brought about by the electron-withdrawing properties of the carbonyl groups.

Fluorocarbonyl derivatives of transition metals are rare but the trends found in this work suggest that in rigorously aprotic solvents, fluoro derivatives may frequently be found to be the most stable halocarbonyl species.

Acknowledgment.—The author wishes to thank G. V. Johnson and R. E. Miller for experimental assistance and Professor Jack Halpern for helpful discussions.

(13) R. G. Pearson, *J. Amer. Chem. Soc.*, **85**, 3433 (1963).

(14) Attempts were made to determine the equilibrium constants of these systems accurately for more direct comparison with the iridium(I) systems but apparently reactions of the type $IrX_4(CO)_2^-$ + Y^- → $IrX_2(CO)_2^-$ + YX_2^- become important with relatively small excesses of halide in these systems. The overlap of electronic spectra of the resulting mixture of iridium(I) and iridium(III) complexes made quantification difficult. A similar reaction in which halide ion reduces $Pt(CN)_4X_2^-$ ions has been reported previously: A. J. Poë and D. H. Vaughan, *Inorg. Chim. Acta*, **2**, 159 (1968).

(15) Ambivalent acceptor is used here to mean an acceptor displaying *simultaneous* hard and soft acceptor behavior.

J. inorg. nucl. Chem., 1971, Vol. 33, pp. 413 to 419.   Pergamon Press.   Printed in Great Britain

THE HARD AND SOFT ACID–BASE PRINCIPLE AND METAL ION ASSISTED LIGAND SUBSTITUTION PROCESSES

MARK M. JONES and HOWELL R. CLARK

Department of Chemistry, Vanderbilt University, Nashville, Tenn. 37212

(*Received* 17 *July* 1970)

Abstract—The hard and soft acid–base principle is shown to be a reliable guide of considerable generality in the correlation of metal ion assisted and metal ion catalyzed ligand substitution reactions of the type

$$M_{(1)}\text{–}X + M_{(2)}\text{–}L \xrightarrow{L} M_{(1)}\text{–}L + M_{(2)}\text{–}X.$$

Numerous examples of reactions of this type which have been reported in the literature are shown to be easily explained on the basis of a rearrangement to a more effective matching up of the hard and soft properties of the metal ions and ligands involved. Use of this principle also leads to the expectation that a very large number of such reactions should be possible, and some previously reported examples are presented to show the predictive advantages of such an approach.

THE USE of the principle of hard and soft acids and bases has been shown by Pearson and his colleagues to lead to a very effective ordering of a wide range of inorganic and organic reactions[1]. Subsequently, this principle has been used by others to correlate thermodynamic parameters[2] and other phenomena[3]. While some details of the principle, such as the terminology, have been subjected to criticism, the utility of the ideas it summarizes has been established. For the present purposes it is only necessary to note that Pearson has introduced the notion that the *rates* of displacement reactions should correlate with the acid–base characteristics of the central atom and the exchanged groups. This has been cast in a more definite form for organic reactions by Saville[1c], who has presented this principle in the form of two rules which give the directions in which acid–base reactions can be expected to proceed. These can be summarized as

Rule 1

$$Z \quad A\text{------}X \quad E \longrightarrow Z\text{------}A + X\text{------}E$$

Z	A	X	E	Z	A	X	E
hard	hard	soft	soft	hard	hard	soft	soft
base	acid	base	acid	base	acid	base	acid

1. (a) R. G. Pearson, *J. Am. chem. Soc.* **85**, 3533 (1963); *idem. Science, Lond.* **151**, 172 (1966); *idem. Chemistry in Britain* **3**, 103 (1967); *idem. J. Chem. Educ.* **45**, 581, 643 (1968); (b) R. G. Pearson and J. Songstad, *J. Am. chem. Soc.* **89**, 1827 (1967); (c) B. Saville, *Angew. Chem.* **6**, 928 (1967).
2. S. Ahrland, *Structure and Bonding* **5**, 118 (1968).
3. *Symposium: Structure and Bonding* **1**, 207 (1966).

413

Rule 2

$$Z \quad A\text{---}\!X \quad E \longrightarrow Z\text{---}\!A + X\text{---}\!E$$

| soft | soft | hard | hard | soft | soft | hard | hard |
| base | acid | base | acid | base | acid | base | acid |

These rules suggest that this same principle can function as a guide to metal ion assisted and metal ion catalyzed ligand substitution processes. Pearson's viewpoint on this, which emphasizes the rate aspects of such reactions, implies that the *rates* of such reactions are also governed by the nature of the acid–base interactions involved. This leads naturally to the examination of this hypothesis to see if it does, in fact, correlate the available data on metal ion assisted and metal ion catalyzed ligand substitution processes.

The use of these different categories of acids and bases to explain the interaction in metal catalyzed ligand substitution processes is possible even if one ignores their designations as "hard" or "soft". This is because the categories have an extensive empirical basis quite independent of any particular theory which interprets acid–base behavior in more detailed terms.

The ability of one metal ion to influence the course of the substitution reactions of a complex of another metal ion has been recognized, somewhat vaguely, for many years. It was realized over 60 years ago that certain cobalt(III) complexes react with silver nitrate solutions such that both ionic and coordinated chloride are rapidly and completely precipitated as silver chloride. For example, aqueous solutions of both $[Co(NH_3)_3Cl_3]$ and $[Co(NH_3)_3(H_2O)Cl_2]Cl$ react immediately with silver nitrate solution to give a silver chloride precipitate which contains all their chloride[4]. Subsequently it was found that the reaction of solutions of $[Cr(H_2O)_4Cl_2]Cl \cdot 6H_2O$ with silver nitrate solutions gave more silver chloride than could be accounted for by the uncoordinated chloride alone[5]. Since this reaction with Ag(I) was proposed as a test to differentiate between ionic and coordinated chloride, the results were attributed to the hydrolysis of the coordinated chloride. In retrospect, it seems that at least some of the results were the consequence of a silver(I) promoted hydrolysis[6], especially in the case of $[Co(NH_3)_3(OH_2)Cl_2]Cl$, where all of the chloride was immediately precipitated as silver chloride.

The uncertain results obtained in the use of silver nitrate solution to distinguish between coordinated and ionic halides is seen very clearly in early studies on $[Cr(H_2O)_4Cl_2]Cl$ and $[Cr(H_2O)_4Br_2]Br[7]$. Here it is found that the amount of silver halide precipitated "immediately" increased as the amount of silver nitrate or silver perchlorate used in the test was increased. Furthermore, the amount of coordinated bromide lost was considerably greater than that of coordinated chloride. This is readily explainable by the "softer" nature of bromide and the greater K_{sp} of silver bromide.

4. S. M. Jorgensen, *Z. anorg. allg. Chem.* **5**, 188 (1894).
5. A. Werner and A. Gubser, *Chem. Ber.* **39**, 1828 (1906).
6. G. C. Lalor and D. S. Rustad, *J. inorg. nucl. Chem.* **31**, 3219 (1969); P. J. Elving and B. Zemel, *J. Am. chem. Soc.* **79**, 5855 (1957).
7. R. F. Weinland and A. Koch, *Z. anorg. allg. Chem.* **39**, 298 (1904).

In subsequent years various reactions of this type have been studied in more detail, but they have not previously been considered to constitute a category of sufficient homogeneity to be examined as a class.

A relatively few reactions of this category have received rather detailed scrutiny largely because they furnish test cases convenient for the examination of ionic strength effects on reaction rates or for testing theories of octahedral substitution. The first of these on which a kinetic study was carried out is the reaction[8]

$$Co(NH_3)_5Cl^{2+} + Hg^{2+} \longrightarrow Co(NH_3)_5(H_2O)^{3+} + HgCl^+.$$

Subsequently, Brønsted and Livingston[9] made a detailed kinetic study of the reaction

$$Co(NH_3)_5Br^{2+} + Hg^{2+} \longrightarrow Co(NH_3)_5(H_2O)^{3+} + HgBr^+$$

as an example of a reaction of two bipositive ions. The reaction has been examined for the same purpose more recently[10].

As will be documented presently, there are a reasonable number of such reactions. When they are examined for an underlying principle governing their occurrence, it soon becomes apparent that they fall very neatly into the categories anticipated from the principle of hard and soft acids and bases. This principle leads to the formulation of all reactions of this type as specific examples of the reaction

$$M_{(1)} - X + M_{(2)} - L \longrightarrow M_{(1)} - L + M_{(2)} - X$$

where $M_{(1)}$ and $M_{(2)}$ are Lewis acids and X and L Lewis bases, and $M_{(2)}$ is present in a labile complex. For the reaction to proceed spontaneously and rapidly, the principle of hard and soft acids and bases requires that the interactions in the products be more closely matched than in the reactants insofar as their hard and soft nature is concerned. To be more specific, if $M_{(1)}$ and L are hard and $M_{(2)}$ and X are soft, the reaction can be expected to be reasonably rapid if $M_{(2)}$ forms labile complexes. Reactions of this sort are metal catalyzed or, more accurately, metal *assisted* substitution reactions. The fact that a variety of these are known leads to the following principle:

Ligand substitution reactions in which a soft ligand is displaced from a hard acid will be assisted (catalyzed) by the addition of a soft acid coordinated to a harder base. Those in which a hard ligand is to be displaced from a soft acid will be assisted (catalyzed) by the addition of a hard acid coordinated to a softer base. In either case the assisting acid cation must be present in a labile complex.

While this principle implies a mechanism in which the assisting cation $M_{(2)}$ forms a weak bond with X prior to its removal from $M_{(1)}$, the overall process may

8. J. N. Brønsted and C. E. Teeter, *J. phys. Chem.* **28**, 583 (1924).
9. J. N. Brønsted and R. Livingston, *J. Am. chem. Soc.* **49**, 435 (1927).
10. A. R. Olsen and T. Simonson, *J. chem. Phys.* **17**, 1167 (1949).

Table 1. Metal ion assisted substitution reactions

Reactant	Product	Metal ion assisting	Species removed	Ref.
$Co(NH_3)_5Cl^{2+}$	$Co(NH_3)_5(OH_2)^{3+}$	Hg^{2+}	Cl^-	[8]
$Co(NH_3)_5Br^{2+}$	$Co(NH_3)_5(OH_2)^{3+}$	Hg^{2+}	Br^-	[9, 10]
$Co(NH_3)_5I^{2+}$	$Co(NH_3)_5(OH_2)^{3+}$	Ag^+	I^-	[6]
$Co(NH_3)_5X^{2+}$	$Co(NH_3)_5(OH_2)^{3+}$	Ag_+, Tl^{3+}, Hg^{2+}	X^-	[11]
$trans$-$Co(trien)Cl_2^+$	$trans$-$Co(trien)(OH_2)Cl^{2+}$	Hg^{2+}	Cl^-	[12, 14]
$trans$-$Co(en)_2Cl_2^+$	$trans$-$Co(en)_2(H_2O)Cl^{2+}$	Hg^{2+}	Cl^-	[13, 14]
$(+)Co(en)_2Cl_2^+$	{70% $(+)[Co(en)_2(H_2O)Cl]^{2+}$ 30% $trans$-$[Co(en)_2(H_2O)Cl]^{2+}$	Hg^{2+}	Cl^-	[14]
$(+)[Co(trien)Cl_2]^+$	$(+)[Co(trien)(H_2O)Cl]^{2+}$	Hg^{2+}	Cl^-	[14]
$(+)[Co(trien)Cl_2]^+$	$(+)[Co(trien)(H_2O)Cl]^{2+}$	Hg^{2+}	Cl^-	[14]
cis-$[Co(en)_2Cl(H_2O)]^{2+}$	cis-$[Co(en)_2(H_2O)_2]^{3+}$	Hg^{2+}	Cl^-	[15]
1-cis-$[Co(en)_2Cl(H_2O)]^{2+}$	1-cis-$[Co(en)_2(H_2O)_2]^{3+}$	Hg^{2+}	Cl^-	[15]
cis-$[Co(en)_2Br(H_2O)]^{2+}$	cis-$[Co(en)_2(H_2O)_2]^{3+}$	Hg^{2+}	Br^-	[15]
$trans$-$[Co(en)_2(N_3)_2]^+$	$trans$-$[Co(en)_2(N_3)(H_2O)]^+$	Hg^{2+}	N_3^-	[15]
cis and $trans$-$[CoA_4XCl]^{n+}$	various $[CoA_4X(OH_2)]^{n+1}$	Hg^{2+}	Cl^-	[16]
$AsF_5(OH)^-$	H_3AsO_4	Fe^{3+}	F^-	[17]
$Fe(CN)_6^{2-}$	$Fe(CN)_5(OH_2)^-$	Hg^{2+}	CN^-	[18]
$Fe(CN)_6^{2-}$	$Fe(CN)_5(OH_2)^-$	$Pt^{4+} Au^{3+}$	CN^-	[19]
cis-$[Co(en)_2(RNH_2)Cl]^{2+}$	cis-$[Co(en)_2(RNH_2)(H_2O)]^{3+}$	Hg_2^{2+}, Hg^{2+}	CL^-	[20]
cis-$[Co(en)_2(NH_3)Br]^{2+}$	cis-$[Co(en)_2(NH_3)(H_2O)]^{3+}$	Hg^{2+}	Br^-	[20]
$Co(HEDTA)Cl^-$	$Co(EDTA)^-$	17 cations mostly divalent	Cl^-	[21]
$Co(HEDTA)Br^-$	$Co(EDTA)^-$	Pb^{2+}	Br^-	[22]
$Co(HEDTA)Br^-$	$Co(EDTA)^-$	Ag^+, Hg^{2+}	Br^-	[23]
$Co(HEDTA)Cl^-$	$Co(EDTA)^-$	18 cations mostly trivalent	Cl^-	[24]
$Cr(H_2O)_5Cl^{2+}$	$Cr(H_2O)_6^{3+}$	Ag^+	Cl^-	[25]
$Cr(H_2O)_4Cl_2^+$	$Cr(H_2O)_6^{3+}$	Ag^+	Cl^-	[25]
$Cr(H_2O)_5Cl^{2+}$	$Cr(H_2O)_6^{3+}$	Hg^{2+}	Cl^-	[26]
$Cr(NH_3)_5Cl^{2+}$	$Cr(NH_3)_5(H_2O)^{3+}$	Hg^{2+}	Cl^-	[27]
$C_6H_5CH_2F$	$C_6H_5CH_2OH$	$Th^{4+}, Zr^{4+}, Al^{3+}$	F^-	[28]
AsF_6^-	H_3AsO_4	Th^{4+}, Al^{3+}	F^-	[29]

Reactant	Product	Acid	Base	Ref.
BF_4^-	H_3BO_3	$Th^{4+}, Zr^{4+}, Al^{3+}, Be^{2+}, Ti^{4+}$	F^-	[29]
PF_6^-	H_3PO_4	$Th^{4+}, Al^{3+}, Zr^{4+}$	F^-	[29]
R_3COH	RCl	Zn^{2+}	OH^-	[32]
$[Co(en)_2(\text{amino acid ester})X]^{2+}$	$[Co(en)_2(\text{amino acid ester})(H_2O)]^{3+}$	Hg^{2+}	Cl^-, Br^-	[34]
R_2CHOH	R_2CHCl	Zn^{2+}	OH^-	[33]
$(CH_3)_3CF$	$(CH_3)_3C(OH)$	Hard Acids	F^-	[30]
$(CH_3)_2CHO{-}\overset{O}{\underset{CH_3}{P}}{-}F$	$(CH_3)_2CHO{-}\overset{O}{\underset{CH_3}{P}}{-}OH$	Complexes of Cu^{2+} and hard acid cations	F^-	[31]
$(CH_3)_2CO{-}\overset{O}{\underset{H}{P}}{-}F$	$(CH_3)_2CO{-}\overset{O}{\underset{H}{P}}{-}OH$	Cu(II) chelates	F^-	[32]
cis and $trans\text{-}[Cr(H_2O)_4Cl_2]^+$	$Cr(H_2O)_6^{3+}$	Hg^{2+}	Cl^-	[35]
$Co(NH_3)_5Br_2^+$	$Co(NH_3)_5(H_2O)^{3+}$	HgS, AgBr	Br^-	[36]
$H_2NOCCH_2CH_2CONH_2$ and other diamidines	$HOOCCH_2CH_2COOH$, etc.	$Zr^{4+}, Th^{4+}, Ce^{4+}$	NH_3	[37]
$C_6H_5CH_2Cl$	$C_6H_5CH_2OH$	Hg^{2+}	Cl^-	[38]
CH_3CN	CH_3COOH	HgO	NH_3	[39]
$RSCH_2COOR$	ROH	Ag^+	$ROOCCH_2S^-$	[40]
$R{-}\overset{O}{\underset{RO}{P}}{-}SEt$	$R{-}\overset{O}{\underset{RO}{P}}{-}F$	AgF	EtS^-	[41]
$R{-}\overset{O}{\underset{RO}{P}}{-}SR$	$RO{-}\overset{O}{\underset{RO}{P}}{-}OCH_3$	Ag^+ in CH_3OH	RS^-	[42]
Cobalt(III) corrinoid	aquo corrinoid	Ag^+	CN^-	[43]
$CH_2{=}CHCl$	$CH_2{=}CHOAc$	Pd(II)	Cl	[44]
Aromatic halide	Aromatic ether	Cu(I)	halide	[45]
Aromatic iodides	aralkyls	Cu(I)	iodide	[46]
Aromatic halides	various	Cu(I)	halide	[47]

be one in which an intermediate of lower coordination number is generated (as in an S_N1 type process). From the wide range of species which undergo such metal assisted processes it is highly probable that the metal ion can furnish assistance in either type of mechanism.

Examples

A listing of reported examples of this type is presented in Table 1. It becomes apparent that the scope of the principle is broader than might have been anticipated and is some interest as a guide to the development of new processes of this sort, including a large number of reactions which are not hydrolyses. While data currently available have been obtained for the most part in aqueous media, a few

Table 1 References

11. F. A. Posey and H. Taube, *J. Am. chem. Soc.* **79**, 255 (1957).
12. A. M. Sargeson and G. H. Searle, *Nature, Lond.* **200**, 356 (1963).
13. A. M. Sargeson, *Austral. J. Chem.* **16**, 352 (1963).
14. A. M. Sargeson, *Austral. J. Chem.* **17**, 385 (1964).
15. D. A. Loesliger and H. A. Taube, *Inorg. Chem.* **5**, 1376 (1966).
16. C. Bifano and R. G. Linck, *Inorg. Chem.* **7**, 908 (1968).
17. W. L. Johnson and M. M. Jones, *Inorg. Chem.* **5**, 1345 (1966).
18. S. Asperger, I. Murati and D. Pavlovic, *J. chem. Soc.* A, 2044 (1969).
19. S. Asperger and D. Pavlovic, *J. chem. Soc.* 1449 (1955).
20. S. C. Chan and S. F. Chan, *J. chem. Soc.* A, 202 (1969).
21. R. Dyke and W. C. E. Higginson, *J. chem. Soc.* 2788 (1963).
22. W. C. E. Higginson and M. P. Hill, *J. chem. Soc.* 1620 (1959).
23. G. Schwarzerbach, *Helv. chim. Acta* **32**, 845 (1949).
24. S. P. Tanner and W. C. E. Higginson, *J. chem. Soc.* A, 1164 (1969).
25. P. J. Elving and B. Zemel, *J. Am. chem. Soc.* **79**, 5855 (1957).
26. J. H. Espenson and J. P. Birk, *Inorg. Chem.* **4**, 527 (1965).
27. J. H. Espenson and S. R. Hubbard, *Inorg. Chem.* **5**, 686 (1966).
28. H. R. Clark and M. M. Jones, *J. Am. chem. Soc.* **91**, 4302 (1969).
29. H. R. Clark and M. M. Jones, *J. Am. chem. Soc.* **92**, 816 (1970).
30. E. S. Rudakov, I. V. Kozhevnikov and V. V. Zamashikov, *Reakts. Sposobnost Org. Soedin* **6**, 573 (1969); *C.A.* **72**, 25406m.
31. R. C. Courtney, R. L. Gustafson, S. J. Westerback, H. Hyytiainen, S. C. Chaberek, Jr. and A. E. Martell, *J. Am. chem. Soc.* **79**, 3030 (1957).
32. T Wagner-Jauregg, B. E. Hackley, Jr., T. A. Lies, O. O. Owens and R. Proper, *J. Am. chem. Soc.* **52**, 802 (1930).
33. H. J. Lucas, *J. Am. chem. Soc.* **52**, 802 (1930).
34. M. D. Alexander and D. H. Busch, *J. Am. chem. Soc.* **88**, 1130 (1966).
35. J. P. Birk, *Inorg. Chem.* **9**, 735 (1970).
36. M. D. Archer and M. Spiro, *J. chem. Soc.* **68**, 73, 78 (1970).
37. E. Bamann, H. Trapmann and H. Muenstermann, *Arch. Pharm.* **296**, 47 (1963).
38. S. Koshy and R. Anantaraman, *J. Am. chem. Soc.* **82**, 1574 (1960).
39. G. Travagli, *Gazz. Chim. Ital.* **87**, 682 (1957); *C.A.* **52**, 1914h.
40. L. Murr and G. Santiago, *J. Am. chem. Soc.* **88**, 1827 (1966).
41. B. Saville, *J. chem. Soc.* 1624 (1961); *ibid.* 4062 (1962).
42. M. Green and R. F. Hudson, *J. chem. Soc.* 3883 (1963).
43. H. A. O. Hill, J. M. Pratt, F. R. Williams and R. J. P. Williams, *Chem. Comm.* 341 (1970).
44. D. G. Brady, *Chem. Comm.* 434 (1970).
45. R. G. R. Bacon and O. J. Stewart, *J. chem. Soc.* 4953 (1965).
46. V. C. R. McLoughlin and J. Thrower, *Tetrahedron* **25**, 5921 (1969).
47. R. G. R. Bacon and H. A. O. Hill, *Q. Rev. chem. Soc.* **19**, 95 (1965).

instances are cited which show that reactions of this type are not restricted to water as a solvent.

In some cases the hydrolysis of a cation may prevent the predicted order of effectiveness of a series of cations from being observed. For example, Zr(IV) has a greater tendency to complex with fluoride than Al(III), but Zr(IV) is the poorer catalyst for fluoride hydrolysis under conditions of low acidity because of its greater tendency to undergo hydrolysis.

The examples in Table 1 well illustrate that the HSAB principle does serve as a reliable guide in the correlation of data presently available on metal-ion assisted ligand substitution processes. It must next be determined if this same principle can serve as a guide to the formulation of conditions appropriate to metal ion assisted substitution processes in systems where they have not been observed previously. The most attractive potential use of this kind of reaction is in the catalysis of substitution reactions which do not introduce the elements of water or other solvent. An indication of the potential scope of such processes can be seen in the Lucas reaction[33]. This reaction is used to replace the hydroxy group of an alcohol with a chloride ion, but it would seem to be potentially capable of replacing hydroxy groups on atoms other than carbon with chlorides.

By implication, we are assuming that these reactions occur by an S_E type mechanism in which the critical step is the formation of a weak bond between the catalyst metal ion and a bound halide or other group. If this is the case there should be some relation between the experimental rate constants and the formation constants corresponding to the catalyst metal complexes produced by the reactions. While such a correlation has been demonstrated to be valid in a few cases[24, 48] it is most apparent for good catalysts and quite poor where the metal ions possess only marginal catalytic activity. This is due to the fact that metal salts can affect the rates of such reactions by means other than direct interaction with the leaving group.

There is every reason to believe that the selectivity of metal–ligand reactivity can be utilized in a very large number of reactions of the type described in this paper.

Reprinted from *Inorganica Chimica Acta Reviews*, **3**, 75–79 (1969)

30

THE EFFECTS OF LIGAND AND METAL HARDNESS ON THE FORMATION OF HYDRIDO– AND ORGANO–METALLIC COMPLEXES BY THE β-INTERACTION

R. J. Cross

Chemistry Department, University of Glasgow, Glasgow, W.2., United Kingdom

CONTENTS

I. INTRODUCTION

The interactions between a transition metal and the β-atom of a ligand are well known, and the phenomena are often referred to as β-interactions.[1] The effects of the interactions are numerous but one is of particular importance and occurs in many catalytic and synthetic reactions. This is the conversion of the transition-metal complex to a hydride by the transfer of a hydrogen atom from the β-atom of the ligand to the metal. The formation of the hydride is often accompanied by the expulsion of an unsaturated compound, the remainder of the ligand. The reactions can be summarised by the general equation

$$L_xM-X-CH_r-R \rightleftharpoons L_xMH+RCHX$$
$$\quad\quad\alpha\quad\beta$$

Reactions of this type are particularly numerous amongst complexes of the heavier platinum metals, and have been observed with a diversity of ligand arrangements, L_x, ususally including some π-bonding moeties. An obvious requirement for the process to proceed is the ability to form a stable unsaturated leaving molecule, RCHX. For example, this can be an aldehyde when the donor atom X is oxygen, or an olefin when X is CH_2. Beyond this, conditions favouring hydride formation are not understood. Where the complex $(Et_3P)_2PtCl(OEt)$ cannot be isolated due to very rapid acetaldehyde elimination, the analogous $(Et_3P)_2Pt(GePh_3)(OEt)$ is a stable compound and cannot be converted to the hydride.

A recent paper[2] discusses the stabilisation of complexes related to those involved in these reactions by alterations in the hardness of the metal atom. This prompts the idea that the β—transfer reactions might be classified according to the compatibility of metal and ligand with respect to the hard and soft acid and base theory.[3] The object of this review is to examine the relevant information to see if it is compatible with this hypothesis. Most examples are taken from the chemistry of the nickel group elements, but some relevant material from other groups is included. The examination is extended to the formation of organometallic compounds by similar reactions, and to the formation and reduction of nitrogenyl complexes.

II. HARDNESS OF THE METAL

One of the most convenient methods for the preparation of platinum hydrides[4] is the reaction between ethanolic potassium hydroxide and a platinum halide complex.[5]

The intermediate ethoxyplatinum complex has not been isolated, but deuterium studies proved that the source of the hydrogen was the β-carbon atom. Acetone was isolated instead of acetaldehyde when isopropanol replaced ethanol. This reaction has been applied to many other complexes. The three halogen atoms of $(R_3P)_3IrCl_3$ can be progressively replaced by hydrogen by refluxing in basic alcohols.[6] Long reflux times with high-boiling alcohols are necessary to replace the second and third chlorine atoms. Phosphine-chloro-complexes of rhodium,[7] ruthenium,[7] and osmium[8] have been reduced in a similar manner. Carbon monoxide often appears in the coordination sphere of these metals after reduction, and it seems that this arises from further reaction of the aldehyde biproducts while they remain coordinated to the metal. In none of these examples was the intermediate alkoxy compound detected.

The function of the base in this process appears to be simply to remove HCl. Not surprisingly, a number of the above reactions can be made to proceed without added base. Organic bases such as Et₃N can replace the KOH with no reduction in yield.[9]

The metal atoms in the complexes discussed so far are all soft acids. Of the ligands involved, RO⁻ is a hard base, where H⁻ is soft. Aldehydes and ketones are also softer than RO⁻ and ROH due to their π-bonding. The reactions, then, can be regarded as unfavourable soft-hard combinations rearranging to give soft-soft complexes.

$$M-O-R \longrightarrow M-H+R_2'C=O$$
$$\text{soft-hard} \qquad\qquad \text{soft-soft (soft)}$$

If M could be hardened, the reaction might become less favourable, as not only would the alkoxide be more stable, the metal hydride would be destablised. One way of testing this is to replace the metal M by other, harder ones. For example, the place of platinum can be taken by palladium and nickel. Although all three metals are normally regarded as soft (Ni[II] is intermediate in its behaviour to bases),[3] the order of hardness is expected to increase in the series Pt < Pd < Ni. Unfortunately it is not possible at present to compare the effects of these metals in exactly analogous complexes, and variations in the ligands can also affect the hardness (see section III). If we assume that the ligand effects are of secondary importance, however, a rough comparison is possible.

Ethanol reacts rapidly at room temperature with bistriethylphosphinedimethylpalladium to eliminate methane.[10]

$$(Et_3P)_2PdMe_2 + EtOH \rightarrow (Et_3P)_2PdMe(OEt) + CH_4$$

The ethoxy derivative formed then decomposes over about 3 hours. The authors assumed a direct radical reaction, but the experimental evidence is compatible with preliminary elimination of CH₃CHO followed by decomposition of (Et₃P)₂PdHMe. No phosphinealkoxynickel complexes are known (nickel hydrides have not been prepared by this route) but Ni(OMe)₂ has been isolated and shows no signs of decomposition.[11] Bridging ligands and octahedrally coordinated nickel appear to be present.

Platinum(IV), which may be comparable in hardness to nickel(II), has many complexes with oxygen-bonded ligands.[12]

III. INFLUENCE OF THE OTHER LIGANDS

The hardness of a metal atom is modified by the ligands in its coordination sphere.[3] In general, hard ligands increase the hardness of the metal. In square-planar platinum(II) complexes, electronic effect of ligands are strongly transmitted stereospecifically towards the trans groups. This is known as the trans-effect.[18] A group of high trans-influence weakens the bonding to the ligand lying opposite it. The effect is often observed as a lengthening of the trans Pt–X

bond, or a decrease in ν(Pt–X) in the i.r. spectrum. It has also been realised for some time that high trans-effect ligands increase the apparent hardness of the metal for the trans ligand. This appears to have a pronounced effect on the reactions under consideration.

The most striking example of this is the isolation of a series of alkoxyplatinum(II) complexes, trans–(Et₃P)₂Pt(GePh₃)OR.[14] Complexes with R = H, Me, Et and i–Pr have been obtained and show no tendency to form the corresponding hydrides. Indeed, the reverse reaction seems favourable.

$$(Et_3P)_2PtH(GePh_3) + CH_3CHO \rightarrow (Et_3P)_2Pt(OEt)(GePh_3) \; ^{15}$$

The Ph₃Ge group has a very large trans–influence, and decreases ν(Pt–H) by about 150 cm⁻¹ to 2050 cm⁻¹ in trans–(Et₃P)₂PtH(GePh₃), compared to trans–(Et₃P)₂PtHCl. Using this criterion, it might be expected that other hydridoplatinum species with similar or lower Pt–H stretching frequencies may also form stable alkoxides. An obvious candidate is the bridging binuclear complex[16]

with ν(Pt–H) at 2005 cm⁻¹. It is interesting to note that attempts to prepare this hydride by the action of alcohol and base on the corresponding chloride (Et₃P)₂Pt₂(PPh₂)₂Cl₂ have met with failure.[9,16]

Chatt and Heaton[2] have discussed this trans hardening effect in relation to the hydroxy compounds shown below, for which two isomers exist, probably due to the orientation of the Pt–P–O–Pt bridges.

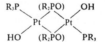

That the bridging phosphinato groups have a very high trans-effect is shown by the low ν(Pt–Cl) of trans chloride ligands. A characteristic of the alkoxy derivatives described above is the facile exchange of the OR⁻ ligand.

$$(Et_3P)_2Pt(GePh_3)OH + ROH \rightarrow (Et_3P)_2Pt(GePh_3)OR + H_2O$$

Similar reactions with the phosphinato hydroxy-complexes may produce more examples of alkoxy-platinum complexes.

Further examples of the lability of oxygen-platinum(II) complexes is provided by the phenoxide compounds (Et₃P)₂PtCl(OPh) and (Et₃P)₂Pt(OPh)₂.[9] They are prepared from cis–(Et₃P)₂PtCl₂ and sodium phenoxide, and the absence of hydrogen on the β-atom prevents their decomposition. Both appear to be cis complexes so by analogy to the above compounds the high trans-influence of Et₃P might further stabilise these hard ligands. The facile exchange of the RO⁻ ligands in these cases leads to decompositions.

$$(Et_3P)_2PtCl(OPh) + EtOH \rightarrow (Et_3P)_2PtHCl + CH_3CHO + PhOH$$

Inorganica Chimica Acta

It is relevant to mention here that thiolate complexes $(R_3P)_2M(SR')_2$ have been isolated for M = Ni, Pd[17] and Pt.[18] All of these are stable in the presence of ethanol, and thus appear to be more stable than their corresponding alkoxides. This reflects the overall soft nature of these metals.

The effects discussed in Section III may invalidate the comparison between Ni, Pd and Pt in Section II. It must be remembered, however, that the *trans*-effect is not transmitted through palladium and nickel to such a great extent as in platinum, so the trend observed may be genuine and general.

IV. HARDNESS OF THE LIGAND

As the ligand involved in the β-atom transfer becomes more compatible with the metal atom, the tendency towards this type of rearrangement diminishes. This is illustrated by a comparison of $(Et_3P)_2$-PtCl(OEt) and $(Et_3P)_2PtCl(CH_2CH_3)$. In the former case only the corresponding hydride can be isolated, whereas its formation from the latter compound is completely reversible.[19]

$$trans-(Et_3P)_2PtCl(C_2H_5) \underset{95°/40° \text{ atm.}}{\overset{180°}{\rightleftharpoons}}$$

$$trans-(Et_3P)_2PtHCl + C_2H_4$$

An intermediate olefin complex appears to be involved and this prevents proof by deuteration studies that a β-transfer is responsible for the reactions.

$$(Et_3P)_2PtClEt \rightleftharpoons (Et_3P)_2PtHCl \cdot C_2H_4 \longrightarrow (Et_3P)_2PtHCl + C_2H_4$$

Alkyl groups are soft compared to alkoxy groups, and their extra stability can be explained in terms of a more compatible soft-soft combination.

As the order of softness is O < N < C in ligands of similar structure, nitrogen donors are expected to show intermediate behaviour with Pt[II], and this appears to be the case. On the one hand the reaction between $(Et_3P)_2PtCl_2$ and NaNEt₂ does produce some platinum hydride.[9] On the other, a number of platinum(II) nitrogen derivatives have been isolated. These include $(Et_3P)_2PtCl(NHR)$ (R = PhCO and tol. SO₂),[20] $(Ph_3P)_2PtH(NC_4H_4O_2)$ (from $(Ph_3P)_4Pt$ and succinimide)[21] and Pt(serinate)₂.[22]

Under rigorous conditions, triethylamine converts platinum dihalides to hydrido-halides.[9]

$$cis-(Et_3P)_2PtCl_2 \xrightarrow[\text{sealed tube}]{Et_3N, 180°} trans-(Et_3P)_2PtHCl$$

The nitrogen biproducts have not been isolated.

Numerous nitrogen derivatives of the harder platinum(IV) are known.[12]

V. RELATED LIGAND SYSTEMS

1. *Formates.* Hot formic acid reduces $cis-(Et_3P)_2$-PtCl₂ to the hydridochloride in a manner similar to ethanol.[5]

$$(Et_3P)_2PtCl_2 + HCOOH \rightarrow (Et_3P)_2PtHCl + HCl + CO_2$$

A probable intermediate is a platinum formate, which could eliminate CO₂ by hydrogen transfer from the β-atom (again carbon). The reaction of formic acid and Li₂PtCl₄ in the presence of 1-octene to give the chloride-bridged $(CO)_2Pt_2Cl_2(C_8H_{17})_2$ may be a similar case.[13] The carbon monoxide could be produced from the formic acid or dimethylformamide solvent.

In a reaction involving an osmium-carbonylphosphine complex, the formate intermediate was isolated.[24]

$$Os(CO)_3(PPh_3)_2 \xrightarrow{HCOOH \text{ aq.}}$$

$$Os(HCO_2)_2(CO)(PPh_3)_2 \xrightarrow[\text{benzene}]{\text{reflux in}}$$

$$OsH_2(CO)(PPh_3)_2 + 2CO_2$$

Furthermore, the reverse reaction has been demonstrated with a cobalt complex.[25]

$$(Ph_3P)_3CoH(N_2) + CO_2 \xrightarrow{1 \text{ atm.}} (Ph_3P)_3Co(CO_2H) + N_2$$

The greater hardness of Co compared to Os and Pt might account for this, but significant comparisons of these different systems are difficult. Isostructural formates $M(HCO_2)_2 \cdot 2H_2O$ (M = Co, Ni, Cu) with bridging formate ligands are known,[26] and similar iron(II) and rhodium(II) compounds have been isolated.

2. *Hydrazine.* Another very convenient method for the preparation of platinum(II) hydrides, the reaction of dihalide complexes with hydrazine, was originally formulated thus.[5]

$$(Et_3P)_2PtCl_2 \xrightarrow{N_2H_4} [(Et_3P)_2Pt(N_2H_4)Cl]Cl$$
$$\downarrow N_2H_4$$
$$(Et_3P)_2PtHCl + NH_4Cl + NH_3 + N_2$$

The amount of nitrogen evolved in experiments using one equivalent of N₂H₄ appeared to be too great, however,[9] and the following course of reactions seemed plausible by comparison with the systems examined above.

$$-Pt-Cl + N_2H_4 \xrightarrow{\text{base}} -Pt-NH-NH_2 + HCl \xrightarrow{\text{fast}}$$

$$-Pt-N=NH + H_2 \xrightarrow{\text{slow}} -Pt-H + N_2$$

The isolation of the bridged complex shown below adds support to this reaction mechanism.[27] It was produced in the reaction between hydrazine and $(Ph_3P)_2PtCl_2$.

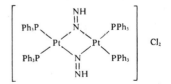

X-ray crystallographic analysis indicated the presence of bridging amine groups as well as the dehydrodiimide complex, so the reaction sequence is not as simple as that shown above.

Nevertheless the reaction appears to proceed at least in part by hydrogen transfer from the β-atom (nitrogen) to platinum. Both the dehydrodiimide and hydride are softer ligands than hydrazine, and should be more compatible with platinum(II). When Pt^{II} is replaced by a harder metal atom the process becomes less favourable. Many hydrazine complexes of platinum(IV) are known,[12] and hydrazine reacts with $(R_3P)_2NiX_2$ to replace the phosphines, producing $(N_2H_4)_2NiX_2$.[28] Only prolonged refluxing with excess hydrazine causes reduction (to nickel metal).

VI. FORMATION OF ORGANOMETALLIC COMPLEXES

Organic ligands R$^-$ are similar in softness to H$^-$ and the same principles might be expected to apply to the transfer of organic moeties from the β-ligand-atom to the metal. In fact such reactions are quite rare, and all of them involve nitrogen as the ligand donor-atom.

Arylplatinum compounds can conveniently be prepared by � route shown below[29] which involves the preparation of arylazoplatinum intermediates. The reaction involves the transfer of the phenyl group from the β-nitrogen atom to platinum. The arylazo intermediates can be reduced to phenylhydrazine by molecular hydrogen.[30]

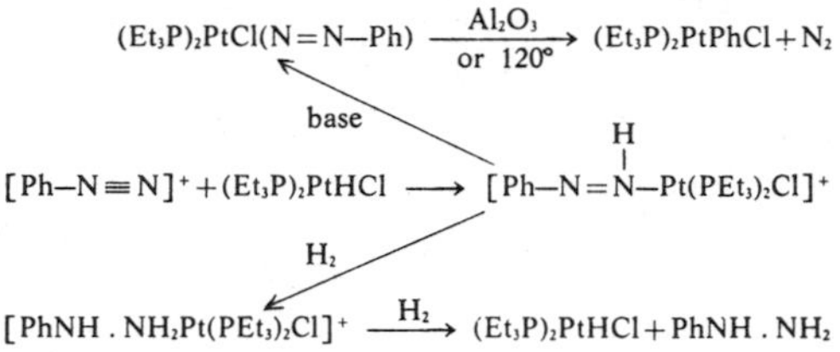

$$(Et_3P)_2PtCl(N=N-Ph) \xrightarrow[\text{or } 120°]{Al_2O_3} (Et_3P)_2PtPhCl + N_2$$

base

$$[Ph-N\equiv N]^+ + (Et_3P)_2PtHCl \longrightarrow [Ph-N=N-\overset{H}{\underset{|}{N}}-Pt(PEt_3)_2Cl]^+$$

H$_2$

$$[PhNH \cdot NH_2Pt(PEt_3)_2Cl]^+ \xrightarrow{H_2} (Et_3P)_2PtHCl + PhNH \cdot NH_2$$

The reaction between phenylhydrazine and $(Et_3P)_2$-

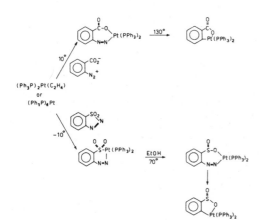

PtCl$_2$ results in the formation of $(Et_3P)_2PtClPh$,[9] although the yield is less than 50%. Arylazo intermediates are probably involved here, also.

Arylazomolybdenum complexes have been isolated,[31] but these have different properties to the platinum derivatives and the arylazo group here seems to function as a 3-electron donor like NO. No tendency to eliminate N$_2$ has been observed up to 100°.

Two further examples which involve N$_2$ elimination and the formation of Pt–C bonds are shown below.[32,33] It is of interest that neither of these reaction products could be made to eliminate CO$_2$ or SO$_2$.

A reaction which may be related is that between Vaska's iridium complex and benzoyl azide.[34] A β-interaction might be involved in this also.

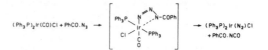

Apart from observing that the mechanisms of these reactions may be the same as in the hydrazine reductions, no further information relevant to the present discussion can be obtained because corresponding reactions with hard-acid complexes are lacking.

A class of reactions included in the general term of β-interactions, but of which the relationship to the β-atom transfer reactions is obscure, is worthy of mention. Azobenzene[35,36,37] and several tertiary amines[38,39] react under mild conditions with various metal complexes to produce σ-bonded metal-carbon species. Some examples are shown below.

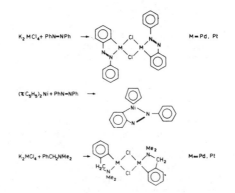

The mechanisms are not understood. The interaction with the metal may include a preliminary transfer of a proton, or direct attack at the ortho carbon atom. The result, however, is the formation of a soft (σ-carbon) ligand. As all the metal atoms involved (zerovalent iron and nickel, divalent palladium and platinum) are soft the reactions can again be classed as unfavourable soft-hard combinations converting to more stable soft-soft. A number of related « ortho » and « β » interactions involving tertiary phosphines are also known.[40,41,42]

Inorganica Chimica Acta

VII. MOLECULAR NITROGEN COMPLEXES

The number of known, well-defined nitrogenyl complexes is growing rapidly. Several complexes have been prepared by the action of hydrazine on metal halides.[43,44,45] These involve the cations Ru-$(NH_3)_5N_2^{2+}$ and $Os(NH_3)_5N_2^{2+}$. The mechanism of the formation of these nitrogenyl compounds is not known but the overall process is analogous to the previously mentioned hydrazine reductions. Molecular nitrogen (isoelectronic with CO) is a soft base, where hydrazine is hard. Both Ru^{II} and Os^{II} are soft acids. Significantly, the osmium nitrogenyl complex is more stable than the ruthenium analogue. Other nitrogenyl complexes, prepared from molecular nitrogen, are all best regarded as soft-soft combinations.[45,46,47,48]

None of these well-defined complexes has yet been successfully reduced, although many attempts have been made. This would represent a reversal of the processes discussed, involving soft-soft acid-base complexes being converted to soft-hard ones. Those systems in which reduction of nitrogen to ammonia has been achieved all involve hard-acid, high-valency metals.[45] The nature of the intermediates in these processes is obscure, but the type of reactions involved are illustrated by two recent publications describing the use of titanium compounds in nitrogen fixation.[49,50] Titanium(IV) is chemically or electrolytically reduced to Ti^{II} and this (softer) species incorporates the nitrogen. The conversion of the nitrogen to ammonia appears to involve either reoxidation to Ti^{IV}, or transfer to another hard-acid centre such as Al^{III}.

VIII. CONCLUSIONS

The above reactions which produce metal hydrides, organometallics and nitrogenyl complexes can all be considered as ligand replacements or ligand rearrangements leading to more compatible (soft-soft) combinations. The evidence suggests that the controlling feature may be the apparent metal hardness, but is insufficient to prove this point conclusively at present. For example the comparisons between nickel, palladium and platinum compounds may be invalid as the ligand arrangements are not identical in each case. That this can profoundly affect the reactions has been shown. Indeed it may be that the ligand combinations produce a bigger effect on these reactions (and on the metal hardness) than the change in metal.

It is very probable that these β-atom interactions are controlled by a number of influences, and the effects of each may vary from case to case. For example, the availability of d-orbitals and extra coordination sites at the transition state may control some of the transfers. *Trans*-effect studies on square-planar platinum(II) compounds indicate that more than one process is involved in stabilising hard-ligand complexes. The hardening properties of some high *trans*-influence, σ-bond-weakening ligands have been discussed, but both oxygen[51] and nitrogen[52]

complexes of Pt^{II} are known with olefins in the *trans* position. It is known[13] that the *trans*-effect of olefins is purely a π-bonding effect.

The advantages of the concept of hardness and softness in these examples may well lie in its vagueness. The concept encompasses a large number of considerations of the type mentioned above, including both σ- and π-effects. Because of this an overall classification of these reactions as outlined above might be possible, where other approaches fail.

IX. REFERENCES

(1) M. L. H. Green, « *Organometallic Compounds* » Vol. 2., 3rd. Ed Methuen 1967, p. 211.
(2) J. Chatt and B. T. Heaton, *J. Chem. Soc.* (A), 2745 (1968).
(3) R. G. Pearson, *Chemistry in Britain*, 3, 103 (1967).
(4) R. J. Cross, Organometallic Chem. Rev., 2, 97 (1967).
(5) J. Chatt and B. L. Shaw, *J. Chem. Soc.*, 5075 (1962).
(6) J. Chatt, R. S. Coffey and B. L. Shaw, *J. Chem. Soc.*, 7391 (1965).
(7) J. Chatt and B. L. Shaw, *Chem. and Ind.* (London), 931 (1960).
(8) J. Chatt and B. L. Shaw, *Chem. and Ind.* (London), 290 (1961).
(9) R. J. Cross, unpublished observations.
(10) G. H. Coates and G. Calvin, *J. Chem. Soc.*, 2008 (1960).
(11) R. W. Adams, E. Bishop, R. L. Martin and G. Winter, *Aust. J. Chem.*, 19, 207 (1966).
(12) See, for example, Gmelin's Handbuch der anorganischen Chemie, No. 68, Platinum.
(13) F. Basolo and R. G. Pierson, *Progress in Inorganic Chemistry*, Vol. 4, New York, Interscience, 1962.
(14) R. J. Cross and F. Glockling, *J. Chem. Soc.*, 5422 (1965).
(15) R. J. Cross and F. Glockling, unpublished observations.
(16) J. Chatt and J. M. Davidson, *J. Chem. Soc.*, 2433 (1964).
(17) R. G. Hayter and F. S. Humiec, *J. Inorg. Nucl. Chem.*, 26, 807 (1964).
(18) P. Klason and J. Wanselin, *J. prakt. Chem.*, 67, 41 (1903).
(19) J. Chatt, R. S. Coffey, A. Gough and D. T. Thomson, *J. Chem. Soc.* (A), 190 (1968).
(20) W. Beck, M. Bauder, W. P. Fehlhammer, P. Pöllmann and H. Schäckl, *Inorg. Nucl. Chem. Lett.*, 4, 143 (1968).
(21) D. M. Roundhill, *Chem Comm.*, 567 (1969).
(22) L. M. Volshtein and T. R. Lastushkina, Russian *J. Inorg. Chem.*, 14, 246 (1969).
(23) E. Lodewijk and D. Wright, *J. Chem. Soc.* (A), 119 (1968).
(24) K. R. Laing and W. R. Roper, *J. Chem. Soc.* (A), 1889 (1969).
(25) L. S. Pu, A. Yamamoto and S. Ikeda, *J. Amer. Chem. Soc.*, 90, 3896 (1968).
(26) C. Oldham, *Progress in Inorganic Chemistry*, Vol. 10, New York, Interscience (1968).
(27) G. C. Dobinson, R. Mason, G. B. Robertson, R. Ugo, F. Conti, D. Morelli, S. Cenini and F. Bonati, *Chem. Comm.*, 1967, 739.
(28) G. Mattogno, A. Monaci and F. Tarli, *Inorg. Nucl. Chem. Lett.*, 4, 315 (1968).
(29) G. W. Parshall, *J. Amer. Chem. Soc.*, 87, 2133 (1965).
(30) G. W. Parshall, *J. Amer. Chem. Soc.*, 89, 1822 (1967).
(31) R. B. King and M. B. Bisnette, *Inorg. Chem.*, 5, 300 (1965).
(32) C. D. Cook and G. S. Jauhal, *J. Amer. Chem. Soc.*, 90, 1465 (1968).
(33) T. L. Gilchrist, F. J. Graveling and C. W. Rees, *Chem. Comm.*, 821 (1968).
(34) J. P. Collman, M. Kubota, F. D. Vastine, J. Y. Sun and J. W. Kang, *J. Amer. Chem. Soc.*, 90, 5430 (1968).
(35) A. C. Cope and R. W. Siekman, *J. Amer. Chem. Soc.*, 87, 3272 (1965).
(36) J. P. Kleiman and M. Dubeck, *J. Amer. Chem. Soc.*, 85, 1544 (1963).
(37) M. M. Bagga, P. L. Pauson, F. J. Preston and R. I. Reed, *Chem. Comm.*, 543 (1965).
(38) A. C. Cope and E. C. Friedrich, *J. Amer. Chem. Soc.*, 90, 909 (1968).
(39) J. M. Klegman and A. C. Cope, *J. Organometallic Chem.*, 16, 309 (1969).
(40) J. Chatt and J. M. Davidson, *J. Chem. Soc.*, 843 (1965).
(41) M. A. Bennett and D. L. Milner, *Chem. Comm.*, 581 (1967).
(42) G. W. Parshall, W. H. Knoth and R. A. Schunn, *J. Amer. Chem. Soc.*, 91, 4990 (1969).
(43) A. D. Allan, F. Bottomley, R. O. Harris, V. P. Reinsalu and C. V. Senoff, *J. Amer. Chem. Soc.*, 89, 5595 (1967).
(44) A. D. Allen and J. R. Stephens, *Chem. Comm.* 1147 (1967).
(45) R. Murray and D. C. Smith, *Coordination Chem. Rev.*, 3, 429 (1968).
(46) A. Sacco and M. Aresta, *Chem. Comm.*, 1223 (1968).
(47) G. M. Bancroft, M. J. Mays and B. E. Praier, *Chem. Comm.*, 585 (1969).
(48) J. Chatt, J. R. Dilworth and G. J. Leigh, *Chem. Comm.*, 687 (1969).
(49) E. E. van Tamelen, R. B. Fechter, S. W. Schneller, G. Boche, R. H. Greeley and B. Åkermark, *J. Amer. Chem. Soc.*, 91, 1551 (1969).
(50) E. E. van Tamelen and D. A. Seeley, *J. Amer. Chem. Soc.*, 91, 5194 (1969).
(51) D. G. McMane and D. S. Martin, Jr., *Inorg. Chem.*, 7, 1169 (1968).
(52) D. V. Claridge and L. M. Venanzi, *J. Chem. Soc.*, 3419 (1964).

[Reprinted from the Journal of the American Chemical Society, **87**, 241 (1965).]
Copyright 1965 by the American Chemical Society and reprinted by permission of the copyright owner.

31

Nucleophilic Constants and Substrate Discrimination Factors for Substitution Reactions of Platinum(II) Complexes

Umberto Belluco, Lucio Cattalini, Fred Basolo, Ralph G. Pearson, and Aldo Turco

*Contribution from the Chemical Laboratory of Northwestern University,
Evanston, Illinois, and the Istituto Chimica Generale, Universita di Padova,
Padova, Italy. Received July 15, 1964*

The kinetics for the substitution reactions of six different platinum complexes of the type trans-[PtL$_2$Cl$_2$] with eighteen different nucleophilic reagents are reported. A set of nucleophilic reactivity constants, n$_{Pt}$, are defined for these reagents by selecting trans-[Pt(py)$_2$Cl$_2$] as a standard. Other platinum complexes then obey the equation log k$_Y$ = sn$_{Pt}$ + log k$_S$, where k$_Y$ is a second-order rate constant for the nucleophile Y, s is a nucleophilic discrimination factor, and k$_S$ is an "intrinsic" reactivity measure which is equal to the first-order rate constant for the reaction in which the solvent is the nucleophile. Complexes with a low intrinsic reactivity have a large value of s, and complexes which are highly reactive show little discrimination. Strong π-bonding ligands, in the cis position, produce low values of k$_S$ but large values of s.

Introduction

The chemical literature contains a large amount of data on rates of bimolecular, nucleophilic substitution reactions.

$$Y + M-X \longrightarrow M-Y + X \qquad (1)$$

Here M is an electrophilic atom, which will be a metal in the cases of interest in this paper, Y and X are nucleophilic atoms or groups also called ligands, and M has other inert groups or ligands attached to it. The mechanism is the familiar S$_N$2 process in which a transition state, or possibly a reactive intermediate, of increased coordination number is formed with both Y and X bound to M. The substrate atoms (M), for which the most data are available,[1] include hydrogen, boron, carbon, nitrogen, oxygen, phosphorus, sulfur, platinum(II), and, more recently, palladium(II)[2] and gold-(III).[3]

A number of attempts have been made to correlate the rates of eq. 1 with other kinetic and thermodynamic data as Y is varied. These are all based on the principle of linear free-energy relationships.[4] Some of these include the Brønsted relationship,[5] the Swain and Scott equation,[6] and the Edwards equations.[7] The latter are particularly interesting in that the oxidation potential of the nucleophile is taken as one of the parameters.

Recent discussions have emphasized the fact that no one scale of nucleophilic reactivity exists but that the substrate M–X determines the order of effectiveness of various Y groups.[1,8] In particular a simple, useful rule may be formulated. If we define a "hard" acid or base as one of low polarizability and a "soft" acid or base as one of high polarizability, then hard acids combine best with hard bases and soft acids combine best with soft bases.[8] The nucleophilic–electrophilic interaction in the transition state of reaction 1 is, of course, a generalized acid–base reaction. Thus, polarizable metal atoms such as platinum(II) will react rapidly with easily polarizable reagents (which are also those of more positive oxidation potential).

While there is ample evidence to substantiate the above statement,[8,9] there is relatively little evidence as to how changes in the nature of the leaving group X and the inert ligands changes the selectivity of platinum complexes toward a series of nucleophilic reagents. A previous conclusion[9a] that increased negative charge on the complex increases the relative reactivity toward polarizable reagents must be viewed with some caution. The conclusion was based largely on a comparison of rates of reaction with chloride ion and nitrite ion. We

(1) For a review and references, see J. O. Edwards and R. G. Pearson, *J. Am. Chem. Soc.*, **84**, 16 (1962).

(2) (a) R. G. Pearson and D. A. Johnson, *ibid.*, **86**, 3983 (1964); (b) W. H. Baddley, Doctoral Thesis, Northwestern University, Evanston, Ill., 1964.

(3) W. H. Baddley and F. Basolo, *Inorg. Chem.*, **3**, 1087 (1964).

(4) For a general discussion and for details of the various equations proposed, see J. E. Leffler and E. Grunwald, "Rates and Equilibria of Organic Reactions," John Wiley and Sons, Inc., New York, N. Y., 1963, Chapter 9.

(5) J. N. Brønsted, *J. Am. Chem. Soc.*, **51**, 428 (1929).

(6) C. G. Swain and C. B. Scott, *ibid.*, **75**, 141 (1953).

(7) J. O. Edwards, *ibid.*, **76**, 1540 (1954); **78**, 1819 (1956); see also D. H. McDaniel and A. Yingst, *ibid.*, **86**, 1334 (1964).

(8) R. G. Pearson, *ibid.*, **85**, 3533 (1963).

(9) (a) F. Basolo and R. G. Pearson, *Progr. Inorg. Chem.*, **4**, 388 (1962); (b) H. B. Gray, *J. Am. Chem. Soc.*, **84**, 1548 (1962); (c) H. B. Gray and R. J. Olcott, *Inorg. Chem.*, **1**, 481 (1962).

have now found that nitrous acid in small amounts has a pronounced effect on the reactions of some platinum-(II) complexes with nitrite ion. This subject will be reported on in detail at a later date. The rate data of the present paper with nitrite ion were obtained in alkaline solution in order to prevent the presence of HNO_2.

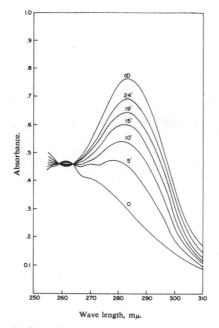

Figure 1. Spectral changes during the reaction in methanol at 30° of trans-[Pt(py)$_2$Cl$_2$] (zero time) with I⁻ to yield trans-[Pt-(py)$_2$I$_2$] (infinite time); complex concentration 5×10^{-5} M.

In this work we report the results of an extensive kinetic study of the substitution reactions of a number of platinum(II) complexes of the type trans-[PtL$_2$Cl$_2$] with a variety of nucleophiles. The solvent was usually methyl alcohol and the temperature 30°. The object was, first, to try to order the nucleophiles in terms of relative reactivity and to see what properties made a reagent good, and, second, to observe the effect of changing the substrate platinum complex on the relative reactivities. Finally, it was hoped that a general, quantitative correlation of the rate constants might be found.

Experimental

Materials. The Pt(II) complexes used in this investigation were all known compounds and were prepared by the methods reported in the literature. Each was characterized by means of a platinum analysis by burning the compound to yield a platinum residue. References to the methods of preparation are as follows[10]: trans-[Pt(py)$_2$Cl$_2$],[11] trans-[Pt(pip)$_2$Cl$_2$],[12]

(10) The symbols used are py, pyridine; pip, piperidine; Et, ethyl; s-Bu, sec-butyl; en, ethylenediamine; and dien, diethylenetriamine.

trans-[Pt(PEt$_3$)$_2$Cl$_2$],[13] trans-[Pt(AsEt$_3$)$_2$Cl$_2$],[13] trans-[Pt(S-(s-Bu)$_2$)$_2$Cl$_2$],[14] trans-[Pt(SeEt$_2$)$_2$Cl$_2$],[15] [Pt(en)-Cl$_2$],[16] and [Pt(dien)Cl]Cl.[17]

The solvent methanol was purified by distillation after refluxing over Mg(OCH$_3$)$_2$ to remove water. The other chemicals were all reagent grade commercial materials and were used without further purification.

Kinetics. The rates of reaction were followed spectrophotometrically by measuring changes in optical density over a period of time at some selected wave length in the ultraviolet region. The instruments used were either a Beckman DK-2 or a Beckman DU with appropriate attachments to maintain the reaction mixture, contained in a 1-cm. quartz cell, at constant ($\pm 0.1°$) temperature. Each system was characterized at least once by recording the spectral change in the region 230–310 mµ during the reaction. In most cases, these spectra (Figure 1) showed well-defined isosbestic points which are indicative of only two absorbing species, the substrate and the product. This suggests that the reactions proceed by a slow, rate-determining first step (2) followed by a rapid second step (3). That this supposition is correct is supported by

$$trans\text{-}[PtL_2Cl_2] + Y \xrightarrow{k_{obsd}} trans\text{-}[PtL_2ClY] + Cl^- \quad (2)$$

$$trans\text{-}[PtL_2ClY] + Y \xrightarrow{fast} trans\text{-}[PtL_2(Y)_2] + Cl^- \quad (3)$$

the isolation and characterization of several of the reaction products. Such a result is understandable, realizing that the *trans* effects[9a] of most of the nucleophiles Y that were used are greater than the *trans* effect of Cl⁻. This would not be the case for the reactions of [Pt-(en)Cl$_2$] or for the reactions of the nucleophiles NH$_3$, py, NH$_2$OH, and NH$_2$NH$_2$. Nevertheless, for these reactions the experimental infinite time optical density was used, and the kinetic plots were linear for at least two half-lives. The method used to investigate the exchange of chloride ion is that described earlier.[18]

The kinetic studies were performed with an excess of reagent present to provide pseudo-first-order conditions. The substrate concentration varied between 5×10^{-5} and 1×10^{-3} M and the reagent concentrations from about 5×10^{-3} to 10^{-1} M. The data obtained were plotted according to the usual first-order rate law. Excellent linear plots were obtained, and the values of the pseudo-first-order rate constants, k_{obsd}, were generally reproducible to better than 5%. Some curvature was obtained when an insufficient excess of reagent was used. In these few instances the values of k_{obsd} were estimated by considering only the initial points in the curves.

The solvent used was generally methanol, but in some cases it was necessary to use methanol–water (80:20

(11) G. B. Kauffman, *Inorg. Syn.*, 7, 251 (1963).
(12) J. Chatt, L. A. Duncanson, and L. M. Venanzi, *J. Chem. Soc.*, 4461 (1955).
(13) K. A. Jensen, *Z. anorg. allgem. Chem.*, **229**, 225 (1936).
(14) K. A. Jensen, *ibid.*, **225**, 115 (1935).
(15) K. A. Jensen, *ibid.*, **225**, 94 (1935).
(16) F. Basolo, J. C. Bailar, Jr., and B. R. Tarr, *J. Am. Chem. Soc.*, 72, 2433 (1950).
(17) F. G. Mann, *J. Chem. Soc.*, 466 (1934).
(18) A. Belluco, L. Cattalini, and A. Turco, *J. Am. Chem. Soc.*, 86, 226 (1964).

Table I. Rates of Reaction[a] for Some Platinum(II) Complexes with Different Nucleophiles in Methanol at 30° and $\mu = 0.1$

Y	trans-[Pt(py)$_2$Cl$_2$]	trans-[Pt(pip)$_2$Cl$_2$]	trans-[Pt(AsEt$_3$)$_2$Cl$_2$]	trans-[Pt(PEt$_3$)$_2$Cl$_2$]	trans-[Pt(SeEt$_2$)$_2$Cl$_2$]	trans-[Pt(S(s-Bu)$_2$)$_2$Cl$_2$][b]	[Pt(en)Cl$_2$][c]
CH$_3$OH[d]	1×10^{-5}	1.2×10^{-5}	[e]	[e]	2×10^{-5}	1×10^{-5}	5×10^{-5}[f]
CH$_3$O$^-$	≤0.1	≤0.1	...	...	≤0.1	≤0.1	...
^{36}Cl$^-$	0.45[g]	0.925	0.69	0.029	...	0.074	...
NH$_3$	0.47	0.6	...	...	...	...	0.5
C$_5$H$_5$N	0.55	...	...	...	...	...	...
NO$_2$$^-$	0.68	2.04	0.1	0.027	1.11	...	...
N$_3$$^-$	1.55	5.30	0.8	0.2	7.5	...	1
NH$_2$OH	2.9	...	...	...	...	...	...
H$_2$NNH$_2$	2.93	...	...	...	...	...	...
Br$^-$	3.7	6.16	1.63	0.93	6.35	0.21	1.7
C$_6$H$_5$SH	5.7	...	...	...	...	...	...
SO$_3$$^{2-}$	250[g]	400	...	...	...	...	...
I$^-$	107	...	650	236	1100	1.94	22
SCN$^-$	180	399	565	371	675	2.45	40
SeCN$^-$	5150	3310	12,300 (ext.)	6950	13,500 (ext.)	...	...
C$_6$H$_5$S$^-$	6000	...	...	...	...	...	...
S=C(NH$_2$)$_2$	6000	3500	...	...	22,900 (ext.)	...	170
S$_2$O$_3$$^{2-}$	9000	...	...	...	...	...	...

[a] Values of $10^3 k_Y$ in M^{-1} sec.$^{-1}$ and k_S in sec.$^{-1}$. [b] 55°. [c] 35°, water solvent. [d] Values of k_S in sec.$^{-1}$. [e] Value was too small to measure. [f] Value for H$_2$O. [g] Estimated from the data on trans-[Pt(pip)$_2$Cl$_2$].

v./v.) in order to be able to dissol e sufficient reagent. Duplicate runs for the reaction of trans-[Pt(py)$_2$Cl$_2$] with I$^-$ and with thiourea show that the rates of reaction in these two solvents are almost the same. Whenever possible, the ionic strength was maintained at 0.1 by the addition of either LiClO$_4$ or LiNO$_3$. Difficulties with precipitation or with low transmission of light prevented maintenance of the ionic strength in about one-fourth of the runs. However, the effect of ionic strength is small judging from a study of the rates of reaction of trans-[Pt(py)$_2$Cl$_2$] with I$^-$ at various ionic strengths. Reactions with NO$_2$$^-$ were carried out with added OH$^-$ (the concentration of OH$^-$ being $^1/_{10}$ that of NO$_2$$^-$) to prevent the formation of HNO$_2$.

Results

The rates of the reactions investigated follow the two-term rate law (4) that is well recognized for substitution reactions of platinum(II) complexes.[9] The rate constant k_1 is for the solvent path, and k_2 is for the

$$\text{rate} = k_1[\text{complex}] + k_2[\text{complex}][\text{reagent}] \qquad (4)$$

direct reagent path. For reaction 2, an excess of reagent was used, and the pseudo-first-order rate constant, k_{obsd}, was determined. This experimental rate constant is related to k_1 and k_2 as given by eq. 5. For

$$k_{obsd} = k_1 + k_2[\text{Y}] \qquad (5)$$

each reaction, the values of k_{obsd} were determined at about five different concentrations of nucleophile, Y. Linear plots (Figure 2) of k_{obsd} vs. [Y] were obtained as is predicted by eq. 5. The extrapolated values of k_1 (at zero [Y]) and the calculated values of k_2 (slopes of the plots) are reported in Table I.

Discussion

The assumption will be made that all the substitution reactions reported in this paper proceed by an S$_N$2 mechanism. The evidence for this assumption

is ample and has been summarized elsewhere.[19] The experimental rate law contained in eq. 4 and 5 may now be rewritten as

$$k_{obsd} = k_S + k_Y[\text{Y}] \qquad (6)$$

that is, k_1 is interpreted as a pseudo-first-order constant for a bimolecular nucleophilic reaction of the solvent,

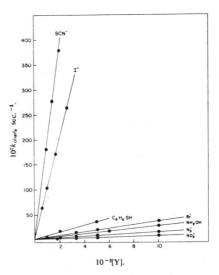

Figure 2. Rates of reaction of trans-[Pt(py)$_2$Cl$_2$] in methanol at 30° as a function of the concentrations of different nucleophiles.

k_S, and k_2 as a bimolecular constant for the nucleophilic reagent, k_Y. The methanol complex is known to

(19) F. Basolo and R. G. Pearson, "Mechanisms of Inorganic Reactions," John Wiley and Sons, Inc., New York, N. Y., 1958, Chapter 4; see also ref. 9.

react rapidly with Y so that the product is the same for both the k_S and k_Y paths.

$$trans\text{-}[PtL_2ClCH_3OH] + Y \xrightarrow{fast}$$

$$trans\text{-}[PtL_2ClY] + CH_3OH \quad (7)$$

Quantitative evaluation of the nucleophilic properties of various reagents generally brings in their basicities toward the proton and a characteristic which may be loosely defined as polarizability or electronegativity. The nature of the electrophilic substrate determines which of the properties makes the greatest contribution. It is already known[1] that platinum(II) forms substrates in which basicity is of little importance. This is borne out by the present work as is shown by the kinetic data of Table I and the pK_a values of the nucleophiles listed in Table II.

Table II. Some Properties of Different Nucleophiles

Y	E° [a]	pK_a[a]	$pK_{CH_3Hg^+}$[b]	n_{Pt}
H_2O	-2.60	-1.7	...	0^c
C_5H_5N	-1.40	5.3	4.8	1.74
Cl^-	-1.36	(-4)	5.45	1.65^d
NH_3	-0.76	9.5	8.4	1.67
Br^-	-1.09	(-7)	6.7	2.79
N_3^-	-1.02	4.7	...	2.19
OH^-	-0.95	15.7	9.5	$\leq 1^c$
NO_2^-	-0.87	3.4	...	1.83
SCN^-	-0.77	(-0.7)	6.1	4.26
I^-	-0.54	(-10)	8.7	4.03
$SeCN^-$	$(>-0.54)^e$	...	...	5.71
SO_3^{2-}	-0.03	9.1	8.16	4.40^d
$S_2O_3^{2-}$	$+0.03^f$	1.9	10.95	5.95
$C_6H_5S^-$	$(+0.3)$	(7)	(19.7)	5.78
$S=C(NH_2)_2$	$+0.42$	0.4	...	5.78
C_6H_5SH	...	...	...	2.75
NH_2OH	...	9.8^g	...	2.46
H_2NNH_2	...	7.9^g	...	2.47

[a] Data from ref. 7. Values of E° are for $Y(aq) = 0.5(Y-Y)(aq) + e(aq)$ [b] R. B. Simpson, *J. Am. Chem. Soc.*, **83**, 4711 (1961); M. Schellenberg and G. Schwarzenbach, Proceedings of the Seventh International Conference on Coordination Chemistry, Stockholm and Uppsala, Sweden, June 25–29, 1962, p. 157. [c] Value for CH_3OH and CH_3O^-. [d] Estimated from the rates of reaction of $trans\text{-}[Pt(pip)_2Cl_2]$. [e] T. Moeller, "Inorganic Chemistry," John Wiley and Sons, Inc., New York, N. Y., 1959, p. 466. [f] B. V. Ptitsyn and V. A. Kozlov, *Zh. Anal. Khim.*, **2**, 259 (1947). [g] F. A. Cotton and G. Wilkinson, "Advanced Inorganic Chemistry," Interscience Division of John Wiley and Sons, Inc., New York, N. Y., 1962, pp. 250–253.

Polarizability, on the other hand, seems to play a major role, and the only difficulty is to find a suitable set of numbers to assess quantitatively the polarizability or electronegativity effect. Edwards[7] has had success with the use of electrode potentials to estimate nucleophilic reactivities. It is very desirable to be able to use such extrakinetic data to predict rates of reaction. The disadvantage is that E° values are not known for many common reagents, since the coupled product Y–Y is often incapable of existence in reaction 8. Table II gives the E° values listed by Edwards[7]

$$Y(aq) = 0.5(Y-Y)(aq) + e(aq) \quad (8)$$

for a number of cases. Several of these, such as for NH_3, N_3^-, and NO_2^-, are only estimated. Figure 3

shows a plot of log k_Y values *vs.* E° values for the case of $trans\text{-}[Pt(py)_2Cl_2]$. There is indeed a rough correlation, similar to what has been noted before.[1] It is possible that some of the deviations are due to incorrect E° values.[20] The other platinum(II) complexes give results similar to Figure 3 when plotted.

A number of attempts were made to find methods of correlating the rate data of Table I. One was to use the value of λ_{max} for the charge-transfer spectra of the complexes $[Co(NH_3)_5X]^{2+}$ where $X = Cl^-$, Br^-, I^-, N_3^-, NO_2^-, and NH_3. Such charge-transfer bands are a measure of the ease of transfer of an electron to the metal from the ligand.[21] While some correlation existed, no quantitative relationship could be found. This was also the case for the attempted correlation of rates with $E^{\circ} - 0.5D$, where D is the bond-dissociation energy of Y–Y in electron volts.

Since platinum(II) is a typical soft acid, it was thought that perhaps a series of equilibrium constants for some other, standard, soft acid could be used as a reference. This would parallel the use of pK_a values to correlate nucleophilic reactivity in the case of typical hard acid substrates (the Brønsted relationship). Table II shows a number of pK data for the formation constants of various bases with CH_3Hg^+, a soft acid (eq. 9). A log k_Y–log $K_{CH_3Hg^+}$ plot shows about

$$CH_3Hg^+ + Y^- \longrightarrow CH_3HgY \quad (9)$$

as good a correlation as does Figure 3 for $trans\text{-}[Pt(py)_2Cl_2]$ and the other platinum complexes studied. Hydroxide ion, or methoxide ion in methanol, is badly off again since at equilibrium it is bound fairly strongly to CH_3Hg^+, whereas it was not found to be a nucleophilic reagent for any of the platinum complexes. It is also known that hydroxide ion is bound strongly at equilibrium for platinum(II) complexes.[22]

Since none of the above attempts was satisfactory, it was finally decided to use the rate constants for the complex $trans\text{-}[Pt(py)_2Cl_2]$ with various nucleophiles as standards. This complex was selected because data were available for the largest number of reagents. The procedure is essentially that of Swain and Scott[6] except that a different reference substrate is used. Methyl bromide, the reference used by Swain and Scott, gives constants which are correlated well by oxidation–reduction potentials.[7] Hydroxide ion would be assigned a fairly large nucleophilic constant in disagreement with the results for platinum(II).

We define a set of nucleophilic reactivity constants, n_{Pt}, by the equation

$$\log \left(\frac{k_Y}{k_S} \right)_0 = n_{Pt} \quad (10)$$

where k_Y and k_S refer to the rate constants for the reaction of $trans\text{-}[Pt(py)_2Cl_2]$ in methanol at 30°. Table

(20) J. O. Edwards, "Inorganic Reaction Mechanisms," W. A. Benjamin, Inc., New York, N. Y., 1964, p. 54, has given a corrected value of E° for NH_3 which fits the rate data much better. This value does not come from oxidation–reduction data but from other rate data, however.

(21) C. K. Jørgensen, "Orbitals in Atoms and Molecules," Academic Press, New York, N. Y., 1962, Chapter 7.

(22) Reference 9a, p. 401; I. Leden and J. Chatt, *J. Chem. Soc.*, 2371 (1956); T. S. Elleman, J. W. Reishus, and D. S. Martin, Jr., *J. Am. Chem. Soc.*, **80**, 536 (1958).

II gives values of n_{Pt} for a number of nucleophilic reagents. A plot of log k_Y for other platinum complexes against n_{Pt} gives reasonably good straight lines in all cases. Figure 4 provides examples. This suggests the linear free energy relationship

$$\log k_Y = sn_{Pt} + \log k_S \qquad (11)$$

The intercepts of plots such as Figure 4 are fairly close to the values of k_S for each substrate in those cases where k_S was measurable. For the PEt$_3$ and AsEt$_3$ systems, the intercepts give predicted values of k_S which are small, in agreement with the observation that the rates are too slow to measure (Table I). It is also convenient that methoxide ion no longer is a problem, since its rate constant is immeasurably small for all complexes.

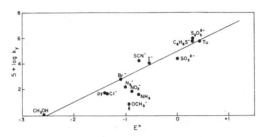

Figure 3. Rates of reaction of *trans*-[Pt(py)$_2$Cl$_2$] in methanol at 30° with different nucleophiles as a function of their $E°$ values.

The constant s is dependent on the nature of the complex. It is a nucleophilic discrimination factor. A large value of s means that the complex is very sensitive to changes in the nature of the nucleophilic reagent. Table III gives values of s and log k_S for several

Table III.[a] Nucleophilic Discrimination Parameters, s, and Intercepts, log k_S, for Several Platinum(II) Complexes

Complex	log k_S[b]	Std. dev. of log k_S	s	Std. dev. of s
trans-[Pt(PEt$_3$)$_2$Cl$_2$][c]	−6.83 (−)[d]	0.35	1.43	0.10
trans-[Pt(AsEt$_3$)$_2$Cl$_2$][c]	−5.75 (−)[d]	0.42	1.25	0.12
trans-[Pt(S(s-Bu)$_2$)$_2$Cl$_2$][e]	−4.95 (−5.00)	0.05	0.57	0.02
trans-[Pt(SeEt$_2$)$_2$Cl$_2$][c]	−4.67 (−4.70)	0.19	1.05	0.05
trans-[Pt(pip)$_2$Cl$_2$][c]	−4.56 (−4.92)	0.13	0.91	0.04
[Pt(en)Cl$_2$][f]	−4.33 (−4.30)	0.09	0.64	0.03
[Pt(dien)Br]$^+$ [g,h]	−4.06 (−3.92)	0.21	0.75	0.07
[Pt(dien)Cl]$^+$ [f,i]	−3.61 (−3.70)	0.25	0.65	0.06
[Pt(dien)H$_2$O]$^{2+}$ [g,j]	−0.44[k]	0.37	0.44	0.12

[a] The intercepts, log k_S, and slopes, s, for different nucleophiles, Y, were determined by the method of least squares using eq. 11. [b] Values in parentheses are the experimental values taken from Table I or from the literature. [c] Methanol solution at 30°. [d] Value was too small to determine experimentally. [e] Methanol solution at 55°. [f] Water solution at 35°. [g] Water solution at 25°. [h] From ref. 9b; eight nucleophiles. [i] From ref. 2b; four nucleophiles. [j] From ref. 9c; five nucleophiles. [k] This should equal the rate of water exchange, which is not known.

platinum(II) complexes. These are found from the slopes and intercepts of straight lines such as Figure 4 by the use of a least-squares analysis. Included are

not only the results in methanol but a few cases studied in water as the solvent. Some of these data are taken from the literature.

It can be seen that quite a reasonable spread of s values is found, from 0.44 to 1.43. Of course, two different solvents are involved and temperatures ranging from 25 to 55°, and s is expected to be a function of both temperature and solvent. The difference between water and methanol is not apt to be large, however, and only the data for the *sec*-butyl sulfide complex are at a temperature more than 5° different from 30°.

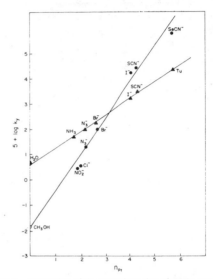

Figure 4. Correlation of the rates of reaction of Pt(II) complexes with the standard *trans*-[Pt(py)$_2$Cl$_2$] for different nucleophiles: ●, *trans*-[Pt(PEt$_3$)$_2$Cl$_2$] in methanol at 30°; ▲, [Pt(en)Cl$_2$] in water at 35°.

Ignoring this last case, there is a strong inverse correlation between s and k_S. A large value of s goes with a very small value of k_S.[23] Such a result is entirely reasonable if we interpret k_S as a measure of the "intrinsic" reactivity of the platinum complex; that is, it is the rate constant for the poorest nucleophilic reagent whose effect can be measured in any solution.[24] With such a poor nucleophile, the greatest burden is put on the complex to reach the activated complex for reaction. Now a complex of high intrinsic reactivity, such as [Pt(dien)H$_2$O]$^{2+}$, will not be very discriminating in reacting with different nucleophiles, and s will be small.[25]

The arsine and phosphine complexes show the largest discrimination between various reagents and the lowest intrinsic reactivity. It is probably sig-

(23) This statement may not always apply for reactions involving different leaving groups. Thus, the s values for [Pt(dien)Cl]$^+$ and [Pt(dien)H$_2$O]$^{2+}$ differ less than might be expected on the basis of their large difference in k_S.
(24) If any nucleophile, such as ClO$_4$, has a rate constant less than the solvent, its kinetic effect will never be measurable.
(25) The limiting case of such a reactive species with low discrimination would be a high energy species of reduced coordination number, such as the carbonium ion in organic chemistry. Compare the discussion in C. G. Swain, C. B. Scott, and K. H. Lohmann, *J. Am. Chem. Soc.*, 75, 135 (1953).

Belluco, Cattalini, Basolo, Pearson, Turco / Platinum(II) Complexes   245

nificant that these two complexes are generally believed to be capable of the largest amount of metal-to-ligand π-bonding. Such π-bonding is probably small in the ground state but becomes quite large in the transition state.[26] It is a variable which can respond to satisfy a demand. Thus, if the more reactive reagents donate electrons more strongly to the platinum atom in the transition state, then more π-bonding can occur to remove excess charge from the metal. It should be recalled that the primary bonding interaction between soft acids and soft bases is believed to be covalent in character.[8]

(26) L. E. Orgel, *J. Inorg. Nucl. Chem.*, **2**, 137 (1956); D. M. Adams, J. Chatt, J. Gerratt, and A. D. Westland, *J. Chem. Soc.*, 734 (1964); C. K. Jørgensen, "Inorganic Complexes," Academic Press, New York, N. Y., 1963, p. 7 and Chapter 9.

One final point of interest is that n_{Pt} for NH_2NH_2 and NH_2OH is appreciably larger than for NH_3, pyridine, NO_2^-, and N_3^-. This suggests that the α-effect[1] is responsible for their greater reactivity compared to other nucleophiles in which the donor atom is nitrogen.

Acknowledgments. The research performed at Northwestern University was supported by the U. S. Atomic Energy Commission, COO-1087-79, and that performed at the University of Padova was supported by a NATO grant. U. B. wishes to thank the Institute of General Chemistry at the University of Padova for a leave of absence, the Italian Council for Research (C.N.R Rome) for a Fellowship, and the Fulbright program for a travel grant.

328

32

CONTRIBUTION FROM THE ISTITUTO DI CHIMICA GENERALE,
UNIVERSITA' DI PADOVA, PADUA, ITALY

Monohalide Displacements of *trans*-[Pt(P(C₂H₅)₃)₂Cl₂] in Dipolar Aprotic Solvents

BY UMBERTO BELLUCO, MARIO MARTELLI, AND ANGELO ORIO

Received September 30, 1965

Kinetic data for monohalide displacement of *trans*-[Pt(P(C₂H₅)₃)₂Cl₂] with various nucleophiles in acetone and dimethyl sulfoxide are reported. From a comparison of these rate data with the rates of the corresponding reactions in methanol, it appears that the nucleophilic reactivity order remains the same regardless of the nature of the solvent. The reactivity order parallels the polarizability order of the entering groups. Saturated carbon substrates present a reverse reactivity order as well as a large shift of the reactivity of halide ions in going from protic to dipolar aprotic solvents, depending on the anion solvation. However, in the SN2 reactions at soft centers, which generally enhance the polarizability of the nucleophiles, the nucleophilic order and the reactivity are practically unaffected when the solvent is changed. In accordance with the reagent solvation, the nucleophilic discrimination factor of *trans*-[Pt(P(C₂H₅)₃)₂Cl₂] is greater in protic than in dipolar aprotic solvents.

Introduction

Most of the rate measurements on platinum(II) complexes refer to reactions in hydroxylic solvents such as methanol and water. In these solvents the order of reactivity of the entering groups is found to be the same for different Pt(II) substrates.[1,2] In dipolar aprotic solvents, the only available data concern chloride exchange of *trans*-[Pt(py)₂Cl₂].[3]

The most important factors recognized to influence the nucleophilic reactivity toward saturated carbon (for which a large amount of data is available) are the following: geometrical factors including steric hindrance[4] or steric acceleration,[5] ion aggregation,[6] M basicity,[7] M–Y bond strength,[8] polarizability, and H basicity.[9] Moreover, it has been recently pointed out

(1) For a review and references see: (a) F. Basolo and R. G. Pearson, *Progr. Inorg. Chem.*, **4**, 388 (1962); (b) W. H. Baddley and F. Basolo, *Inorg. Chem.*, **3**, 1087 (1964); (c) U. Belluco, L. Cattalini, F. Basolo, R. G. Pearson, and A. Turco, *J. Am. Chem. Soc.*, **87**, 241 (1965).

(2) F. Basolo, "Mechanisms of Inorganic Reactions," Summer Symposium of Inorganic Chemistry of the American Chemical Society, Kansas, 1964, p 82.

(3) R. G. Pearson, H. B. Gray, and F. Basolo, *J. Am. Chem. Soc.*, **82**, 787 (1960).

(4) H. C. Brown, *J. Chem. Soc.*, 1248 (1956).

(5) J. F. Bunnett and T. Okamoto, *J. Am. Chem. Soc.*, **78**, 5363 (1956).

(6) S. Winstein, L. G. Savedoff, S. Smith, I. D. R. Stevens, and J. S. Gall, *Tetrahedron Letters*, **9**, 24 (1960).

(7) A. J. Parker, *Proc. Chem. Soc.*, 371 (1961).

(8) R. F. Hudson, *Chimia* (Milan), **16**, 173 (1962).

(9) (a) J. O. Edwards and R. G. Pearson, *J. Am. Chem. Soc.*, **84**, 16 (1962); (b) J. O. Edwards, *ibid.*, **76**, 1540 (1954). For a review and references see also: J. O. Edwards, "Inorganic Reaction Mechanisms," W. A. Benjamin, New York, N. Y., 1964, p 51; J. F. Bunnett, *Ann. Rev. Phys. Chem.*, 271 (1963); W. P. Jencks and J. Carriuolo, *J. Am. Chem. Soc.*, **82**, 1778 (1960).

that the order of nucleophilic reactivity is largely affected by the solvent.[10] Some of these factors are expected to be more important than others, and two among these appear to be generally predominant: the energy required to remove the fraction of an electron from Y^- in solution and the energy corresponding to the fraction of desolvation at the transition state.[8, 10c]

In the case of bimolecular reactions of low-spin d^8 systems, the entering group reactivity order was identified with the polarizability order of the reagents.[1,2] However, in the case of reactions at soft centers, such as Pt(II), no attempts have been made, up to now, to distinguish between the solvation of the entering group and the contribution of polarizability to the nucleophile reactivity.

In this work we report the rates of substitutions of *trans*-[Pt(P(C₂H₅)₃)₂Cl₂] with Cl^-, Br^-, I^-, SCN^-, and thiourea in acetone and in dimethyl sulfoxide. The rate data are compared with the corresponding data of the same substrate in methanol.

Experimental Section

Materials.—The complexes *trans*-[Pt(P(C₂H₅)₃)₂Y₂] (Y = Cl, Br, I, SCN) were prepared by the method of Jensen.[11] All other materials used were reagent grade.

Acetone was refluxed with potassium permanganate, dried with potassium carbonate, and fractionated, bp 56.5°; dimethyl sulfoxide (DMSO) was fractionated under reduced pressure in a nitrogen atmosphere.

Kinetics.—The chloride reactions were carried out with two different methods; the first method was described in a previous paper.[12]

With the second method, the solution of the *trans*-[Pt(P-(C₂H₅)₃)₂Cl₂] labeled with $^{36}Cl^-$ and that of LiCl or (C₄H₉)₄NCl were mixed in the required volumes and thermostated at 25 ± 0.1°. Aliquots of the mixture (2 ml) were removed at time intervals, and the complex was separated from uncomplexed chloride by making use of an anion-exchange resin, Dowex 1X4 (200–400 mesh), nitrate form.

The resin was washed with several portions of acetone, and the washings were added to the original effluent. The resulting solution was concentrated to a known volume (10.5 ml). The solution activity of the liquid samples (10 ml) was determined by means of an immersion-type Geiger–Müller counter. The reliability of this separation method had been tested by passing through the resin an acetone solution of Li^{36}Cl; the radioactivity was absent in the effluent. It was also found that the radioactivity of the complex was unchanged by passing through the same resin. The time necessary for the separation was negligible as compared to the half-time of the reaction.

The kinetics were analyzed by means of the well-known McKay law for simple exchange reactions involving one atom of each reactant and a negligible kinetic isotope effect.[13]

The rate constants obtained with both the methods used were reproducible to within 5%.

Other reactions of *trans*-[Pt(P(C₂H₅)₃)₂Cl₂] with Br^-, I^- (lithium or tetrabutylammonium salts), SCN^- (potassium salt), and thiourea were followed spectrophotometrically in the ultraviolet region by using both a Beckman DK-2A recording appara-

tus and a Beckman DU with a cell compartment thermostated at 25 ± 0.1°. The details of the experimental procedure have been given in a previous paper.[12] For these reactions, excess reagent was used in order to avoid unfavorable equilibrium. Thus, pseudo-first-order rate constants, k_{obsd} (sec^{-1}), were determined. The rate data are listed in Table I.

TABLE I

PSEUDO-FIRST-ORDER RATE CONSTANTS FOR THE REACTIONS:

$$\text{trans-[Pt(P(C₂H₅)₃)₂Cl₂]} + Y^- \xrightarrow{k_{obsd}} \text{trans-[Pt(P(C₂H₅)₃)₂YCl]} + Cl^- \text{ AT 25° IN ACETONE}^a$$

Nucleophile Y	Nucleophile concn., 10^2 M	$10^2 k_{obsd}$, sec^{-1}	$10^4 K^b$
Cl$^-$ (LiCl)	4	0.048	0.02
	1.36	0.016	
	0.68	0.009	
	0.17	0.002	
((n-C₄H₉)₄NCl)	17.5	0.048	22.8
	8.75	0.029	
	4.37	0.017	
	1.05	0.007	
	0.10	0.001	
Br$^-$ (LiBr)	16.4	1.53	5.22
	13.5	1.28	
	6.7	0.67	
	4.0	0.42	
	1.6	0.16	
((n-C₄H₉)₄NBr)	16	1.8	32.9
	13	1.49	
	8	0.96	
	4	0.51	
	2	0.25	
I$^-$ (LiI)	6.4	11.1	69
	3.75	6.7	
	0.65	1.3	
	0.37	0.63	
((n-C₄H₉)₄NI)	6.28	14.45	64.8
	4	10.00	
	3.14	7.00	
	2	4.60	
	1	2.34	
SCN$^-$ (KSCN)	7.8	9.4	38.3
	3.9	4.6	
	1.0	1.46	
	0.5	0.79	
Thiourea	0.25	97.5	
	0.2	80.0	
	0.12	50.5	

^{a} Complex concentration is about 2×10^{-4} M. ^{b} The value of ion-pair dissociation constants, K, are taken from ref 6.

The reactions proceed by a slow rate-determining first step (1), followed by a rapid second step (2),

$$\text{trans-[Pt(P(C₂H₅)₃)₂Cl₂]} + Y^- \xrightarrow{k_{obsd}} \text{trans-[Pt(P(C₂H₅)₃)₂ClY]} + Cl^- \quad (1)$$

$$\text{trans-[Pt(P(C₂H₅)₃)₂YCl]} + Y^- \xrightarrow{fast} \text{trans-[Pt(P(C₂H₅)₃)₂Y₂]} + Cl^- \quad (2)$$

Results and Discussion

The data in Table I show that k_{obsd} values depend on the concentration and nature of the reagent.

In the substitutions of Pt(II) complexes in hydroxylic solvents it has always been found that the reactions obey the rate law: $k_{obsd} = k_1 + k_2[Y:]$, where k_1 is a pseudo-first-order rate constant for the reaction involving the solvent as entering group and k_2 is a second-order rate constant for the direct reaction with $Y:$.

(10) See ref 6 and: (a) A. J. Parker, *J. Chem. Soc.*, 4398 (1961); (b) E. S. Gould "Mechanism and Structure in Organic Chemistry," Holt, Rinehart, and Winston, New York, N. Y., 1959. p 260; (c) C. A. Bunton, "Nucleophilic Substitution at a Saturated Carbon Atom," Elsevier Publishing Co., New York, N. Y., 1963.

(11) K. A. Jensen, *Z. Anorg. Allgem. Chem.*, **229**, 225 (1936).

(12) U. Belluco, L. Cattalini, and A. Turco, *J. Am. Chem. Soc.*, **86**, 226 (1964).

(13) G. Friedlander and J. Kennedy, "Nuclear and Radiochemistry," John Wiley and Sons, Inc., New York, N. Y., 1955, p 315.

Plots of k_{obsd} vs. [Y:] give straight lines with k_1 as intercept and k_2 as slope.[14]

In the case of *trans*-[Pt(P(C₂H₅)₃)₂Cl₂] reacting with different nucleophiles in DMSO, linear plots of k_{obsd} vs. the reagent concentrations were also obtained in accordance with the complete dissociation of the examined salts in this solvent.[15] The extrapolated value of k_1 in this solvent is about 2×10^{-5} sec^{-1}. Complex concentration is about 10^{-4} M. For each entering group, at least four concentrations have been examined, in the range 10^{-3} to 10^{-1} M.

However, for the reactions of *trans*-[Pt(P(C₂H₅)₃)₂Cl₂] in acetone the same plots give smooth curves passing through the origin with the initial slope being equal to k_2. Only in the reactions with Cl$^-$ was a nonzero intercept found. The estimated k_1 value for the *trans*-[Pt(P(C₂H₅)₃)₂Cl₂] in acetone is less than 10^{-6} sec^{-1}, and thus the maximum possible contribution of k_1 to the lowest reaction rate in acetone of the examined reagents is then less than 1% (except for Cl$^-$). Consequently, it is possible to apply the treatment of a simple bimolecular reaction after correction of k_{obsd} for the k_1 value.

In Figure 1, the k_{obsd} values of the *trans*-[Pt(P(C₂H₅)₃)₂Cl₂] reaction with bromide in acetone are reported as a function of the initial concentration of the reagents.

It can be seen that the apparent second-order rate constants (slopes of the plots) decrease by increasing the salt concentration. This is interpreted as being due to the occurrence in acetone of free and paired ions having different reactivities toward the substrate. Such a smooth and orderly variation in the apparent second-order rate constants, as shown in Figure 1, has been observed in bimolecular substitutions of alkyl halides with halide ions in liquid sulfur dioxide.[16] The same behavior was also observed in the substitutions of toluenesulfonates in dimethylformamide and was attributed to an association of salts to ion pairs and not to a salt effect or to the ionic strength effect.[17] It is worth noting that salt effects are generally absent in the reaction of ions with neutral molecules;[18] this trait has also been noted in reactions of platinum(II) complexes.[1c, 14]

In this present work, constant ionic strength was avoided since the addition of any salt would exert a buffering effect on the dissociation of the ion pairs.

The rate constants for the bimolecular reaction of free anions were estimated by Acree's analysis,[19] which ascribes different reactivities to free ions (k_2) and paired ions (k_p); *i.e.*, $k_{obsd} = \alpha k_2 C + (1 - \alpha)k_p C$ (C is the initial reagent concentration and α is the degree of

(14) (a) H. B. Gray, *J. Am. Chem. Soc.*, **84**, 1548 (1962); (b) H. B. Gray and R. J. Olcott, *Inorg. Chem.*, **1**, 481 (1962). See also ref 1a and 1c.

(15) P. G. Sears, G. R. Lester, and L. R. Dawson, *J. Phys. Chem.*, **60**, 1433 (1956).

(16) N. N. Lichtin and K. N. Rao, *J. Am. Chem. Soc.*, **83**, 2417 (1961).

(17) W. M. Weaver and J. D. Hutchison, *ibid.*, **86**, 261 (1964).

(18) E. A. Moelwyn-Hughes, "The Kinetics of Reactions in Solution," Oxford University Press, London, 1942, pp 128–130.

(19) (a) H. C. Robertson and S. F. Acree, *J. Am Chem. Soc.*, **37**, 1902 (1915); (b) J. H. Shroder and S. F. Acree, *J. Chem. Soc.*, 2582 (1914); (c) E. K. Marshal and S. F. Acree, *J. Phys. Chem.*, **19**, 589 (1915).

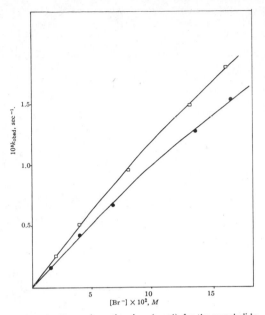

Figure 1.—Rates of reaction, k_{obsd} (sec^{-1}), for the monohalide displacements of *trans*-[Pt(P(C₂H₅)₃)₂Cl₂] at 25° in acetone as a function of the concentration of Br$^-$ salts: LiBr, ●; (*n*-C₄H₉)₄-NBr, □.

dissociation of the salts). The validity of this analysis is supported by the fact that the k_2 values obtained by this treatment are independent of the nature of the cation. In fact, the lithium and tetrabutylammonium salts lead to the same values of k_2 for the free ions, as shown in Figure 2. The small amount of reactivity data for ion pairs (k_p) or higher aggregates and the lack of unequivocal information regarding their structure in the examined solvents make it difficult to discuss the reactivity of such reagent species.

In bimolecular nucleophilic substitutions toward some saturated carbon substrates, the nucleophilicity order found in protic solvents is I$^-$ > Br$^-$ > Cl$^-$.[6, 20] Winstein, *et al.*,[6] and, more recently, Weaver and Hutchison,[17] found that the halide reactivity order for the reaction toward alkyl halides or toluenesulfonates is the exact reverse in acetone and dimethylformamide. Thus, in dimethylformamide the relative rates of reaction at 0° of lithium halides with methyl tosylate after ion-pairing correction are[17] Cl$^-$ (9.1) > Br$^-$ (3.4) > I$^-$ (1.0). An opposite order occurs for sodium halides reacting with ethyl tosylate in aqueous dioxane at 50° where the order is[21] Cl$^-$ (0.14) < Br$^-$ (0.32) < I$^-$ (1.0).

In anhydrous acetone the reactivity order at 25° of lithium halides reacting with *n*-butyl *p*-bromobenzene-sulfonate is[6] Cl$^-$ (18.3) > Br$^-$ (4.3) > I$^-$ (1.0). It is interesting to point out that, in respect to Pt(II) reac-

(20) (a) A. J. Parker, *Quart. Rev.* (London), **16**, 163 (1962); (b) J. F. Bunnett, *Ann. Rev. Phys. Chem.*, 271 (1963).

(21) H. R. McCleary and L. P. Hammett, *J. Am. Chem. Soc.*, **63**, 2254 (1941).

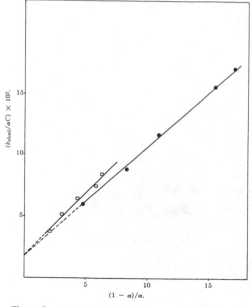

Figure 2.—Acree's analysis for the reaction of *trans*-[Pt-(P(C_6H_5)$_3$)$_2$Cl$_2$] at 25° in acetone with LiBr (●) and (*n*-C_4H_9)$_4$-NBr (□).

TABLE III

SECOND-ORDER RATE CONSTANTS OF FREE IONS FOR THE MONOHALIDE DISPLACEMENTS OF *trans*-[Pt(P(C_2H_5)$_3$)$_2$Cl$_2$] IN VARIOUS SOLVENTS

Nucleophile Y	Acetone[a] $10^3 k_2$, M^{-1} sec^{-1}	Dimethyl sulfoxide[a] $10^3 k_2$, M^{-1} sec^{-1}	Methanol[b] $10^3 k_2$, M^{-1} sec^{-1}	$n°_{Pt}$[c]
Cl$^-$	1.5	0.7	0.029	3.04
Br$^-$	16	5.8	0.93	4.18
I$^-$	230	160	263	5.42
SCN$^-$	180	246	371	5.65
Thiourea	40,000	15,200		7.17
Acetone[d]	~10^{-6}	...	...	
DMSO[d]	...	2×10^{-5}	...	
CH_3OH[d]	...	...	~10^{-7}	0

[a] Temperature 25°.   [b] Data from ref 1c. Temperature 30°.
[c] Nucleophilic reactivity constants, $n°_{Pt} = \log (k_2/k°_2)_0$, for the reaction of *trans*-[Pt(py)$_2$Cl$_2$] in methanol at 30° with different nucleophiles (see ref 23).   [d] Values of k_1 in sec^{-1}.

the same in protic and aprotic solvents. It clearly appears that the nucleophilic reactivity parallels the polarizability of the reagents.

As shown in Table IV, the difference of reactivities on Pt(II) displacements in the two classes of solvents is that the nucleophilic discrimination factor[23] is greater in protic than in aprotic solvents, as a consequence of the relative reagent solvations in the considered solvents.[20a] Of course, the transition state structure for these reactions is assumed to be the same in different solvents.

TABLE IV

NUCLEOPHILIC DISCRIMINATION FACTORS OF *trans*-[Pt(P(C_2H_5)$_3$)$_2$Cl$_2$] IN DIFFERENT SOLVENTS AT 25°

Solvents	Nucleophilic discrimination factors
Methanol[a]	1.43
Acetone	1.04
Dimethyl sulfoxide	1.06

[a] Data at 30° (ref 1c).

It is of interest to note that no relevant solvent effect was found on the leaving group in reactions of some Pt(II) complexes.[3, 24]

Moreover, in the present case the possibility of a strong interaction between Pt(II) and the dipolar solvent in the rate-determining step is reduced, owing to the presence of two strong π-bonding ligands coordinated to the metal ion.[3]

It must be pointed out that differentiation of the reaction rates of single reagents and, consequently, the study of the solvent effect have been possible with the complex chosen because its nucleophilic discrimination is sufficiently high in all of the solvents used.

tions, in S_N2 reactions of methyl chloride[22] there is a larger shift (about 100-fold) in the reactivity of halide ions as the solvent changes from water to acetone (Table II).

TABLE II

SOLVENT EFFECT ON S_N2 REACTIONS OF *trans*-[Pt(P(C_2H_5)$_3$)$_2$Cl$_2$] AND OF CH_3Cl

Nucleophile	(CH$_3$)$_2$CO (25°) $10^3 k_2$, M^{-1} sec^{-1} Pt(II)	CH$_3$OH (30°) $10^3 k_2$, M^{-1} sec^{-1} Pt(II)	k_2(CH$_3$)$_2$CO/ k_2(H$_2$O) (25°), CH$_3$Cl
Cl$^-$	1.5	0.029	2500
Br$^-$	16	0.93	200
I$^-$	230	263	80

The observed effects of solvents on rates of the S_N2 reactions of methyl chloride were rationalized in terms of solvation differences of the anionic reagents. In protic solvents the retarding effect in the rate of reaction of small anions with high charge density is due to their solvation which is greater than that for the large polarizable ions with lower charge density. In dipolar aprotic solvents, such as acetone and dimethylformamide, the observed reactivity order is expected, owing to the smaller chloride solvation with respect to that of the iodide ion.

In most systems, as quoted, "the solvation is one of the most important factors controlling nucleophilicity, and hence the two factors affecting nucleophilic order cannot be separated."[8]

For reactions of *trans*-[Pt(P(C_2H_5)$_3$)$_2$Cl$_2$] the rate data in Table III show that the reactivity order of free ions is

(22) A. J. Parker, *J. Chem. Soc.*, 1328 (1961), and references therein.

(23) The nucleophilic discrimination factors (*s*) are calculated by use of the least-squares analysis on the linear free energy relationship: log k_2 = $sn°_{Pt}$ + log $k_2°$. The nucleophilic reactivity constants, $n°_{Pt}$, are defined as log $(k_2/k_2°)_0$ for the reaction of each nucleophile Y: with the *trans*-[Pt(py)$_2$-Cl$_2$] in methanol at 30° selected as standard conditions. k_2, M^{-1} sec^{-1}, is the second-order rate constant for the nucleophile Y: and $k_2°$, M^{-1} sec^{-1}, is the second-order constant for the reaction in which the methanol is the entering group. The $k_2°$ value was computed from the pseudo-first-order rate constant k_1, sec^{-1}, of *trans*-[Pt(py)$_2$Cl$_2$] in CH$_3$OH at 30° by simply dividing by the "concentration" of the solvent. See ref. 1c; U. Belluco, "Coordination Chemistry Reviews," Elsevier Publishing Co., New York, N. Y., in press; ref 14b; R. J. Mawby, F. Basolo, and R. G. Pearson, *J. Am. Chem. Soc.*, **86**, 3994 (1964).

(24) U. Belluco, unpublished data.

One general conclusion which can be drawn is that, when the electrophilic center enhances the contribution of the polarizability of the reagent in S_N2 reactions, then the nucleophilic order and the reactivity are little influenced by the nature of the solvent. In other words, in the reactions at soft centers such as Pt(II), in which the polarizability of the nucleophile exerts an important role, the nature of the solvent becomes relatively less important.

In these reactions, it is therefore possible to distinguish the contribution of solvation from that of the reagent polarizability, owing to the fact that the solvation effect is a relatively small one in comparison to the contribution from the polarizability of the reagents.

In most cases of S_N2 reactions at saturated carbon centers, the extent of bond formation realized at the transition state is relatively small.[8] In the case of displacements on Pt(II) complexes, however, one can conclude that bond formation at the transition state is the driving force of the reaction,[25] whereas solvation exerts a secondary effect. This assumption is supported by the fact that stable five-coordinated complexes of d^8 ions have recently been prepared.[26] Furthermore, Pt(II) reactions are accompanied by relatively small activation enthalpies and by rather negative values of activation entropies; this is consistent with a net increase in bonding at the transition state.[27]

Acknowledgments.—This work was supported by the Italian Consiglio Nazionale delle Ricerche (CNR, Rome).

(25) For a discussion on the platinum–nucleophile bond see R. G. Pearson, *J. Am. Chem. Soc.*, **85**, 3533 (1963), and ref 1a and 9a.

(26) For a recent review see L. M. Venanzi, *Angew. Chem., Intern. Ed. Engl.*, **3**, 453 (1964); see also R. O. Cramer, R. V. Lindsey, C. T. Prewitt, and V. G. Stolberg, *J. Am. Chem. Soc.*, **87**, 658 (1965); C. M. Harris, R. S. Nyholm, and D. J. Philips, *J. Chem. Soc.*, 4379 (1963); C. M. Harris and R. S. Nyholm, *ibid.*, 63 (1957); J. W. Collier, F. G. Mann, D. G. Watson, and H. R. Watson, *ibid.*, 1803 (1964); P. C. Westland, *ibid.*, 3060 (1965).

(27) U. Belluco, R. Ettorre, F. Basolo, R. G. Pearson, and A. Turco, *Inorg. Chem.*, **5**, 591 (1966).

33

Ligand Substitution Catalysis *via* Hard Acid–Hard Base Interaction

Howell R. Clark and Mark M. Jones

Contribution from the Department of Chemistry, Vanderbilt University, Nashville, Tennessee 37203. Received September 6, 1969

Abstract: Spectrophotometric and potentiometric methods have been used to obtain quantitative rate data hich show that certain cations accelerate the hydrolytic displacement of fluoride from a variety of anionic fluoro complexes. Typical hard acids, such as Be(II), Al(III), Zr(IV), and Th(IV), catalyze the hydrolysis of $AsF_5(OH)^-$, PF_6^-, BF_4^-, and AsF_6^-. The extent of catalysis varies with the catalyst: total fluoride ratio, the pH of the hydrolysis media, and the stability of the catalyst ion–fluoro complex. Catalyzed hydrolyses are first order in substrate only when the ratio of catalyst to substrate is large. Direct interaction between bound fluoride and the catalyst ion rather than promotion of an acid catalyzed reaction is shown by the extensive hydrolysis of PF_6^- and A F_6^- under conditions of acidity (pH 2) where the uncatalyzed hydrolysis has a half-life of several months. These results show that the hard and soft acid–base theory can be used to select catalysts for replacement of ligands by basing the selection on the hard or soft nature of the ligand to be replaced.

The hard and soft acid–base theory[1] has contributed significantly to the understanding of acid–base interactions and has become a useful tool for predicting the acid–base behavior of species in many types of reaction. Previously[2] it has been shown that soft acids

can catalyze the replacement of soft bases present as ligands within the coordination sphere of a metal. The objective of the present study was to determine the validity of the generalization of this phenomenon to systems of hard acids and hard bases. Specifically the authors wished to determine if hard acids *in general* were able to catalyze the hydrolytic displacement of complexed fluoride. The hydrogen ion catalysis of

(1) R. G. Pearson, *J. Amer. Chem. Soc.*, **85**, 3533 (1963).

(2) (a) C. Bifano and R. G. Linck, *Inorg. Chem.*, **7**, 908 (1968), and references therein; (b) S. P. Tanner and W. C. E. Higginson, *J. Chem. Soc., A*, 1164 (1969).

such reactions is now well documented[3] and is found both in reactions of the sort

$$[Co(NH_3)_5F]^{2+} + H_2O + H^+ \longrightarrow [Co(NH_3)_5(H_2O)]^{3+} + HF$$

and with fluoro anions as in

$$PF_6^- + H^+ + 4H_2O \longrightarrow H_3PO_4 + 6HF$$

The small body of published work on metal chelate catalyzed fluoride hydrolysis gives some indication of a general interaction between chelate derivatives of hard acid cations and complexed fluoride. The acid hydrolysis of $AsF_5(OH)^-$ has been shown to be catalyzed by a number of iron(III) complexes.[4,5] Furthermore, the hydrolysis of the P-F bond in diisopropyl fluorophosphate[6] and isopropyl methylphosphonofluoridate (Sarin)[7] is catalyzed by chelates of copper(II), zirconium(IV), thorium(IV), and uranium(VI), all in nearly neutral media. Because of interest in chelated and unchelated metal ions, both have been examined as catalysts. When chelates are used, their decolorization by released fluoride can be used to follow the course of the reaction. When aquo ions are used as catalysts, a potentiometric method, adapted from that recently used in this laboratory for studying the acid hydrolysis of PF_6^-,[3b] was used. This method was used in all studies on the metal ion catalyzed hydrolysis of PF_6^- and BF_4^-. This technique offers considerable flexibility in the choice of experimental conditions.

Experimental Section

Materials. $KAsF_5(OH)$ was prepared by literature procedures[8,9] and analyzed for arsenic by an iodimetric procedure following complete alkaline hydrolysis of the complex.[8] *Anal.* Calcd for $KAsF_5(OH)$: As, 33.14. Found: As, 33.26.

KPF_6 and $KAsF_6$ were purchased from the Ozark-Mahoning Co. of Tulsa, Okla. These products were recrystallized from aqueous potassium hydroxide and washed, following suction filtration, with portions of ice water, ethanol, and ether in that order. The purified samples were dried at 100° for 2 hr and stored in a desiccator. Neither material gave evidence of ionic fluoride prior to hydrolysis.

$NaBF_4$ was purchased from Alfa Inorganics, Inc., Beverly, Mass., and was purified by addition of ethanol to a small volume of cold saturated aqueous solution until a slight turbidity appeared, whereupon it was quickly chilled until a considerable quantity of solid precipitated. After collection on a filter the solid was redissolved in the minimum volume of cold water and the precipitation and chilling steps repeated. The filtered solid from the second precipitation was washed successively with portions of ethanol and ether and dried at 100° for 2 hr before storing in a desiccator. The recrystallized material still contained 0.8% of the total fluoride in the ionic condition, for which a correction was made in treating rate data.

The samples were analyzed after ten or more half-lives to ascertain total fluoride release in most cases. Recoveries ranged from 98 to 102% of theoretical fluoride for all systems except as noted.

(3) See, for example: (a) M. Anbar and S. Guttmann, *J. Phys. Chem.*, **64**, 1896 (1960); (b) A. E. Gebala and M. M. Jones, *J. Inorg. Nucl. Chem.*, **31**, 771 (1969); (c) L. N. Devonshire and H. H. Rowley, *Inorg. Chem.*, **1**, 680 (1962).

(4) W. L. Johnson and M. M. Jones, *ibid.*, **5**, 1345 (1966).

(5) G. Slate and M. M. Jones, unpublished results.

(6) T. Wagner-Jauregg, B. E. Hackley, Jr., T. A. Lies, O. O. Owens, and R. Proper, *J. Amer. Chem. Soc.*, **77**, 922 (1955).

(7) R. C. Courtney, R. L. Gustafson, S. J. Westerback, H. Hytiainen, S. C. Chaberek, Jr., and A. E. Martell, *ibid.*, **79**, 3030 (1957).

(8) H. M. Dess and R. W. Parry, *ibid.*, **79**, 1589 (1957).

(9) L. Kolditz and W. Rohnsch, *Z. Anorg. Allg. Chem.*, **293**, 168 (1957).

Chelating dyes used in the spectrophotometric study of the $AsF_5(OH)^-$ ion were obtained commercially and used throughout the study. Concentrations of acids used as hydrolysis media were checked by titration against standard sodium hydroxide.

Stock indicator solutions, except in instances where rate comparisons were sought from varying the metal:indicator ratio, contained equimolar quantities of the dye and metal ion (usually 10^{-3} *M* in each) dissolved in the same medium to be used for the hydrolysis reaction. Aliquots supplying the desired indicator concentration were then diluted to volume (100 ml) prior to a kinetic run. All metal-indicator solutions were stable as evidenced by a negligible change in absorbance upon aging several days.

Rate Measurements. Spectrophotometric. The procedure used was basically that described in ref 4. The present metal-indicator systems have higher molar absorptivities and are more sensitive to fluoride than is the Fe(III)-Ferron chelate previously used.[4] The amount of solid complex per run was reduced to 0.01-0.02 mmol and was weighed on a microbalance.

Calibration with standard sodium fluoride was necessary for each level of indicator concentration used, as the absorbance of none of the systems follows Beer's law with respect to fluoride. Absorbance readings were made with the Beckman Model B spectrophotometer zeroed on a solution containing an equimolar concentration of the indicator dye in the absence of metal ions. The wavelengths (mμ) used are as follows: Al(III)-Alizarin Red S, 490; Be(II)-Eriochrome Cyanine R, 550; and Zr(IV)-Xylenol Orange, 590 (a shoulder wavelength, the use of which was necessary to obtain an adequate fluoride capacity for study due to the high molar absorptivity of this complex).

Potentiometric. All kinetic runs were made in wide-mouth plastic bottles with plastic screw-type caps. A weighed sample (0.1 mmol) of the fluoro complex salt was added to a previously thermostated bottle containing 100 ml of the hydrochloric acid-metal ion solution selected for the hydrolysis reaction medium. A timer was started and the bottle shaken briefly. At intervals, 5-ml aliquots were withdrawn and immediately discharged into 20 ml of 0.5 *M* sodium citrate to quench the reaction and to partially displace the fluoride coordinated by the catalyst ion. The resulting solution was then diluted to 100 ml with an acetate buffer containing 0.5 *M* sodium acetate and 0.5 *M* acetic acid in such proportion to establish the pH of the final solution at 5.00 ± 0.1. The diluent was 1.1 *M* sodium acetate for the runs in 6 *M* hydrochloric acid. The diluted solutions were analyzed by measuring the potential attained by the fluoride electrode *vs.* the sce and comparison with a calibration curve. The fluoride electrode used was an Orion Model 94-09 in conjunction with a Beckman Research Model pH meter, which allowed measurement of potential to ±0.1 mV.

Separate calibration data were required for each change in acid and/or catalyst concentration. Shifts toward higher potential occurred as metal ion concentration was increased. These calibration curves were constructed on single cycle semilog paper from potentials attained by the fluoride electrode in solutions of known fluoride concentration. Calibration solutions were identical in composition and pH with the analysis solutions from the pertinent kinetic run except for substitution of a quantity of standard sodium fluoride solution for the sample.

For runs in low acid and high catalyst ion concentrations, *i.e.* for data to construct Figure 1, the 5-ml aliquots of hydrolysate were initially discharged into 10 ml of 0.5 *M* sodium citrate containing 5 ml of hydrochloric acid, since complexation of the catalyst ion was more rapid and repeatable if initiated in acidic solution. Additional 0.5 *M* sodium citrate was then added to produce a [citrate]: [metal ion] ratio of approximately 10 in the case of Al(III) and 30 with Th(IV). The concentration of hydrochloric acid was chosen so as to yield an analysis solution having a pH near 5.0 when all other components were added and the volume adjusted to 100 ml with water. For example, in runs using 0.12 *M* Th(IV) the hydrolysate aliquots were initially added to 10 ml of 0.5 *M* sodium citrate and 5 ml of 3.0 *M* HCl, then 30 ml of 0.5 *M* sodium citrate and 50 ml of water were added to give 100 ml of solution having a pH of 5.0.

Response time for the fluoride electrode in media containing large concentrations of catalyst ion and citrate (relative to fluoride) is much longer than in media without these interferences. Several minutes were sometimes required for the meter to attain a stable null reading, particularly when a solution was being analyzed that had lower fluoride than the one just preceding. Daily calibration checks were required for runs of long duration when large catalyst concentrations were being used.

The potentiometric method is particularly applicable to the study of fluoro species whose hydrolysis reactions are quenched by buffering in the pH 5 region. Substances which continue to react at this pH, *e.g.*, $AsF_5(OH)^-$, require some procedural modifications even in the absence of catalyst ions and can only be studied with catalyst if the electrode potential stabilizes rapidly, *i.e.*, rapid equilibration between citrate and the catalyst species. Electrode potential readings must be taken as quickly as good mixing is achieved and the meter has attained initial equilibrium. Higher fluoride concentrations (*e.g.*, $10^{-4}-10^{-3}$ M) expedite rapid electrode equilibration. The more active catalyst ions (*e.g.*, Zr(IV)) apparently do not release coordinated fluoride instantly in the presence of excess citrate, thereby causing a delayed electrode response which in turn restricts the value of rate data for metal ion catalyzed hydrolyses of substances which have significant reaction rates in weakly acidic media.

Results

The conditions for, and rate data from, a typical kinetic run are summarized in Table I, together with the relevant calibration points.

Table I. The Al(III) Catalyzed Hydrolysis of BF_4^- [a]

Time, min	mV	[F⁻] × 10³ (sample basis) Total	Net	[(F∞ − F_t)] × 10³
0				3.978
1.0	120.8	0.520	0.487	3.491
2.0	106.1	0.960	0.927	3.051
3.0	98.0	1.346	1.313	2.665
5.0	89.1	1.952	1.919	2.059
7.0	84.1	2.44	2.407	1.571
10.0	79.8	2.94	2.907	1.071
14.0	76.6	3.40	3.367	0.611
∞	72.5	4.06	4.027	

Calibration

[F⁻] × 10⁴	mV
0.25	121.8
0.50	104.9
1.0	88.5
2.0	72.9

[a] $[H^+] = 0.1$, $[Al(III)] = 0.04$, $[NaBF_4] = 1.003 \times 10^{-3}$ M.

In this run the tetrafluoroborate sample contained 4.011×10^{-3} M total fluoride of which 0.033×10^{-3} M was initially free and 3.978×10^{-3} M present in BF_4^-. For analysis 5-ml aliquots were added to 20-ml portions of 0.5 M sodium citrate and diluted to 100 ml with a solution containing equal volumes of 0.5 M sodium acetate and 0.5 M acetic acid. The pH of this solution was 5.02 and the fluoride electrode readings were obtained on it.

K[AsF₅(OH)]. The hydrolysis of this complex is catalyzed by each of the following or complexes thereof: Be(II), Al(III), Zr(IV), and Ti(IV). The first three were studied as colored chelates with Eriochrome Cyanine R (ECR), Alizarin Red S (ARS), and Xylenol Orange (XO), respectively. No colored complex was found for Ti(IV), which would equilibrate rapidly with free fluoride, an essential for the metal–indicator method. Fortunately the equilibration between citrate ion and Ti(IV) from an aliquot of hydrolysate is sufficiently rapid to permit use of the potentiometric method to measure the catalytic effect of Ti(IV).

The hydrolysis of the $AsF_5(OH)^-$ ion in the absence of the metal ion catalyst and at constant acidity conforms to a first-order rate law with respect to the complex.[10] When the catalytic effect from metal ions is large, as was the case for all ions studied except Be(II), the reaction is no longer simple first order, but shows decreasing rates as the reaction progresses. This behavior was general for all anionic fluoro species studied when the catalyst concentration was of similar magnitude to that of the fluoro complex.

For the catalytic reactions of $AsF_5(OH)^-$, rate constants valid for 50% or more of the reaction could be obtained only in the case of mild catalysis by the Be(II)–ECR complex. However, a comparison of typical times required to release half of the total fluoride reflects the magnitude of catalysis among the various systems. Such a comparison is summarized in Table II.

The rates of hydrolysis of the $AsF_5(OH)^-$ ion in 0.10 and 1.0 M HCl as shown in Table II suggest that acid catalysis is operative. This is in agreement with the complete acid hydrolysis pattern for this complex recently resolved in this laboratory.[10]

Half-life data in Table II for the Al(III)–ARS system doped with fluoride ion demonstrate that catalytic activity decreases sharply with consumption of fluoride ion by the catalyst.

KPF₆. The hydrolysis of the PF_6^- ion in 1.0 M HCl is markedly catalyzed by Th(IV) and Zr(IV). The activity of the latter persists even in 6.0 M HCl. Considerably less catalytic effect is shown by Al(III) and Be(II). The reactions are first order with respect to PF_6^- only when a large excess of the catalyst ion is used, *i.e.*, pseudo-first-order catalyst concentrations. Under these conditions only the MF^{n-1} species can form in significant amount.

Typical half-life data for the uncatalyzed hydrolysis and metal ion catalyzed hydrolyses under nonpseudo-first-order conditions are shown in Table III. Table IV presents kinetic data collected using psuedo-first-order catalyst concentrations.

Precipitate formation in the hydrolysate during late stages was observed for Zr(IV)-catalyzed reactions when excess catalyst was used. This was most likely a phosphate salt, since calibration solutions of the same composition except for PF_6^- showed no tendency to give a precipitate. No precipitate was noted for either Th(IV)- or Al(III)-catalyzed reactions. An attempt to use zirconium(IV) sulfate as catalyst resulted in early precipitate formation. Solutions of 0.1 M zirconium(IV) sulfate in 1.0 M HCl exhibited considerable aging just from standing several days at room temperature.

Data reported in Tables III and IV came from media having hydronium ion concentration in great excess over that of the catalyst ion. There was some question as to whether the catalyst ions were interacting directly with bound fluoride or were merely assisting in the acid hydrolysis by facilitating removal of hydrogen fluoride. Accordingly, the hydrolysis of the PF_6^- ion was run in the presence of varying excess amounts of Al(III) and Th(IV) with the acidity adjusted to approximately 0.01 M. At this acidity the rate contribution from acid catalysis could not be significant. A measurable rate under these conditions together with an increase in rate

(10) A. E. Gebala and M. M. Jones, *J. Inorg. Nucl. Chem.*, in press.

Table II. Comparison of Half-Lives for the Hydrolysis of Hydroxopentafluoroarsenate(V) at 25°

Catalyst	[Catalyst]/ [total fluoride]	Reaction medium	$t_{1/2}$, min
None		Acetate buffer[e] pH 4.0	101[f]
Be(II)–ECR (1:1)	0.350[a]	Acetate buffer[e] pH 4.0	80
	0.212[a]	Acetate buffer[e] pH 4.0	87.5
Be(II)–ECR (2:1)	0.418[a]	Acetate buffer[e] pH 4.0	69.0
	0.346[a]	Acetate buffer[e] pH 4.0	72.5
Al(III)–ARS (1:1)	0.500[b]	Acetate buffer[e] pH 4.0	5.2
	0.247[b]	Acetate buffer[e] pH 4.0	10.7
Al(III)–ARS (2:1)	0.555[b]	Acetate buffer[e] pH 4.0	3.6
	0.313[b]	Acetate buffer[e] pH 4.0	6.8
Al(III)–F⁻–ARS (2:2:1)	0.492[b]	Acetate buffer[e] pH 4.0	6.7
	0.290[b]	Acetate buffer[e] pH 4.0	17.1
None		1.0 M HCl	84
Zr(IV)–XO (1:1)	0.450[c]	1.0 M HCl	5.2
	0.385[c]	1.0 M HCl	6.4
	0.300[c]	1.0 M HCl	9.2
	0.227[c]	1.0 M HCl	13.6
None		0.10 M HCl	198
Ti(IV)[g]	0.50[d]	0.10 M HCl	50
	0.30[d]	0.10 M HCl	80

[a] Initial [KAsF$_5$(OH)] (10⁴), 1.1–1.2. [b] Initial [KAsF$_5$(OH)] (10⁴), 1.6–1.8. [c] Initial [KAsF$_5$(OH)] (10⁴), 1.25–1.35. [d] Initial [KAsF$_5$-(OH)] (10⁴), 20. [e] Prepared by adding 0.5 M NaC$_2$H$_3$O$_2$ to 0.5 M HC$_2$H$_3$O$_2$ to a reading of 4.0 on a pH meter. [f] Calculated from the extrapolated value: $k_0 = 6.84 \times 10^{-3}$ min⁻¹. This value was obtained from a plot of k_{obsd} vs. [catalyst] for the Be(II)–ECR (1:1) system using five [catalyst] values ranging from 0.00006 to 0.0002 M. [g] Potentiometric method used with fluoride electrode potentials read rapidly in a pH 4.0 buffer.

Table III. Comparison of Half-Lives for the Catalyzed and Uncatalyzed Hydrolysis of Hexafluorophosphate(V) at 25°

Catalyst	[Metal ion] total/ [fluoride] total[a]	Reaction medium, M HCl	$t_{1/2}$
None		1.0	1565 hr
Zr(IV)	0.25	1.0	718 hr
	0.50		180 hr
	1.0		78 hr
Th(IV)	0.50	1.0	676 hr
	1.0		524 hr
	1.5		213 hr
Al(III)	0.50	1.0	1290 hr
	2.0		767 hr
Be(II)	0.50	1.0	1370 hr
	1.5		1050 hr
None		6.0	1135 min[b]
Zr(IV)	0.50	6.0	298 min
	1.0		118 min
	3.0		44 min

[a] Initial molarity of KPF$_6$ was 0.001 in all cases. [b] Reference 3b.

Table IV. Experimental Rate Constants from the Metal Ion Catalyzed Hydrolysis of Hexafluorophosphate(V) at 25°

Catalyst[a]	Reaction medium, M HCl	$10^4 k_{obsd}$, min⁻¹	k_{cat}[b]/k_0
None	1.0	0.0764	
	2.0	0.214[c]	
	4.0	1.03[c]	
	6.0	6.10	
Zr(IV)	1.0	3.98	51.1
	2.0	16.1	74.3
	4.0	85.8	82.4
	6.0	410	66.2
Th(IV)	1.0	3.19	40.8
	2.0	3.92	17.3
	4.0	8.84	7.6
Al(III)	1.0	0.451	4.9
	2.0	0.745	2.5

[a] The mole ratio of catalyst to total fluoride was 10:1 for all catalysts used; initial molarity of KPF$_6$ was 0.001 in each case. [b] $k_{cat} = k_{obsd} - k_0$. [c] Reference 3b.

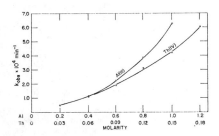

Figure 1. A plot of observed rate constants vs. molar concentration of catalyst for the catalyzed hydrolysis of PF$_6$⁻. The upper abscissa represents the concentration of AlCl$_3 \cdot$6H$_2$O and the lower abscissa represents that of Th(NO$_3$)$_4 \cdot$4H$_2$O. Ionic strength was maintained at 6.0 M for all runs with Al(III) and at 2.0 M for those with Th(IV). The initial pH was 2.0 for all runs. The k_{obsd} values plotted are valid only for approximately the first half-life (see Discussion section).

occurs is apparent from the results of these runs shown in Figure 1.

NaBF₄. Table V summarizes the data for the effect of metal cations on the rate of acid hydrolysis of the BF$_4$⁻ ion. Significant rate increases are apparent for all the hard acid cations examined, but that of Th(IV) exceeds all others. Pseudo-first-order conditions with respect to catalyst were used in all cases. At constant acidity the rate law is

$$-d[BF_4^-]/dt = k_{obsd}[BF_4^-]$$

Rate constants reported in Table V were evaluated by the slope of the first-order plots. Such plots were typically linear through three half-lives. In some cases the reaction was quite rapid, and although good agreement was consistently observed for replicate runs, the rate constants are shown as approximate because of inherent effects on accuracy due to the rapid rate.

KAsF₆. Displacement of fluoride from the As–F bond in this complex was accomplished only by use of high catalyst concentration and reflux temperature. As such, a quantitative rate study was not feasible. Significant, however, is the fact that the hydrolysis is catalyzed by hard acid cations. After refluxing a 0.001

as the catalyst concentration increased would point convincingly to a direct interaction between catalyst species and the fluoro complex. That this indeed

Clark, Jones / Ligand Substitution Catalysis via Hard Acid–Hard Base Interaction

Table V. Experimental Rate Constants for the Metal Ion Catalyzed Hydrolysis of Tetrafluoroborate at 25°

Catalyst[a]	Reaction medium, [H+]	$10^3 k_{obsd}$, min^{-1}	k_{cat}/k_0
None	1.0	7.17[b]	
	0.1	0.830[b]	
	0.01	0.013[b]	
Th(IV)	1.0	~1000	~140
	0.1	~1300	~1,600
	0.01	~1500	~1.15 × 10⁵
Zr(IV)	1.0	800	111
	0.1	160	192
	0.01	5.77[c]	445
Al(III)	1.0	95	12.2
	0.1	133	159
	0.01	157	12,100
Ti(IV)	1.0	39.6	4.52
	0.1	32.0[d]	37.5
Be(II)	1.0	57.6	7.02
	0.1	63.0	75.0
	0.01	71.0	5,460
Al(III)	0.1[e]	124	149
	0.01[e]	114	8,780
	0.001[e]	110	

[a] The mole ratio of catalyst to total fluoride was 10:1 in each case; initial molarity of NaBF$_4$ was 0.001. [b] Experimentally determined using identical conditions for fluoride analysis in runs with catalyst except for omission of catalyst. Reference 3a gives $t_{1/2} = 98$ min for 1.0 M HCl and 850 min for 0.1 M HCl. Above values correspond to 96.5 min for 1 M HCl and 830 min for 0.1 M HCl. [c] A NaOAc–HCl buffer was used to keep the Zr(IV) in solution. Complexation of Zr(IV) by OAc$^-$ decreased the catalytic activity. [d] Freshly diluted from a more concentrated Ti(IV) solution in 1 M HCl. Aging rapidly decreased catalytic activity due to hydrolysis of Ti(IV). [e] NaCl added to adjust ionic strength to 0.26.

M solution of the complex in 1.0 M Al(III) at pH 2 for 80 hr, the hydrolysis was 40% complete. A similar experiment using 0.18 M Th(IV) yielded 29% hydrolysis after 98 hr. Refluxing the same amount of complex for 80 hr in 1.0 M Cd(II) at identical pH produced no detectable hydrolysis, thus indicating that the catalyst role demands a hard acid.

Discussion

Data presented in Tables II–V apparently represent the first reported quantitative evidence for interaction between hard acid metal cations and fluoride ion bound in stable anionic complexes in strongly acidic media. In addition, these data are probably the first involving metal ions functioning as catalysts for fluoride displacement in the absence of seriously competing ligands. Certainly one of the most significant aspects of the study has been the development of a simple experimental technique for the production of quantitative data on fluoride systems that appear to be well within the usual limitations of kinetic precision and accuracy. While no data are available for comparison for the catalyzed reactions, a comparison of our uncatalyzed rate data for tetrafluoroborate using the fluoride electrode and that of Anbar and Guttmann[3a] using a counting technique shows agreement within approximately 2%. Acceptable agreement has been established in a related study with acid hydrolysis rate constants of benzyl fluoride determined by Swain and

Spalding[11] using a steam distillation for fluoride separation followed by a titrimetric analysis. Still another favorable comparison has been established with certain rate constants for the acid hydrolysis of monofluorophosphate reported by Devonshire and Rowley.[3c]

The fluoride electrode technique as described herein has limitations which restrict its generality. One has been stated previously in that the uncatalyzed hydrolysis rate must be negligible in the pH 5 region, which is nearly optimum for measuring the electrode potential. This difficulty can usually be circumvented in the study of hydrolysis uncatalyzed by metal ions but in general cannot be managed otherwise. A second limitation seems likely in dealing with fluorometalate complexes and the necessity for using a large excess of an auxiliary complexing agent (such as the citrate ion) to generate free fluoride ion, the only form to which the electrode responds. Interaction between the auxiliary agent and the fluorometalate species could possibly lead to erroneous potentials. Systems of this type have not been examined from a metal ion catalyzed standpoint in this laboratory.

From an examination of the results given here it appears probable that *hard acid–hard base interaction will generally accelerate fluoride displacement*. This work has demonstrated a reasonable generality for catalyzed displacement of fluoride from the As–F, P–F, and B–F linkages. Another study in progress indicates that certain C–F bonds readily yield to metal ion catalysis. Exceptions are foreseeable for metal ion catalysis as have been noted for acid catalysis, and the two processes will not necessarily parallel. Chelates can arise in these reactions which stabilize, rather than accelerate, the hydrolysis of the species.

Possible inhibitive behavior from catalyst ions was noted in collecting the data shown in Figure 1, particularly for Al(III). Plots of log [PF$_6^-$] vs. time for these systems remain linear only for approximately one half-life. Beyond this point the rate diminishes and becomes extremely slow after 80–85% of reaction. This indicates that the final oxygenated intermediate, the PO$_3$F^{2-} ion, resists hydrolysis under these conditions and that the rate decrease is caused by a buildup of the PO$_3$F^{2-} species. Gustafson and Martell[12] have proposed a four-membered chelate intermediate for the hydrolysis of Sarin, and Courtney, *et al.*,[7] have done likewise for the hydrolysis of diisopropylfluorophosphate. Comparison of structural features of these compounds with those of the PO$_3$F^{2-} ion suggest an even greater possibility of chelation for the latter. Separate experiments have shown that Al(III) and Th(IV), as well as other ions, *inhibit* the hydrolysis of the PO$_3$F^{2-} ion in solutions with [H+] in the 0.1 M region. High acidity affords an acid-catalyzed route for the hydrolysis,[3c] which counteracts the inhibitive effect from metal ions. This accounts for rate decreases not being observed early in the reactions leading to data presented in Table IV.

The extent of catalysis is at least qualitatively related to the catalyst species employed and the acidity of the hydrolysis media. The change in greatest catalytic efficacy from Zr(IV) for the PF$_6^-$ ion to Th(IV) for the

(11) C. G. Swain and R. E. J. Spalding, *J. Amer. Chem. Soc.*, **82**, 6104 (1960).
(12) R. L. Gustafson and A. E. Martell, *ibid.*, **84**, 2309 (1962).

BF$_4^-$ ion suggests that relative fluoride affinity[13] may be an imperfect criterion for predicting catalytic effectiveness. Obviously the catalysis is not restricted completely to metal ion–fluoride interaction but may be influenced by metal ion interaction with the oxygenated hydrolysis product as well. This could be a factor in the efficiency of Zr(IV) for accelerating fluoride displacement from the PF$_6^-$ ion. The results of this study have clearly confirmed that catalytic activity of a given hard acid is greatest in the absence of competing ligands. In addition, there is a definite link between catalytic activity and hydrolytic properties of the catalyst ion on the one hand and the stability of the fluoro complex resulting from catalytic action on the other. It is probable that all fluoro species involving the catalyst are collectively subject to acid hydrolysis. On that assumption, the range of acidity wherein a catalyst ion is active is broadly defined by the most basic environment in which its electrophilic nature has not been significantly diminished by hydroxyl ion attack and the other extreme as the most acidic environment in which the various fluoro complex species escape disruption *via* production of HF.

There is marked variation among the catalyst ions considered in this work with respect to the activity limits just elaborated. For instance, Zr(IV) is seen from Tables III and IV to be most active in rather acidic media, peaking in the 2–4 M acid region in the case of PF$_6^-$. On the other hand, Al(III) shows greater activity with decreasing acidity, both with the PF$_6^-$ ion (Table IV and Figure 1) and the BF$_4^-$ ion (Table V) hydrolyses. The activity of Ti(IV) is apparently confined to a narrow range of acidity. No acceleration of the rate of the AsF$_5$(OH)$^-$ ion hydrolysis was observed on adding Ti(IV) to a 1 M HCl hydrolysis medium, whereas its use in 0.1 M HCl resulted in a significant increase in rate as shown in Table II.

The complexity of kinetic behavior arising from the use of catalyst concentrations of the same order of magnitude as those of the total fluoride may be rationalized by assigning a diminishing catalytic efficiency of catalyst species as fluoride is coordinated. In this manner, the observed first-order rate may be represented as the uncatalyzed rate plus a summation of catalytic contributions to the rate from various fluoro complexes of the catalyst, namely

$$k_{obsd} = k_0 + \sum_{n=0}^{n=N} k_n [MF_n]^{y-n}$$

where an active catalyst ion, M^{y+}, having a maximum coordination number N, has a greater catalytic effect than the MF^{y-1} species, which in turn is more active than the MF$_2^{y-2}$ species, etc. The reasonableness of this rationale arises from the experimentally observed decreasing rates in all cases involving [catalyst]:[total fluoride] ratios of near unity when treated by first-order kinetics. Furthermore, it is evident from Table II, where the catalyst was introduced as equimolar amounts of Al(III) and fluoride, that sharply diminished catalysis is found in the presence of fluoride. A final piece of evidence may be cited from hydrolyses using

pseudo-first-order catalyst concentrations. By using a tenfold excess of catalyst, both the PF$_6^-$ and BF$_4^-$ systems yielded linear first-order plots for at least two half-lives, thus indicating a constancy of rate only when the MF^{y-1} species is the exclusive product in significant amount.

Displacement of fluoride from the As–F bond in the AsF$_6^-$ ion proceeds at a measurable rate only when excess catalyst concentration and reflux temperature are used. This correlates with the extreme resistance of this complex to acid hydrolysis.[14]

The hydrolytic susceptibility of the As–F bond in AsF$_5$(OH)$^-$ compared to that in AsF$_6^-$ is not fully understood. One of the more plausible rationales is that of Kolditz and Rohnsch,[9] who suggest that the AsF$_5$(OH)$^-$ ion probably has octahedral asymmetry arising from the As–O bond being longer than the As–F bond, thereby facilitating electrophilic attack on the fluoride opposite the hydroxo group. The extra fluorine, together with the increased symmetry, makes HAsF$_6$ a much stronger acid than HAsF$_5$(OH), thereby facilitating protonation in the latter case.

The clarification of the mechanistic role of metal ions in fluoride displacement was not an immediate objective of this work; however, the results obtained do provide information on this. Most of the data have been derived from systems wherein the hydronium ion concentration greatly exceeded that of the metal ion catalyst. This was done, first of all, for comparisons of catalyzed *vs.* uncatalyzed rates and, secondly, to exclude basic hydrolysis of the catalyst species. The role of the catalyst under these conditions might be construed to exclude direct interaction with the fluoro species and be limited to assistance in the expulsion of HF from the protonated intermediate. The results with the AsF$_6^-$ ion, the data for the PF$_6^-$ ion shown in Figure 1, and the final pieces of data for the BF$_4^-$ ion in Table V prove conclusively that direct coordination is operative. All of these runs involved acid concentrations too small to contribute significantly to the rate. Only the catalyst ions functioning independently of acid hydrolysis could have brought about the large rate increases observed.

A simple interpretation of the process would thus incorporate coordination of the aquated catalyst ion to one or more fluorine atoms, thereby creating a species susceptible to nucleophilic attack, presumably by a water molecule. Indeed, the metal ion interaction must parallel that of hydronium ion to some degree. Multiple coordination is possible, however, from a single metal ion, and the electrophilic nature of the intermediate is apparently much greater than from protonation. This presumption becomes more reasonable when one considers the relative catalytic efficacy of 1.0 M HCl with that of 1.0 M AlCl$_3$. In the latter solution the initial decomposition of the PF$_6^-$ ion is more than 80 times faster than the first; it has an accelerative effect almost identical with that of 6.0 M HCl on the rate of fluoride release from the PF$_6^-$ ion. The greater polarizing effect of Th(IV) over Al(III) on the incipient fluoride ion is manifested in the fact that 0.18 M Th(IV) is essentially equivalent in accelerative effect to 1.0 M

(13) S. Ahrland, J. Chatt, and N. R. Davies, *Quart. Rev.* (London), 12, 265 (1958).

(14) W. L. Lockhart, D. O. Johnston, and M. M. Jones, *J. Inorg. Nucl. Chem.*, 31, 407 (1969).

Clark, Jones / *Ligand Substitution Catalysis via Hard Acid–Hard Base Interaction*

Al(III). Other investigations into the scope and mechanistic aspects of metal ion catalyzed hydrolysis of fluorine-containing species are in progress. Presently there is optimism regarding application to certain C–F linkages. This type of interaction has practical consequence in addition to the theoretical application of the hard and soft acid–base theory to metal-catalyzed substitutions in a number of organic reactions.

Acknowledgment. Financial support for this work from a National Science Foundation Science Faculty Predoctoral Fellowship and the U. S. Atomic Energy Commission is gratefully acknowledged.

34

Formation of Metal Ion–Nitrene Complexes

Sir:

The production of electron-deficient nitrene species *via* the action of ultraviolet light or acid on organic azides is well established.[1,2] The stabilization of nitrenes by coordination to transition metals has been hypothesized,[3] and recently a copper–nitrene intermediate has been proposed in the copper-catalyzed decomposition of benzenesulfonyl azide.[4] However, as yet no well-established case of a metal–nitrene complex has been described. We wish now to report the formation of nitrene coordinated to ruthenium ion *via* the action of acid on ruthenium(III) azide complexes.

The addition of aqueous acid to ruthenium(III) azides causes the vigorous evolution of gas. Mass spectral analysis showed this to be greater than 99% nitrogen. This behavior is markedly different from that of most transition metal azides with acid. In general, acid merely catalyzes the displacement of coordinated azide as hydrazoic acid. For the action of H_2SO_4 on $[Ru(NH_3)_5N_3](N_3)_2$, the products have been characterized from their ultraviolet spectra. Two intense bands were produced at 220 and 262 mμ (Figure 1). Their wavelengths and intensities agree well with those reported for $[Ru(NH_3)_5N_2]^{2+}$ (A) and the dimer $[(NH_3)_5Ru-N_2-Ru(NH_3)_5]^{4+}$ (B), respectively.[5] When reaction solutions were made slightly basic the band at 262 mμ gradually decreased in intensity and the band at 220 mμ increased. A sharp isosbestic point was observed at 240 mμ. This latter behavior confirms the assignment of products since Taube, *et al.*, have previously shown that $[(NH_3)_5Ru-N_2-Ru(NH_3)_5]^{4+}$ dissociates in weakly basic media into $[Ru(NH_3)_5N_2]^{2+}$ (A) and $[Ru(NH_3)_5OH]^{+}$.[5] Samples isolated from acid solution (as tetraphenylborate salts) at the end of gas evolution were diamagnetic and exhibited strong infrared bands in the region (2100–2200 cm^{-1}) associated

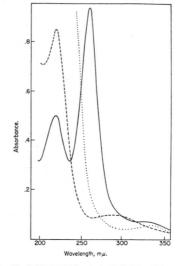

Figure 1. $[Ru(NH_3)_5N_3](N_3)_2 + 1\ N\ H_2SO_4,\ 7.8 \times 10^{-5}\ M$ (———); $[Ru(NH_3)_5N_3](N_3)_2 + H_2O,\ 9.9 \times 10^{-5}\ M$ (------); $[Ru(NH_3)_5N_3](N_3)_2 + 1\ N\ H_2SO_4,\ 7.8 \times 10^{-5}\ M,\ + 1.90 \times 10^{-4}\ M$ thiourea (·······).

with ruthenium(II)–nitrogen complexes. The analyses of these solids corresponded to mixtures of A and B.

The results of product composition studies on the acid reaction of $[Ru(NH_3)_5N_3](N_3)_2$ are summarized in Table I. In 0.80 M H_2SO_4, reaction is complete within 5 min. This rapid appearance of $[(NH_3)_5Ru-N_2-Ru(NH_3)_5]^{4+}$ (B) eliminates the possibility that it is produced by the combination of $[Ru(NH_3)_5N_2]^{2+}$ and $[Ru(NH_3)_5(H_2O)]^{2+}$. The rate of the latter reaction has been measured by Itzkovitch and Page.[6] Under the conditions employed here, such combination would require several hours.

The vigorous evolution of nitrogen and the rapid formation of nitrogen-bridged dimers are very reminiscent of the behavior of organic azides when treated with acid or ultraviolet radiation.

$$Ph-N_3 \xrightarrow{h\nu} Ph-N + N_2$$

$$Ph-N + Ph-N \longrightarrow Ph-N{=}N-Ph$$

We believe a similar pathway involving a metalated nitrene (C) is present in the acid reaction of $[Ru(NH_3)_5N_3]^{2+}$ (eq 1). Strong support for the presence

$$[(NH_3)_5Ru-N{=}N{=}N]^{2+} \xrightarrow{H^+} [(NH_3)_5Ru-\overset{H}{N}-N{=}N]^{3+} \longrightarrow$$

$$[(NH_3)_5Ru-NH]^{3+} + N_2$$
$$C$$

$$2[(NH_3)_5Ru-NH]^{3+} \longrightarrow$$
$$C$$

$$[(NH_3)_5Ru-\overset{H}{N}{=}\overset{H}{N}-Ru(NH_3)_5]^{6+}$$

$$\downarrow$$

$$[(NH_3)_5Ru-N_2-Ru(NH_3)_5]^{4+} + 2H^+ \quad (1)$$
$$B$$

of the nitrene intermediate (C) was obtained from trapping experiments. In the presence of small

Table I. Reaction of H_2SO_4 on $[Ru(NH_3)_5N_3](N_3)_2$ at Room Temperature^a

Complex concn, M	H_2SO_4 concn, M	Per cent yield A	Per cent yield B	Total per cent yield (A + B)
9.5×10^{-5}	Water	85	0	85
9.5×10^{-5}	0.05	102	8	110
9.5×10^{-5}	0.15	53	45	98
9.5×10^{-5}	0.40	47	52	99
9.5×10^{-5}	0.80	40	67	107
9.5×10^{-5}	4.0	30	65	95
9.5×10^{-5}	0.80	40	67	107
9.9×10^{-4}	0.80	45	50	95
1.1×10^{-2}	0.80	93	8	101
6.5×10^{-2}	0.80	95	1	95

^{a} The most significant features apparent from Table I are: (i) the production of large amounts of $[(NH_3)_5Ru-N_2-Ru(NH_3)_5]^{4+}$ (B), (ii) the absence of dimer (B) formation unless acid is present, and (iii) the marked decrease in dimer (B) yield with increasing concentration of $[Ru(NH_3)_5N_3](N_3)_2$.

(1) L. Horner and A. Christman, *Angew. Chem. Intern. Ed. Engl.*, **2**, 599 (1963).
(2) W. Lwowski, *ibid.*, **6**, 5899 (1968).
(3) R. Gleiter and R. Hoffman, *Tetrahedron*, **24**, 5899 (1969).
(4) H. Kwart and A. Kahn, *J. Am. Chem. Soc.*, **89**, 1950 (1967).
(5) D. F. Harrison, E. Weissberger, and H. Taube, *Science*, **159**, 320 (1967).
(6) I. J. Itzkovitch and J. A. Page, *Can. J. Chem.*, **46**, 2743 (1968).

Communications to the Editor

amounts of thiourea, diethyl sulfide, or I⁻, the reaction of acid on [Ru(NH₃)₅N₃](N₃)₂ generated no dimer (B) (see Figure 1). This behavior may be readily rationalized in terms of the Lewis acid character of $[(NH_3)_5Ru-NH]^{3+}$ (C). Coordinated nitrene, like the similar carbene, is expected to be a soft acid.[7] Combination of $(NH_3)_5Ru-NH$ with itself to produce the dimer (B) is quenched by the more rapid reaction of the nitrene with the soft bases thiourea, $(C_2H_5)_2S$, or I⁻. On the other hand, reagents such as Cl⁻, PF_6^-, CH_3CN, or dimethyl sulfoxide had no observable effect on the course of the acid reaction.

The decrease in yield of the dimer (B) with increasing initial concentration of $[Ru(NH_3)_5N_3](N_3)_2$ (Table I) suggests that $[Ru(NH_3)_5N_3]^{2+}$ may also function as an effective trap for the nitrene (C) (eq 2). A combina-

$$[(NH_3)_5Ru-N=N=N]^{2+} + [HN-Ru(NH_3)_5]^{3+} \longrightarrow$$

$$[(NH_3)_5Ru-N=N\overset{|}{-}N \cdots\cdots \overset{H}{\overset{|}{N}}-Ru(NH_3)_5]^{5+}$$

$$\swarrow$$

$$2[Ru(NH_3)_5N_2]^{2+} + H^+ \qquad (2)$$
$$A$$

tion of eq 1 and 2 explains the production of both A and B in the acid reaction of $[Ru(NH_3)_5N_3]^{2+}$.

The behavior of cis-$[Ru(en)_2(N_3)_2]PF_6$ and cis-$[Ru(trien)(N_3)_2]PF_6$ in acid is very similar to that of $[Ru(NH_3)_5N_3](N_3)_2$. The final spectra again exhibited two intense bands at 221 and 264 mμ. The band at 221 mμ agrees well with the known spectrum of cis-$[Ru(en)_2N_2H_2O]^{2+}$ (A′).[8] The bands at 264 mμ have been assigned to the dimers $[(A-A)_2H_2ORu-N_2-RuH_2O-(A-A)_2]^{4+}$ (B′) (A-A = en, 0.5 trien). The latter compounds have been independently prepared in situ, and their spectra ($\epsilon_{264} = 48,000$) agree closely with that reported for $[(NH_3)_5Ru-N_2-Ru(NH_3)_5]^{4+}$.[5] The presence of small amounts of thiourea or I⁻ in the acid reactions again eliminated the formation of the dimers

(B′), supporting the presence of reactive nitrene intermediates.

Attempts to observe an esr signal for the postulated nitrene intermediates have been unsuccessful. Solutions of cis-$[Ru(en)_2(N_3)_2]PF_6$ (0.038 M) in H_2SO_4 (0.80 M) were frozen in liquid nitrogen after various reaction times, and their esr spectra recorded. All resonances observed in the region 1000–8000 G have been assigned to the unpaired electron of ruthenium(III). These bands decreased with increasing reaction time due to the formation of ruthenium(II). The failure to observe a nitrene resonance suggests it may be present in a singlet electronic state, though it may also be due to a low nitrene concentration.

A search has been made for similar acid-catalyzed behavior among other transition metal azide complexes. Azide complexes of Co(III), Rh(III), Pt(II), Pd(II), and Au(III) generated no gas when dissolved in 4 M H_2SO_4 at room temperature. However, similar behavior has been found in these laboratories for the Ir(III) complex trans-$[Ir(en)_2(N_3)_2]PF_6$.[9] Ultraviolet irradiation of metal azide complexes appears to be an alternative source of metalated nitrenes.[10] Compounds such as $ReCl_3(P(C_2H_5)_2C_6H_5)_2NC_6H_4X$ are known and their structures have been determined.[11] While these may formally be considered as arylnitrenes coordinated to metal ions, they are very stable and are best formulated as arylimino complexes.[12]

Acknowledgment. This work was supported by the National Science Foundation, Grant 6341X to Northwestern University.

(7) R. G. Pearson and J. Songstad, *J. Am. Chem. Soc.*, **89**, 1827 (1967).
(8) L. Kane-Maguire, P. S. Sheridan, F. Basolo, and R. G. Pearson, *ibid.*, **90**, 5295 (1968).
(9) R. Bauer, unpublished results.
(10) B. Hoffman, private communication.
(11) D. Bright and J. A. Ibers, *Inorg. Chem.*, **7**, 1099 (1968).
(12) J. Chatt, J. D. Garforth, N. P. Johnson, and G. A. Rowe, *J. Chem. Soc.*, 1012 (1964).

Leon A. P. Kane-Maguire, Fred Basolo, Ralph G. Pearson
Department of Chemistry, Northwestern University
Evanston, Illinois 60201
Received May 14, 1969

[Reprinted from Inorganic Chemistry, **7**, 1475 (1968).]

35

Reactivity Patterns in Inner- and Outer-Sphere Reductions of Halogenopentaamminecobalt(III) Complexes[1]

Sir:

The reductions of halogenopentaamminecobalt(III) complexes by various reducing agents have been extensively investigated. For all of the reducing agents studied the rate constants vary monotonically as the halide is changed from fluoride to iodide. The reactivity order has come to be known as "normal" or "inverse" depending on whether the rate increases or decreases with increasing atomic number of the halogen. The normal reactivity order (F < Cl < Br < I) is observed for the inner-sphere reductions by chromium(II)[2] and pentacyanocobaltate(II);[3] for the outer-

(1) This work was supported by the National Science Foundation under Grant GP-6528.

(2) J. P. Candlin and J. Halpern, *Inorg. Chem.*, **4**, 766 (1965).

(3) J. P. Candlin, J. Halpern, and S. Nakamura, *J. Am. Chem. Soc.*, **85**, 2517 (1963).

sphere reductions, by tris(bipyridyl)chromium(II)[4] and hexaammineruthenium(II),[5] as well as for the reductions by vanadium(II)[4,6] which proceed by an unknown mechanism. In contrast, the reductions by europium(II)[4] and iron(II)[6,7] which also proceed by unknown mechanisms obey the inverse order (F > Cl > Br > I).

It was noted[2,4] that the latter reactivity order parallels the order of the thermodynamic stabilities of the europium(III) and iron(III) halide complexes. On this basis, it was suggested[2,4] that the inverse order is determined by the driving force for reaction and, therefore, that the inner-sphere mechanism obtains for the europium(II) and iron(II) reductions. The assumptions involved in correlating the reactivity sequence with the standard free energy change for the reactions have been explicitly stated.[6] In this context, it is noteworthy that the reversal in reactivity in the reactions of chromium(II) and iron(II) with the halogenopentaamminecobalt(III) complexes has been predicted by using the Marcus equation.[8]

It is apparent that trying to understand the reactivity orders described above poses a challenging problem, but so far only some of the factors that need to be taken into consideration have been enumerated.[8-10] Moreover, if we accept, as seems likely, that the iron-(II) and europium(II) reductions proceed *via* halide-bridged transition states, then it is clear that the reactivity orders are not useful indirect criteria for distinguishing between inner- and outer-sphere mechanisms.

In the present note we wish to examine the reductions of halogenopentaamminecobalt(III) complexes on the basis of formal calculations of equilibrium constants involving transition states.[11] It must be noted that in performing these calculations, it is assumed that the transmission coefficient κ, in the transition-state theory expression $k = \kappa(RT/Nh)K^{\ddagger}$, is unity. Viewed in this manner, the reactions under consideration fall into three categories: (1) inner-sphere reductions by class a or hard metal ions, (2) outer-sphere reductions, and (3) inner-sphere reductions by class b or soft metal ions.

The equilibrium constant for the reaction of the transition state $[(NH_3)_5CoFCr^{4+}]^{\ddagger}$ with iodide ion to produce the transition state $[(NH_3)_5CoICr^{4+}]^{\ddagger}$ and fluoride ion (eq 5) can be obtained by appropriate combination of reactions 1–4. The value of Q_5 (subject

$$Co(NH_3)_5F^{2+} + Cr^{2+} \rightleftarrows [(NH_3)_5CoFCr^{4+}]^{\ddagger} \qquad k_1 \quad (1)$$

(4) J. P. Candlin, J. Halpern, and D. L. Trimm, *J. Am. Chem. Soc.*, **86**, 1019 (1964).
(5) J. F. Endicott and H. Taube, *ibid.*, **86**, 1686 (1964).
(6) H. Diebler and H. Taube, *Inorg. Chem.*, **4**, 1029 (1965).
(7) J. H. Espenson, *ibid.*, **4**, 121 (1965).
(8) N. Sutin, *Ann. Rev. Phys. Chem.*, **17**, 119 (1966).
(9) H. Taube, Welch Conference, "Modern Inorganic Chemistry," Houston, Texas, 1962.
(10) F. Basolo and R. G. Pearson, "Mechanisms of Inorganic Reactions," 2nd ed, John Wiley and Sons, Inc., New York, N. Y., 1967, Chapter 6.
(11) Dr. T. W. Newton (private communication to J. H. Espenson, *Inorg. Chem.*, **4**, 1025 (1965)) has previously carried out calculations of acid dissociation constants of transition states. This type of calculation can be generalized to many other formal reactions of transition states: T. W. Newton and F. B. Baker, *J. Phys. Chem.*, **67**, 1425 (1963); D. E. Pennington and A. Haim, to be submitted for publication.

$$Co(NH_3)_5I^{2+} + Cr^{2+} \rightleftarrows [(NH_3)_5CoICr^{4+}]^{\ddagger} \qquad k_2 \quad (2)$$

$$Co(NH_3)_5OH_2^{3+} + F^- \rightleftarrows Co(NH_3)_5F^{2+} + H_2O \qquad Q_3 \quad (3)$$

$$Co(NH_3)_5OH_2^{3+} + I^- \rightleftarrows Co(NH_3)_5I^{2+} + H_2O \qquad Q_4 \quad (4)$$

$$[(NH_3)_5CoFCr^{4+}] + I^- \rightleftarrows [(NH_3)_5CoICr^{4+}]^{\ddagger} + F^-$$
$$Q_5 = k_2Q_4/k_1Q_3 \quad (5)$$

to the validity of the above assumption) gives a measure of the relative stabilities of the transition states with fluoride and iodide bridges, and it is seen that this formal calculation effectively provides a correction of the rate constant ratio k_2/k_1 for the difference in free energies of the ground states.

The results of our calculations are presented in Table I and depicted schematically in Figure 1. Some com-

TABLE I

FORMAL EQUILIBRIUM CONSTANTS AT 25° FOR SUBSTITUTION OF HALIDE IONS IN VARIOUS TRANSITION STATES[a]

Reaction	Q
$[(NH_3)_5CoFCr^{4+}]^{\ddagger} + I^- \rightleftarrows$ $[(NH_3)_5CoICr^{4+}]^{\ddagger} + F^-$	6.4×10^{-2}
$[(NH_3)_5CoFV^{4+}]^{\ddagger} + Br^- \rightleftarrows$ $[(NH_3)_5CoBrV^{4+}]^{\ddagger} + F^-$	7.3×10^{-2}
$[(NH_3)_5CoFFe^{4+}]^{\ddagger} + Br^- \rightleftarrows$ $[(NH_3)_5CoBrFe^{4+}]^{\ddagger} + F^-$	1.6×10^{-3}
$[(NH_3)_5CoFEu^{4+}]^{\ddagger} + I^- \rightleftarrows$ $[(NH_3)_5CoIEu^{4+}]^{\ddagger} + F^-$	2.2×10^{-5}
$[(NH_3)_5CoClRu(NH_3)_6^{4+}]^{\ddagger} + I^- \rightleftarrows$ $[(NH_3)_5CoIRu(NH_3)_6^{4+}]^{\ddagger} + Cl^-$	2.5×10^2
$[(NH_3)_5CoFCr(bipy)_3^{4+}]^{\ddagger} + Br^- \rightleftarrows$ $[(NH_3)_5CoBrCr(bipy)_3^{4+}]^{\ddagger} + F^-$	3.9×10^2
$[(NH_3)_5CoFCo(CN)_5^-]^{\ddagger} + Br^- \rightleftarrows$ $[(NH_3)_5CoBrCo(CN)_5^-]^{\ddagger} + F^-$	$>1.6 \times 10^4$

[a] Rate constants used in calculations taken from ref 2–7. Equilibrium constants taken from: D. A. Buckingham, I. I. Olsen, A. M. Sargeson, and H. Satrapa, *Inorg. Chem.*, **6**, 1027 (1967); C. H. Langford, *ibid.*, **4**, 265 (1965). The equilibrium constants at 25° for the formation of $Co(NH_3)_5X^{2+}$ (X = F, Cl, Br, I) from $Co(NH_3)_5OH_2^{3+}$ and the appropriate halide are 25, 1.11, 0.35, and 0.12 M^{-1} for F, Cl, Br, and I, respectively.

ments regarding the calculations in Table I are pertinent. For the vanadium(II) reductions, we chose to calculate the equilibrium constant for substitution of fluoride by bromide rather than by iodide. The rate constants for the reductions of $Co(NH_3)_5F^{2+}$, $Co(NH_3)_5$-Cl^{2+}, and $Co(NH_3)_5Br^{2+}$ by vanadium(II) fall in the range 5–15 M^{-1} sec^{-1}, characteristic of inner-sphere reductions,[12-14] whereas the rate constant for Co-$(NH_3)_5I^{2+}$ is 1.2×10^2 M^{-1} sec^{-1}, a value that suggests the operation, at least in part, of an outer-sphere mechanism.[12] The reductions by iron(II) and europium(II) are formulated as inner-sphere reactions. Direct evidence for bridged transition states in the reduction of a variety of cobalt(III) complexes by iron(II) has been previously presented.[15] The mechanism is unknown for europium(II), but, as will be seen below, the present comparisons suggest an inner-sphere mechanism. When the rate constants for reductions of the fluoro- or iodopentaamminecobalt(III) complexes have not been

(12) B. R. Baker, M. Orhanovic, and N. Sutin, *J. Am. Chem. Soc.*, **89**, 722 (1967).
(13) J. H. Espenson, *ibid.*, **89**, 1276 (1967).
(14) H. J. Price and H. Taube, *Inorg. Chem.*, **7**, 1 (1968).
(15) A. Haim and N. Sutin, *J. Am. Chem. Soc.*, **88**, 5343 (1966).

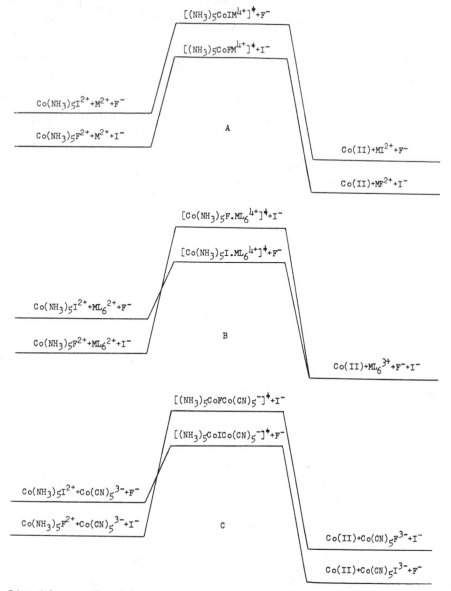

Figure 1.—Schematic free energy diagram of reactants, transition states, and products for three categories of reactions: A, $Co(NH_3)_5X^{2+}$ + M^{2+} (M = Cr, V, Fe, Eu); B, $Co(NH_3)_5X^{2+}$ + ML_6^{2+} (ML_6 = $Ru(NH_3)_6$, $Cr(bipy)_3$); C, $Co(NH_3)_5X^{2+}$ + $Co(CN)_5^{3-}$.

reported, the calculations are carried out for the corresponding chloride or bromide complexes, respectively. Equilibrium constants involving the intermediate halides are not included in Table I. However, in all cases the equilibrium constants vary monotonically as one goes through the sequence Cl, Br, I.

An examination of the values of Q listed in Table I reveals that, for chromium(II), iron(II), europium-(II), and vanadium(II), the substitution of a bridging fluoride by a bridging iodide (or bromide) is an *un-favorable* process. Therefore, using the values of Q as a measure of stability, it is seen that the stability order of the transition states is F > Cl > Br > I and is the same as the reactivity order for iron(II) and europium(II), but opposite to that for chromium(II) and vanadium(II). It is noteworthy that when one compares reactivities, these four reducing agents fall into two classes. However, when one compares the stabilities of the transition states, all of these reducing agents exhibit a common sequence.

For the outer-sphere reductions by $Cr(bipy)_3^{2+}$ and $Ru(NH_3)_6^{2+}$, the substitution of a coordinated fluoride (or chloride) by bromide (or iodide) is a *favorable* process, and, therefore, the stability order of transition states $F < Cl < Br < I$ is the same as the reactivity order. Finally, for the inner-sphere reductant $Co(CN)_5^{3-}$, substitution of fluoride by bromide is favorable, and the stability order for the transition states ($F < Cl < Br < I$) is the same as the reactivity order.

The differences between the three categories are displayed in Figure 1. For the reductions by the class a[16] or hard[17] metal centers, the stability orders for reactants, transition states, and products follow the sequence $F > Cl > Br > I$. For the outer-sphere reactions, there is a reversal in stability order as one goes from reactants ($F > Cl > Br > I$) to transition states ($F < Cl < Br < I$). Finally, for the inner-sphere reductions by the soft[18] center $Co(CN)_5^{3-}$, the stability order of the transition states and the products ($F < Cl < Br < I$) is the reverse of the stability order of the reactants.

Two other notable classes of reactions that follow the normal reactivity order are the chromium(II)-catalyzed dissociation of halogenopentaamminechromium(III) complexes[19] and the exchange of chromium atoms between halogenopentaaquochromium(III) complexes and chromium(II).[20] Unfortunately, no thermodynamic data are available for the pentaammine series. However, equilibrium constants have been reported for the formation of $Cr(H_2O)_5F^{2+}$ and $Cr(H_2O)_5Cl^{2+}$.[21,22] Combining these values with the rate constants for the $Cr(H_2O)_5F^{2+}-Cr^{2+}$ and $Cr(H_2O)_5Cl^{2+}-Cr^{2+}$ reactions,[20] we arrive at a value of $Q_0 \sim 1.5 \times 10^{-2}$ for

$$[(H_2O)_5CrFCr^{4+}] \ddagger + Cl^- \rightleftarrows [(H_2O)_5CrClCr^{4+}] \ddagger + F^- \quad (6)$$

The substitution of a bridging fluoride by a bridging chloride is an unfavorable reaction, and we conclude that although the reactivity order is $Cl > F$, the stabil-

ity order of the transition states is $F > Cl$. It is noteworthy that this inner-sphere system displays a behavior entirely analogous to that of the other inner-sphere systems $Co(NH_3)_5X^{2+} + Cr^{2+}$, Fe^{2+}, V^{2+}, and Eu^{2+}, in spite of the fact that the standard free energy change for the exchange reactions is zero.

The exchange reactions between iron(II) and iron(III) catalyzed by fluoride and chloride ions exhibit the inverse reactivity order.[8] It has been shown that part of the $FeCl^{2+}-Fe^{2+}$ reaction proceeds *via* an inner-sphere mechanism.[23] For the fluoride-catalyzed reaction, FeF^{2+} has been shown to be the reactive species,[24] and it appears reasonable to assume that an inner-sphere mechanism obtains. The equilibrium constant for reaction 7 calculated from the known rate[25]

$$[FeFFe^{4+}] \ddagger + Cl^- \rightleftarrows [FeClFe^{4+}] \ddagger + F^- \quad (7)$$

and equilibrium[8] constants is $Q_7 \sim 10^{-5}$. The reaction is unfavorable, and again the transition states exhibit the stability order $F > Cl$. It is noteworthy, as shown above for the reductions of $Co(NH_3)_5X^{2+}$ by europium(II) and iron(II), on one hand, and by chromium(II) and vanadium(II), on the other hand, that there is a reversal in the reactivities of fluoride and chloride complexes in going from the iron to the chromium-exchange systems. However, when one compares the stabilities of the transition states, the two systems conform to the stability order $F > Cl$.

On the basis of the systematization presented above, it seems possible that a comparison of the stabilities of the transition states, rather than the reactivity order for the series F, Cl, Br, and I, could provide a useful indirect criterion for distinguishing between inner- and outer-sphere mechanisms.

Acknowledgment.—The author is grateful to Dr. N. Sutin for illuminating comments and criticisms.

(16) S. Ahrland, J. Watt, and N. R. Davies, *Quart. Rev.* (London), **12**, 265 (1958).
(17) R. G. Pearson, *J. Am. Chem. Soc.*, **85**, 3533 (1963).
(18) R. Grassi, A. Haim, and W. K. Wilmarth, *Inorg. Chem.*, **6**, 237 (1967)
(19) A. E. Ogard and H. Taube, *J. Am. Chem. Soc.*, **80**, 1084 (1958).
(20) D. L. Ball and E. L. King, *ibid.*, **80**, 1091 (1958).
(21) T. W. Swaddle and E. L. King, *Inorg. Chem.*, **4**, 532 (1965).
(22) C. F. Hale and E. L. King, *J. Phys. Chem.*, **71**, 1779 (1967).

(23) R. J. Campion, T. J. Conocchioli, and N. Sutin, *J. Am. Chem. Soc.*, **86**, 4591 (1964).
(24) J. Menashi, S. Fukushima, and W. L. Reynolds, *Inorg. Chem.*, **3**, 1242 (1964).
(25) N. Sutin, *Ann. Rev. Nucl. Sci.*, **12**, 285 (1962).
(26) Fellow of the Alfred P. Sloan Foundation, 1965–1968.

DEPARTMENT OF CHEMISTRY ALBERT HAIM[26]
STATE UNIVERSITY OF NEW YORK
STONY BROOK, NEW YORK 11790
RECEIVED JANUARY 30, 1968

[Reprinted from the Journal of the American Chemical Society, 94, 856 (1972).]

36

856

Homolytic and Ionic Mechanisms in the Ligand-Transfer Oxidation of Alkyl Radicals by Copper(II) Halides and Pseudohalides

C. L. Jenkins and J. K. Kochi*

Contribution from the Department of Chemistry, Indiana University, Bloomington, Indiana 47401. Received April 16, 1971

Abstract: The facile oxidation of alkyl radicals by various copper(II) halides and pseudohalides involves the transfer of the ligand from the copper(II) oxidant to the alkyl radicals. Two simultaneous processes are described in which atom transfer occurs primarily *via* a homolytic transition state involving minimal polar character. The competing oxidative substitution proceeds *via* cationic intermediates, and an alkylcopper species similar to those described in oxidative solvolysis during electron-transfer oxidation is proposed. Atom transfer usually prevails during ligand-transfer oxidation with most alkyl radicals, but the cationic route becomes important when alkyl radicals such as cyclobutyl and β-anisylethyl, which afford stabilized carbonium ions, undergo ligand-transfer oxidation. The cationic route can also be promoted by use of dioxane as solvent. The scrambling of carbon atoms in cyclobutyl radicals and isotopic labeling in β-anisylethyl radicals are used as probes for cationic intermediates. In a more general sense, ligand-transfer and electron-transfer categories are superseded by a unified mechanism for the oxidation and reduction of free radicals by metal complexes based on the hard and soft acid–base classification.

A variety of alkyl radicals R· are readily oxidized by copper(II) halides (Cl, Br, I) and pseudohalides (SCN, N$_3$, CN) by a process which has been described as *ligand transfer*.[1] The relevant reaction involves the transfer of the ligand associated with the metal [copper(II)] to the free radical and concomitant reduction of the metal as given in eq 1. These reactions proceed at

$$R\cdot + Cu^{II}X_2 \longrightarrow R-X + Cu^IX \qquad (1)$$

$$X = Cl, Br, I, SCN, N_3, CN$$

rates which approach the diffusion-controlled limit, having second-order rate constants of 3.6×10^8, 1.1×10^9, and 4.3×10^9 M^{-1} sec^{-1} at 25° for the transfer of thiocyanate, chloride, and bromide, respectively. Iodide, azide, and cyanide can also be efficiently transferred from copper(II) complexes to alkyl radicals. Thus, they can and do form the vital link in a number of catalytic reactions involving copper.[2]

In this report we wish to examine the mechanism of the *ligand-transfer* oxidation of alkyl radicals by copper(II) halides and pseudohalides in the light of our knowledge of the electron-transfer process described in the foregoing paper.[3] The role of carbonium ions as intermediates and the effect of the solvent in ligand-transfer oxidation of alkyl radicals are of particular importance in probing for the mechanism of this interesting process. The oxidations of homoallylic (cyclobutyl, cyclopropylmethyl, and allylcarbinyl) radicals are examined in detail, and factors leading to cationic rearrangement delineated. Finally, deuterium labeling in β-phenethyl and β-anisylethyl radicals establishes a duality of mechanisms, one path homolytic and the other heterolytic in which carbonium ions participate.

The catalytic decomposition of diacyl peroxides by copper complexes represents a method generally adaptable for the production of a variety of alkyl radicals

for ligand-transfer studies.[1] For example, the catalytic reaction of propionyl peroxide and copper(II) chloride has been shown to proceed *via* a chain reaction involving the copper(I)–copper(II) redox cycle (X = Cl). The kinetic chain length in these catalytic re-

$$(CH_3CH_2CO_2)_2 + Cu^IX \longrightarrow CH_3CH_2\cdot + CO_2 + \\ CH_3CH_2CO_2Cu^{II}X \qquad (2)$$

$$CH_3CH_2CO_2Cu^{II}X + X^- \rightleftharpoons CH_3CH_2CO_2^- + Cu^{II}X_2 \qquad (3)$$

$$CH_3CH_2\cdot + Cu^{II}X_2 \longrightarrow CH_3CH_2X + Cu^IX, \text{ etc.} \qquad (4)$$

actions is in excess of 20. Other alkyl radicals used in this study were produced by the catalytic decomposition of the appropriate diacyl peroxide.

Results

Oxidation of Neopentyl Radicals by Copper(II) Chloride. Neopentyl radicals were derived in the catalytic decomposition of *tert*-butylacetyl peroxide. The oxidation of neopentyl radicals by copper(II) chloride afforded neopentyl chloride exclusively as shown in Table I. Control experiments showed that *tert*-amyl chloride which was deliberately added prior to the initiation of the reaction could be recovered quantitatively.

Oxidation of Homoallylic Radicals by Copper(II) Halides and Pseudohalides. The allylcarbinyl, cyclobutyl, and cyclopropylmethyl species are designated here as homoallylic radicals. Each of these radicals was derived separately in acetonitrile from the corresponding diacyl peroxide by the catalytic method described above. Independent studies have shown that cyclobutyl and allylcarbinyl radicals did not undergo rearrangement under these conditions.[4] The isomerization of cyclopropylmethyl radical to allylcarbinyl radical, on the other hand, proceeds with a first-order rate constant[5] of at least 1×10^8 sec^{-1}.

(1) (a) J. K. Kochi and R. V. Subramanian, *J. Amer. Chem. Soc.*, **87**, 1508 (1965); (b) C. L. Jenkins and J. K. Kochi, *J. Org. Chem.*, **36**, 3095, 3103 (1971).

(2) J. K. Kochi, *Rec. Chem. Progr.*, **27**, 207 (1966).

(3) C. L. Jenkins and J. K. Kochi, *J. Amer. Chem. Soc.*, **94**, 843 (1972).

(4) J. K. Kochi and A. Bemis, *ibid.*, **90**, 4038 (1968).

(5) Estimated from the rate of isocholesteryl to cholesteryl rearrangement [D. J. Carlsson and K. U. Ingold, *ibid.*, **90**, 7047 (1968)].

Table I. Reaction of Neopentyl Radicals with Copper(II) Chloride in Acetonitrile at 25°

		Reagents, M		Products, mol%		
CuCl$_2$	LiCl	CuCl	*tert*-Butyl acetyl peroxide	CO$_2$	⤳Cl	⤳Cl
0.040	0.080	0.010	0.040	97	96	0
0.040	0.080	0.020	0.040	99	85	24[a]

[a] Control experiment, 0.24 mmol (24 mol% based on peroxide) was added prior to reaction.

Table II. Oxidation of Homoallylic Radicals by Copper(II) Bromide[a]

(RCO$_2$)$_2$, M	CuBr$_2$, M	CO$_2$	⤳Br	◁—Br	⤳Br
☐	0.100	98	0	97	0
⤳	0.100	95	90	0	0
▷—	0.050	99	30	2.8	65
▷—·	0.150	100	0.8	4.6	85

[a] Reactions carried out with 0.06 M peroxide contained in 20 ml of propionitrile at 0°.

The oxidation of allylcarbinyl radicals by copper(II) thiocyanate proceeded without rearrangement. A similar treatment of cyclobutyl radicals also produced mainly cyclobutyl thiocyanate, but small amounts of the isomeric allylcarinyl and cyclopropylmethyl thiocyanates were also produced as shown in Table III. The oxidation of cyclopropylmethyl radicals by copper(II) thiocyanate afforded a mixture consisting largely of cyclopropylmethyl and allylcarbinyl thiocyanate, the relative amounts of which were directly determined by the concentration of copper(II) thiocyanate. A quantitative determination of the latter by use of eq 5

Table III. Oxidation of Homoallylic Radicals by Copper(II) Thiocyanate[a]

(RCO$_2$)$_2$	M	Cu^{II}(SCN)$_2$, M	KSCN, M	CO$_2$	⤳SCN	◁—SCN	⤳SCN
☐	(0.040)	0.040	0.200	100	1.0	99	3.8
		0.020	0.080	100	1.7	95	5.1
⤳·	(0.040)	0.080	0.200	100	99	0	0
		0.120	0.280	101	99	0	0
▷—·	(0.040)	0.040	0.120	98	79	1.2	12
		0.120	0.280	96	64	2.5	25

[a] Reaction carried out with 0.040 M peroxide contained in 25 ml of acetonitrile at 0°.

The oxidation of cyclobutyl and allylcarbinyl radicals with copper(II) *bromide* occurred with no rearrangement. Some representative results are given in Table II. The oxidation of cyclopropylmethyl radical by copper(II) afforded predominately cyclopropylmethyl bromide. The smaller amounts of allylcarbinyl bromide were attributable to the isomerization of the radical, the competition from which could be largely overwhelmed by increasing the concentration of copper(II) bromide. A kinetic scheme presented below was based

$$ \triangleright\!\!-\!\cdot \; + \; Cu^{II}X_2 \; \xrightarrow{k_L} \; \diagdown\!\!\diagup X \; + \; Cu^{I}X, \text{etc.} $$

$$ k_r \big\downarrow $$

$$ \diagup\!\!-\!\cdot \; + \; Cu^{II}X_2 \; \xrightarrow{k_L} \; \diagdown\!\!\diagup X \; + \; Cu^{I}X, \text{etc.} $$

on a knowledge of this competing first-order rate constant for isomerization. If we assume that k_L for allylcarbinyl and cyclopropylmethyl radicals are approximately the same and the isomerization is irreversible, the relative yields of products may be expressed as

$$ \frac{\diagdown\!\!\diagup^X}{\triangleright\!\!\diagup X} \; = \; \frac{k_r}{k_L} \frac{1}{[Cu^{II}X_2]} \tag{5} $$

A value of $k_L(Br) = 4.3 \times 10^9\ M^{-1}\ sec^{-1}$ at 25° was obtained.[1] No isomerization of any of the homoallylic bromides occurred under these reaction conditions.

led to a value of the second-order rate constant k_L for thiocyanate transfer of $3.6 \times 10^8\ M^{-1}\ sec^{-1}$. Small but discrete amounts of cyclobutyl thiocyanate were also formed. Scrutiny of the reaction showed, however, that no homoallylic isothiocyanates were formed or isomerized under the conditions of the reaction.

The oxidation of cyclobutyl and allylcarbinyl radicals by copper(II) *chloride*, in contrast to copper(II) bromide or copper(II) thiocyanate, afforded a mixture of homoallylic chlorides representing extensive rearrangement of the structure of the parent radical. *The composition of each mixture was independent of the concentration of the copper(II) chloride.* Some typical results are summarized in Table IV. The composition of the products resulting from the oxidation of cyclopropylmethyl radical was highly dependent on the concentration of copper(II) chloride, and at very low concentrations of copper(II) the isomeric mixture was the same as that obtained from the oxidation of allylcarbinyl radical. It could be shown from studies[1b] carried out at higher concentrations of copper(II) chloride that the oxidation of cyclopropylmethyl radicals also produced the unique mixture of homoallylic chlorides given in eq 8, the composition of which was also independent of the concentration of copper(II) chloride. Kinetic analysis of the type described for bromide and thiocyanate led to a second-order rate constant for chlorine transfer of $k_L(Cl) = 1.1 \times 10^9\ M^{-1}\ sec^{-1}$.

Jenkins, Kochi / Ligand-Transfer Oxidation of Alkyl Radicals

Table IV. Oxidation of Homoallylic Radicals by Copper(II) Chloride[a]

(RCO$_2$)$_2$	Cu^{II}Cl$_2$	Cu^{I}Cl[b]	Products, mol % CO$_2$	C$_4$H$_7$Cl	Distribution, % ⋁—Cl	◇—Cl	⋀—Cl
☐	0.020	0.013	95	91	4	71	25
	0.230	0.057	100	93	5	70	25
⌐	0.040	0.013	89	89	81	6	13
	0.120	0.013	95	92	82	5	13
▷	0	0.020	95	65	80	6	14
	0.240	0.010	97	88	29	9	62

[a] Reaction carried out with 0.040 M peroxide in acetonitrile at 0°. [b] Initiation by Cu^{I}Cl with an equimolar amount of lithium chloride added.

Table V. Oxidation of Cyclobutyl Radicals by Copper(II) Chloride in Nonaqueous Solutions[a]

Solvent	Reactants CuCl$_2$, M	LiCl, M	Products CO$_2$, %	C$_4$H$_7$Cl, %	Distribution, % ⋁—Cl	◇—Cl	⋀—Cl
Pyridine[b]	0.043	0	96	58	0	100	0
HMPA[c]	0.052	0	99	100	1	86	13
DMSO[d]	0.040	0	103	57	3	83	14
	0.040	0.106	102	56	1	95	4
	0.080	0	98	63	2	84	14
	0.080	0.213	99	87	2	92	6
	0.120	0.320	98	79	0.5	96	3.5
DMF[e]	0.050	0	95	58	2	68	20
HOAc[f]	0.044	0.480	100	71	1	72	24
CH$_3$CN	0.060	0	99	90	3	70	27

[a] In solutions containing 0.04 M cyclobutanecarbonyl peroxide at 0°. Reactions initiated by 0.01 M copper(I) chloride. [b] With 2.5 vol % acetonitrile. [c] Hexamethylphosphoric triamide with 4% acetonitrile. [d] Dimethyl sulfoxide with 4 vol % acetonitrile. [e] Dimethylformamide with 4% acetonitrile. [f] With 4% acetonitrile.

$$[\text{⋁—Cl} + \text{◇—Cl} + \text{⋀—Cl}] + Cu^I Cl$$

⌐ + Cu^{II}Cl$_2$ ⟶ 81% 6% 13% (6)

☐ + Cu^{II}Cl$_2$ ⟶ 4% 70% 26% (7)

▷ + Cu^{II}Cl$_2$ ⟶ 9% 2% 89% (8)

The oxidation of each homoallylic radical by copper-(II) chloride in acetonitrile thus produced a *unique* mixture of homoallylic chlorides characteristic of that radical, and the composition of this mixture was independent of the concentration of copper(II) chloride. Moreover, the predominant isomer in each mixture was that resulting from the intact structure of the precursor radical. Careful study proved that no rearrangement of the products occurred subsequent to their formation. The observed distribution of isomeric homoallylic chlorides shown in eq 6–8 thus resulted from a kinetically controlled oxidation process.

Effect of Solvent on the Oxidation of Cyclobutyl Chloride by Copper(II) Chloride. Since the foregoing studies showed that oxidation of cyclobutyl radicals in acetonitrile produced a mixture of homoallylic chlorides, the medium was varied in order to determine the effect of solvent on the distribution of homoallylic chlorides and its relationship to the mechanism. In fact, cyclobutyl chloride was the only product of oxidation of cyclobutyl radicals by copper(II) chloride when the reaction was carried out in pyridine (Table V). The reaction was slightly less selective in dimethyl sulfoxide (DMSO) and hexamethylphosphoric triamide (HMPA). Acetic acid, acetonitrile, and N,N-dimethylformamide, despite their different chemical and physical properties, induced approximately the same amount of rearrangement. The stability of cyclobutyl chloride

to isomerization was established in each solvent under these conditions.

The specificity of the oxidation of cyclobutyl radicals by copper(II) chloride was enhanced by the addition of lithium chloride. Thus, the selectivity in the formation of cyclobutyl chloride was raised from 83% in the presence of 0.040 M copper(II) chloride to 95% on the addition of 0.106 M lithium chloride in DMSO solutions.

The specificity could also be deliberately *lowered* by the use of 1,4-dioxane or 1,2-dimethoxyethane as the solvent. Copper(II) chloride by itself was insoluble in dioxane and small amounts of water (>2 wt %) were required for dissolution. However, the presence of water in varying amounts caused very little change in the distribution of homoallylic chlorides produced in aqueous dioxane solutions as shown in Table VI. Water alone could not have been responsible for these changes since it had little effect in acetonitrile solutions in amounts up to 16 wt %. At the upper limit of water studied, a mixture of both homoallylic chlorides as well as alcohols was produced. Significantly, the distribution of isomeric homoallylic *alcohols* obtained under these conditions was essentially the same as the composition of the homoallylic *chlorides* produced in dioxane. Furthermore, the mixture of homoallylic chlorides obtained in these aqueous solutions had a slightly higher component of rearranged chlorides than that obtained in pure acetonitrile (Table VI).

Copper(II) Chloride Species in Solution. The absorption spectrum of copper(II) chloride dissolved in various media was examined. The principal absorption bands in the visible region are listed in Table VII. Howald, *et al.*,[6] examined the spectra of mixtures of

(6) R. P. Eswein, E. S. Howald, R. A. Howald, and D. P. Keeton,

Table VI. Oxidation of Cyclobutyl Radicals by Copper(II) Chloride in Ethereal or Partially Aqueous Solutions[a]

Solvent	H₂O, wt %	CuCl₂, M	LiCl, M	CO₂	C₄H₇Cl	⌇⌇Cl	◇–Cl	▷⌇Cl
Dioxane	2[b]	0.041	0	91		6	55	39
	4[b]	0.045	0	93	74	6	51	43
	4[b]	0.042	0.144	93	82	6	59	35
	4[c]	0.045	0	90		6	52	42
	8[c]	0.044	0	88		6	58	41
	12[c]	0.044	0	87		6	54	40
	16[c]	0.055	0	85		6	54	40
CH₂OCH₃	2[b]	0.049	0	98	84	3	59	38
CH₂OCH₃	2[b]	0.052	0	83	62	3	57	40
CH₃CN	0	0.060	0	99	90	3	70	27
	2	0.040	0	99	84	3	70	27
	4	0.045	0	99	79	3	68	29
	16	0.049	0	101	64	3	63	34
					18[d]	3[d]	52[d]	45[d]

[a] In solutions containing 0.04 M cyclobutanecarbonyl peroxide at 0°. Reactions initiated with 0.01 M copper(I) chloride. [b] In addition to 2 vol % acetonitrile. [c] In addition to 4 vol % acetonitrile. [d] Alcohols.

Table VII. Visible Spectra of Copper(II) Chloride in Various Solvents

Solvent	Added ligand	λ_max, nm	ε,[a] M⁻¹ cm⁻¹	Principal species
CH₃CN		750	55	{ CuCl₂
		462	508	CuCl₃⁻ }
		312	3620	
CH₃CN	LiCl[b]	465	1580	{ CuCl₃⁻
		315	3750	CuCl₄²⁻ }
HMPA		750	19	{ CuCl₂
		465 (s)	51	CuCl₃⁻ }
		360 (s)	600	
		340	1920	
DMSO		300	3820	CuCl₂
DMSO	LiCl[c]	405	1330	CuCl₄,₃,₂²⁻,⁻,⁰
		300	3170	
Dioxane[d]		750	38.4	CuL₆²⁺+2Cl⁻[e]

[a] Apparent extinction coefficient (*i.e.* absorbance divided by the molar concentration of copper). [b] 2 equiv of LiCl/CuCl₂. [c] Excess. [d] Containing 4 wt % water. [e] L = dioxane, water.

copper(II) chloride and lithium chloride in acetonitrile and acetic acid solutions and concluded that the appearance of bands at 375 and 450 nm was due to the tetrachlorocuprate ion. The absorption bands of various analogous chlorocopper(II) complexes have also been described in other solvents. It is apparent from data presented in Table VII that solutions of copper(II) chloride in hexamethylphosphoric triamide, dimethylformamide, pyridine, and dimethyl sulfoxide also consisted predominantly of chlorocopper(II) species.[7] Furthermore, the addition of lithium chloride to a solution of copper(II) chloride promoted the formation of the more highly chlorinated copper(II) species, in accord with expectations based on the mass-action principle.

On the other hand, it was immediately apparent from visual observations that a solution of copper(II) chloride in dioxane consisted of species which were significantly different from the chlorocopper(II) complexes present in those media described above. For example, the latter consisted of dark yellow to green solutions,

whereas solutions of copper(II) chloride in either dioxane or dimethoxyethane were almost *colorless*. The difference was further apparent in the visible absorption spectra shown in Figure 1. The similarity of the absorption spectrum of copper(II) chloride in dioxane to that of copper(II) perchlorate in water,[8] as well as the

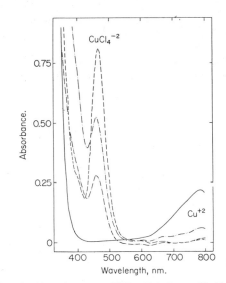

Figure 1. Absorption spectra of 3.95×10^{-3} M copper(II) chloride in 96 vol % dioxane–water (——). Comparison with 4.70×10^{-4} M copper(II) chloride (— · —) and 9.4×10^{-4} M copper(II) chloride (· — ·) in acetonitrile. Solution of 4.75×10^{-4} M copper(II) chloride and 1.00×10^{-3} M lithium chloride (- - -).

absence of charge-transfer bands[7] (associated with electronic transitions from the chloride ligands to copper), indicates that these are outer sphere complexes of copper(II) chloride.[9] Solvolysis shown in eq 9 was only

J. Inorg. Nucl. Chem., **29,** 437 (1967); see also R. D. Willet, *et al., Inorg. Chem.,* **6,** 1666, 1885 (1967); C. Furlani and G. Morpurgo, *Theor. Chim. Acta,* **1,** 102 (1963).

(7) W. E. Hatfield and R. Whyman, *Transition Metal Chem.,* **5,** 47 (1969); F. Gutmann, *Coord. Chem. Rec.,* **2,** 239 (1967); D. W. Meek, W. E. Hatfield, R. S. Drago, and T. S. Piper, *Inorg. Chem.,* **3,** 841, 1637 (1964); D. Culpin, P. Day, R. Edwards, and R. J. P. Williams, *Chem. Commun.,* 450 (1965)

(8) F. A. Cotton and G. Wilkinson, "Advanced Inorganic Chemistry," second ed, Interscience, New York, N. Y., 1966, p 906.

(9) V. E. Mironow, Yu. A. Makashev, and I. Ya. Mavrina, *Russ. J. Inorg. Chem.,* **14**(5), 746 (1969); M. W. Andreeva and V. G. Khaldin, *ibid.,* **14**(5), 626 (1969).

Jenkins, Kochi / Ligand-Transfer Oxidation of Alkyl Radicals

Table VIII. Oxidation of β-Phenethyl and β-Anisylethyl Radicals by Copper(II) Chloride and Bromide.   Effect of Solvents[a]

Radical	CuX$_2$	Solvent	Product[b,f]		Rearrangement, %
C$_6$H$_5$CD$_2$CH$_2$·	CuCl$_2$	CH$_3$CN	[benzene ring]—CD$_2$CH$_2$Cl 100%		0
	CuCl$_2$	Dioxane[c]	100%		0
	CuBr$_2$	CH$_3$CN	[benzene ring]—CD$_2$CH$_2$Br 100%		0
CH$_3$OC$_6$H$_4$CH$_2$CD$_2$·			C(–Ar)D$_2$CH$_2$Cl	C(–Ar)H$_2$CD$_2$Cl	
	CuCl$_2$	CH$_3$CN	28 (25)	72 (75)	54
	CuCl$_2$	Dioxane[d]	47 (46)	53 (54)	94
			C(–Ar)D$_2$CH$_2$Br	C(–Ar)H$_2$CD$_2$Br	
	CuBr$_2$	CH$_3$CN	6	94	12
	CuBr$_2$	Dioxane[e]	9	91	18

[a] Radicals obtained from copper-catalyzed decomposition of the corresponding hydrocinnamoyl peroxide (0.04 M) with 0.045 M copper(II) halide and 0.01 M copper(I) halide at 0°.   [b] Absolute yields of aralkyl halides high but not determined.   Values given are relative purity after separation by preparative gas chromatography and analysis by pmr.   Values in parentheses determined by d(deuterium)mr.   [c] Containing 3 wt % water and 3 wt % acetonitrile.   [d] Containing 3 wt % water and 7 wt % acetonitrile.   [e] Containing 4 wt % water and 1 wt % acetonitrile.   [f] Ar = p-CH$_3$OC$_6$H$_4$–.

$$Cu^{II}Cl_2 + 6L \rightleftharpoons Cu^{II}L_6{}^{2+} + 2Cl^- \qquad (9)$$
$$L = \text{dioxane, water}$$

slightly reversed by anation in the presence of excess chloride.   Furthermore, hydrolysis was much less important in aqueous acetonitrile solutions than in aqueous dioxane solutions.

Oxidation of Arylethyl Radicals by Copper(II) Chloride and Bromide.   Phenonium Ions as Intermediates. Phenethyl and anisylethyl radicals do not isomerize under these conditions.[4,10]   The corresponding cations, on the other hand, are known to undergo extensive rearrangement to equilibrate the α- and β-carbon atoms.[11] These aralkyl radicals have been successfully employed to probe for carbonium ion intermediates in the previous study of electron-transfer oxidation by copper(II) actate.[12]

The oxidation of phenethyl and anisylethyl radicals was examined in this study with copper(II) chloride and bromide as oxidants in acetonitrile and in dioxane.   The oxidation of β,β-dideuterio-β-phenethyl radical by either copper(II) chloride or bromide afforded no evidence of rearranged products being formed in acetonitrile or in dioxane.   Excellent yields of only β,β-dideuterio-β-phenethyl chloride and bromide were obtained in the presence of copper(II) chloride and bromide, respectively, as shown in Table VIII.

On the other hand, the oxidation of α,α-dideuterio-β-anisylethyl radicals under the same conditions gave significant amounts of products of rearrangement. Thus, the oxidation of α,α-dideuterio-β-anisylethyl radical by copper(II) chloride in acetonitrile afforded a mixture of dideuterio-β-anisylethyl chlorides containing 26% deuterium label in the β position.   The same oxidation in dioxane produced 94% rearrangement.

Analysis of the isomeric mixture by proton magnetic resonance spectroscopy gave the same results as those obtained from the examination of the deuterium magnetic resonance spectra.   Oxidation of α,α-dideuterio-β-anisylethyl radicals by copper(II) bromide produced only 12% rearrangement in acetonitrile.   Slightly more rearrangement (18%) was observed when the oxidation with copper(II) bromide was carried out in dioxane solutions.

Rearrangement of the β-arylethyl moiety during oxidation thus depended on both the copper(II) halide as well as the solvent, since β-phenethyl radical afforded no products of rearrangement.   The oxidation of β-anisylethyl radical, however, led to a 12% scrambling of the α and β carbon atoms during oxidation by copper bromide in acetonitrile and 18% scrambling when the same oxidation was carried out in dioxane.   Similarly, the equilibration of the side chain was extensive (52%) when the oxidation of β-anisylethyl radical was carried out by copper(II) chloride in acetonitrile and almost complete (94%) when carried out in dioxane.

Discussion

The ligand-transfer oxidation of alkyl radicals by metal complexes presents several interesting mechanistic questions.   Thus, the reactivity of alkyl radicals in atom-transfer processes, particularly those involving hydrogen and halogen, are well-known phenomena in the area of free-radical chemistry.[13]   The importance of polar effects in the transition states of some atom-transfer processes has been described, but carbonium ions are not intermediates in these processes.[14]   On the other hand, the role of copper(II) in wholly inorganic redox processes[15] can be described both in terms of inner sphere and outer sphere mechanisms.[16]   The first ques-

(10) R. Friedlina, *Advan. Free-Radical Chem.*, **1**, 215 (1965).

(11) C. C. Lee, G. Slater, and J. Spinks, *Can. J. Chem.*, **35**, 1417 (1957); W. H. Saunders, S. Asperger, and D. Edison, *J. Amer. Chem. Soc.*, **80**, 2423 (1958); L. Eberson, J. Petrovich, R. Baird, D. Dyckes, and S. Winstein, *ibid.*, **87**, 3504 (1965); J. Nordlander and W. Deadman, *ibid.*, **90**, 1590 (1968); B. G. van Leuwen and R. J. Ouellette, *ibid.*, **90**, 7056, 7061 (1968); C. C. Lee and R. Tewari, *Can. J. Chem.*, **45**, 2256 (1967); R. Jablonski and E. Snyder, *Tetrahedron Lett.*, 1103 (1968); Y. Yukawa, *et al.*, *ibid.*, 847 (1971); A. Laurent and R. Tardivel, *C. R. Acad. Sci., Ser. C*, **272**, 8 (1971).

(12) J. K. Kochi, A. Bemis, and C. L. Jenkins, *J. Amer. Chem. Soc.*, **90**, 4616 (1968).

(13) W. A. Pryor, "Free Radicals," McGraw-Hill, New York, N. Y., 1966, p 149ff.; C. Walling, "Free Radicals in Solution," Wiley, New York, N. Y., 1957.

(14) E. S. Huyser, "Free Radical Chain Reactions," Wiley-Interscience, New York, N. Y., 1970, p 77ff.

(15) F. Basolo and R. G. Pearson, "Mechanisms of Inorganic Reactions," Wiley, New York, N. Y., 1967, p 543ff.; O. J. Parker and J. H. Espenson, *J. Amer. Chem. Soc.*, **91**, 1313, 1968 (1969); **89**, 5730 (1968); *Inorg. Chem.*, **7**, 1619 (1968).

(16) H. Taube, *Chem. Rev.*, **50**, 69 (1950); *Advan. Inorg. Radiochem.*, **1**, 1 (1959); A. G. Sykes, *ibid.*, **10**, 394 (1967).

tion we pose, therefore, is whether carbonium ions are intermediates in the ligand-transfer oxidation of alkyl radicals by copper(II) chloride.

Carbonium Ions as Intermediates in Ligand Transfer. The facile oxidation of alkyl radicals by various copper-(II) halides and pseudohalides involves the transfer of a ligand from the copper(II) oxidant to the alkyl radical. *Free* carbonium ions as such are not intermediates since the transfer of chloride, bromide, iodide, thiocyanate, azide, and cyanide can be effected in protic media such as acetic acid without direct intervention of the external nucleophile.[1] However, cationoid intermediates are not necessarily precluded, since the subtleties of reactions occurring from ion pairs can exclude direct participation of solvent, as studies of the solvolysis of alkyl derivatives have shown.[17]

A much more sensitive probe for the participation of carbonium ions along the reaction coordinate is provided by the study of alkyl moieties prone to cationic rearrangement. Three such systems have been employed in this study: neopentyl, homoallylic, and β-arylethyl radicals. A comparison of these alkyl radicals shows that the contribution from a cationic pathway varies with both the alkyl moiety as well as the copper(II) oxidant. Thus, ligand-transfer oxidation of neopentyl radical occurs with *no evidence of rearrangement*, since *tert*-amyl derivatives are absent.[4] On the other hand, homoallylic systems represented by cyclobutyl, allylcarbinyl, and cyclopropylmethyl are more susceptible to cationic rearrangement.[18] The predominant product obtained from the oxidation of each homoallylic radical by copper(II) chloride in acetonitrile was that isomeric chloride resulting from the radical with the structure intact. That is, cyclobutyl radical afforded principally cyclobutyl chloride, allylcarbinyl radical gave largely allylcarbinyl chloride. Even cyclopropylmethyl radical produced mainly cyclopropylmethyl chloride, provided the copper(II) chloride concentration was sufficiently high. In every case, the by-product from the oxidation of each isomeric radical was a mixture of the other two homoallylic chlorides.

Oxidation of each homoallylic radical by copper(II) bromide and by copper(II) thiocyanate gave similar results, although the amount of rearrangement varied appreciably with the copper(II) oxidant. The approximate extent of rearrangement accompanying ligand-transfer oxidation of each homoallylic radical by various copper(II) complexes in acetonitrile is summarized in Table IX.

Homoallylic cations are implicated as the precursors for the rearranged products, since the distributions of the isomeric by-products from the ligand-transfer oxidations of all of the homoallylic radicals shown in Tables II–IV are characteristic of the results obtained in solvolytic studies of these systems.[18,19] Furthermore,

(17) (a) S. Winstein, B. Appel, R. Baker, and A. Diaz, *Chem. Soc., Spec. Publ.*, No. 19, 109 (1965); (b) S. G. Smith and J. P. Petrovich, *J. Org. Chem.*, 30, 2882 (1965); H. L. Goering, *et al.*, *J. Amer. Chem. Soc.*, 92, 7401 (1970); 93, 1224 (1971); A. Ceccon, A. Fava, and I. Papa, *ibid.*, 91, 5547 (1969); L. A. Spurlock and W. G. Cox, *ibid.*, 93, 146 (1971).

(18) R. Breslow, "Molecular Rearrangements," Part I, P. de Mayo, Ed., Interscience, New York, N. Y., 1963, p 254ff; G. A. Olah and P. v. R. Schleyer, "Carbonium Ions," Wiley, New York, N. Y., 1969.

(19) *Cf.* M. Hanack and A. J. Schneider, *Angew. Chem., Int. Ed. Engl.*, 6, 666 (1967); R. Moss and F. Shulman, *Tetrahedron*, 24, 2881 (1968); *J. Amer. Chem. Soc.*, 90, 2731 (1968); Z. Majerski, S. Borcic, and D. E. Sunko, *Tetrahedron*, 25, 301 (1969); K. L. Servis and J. D. Roberts,

Table IX. Rearrangement Resulting from Ligand-Transfer Oxidation of Homoallylic and Neopentyl Radicals by Copper(II) Complexes[a]

$Cu^{II}X_2$	Relative rate[b]	—Homoallylic radical, %—			Neopentyl radical, %
$Cu^{II}Br_2$	12	0	0	<1	
$Cu^{II}Cl_2$	3	20	50	15	0
$Cu^{II}(NCS)_2$	1	0	5	3	

[a] The extent of arrangement extrapolated from Tables II-VI.
[b] $[Cu(NCS)_2] = 1.0$.

any isomerization of the alkyl radical precursors has been taken into account in these results.

The contribution of a polar substituent effect in promoting the cationic pathway is obtained by comparing the ligand-transfer oxidation of β-phenethyl and β-anisylethyl radicals. Isotopic labeling allows the examination of carbonium ion intermediates in this system, since complete equilibration of the α and β carbons invariably results from a phenonium ion intermediate.[12] Solvolysis studies show that an even greater driving force for cation formation in the form of the bridged

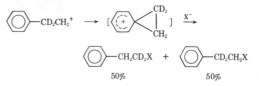

anisonium ion is provided by the presence of a *p*-methoxy group.[20] Rearrangement accompanying ligand transfer is summarized in Table X, which shows that

Table X. Contribution from the Cationic Path in the Ligand-Transfer Oxidation of β-Arylethyl Radicals in Acetonitrile

$Cu^{II}X_2$	Rearrangement, %	
	$C_6H_5CD_2CH_2\cdot$	$CH_3OC_6H_4CH_2CD_2\cdot$
$Cu^{II}Cl_2$	0	54
$Cu^{II}Br_2$	0	12

a significantly greater driving force for the cationic path is, indeed, derived from the oxidation of β-anisylethyl radical compared to β-phenethyl radical.

Dual Mechanisms in Ligand-Transfer Oxidation. These studies show that two modes of oxidation occur during ligand-transfer oxidation of alkyl radicals by copper(II) halides and pseudohalides. One pathway involves *direct conversion* of the alkyl radical (R·) to the substitution product (RX). The other more *indirect route* generates a carbonium ion of sufficient integrity somewhere along the reaction coordinate to undergo complete equilibration of the α- and β-carbon

J. Amer. Chem. Soc., 87, 1331 (1965); P. v. R. Schleyer and G. W. Van Dine, *ibid.*, 88, 2321 (1966); K. L. Servis and J. D. Roberts, *Tetrahedron Lett.*, 1369 (1967); H. G. Richey and J. M. Richey, *J. Amer. Chem. Soc.*, 88, 4971 (1966). A common intermediate is not necessarily implied. See I. Lillien and L. Handloser, *ibid.*, 93, 1682 (1971); *J. Org. Chem.*, 33, 3841 (1968); 34, 3658 (1969); P. v. R. Schleyer and Z. Majerski, *J. Amer. Chem. Soc.*, 93, 665 (1971); D. D. Roberts, *J. Org. Chem.*, 36, 1913 (1971).

(20) W. H. Saunders, S. Asperger, and D. Edison, *J. Amer. Chem. Soc.*, 80, 2423 (1958); G. A. Olah, M. Comisarow, E. Namenworth, and B. Ramsey, *ibid.*, 89, 711 (1967).

Jenkins, Kochi | Ligand-Transfer Oxidation of Alkyl Radicals

862

atoms of the β-anisylethyl moiety and extensive scrambling of the carbon atoms in the homoallylic precursors. Stabilization of the alkyl cation in these systems plays an important role in the indirect route, since the neopentyl and β-phenethyl analogs show no tendency to afford carbonium ion intermediates under the same conditions.

The results of ligand-transfer oxidation of the series of neopentyl, homoallylic, and β-arylethyl radicals given here are in sharp contrast to the electron-transfer oxidation of the same radicals previously examined under comparable conditions. In every example pertaining to the electron-transfer oxidation of these alkyl radicals, *oxidative solvolysis invariably led to complete cationic rearrangement of the alkyl moiety.* Thus, electron-transfer oxidation of neopentyl radicals afforded only products derived from the *tert*-amyl cation.[4] Similarly, oxidative substitution of the β-phenethyl radical by copper(II) acetate led to complete equilibration of the α- and β-carbon atoms.[12] Finally, the oxidation of cyclobutyl radicals under the same conditions produced the same mixture of extensively rearranged homoallylic acetates derived by other cationic processes.[4]

The partial cationic rearrangement observed during the ligand-transfer oxidation of certain alkyl radicals and the absence of rearrangement in others indicate that at least two processes are operative (*cf.* Tables IX and X). The extent of rearrangement, furthermore, is invariant with the concentration of the copper(II) halide or pseudohalide, but highly dependent on the structure of, and the ligand associated with, the copper(II) nucleus. These results are inconsistent with competition from a kinetically first-order process. We propose that the *direct* and *indirect* pathways in ligand-transfer oxidation are represented by independent simultaneous second-order processes. The direct process is described as *atom transfer* and the indirect route is represented as *oxidative substitution.* These mechanisms will be elaborated further.

The Role of Oxidative Substitution in Ligand-Transfer Processes. Ligand-transfer oxidation proceeding *via* a pathway represented as *oxidative substitution* is given by eq 10 and 11, in which copper(II) chloride is used

$$R \cdot + CuCl_2 \rightleftharpoons \underset{I}{RCuCl_2} \qquad (10)^{21}$$

$$I \rightleftharpoons RCuCl^+Cl^- \longrightarrow R-Cl + Cu^ICl \qquad (11)$$

for illustrative purposes. The formation of a carbonium ion intermediate by this route is accommodated by a metastable alkylcopper species I akin to that presented earlier in connection with electron-transfer processes (see the preceding paper).[3]

According to this formulation, the formation of carbonium ion intermediates leading to ligand transfer is promoted by ionization of the alkylcopper intermediate I. Evidence for ionization of copper(II) complexes is obtained from an examination of the absorption spectra. Thus, the visible absorption spectrum of copper(II) chloride shows that it exists in dioxane as an outer sphere complex (see Figure 1). In strong contrast, inner sphere chlorocopper(II) species are prevalent in acetonitrile.[22]

(21) All of the coordination around the copper nucleus is not included explicitly, unless required for the discussion.

$$CuCl_2 \overset{dioxane}{\rightleftharpoons} Cu^{2+} 2Cl^- \qquad (12)$$

$$CuCl_2 \overset{CH_3CN}{\rightleftharpoons} Cu^{2+} 2Cl^- \qquad (13)$$

The formation constants of chlorocopper(II) have not as yet been measured in dioxane. However, aqueous solutions form a reasonable approximation since the visible absorption spectrum of copper(II) chloride is virtually the same in dioxane as in water. The formation constants of inner sphere chlorocopper(II) species also differ markedly in acetonitrile and in water as shown in Table XI.

Table XI. Formation Constants of Inner Sphere Chlorocopper(II) Species in Acetonitrile, Water, and Methanol

Equilibrium		$-K_i$ formation constant[a]		
		CH_3CN	CH_3OH	H_2O
$Cu^{II} + Cl^- \rightleftharpoons Cu^{II}Cl^+$	$K_1 = 10^{9.7}$	$10^{4.2}$	1.0
$Cu^{II}Cl^+ + Cl^- \rightleftharpoons Cu^{II}Cl_2$	$K_2 = 10^{7.9}$	$10^{6.5}$	0.2
$Cu^{II}Cl_2 + Cl^- \rightleftharpoons Cu^{II}Cl_3^-$	$K_3 = 10^{7.1}$		0.04
$Cu^{II}Cl_3^- + Cl^- \rightleftharpoons Cu^{II}Cl_4^{2-}$	$K_4 = 10^{3.7}$		0.01

[a] From ref 22a, d, and e.

The ionization of the metastable chloroalkylcopper intermediate I should be promoted by dioxane much like the chlorocopper(II) species themselves. Copper(II) complexes are, in general, highly substitution labile

$$RCuCl_2 \overset{dioxane}{\rightleftharpoons} RCuCl^+Cl^- \qquad (14)$$

and anation should not place kinetic restraints on the reactivity of the alkylcopper species. The coordinatively unsaturated cationic alkylcopper species is subject to heterolysis of the alkylcopper bond in a manner reminiscent of the previously described oxidative substitution in electron-transfer oxidation[3] and demercuration of alkylmercurinium cations.[23]

$$RCuCl^+ \longrightarrow R^+Cu^ICl, \text{ etc.} \qquad (15)$$

Extensive cationic rearrangement of the cyclobutyl moiety during ligand-transfer oxidation of cyclobutyl radicals by copper(II) chloride in dioxane is consistent with this formulation. A qualitative comparison of the importance of the cationic (oxidative substitution) route in acetonitrile and dioxane is presented in Table XII. The same phenomenon can be evaluated quan-

Table XII. Promotion of Oxidative Substitution by Solvent in Ligand–Transfer Oxidations

Radical	$Cu^{II}X_2$	Solvent	Cationic rearrangement, %
□	$Cu^{II}Cl_2$	CH_3CN	~50
	$Cu^{II}Br_2$	CH_3CN	0
	$Cu^{II}Cl_2$	Dioxane	~100
$CH_3OC_6H_4CH_2CD_2 \cdot$	$Cu^{II}Cl_2$	CH_3CN	54
	$Cu^{II}Cl_2$	Dioxane	94
	$Cu^{II}Br_2$	CH_3CN	12
	$Cu^{II}Br_2$	Dioxane	18

(22) (a) S. E. Manahan and R. T. Iwamoto, *Inorg. Chem.*, **4**, 1409 (1965); (b) see also C. C. Hinckley, *ibid.*, **7**, 396 (1968); (c) R. Meyers and R. D. Willet, *J. Inorg. Nucl. Chem.*, **29**, 1546 (1967); (d) J. Bjerrum, *Kem. Maanedsbl. Nord. Handelsblad Kem. Ind.*, **26**, 24 (1945); (e) S. E. Manahan and R. T. Iwamoto, *J. Electroanal. Chem.*, **13**, 411 (1967).

(23) F. R. Jensen, R. Ouellette, G. K. Knutson, D. Babbe, and R. Hartgerink, *Trans. N. Y. Acad. Sci.*, **30** (2), 751 (1968).

Table XIII. Rates of Ligand-Transfer Oxidation by Copper(II) Halides and Pseudohalides

Radical	Copper(II)	k_r/k_L	$k_L{}^a$	k_L (relative)[b]
5-Hexenyl	$Cu^{II}(NCS)_2$	3.9×10^{-4}	2.6×10^8	
Cyclopropylmethyl	$Cu^{II}(NCS)_2$	2.7×10^{-1}	3.6×10^8	1
5-Hexenyl	$Cu^{II}Cl_2$	$<4 \times 10^{-4}$	$>2 \times 10^8$	
Cyclopropylmethyl	$Cu^{II}Cl_2$	9.2×10^{-2}	1.1×10^9	3
5-Hexenyl	$Cu^{II}Br_2$	$<4 \times 10^{-4}$	$>2 \times 10^8$	
Cyclopropylmethyl	$Cu^{II}Br_2$	2.3×10^{-2}	4.3×10^9	12

[a] At 25° assuming $k_r = 1 \times 10^5$ sec^{-1} for 5-hexenyl radical and 1×10^8 sec^{-1} for cyclopropylmethyl radical in units of M^{-1} sec^{-1}. [b] $k[Cu(NCS)_2] = 1.0$.

titatively by examining the ligand-transfer oxidation of isotopically labeled β-anisylethyl radical. The magnitude (both absolute and relative) of the solvent effect shown in Table XII is also more marked with copper(II) chloride than bromide.

The similarity between the pathways involving oxidative substitution in ligand transfer and oxidative solvolysis in electron transfer is shown by the oxidation of cyclobutyl radicals with copper(II) chloride in aqueous acetonitrile (Table VI). The homoallylic alcohols formed under these conditions represent oxidative solvolysis. The composition of this mixture of isomeric alcohols is strikingly similar to the mixture of homoallylic chlorides obtained from the ligand-transfer oxidation of cyclobutyl radicals by copper(II) chloride in dioxane. The transition states for the reaction of the cyclobutyl moiety with water or chloride during these oxidative processes are, no doubt, related to similar alkylcopper species.

The Role of Atom Transfer in Ligand-Transfer Processes. The alternate route available in ligand-transfer oxidation is represented as *atom transfer* in eq 16. The transition state for atom transfer is free

$$R \cdot + Cu^{II}Cl_2 \longrightarrow [R \cdots Cl \cdots CuCl] \ddagger \longrightarrow R–Cl + Cu^ICl \quad (16)$$

radical in nature, and polar effects are small. Thus, even alkyl radicals with electron-withdrawing α substituents such as cyano, carbonyl, and halogen are readily oxidized by copper(II) chloride.[24a] The same radicals are, by and large, inert to electron-transfer oxidants.[24b] Furthermore, the microscopic reverse process is represented by the direct transfer of a halogen atom from an alkyl halide to chromium(II),[25] and

$$R–X + Cr^{II} \longrightarrow R \cdot + Cr^{III}X \quad (17)$$

a kinetic study of the reduction of substituted benzyl halides shows no polar effect.[26] The atom-transfer mechanism is, thus, an example of a more general inner sphere process involving a bridge-activated complex. A classic example of the latter is represented in eq 18.[16,27]

$$(NH_3)_5Co^{III}Cl + Cr^{II} \longrightarrow [(NH_3)_5Co \cdots Cl \cdots Cr] \ddagger \longrightarrow$$
$$(NH_3)_5Co^{II} + Cr^{III}Cl, \text{ etc.} \quad (18)$$

An atom-transfer mechanism, of course, also bears a direct relationship to chain-transfer reactions of alkyl radicals with a variety of halogen compounds.[13,14,28]

(24) (a) J. K. Kochi and F. F. Rust, *J. Amer. Chem. Soc.*, **84**, 3946 (1962); J. Kumamoto, H. E. DelaMare, and F. F. Rust, *ibid.*, **82**, 1935 (1960); C. H. Bamford, A. Jenkins, and R. Johnston, *Proc. Roy. Soc.*, *Ser. A*, **239**, 214 (1957); (b) J. K. Kochi and D. M. Mog, *J. Amer. Chem. Soc.*, **87**, 522 (1965).
(25) J. K. Kochi and J. W. Powers, *ibid.*, **92**, 137 (1970).
(26) J. K. Kochi and D. D. Davis, *ibid.*, **86**, 5264 (1964).
(27) H. Taube and H. Meyers, *ibid.*, **76**, 2103 (1954); H. Taube, H. Meyers, and R. C. Rich, *ibid.*, **75**, 4118 (1953).
(28) D. F. DeTar and D. V. Wells, *ibid.*, **82**, 5839 (1960); L. O. Moore, *J. Phys. Chem.*, **75**, 2075 (1971).

Rates of Ligand-Transfer Oxidation of Alkyl Radicals. The ligand-transfer oxidations of alkyl radicals by copper(II) halides and pseudohalides are characterized by their extremely high rates. Competitive kinetic studies based on the isomerization of 5-hexenyl and cyclopropylmethyl radicals provide results which are in remarkable agreement with one another,[1] considering a variation in rates of a factor of 10^3. The second-order rate constants, k_L, for the ligand-transfer oxidation of copper(II) radicals by copper(II) thiocyanate, chloride, and bromide are summarized in Table XIII.

The values of k_L, however, represent a composite of a number of individual rate constants k_i for various copper(II) species: $k_L[CuCl_2] = k_1[CuCl^+] + k_2[CuCl_2] + k_3[CuCl_3^-] + k_4[CuCl_4^{2-}]$. For example, the various chlorocopper(II) species represented in Table XI are present in appreciable amounts in acetonitrile solutions. At this juncture it is impossible to separate the specific rate constant k_i by which each of these chlorocopper(II) species oxidizes an alkyl radical. If the composite value of k_L listed in Table XIII includes all of these species, it is reasonable to expect that some values of k_i (particularly for $CuCl_4^{2-}$) may actually be at the diffusion-controlled limit since even the value of k_e for oxidative elimination by copper(II) acetate is in the range of 10^6–10^7 M^{-1} sec^{-1}.[12,29]

Summary

Atom Transfer *vs.* Oxidative Substitution in Ligand-Transfer Processes. In ligand-transfer oxidations, the atom-transfer mechanism (eq 16) is generally the energetically more favorable process and usually represents the major course of reaction. The alternative pathway involving oxidative substitution (eq 10 and 11) by virtue of the formation of carbonium ion intermediates is more applicable to alkyl radicals capable of forming stabilized carbonium ions. Although solvents such as dioxane promote oxidative substitution, the effect is apparent in the ligand-transfer oxidation of only radicals such as cyclobutyl and β-anisylethyl, in which the competition between atom transfer and oxidative substitution is already delicately balanced. Thus, oxidative substitution cannot be promoted in neopentyl and phenethyl radicals by solvent changes, and only atom transfer prevails.

For a given alkyl radical, the relative importance of atom transfer and oxidative substitution pathways also depends on the ligand involved. For example, the ligand-transfer oxidation of cyclobutyl radicals by copper(II) bromide occurs only by atom transfer (Table IX). On the other hand, as much as 50% of the oxidation of cyclobutyl radicals by copper(II) chloride takes place by oxidative substitution. Copper(II) thiocya-

(29) J. K. Kochi and R. V. Subramanian, *J. Amer. Chem. Soc.*, **87**, 4855 (1965).

Jenkins, Kochi | Ligand-Transfer Oxidation of Alkyl Radicals

nate occupies an intermediate position. The same general trend of $Cu^{II}Br > Cu^{II}NCS > Cu^{II}Cl$ in atom-transfer processes is also established with cyclopropylmethyl and allylcarbinyl radicals.

The competition between atom transfer and oxidative substitution is summarized in Scheme I.

Scheme I. Mechanisms of the Ligand-Transfer Oxidation

Atom transfer (direct)

$$R \cdot + Cu^{II}X_2 \longrightarrow [R\text{---}X\text{---}CuX]^+ \longrightarrow R\text{-}X + Cu^{I}X$$

Oxidative substitution (alkylcopper intermediate)

$$R \cdot + Cu^{II}X_2 \rightleftharpoons RCuX_2$$
$$RCuX_2 \rightleftharpoons RCuX^+X^- \rightleftharpoons R^+ Cu^{I}X_2^-$$
$$R^+Cu^{I}X_2^- \longrightarrow R\text{-}X + Cu^{I}X$$

A Unified View of Ligand-Transfer and Electron-Transfer Oxidations of Alkyl Radicals. In a more general sense, the oxidation of alkyl radicals by copper(II) oxidants transcends the relatively arbitrary classification into electron-transfer and ligand-transfer categories, which are based largely on stoichiometry. For example, a distinction between the analogous metastable alkylcopper intermediates leading to (1) oxidative solvolysis in electron transfer and (2) oxidative substitution during ligand transfer cannot be clearly made. In each, the ionization of the ligand plays an important role in the formation of carbonium ion intermediates.

A more unified and general classification of the mechanisms of oxidation of free radicals by metal complexes may be made by considering the position at which the metal complex is attacked by the free radical. Within the present context, for example, the reaction of alkyl radicals with copper(II) complexes by attack on the ligand leads to atom transfer, a process which is largely free radical or homolytic in nature. On the other hand, attachment of the alkyl radical to the copper(II) nucleus results in an alkylcopper intermediate which can be subsequently partitioned in a variety of ways including (1) oxidative elimination to afford alkenes, (2) oxidative substitution to form an alkyl derivative from the ligand,[30] and (3) oxidative solvolysis by participation of the solvent.

Central to this mechanistic classification is the ability of the free radical to discriminate among several sites on the metal complex. We propose as a working hypothesis for further study that such a distinction be based on the hard and soft acid–base classification proposed by Pearson and others.[31]

Ligands and alkyl radicals are listed in Table XIV qualitatively in *order of descending hardness.* The horizontal dashed line represents the border region above which alkyl radicals react with copper(II) complexes by attachment to the copper nucleus. Below the dashed line, reaction occurs primarily on the ligand and an atom transfer (inner sphere) mechanism pre-

Table XIV. Hard and Soft Acid–Base Classification of the Oxidation of Alkyl Radicals by Cupric Complexes

	$Cu^{II}(X_2)(L)_4$		Alkyl
	X^-(anion)	L(neutral)	radical
Perchlorato	ClO_4^-	Water	Methyl
Triflato	$CF_3SO_3^-$	Acetonitrile	Ethyl
Fluoroborato	BF_4^-	1,4-Dioxane	*n*-Butyl
Trifluoroacetato	$CF_3CO_2^-$	Pyridine	Isopropyl
Acetato	$CH_3CO_2^-$	2,2'-Bipyridine	*sec*-Butyl
Chloro	Cl^-	Phenanthroline	*tert*-Butyl
Thiocyanato	SCN^-	Acetic acid	Benzyl
Bromo	Br^-	Dimethyl sulfoxide	Allyl
Iodo	I^-	Dimethylformamide	

vails. Attack on ligand and attachment to copper are competitive with the chloro ligand which lies in the borderline region.

Alkyl radicals can be similarly placed on a hardness scale, which increases from benzyl and allyl radicals to tertiary, secondary, and primary alkyl radicals and finally to methyl radical itself. The mode of reaction between a given alkyl radical and a copper(II) complex would then be determined by their relative positions on these scales. The extension of these concepts to the oxidation and reduction of other free radicals and metal systems would be desirable.

Experimental Section

Materials. Reagents and chemicals used in the previous study have been described[1] and others are given below.

Copper(II) Thiocyanate. Cupric sulfate pentahydrate (0.5 mol) was dissolved in 750 ml of water and potassium thiocyanate (1.0 mol) was added slowly with rapid stirring under an argon blanket. The black precipitate was filtered and washed twice with ethanol and three times with ethyl ether. The material was then dried overnight in a vacuum oven at 50° (20 mm) giving a 63% yield of $Cu(NCS)_2$. The copper content was determined by electrodeposition. *Anal.* Calcd for $Cu(NCS)_2$: Cu, 35.4. Found: Cu, 37.4.

Solutions of copper(I) chloride, bromide, and acetate in acetonitrile were prepared by simply allowing the copper(II) salt and copper metal to react in a closed vessel fitted with a rubber septum. The resulting copper(I) solution was standardized by reaction with an excess of acidic ferric chloride solution and then titrating the ferrous produced with cerium(IV) to a ferrous phenanthroline end point.

Homoallylic Halides. Samples of allylcarbinyl, cyclobutyl, and cyclopropylcarbinyl chlorides and bromides were generously donated by Dr. H. Lin.[32]

Allylcarbinyl Thiocyanate. To 15.6 g of toluenesulfonyl chloride (0.083 mol) in 30 ml of pyridine was added 4 g of allylcarbinyl alcohol at 0°. The reaction was stirred for 3 hr and then poured into an ice–water mixture and extracted with methylene chloride. After, the extract was washed successively with dilute sulfuric acid, sodium bicarbonate, and water. After drying, the solvent was removed at room temperature by rotary evaporation to yield 10.2 g of yellow oil (allylcarbinyl tosylate). This yellow oil (10.1 g) was added to 15 g of potassium thiocyanate in 200 ml of reagent grade acetone and refluxed for 2 hr. After the work-up described for 5-hexenyl thiocyanate,[1b] distillation yielded 1.5 g of allylcarbinyl thiocyanate; bp 82–83° (25 mm). The product was confirmed by its infrared and pmr spectra.

Cyclopropylcarbinyl Thiocyanate.[33] By the procedure described for allylcarbinyl thiocyanate, 10 g of cyclopropylcarbinyl alcohol reacted with 40 g of toluenesulfonyl chloride to yield 27 g of cyclopropylcarbinyl tosylate. Reaction of 20 g of cyclopropylcarbinyl tosylate with 43 g of potassium thiocyanate was carried out at room temperature. After the work-up (see 5-hexenyl thiocyanate), distillation resulted in a mixture boiling over a wide range (45–90°

(30) Oxidative substitution in the ligand-transfer oxidation of alkyl radicals by copper(II) halides and pseudohalides is akin to a similar process for the formation of methyl acetate from the electron-transfer oxidation of methyl radicals by copper(II) acetate (see preceding paper). The relationships among oxidative solvolysis and oxidative displacement in electron transfer and oxidative substitution in ligand transfer are intertwined, but not yet completely clear.

(31) R. G. Pearson, *Science,* **151,** 172 (1966); *J. Amer. Chem. Soc.,* **85,** 3533 (1963); S. Arhland, J. Chatt, and N. R. Davies, *Quart. Rev., Chem. Soc.,* **12,** 265 (1958); M. M. Jones and H. R. Clark, *J. Inorg. Nucl. Chem.,* **33,** 413 (1971).

(32) *Cf.* G. A. Olah and C. H. Lin, *J. Amer. Chem. Soc.,* **90,** 6468 (1968).

(33) L. A. Spurlock and P. E. Newallis, *Tetrahedron Lett.,* 303 (1966).

{22 mm)). Approximately 1 g of material boiling at 89–90° (22 mm) was collected, and the pmr spectrum indicated that it was predominately of cyclopropylcarbinyl structure. By means of preparative gc on a 9-ft 20% FFAP column at 100°, 53 mg of material was obtained which was confirmed by pmr and infrared spectroscopy to be cyclopropylcarbinyl thiocyanate.

Cyclobutyl Thiocyanate. Cyclobutanol (10.5 g) was prepared from cyclopropylcarbinol and allowed to react with 40 g of toluenesulfonyl chloride in 80 ml of pyridine at 0° by the procedure described for allylcarbinyl thiocyanate to afford 31 g of cyclobutyl tosylate. To 26 g of potassium thiocyanate in 150 ml of reagent grade acetone 26 g of cyclobutyl tosylate was added and refluxed for 12 hr. After the work-up (see 5-hexenyl thiocyanate), 5 g of crude material was obtained. Analysis by gas chromatography (11-ft XF1150 at 150°) gave the following composition of products: allylcarbinyl isothiocyanate, 1.4%, cyclobutyl isothiocyanate, 7.5%, cyclopropylmethyl isothiocyanate, 8.3%, allylcarbinyl thiocyanate, 14.5%, cyclobutyl thiocyanate, 23.5%, cyclopropylmethyl thiocyanate, 44.5%. A pure sample of cyclobutyl thiocyanate was obtained by preparative gas chromatography on a 9-ft XF1150 column at 130°. The structure was confirmed by infrared and pmr analysis.

Homoallylic Isothiocyanates. Each of the homoallylic isothiocyanates was prepared by reaction of the corresponding amine with carbon disulfide by the method of Hodgkins and Ettlinger.[34] Cyclobutylamine was generously donated by Dr. L. Friedman. Each of the amines gave isomerically pure material by gas chromatography, pmr, and infrared analysis.

Hydrocinnamic-β,β-d_2 Acid. Ethyl benzoate (0.225 mol) was reduced with 0.119 mol of lithium aluminum deuteride (E. Merck AG, 99% min D) in ethyl ether to yield 0.14 mol (62% yield) of benzyl-α,α-d_2 tosylate via the sodium salt.[35] Pmr spectrum of the tosylate showed no resonances at τ 5 characteristic of the benzyl protons. Sodium diethylmalonate (0.15 mol) was benzylated in 70 vol % dimethyl sulfoxide-ethanol by addition of benzyl tosylate (0.13 mol) to a solution containing a 200% excess of diethyl malonate. After work-up, distillation yielded 26 g of diethyl benzylmalonate-d_2 boiling at 108–113° (1 mm). It was then saponified by treating diethyl benzylmalonate with 2.2 equiv of aqueous potassium hydroxide at 100° for 3 hr, acidified with dilute sulfuric acid, and decarboxylated by heating at 100° for 2 hr. The acid was then extracted with pentane and on crystallization afforded 8 g of hydrocinnamic-β,β acid, mp 47.8–48.2° [lit.[37] mp 48.6°]. The pmr spectrum in carbon tetrachloride showed aromatic resonance (τ 2.9) and the broadened α-methylene protons (τ 7.4). The integrated intensities of aromatic to methylene were 5:2 and no resonances for the benzyl protons. The acid was also analyzed for deuterium by combustion followed by the falling drop method.[37] *Anal.* Calcd for $C_9H_8D_2O$: C, 71.0; H, 5.3; D, 2.7. Found: C, 71.5; H, 5.5; D, 2.6; 1.96, 1.99 atoms of D/molecule. *p*-Methoxyhydrocinnamic-α,α-d_2 acid was described previously.[4]

Preparation of Diacyl Peroxides. A solution of 1.1 equiv of pyridine and 50 ml of diethyl ether was cooled to −10°, and 0.55 equiv of 30% hydrogen peroxide was added portionwise so that the temperature did not exceed 0°. The solution was stirred rapidly and the acid chloride (1.0 equiv) added dropwise, maintaining the temperature between −5° and −10°. The reaction was stirred an

additional 2 hr at 0°, after which it was neutralized with a small amount of chilled 10% sulfuric acid (throughout the work-up the solution is *not allowed to warm above 10°*). The reaction was diluted with ether and the peroxide extracted after flooding with excess ice-water. The solution was diluted further with pentane, and washed with chilled 10% sulfuric acid, 10% sodium bicarbonate, and distilled water. The pentane-ether extract was dried and concentrated on a rotary evaporator with the aid of a water aspirator. Final traces of solvent were removed by use of a vacuum pump. (The flask was always maintained in an ice bath throughout the evaporation.) The diacyl peroxide was also made directly from the carboxylic acid and anhydrous hydrogen peroxide in the presence of dicyclohexylcarbodiimide by the method of Greene and Kazan.[38] Diacyl peroxides were analyzed by infrared spectroscopy and showed characteristic doublets at 1780 and 1800 cm^{-1}. The purity was further determined by iodometric titration using ferric chloride (0.02% solution) as a catalyst. Peroxides were not used unless their purity exceeded 98%.

The General Procedure for Reaction. The copper(II) complex and any other additives were dissolved in the appropriate solvent contained in a 125-ml Erlenmeyer flask. The diacyl peroxide was then added from a standard solution by means of a glass pipet. The flask was capped with a rubber septum and flushed with a stream of helium for approximately 15 min to remove any oxygen from the system. The appropriate copper(I) species in solution was then added by means of a hypodermic syringe through the rubber septum to initiate the decomposition. The reactions were then run until carbon dioxide was liberated completely.

General Reaction Procedure for Copper(II) Bromide and Thiocyanate. The copper(II) complex and additives, if any, were dissolved in the appropriate solvent. The 125-ml Erlenmeyer flask was then capped with a rubber septum and the solution flushed with helium for approximately 15 min to remove oxygen. A solution of the diacyl peroxide was added by means of a hypodermic syringe. The reaction was then run until carbon dioxide was no longer evolved.

Analysis. All gaseous and liquid products were analyzed by standard gas chromatographic procedures using the internal standard method. In all cases standard solutions or standard gas samples were prepared, and calibrations of products against the internal standard were determined under reaction conditions. Carbon dioxide, methane, ethane, and ethylene were determined on a 2-ft Porapax Q (150–200 mesh) column at room temperature by gas chromatography using thermal conductivity detectors. Gaseous hydrocarbons were determined on a 15-ft column of 15% Dowtherm A on firebrick or a 20-ft column of 20% silver nitrate and 25% benzonitrile on Chromosorb W. The analyses for liquid products were performed by removing an aliquot of the reaction mixture and an aliquot of the internal standard solution to a separatory funnel. Excess water was added and the organic products were extracted with an appropriate solvent, usually ethyl ether or pentane. The extract was then washed several times with water and 10% sodium bicarbonate to remove the solvents (usually acetonitrile). In those cases where the products were soluble in water the reaction mixture was analyzed directly by adding an aliquot of the internal standard to an aliquot of the reaction mixture. Possible interference in the analysis from the presence of copper salts was carefully checked.

Acknowledgment. We wish to thank the National Science Foundation for generous financial support of this work.

(34) J. Hodgkins and M. Ettlinger, *J. Org. Chem.*, **20**, 404 (1955).

(35) J. K. Kochi and G. S. Hammond, *J. Amer. Chem. Soc.*, **75**, 3433 (1953).

(36) R. Weast, Ed., "Handbook of Physics and Chemistry," 47th ed, Chemical Rubber Publishing Co., Cleveland, Ohio, 1966.

(37) Analysis performed by J. Nemeth, University of Illinois.

(38) F. D. Greene and J. Kazan, *J. Org. Chem.*, **28**, 2168 (1963).

Jenkins, Kochi / *Ligand-Transfer Oxidation of Alkyl Radicals*

Part VI

Editor's Comments on Papers 37–45

We come next to a series of papers which illustrate the use of the HSAB principle in organic chemistry. The initial paper is by Hudson, who was one of the first to apply HSAB to organic chemistry. The second paper is by Saville and gives his very important rules for the selection of catalysts for nucleophilic and electrophilic substitutions. Indeed more than catalysis is considered, since all multicenter chemical reactions involving interconnected acid–base reactions can be included.

The next two contributions are from Songstad and myself. We point out, quoting some earlier results by Hine, that symbiosis is particularly important in organic chemistry. We also stressed the novel symbiosis that exists between the entering and leaving group in an S_N2 reaction.

The paper by Trahanovsky and Doyle complains, however, that the HSAB principle failed when applied to a particular case. In their example a competition existed between acetic acid (a hard base) and the olefinic double bond (a soft base). The prediction from symbiosis is that hex-5-enyl iodide should give more olefin attack, leading to cyclic product, than hex-5-enyl tosylate. Unfortunately the opposite result was found, a clear failure of HSAB.

In extenuation I can only say that the next paper on nucleophilic reactivity mentions that the double bond of allyl alcohol is an immeasurably weak reagent towards methyl iodide. It is likely, therefore, that the acetolysis of hex-5-enyl species involves some ion-pair formation as well as direct acetolysis. Attack on the ion-pair by the olefinic double bond could lead to cyclization. Failure of the HSAB principle, in other words, may be an indication that the reaction is more complex than imagined. If the reaction is not simple S_N2, there is no reason for symbiosis to exist.

The next papers in this section deal with nucleophilic reactivity. The article by Teichmann and Hilgetag is concerned with the behavior of the ambident thiophosphoryl group. The two papers by Kice and his co-workers deal with the relative behavior of sulfenyl (RS), sulfinyl (RSO), and sulfonyl (RSO_2) sulfur towards various nucleophiles. It is found that they form a series of electrophiles of increasing hardness, as expected from HSAB.

Coordination Chemistry Reviews
Elsevier Publishing Company, Amsterdam
Printed in the Netherlands.

37

THE CONCEPT OF HARD AND SOFT ACIDS AND BASES AND NUCLEOPHILIC DISPLACEMENT REACTIONS

R. F. HUDSON

Cyanamid European Research Institute, Cologny–Geneva (Switzerland)

If we consider a simple ionic equilibrium,

$$N^{(-)} + M^{+} \rightleftharpoons MN$$

the free energy change ΔF is given by

$$\Delta F = -RT \ln K = E_N^s - D_{MN} - E_M^s \tag{1}$$

where E_N^s is the energy required (including entropy of solvation) to remove an electron from N^- (in solution), E_M^s the gain in energy on adding an electron to M^+ (in solution), and D_{MN} the dissociation energy of compound MN. The order of equilibrium constants for the combination of a series of ligands with a given metal M is therefore determined by the changes in E_N^s and D_{MN}.

In general, when a series of ligands from the same group of the Periodic Table is considered *e.g.*

$$F^{(-)}, \quad Cl^{(-)}, \quad Br^{(-)}, \quad I^{(-)}$$
$$RO^{(-)} \quad RS^{(-)} \quad RSe^{(-)} \quad RTe^{(-)}$$
$$R_3N \quad R_3P \quad R_3As$$

increases in D_{MN} (ΔD_{MN}) are accompanied by (approximately linear) increases in E_N^s (ΔE_N^s). This follows[1] from the tendency of D_{MN} to increase with D_{H-N} provided that N is more electronegative than M, *i.e.*

$$D_{MN} = k D_{H-N} + \text{const.}$$

as shown by data in Fig. 1. Also E_N^s for a given series increases (almost linearly) with D_{HN}, *i.e.*

$$D_{HN} = k' E_N^s + \text{const.} \quad (k' \text{ is the slope of Fig.2)}$$

as shown in Fig. 2. Consequently for each series of ligands we have

$$D_{MN} = k k' E_N^s + \text{const.}$$

and

$$RT \ln \frac{K_1}{K_2} = (k k' - 1) \Delta E_N^s \tag{2}$$

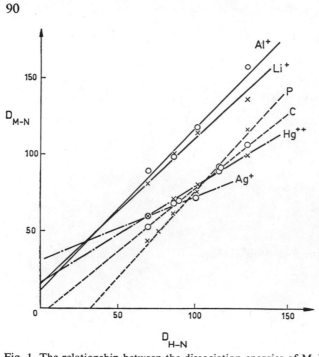

Fig. 1. The relationship between the dissociation energies of M–N bonds, where M is Al⁺, Li⁺, P^{3+}, C, Hg^{2+}, and Ag⁺, and the corresponding bonds formed by the ligand N and hydrogen. N = halogen.

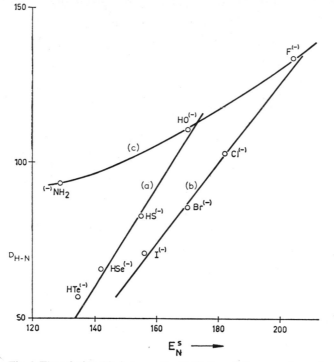

Fig. 2. The relationship between the H–N dissociation energies and solution electron affinities, E_N^s.

Thus in general we have two orders depending on the magnitude of k (which depends on the nature of M) and k' (which depends on the ligand series considered), *i.e.*

A. Order	B. Order
$(kk' > 1, i.e.\ \Delta D_{MN} > \Delta E_N^s)$	$(kk' < 1, i.e.\ \Delta D_{MN} < \Delta E_N^s)$
$F^- > Cl^- > Br^- > I^-$	$I^- > Br^- > Cl^- > F^-$
$RO^- > RS^-$	$RS^- > RO^-$

Order A is characteristic of a hard metal since it tends to combine preferentially with hard bases (*e.g.* F^-, OR^-), order B being characteristic of a soft metal[2]. Since the constant k' is similar for these two ligand series, these orders will define soft and hard behaviour, in the same way as A and B metals are defined in the Chatt-Ahrland-Davis classification[3]. A given order is however not usually obtained if the two series are combined, since the $D_{H-N} - E_N$ lines for the two series do not coincide (Fig. 2), as shown by the following data:

1) $F < Cl < Br < I < OH < RS$ $\quad$ $MeHg^+$ equilibria[4]

2) $F < Cl < Br < OH < I < RS$ $\quad$ S_N2 displacement on carbon[5]

3) $F < OH < Cl < Br < I < RS$ $\quad$ Ag^+ equilibria[4]
$\qquad\qquad\qquad\qquad\qquad\qquad$ Pt^{II} displacements[6]

These orders differ in the relative position of OH^- since bond energy changes are more important for this ligand (*vide infra*).

For a wider comparison, the soft–hard classification will depend on the particular series of ligands chosen (*i.e.* on the magnitude of k'). Referring to Fig. 2, it is noted that kk' may be less than 1 when ligand series (c) is considered, whereas $kk' > 1$ for series (a) and (b). For example, for proton neutralisation (in water) we have

but $\quad$ $F^- > Cl^- > Br^- > I^-$ and $RO^- > RS^-$ *i.e.* "hard" orders

$R_3C^- \gg F^-$ *i.e.* "soft" order.

The order $F^- > R_3C^-$ can be observed only for combination with very hard acids (*e.g.* Al^{3+}) where bonding is largely electrostatic. That R_3C^- is very soft is shown by the preferential reaction at soft electrophilic centres[7], *e.g.*

$$CH_3CH{=}CH{-}CH_3 \xleftarrow{R_3C^-} CH_3{-}CH{-}CH{-}CH_3 \xrightarrow{OH^-} CH_2{=}CH{-}CH{=}CH_2$$
$$+ Br_2 \qquad\qquad\qquad \overset{|}{Br}\ \overset{|}{Br} \qquad\qquad\qquad + 2\,HBr$$

The significant feature* is that

$$\left[\frac{k_{R_3C^-}}{k_{HO^-}}\right]_{Br} > \left[\frac{k_{R_3C^-}}{k_{HO^-}}\right]_H$$

showing R_3C^- to be a very soft nucleophile relative to HO^-.

---

* There are unfortunately exceptions to this soft and hard definition.

It is to be noted that H^+ is not the hardest acid, since in the complex or compound it is less electron deficient than a proton. A small electron density produces considerable soft character (*e.g.*, the hydride ion is a very soft base).

Kinetic Processes[8]

In a displacement reaction, the nucleophilic order will depend in a similar manner on the relative values of ΔD_{NM} and $\Delta E_N{}^s$ and also on the extent of bond formation between the nucleophile N^- and electrophilic centre X. Consider the following displacement reaction involving a transfer of charge of *ze* from N^- to the electrophile XY,

$$N^- + XY \rightleftharpoons \overset{(1-z)-}{N} --- \overset{(\Delta-z)+}{X} --- \overset{\Delta-}{Y} \rightarrow NX + Y^-$$

The rate constant *k* is given by

$$-RT \ln k \equiv \Delta F^* = \alpha E_N{}^s - \beta D_{XN} + f(XY) \tag{3}$$

For a given electrophile, we may assume that the change in energy, $f(XY)$ is constant. Now $\alpha \rightarrow 1$, $\beta \rightarrow 1$, when $z \rightarrow 1$ and $\alpha \rightarrow 0$, $\beta \rightarrow 0$ when $z \rightarrow 0$. For intermediate values, we consider that $\alpha > \beta$ since the D_{MN} term includes inter electronic repulsions in the NXY bond, which are very important in transition states.

These considerations lead to the following situations:
a) When $\alpha \rightarrow 1$, the rate order will follow the thermodynamic order as given by equation (1). This may be the hard or soft order depending on the values of D_{XN}.
b) When $\alpha \rightarrow 0$, we have zero selectivity (limiting S_N1).
c) For small values of α ($\alpha < 0.5$), the βD_{XN} term is relatively small, and the rate changes follow $E_N{}^s$, *i.e.* the order for soft acids.

We reach the important conclusion therefore that the soft or hard behaviour is determined not only by the values of $E_N{}^s$ and D_{XN}, but also by the extent of bond formation in the transition state. This is shown by the rate orders frequently observed for acylation (I) and alkylation (II).

$$\text{RCOY} + N^- \rightleftharpoons R-\overset{\overset{O^-}{|}}{\underset{\underset{Y}{|}}{C}}-N \qquad\qquad \text{RCH}_2Y + N^- \rightleftharpoons N^{(\Delta-)} \cdots \overset{\overset{R}{|}}{\underset{\overset{}{H\ \ H}}{C}} \cdots Y^{(\Delta-)}$$

$$\text{(I)} \qquad\qquad\qquad\qquad\qquad\qquad \text{(II)}$$

$$F^- > Br^-, I^- \qquad\quad I^- > Br^- > Cl^- > F^-$$

$$RO^- > RS^- \qquad\qquad RS^- \gg RO^-$$

With this added proviso, electrophiles can be divided into soft and hard as shown by the following examples.

Soft electrophiles
$R_1R_2R_3CY$, R_1R_2PY, RSY, Br–Y
H–Y ($\alpha \to 0$)

Hard electrophiles
R_1R_2POY, RCOY, RSO_2Y
HY ($\alpha \to 1$)

Since the nucleophilic order depends on the degree of bonding in the transition state (by changing k of equation 2), a common order of increasing nucleophile or ligand softness cannot be established. Variations in different orders will follow the kinetic equation given above. As an example we compare, in Fig. 3, the stability constants for combination with[4] CH_3Hg^+ with the rate constants for displacement at a complex[6] of Pt^{II} by a series of nucleophiles.

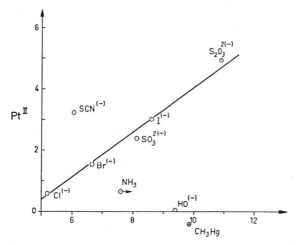

Fig. 3. The relationship between the equilibrium constant for the addition of ligand N to the CH_3Hg^+ cation and the rate of displacement on Pt^{II} complexes[6].

Basic ligands (RO^- and RNH_2) have a relatively greater affinity for $MeHg^+$ than for Pt^{II} since the bond is fully formed in the first case. The β term of equation (3) is therefore relatively small compared with the corresponding term in equation (1). On the other hand, the thiocyanate ion has a greater affinity for Pt^{II} than for CH_3Hg^+, probably because conjugation

$$\ominus_{\overset{\frown}{N}=C\overset{\frown}{=}S:} \xrightarrow{\hspace{1cm}} XY$$

reduces the activation energy of the kinetic process. A similar high reactivity is observed in S_N2 displacements on carbon[5].

Ambient ions[9]

The reactions of ambient ions also provide exceptions to the simple classification of hard and soft acids and bases. In the following reactions,

$$CH_3CONCS \xleftarrow{CH_3COX} NCS^- \xrightarrow{CH_3X} CH_3SCN$$

Coordin. Chem. Rev., 1 (1966) 89–94

$$(RO)_2-\overset{\overset{O}{\|}}{P}-O-\overset{\overset{S}{\|}}{P}(OR)_2 \xleftarrow{(RO_2POX)} (RO)_2P\overset{\diagup O}{\diagdown_{S^-}} \xrightarrow{CH_3X} (RO)_2P\overset{\diagup O}{\diagdown_{SCH_3}}$$

the soft electrophile reacts with the soft sulphur atom and the hard electrophile with the hard nitrogen or oxygen atom in agreement with the general principle that soft electrophiles tend to react preferentially with soft nucleophiles and hard with hard. A similar rule was advanced by Kornblum et al.[9] for the reactions of nitrite and enolate ions with alkyl halides.

There are however many examples where the simple rule is contradicted[10], e.g.

$$(1)\ (RO)_2PO^- + (RO)_2P(O)X \rightarrow (RO)_2\overset{\overset{O}{\|}}{P}-\overset{\overset{O}{\|}}{P}(OR)_2$$

$$(2)\ (RO)_2PO^- + (RO)_2PX \rightarrow (RO)_2P-O-P(OR)_2$$

The preferential reaction of soft phosphorus in $(RO)_2P-O^-$ with hard P in $(RO)_2POCl$, is determined by the high P=O bond energy in the transition state (which resembles the addition intermediate), hence the thermodynamically stable product is obtained. It is possible, however, that the reaction leading to the isomer is reversible, in which case the thermodynamic product is always formed. In reaction(2) the transition state is closer to the reactants, and the greater charge on oxygen relative to phosphorus promotes reaction on the former (particularly in the aprotic or non-polar solvents used).

We conclude therefore that the interpretation of ambient ion action requires a knowledge of the nature of the transition state (in addition to the extent of reversibility of the competing reactions), and that a simple soft and hard rule, although frequently obeyed, is not a general one for this kind of reactant.

REFERENCES

1 L. A. ERREDE, J. Phys. Chem., 64 (1960) 1031; R. S. NEALE, J. Phys. Chem., 68 (1964) 143.
2 R. G. PEARSON, J. Am. Chem. Soc., 85 (1963) 3533.
3 S. AHRLAND, J. CHATT AND N. R. DAVIES, Quart. Rev. (London), 12 (1958) 265.
4 G. SCHWARZENBACH AND M. SCHELLENBERG, Helv. Chim. Acta, 48 (1965) 28.
5 C. G. SWAIN AND C. B. SCOTT, J. Am. Chem. Soc., 75 (1953) 141.
6 U. BELLUCO, L. CATTALINI, F. BASOLO, R. G. PEARSON AND A. TURCO, J. Am. Chem. Soc., 87 (1965) 241.
7 R. F. HUDSON, Chem. Eng. News, (May 31, 1965) 90.
8 R. F. HUDSON, Chimia, 16 (1962) 173; Chim. Ind. (Milan), 46 (1964) 1177.
9 N. KORNBLUM, R. A. SMILEY, R. K. BLACKWOOD AND D. C. IFFLAND, J. Am. Chem. Soc., 77 (1955) 6269.
10 R. F. HUDSON, Structure and Mechanism in Organo-Phosphorus Chemistry, Academic Press, London, 1965, p. 125.

38

The Concept of Hard and Soft Acids and Bases as Applied to Multi-Center Chemical Reactions [**]

BY B. SAVILLE [*]

Many chemical bonds of differing types and strengths have recently been regarded by Pearson [1] as representing partnerships between (Lewis) acids and (Lewis) bases. Most acceptor molecules or ions (acids) can be placed in one or other of two categories, graphically termed "Hard" and "Soft". There are also two broad categories of donor molecules or ions (bases) which can also be termed Hard and Soft. On the whole, strong chemical bonds are partnerships between either a Hard base and a Hard acid or a Soft base and Soft acid, whereas weaker bond types most usually result in cases of either Hard base-Soft acid or Soft base-Hard acid interactions. The present paper shows how this concept of acidity and basicity can be applied in the interpretation of multi-center chemical reactions involving interconnected acid-base relationships. In particular, four-center substitutions and additions involving cooperative attack by nucleophiles and electrophiles at various chemical bonds have been examined, and a conclusion is reached that especially reactive patterns of reactants can be developed if the substrates contain bonds between either a hard acid and a soft base, or a soft acid and a hard base. Indeed, the arguments can be elaborated to provide two distinct Rules which should be of interest in the interpretation of metal-ion assisted reactions and in the design of novel syntheses.

1. Introduction

In 1963 *Pearson* published a notable paper [1] on the concept of hard and soft acids and bases. In essence, the idea that *Pearson* succeeded in conveying so con-

vincingly is that practically any type of chemical bond one cares to choose (be it either a strong σ or π bond of an organic molecule, a ligand-metal interaction, a bond between charge-transfer complex partners, or even a solvation bond) can be thought of as an acid bound to a base, this being true as long as one is prepared to adopt the view that there are two general classes of acids and two general classes of bases.

Pearson brought out powerfully the generalization that some bases, which can be termed hard, tend to form their strongest bonds with a fairly definite class of acids which he also called hard. On the other hand, hard bases interact rather weakly with a second class of acids which were termed soft. However, the latter

[*] Prof. B. Saville
Natural Rubber Producers' Research Association
Welwyn (England)
Present address:
Dept. of Chemistry, Oregon State University
Corvallis, Oregon 97331 (U.S.A.)
[**] Presented in part as lectures at Imperial College (London), Geneva (Cyanamid European Research Institute), and St. Andrews (Scotland).
[1] *R. G. Pearson*, J. Amer. chem. Soc. *85*, 3553 (1963); Science (Washington) *151*, 172 (1966).

form their strongest bonds with bases drawn from a second class, also labeled soft. The soft bases generally (but not always) fail to give strong bonds with hard acids.

acids   bases
hard ⇄ hard
soft ⇄ soft

How can one determine whether a molecule or an ion is hard or soft?

At an international symposium on this subject [2] it was generally agreed that hardness has to do with bonds which have high ionic character, whereas softness relates more directly to the current picture of covalent bonds. Thus, factors such as small atomic radius, high effective nuclear charge, low polarizability, and low ease of oxidation confer hard character. We find among hard acids the proton, cations of small size (Li$^+$, Be^{2+}), and cations or molecules in which substituents induce a high positive charge on the central atom (Mg^{2+}, Fe^{3+}, Co^{3+}, Th^{4+}, BF$_3$, AlCl$_3$, SiF$_4$).

These species give strong bonds with a restricted range of hard bases which are drawn from N, O, and F as donor atoms. It is interesting that carbanionoid systems (R$_3$C$^\ominus$ or R$_2$C$\overset{\delta\ominus}{=\!=\!=}$), also containing a first row element, are not very hard as bases, especially in cases where the 2p electron density is low, as in an olefin. Indeed, saturated carbon as a base seems to be of hard-soft borderline character but falling, on the whole, into the wider group of soft bases.

Outstanding examples of soft bases are the hydride ion and ions derived from P(III), As(III), S(II), Se(II) (Cl(I)), Br(I), I(I), which form bonds of greater covalent character with soft acids.

Soft acids are usually characterized by low positive charge, large size, filled outer orbitals, and high polarizability and include derivatives of saturated C(IV), O(II) (surprisingly), S(II), P(III), Cl, Br, I (as cations or atoms) Hg(I), Hg(II), Ag(I), B(Alk)$_3$, Pt(II), Pt(IV), Pd(IV), as well as some electron-deficient π systems such as those in polynitroaromatics and tetrahalogenoquinones. A complete list and some explanation of these class divisions is provided in *Pearson*'s paper [1].

Although R$_3$C$^\ominus$ is not a very hard base it can still bond well with hard acid centers on account of its high charge density; nevertheless, a polarized olefin

$$\overset{}{\underset{}{>}}C=C\overset{}{\underset{}{<}} \longleftrightarrow \overset{\delta\oplus}{>}C\!\!-\!\!\!-\!\!\!-\!\!C\overset{\delta\ominus}{<}$$

is most decidedly a soft base [3]. A carbon radical is soft both as an acid and a base, whereas a carbonium ion must be regarded as a hard acid. Saturated carbon centers as acids (*e.g.* in nucleophilic substitutions) are definitely soft. Carbonyl and thiocarbonyl carbon centers which, as $>$C$^\oplus$–O$^\ominus$ and $>$C$^\oplus$–S$^\ominus$, resemble

carbonium centers are hard like the proton [4,5], whereas aromatic carbon both as a donor and acceptor tends to be soft.

It will be seen that the above statements are in accord with the following well-known distinctions in simple organic reactions:

(i) It is much easier to make a soft acid such as Br$^+$ or Ag$^+$ react with an olefin than to convert the latter into a carbonium ion with a proton source (hard acid).

(ii) Carbon radicals add to soft oxygen π bonds or attack halogen compounds (soft acids) with great ease, or combine with themselves, but abstract hydrogen (hard acid) from activated methylene groups only slowly.

(iii) The t-butyl cation (CH$_3$)$_3$C$^\ominus$ in aqueous solution reacts preferentially with hard solvent molecules but not with added soft thio nucleophiles (bases), such as S=C(NH$_2$)$_2$ or S$_2$O$_3^{2\ominus}$.

(iv) The reagents generally used in attacking carbonyl groups, eliminating water, or displacing attached basic groups are HO$^\ominus$, RO$^\ominus$, HOO$^\ominus$, and a wide range of reagents containing amino groups; the latter are hard bases. Iodide ion and thiolate ions (soft bases) are less suitable reagents.

We shall now see how usefully the concept can be applied to aspects of chemical reactivity. We shall be concerned not only with equilibrium situations, but also with kinetic aspects – indeed, with hardness and softness seen in regard to the rate of interaction of acidic and basic centers. In the majority of situations it seems that the hard-soft disctinctions based on equilibrium properties may be applied to rate processes, but, as *Pearson* points out, acids and bases tend to exhibit soft character more in kinetic investigations than in equilibrium investigations [1]. This follows from the fact that bonds are longer in transition state configurations. Softness can in part be due to overlap between diffuse orbitals and thus can contribute to bonding at an earlier point along a reaction coordinate.

Notwithstanding such possible complications, the author is of the opinion that the concept may find major application when reactions are considered in which two or more pairs of atoms are at some stage involved in bond making and bond breaking processes, *i.e.* in the field of multi-center reactions. This article is devoted to developing a general picture of four-center reactions.

2. Four-Center Substitutions and Additions

Bimolecular nucleophilic substitution can be represented as follows:

$$Z: + \, A\text{--}X \; \rightarrow \; [Z\text{--}A]^\oplus \, [:X]^\ominus \qquad (1)$$

where Z: is a nucleophile which always carries an unshared pair of outer-shell electrons on the attacking atom; A is the electrophilic center, *i.e.* one which, after departure of the nucleophilic leaving group :X,

[2] For a summary of papers, see Chem. and Engng. News *43*, No. 22, p. 90 (1965).

[3] *J. Chatt*, J. inorg. Nuclear Chem. *8*, 515 (1958).

[4] *J. O. Edwards* and *R. G. Pearson*, J. Amer. chem. Soc. *84*, 16 (1962).

[5] *R. G. Pearson*, *D. N. Edgington*, and *F. Basolo*, J. Amer. chem. Soc. *84*, 3233 (1962).

Angew. Chem. internat. Edit. / Vol. 6 (1967) / No. 11

929

366

can present a vacant orbital to the electron-pair of Z: for bonding purposes. Thus, as reaction proceeds, the center A changes its partnership with :X for a more satisfactory one with Z:.

Let us consider now the factors that influence the bimolecular substitution system. Clearly, the incipient development of a strong bond between Z: and A (particularly if this strength is maintained at the strained-bond lengths in transition states) will be of key importance in determining reaction rate, and may be of dominant importance if the new Z–A bond is nearly completely formed before the old A–X bond breaks. On the other hand, if in the transition state considerable stretching of the A–X bond has already occurred, the electron pair of the A–X σ bond begins to reside on X. The easier the pair can be accommodated on X the more rapidly will substitution proceed. Generally, the bonding pair of A–X is finally delocalized and stabilized if X is an electron-deficient π system spread over atoms of high effective positive charge. Thus *(1)* and *(2)* are two of the best leaving-groups because of the charge spreading contributions of the type *(3)* and *(4)*. By contrast, $-\ddot{O}H$ and $-\ddot{O}R$

$$(1)\ \ -\underset{\underset{O}{\overset{\|}{}}}{\overset{\overset{O}{\|}}{\ddot{O}}}\text{-S-R} \qquad -\underset{\underset{O}{\overset{\|}{}}}{\overset{\overset{O}{\|}}{\ddot{O}}}\text{-Cl=O}\ \ (2)$$

$$(3)\ \ -\underset{\underset{O}{\overset{\|}{}}}{\overset{\overset{O^{\ominus}}{|}}{\overset{\oplus}{O}}}\text{=S-R} \qquad -\underset{\underset{O}{\overset{\|}{}}}{\overset{\overset{O^{\ominus}}{|}}{\overset{\oplus}{O}}}\text{=Cl=O}\ \ (4)$$

are much poorer leaving groups because the 2p electron pairs reside firmly on one oxygen atom.

The interaction of the developing species [:X]$^{\ominus}$ with the solvent can also be a stabilizing factor; however, the effect is complicated since it is often impossible to arrange for a solvent to bond well with [:X]$^{\ominus}$ and not with :Z. *Pearson*[1] has discussed *Parker*'s view [6] that solvation reduces basicity of small (hard) anions whilst having little effect on basicity of large (soft) anions. Selective assistance of substitutions is therefore possible on this basis.

A greatly increased driving force for substitution (1) above can be provided by increasing the effective charge of the nucleophilic atom in the leaving group, a condition that can be achieved if :X carries at least two pairs of free p electrons. In this case, A–X (more correctly A–$\ddot{X}$) can coordinate with an added electrophilic acceptor molecule or ion E.

$$A\text{-}\ddot{X} + E \ \rightleftharpoons\ A\text{-}\overset{\oplus}{X}\text{-}\overset{\ominus}{E} \qquad (2)$$
$$(5)$$

A new reactive substrate *(5)* will develop in which X carries a higher effective nuclear charge than it does in A–$\ddot{X}$ bond. The higher charge will now assist X in accepting and stabilizing the electron pair of the A–X bond in A–X$^{\oplus}$–E$^{\ominus}$ which we wish to break.

[6] *A. J. Parker*, Quart. Rev. (Chem. Soc., London) *14*, 163 (1962); *J. Miller* and *A. J. Parker*, J. Amer. chem. Soc. *83*, 117 (1961).

Adding the electrophile E thus causes a rate-enhancement of substitution (1). The rate constant k_{Ass} for the four-center electrophilic substitution (3),

$$Z\text{:}\curvearrowright A\underset{}{\overset{\frown}{\text{-}}}\overset{\oplus}{X}\text{-}\overset{\ominus}{E} \ \xrightarrow{\ k_{Ass}\ }\ [Z\text{-}A]^{\oplus}[\ :X\text{-}E]^{\ominus} \qquad (3)$$

should be greater than the second-order rate constant, k, of the normal substitution (1). Also, if reaction (1) were reversible, the final equilibrium position would be displaced well to the right in the presence of E. The species E would also serve to reduce the nucleophilicity of [:X]$^{\ominus}$.

This rather general statement is based on accepted orders of leaving group activity (R$_2$O$^{\oplus}$, RHO$^{\oplus}$– $\gg$ R$\ddot{O}$–, R$_3$N$^{\oplus}$– $\gg$ R$_2\ddot{N}$–, R$_2\overset{\oplus}{S}$– $\gg$ R$\ddot{S}$–) given in text-books dealing with reaction mechanisms [7].

Although the above theme of electrophilic assistance of a bimolecular substitution has been developed having regard to a process in whose transition state considerable A–$\ddot{X}$ bond stretching has occurred, it should be applicable also to the case where formation of a Z–A bond is well developed before the A–$\ddot{X}$ bond begins to break. This follows from the idea that for Z–A bond formation to precede A–$\ddot{X}$ bond breaking an initial vacant orbital must be available in A to receive the nucleophile Z:. However, if X carries an electron pair this will be involved in π bonding with the vacant orbital of A; thus the effect of adding E would be to uncouple the π bond, leaving the orbital on A vacant and ready to receive Z: more easily than before.

We therefore come to visualize a general form of four-center reaction (4) involving the cooperative action of a nucleophile (Z:) and an electrophile (E) on a substrate (A–$\ddot{X}$) [*].

The arrows can be numbered to indicate the timing of the covalency changes. Equation (5) represents an electrophilically assisted nucleophilic substitution.

A nucleophilically assisted electrophilic substitution, in which the need for an unshared electron-pair on X is removed but in which a vacant orbital on A is mandatory, is depicted in eq. (6).

There will also be situations in which Z: and E are neighboring atoms or groups in the same molecule (eq. (7)).

$$Z\text{:}\curvearrowright A\underset{}{\overset{\frown}{\text{-}}}\overset{\overset{\displaystyle\ddot{}}{}}{X}\overset{\downarrow}{}E \ \longrightarrow\ [Z\text{-}A]^{\oplus}[\ :X\text{-}E]^{\ominus} \qquad (4)$$

$$Z\text{:}\overset{2}{\curvearrowright} A\underset{3}{\overset{\frown}{\text{-}}}\overset{\overset{1}{}}{X}\overset{\downarrow}{}E \ \longrightarrow\ [Z\text{-}A]^{\oplus}[\ :X\text{-}E]^{\ominus} \qquad (5)$$

$$Z\text{:}\overset{1}{\curvearrowright} A\underset{2}{\overset{\frown}{\text{-}}}X\overset{3}{\longrightarrow}E \ \longrightarrow\ [Z\text{-}A]^{\oplus}[\ X\text{-}E]^{\ominus} \qquad (6)$$

$$\overset{Z}{\underset{A}{}}\overset{\overset{\displaystyle E}{|}}{\underset{\underset{\displaystyle\ddot{X}}{|}}{}} \ \longrightarrow\ [\ :Z\text{-}A]^{\oplus}[\ :X\text{-}E]^{\ominus} \qquad (7)$$

$$Z\text{:}\curvearrowright A\overset{\frown}{\text{-}}X'\curvearrowright E \ \longrightarrow\ \overset{\ominus}{Z}\text{-}A\text{-}X'\text{-}E\ \cdot \qquad (8)$$

[7] *C.K.Ingold*: Structure and Mechanism in Organic Chemistry. Bell and Sons, London 1953, p. 338ff. *E. S. Gould*: Mechanism and Structure in Organic Chemistry. Holt, Rinehart and Winston, New York 1959, p. 261.

[*] *R. E. Dessy* [83] has approached the subject of multicenter substitutions from the standpoint of the conventional electrophilic substitution but has arrived at much the same picture.

930

Angew. Chem. internat. Edit. / *Vol. 6 (1967)* / *No. 11*

367

These pictures of four-center substitutions in which A−X σ bonds become broken can be easily extended to the case of electrophilically assisted nucleophilic additions to π electron systems, in which π bonds are opened rather than σ bonds. Thus, if we consider A=X′ as substrate in place of A−Ẍ, then a four-center process can be written as eq. (8).

The rate of this process should again be greater than that of the corresponding process in the absence of the electrophile, E.

3. Two Rules Concerning Choice of Reactants

The above four-center reactions can be regarded as (i) formation of acid-base pairs and (ii) destruction of other pairs in the reactants (Scheme 1).

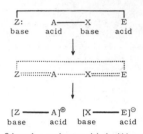

Scheme 1. —— denotes original acid-base pairs of reactants; ······· acid-base pairs being destroyed; :::::: developing acid-base pairs; —— final acid-base interactions of the products.

How can the concept of hard and soft acids and bases be applied to the choice of reactants in four-center reactions? We believe that in order for reactions to proceed easily to afford thermodynamically well-stabilized products, the acid-base pairs [*] of the products Z−A and Ẍ−E must be bonded much more firmly than the acid-base pairs Z−E and A−Ẍ [**] present in the reactants. This condition will be best fulfilled when the substrate bond A−Ẍ possesses hard-soft dissymmetry, i.e. when A is hard and Ẍ is soft (1st case) or when A is soft and Ẍ is hard (2nd case).

For the first case, bearing in mind that it is important that Z−A, when developed, should be a strong bond it will be appropriate to make Z: a hard base. Also, as Ẍ is to bond strongly with E the latter should be a soft acid. A rational four-center reaction pattern can be written:

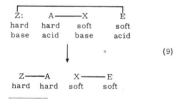

(9)

[*] From this point we omit overall charges in our representations of acid-base partnerships.
[**] Alternatively, here read the A = X′ π bond in the case of substrates undergoing nucleophilic addition.

It is only when the A−X bond possesses hard-soft dissymmetry that is it possible for the preferred Z: and E reagents to have a low mutual interaction energy, as required by the earlier deduction.

Equation (9) can now be formulated as the "first rule" for reagent choice in achieving particularly easy four-center substitutions. This is stated as follows:

If the electrophilic center A of a substrate A−Ẍ is a hard acid, then Ẍ should be a soft base. A−Ẍ will then react most easily with a nucleophile Z:, which is a hard base, together with an electrophile E, which is a soft acid.

In the second case of a substrate possessing hard-soft dissymmetry a second, complementary rule can be derived (eq. (10)).

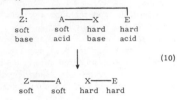

(10)

Here, the weak Z——E and A——X bonds of the reactants are replaced by strong Z−A (soft-soft) and X−E (hard-hard) bonds in the products.

The above rules should apply equally well to the variants of four-center substitution (equations (4)−(7)) and also to four-center assisted nucleophilic additions eq. (8). (In the latter case, one has to decide whether the atoms terminating an opened π bond are hard or soft.)

If A−X is initially either a hard-hard or soft-soft partnership, Z: and E must be either both hard or both soft. In other words, Z: and E are no longer co-operative reagents, and one cannot expect easy forward reactions under such conditions.

It is now of interest to consider a selection of known four-center reactions to see if reagents used in these cases do indeed conform to the rules proposed above. To begin, we shall consider the first rule and note that here the electrophilic center is to be a hard acid. Examples of hard acid centers include those based on boron, carbon in R−CO− and R−CS−, phosphorus in $R_2PO−$, sulfur in $R−SO_2−$ Al(III), Si(IV), Sn(IV), and of course $H^\oplus$.

These provide a wide range of substrates when coupled to soft leaving-group centers Ẍ, such as RS−, $R_2P−$, halide, $H^\ominus$, and $R_3C^\ominus$.

4. Examples of the Operation of the First Rule

4.1. Cleavage of Aryl-Boron Bonds

In the bromination of an aromatic compound it is generally assumed that the π electron system of the benzene ring attacks a bromine molecule (suitably activated in the sense $Br^{\delta\oplus}\cdots Br^{\delta\ominus}$) to provide an

931

368

intermediate *(6)* which then loses a proton in affording the observed substitution product:

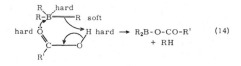

It appears that a similar mechanism applies in the reaction of bromine with aryl-boron bonds. The ease of this process should be considerably enhanced if the vacant orbital on boron is first filled by the electrons of a suitable nucleophilic base (Z) to provide a four-center transition state, for example in the cleavage of benzeneboronic acid by means of bromine.

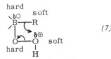

In acetic acid, the rate of this reaction is proportional to the product of concentration of substrate, bromine, and the added bases $OAc^\ominus$, or H_2O [8]. As the first rule predicts, the hard boron center responds to hard oxygenated bases, but not at all to soft $Br^\ominus$ species.

Clearly, also, the soft aromatic leaving group bonds well with the soft electrophilic $Br^\ominus$ center. Thus it seems that equation (12) furnishes excellent confirmation of the first rule:

Z:	A	X	E
oxygen base	boron center [*]	benzene nucleus	bromine (electrophilic acid)
hard	hard	soft	soft

If the first rule is to be useful, it must provide predictive information. It is therefore suggested that hypohalous acids, peracids, mercuric acetate, or bromine fluoride, *i.e.* reagents which all combine a soft electrophilic acid (E) [$Hal^{\delta\oplus}$, $HO^{\delta\oplus}$, $Hg^\oplus X$, $Br^{\delta\oplus}$] with a hard nucleophilic base [$HO^\ominus$, $RCOO^\ominus$, $F^\ominus$], would also be effective in cleaving aryl-boron bonds [**].

4.2. Cleavage of Boron Alkyls

The hydrolysis of trialkyl borons, even when catalysed by strong mineral acids, is remarkably slow [9,10]. This can be ascribed to the difficulty in producing the unusual "boronium" state (two vacant 2p orbitals on boron) demanded by the protolysis mechanism (eq. (13)).

$$\overset{|}{\underset{R}{B}} \overset{}{\curvearrowleft} H^\oplus \longrightarrow \overset{|}{\underset{}{B}}\oplus + RH \qquad (13)$$

[8] *H. G. Kuivila* and *E. K. Easterbrook*, J. Amer. chem. Soc. *73*, 4629 (1951).

[*] Boron is a hard center in BF_3 (cf. $BF_4^\ominus$) but a soft one in BH_3 (cf. $BH_4^\ominus$) and BPh_3 (cf. $BPh_4^\ominus$). Hard ligands, such as $OH^\ominus$, confer hardness to boron.

[**] *A. D. Ainley* and *F. Challenger* (J. chem. Soc. (London) *1930*, 2171) have shown that hypoiodite ion, neutral hydrogen peroxide, and aqueous mercuric bromide are effective in cleaving aryl-boron bonds. Zinc and cadmium halides also serve as hydrolysis catalysts.

[9] *D. Ulmschneider* and *J. Goubeau*, Chem. Ber. *90*, 2733 (1957).

[10] *J. R. Johnson, H. R. Snyder,* and *M. G. Van Campen jr.*, J. Amer. chem. Soc. *60*, 115 (1938).

Even though the developing electron-deficiency at boron in such a process could be offset by prior co-ordination with a base, the bases available in solutions of strong mineral acids will be weak and relatively ineffectual. On the other hand, conventionally weak acids such as propionic acid, which supply a relatively potent hard carboxylate base, are far better reagents for the protolysis of boron alkyls [11-13]. The reaction can be visualized as a four-center substitution:

$$\begin{array}{c} R \quad hard \\ R-B{\cdots}R \ soft \\ hard \ O \qquad H \ hard \\ \diagdown C \diagup \\ | \\ R' \end{array} \longrightarrow R_2B\text{-}O\text{-}CO\text{-}R' \atop + RH \qquad (14)$$

However, equation (14) does not conform with the requirements of the first rule; the hard proton has to be changed for a soft acid center *e.g.*, Hal^+, RS^+, or $HO^\oplus$ if rate enhancement is to occur. Indeed, perbenzoic acid (possessing the soft, electrophilic OH center) reacts with tri-n-butylborane at 0 °C in chloroform [1,17,57].

The use of alkaline hydrogen peroxide [14,15] as a general reagent for cleaving organoboranes is a similar example. The hard, electron-dense, oxygen base of hydroperoxide anion can coordinate with the vacant boron orbital to give an intermediate *(7)* (cf. [16]),

$$\begin{array}{c} hard \\ _{\delta\ominus O}\quad soft \\ \overset{\diagdown}{B}\text{---}R \\ \diagup \quad \diagdown \\ O \quad _{\delta\oplus O}\ soft \\ hard \ | \\ H \end{array} \qquad (7)$$

in which the soft alkyl group has enhanced nucleophilicity and, moreover, finds itself in close proximity to a soft electrophilic $OH^{\delta\ominus}$ center. Other reagents which could be suggested for the cleavages of boron alkyls, on the basis of the first rule, are hypochlorite ion and chloramine [*].

[11] *H. Meerwein, G. Hinz, H. Majert,* and *H. Sönke*, J. prakt. Chem. *147*, 226 (1936).

[12] *J. Coubeau, R. Epple, D. Ulmschneider,* and *H. Lehmann*, Angew. Chem. *67*, 710 (1955).

[13] *H. C. Brown* and *K. Murray*, J. Amer. chem. Soc. *81*, 4108 (1959).

[14] *H. C. Brown*: Hydroboration. Benjamin, New York 1962, p. 69.

[15] *H. C. Brown* and *B. C. Subba Rao*, J. Amer. chem. Soc. *78*, 5694 (1956).

[15a] *H. C. Brown. N. C. Herbert,* and *C. H. Snyder*, J. Amer. chem. Soc. *83*, 1001 (1961).

[16] *H. G. Kuivila*, J. Amer. chem. Soc. *76*, 870 (1954).

[17] *J. R. Johnson* and *M. G. Van Campen jr.*, J. Amer. chem. Soc. *60*, 121 (1938).

[*] Chloramine indeed cleaves trialkylboranes (*H. C. Brown, W. R. Heydkamp, E. Breuer,* and *W. S. Murphy*, J. Amer. chem. Soc. *86*, 3565 (1964) but in the opposite sense to the first rule, to afford alkylamines. *N*-chlorodialkylamines R_2NCl, however, give alkyl chlorides with trialkylboranes (*J. G. Sharefkin* and *H. D. Banks*, J. org. Chemistry *30*, 4313 (1965)). It is conceivable, in *Brown's* reaction, that the alkyl attachments confer sufficient soft character on boron to attract soft $Cl^\ominus$. The $-NH_2$ group is left positively polarized and soft enough to attack the alkyl group.

932

Angew. Chem. internat. Edit. / Vol. 6 (1967) / No. 11

369

Cl———O$^{\ominus}$ Cl———$\ddot{N}H_2$
soft   hard   soft   hard

H. C. Brown [15a] has also described the reaction of silver, gold, and platinum oxides with triethylboron to furnish noble metal alkyls in accordance with the first rule.

In the intermediate *(7)* the leaving group R (alkyl) carries no lone pairs. On the other hand the center A (boron) has a vacant orbital. It thus seems that bond formation between boron and oxygen occurs first, and is followed by bond formation between R and OH; finally breakage of the R—B and O—O bonds takes place. The reactions are best thought of as nucleophilically assisted electrophilic substitutions.

4.3. Four-Center Systems in the Cleavage of Hydrogen Compounds

This field is less rich than might have been supposed because the conventionally weak acids (in which proton-transfer rates might be measurably slow) are often of the type H$^{\oplus}$ (hard acid) joined to a hard base, and are therefore not examples of an A—X substrate according to the first rule. Conversely, protons joined to soft-base centers (*e.g.* Hal$^{\ominus}$) give strong acids in the conventional sense in which proton transfer is too fast to study. However, we can consider thiols (eq. (15)), selenols, and phosphines as intermediate types.

hard
base $\curvearrowright$H$\overset{\curvearrowleft}{\frown}$SR   →   [base-H]$^{\oplus}$ SR$^{\ominus}$ (15)

It would be of interest to know whether the forward dissociation rates of RSH, RSeH, and R$_2$PH can be enhanced by the presence of a suitable soft acid such as Ag(I), Hg(II), or I$_2$. The fact that equilibria are pushed well to the right in the presence of such agents forms the basis of analytical methods [18]. When the proton (A) is bonded to a very potent soft base (X) such as H$^{\ominus}$ and R$^{\ominus}$, proton transfer becomes slow and the system is suited to detailed kinetic studies. We ought then particularly to see the operation of the first rule in the nucleophilic cleavage of the hydrogen molecule, and in the cleavage of CH bonds in olefins and paraffins.

4.3.1. Cleavage of the Hydrogen Molecule

Halpern [19,20] has described an important form of homogeneous activation of molecular hydrogen (eq. (16)).

Base:$\curvearrowright$H$\overset{\curvearrowleft}{\frown}H\downarrow$M   →   [Base-H]$^{\oplus}$ [H-M]$^{\ominus}$ (16)

Here the preferred bases appear to be of the hard category (carboxylate ions, pyridine derivatives) and the entity M is a soft metal cation, notably Ag(I), Hg(I), or Hg(II), Cu(I), and Pt(II). A large range of reagent combinations were tested but only the above types were found to be effective. The pattern is seen to be completely consistent with the first rule. It is interesting to note that F$^{\ominus}$ (a very hard base accord-

ing to *Pearson*) is particularly effective in promoting the silver ion-assisted heterolysis of the hydrogen molecule [21].

The author is unaware of an example that demonstrates the assisted ionisation of a paraffin in the sense H$\overset{\frown}{}$CR$_3$. Recently, evidence has appeared that olefins form complexes with palladium-group metal centers according to eq. (17) [22].

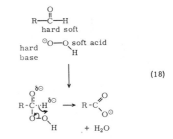

The very soft palladium center can be thought of as assisting H—C ionization by supplying, within the initial π complex, a soft coordination position ready to acquire the electrons of the CH bond.

4.4. Cleavage of Bonds to Carbonyl Centers: Acyl Derivatives of Soft Bases

Pearson has firmly classified the carbonyl group as hard both on kinetic and equilibrium behavior [1]. In the well-known stable carbonyl [*] compounds (*e.g.* carboxylic acids, esters, and amides) the carbonyl center is bonded to hard basic groups (OH, OR, NHR, NR$_2$). Their reactions cannot be predicted in terms of the first rule. The acid fluorides RCOF may also be placed in this category.

However, in derivatives such as RCO—H, RCO—SH, RCO—SR, RCO—SeR, RCO—PR$_2$, and RCO—Hal (Cl, Br, I) the hard carbonyl center R—CO$^{\ominus}$ is bonded to the soft bases H$^{\ominus}$, SH$^{\ominus}$, *etc.*

Hence these acyl compounds should participate in four-center reactions and respond particularly to a combination of a hard nucleophilic base and a soft electrophilic acid.

4.4.1. Reactions of Aldehydes

In the Baeyer-Villiger oxidation [23] aldehydes are oxidized with alkaline hydrogen peroxide:

$$
(18)
$$

[18] *B. Saville*, Analyst *86*, 29 (1961).

[19] *J. Halpern*, Quart. Rev. (chem. Soc., London) *10*, 463 (1956).

[20] *J. Halpern*, Annu. Rev. physic. Chem. *16*, 103 (1965).

[21] *M.T. Beck, I. Gimesi,* and *J. Farkas*, Nature (London) *197*, 73 (1963).

[22] *R. Hüttel, J. Kratzer,* and *M. Bechter*, Chem. Ber. *94*, 766 (1961); *G.W. Parshall* and *G. Wilkinson*, Inorg. Chem. *1*, 896 (1962).

[*] We here use carbonyl not in the sense of a derivative of carbon monoxide but rather to refer to the carbon center of R—CO—X.

[23] *C. H. Hassall*, Org. Reactions *9*, 73 (1957).

A hard oxyanion attacks carbonyl and the leaving soft hydride ion is removed by a sterically well-placed soft electrophilic O(II) center.

In distinction to this process, no report has been made of the conceptually similar reaction (19). Aldehyde-carboxylic exchange does not occur because the electrophilic R'CO center

$$R\text{-}\overset{\overset{O}{\|}}{C}\text{-}H \quad + \quad \overset{H\;\;\;\;O}{\underset{H\;\;\;\;R'}{O\text{-}C}} \nrightarrow R\text{-}\overset{\overset{O}{\|}}{C}\overset{O}{\diagdown} + H\text{-}\overset{\overset{O}{\|}}{C}\text{-}R \quad (19)$$

is hard and will therefore not bond easily with the potential hydride ion. Another example of the oxidation of aldehydes (conforming to the first rule) is the silver ion-assisted reaction in aqueous alkali [24-26].

$$\underset{hard}{R\text{-}\overset{\overset{O}{\|}}{C}\text{-}H} \underset{soft}{\overset{Ag^{\oplus}}{\longrightarrow}} \underset{soft}{} \longrightarrow R\text{-}\overset{\overset{O}{\|}}{C}\overset{O}{\diagdown} + H\text{-}Ag \quad (20)$$

$$\left.\begin{array}{c}{}^{\ominus}OH \\ \text{or } H_2O\end{array}\right\}hard \qquad \downarrow \qquad {}^{1}\!/_{2}\,H_2 + Ag$$

4.4.2. Thioacids and Thioesters

Thioacetic acid, and particularly thioacetamide, precipitate certain soft heavy-metal cations [27-30] as their sulfides. Since the hydrolysis of the carbonyl-sulfur bond is quite slow in the absence of heavy cations [31], and yet precipitation of the heavy metal sulfides can occur nearly instantaneously, it appears that the soft metal electrophile can actively assist hydrolysis (see eq. (21)), in accordance with predictions of the first rule.

$$R\text{-}\overset{\overset{O}{\|}}{C}\text{-}S^{\ominus} + M^{\oplus} \rightleftharpoons R\text{-}\overset{\overset{O}{\|}}{C}\text{-}SM$$

$$R\text{-}\overset{\overset{O}{\|}}{C}\text{-}SM + M^{\oplus} \rightleftharpoons R\text{-}\overset{\overset{O}{\|}}{C}\text{-}\overset{\oplus}{S}M_2$$

$$\underset{hard}{H_2O:\curvearrowright}\underset{\underset{O}{\|}}{\overset{R}{\underset{|}{C}}}\text{-}\overset{\oplus}{\underset{\underset{soft}{}}{S}}\overset{M}{\underset{M}{\diagup}} \longrightarrow H_2\overset{\oplus}{O}\text{-}\overset{\overset{}{}}{C}\text{-}R + M_2S \quad (21)$$
$$\qquad\qquad hard\;\;soft\;soft \qquad\qquad\qquad \overset{\|}{O}$$

It should be possible to add a hard base nucleophile to the above kind of reaction system in order to provide new acylation products at increased rates. In this regard it is known that thiolesters not only undergo

[24] I. A. Pearl, US.-Pat. 2419158 (1947), Sulfite Products Corporation.
[25] A. A. Goldberg and R. P. Linstead, J. chem. Soc. (London) 1928, 2343.
[26] M. Délépine and P. Bonnet, C. R. hebd. Séances Acad. Sci. 149, 39 (1909).
[27] A. R. Armstrong, J. chem. Educat. 37, 413 (1960).
[28] L. M. Andreasov, E. I. Vail, V. A. Kremer, and V. A. Shelikovskii, Z. analyt. Chim. 13, 657 (1958).
[29] G. C. Krijn, C. J. J. Rouws, and G. den Boef, Analytica chim. Acta 23, 186 (1960).
[30] D. F. Bowersox, D. M. Smith, and E. H. Swift, Talanta 2, 142 (1959).
[31] M. Cefola, S. Peter, P. S. Gentile, and A.V. Celiano, Talanta 9, 537 (1962).

934

hydrolysis assisted by the soft acids Ag+, Hg2+, and Pb2+, but also undergo a faster, metal ion-assisted reaction with aniline or with amino groups of amino acids to afford amides in good yield [32,33]:

$$\underset{hard}{C_6H_5\text{-}\overset{..}{N}H_2} \quad \underset{hard\;\;\;soft\;soft}{C_6H_5\text{-}\overset{\overset{O}{\|}}{C}\text{-}SR'\;Pb^{2\oplus}} \rightarrow C_6H_5\text{-}NH\text{-}CO\text{-}C_6H_5$$
$$\qquad\qquad\qquad\qquad\qquad\qquad + H^{\oplus} + RSPb^{\oplus} \quad (22)$$

$$R'SH = HOCH_2\text{-}C(CH_3)_2\text{-}CHOH\text{-}CO\text{-}NH\text{-}(CH_2)_2\text{-}CO\text{-}NH\text{-}(CH_2)_2\text{-}SH$$
(Pantetheine)

4.5. Phosphoryl and Sulfonyl Compounds

In reactions similar to that above, $>$P(O)$-$S bonds can be rapidly cleaved as follows:

(a) A solution of fluoride ions in aqueous bromine [34] converts a phosphorothioic acid into a phosphorofluoridate instantly, whilst in the absence of fluoride

$$\underset{hard\;hard}{F^{\ominus}\overset{\overset{O}{\|}}{\underset{|}{P}}\text{-}\underset{soft}{SH}} \quad \underset{soft}{\overset{\delta\oplus\;\;\;\delta\ominus}{Br\cdots Br}} \rightarrow F\text{-}\overset{\overset{O}{\|}}{\underset{|}{P}} + [SBr_2] + HBr + Br$$
$$\qquad\qquad\qquad\qquad\qquad\qquad\qquad \downarrow H_2O,\,Br_2 \qquad (23)$$
$$\qquad\qquad\qquad\qquad\qquad\qquad\qquad H_2SO_4$$

ions overall oxidative hydrolytic decomposition to a phosphoric acid and sulfate occurs with near equal ease (eq. (23)).

(b) Aqueous bromine or chlorine converts thiophosphate esters into acids and alkanesulfonyl halides [35,36]:

$$\underset{hard\;\;soft}{H_2O:\curvearrowright\overset{\overset{O}{\|}}{\underset{|}{P}}\text{-}\underset{soft}{SR}} \quad \underset{soft}{Br_2} \longrightarrow H_2\overset{\oplus}{O}\text{-}\overset{\overset{O}{\|}}{\underset{|}{P}} + RSBr$$
$$\qquad\qquad\qquad\qquad\qquad\qquad + Br^{\ominus} \qquad \downarrow H_2O,\,Br_2 \quad (24)$$
$$\qquad\qquad\qquad\qquad\qquad\qquad\qquad RSO_2Br$$

and mechanistically similar examples occur with thiolesters etc. [37].

In this reaction, the formation of alkyl bromides as well as sulfonyl bromides can be ascribed to high concentrations of bromide ions developing in preparative-scale reactions. The bromide ions (soft) prefer to attack the soft alkyl groups of the presumed S-bromosulfonium intermediate in preference to the hard phosphonyl center

$$\underset{}{\overset{\oplus}{Br\text{-}\underset{\underset{Br^{\ominus}\;soft}{\curvearrowleft}}{\underset{CH_2R\;soft}{S}}}\overset{\overset{O}{\|}}{\overset{P}{\diagup}\;hard}} \longrightarrow \overset{\overset{O}{\|}}{\underset{|}{P}}\text{-}SBr + BrCH_2R \quad (25)$$

[32] R. Schwyzer and C. Hurlimann, Helv. chim. Acta 37, 155 (1954).
[33] R. Schwyzer, US.-Pat. 2786048 (1957), Ciba Pharmaceutical Products, Inc.
[34] B. Saville, unpublished.
[35] B. Saville, Chem. and Ind. 1956, 660.
[36] C. J. M. Stirling, J. chem. Soc. (London) 1957, 3597.
[37] I. B. Douglas and T. B. Johnson, J. Amer. chem. Soc. 60, 1486 (1938).

Angew. Chem. internat. Edit. / Vol. 6 (1967) / No. 11

371

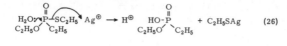

$$H^{\oplus} \qquad HO\text{-}\overset{O}{\underset{C_2H_5O}{\overset{\|}{P}}}C_2H_5 \; + \; C_2H_5SAg \qquad (26)$$

$$\qquad F\text{-}\overset{O}{\underset{C_2H_5O}{\overset{\|}{P}}}C_2H_5 \; + \; C_2H_5SAg \qquad (27)$$

(c) A solution of fluoride and silver ions in water reacts with O,S-diethylethylthiophosphonate [38] rapidly giving the products of hydrolysis and substitution by fluoride ion (see equations (26) and (27)).

In the case of reactions (26) and (27), which proceed in accordance with the first rule, a detailed kinetic analysis [38] has revealed that the electrophile ($Ag^{\oplus}$)-modified substrate (that is susceptible to rapid attack by hard H_2O or $F^{\ominus}$ reagents) probably contains two silver ions per molecule (see also [39]).

This contrasts with monoprotonation of oxygen or nitrogen derivatives in proton catalysed hydrolyses of esters or amides. The coordination of more than one silver ion to one atom of sulfur is most plausibly explained by the postulate that, in a silver-sulfur bond, metal to ligand electronic feed-back involving $d_\pi-d_\pi$ bonding has the effect of leaving the positivity of the sulfur atom only slightly enhanced. The sulfur is then able to use its remaining pair of 3p electrons to coordinate with yet another heavy-metal ion until high positive charge on sulfur precludes further coordination, and, moreover, produces a good leaving group (from the phosphonyl center).

Thus the leaving-group activities may be written:

$$-SR \; < \; -\overset{\oplus}{\underset{R}{S}}{}^{\nearrow Ag} \; < \; -\overset{2\oplus}{\underset{Ag}{S}\text{-}R}{}^{\nearrow Ag}$$

It may well be that this kind of multi-coordination is general for soft polarizable bases interacting with soft acids (being formally analogous to the multiplicity of oxidation numbers often characteristic of soft bases), so that many reactions in which assistance derives from soft leaving group/soft acid interaction will be of high kinetic order referred to the assisting acid. Recent work by *Schwarzenbach* and *Schellenberg* [40] on the binding of thio ligands to the methylmercuric cation is in accord with this picture, as is the recognition of species such as $Ag_3I^{2\oplus}$, $Ag_3S^\ominus$, $6\,AgNO_3\cdot AgC\equiv CAg$ [41-43].

Similar activation of cleavage of bonds between sulfur(II) and sulfonyl sulfur — a hard acid center — has been observed

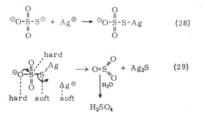

$$\overset{O}{\underset{O}{\overset{\|}{{}^\ominus O\text{-}S\text{-}S^\ominus}}} \; + \; Ag^{\oplus} \; \longrightarrow \; \overset{O}{\underset{O}{\overset{\|}{{}^\ominus O\text{-}S\text{-}S\text{-}Ag}}} \qquad (28)$$

$$\longrightarrow \; O{=}S\overset{O}{\underset{O}{\overset{\diagup}{\diagdown}}} \; + \; Ag_2S \qquad (29)$$

$$\downarrow H_2O$$
$$H_2SO_4$$

in polythionate and thiosulfate chemistry [44]. Thus, in strong aqueous solutions of silver or mercury salts, thiosulfate ion breaks down readily according to eq. (28) and (29).

(d) Phosphorylations of biological importance are often geared to redox processes, and phosphorylated hydroquinone derivatives of the vitamin K, ubiquinone, and the vitamin E groups are known to be important intermediates [45-47].

Model studies [45] have shown the activation of O-phosphorylated quinols by oxidizing agents (air, Br_2, or Ce^{4+} ions) in the sense:

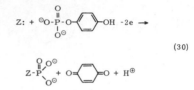

$$(30)$$

$$Z\text{-}\overset{O}{\underset{O^\ominus}{\overset{\|}{P}}}{-}O^\ominus \; + \; O{=}\!\!\!\bigcirc\!\!\!{=}O \; + \; H^\oplus$$

Hard nucleophiles (*e.g.* ADP, water, or inorganic phosphate) attack the hard phosphoryl center whilst the p-dihydroxyaryl system (a soft leaving group) is removed because of interaction with oxidizing agents, all of which can be regarded as soft electrophiles. It has been suggested [45] that certain O,O'-diphosphorylated quinols can become active phosphorylating agents in the presence of quinones. In this case the soft-soft interaction of partners in a charge-transfer complex, as envisaged by *Pearson* [1], becomes the driving force of assisted phosphorylation as depicted by eq. (31).

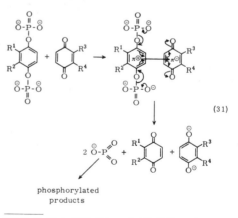

$$(31)$$

phosphorylated products

[38] *B. Saville*, J. chem. Soc. (London) *1961*, 4624.

[39] *B. Saville*, J. chem. Soc. (London) *1962*, 4062.

[40] *G. Schwarzenbach* and *M. Schellenberg*, Helv. chim. Acta *48*, 28 (1965).

[41] *B. Reuter* and *K. Hardel*, Angew. Chem. *72*, 138 (1960); Naturwissenschaften *48*, 161 (1961).

[42] *R. Vestin* and *E. Ralf*, Acta chem. scand. *3*, 101, 107 (1949).

[43] *I. Leden* and *C. Parck*, Acta chem. scand. *10*, 535 (1956).

[44] *J. N. Friend*: Text-book of Inorganic Chemistry, Griffin, London 1931, Vol. 7, part 2, p. 199 ff. and references quoted therein.

[45] *V. M. Clark*, *G. W. Kirby*, and *A. Todd*, Nature (London) *181*, 1131 (1958).

[46] *R. D. Dallam* and *J. F. Taylor*, Federat. Proc. *18*, 826 (1959).

[47] *V. C. Wessels*, Recueil Trav. chim. Pays-Bas *73*, 529 (1954).

Angew. Chem. internat. Edit. / Vol. 6 (1967) / No. 11

935

Hadler [48] has argued differently and suggests the intermediacy of enol phosphates in biological phosphorylations (eq. (32)).

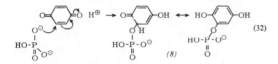

$$(32)$$

However, the oxidative activation of molecules such as *(8)* follows the same general pattern as that proposed in eq. (31).

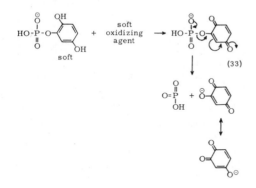

$$(33)$$

4.6. Other Examples of Substitutions Assisted According to the First Rule

The organometallic field contains numerous examples of exchange reactions of the type (34).

$$X-M^1 \rightleftharpoons \begin{matrix} X \\ | \\ M^2 \end{matrix} + \begin{matrix} M^1 \\ | \\ X' \end{matrix} \qquad (34)$$

$$M^2-X'$$

Clearly, the preferred equilibrium positions in these cases should be predictable from the first rule. Equation (35) represents a relevant reaction of organotin compounds [49].

$$\begin{matrix} \text{hard} & \text{soft} \\ (C_4H_9)_3Sn-SR & (9) \\ ROCO-N\cdots\cdots HgX \rightarrow (C_4H_9)_3Sn-NR'-COOR & (35) \\ \quad | & + RSHgX \\ \quad R' & \\ \text{hard} \quad \text{soft} & \\ \text{base} \quad \text{acid} & \end{matrix}$$

Sn(IV) is usually regarded as a hard base [1]; however, we do not know the extent to which the three soft butyl groups confer softness on the tin center. Presumably, electron-attracting substituents on tin facilitate the above type of reaction. One can thus understand why *N*-halogenoamides also react with compounds such as *(9)* to furnish tin amides and sulfenyl

halides. A further example of a multi-center reaction in organotin chemistry [51] is the decomposition of trialkyltin hydrides by acids in dimethylsulfoxide in the presence of basic substances [51] (eq. (36)).

$$\text{Base:} \curvearrowright R_3Sn-H^\curvearrowright H^\oplus \longrightarrow \text{Base}^\oplus SnR_3 + H_2 \qquad (36)$$

The activity of the assisting base is related roughly to its position in the spectrochemical series, which in turn is an approximate measure of hardness. According to the first rule, soft metal cations should be more effective than the proton as the electrophile in the above type of reaction.

4.7. Assisted Nucleophilic Additions

When, as in eq. (37), a π bond can be opened to provide a positively charged hard-acid center with a vacant orbital and simultaneously a negatively charged soft-base center, it is to be expected that such an

$$A=X' \rightarrow \underset{\text{hard} \quad \text{soft}}{\overset{\oplus}{A}-\overset{\ominus}{X}} \qquad (37)$$

opening might best be brought about by the combined action of a hard nucleophile (Z:) and a soft electrophile (E).

Olefins, $R^1R^2C=CR^3R^4$, where neither of the pairs R^1R^2 or R^3R^4 markedly stabilize the carbonium center, and thiocarbonyl compounds such as carbon disulfide, are therefore substrates to be considered in this regard. The reaction of alcoholic mercuric acetate solutions with olefins, and the oxidation of ethylene to a hydroxyethyl ligand under the action of $PdCl_2$ complexes [20,50], are good examples of such opening of olefinic π bonds.

$$\begin{matrix} \text{hard} \quad \text{soft} \\ \delta\oplus\,CH_2\cdots\cdots CH_2\,\delta\ominus \\ HO^\ominus\!-\!Pd\!-\!Cl \rightarrow HO\!-\!CH_2\!-\!CH_2\!-\!Pd^{II}\;Cl \quad (38) \\ \quad | \\ \text{hard} \quad Cl\;\;\text{soft} \qquad\qquad + Cl^\ominus \\ \text{base} \qquad \text{acid} \end{matrix}$$

The reaction of silver carboxylates with carbon disulphide gives acid anhydrides [52] (eq. (39)).

Finally, certain trialkylaluminums, trialkyl borons, and dialkylphosphines are spontaneously inflamma

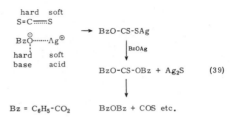

$$Bz = C_6H_5-CO_2 \qquad BzOBz + COS \text{ etc.}$$

[48] *H. I. Hadler*, Experimentia *17*, 268 (1961).

[49] *A. G. Davies* and *G. J. D. Peddle*, Chem. Commun. *5*, 96 (1965).

[50] *P. M. Henry*, J. Amer. chem. Soc. *86*, 3246 (1964).

[51] *R. E. Dessy, F. E. Paulik*, and *T. Hieber*, J. Amer. chem. Soc. *86*, 28 (1964).

[52] *D. Bryce-Smith*, Proc. chem. Soc. (London) *1957*, 20.

936

Angew. Chem. internat. Edit. / Vol. 6 (1967) / No. 11

373

ble [53,54] (eq. (40) and (41)). Could this be a consequence of easy concerted acid-base activity?

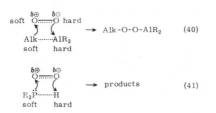

$$\text{Alk-O-O-AlR}_2 \quad (40)$$

$$\rightarrow \text{ products} \quad (41)$$

5. Examples of the Operation of the Second Rule

The saturated carbon center R_3C is a good example of a soft electrophilic center A of a substrate $A-\ddot{X}$ whose reactions should be interpretable in the light of the second rule. This center responds very directly to softness in the attacking nucleophile in the case of simple S_N2 displacements [4,55,56].

As other examples of a soft A center we have bivalent oxygen (as in peroxides and hypohalites) [4,57,58], sulfur [4,59,60], and probably trivalent nitrogen [61,62] in the rare cases of N attached to good leaving groups. The reason for the softness of these atoms as acids has not been made clear but may well concern the overlap of filled 2p or 3p orbitals on the oxygen or sulfur atom with empty p or d orbitals of the approaching soft nucleophile, a π-bonding effect which can strengthen the new bond both in the transition state and the final state (see, e.g., eq. (42)).

$$N\equiv C^{\ominus} + \overset{..}{\text{O}}\text{-OH} \rightarrow \left[N \equiv C \cdots \overset{\delta\oplus}{\text{O}} \cdots \overset{\delta\ominus}{\text{OH}} \right]$$

$$(42)$$

$$\overset{\ominus}{N}=C=\overset{\oplus}{O} + OH^{\ominus} \leftarrow \left[\overset{\delta\ominus}{N}=C \cdots \overset{\delta\oplus}{O} \cdots \overset{\delta\ominus}{OH} \right]$$

Softness in the case of saturated carbon is more difficult to rationalize; however, since most saturated carbon centers possess at least one C—H bond it is possible that in the transition state the σ electrons of this bond are somewhat delocalized across to the vacant orbital of the approaching soft nucleo-

[53] G. E. Coates: Organo-metallic Compounds. Methuen, London 1956, Chapter III.

[54] G. M. Kosolapoff: Organophosphorus Compounds. Chapman and Hall, London 1958, Chapter II.

[55] C. G. Swain and C. B. Scott, J. Amer. chem. Soc. 75, 141 (1953).

[56] J. O. Edwards, J. Amer. chem. Soc. 76, 1540 (1954); 78, 1819 (1956).

[57] J. O. Edwards, paper given at Peroxide Reaction Mechanism Conference. Brown University, Providence, R.I. (USA), June 1960.

[58] M. C. R. Symons, Chem. and Ind. 48, 1480 (1960).

[59] A. J. Parker and N. Kharasch, Chem. Reviews 59, 583(1959).

[60] P. D. Bartlett, A. K. Colter, R. E. Davis, and W. R. Roderick, J. Amer. chem. Soc. 83, 109 (1961).

[61] H. H. Sisler, A. Sarkis, H. S. Ahuja, R. J. Grage, and N. L. Smith, J. Amer. chem. Soc. 81, 2982 (1959).

[62] F. N. Collier jr., H. H. Sisler, J. G. Calvert, and F. R. Hurley, J. Amer. chem. Soc. 81, 6177 (1959).

phile — in a manner similar to the current picture of hyperconjugation involving vacant carbon 2p orbitals.

Positive halogen compounds can also usefully be regarded as $A-\ddot{X}$ derivatives in which A is a soft-acid halogen atom.

5.1. Cleavage of Carbon-Oxygen Bonds

The simple ethers are typical substrates involving soft saturated carbon and hard basic oxygen as leaving groups. The classical method, due to Zeisel [63,64], of cleaving ethers provides a good example of the operation of the second rule (see eq. (43)).

$$I^{\ominus} \xrightarrow{} \overset{|}{C} \xrightarrow{} \overset{|}{\text{O}} \xrightarrow{} H^{\oplus} \rightarrow I - \overset{|}{C} + HOR \quad (43)$$

soft   soft   hard hard
base   acid   base acid

It is significant that if a weak hard nucleophile such as $HSO_4^{\ominus}$, is used in place of $I^{\ominus}$ or $Br^{\ominus}$ (soft) a much slower cleavage takes place [64]. Also, replacement of the hard ether oxygen atom by a soft sulfur atom (thus violating the conditions of the second rule) gives an almost unreactive system [65].

The reaction of halogen acids with n-alkanols to yield alkyl halides, and the reaction of thiourea with n-alkanols and mineral acids (eq. (44)) [66] likewise obey the second rule.

$$\underset{H_2N}{\overset{H_2N}{>}}\!\!\overset{\oplus}{C}\text{-}S^{\ominus} \quad \overset{|}{C}\text{-}\overset{|}{\underset{H}{\overset{H}{O}}} \quad \underset{H_2N}{\overset{H_2N}{>}}\!\!\overset{\oplus}{C}\cdots S\overset{|}{C} + H_2O \quad (44)$$

soft $H^{\oplus}$

(10)

A more recent reagent for the cleavage of ethers is boron tribromide [67,68], in which a hard boron center (E) is present in the same molecule as an incipient soft nucleophilic bromide ion (Z). With this reagent, cleavage (eq. (45)) may occur at temperatures as low as −60 °C. Violation of the second rule pattern by changing from BBr_3 to BF_3 (F is hard) leads to non-reactive addition compounds, at least one of which can be distilled unchanged [69]. The many examples given

soft   hard
acid   base

$$>\!\overset{|}{C}\!\!\overset{R}{\underset{Br-B}{\overset{|}{\underset{Br}{}}}}Br \rightarrow \overset{|}{\underset{Br}{C}} + \overset{R}{\underset{BBr_2}{O}} \quad (45)$$

soft   hard

[63] S. Zeisel, Mh. Chem. 6, 989 (1885).

[64] R. L. Burwell jr., Chem. Reviews 54, 615 (1954).

[65] G. K. Hughes and E. O. P. Thompson, J. Proc. Roy. Soc. New South Wales 83, 269 (1950).

[66] R. L. Frank and P. V. Smith, J. Amer. chem. Soc. 68, 2103 (1946).

[67] J. F. W. McOmie and M. L. Watts, Chem. and Ind. 1963, 1658.

[68] F. L. Benton and T. E. Dillon, J. Amer. chem. Soc. 64, 1128 (1942).

[69] H. C. Brown and R. M. Adams, J. Amer. chem. Soc. 64, 2557 (1942).

Angew. Chem. internat. Edit. / Vol. 6 (1967) / No. 11

937

in *Gerrard*'s papers on the reaction of alcohols with boron halides involving carbon-oxygen bond cleavage would similarly fit into the second category of four-center assisted substitutions [70, 71].

Further examples of a saturated carbon center joined to a hard oxygen undergoing reactions according to the second rule appear in the chemistry of phosphate esters. Lord *Todd* and his co-workers [72, 73] noted that lithium chloride is a valuable reagent for the dealkylation of phosphate and pyrophosphate esters (eq. (46)).

$$
\underset{\substack{\text{soft}}}{Cl^{\ominus}}\overset{\frown}{}CH_2\underset{\substack{|\\C_6H_5\ Li^{\oplus}\\ \text{hard}}}{\overset{\overset{\displaystyle O}{\parallel}}{O-P}}\longrightarrow ClCH_2\text{-}C_6H_5 + LiO\text{-}\overset{\overset{\displaystyle O}{\parallel}}{P}\diagdown \tag{46}
$$

Since $Cl^{\ominus}$ is not an outstandingly potent soft nucleophile it seems possible that the lithium ion (which *Pearson* clearly classifies as hard [11]) may render electrophilic assistance.

Lithium salts are frequently used in studies of salt effects in organic reactions. The question now occurs whether catalysis by salts, in appropriate cases, should be re-interpreted to include specific electrophilic assistance, in addition to ion-atmosphere effects.

5.2. Substitutions at Oxygen and Sulfur Atoms

Substitutions at O and S centers tend to require soft nucleophilic reagents (Z:) [57, 58] and hence we should again expect to find examples of assistance by hard acids when the leaving-group is hard, as required by the second rule. Such a case is exemplified by the acid-catalysed reaction of hydrogen peroxide with iodide ion [74, 75] (eq. (47) and (48)).

$$
\underset{\substack{\text{soft}\\ \text{base}}}{I^{\ominus}}\overset{\frown}{}\underset{\substack{\text{soft}\\ \text{acid}}}{\overset{H}{O}}\overset{\delta\ominus}{-}\underset{\substack{\text{hard}\\ \text{base}}}{\overset{\delta\ominus}{O}}\underset{\substack{H\\ \text{hard}\\ \text{acid}}}{\overset{\frown}{}H^{\oplus}} \longrightarrow IOH + H_2O \tag{47}
$$

$$
\underset{\substack{\text{soft}\\ \text{base}}}{I^{\ominus}}\quad \underset{\substack{\text{soft}\\ \text{acid}}}{I-\overset{..}{O}}\overset{\frown}{\underset{H}{}}H^{\oplus} \longrightarrow I\text{-}I + H_2O \tag{48}
$$

Indeed, practically every known proton-catalysed reaction of hydrogen peroxide involves attack by a soft nucleophile on oxygen (*e.g.* epoxidation of olefins, oxidation of sulfides to sulfoxides and sulfones). Iodide ion attack on hydrogen peroxide is also strongly catalysed by molybdic acid [76], and, although the

mechanism is unknown, the reaction classifies this acid as hard.

Outside these examples, however, little use seems to have been made of behavior conforming to the second rule in connection with the cleavage of O–O or O–N bonds. It is here considered that soft nucleophiles such as RSR', R_3P:, and R_3As: used in conjunction with BX_3, SnX_4, or AlX_3 could become interesting reagents; for example, equation (49) represents a proposed cleavage of hydroxylamine. (This reaction is

$$
\tag{49}
$$

similar in principle to the reaction of sulfurous acid $[H\overset{..}{O}SO_2{}^{\ominus}$: soft, $H^{\ominus}$: hard] with hydroxylamine [77] [*].)

For substitutions at bivalent sulfur centers, the second rule demands that those centers be attached to hard leaving groups. Hence the sulfenic acids and esters (RS–OH, RS–OR') and sulfenamides (RS–NR'R'') are suitable substrates. The sulfenic esters readily react with soft base/hard acid combinations (*e.g.* HX [X = Hal], HSR) as also do the sulfenamides [78, 79].

Thiols, as well as hydrogen sulfide, generally react with sulfenamides according to equation (50), and with halogen acids according to equations (51) and (52) [78].

$$
\begin{array}{c}
\underset{\substack{\text{soft}\\ \text{base}}}{R\text{-}S}\cdots\cdots\underset{\substack{\text{hard}\\ \text{acid}}}{H} \\[2pt]
\diagdown\overset{\frown}{}R'' \\[2pt]
\underset{\substack{\text{soft}\\ \text{acid}}}{R'\text{-}S}\overset{\frown}{}\underset{\substack{|\\ R''\\ \text{hard}\\ \text{base}}}{N}
\end{array}
\longrightarrow RSSR' + HNR_2'' \tag{50}
$$

Dialkylphosphites and sulfenamides also react by the same principle [80], as also do thiosulfuric acid and sulfenamides [80a].

$$
HI + R'S\text{-}NR_2'' \longrightarrow ISR' + HNR_2'' \tag{51}
$$

$$
I^{\ominus} + ISR'' \underset{-H^{\oplus}}{\overset{+H^{\oplus}}{\rightleftarrows}} I_2 + R'SH \tag{52}
$$

The soft base and hard acid (Z: and E) requirements of the reagents for assistance under the second rule can sometimes be provided through the opening of a suitable π bond. For example, π bond opening of a C=S bond in carbon disulfide leads to the state

[70] *W. Gerrard* and *M. F. Lappert*, J. chem. Soc. (London) *1951*, 1020.

[71] *W. Gerrard* and *M. F. Lappert*, J. chem. Soc. (London) *1952*, 1486.

[72] *V. M. Clark* and *A.Todd*, J. chem. Soc. (London) *1950*, 2031.

[73] *J. Lecocq* and *A.Todd*, J. chem. Soc. (London) *1954*, 2381.

[74] *P. Rumpf*, C. R. hebd. Acad. Sci. *198*, 256 (1934).

[75] *H. A. Liebhafsky* and *A. Mohammad*, J. physic. Chem. *38*, 857 (1934).

[76] *A. I. Vogel*: Quantitative Inorganic Analysis. 3rd Edit., Longmans, London 1961, p. 363.

[77] *D. S. Brackman* and *W. C. E. Higginson*, J. chem. Soc. (London) *1953*, 3896.

[*] Actually, hydroxylamine tends to react in the form $NH_2{}^{\delta\oplus}$ (soft)···$OH^{\delta\ominus}$ (hard) in this case giving sulfamic acid as the main product. However, arsenite (soft nucleophile) combines with the $OH^{\delta\oplus}$ (soft) pole and displaces ammonia.

[78] *P.-L. Hu* and *W. Scheele*, Kautschuk u. Gummi *18*, 290 (1965).

[79] *M. R. Porter* and *B. Saville*, unpublished; see also *W.Scheele* and *J. Helberg*, Kautschuk u. Gummi *15*, S.WT 400, (1962).

[80] *K. A. Petrov*, *N. K. Bliznyuk*, and *V. A. Saostenok*, Ž. Obšč. Chim. *31*, 1361 (1961).

[80a] *O. Foss*, Acta chem. scand. *1*, 307 (1947).

938

Angew. Chem. internat. Edit. / Vol. 6 (1967) / No. 11

375

$S=C^{\oplus}-S^{\ominus}$ in which $C^{\oplus}$ is a hard acid and $S^{\ominus}$ is a soft base. Hence certain "insertion" reactions of carbon disulfide can be interpreted; an example [81] is given in equation (53).

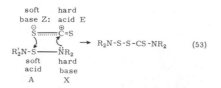

$$\longrightarrow \quad R_2N-S-S-CS-NR_2 \qquad (53)$$

An analogous reaction takes place with tervalent phosphorus [82] (eq. (54)), thus indicating that P(III) in its rather unusual role of an electrophilic center is soft.

$$\longrightarrow \quad R_2P-S-CS-NR_2 \qquad (54)$$

5.3. The Reaction of Methoxycarbonylmercury Complexes

Dessy discovered an analogue of the α-elimination mechanism [83] (equation (55)) which involves the decomposition of

$$\text{Base:} \curvearrowright H \colon\! C \colon\! X \quad \longrightarrow \quad [\text{Base H}]^{\oplus} \cdot \overset{|}{\underset{|}{C}} \cdot \ [:X]^{\ominus} \qquad (55)$$

methoxycarbonylmercury compounds such as *(11)* by acids in the presence of anionic and neutral "assistors" (equation (56)). The "assistor"-bases are of the polarizable soft type

$$R-Hg-CO-OCH_3 + H^{\oplus} \xrightarrow[\text{DMSO}]{Z:} R-Hg^{\oplus}(Z:) + CO + CH_3OH \qquad (56)$$
(11)

(RSH, $I^{\ominus}$, $Br^{\ominus}$, *etc.*). This pseudo-four-center substitution conforms well with the requirements of the second rule.

$$\longrightarrow \quad R-Hg-Z + CO + CH_3OH \qquad (57)$$

6. Final Remarks

How may the "Rules" outlined be best applied in practice? If a given reagent assembly being used in a four-center reaction does not conform to the patterns

[81] E. S. Blake, J. Amer. chem. Soc. 65, 1267 (1943).
[82] R. F. Hudson and J. Searle, personal communication.
[83] R. E. Dessy and F. E. Paulik, J. Amer. chem. Soc. 85, 1812 (1963).

of the first or second rules it should be of interest to study the effect of changing the assembly to conform with one or other of these cooperative patterns. As a final illustration it is noted that the hydrolysis of a thiol ester is weakly catalysed by protons (see equation (58)). In this case the hard proton does not

$$H_2O\colon \quad RCO-SC_2H_5 \quad H^{\oplus} \longrightarrow H_2\overset{\oplus}{O}-COR$$
$$\text{hard} \quad \text{hard} \quad \text{soft} \quad \textit{hard} \quad + \ HSC_2H_5 \qquad (58)$$

"match" with the soft SC_2H_5 group. However, if the catalysing proton is changed for a soft acid, such as silver ion, then the thiol ester is hydrolysed far more rapidly.

It is hoped that the principles arrived at here may further prove useful in interpreting metal-ion assisted

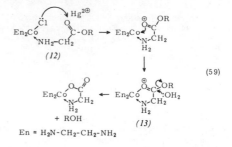

$$(59)$$

$$En = H_2N-CH_2-CH_2-NH_2$$

reactions of biochemical (*e.g.* enzymic processes) and industrial (*e.g.* heterogeneous metal catalysis) importance. In concluding this section mention may be made of the elegant use of the hard-soft principle by *D. H. Busch* [84] in his studies of the cobalt-complex assisted hydrolysis of glycine esters (eq. (59)).

The chloro derivative *(12)* is stable until a chloride ion is removed by Hg^{2+} ions (soft base interacting with soft acid). This leaves a hard vacant acid center on Co^{3+} that can coordinate with the hard oxygen atom of the glycine moiety thereby leaving a hard carboxyl carbon center in *(13)*. This compound is now suitably activated for attack by a hard water molecule.

The Author acknowledges with sincere thanks the encouraging interest shown in these ideas by Professor R. F. Hudson (in particular), by his colleagues Dr. A. A. Watson and Mr. G. M. C. Higgins, and by Professors R. G. Pearson, Sir Ronald Nyholm, F. R. S., G. Wilkinson, F. R. S., and J. I. G. Cadogan.

Received: March 7th, 1966   [A 605 IE]
German version: Angew. Chem. 79, 966 (1967)

[84] D. H. Busch, Chem. Engng. News 43, No. 10, p. 58 (1965).

[Reprinted from the Journal of the American Chemical Society, **89**, 1827 (1967).]
Copyright 1967 by the American Chemical Society and reprinted by permission of the copyright owner.

39

Application of the Principle of Hard and Soft Acids and Bases to Organic Chemistry

Ralph G. Pearson and Jon Songstad[1]

Contribution from the Department of Chemistry, Northwestern University, Evanston, Illinois 60201. Received November 2, 1966

Abstract: The principle of hard and soft acid and bases (HSAB principle) is applied to organic chemistry. Organic molecules are viewed as Lewis acid–base complexes and their relative thermodynamic stability explained in terms of two factors. One is the tendency of intrinsically strong acids to coordinate to the strongest bases. The second is the special stabilization of combinations of hard acids and bases, or soft acids and bases. The symbiotic principle is illustrated, which states that there is an extra stabilization if several soft bases (ligands) or several hard bases cluster about a single acidic atom. The same two principles are applied to rates of nucleophilic and electrophilic substitution reactions in organic chemistry.

Recently[2] a generalization was proposed which makes it possible to correlate a great many phenomena in various areas of chemistry. Use is made of the concept of generalized, or Lewis, acids and bases. The generalization may be called the principle of hard and soft acids and bases (HSAB). It states that hard acids prefer to coordinate to hard bases and soft acids prefer to coordinate to soft bases.

These terms are qualitatively defined in the following ways: soft base—donor atom is of high polarizability, low electronegativity, easily oxidized, and associated with empty, low-lying orbitals; hard base—donor atom is of low polarizability, high electronegativity, hard to oxidize, and associated with empty orbitals of high energy and hence inaccessible; soft acid—the acceptor atom is of low positive charge, large size, and has several easily excited outer electrons; hard acid—acceptor atom is of high positive charge, small size, and does not have easily excited outer electrons

Operationally, acids may be defined by following the procedures of Schwarzenbach[3] and Ahrland, Chatt, and Davies.[4] These workers divided metal ions (which are Lewis acids) into two classes called A and B by Schwarzenbach and a and b by Ahrland, Chatt, and Davies. Hard acids follow the same pattern as class a metal ions, and soft acids show the pattern of class b metal ions. For complexes with different donor atoms, the following sequences of stabilities are found.

hard $\begin{cases} N \gg P > As > Sb \\ O \gg S > Se > Te \\ F > Cl > Br > I \end{cases}$

soft $\begin{cases} N \ll P > As > Sb \\ O \ll S \sim Se \sim Te \\ F < Cl < Br < I \end{cases}$

Soft bases might be operationally defined by considering the equilibrium[5]

$$CH_3Hg^+(aq) + BH^+(aq) \rightleftharpoons CH_3HgB^+(aq) + H^+(aq) \quad (1)$$

If the equilibrium constant for this reaction is much greater than unity, the base B is soft. If it is near unity, or less than unity, the base is hard. The proton is the simplest hard acid and the methylmercury cation is one of the simplest soft acids. Table I contains a listing of hard and soft bases for later reference.

If the equilibrium constants of eq 1 are used to rank a series of bases, the following order of decreasing softness is obtained.

$$I^- > Br^- > Cl^- > S^{2-} > RS^- > CN^- > H_2O > NH_3 \sim F^- > OH^-$$

It turns out that this is not a universal order since a change in one of the reference acids will give a different series. The reason for this may be seen by considering the generalized acid–base exchange reaction.

$$A:B' + A':B \rightleftharpoons A:B + A':B' \quad (2)$$

We expect such a reaction to proceed such that the strongest acid, A, is found coordinated to the strongest base, B. The terms hard and soft do not mean the same as strong and weak. Thus an acid is characterized by at least two properties, its strength and its hardness, or softness; the same is true for a base. It is well known that there is no universal order of acid or base strength; still we recognize that some Lewis acids, such as H⁺, are much stronger than other acids, such as I_2, or that H⁻ is a much stronger base than H_2O. The HSAB principle then states that there is an extra stabilization in A:B if both the acid and base are hard, or if both are soft.[6]

We can usually recognize hardness or softness in a qualitative way by examining an acid or base, particularly the donor or acceptor atoms. The situation may be something like that for the terms solvent polar-

(1) Chemistry Department, Bergen University, Norway. Supported by the Royal Norwegian Council for Scientific and Industrial Research.
(2) R. G. Pearson, *J. Am. Chem. Soc.*, **85**, 3533 (1963); *Science*, **151**, 172 (1966).
(3) G. Schwarzenbach, *Experientia Suppl.*, **5**, 162 (1956); *Advan. Inorg. Chem. Radiochem.*, **3**, 257 (1961).
(4) S. Ahrland, J. Chatt, and N. R. Davies, *Quart. Rev.* (London), **12**, 265 (1958).

(5) G. Schwarzenbach and M. Schellenberg, *Helv. Chim. Acta*, **48**, 28 (1965); G. Schwarzenbach, *Chem. Eng. News*, **43**, 92 (May 31, 1965).
(6) Thus the equilibrium constant for the reaction A + :B $\rightleftharpoons$ A:B might be characterized by an equation such as log $K = S_A S_B + \sigma_A \sigma_B$. The factors S_A and S_B are *strength* factors for the acid and base; σ_A and σ_B are softness factors. For a hard acid or base, σ would be negative; for a soft acid or base, σ would be positive; see R. S. Drago and B. B. Wayland, *J. Am. Chem. Soc.*, **87**, 3571 (1965).

Table I. Classification of Bases

Hard	Soft	Borderline
H_2O, OH^-, F^-, $CH_3CO_2^-$, $PO_4{}^{3-}$, $SO_4{}^{2-}$, $CO_3{}^{2-}$, ClO_4^-, NO_3^-, ROH, RO^-, R_2O, NH_3, RNH_2, N_2H_4	R_2S, RSH, RS^-, I^-, SCN^-, $S_2O_3{}^{2-}$, Br^-, R_3P, R_3As, $(RO)_3P$, CN^-, RNC, CO, C_2H_4, C_6H_6, H^-, R^-	$C_6H_5NH_2$, C_5H_5N, N_3^-, Cl^-, NO_2^-, $SO_3{}^{2-}$

ity or electronegativity. These useful concepts lack a precise definition, or rather several definitions exist to suit various kinds of data.

In spite of this inability to make the rules quantitative at present, we hope in this paper to show that the principle of hard and soft acids and bases is extremely useful. We will take the area of organic chemistry for which a wealth of data exists for which little correlation has been done in terms of hardness and softness concepts.[7] It will turn out that much, but not all, of what we have to say has been noted before and explained in various ways. We wish to show an underlying pattern in all of these phenomena.

It should be stressed that the HSAB principle is not a theory but is a statement about experimental facts. Accordingly an explanation of some observation in terms of hard and soft behavior does not invalidate some other, theoretical explanation. In fact, the various theories which have been put forward[2] to explain the principles of hard and soft acids and bases in general usually include the previous explanations for the particular cases to be discussed in this paper.

Thermodynamic Examples

We will first show how the thermodynamic stabilities of many kinds of organic molecules can be rationalized by the hard–soft concept. This leads to a better understanding of many well-known facts and to a prediction of some results that are probably not well known to most organic chemists.

The method that is used is to mentally break down an organic species into a Lewis acid fragment such as a carbonium ion or acylium ion, and a base fragment such as a carbanion, a hydride ion, or a halide ion. The stability of the molecule is then considered in terms of the acid–base interaction

$$A + :B \longrightarrow A:B \qquad (3)$$

When the acceptor atom of A is carbon, we are talking about what Parker has called carbon basicity.[8] An important paper by Hine and Weimar[9] has recently discussed this subject in a different way.

Some extended results from this paper may be conveniently used to introduce the subject. Table II shows the calculated equilibrium constants[9] for the gas phase reactions

$$CH_3OH(g) + HB(g) \rightleftharpoons CH_3B(g) + H_2O(g) \qquad (4)$$

It may be noted that when HB is a neutral molecule, the equilibrium constants in aqueous solution are not

(7) (a) A recent book by R. F. Hudson, "Structure and Mechanism in Organo-Phosphorus Chemistry," Academic Press Inc., New York, N. Y., 1965, uses the ideas of hard and soft to explain substitution reactions of organophosphorus compounds. (b) B. Saville, *Chem. Eng. News,* **43,** 100 (May 31, 1965); *Angew. Chem.,* in press, has used the concept to discuss catalyzed reactions of organic chemistry.

(8) A. J. Parker, *Proc. Chem. Soc.,* 371 (1961).

(9) J. Hine and R. D. Weimar, Jr., *J. Am. Chem. Soc.,* **87,** 3387 (1965).

Table II. Equilibrium Constants at 25°, K_{eq}, for the Reaction $HB(g) + CH_3OH(g) \rightleftharpoons CH_3B(g) + H_2O(g)$

B	$K_{eq}{}^b$	K_a
H^-	1×10^{21}	$10^{-29\,c}$
CN^-	3×10^{14}	7×10^{-10}
$CH_2COCH_3^-$	2×10^{11}	10^{-20}
CH_3S^-	2×10^9	5×10^{-11}
CH_3^-	3×10^8	$10^{-40\,c}$
I^-	2×10^8	$10^{9.5}$
SH^-	3×10^7	1×10^{-7}
Br^-	1×10^7	10^9
$N_2H_3^-$	6×10^6	1×10^{-8}
Cl^-	3×10^5	10^7
$NO_2^-\,{}^a$	$\sim10^4$	5×10^{-4}
NH_2^-	8×10^3	$10^{-33\,d}$
$C_6H_5O^-$	3×10^3	1×10^{-10}
CH_3O^-	3×10^3	10^{-15}
F^-	$\sim10^{-3}$	1.4×10^{-3}

^{a} Bonding to O in each case. CH_3NO_2 is about 1.5 kcal more stable than CH_3ONO. ^{b} Values of K_{eq} are from ref 9, except for $B=NH_2^-$, $N_2H_3^-$, NO_2^-, CH_3^-, and F^-. See ref 13, 16, and 31 for data; also Technical Notes 270-1 and 270-2, National Bureau of Standards, 1965–1966. ^{c} Estimated. See F. Basolo and R. G. Pearson, "Mechanisms of Inorganic Reactions," John Wiley and Sons, Inc., New York, N. Y., 1958, p 344. ^{d} Estimated from K_a of aniline, phenol, and water. See F. G. Bordwell, "Organic Chemistry," The Macmillan Co., New York, N. Y., 1963, p 867.

different from those in the gas phase by more than a factor of 25 or so for reaction 4. Also the value of $\Delta H°$ is not different from the value of $\Delta G°$, in the gas, by more than a kilocalorie or two. Since we will be discussing large differences, it will be possible to get data reasonably comparable to that in Table II by just knowing gas phase or aqueous heats of formation.

It can be seen that the equilibrium constant for reaction 4 is very large for bases such as CH_3^-, I^-, CH_3S^-, and H^- which are listed as soft bases in Table I. For hard bases such as CH_3O^- and $C_6H_5O^-$, the equilibrium constant is much smaller, and, for F^-, the constant is less than unity. The immediate conclusion is that the methyl carbonium ion is a softer acid than is the proton which is a hard acid. Note that a large equilibrium constant for eq 4 does not mean that CH_3^+ is a stronger acid than H^+ in an intrinsic sense. It means only that CH_3^+ prefers the soft base and H^+ prefers the hard base in a competition of the type given by eq 2. It will be shown in the Appendix that H^+ is intrinsically a much stronger Lewis acid than CH_3^+. There is obviously no correlation between the equilibrium constants for eq 4 and the strength of the bases toward the proton in water. The K_a values for the conjugate acids are given in Table II for each base. Some of these are only estimated values.

The equilibrium constants form an order of decreasing softness as follows.

$$H^- > CN^- > CH_2COCH_3^- > CH_3S^- > CH_3^- > I^- > SH^- >$$

$$Br^- > Cl^- > ONO^- > NH_2^- > CH_3O^- > F^-$$

1829

Where the bases are the same, the order is quite similar to, but not identical with, those for CH_3Hg^+.

The criteria for class b, or soft, behavior are given by displacements such as

$$CH_3OH + HS^- \longrightarrow CH_3SH + OH^- \qquad (5)$$

$$CH_3F + I^- \longrightarrow CH_3I + F^- \qquad (6)$$

in protic solvents such as water or methanol. The data in ref 9 show that displacements such as eq 5 do occur, but just barely. Also literature data on eq 6 show that the equilibrium constant is slightly less than unity.[10] The data of Table II, however, show the very large increase in preference for I^- over F^- when CH_3^+ is compared to the hard proton. The general conclusion is that the methyl carbonium ion is fairly soft but not as extremely soft as CH_3Hg^+, for example. It may be considered as a borderline case.

We can now make the prediction that displacements such as

$$(CH_3)_3N + PH_3 \longrightarrow (CH_3)_3P + NH_3 \qquad (7)$$

$$CH_3NR_3^+ + PR_3 \longrightarrow CH_3PR_3^+ + NR_3 \qquad (8)$$

will occur. The reactions will be relatively independent of the solvent. These are not trivial predictions since amines are always stronger bases (toward the proton) than their corresponding phosphines. One normally accepts the idea that the stronger base is bound more tightly. The concept of soft bases being bound more by the soft methyl carbonium ion produces the opposite prediction.

The heat of reaction 7 is -24.7 kcal/mole so that it goes as predicted. The corresponding reactions in which AsH_3 and SbH_3 are converted to $(CH_3)_3As$ and $(CH_3)_3Sb$ by reaction with $(CH_3)_3N$ are also strongly favored.[11] It is also of interest that a number of displacements of the kind shown in eq 8 can be carried out.[12]

An important point is what happens to the hardness or softness of a carbonium ion as its composition is varied. As an extreme case, we may go to the acylium ion, CH_3CO^+. Table III shows some equilibrium data for the exchange reaction

$$CH_3COOH(g) + HB(g) \rightleftharpoons CH_3COB(g) + H_2O \qquad (9)$$

These were calculated from heats of reaction only. The entropy changes will be small and would not change the conclusions that may be drawn. Some equilibrium constants for reaction 9 in aqueous solution are also available and show the same behavior.[9]

We see that in CH_3CO^+ the great preference that CH_3^+ showed for soft bases has vanished. The equilibrium constant except for alkoxide and amide are less than unity. This simply means that the acylium ion prefers OH^- as a base even more than the proton does. It is a hard Lewis acid. Table III includes the ΔpK_{eq} for reaction 9 and reaction 4. These numbers also give a pattern of decreasing softness from H^- to NH_2^-.

It is not surprising that the acylium ion is much harder than an alkyl carbonium ion. The electronega-

Table III. Equilibrium Constants at 25°, K_{eq}, for the Reaction
$$CH_3COOH(g) + HB(g) \rightleftharpoons CH_3COB(g) + H_2O(g)$$

B	$K_{eq}{}^a$	ΔpK_{eq}
H^-	10^{-5}	26
I^-	10^{-15}	23
$CH_2COCH_3^-$	10^{-10}	21
CH_3^-	10^{-9}	17
SH^-	10^{-6}	13
Cl^-	10^{-6}	11
F^-	10^{-5}	5
CH_3O^-	1	3
NH_2^-	10^4	0

^{a} Calculated from heats of formation only, ignoring the small entropy changes. Data from ref 13 and 14, except for $CH_2COCH_3^-$ which is from J. L. Wood and M. M. Jones, *Inorg. Chem.*, **3**, 1553 (1964).

tive oxygen atom would withdraw charge from the carbon making it a more positive center. In the same way we would expect CF_3^+ to be harder than CH_3^+. This can be verified by examining the reaction[13]

$$CF_3I(g) + CH_3F(g) \longrightarrow CF_4(g) + CH_3I(g) \qquad (10)$$

$$\Delta H^\circ = -18 \text{ kcal/mole at } 25^\circ$$

Thus CF_3^+ prefers to bind F^- and CH_3^+ prefers to bind I^- which justifies the statement that CF_3^+ is harder than CH_3^+.

The effect of replacing H atoms by alkyl groups in the methyl carbonium ion gives an unexpected result. Consider the sequence of reactions[14]

$$CH_3OH(g) + H_2S(g) \longrightarrow CH_3SH(g) + H_2O(g) \qquad (11)$$

$$\Delta H^\circ = -10.4 \text{ kcal/mole at } 25^\circ$$

$$C_2H_5OH(g) + H_2S(g) \longrightarrow C_2H_5SH(g) + H_2O(g) \qquad (12)$$

$$\Delta H^\circ = -7.8 \text{ kcal/mole at } 25^\circ$$

$$i\text{-}C_3H_7OH(g) + H_2S(g) \longrightarrow i\text{-}C_3H_7SH(g) + H_2O(g) \qquad (13)$$

$$\Delta H^\circ = -5.8 \text{ kcal/mole at } 25^\circ$$

$$t\text{-}C_4H_9OH(g) + H_2S(g) \longrightarrow t\text{-}C_4H_9SH(g) + H_2O(g) \qquad (14)$$

$$\Delta H^\circ = -4.2 \text{ kcal/mole at } 25^\circ$$

$$C_6H_5OH(g) + H_2S(g) \longrightarrow C_6H_5SH(g) + H_2O(g) \qquad (15)$$

$$\Delta H^\circ = -3.2 \text{ kcal/mole at } 25^\circ$$

The conclusion is that replacing H atoms in CH_3^+ by methyl groups leads to a progressively harder carbonium ion. Thus $(CH_3)_3C^+$ is harder than CH_3^+.

This result is unexpected because the usual concept of an alkyl group is that it is electron releasing with respect to hydrogen. In fact the reverse must be true. Carbon is a more electronegative element than hydrogen.[15] Replacing H by CH_3, or other alkyl group, must result in a small removal of negative charge from the carbon atom of CH_3^+. Hence it becomes harder. It is also quite reasonable that $C_6H_5^+$ is harder than CH_3^+ since H atoms have been replaced by C atoms. We also note that sp^2 carbon is more electronegative than sp^3 carbon.[15]

(10) R. H. Bathgate and E. A. Moelwyn-Hughes, *J. Chem. Soc.*, 3642 (1959); A. J. Parker, *ibid.*, 1328 (1961).

(11) S. B. Hartley, *et al.*, *Quart. Rev.* (London), **17**, 204 (1963); L. H. Long and J. F. Sackman, *Trans. Faraday Soc.*, **51**, 1062 (1955); **52**, 1201 (1956); F. D. Rossini, *et al.*, National Bureau of Standards Circular 500, U. S. Government Printing Office, Washington, D. C., 1952.

(12) H. Hellman and O. Schumacher, *Ann. Chem.*, **640**, 79 (1961).

(13) Heats of formation are from S. W. Benson, *J. Chem. Educ.*, **502** (1965), and estimates of P. G. Maslov and Yu. P. Maslov, *Khim. i Tekhnol. Topliev i Masel*, **3**, 50 (1958); *Chem. Abstr.*, **53**, 1910 (1958).

(14) Heats of formation of oxygenated compounds from J. H. S. Green, *Quart. Rev.* (London), **15**, 125 (1961); of sulfur compounds, H. Mackle and P. A. G. O'Hare, *Tetrahedron*, **19**, 961 (1963).

(15) The group electronegativity of CH_3 is given as 2.30 compared to 2.20 for H. See H. J. Hinze, M. A. Whitehead, and H. H. Jaffé, *J. Am. Chem. Soc.*, **85**, 148 (1963).

Pearson, Songstad | Hard and Soft Acids and Bases

It has recently been experimentally demonstrated by Laurie and Muenter[16] that a methyl group bonded to saturated carbon is electron withdrawing with respect to hydrogen. This is a pure inductive effect. The common belief that an alkyl group is electron donating with respect to hydrogen is largely based on examples in which an unsaturated carbon is involved. In such cases hyperconjugation will be important and an alkyl group can become a net electron donor. Examples would be (see also the Appendix)

$$H^+CH_2=\overset{\overset{\displaystyle H}{|}}{C}C^-H_2 \longleftrightarrow CH_3\overset{\overset{\displaystyle H}{|}}{C}=CH_2$$

$$H^+CH_2=C(CH_3)_2 \longleftrightarrow CH_3C^+(CH_3)_2$$

$$H^+CH_2=\underset{\underset{\displaystyle OH}{|}}{C}O^- \longleftrightarrow CH_3\underset{\underset{\displaystyle OH}{|}}{C}=O$$

A number of interesting correlations can now be made with the knowledge that hardness decreases in the order

$$C_6H_5^+ > (CH_3)_3C^+ > (CH_3)_2CH^+ > C_2H_5^+ > CH_3^+$$

For example[17]

$$CH_4(g) + H_2O(g) \longrightarrow CH_3OH(g) + H_2(g) \tag{16}$$

$$\Delta H° = 27.6 \text{ kcal/mole at } 25°$$

$$C_2H_6(g) + H_2O(g) \longrightarrow C_2H_5OH(g) + H_2(g) \tag{17}$$

$$\Delta H° = 21.8 \text{ kcal/mole at } 25°$$

$$C_3H_8(g) + H_2O(g) \longrightarrow i\text{-}C_3H_7OH(g) + H_2(g) \tag{18}$$

$$\Delta H° = 17.2 \text{ kcal/mole at } 25°$$

$$i\text{-}C_4H_{10}(g) + H_2O(g) \longrightarrow t\text{-}C_4H_9OH(g) + H_2(g) \tag{19}$$

$$\Delta H° = 15.4 \text{ kcal/mole at } 25°$$

$$C_6H_6(g) + H_2O(g) \longrightarrow C_6H_5OH(g) + H_2(g) \tag{20}$$

$$\Delta H° = 14.9 \text{ kcal/mole at } 25°$$

The increasing tendency for these reactions to occur can be neatly correlated with the fact that hard acids prefer to coordinate to OH⁻ and soft acids prefer H⁻. Isomerizations, such as

$$i\text{-}C_4H_9OH(g) \longrightarrow t\text{-}C_4H_9OH(g) \tag{21}$$

$$\Delta H° = -7.0 \text{ kcal/mole at } 25°$$

find a ready explanation. Further examples will be given under the next heading.

Symbiosis

Jørgensen[18] has pointed out that a common phenomenon occurs in coordination chemistry: soft ligands, or bases, tend to flock together on a central metal atom, and hard ligands tend to flock together. This mutual stabilizing effect was called symbiosis. The same symbiotic phenomenon occurs in organic chemistry. It was first pointed out by Hine,[19] who showed that C–F and C–O bonds in the same molecule tend to reinforce each other. Fluorine and oxygen being hard, and hydrogen (as hydride ion) being soft, the replacement of one hydrogen by F or O would make it easier for the next to be replaced. The reason for

this has already been given in discussing why CH_3CO^+ and CF_3^+ are harder than CH_3^+. Thus piling up soft bases on an acceptor atom makes it soft, and piling up hard bases on an acceptor atom makes it hard. An outstanding example would be BH_3 which is a soft acid, and BF_3 which is a hard acid.

The least stable combinations would be mixtures of hard and soft ligands on one center. This is in agreement with the following cases.

$$2CH_2F_2(g) \longrightarrow CH_4(g) + CF_4(g) \tag{22}$$

$$\Delta H° = -26 \text{ kcal/mole at } 25°$$

$$2CH_2O(g) \longrightarrow CH_4(g) + CO_2(g) \tag{23}$$

$$\Delta H° = -56.5 \text{ kcal/mole at } 25°$$

While thermodynamic data are not available, it is known[20] that the hemithioformal is unstable with

$$2H_2C\overset{\displaystyle SCH_2C_6H_5}{\underset{\displaystyle OH}{}} \overset{H^+}{\longrightarrow} CH_2(OH)_2 + CH_2(SCH_2C_6H_5)_2 \tag{24}$$

respect to disproportionation. A trace of acid is necessary to catalyze the conversion.

More quantitative data[20b] show that it is always more difficult to replace one OR group with SR in $R'_2\text{-}C(OR)_2$ compounds than in R'_3COR compounds; thus the two oxygen atoms together have a symbiotic, stabilizing effect. Equally well one can say that the carbonium ion ROR'_2C^+ is harder than R'_3C^+.

An interesting example of symbiosis is found by comparing an alkyl group as a base, such as CH_3^-, with the H^- ion. Tables II and III show that H^- is softer than CH_3^-. Therefore, the most stable hydrocarbons should contain carbon atoms with the maximum number of C–H bonds or the maximum number of C–C bonds. The poorest combination has equal numbers of bonds to C and H for each carbon. We can predict the direction of the following reactions.

$$2CH_3CH_2CH_3 \longrightarrow CH_4 + (CH_3)_4C \tag{25}$$

$$\Delta H° = -8.0 \text{ kcal/mole at } 25°$$

$$CH_3CH_2CH_2CH_2CH_3 \longrightarrow (CH_3)_4C \tag{26}$$

$$\Delta H° = -4.7 \text{ kcal/mole at } 25°$$

We have by this rule a simple explanation for the extra stability of highly branched hydrocarbons. As we mentioned earlier, explaining something by the concept of hard and soft acids and bases does not necessarily invalidate other, theoretical explanations of the same thing.

There are some limits to the concept of symbiosis. Reaction 27 (cf. eq 10) is endothermic by 50 kcal

$$CI_3F(g) + CF_3I(g) \longrightarrow CF_4(g) + CI_4(g) \tag{27}$$

because of the very positive heat of formation of CI_4.[13] Presumably, steric strain in this molecule overcomes any symbiotic effects. Also in the successive oxidation of methane, the exothermicity increases in

$$CH_4 \overset{O}{\longrightarrow} CH_3OH \overset{O}{\longrightarrow} CH_2(OH)_2 \overset{O}{\longrightarrow} HCOOH \longrightarrow H_2CO_3 \tag{28}$$

(16) V. W. Laurie and J. S. Muenter, *J. Am. Chem. Soc.*, **88**, 2883 (1966).

(17) Heats of formation of hydrocarbons from F. D. Rossini, *et al.*, "Selected Values of Physical and Thermodynamic Properties," Carnegie Press, Pittsburgh, Pa., 1953.

(18) C. K. Jørgensen, *Inorg. Chem.*, **3**, 1201 (1964).

(19) J. Hine, *J. Am. Chem. Soc.*, **85**, 3239 (1963).

(20) (a) H. Böhme and H. P. Teltz, *Ann. Chem.*, **620**, 1 (1959); (b) see also W. P. Jencks, *Progr. Phys. Org. Chem.*, **2**, 104 (1964); G. E. Lienhard and W. P. Jencks, *J. Am. Chem. Soc.*, **88**, 3982 (1966).

each step, until the last. The probable explanation is that the acid HOCO⁺ is much stronger than any of its predecessors. Since H⁻ is also much stronger than OH⁻, there is a larger reluctance for HCOOH to become HOCOOH than expected.

Carbenes

A consideration of carbenes as Lewis acids illustrates the way in which the idea of hard and soft acids and bases can relate information from one area to another. Singlet states of carbenes are, of course, Lewis acids and should react as such. Bond formation to bases will also drive ground triplet states to excited singlet states. Simple considerations make it clear that CH_2 should be a softer Lewis acid than CH_3^+, which is already fairly soft.

We immediately can understand the greater stability of phosphine compared to amine ylides, and of sulfide compared to oxide ylides.[21]

$$(CH_3)_3NCH_2 + P(C_6H_5)_3 \longrightarrow CH_2P(C_6H_5)_3 + N(CH_3)_3 \quad (29)$$

$$H_2OCH_2 + R_2S \longrightarrow CH_2SR_2 + H_2O \quad (30)$$

$$(CH_3)_2S{=}OCH_2 \longrightarrow (CH_3)_2SCH_2 \quad (31)$$
$$\overset{\|}{O}$$

The instability of O ylides may help explain why the reaction

$$CCl_3^- \longrightarrow Cl^- + CCl_2 \quad (32)$$

appears to go by an S$_N$1 mechanism rather than by an S$_N$2 mechanism with water acting as the nucleophile.[22]

The reverse of reaction 32 occurs readily. The rate at which various anions trap dichlorocarbene is given by the expected sequence $I^- \sim Br^- > Cl^- > F^-, NO_3^-, ClO_4^-$.[23] We can predict that the following equilibrium will lie well to the right.

$$CH_2F^- + I^- \rightleftharpoons CH_2I^- + F^- \quad (33)$$

The CH_2I^- is to be regarded as an acid–base complex of the soft acid CH_2 and the soft base I^-.

More important, since CH_3^+ is harder than CH_2, we can predict that the equilibrium

$$CH_3I + CH_2F^- \rightleftharpoons CH_2I^- + CH_3F \quad (34)$$

will also lie to the right. This result is unexpected since it says that an iodine atom is more acid strengthening than a fluorine atom, when attached to the carbon which loses the proton. Classical organic theory predicts just the opposite, the inductive effect of F being assumed greater than that of I.[24]

The data on the effect of halogen substituents on carbanion stability are somewhat incomplete. What there are seem to support the acid strengthening order $I > Br > Cl > F$. Using deuterium exchange as a measure of acidity, Hine found this order for various haloforms, pure and mixed.[25] Bell found $Br > Cl$ for bromoacetone and chloroacetone from the rates of enolization.[26]

(21) W. Kirmse, "Carbene Chemistry," Vol. I, Academic Press Inc., New York, N. Y., 1964.
(22) W. J. Le Noble, *J. Am. Chem. Soc.*, **87**, 2434 (1965).
(23) J. Hine and A. M. Dowell, *ibid.*, **76**, 2688 (1954).
(24) G. Branch and M. Calvin, "The Theory of Organic Chemistry," Prentice-Hall, Inc., New York, N. Y., 1941, Chapter 6.
(25) J. Hine, *et al.*, *J. Am. Chem. Soc.*, **79**, 1406 (1957); **80**, 819 (1958).
(26) R. P. Bell, E. Gelles, and E. Möller, *Proc. Roy. Soc.* (London), **A198**, 310 (1949); R. P. Bell and O. M. Lidwell, *ibid.*, **A176**, 88 (1940).

More clear-cut results are found for the predictions

$$CH_2OR^- + CH_3SR \longrightarrow CH_3OR + CH_2SR^- \quad (35)$$

$$CH_2NR_2 + CH_3PR_3^+ \longrightarrow CH_3NR_2^+ + CH_2PR_3 \quad (36)$$

It is well established[27] that SR is more acid strengthening than OR, and PR₃ is more acid strengthening than NR₃. The factors are as large as 10⁶.

It is pertinent to ask whether the replacement of a hydrogen atom by a halogen atom should increase acid strength at all, since H⁻ is very soft and CH_2 is a soft acid. Consider the equilibrium

$$CH_2F^- + CH_4 \rightleftharpoons CH_3F + CH_3^- \quad (37)$$

This must lie to the left since CH_3^+ is a much *stronger* acid than CH_2, and H⁻ is a much *stronger* base than F⁻. The tendency for the strongest acid to combine with the strongest base, in this case, will outweigh any considerations due to hardness or softness. In the important case of an alkyl substituent

$$CH_3CH_2^- + CH_4 \rightleftharpoons CH_3CH_2 + CH_3^- \quad (38)$$

equilibrium will lie to the right since H⁻ is both a somewhat weaker base and a somewhat softer base than CH_3^-. Thus, CH_2 will coordinate with H⁻ and CH_3^+ with CH_3^-.

A criterion for soft acids that may be used in the absence of other data is the formation of fairly stable complexes with special soft bases, such as carbon monoxide, olefins, and aromatics.[2] Carbenes are well known to form complexes with these bases.[21] It is likely that π complexes are formed with olefins and aromatic molecules, prior to more extensive reaction.[28] Carbene also forms a complex with soft metal atoms, another characteristic of soft Lewis acids.[29]

Meaning of Stability

A great deal of confusion can arise when the term stable is applied to a chemical compound. One must specify whether it is thermodynamic or kinetic stability which is meant, stability to heat, to hydrolysis, etc. The situation is even worse when a rule such as the principle of hard and soft acids is used. The rule implies that there is an extra stabilization of complexes formed from a hard acid and a hard base, or a soft acid and a soft base. It still says very little about stability in an absolute sense. It is quite possible for a compound formed from a hard acid and a soft base to be more stable than one made from a better matched pair. All that is needed is that the first acid and base both be quite strong, say H⁺ and H⁻ combined to form H_2.

Frequently the rule can be used in a comparative sense, to say that one compound is more stable than another. Here also caution is required in order to make a meaningful statement. We will try to give some illustrations of how the principle can be usefully applied.

Consider the pair of compounds, NaH and CuH. The sodium ion is a hard Lewis acid; the cuprous ion is soft. Since hydride ion is also soft, the natural statement to make is that CuH is more stable than NaH. This clashes with the fact that sodium hydride is a well-known, apparently stable substance, and copper hy-

(27) D. J. Cram, "Fundamentals of Carbanion Chemistry," Academic Press Inc., New York, N. Y., 1965, p 71 ff.
(28) G. A. Russell and D. G. Hendry, *J. Org. Chem.*, **28**, 1933 (1963).
(29) F. D. Mango and I. Dvoretsky, *J. Am. Chem. Soc.*, **88**, 1654 (1966).

dride is a little known, apparently very unstable substance. Actually the heat of formation of NaH is negative, while that of CuH is positive.[30] The latter compound is indeed unstable with respect to its elements and is hard to prepare and harder to keep.

$$CuH(s) \longrightarrow Cu(s) + 0.5H_2(g) \qquad (39)$$

$$\Delta H^\circ = -7.9 \text{ kcal/mole at } 25^\circ$$

$$NaH(s) \longrightarrow Na(s) + 0.5H_2(g) \qquad (40)$$

$$\Delta H^\circ = 13.7 \text{ kcal/mole at } 25^\circ$$

The statement about stability based on the hard and soft concept does have meaning in spite of this. It is necessary, however, that an acid–base reaction be chosen as a reference, rather than the oxidation–reduction of eq 39 and 40. Hydrolysis is a good example.[31]

$$NaH(s) + H^+(aq) \longrightarrow Na^+(aq) + H_2(g) \qquad (41)$$

$$\Delta H^\circ = -43.6 \text{ kcal/mole at } 25^\circ$$

$$CuH(s) + H^+(aq) \longrightarrow Cu^+(aq) + H_2(g) \qquad (42)$$

$$\Delta H^\circ = 5.2 \text{ kcal/mole at } 25^\circ$$

The heats show how readily Na^+ coordinates with H_2O, compared to coordinating with H^-, whereas Cu^+ prefers H^- to H_2O.

The same situation exists for organometallic compounds. $Al(CH_3)_3$, $Zn(CH_3)_2$, and $Hg(CH_3)_2$ are decreasingly less stable toward decomposition into their elements. Toward hydrolysis the opposite pattern is found.[32]

$$2Al(CH_3)_3(l) + 3H_2O(l) \longrightarrow Al_2O_3(s) + 6CH_4(g) \qquad (43)$$

$$\Delta H^\circ/6 = -38 \text{ kcal/mole at } 25^\circ$$

$$Zn(CH_3)_2(l) + H_2O(l) \longrightarrow ZnO(s) + 2CH_4(g) \qquad (44)$$

$$\Delta H^\circ/2 = -22.5 \text{ kcal/mole at } 25^\circ$$

$$Hg(CH_3)_2(l) + H_2O(l) \longrightarrow HgO(s) + 2CH_4(g) \qquad (45)$$

$$\Delta H^\circ/2 = 14 \text{ kcal/mole at } 25^\circ$$

This hydrolytic behavior is consistent with increasing softness of the Lewis acids $Al^{3+} < Zn^{2+} < Hg^{2+}$.

Another confusing situation arises when a molecule can be looked at in two or more ways, as far as considering it as an acid–base complex. This is almost always possible in organic chemistry. For example, methane may be considered as $H^+CH_3^-$ or $CH_3^+H^-$. In the first case we have a hard–soft combination, in the second case a soft–soft combination. Is methane to be considered as unstable or stable? The answer is, of course, that a particular acid–base reaction must first be selected. This, in turn, will dictate the necessary formulation for methane. Two possible cases might be

$$\underset{\text{ss}}{CH_4(g)} + \underset{\text{hh}}{CH_3OH(g)} \longrightarrow \underset{\text{sh}}{CH_3OCH_3(g)} + \underset{\text{hs}}{H_2(g)} \qquad (46)$$

$$\Delta H^\circ = 22 \text{ kcal/mole at } 25^\circ$$

$$\underset{\text{hs}}{CH_4(g)} + \underset{\text{sh}}{CH_3OH(g)} \longrightarrow \underset{\text{ss}}{CH_3CH_3(g)} + \underset{\text{hh}}{H_2O(g)} \qquad (47)$$

$$\Delta H^\circ = -12 \text{ kcal/mole at } 25^\circ$$

The first case considers CH_4 as $CH_3^+H^-$, a soft–soft

(30) I. C. Warf, *J. Inorg. Nucl. Chem.*, **28**, 1031 (1966).
(31) Heats of formation of aqueous ions from F. D. Rossini, *et al.*, ref 11.
(32) Heats of formation from ref 31 and from L. H. Long and R. G. W. Norrish, *Phil. Trans., Roy. Soc. London, Ser. A*, **241**, 587 (1949).

combination. The second case considers it to be $H^+CH_3^-$ a hard–soft combination. The other acid–base combinations are also indicated in eq 46 and 47.

Kinetic Applications

The principle of hard and soft acids and bases may also be applied to the rates of nucleophilic and electrophilic substitution reactions. A paper on nucleophilic reactivity was indeed the first to call attention to those properties which were later labeled hard and soft.[33] It was noted that electrophilic centers such as RCO^+, H^+, RSO_2^+, $(RO)_2PO^+$, and $(RO)_2B^+$ reacted rapidly with nucleophiles which were strongly basic to the proton and not very polarizable, such as OH^- and F^-. Other electrophilic centers such as RCH_2^+, R_2P^+, RS^+, Br^+, R_2N^+, RO^+, and Pt^{2+} reacted rapidly with highly polarizable nucleophiles such as I^- and R_3P.

The general rule can easily be stated: hard electrophilic centers (acids) react rapidly with hard nucleophiles (bases), and soft electrophilic centers react rapidly with soft nucleophiles. The rule refers to S_N2- or S_E2-type mechanisms. An additional corollary can also be stated: softness is more important in rate phenomena than in thermodynamic phenomena. That is, a medium soft acid, such as CH_3^+, will be even more reactive to soft bases than the stability of the products would predict. This rule follows partly from experimental observations and partly from the theories which explain preferential hard–hard and soft–soft interactions.[2]

In an S_N2 or $S_N2(\lim)$ mechanism the rate depends as usual on the difference in free energy of the transition state and the reactants

$$B': + A:B \rightleftharpoons B':A:B \longrightarrow B':A + B \qquad (48)$$

The transition state, in $B':A:B$, is just another acid–base complex with an increased coordination number for the electrophilic atom A and somewhat longer bonds. The same considerations that predict the stability of acid–base complexes in general should predict rates of reaction. The increased coordination number in $B':A:B$ puts an increased negative charge on A and makes it softer. Hence softness in B' is more helpful than for equilibria only.

Table IV shows a partial listing of relative nucleophilic reactivities toward methyl iodide in methanol solvent at 25°. The data are given in terms of a

$$B^- + CH_3I \xrightarrow{k_B} CH_3B + I^- \qquad (49)$$

parameter $n^\circ_{CH_3I}$ patterned after that of Swain and Scott,[34] who used CH_3Br as a reference compound and water as a solvent. The rate constant for methanol, k_B, has been converted to second-order units. That is, the first-order rate constant for methanolysis has been divided by 26 M.

$$n^\circ_{CH_3I} = \log(k_B/k_S) \qquad (50)$$

There is a range of nucleophilic reactivity covering nearly 10^{10}. The important feature for the present purpose is that the reactivities do not parallel the basicities toward the proton, a hard reference acid, as given by the pK_a values shown in the table. Also im-

(33) J. O. Edwards and R. G. Pearson, *J. Am. Chem. Soc.*, **84**, 16 (1962).
(34) C. G. Swain and C. B. Scott, *ibid.*, **75**, 141 (1953).

Table IV. Nucleophilic Reactivity Parameters for Reaction with Methyl Iodide in Methanol at 25°

B	$n°_{CH_3I}{}^a$	pK_a
CH₃OH	0.00	~0
Cl⁻	4.37	−7
(CH₃)₂S	5.34	−5.3
NH₃	5.50	9.2
C₆H₅O⁻	5.79	9.9
Br⁻	5.79	−9
CH₃O⁻	6.29	16
(CH₃)₂Se	6.32	. . .
CN⁻	6.70	9.1
(C₂H₅)₃As	7.10	2
I⁻	7.42	−9.5
(C₂H₅)₃P	8.72	8.9
C₆H₅S⁻	9.92	6.5

^{a} $n°_{CH_3I}$ = log (k_B/k_S) where k_S is in second-order units. Data from J. Songstad, to be published in detail.

portant are the reactivity sequences I⁻ > Br⁻ > Cl⁻, R₂Se > R₂S > ROH, and R₃P > R₃As > R₃N. This is typical soft behavior. Furthermore the increased reactivity of the more polarizable nucleophiles is greater than the increased stability of the products, particularly for the halide ions.

Series such as those of Table IV are very sensitive to the solvent, as Parker particularly has emphasized.[35] The reactivity order for the halide ions can be inverted by switching to a dipolar aprotic solvent, such as dimethylformamide. The equilibrium constants are also drastically changed in the same direction. This does not mean that the high reactivity of polarizable reagents is an accident due to a peculiar choice of solvents.

First of all, it is only anions that are seriously affected by changing solvents.[2] If one stays with neutral reagents, the order remains unchanged on switching from protic to dipolar aprotic solvents. For example, the reactivity of phosphines toward ethyl iodide is essentially the same in the solvents methanol, acetone, and nitromethane.[36] That is, the rates are unchanged on changing solvents, whereas halide ion reactivities change[35] by factors of 400 for I⁻ and 3 × 10⁶ for Cl⁻. Also the rates of reaction of tertiary amines with alkyl iodides increase by less than a factor of ten in going from methanol to acetone or nitrobenzene.[37] It is obvious that the reactivity order P >> N is maintained in all solvents used.

Secondly, the inversion in reactivity for the halides occurs only because CH₃⁺ is a borderline soft acid. If a more typically soft substrate, such as Pt²⁺, is taken, the reactivity sequence I > Br > Cl is maintained even in dipolar aprotic solvents.[38] The spread in reactivity is reduced, as expected. Iodide ion is more reactive than chloride ion by a factor of 10⁴ in methanol and by a factor of only 200 in acetone and dimethyl sulfoxide. Solvation effects, while important, are rather predictable perturbations of the basic pattern of nucleophilic reactivity.

The specific stabilization of small, basic anions by protic solvents is itself an example of the HSAB principle.[2] Hydrogen bonding is a hard acid–hard

base interaction. Hence the activity of F⁻ and CH₃O⁻ is reduced most in going from a solvent such as dimethyl sulfoxide to methyl alcohol. Dipolar aprotic solvents are to be considered as fairly soft compared to water and alcohol.

If the substrate taken is an acyl derivative such as an ester or an acyl halide, the pattern of nucleophilic reactivity is completely changed in line with the hardness of the carbonyl function, RCO⁺. Except for the perturbation known as the *alpha effect*,[33] rates vary fairly regularly in accord with the proton basicity of the nucleophile.[39] Soft bases which are not also strong bases toward the proton, are quite ineffective. The Bronsted relationship often holds between the rate constant and K_a of the conjugate acid.[40]

The prediction can be made that alkyl halides such as t-C₄H₉Cl and CH₃OCH₂Cl should be less sensitive to polarizability in the nucleophile than is CH₃I or CH₃Cl. This follows from the previous discussion that shows (CH₃)₃C⁺ and HOCH₂⁺ to be harder than CH₃⁺. It is well known that sensitivity to the nature of the nucleophilic reagent decreases steadily in the series CH₃X > C₂H₅X > i-C₃H₇X > t-C₄H₉X. This includes sensitivity to the hard bases such as hydroxide ion and various oxygen donor solvents.[42] Nevertheless, in agreement with the above prediction, the falloff in relative rates in the series CH₃X > C₂H₅X > i-C₃H₇X is considerably greater for soft nucleophiles such as S₂O₃²⁻, R₃P, and I⁻ than for hard oxygen donor bases.[43]

There also are some interesting results from studies with ambient nucleophiles.[44] As pointed out earlier, such nucleophiles usually have one donor atom softer than the other.[33] The mode of action can then be predicted from the hardness or softness of the electrophilic substrate. In the case of thiocyanate ion we have

$$CH_3I + SCN^- \longrightarrow CH_3SCN + I^- \quad (51)$$

$$CH_3COCl + NCS^- \longrightarrow CH_3CONCS + Cl^- \quad (52)$$

It is of interest that t-C₄H₉Cl reacts with the hard oxygen atom of NO₂⁻, while the softer CH₃I reacts with the softer nitrogen atom.[44] Also enolate ions react at the soft carbon atom with CH₃I and at the hard oxygen atom with CH₃OCH₂Cl.[45] Hudson has discussed a number of other reactions of ambient nucleophiles from the viewpoint of hardness and softness.[46] The expected results are usually found, though there are some ambiguous cases. The phosphonate anions, (RO)₂PO⁻, have a fairly soft center in the P^III atom and a hard center in the oxygen atom. Nevertheless, the

(35) A. J. Parker, *Quart. Rev.* (London), **16**, 163 (1962); A. J. Parker, *et al.*, *J. Am. Chem. Soc.*, **88**, 1911 (1966); *J. Chem. Soc., Inorg.*, 152 (1966).
(36) W. A. Henderson. Jr.. and S. A. Buckler, *J. Am. Chem. Soc.*, **82**, 5794 (1960).
(37) R. G. Pearson, *J. Chem. Phys.*, **20**, 1478 (1952).
(38) U. Belluco, M. Martelli, and A. Orio, *Inorg. Chem.*, **5**, 592 (1966).

(39) M. L. Bender, *Chem. Rev.*, **60**, 53 (1960); W. P. Jencks and J. Carriulo, *J. Am. Chem. Soc.*, **82**, 1778 (1960); M. Green and R. F. Hudson, *J. Chem. Soc.*, 1055 (1962); T. C. Bruice and S. J. Benkovic, *J. Am. Chem. Soc.*, **86**, 418 (1964).
(40) A Brønsted relationship for a series of bases in which the donor atom is constant does not in itself prove that proton basicity is important. Substituents may be changing the electron density at the donor atom, making it harder or easier.[41]
(41) E. Thorsteinson and F. Basolo, *J. Am. Chem. Soc.*, **88**, 3929 (1966).
(42) See C. G. Swain, R. B. Mosely, and D. E. Brown, *ibid.*, **77**, 3731 (1955), for example.
(43) P. M. Dunbar and L. P. Hammett, *ibid.*, **72**, 109 (1950), for S₂O₃²⁻; ref 36 for R₃P; and C. K. Ingold, *Quart. Rev.* (London), **11**, 1 (1957), for I⁻ and OR⁻.
(44) N. Kornblum, R. A. Smiley, R. K. Blackwood, and D. C. Iffand, *J. Am. Chem. Soc.*, **77**, 6269 (1955).
(45) J. L. Simonsen and R. Storey, *J. Chem. Soc.*, **95**, 2106 (1909).
(46) Reference 7a, Chapter 4; R. F. Hudson, in "Organic Reaction Mechanisms," Special Publication No. 19, The Chemical Society, London, 1965, p 93 ff.

Pearson, Songstad / *Hard and Soft Acids and Bases*

proton is known to bind exclusively to the phosphorus atom.[47] This atom is much more basic than the oxygen atom, even to hard Lewis acids. It is not too surprising that reactions such as

$$(R'O)_2PO^- + (RO)_2\overset{O}{\overset{\|}{P}}X \longrightarrow (R'O)_2\overset{O}{\overset{\|}{P}}\overset{O}{\overset{\|}{--P}}(OR)_2 \quad (53)$$

$$(R'O)_2PO^- + (RO)_2\overset{O}{\overset{\|}{P}} \longrightarrow (R'O)_2POP(OR)_2 \quad (54)$$

can both occur since opposing tendencies exist. With alkyl halides reaction normally occurs at the P atom.

An even more common case than that of ambident nucleophiles is that of molecules with multiple electrophilic centers. Most organic compounds fall into this category. Even as simple a molecule as CH_3Br has three centers, the C, H, and Br atoms. The HSAB principle can be of great value in understanding the mode of attack of various nucleophiles. For example, we have the following reactions of i-C_3H_7Br.[48]

$$i\text{-}C_3H_7Br + C_2H_5O^- \longrightarrow \underset{80\%}{CH_3CH{=}CH_2} \quad (55)$$

$$i\text{-}C_3H_7Br + H\bar{C}(COOC_2H_5)_2 \longrightarrow \underset{80\%}{i\text{-}C_3H_7CH(COOC_2H_5)_2} \quad (56)$$

The hard ethoxide ion attacks the proton, giving elimination, and the soft malonate anion attacks tetrahedral carbon with displacement of bromide ion. It should be noted that the proton basicities of the two anions are virtually identical.[49]

Similarly soft nucleophiles attack the alkyl group of phosphate esters while hydroxide ion and fluoride ion attack the hard phosphorus(V) site.[50] Many other examples can be thought of, some of which are discussed by Hudson.[51]

$$(RO)_3P{=}O + SCN^- \longrightarrow RSCN + (RO)_2\overset{O}{\overset{\|}{P}}{--}O^- \quad (57)$$

$$(RO)_3P{=}O + {}^*OH^- \longrightarrow (RO)_2\overset{O}{\overset{\|}{P}}{-}{}^*O^- + ROH \quad (58)$$

An extreme case is afforded by the α-halo ketones, which present no less than five electrophilic sites. Hard bases attack the hard carbonyl carbon, or the proton. Medium soft bases, such as pyridine or RS^-, attack the tetrahedral carbon to give displacement. The softest bases such as I^- and R_3P will cause dehalogenation of iodo and bromo ketones. Recently nucleophilic attack at the oxygen atom of a halo ketone by an alkyl phosphite has also been demonstrated.[52] Note that the oxygen atom of a carbonyl group is a hard basic site, but is a soft acid site. The same situation is found in hydrogen peroxide, which is composed of soft OH^+ combined with hard OH^-. Only soft bases will cause nucleophilic displacements at peroxide oxygen.[53]

Saville[7b] has used the HSAB principle as a guide for selecting catalysts in acid–base substitution reactions. He points out that the reactions of a hard–soft acid–base complex are the easiest to catalyze. The selection

of the catalyst is given by two rules depending on the substrate.

N:	A:B	E	rule 1
h	h s	s	
N:	A:B	E	rule 2
s	s h	h	

The selection of both the substituting reagent (E or N) and the nucleophilic or electrophilic catalyst (N or E) depends on a match of the hard or soft properties of each part of the acid–base complex. An illustrative case or two is self-explanatory.

$$\underset{h}{H_2O} + \underset{hs}{R\overset{O}{\overset{\|}{C}}SR'} + \underset{s}{Br_2} \longrightarrow RCOH_2^+ + R'SBr + Br^- \quad (59)$$

$$\underset{s}{I^-} + \underset{sh}{R'OR} + \underset{h}{H^+} \longrightarrow ROH + R'I \quad (60)$$

Symbiotic Effects on Rates

Inasmuch as the transition state for an S_N2 reaction may be regarded as an acid–base complex, according to eq 48, there should be symbiotic effects on rates of nucleophilic displacement reactions. That is, a grouping of either several hard ligands or several soft ligands on the central carbon atom should stabilize the transition state and cause an increased rate of reaction. The ligands include the entering and leaving group as well as the three groups, L, in the trigonal plane of the activated complex.

In fact just such an effect has been pointed out by Bunnett.[54] Large ratios for $k_{C_6H_5S^-}/k_{CH_3O^-}$ and $k_{I^-}/k_{CH_3O^-}$ are found when these reagents are B' and either B or L become progressively heavier halides. A number of other examples with other nucleophiles have also been listed.[54,55]

In the case of aromatic nucleophilic substitution, it is usually found that the rate ratio k_{ArF}/k_{ArI} is large for hard nucleophiles such as CH_3O^- and N_3^-, and low for soft nucleophiles such as Br^-, I^-, SCN^-, and $C_6H_5S^-$. This is the predicted result. However, CH_3S^- is definitely anomalous in giving a large rate ratio.[56]

For displacements on tetrahedral carbon there is little doubt that the symbiotic effect is real and seems to be quite general. In order to avoid complications due to steric hindrance, it is best to consider only variations in B and B'. The rule is that higher rates are found when B and B' are both hard or both soft. It is usually necessary to look at ratios of rates to see the effect. It is also necessary to compare rates in the same, or similar solvents.

Tables V and VI show relative rate data for some organic halides. The general pattern is for higher rate ratios for the heavier halides as the softness of the nucleophile increases. The common statement[57] that

(47) G. O. Doak and L. D. Freedman, *Chem. Rev.*, **61**, 31 (1961).
(48) F. G. Bordwell, "Organic Chemistry," The Macmillan Co., New York, N. Y., 1963, p 218.
(49) R. G. Pearson, *J. Am. Chem. Soc.*, **71**, 2212 (1949).
(50) D. C. Harper and R. F. Hudson, *J. Chem. Soc.*, 1356 (1958).
(51) See ref 46, also Chapter 7 of ref 7a; E. Hünig, *Angew. Chem. Intern. Ed. Engl.*, **3**, 548 (1964).
(52) B. Miller, *J. Am. Chem. Soc.*, **88**, 1814 (1966).
(53) J. O. Edwards, "Peroxide Reaction Mechanisms," Interscience Publishers, Inc., New York, N. Y., 1962.

(54) J. F. Bunnett, *J. Am. Chem. Soc.*, **79**, 5970 (1957).
(55) J. F. Bunnett and J. D. Reinheimer, *ibid.*, **84**, 3284 (1962); A. J. Sisti and S. Lowell, *J. Org. Chem.*, **29**, 1635 (1964).
(56) J. Miller and K. W. Wong, *J. Chem. Soc.*, 5454 (1965); K. C. Ho, J. Miller, and K. W. Wong, *ibid.*, *Org.*, 310 (1966).
(57) C. K. Ingold, "Structure and Mechanism in Organic Chemistry." Cornell University Press, Ithaca, N. Y., 1953, p 339; A. Streitwieser. Jr., *Chem. Rev.*, **56**, 601 (1956).

Table V. Reactivity Ratios for Methyl Halides in Water at 25°

$$CH_3X + B^- \longrightarrow CH_3B + X^-$$

Nucleophile, B	k_{CH_3I}/k_{CH_3Cl}	k_{CH_3I}/k_{CH_3F}	Ref
H_2O	13	1×10^2	a
F^-	5.5	...	b
OH^-	10	1×10^2	a
Cl^-	...	1×10^3	b,c
Br^-	86	...	b–d
I^-	24	2.4×10^4	a,e
$S_2O_3^{2-}$	31	...	f

[a] I. Fells and E. A. Moelwyn-Hughes, *J. Chem. Soc.*, 398 (1959). [b] R. H. Bathgate and E. A. Moelwyn-Hughes, *ibid.*, 2647 (1959). [c] R. B. Fahim, Thesis, Cambridge, 1954; H. G. Holland, Thesis, Cambridge, 1954. [d] E. A. Moelwyn-Hughes, *J. Chem. Soc.*, 779 (1938). [e] E. R. Swart and H. LeRoux, *ibid.*, 409 (1957). [f] E. A. Moelwyn-Hughes, *Trans. Faraday Soc.*, **37**, 279 (1941).

Table VI. Reactivity Ratios for Organic Halides at 25°

Reactants	Solvent	k_{Br}/k_{Cl}	k_I/k_{Cl}	Ref
$^-OCH_2CH_2X$	H_2O	59	54	a
$NH_2CH_2CH_2X$	H_2O	70	...	b
$C_6H_5CHSO_2^- \\ CHC_6H_5X$	CH_3OH	280	670	c
$C_6H_5COCH_2X +$ piperidine	CH_3OH	108	114	d
$C_6H_5COCH_2X +$ aniline	$C_2H_5OH–H_2O$	97	99	e
$C_6H_5COCH_2X +$ thiourea	CH_3OH	182	341	d

[a] C. L. McCabe and J. C. Warner, *J. Am. Chem. Soc.*, **70**, 4031 (1948). [b] H. Freundlich and G. Salomon, *Z. Physik. Chem.* (Leipzig), **166**, 161 (1933). [c] J. M. Williams, Ph.D. Thesis, Northwestern University, Evanston, Ill., 1966. [d] W. T. Brannen, Ph.D. Thesis, Northwestern University, Evanston, Ill., 1962. [e] J. W. Baker, *J. Chem. Soc.*, 1148 (1932).

the relative reactivities for alkyl halides, RCl : RBr : RI = 1 : 50 : 100, is valid for alcohol, alkoxide ions, and amines in alcoholic solvents. When softer nucleophiles are used, the rate ratios may become very much larger, even in alcohol solvents.[54,58] In dipolar aprotic solvents, the ratios become higher still.[35,36]

Figure 1 shows a plot of Swain and Scott's n values[34] plotted against some $n°_{CH_3I}$ values defined in eq 50. Since the n values are defined in an analogous way except that CH_3Br is the substrate and water is the solvent, we could call them $n°_{CH_3Br}$ values to be consistent. The slope of Figure 1 is 1.4. A slope greater than unity means that CH_3I reacts relatively faster with the better nucleophiles. These in turn are the softer nucleophiles, in general, so the prediction of a higher rate when both B′ and B are very soft is borne out. A plot of $n°_{CH_3I}$ values in methanol against $n°_{CH_3I}$ values in water gives a slope very nearly equal to 1, perhaps 1.05. The solvent correction is accordingly small.

A comparison of tosylate or sulfate, which are hard, and bromide or iodide, which are soft, as leaving groups also shows the symbiotic effect. For instance, sulfates and tosylates react with enols to give largely O alkylation[59]

$$RC{-}CR_2^- + R'OTs \longrightarrow RC{=}CR_2 + TsO^-$$

with structure showing O and R'O groups (61)

(58) B. O. Coniglio, *et al.*, *J. Chem. Soc., Org.*, 152 (1966).
(59) (a) G. J. Heiszwolf and H. Kloosterziel, *Chem. Commun.*, 51 (1966); (b) W. S. Johnson, *et al.*, *J. Am. Chem. Soc.*, **84**, 2181 (1962).

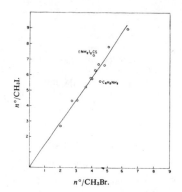

Figure 1. Relative reactivities toward CH_3I in methanol plotted against relative reactivities toward CH_3Br in water. Nucleophiles are (in ascending order) CH_3OH or H_2O, F^-, OAc^-, Cl^-, pyridine, aniline, N_3^-, Br^-, CH_3O^- or OH^-, SCN^-, CN^-, thiourea, I^-, and $S_2O_3^{2-}$.

but bromides and iodides react to give largely C alkylation.[60]

$$RC{=}CR_2^- + R'I \longrightarrow RCCR_2R' + I^- \qquad (62)$$

The rate ratio for alkyl tosylates and alkyl bromides is a function of the nucleophilic reagent used. If a hard nucleophile is used, the ratio, k_{OTs}/k_{Br}, is large. If a soft nucleophile is used, the ratio is small. Some representative data are shown.[61]

		k_{TsO}/k_{Br}
$C_2H_5X \xrightarrow{C_2H_5OH} C_2H_5OC_2H_5$	SN2	15
$CH_3X \xrightarrow{C_2H_5O^-} CH_3OC_2H_5$	SN2	5
$CH_3X \xrightarrow{RS^-} CH_3SR$	SN2	0.3
$n\text{-}C_4H_9X \xrightarrow{I^-} n\text{-}C_4H_9I$	SN2	0.3
$t\text{-}C_4H_9X \xrightarrow{CH_3CN} (CH_3)_2C{=}CH_2$	E1	2000
$C_6H_5CH_2CH_2X \xrightarrow{C_2H_5O^-} C_6H_5CH{=}CH_2$	E2	0.1

The low ratios for RS^- and I^- show the symbiotic stabilization of the transition state when the leaving group is bromide ion. The high ratios for C_2H_5OH and $C_2H_5O^-$ show the same phenomenon when the leaving group is tosylate. The last two examples show a remarkable difference between an E1 and an E2 elimination. The low ratio in the latter case must be a consequence of the polarizable double bond (a soft base) being involved in the transition state.

Acknowledgment. The authors are indebted to a number of persons, especially Professor J. A. Marshall, for helpful discussions. The work was supported by the U. S. Atomic Energy Commission under Grant No. At(11-1)-1087.

Appendix

While it is not possible to set up an absolute order of acid strengths, it is meaningful to say that one Lewis acid is stronger than another for virtually every refer-

(60) Reference 57b; D. Caine, *J. Org. Chem.*, **29**, 1868 (1964); G. Brieger and W. M. Pelletier, *Tetrahedron Letters*, 3555 (1965).
(61) H. M. R. Hoffman, *J. Chem. Soc.*, 6753, 6762 (1965).

Pearson, Songstad | Hard and Soft Acids and Bases

.ence base. This can be demonstrated by comparing the methyl carbonium ion with the proton. The heats of formation of H^+ (367 kcal) and of CH_3^+ (262 kcal) at 25° are needed.[62] We have the following gas phase reactions.

$$CH_3^+(g) + H_2(g) \longrightarrow CH_4(g) + H^+(g) \qquad (63)$$

$$\Delta H° = 87 \text{ kcal/mole at } 25°$$

$$CH_3^+(g) + HI(g) \longrightarrow CH_3I(g) + H^+(g) \qquad (64)$$

$$\Delta H° = 104 \text{ kcal/mole at } 25°$$

$$CH_3^+(g) + H_2O(g) \longrightarrow CH_3OH(g) + H^+(g) \qquad (65)$$

$$\Delta H° = 115 \text{ kcal/mole at } 25°$$

$$CH_3^+(g) + H_3O^+(g) \longrightarrow CH_3OH_2^+(g) + H^+(g) \qquad (66)$$

$$\Delta H° = 100 \text{ kcal/mole at } 25°$$

For the last equation we need the result that the proton affinity of methanol exceeds that of water by 15 kcal.[63]

(62) D. D. Wagman, *et al.*, Technical Notes 270-1 and 270-2, National Bureau of Standards, Washington, D. C., Oct 1965, May 1966.
(63) M. S. B. Munson, *J. Am. Chem. Soc.*, **87**, 2305 (1965).

$$CH_3OH(g) + H_3O^+(g) \longrightarrow CH_3OH_2^+(g) + H_2O(g) \qquad (67)$$

$$\Delta H° = -15 \text{ kcal/mole at } 25°$$

We have the result that H^+ is a stronger acid than CH_3^+ for all possible combinations of a strong, soft base (H^-), a weak, soft base (I^-), a strong, hard base (OH^-), and a weak, hard base (H_2O). These gas phase reactions are the best measure of intrinsic acid strength.

In passing, it may be noted that the effect of the methyl group, when replacing hydrogen, is to always increase the proton affinity of a base by about 15 kcal.[63] This is usually considered to be a base strengthening inductive effect, as in eq 67. It is more consistent to say that eq 67 is exothermic because H^+, which is strong, prefers OH^-, which is strong. The weaker CH_3^+ is then left with the weaker base H_2O. In the same way CH_3NH_2 is a stronger base than NH_3

$$CH_3NH_2 + NH_4^+ \longrightarrow CH_3NH_3^+ + NH_3 \qquad (68)$$

because H^+ binds NH_2^- strongly and the CH_3^+ is left with the weaker base NH_3. As mentioned earlier, the methyl group can be electron donating by hyperconjugation but appears to have an electron-withdrawing inductive effect.

[Reprinted from the Journal of Organic Chemistry, **32**, 2899 (1967).]
Copyright 1967 by the American Chemical Society and reprinted by permission of the copyright owner.

40

Symbiotic Effects in Nucleophilic Displacement Reactions on Carbon

Ralph G. Pearson and Jon Songstad[1]

Department of Chemistry, Northwestern University, Evanston, Illinois 60201

Received February 2, 1967

Inasmuch as the transition state for an S_N2 reaction may be regarded as an acid–base complex, there should be symbiotic effects on the rates of nucleophilic displacement reactions;[2] that is, a grouping of either several hard[3] bases or several soft[3] bases around the central carbon atom should stabilize the transition state and cause an increased rate of reaction. The bases include the entering and leaving groups, B and B′, as well as the three groups, L, in the trigonal plane of the activated complex

Such an effect has been pointed out by Bunnett,[4] although there has been some dispute.[5] If we concentrate on the entering and leaving groups to minimize the effects of steric hindrance, then there is no doubt that symbiosis is important and dominates the so-called leaving-group effect. It is necessary, of course, to examine rates in a fixed solvent. A comparison of the relative rates of reaction of alkyl tosylates with alkyl iodides provides the clearest example. The tosylate group is very hard, being an oxygen donor, whereas the iodide ion is quite soft. Consequently tosylates should react rapidly with hard bases and iodides should react rapidly with soft bases. Much data in the literature show that this prediction is verified when applied to bromides rather than iodides.[6]

Table I shows rate constants which we have obtained for reaction 1 in methanol at 25°. These data may

$$B + CH_3OTs \longrightarrow CH_3B + OTs^- \qquad (1)$$

be compared with the corresponding rate constants for reaction with CH_3I in the same solvent and at the

same temperature.[7] The ratios k_{TsO}/k_I strikingly confirm the prediction in that the ratio is large for hard nucleophiles, *e.g.*, OCH_3^-, CH_3OH, $(C_2H_5)_3N$, and Cl^-, and small for soft nucleophiles, *e.g.*, $C_6H_5S^-$, I^-, $SeCN^-$, and $(C_6H_5)_3P$. In the same way we find that the relative reactivities of RCl, RBr, and RI are functions of the softness of the nucleophile with which they react. The often-quoted[8] ratios of 1:50:100 are valid for alcohol, alkoxide ion, and amines in alcoholic solvents. When soft nucleophiles are used, the ratios become much higher in the same solvents.[9]

Table I

RATE CONSTANTS AT 25° IN METHANOL FOR THE REACTION
$$B + CH_3OTs \longrightarrow CH_3B + TsO^-$$

Nucleophile B	k, M^{-1} sec^{-1}	k_{TsO}/k_I
$C_6H_5S^-$	1.42×10^{-1}	0.13
$SeCN^-$	2.15×10^{-3}	0.23
CH_3O^-	$1.16 \times 10^{-3\,a}$	4.6
$CS(NH_2)_2$	5.60×10^{-4}	0.23
I^-	4.42×10^{-4}	0.13
$(C_6H_5)_3P$	2.3×10^{-4}	0.18
SCN^-	$1.59 \times 10^{-4\,b}$	0.28
$(C_2H_5)_3N$	5.65×10^{-4}	0.95
Br^-	5.75×10^{-5}	0.72
Cl^-	8.5×10^{-6}	2.8
CH_3OH	$2.50 \times 10^{-8\,c}$	210

a A value of 1.17×10^{-3} is given in E. R. Thornton, "Solvolysis Mechanisms," Ronald Press, New York, N. Y., 1964, p 164. *b* Rate constants for SCN^-, N_3^-, OH^-, and $S_2O_3^{2-}$ in water are given by R. E. Davis, *J. Am. Chem. Soc.*, **87**, 3011 (1965). *c* Calculated as a second-order reaction by dividing by 27. The first-order constant of 6.6×10^{-6} sec^{-1} agrees with that of R. E. Robertson, *Can. J. Chem.*, **31**, 589 (1953).

The symbiotic effect is useful in understanding the behavior of ambident nucleophiles as a function of the leaving group. Typically an ambident base, such as an enolate anion, has a softer nucleophilic center (the carbon atom) and a harder center (the oxygen atom). We expect alkyl sulfates and tosylates to react with enolate ions to give largely O alkylation,[10] whereas bromides and iodides react to give largely C alkylation.[11] It is also well known that the percentage of C alkylation *vs.* O alkylation rises steadily as one changes

(1) Chemistry Department, Bergen University, Norway. Supported by the Royal Norwegian Council for Scientific and Industrial Research.

(2) C. K. Jørgensen, *Inorg. Chem.*, **3**, 1201 (1964).

(3) R. G. Pearson, *J. Am. Chem. Soc.*, **85**, 3533 (1963); *Science*, **151**, 172 (1966).

(4) J. F. Bunnett, *ibid.*, **79**, 5970 (1957); J. F. Bunnett and J. D. Reinheimer, *ibid.*, **84**, 3284 (1962).

(5) A. J. Sisti and S. Lowell, *J. Org. Chem.*, **29**, 1635 (1964); K. C. Ho, ᵀ. Miller, and K. W. Wong, *J. Chem. Soc.*, 310B (1966).

(6) H. M. R. Hoffman, *ibid.*, 6753, 6762 (1965).

(7) R. G. Pearson, H. Sobel, and J. Songstad, submitted for publication.

(8) C. K. Ingold, "Structure and Mechanism in Organic Chemistry," Cornell University Press, Ithaca, N. Y., 1953, p 339; A. Streitwieser, Jr., *Chem. Rev.*, **56**, 601 (1956).

(9) R. G. Pearson and J. Songstad, *J. Am. Chem. Soc.*, **89**, 1827 (1967).

(10) G. J. Heiszwolf and H. Kloosterziel, *Chem. Commun.*, **No. 2**, 51 (1966); W. S. Johnson, *et al.*, *J. Am. Chem. Soc.*, **84**, 2181 (1962).

(11) D. Caine, *J. Org. Chem.*, **29**, 1868 (1964); G. Brieger and W. M. Pelletie, *Tetrahedron Letters*, No. 40, 3555 (1965).

from RCl to RBr to RI.[12] This is the order of increasing softness of the leaving group. In conclusion, it should be emphasized that explaining phenomena of organic chemistry by means of the principle of hard and soft acids and bases[3,9] does not necessarily invalidate other, more fundamental explanations. The principle is supposed to be correlative and phenomenological in nature, rather than theoretical.

Experimental Section

Methyl tosylate was made according to the literature[13] and was carefully purified by distillation and crystallization. Titration of the acid formed by complete hydrolysis indicated 100 ± 0.4% purity. Baker Analyzed methanol was refluxed for 5 hr with magnesium methoxide and distilled under nitrogen free from moisture and carbon dioxide. The first 15% of the distillate was discarded and the middle portion was stored under nitrogen for use. The purification of the nucleophiles will be described elsewhere.[7] The rates of reaction with bromide, iodide, and thiocyanate ions were determined by titration of liberated acid due to solvolysis and, simultaneously, following the concentration of these ions by silver nitrate titrations using eosin as indicator. In the case of chloride ions, potassium chromate was used as indicator after neutralizing the sample with NaHCO₃. The rate with SeCN⁻ was determined by potentiometric silver nitrate titration. The rate of solvolysis was followed by titrating the hydrogen ion produced with base. All titrations with base were carried out under nitrogen.

The reaction with triethylamine was studied in a system containing triethylammonium ion as well in order to minimize the reaction with hydroxide ion. The ratio of amine to its conjugate acid was in the range 5:1–10:1. The rate was found by titration of the remaining amine with standard acid. Reactions with triphenylphosphine, thiourea, and thiophenoxide were studied spectrophotometrically at wavelengths of 280, 250, and 290 mμ, respectively, using a Cary spectrophotometer. The optical

(12) N. Kornblum, R. E. Mickel, and R. C. Kerber, *J. Am. Chem. Soc.*, **88**, 5661 (1966).

(13) F. Muth in "Methoden der Organischen Chemie," Vol. 9, Houben-Weyl, Thieme, Stuttgart, 1955, p 659.

(14) A. A. Frost and R. G. Pearson, "Kinetics and Mechanism," John Wiley and Sons, Inc., New York, N. Y., 1961, p 165.

density at infinite time was found from the spectra of methyl-isothiuronium tosylate or methyltriphenylphosphonium tosylate for the first two cases. These spectra were the same as those of the kinetic runs after 15–20 half-lives. Since the solvolysis rate for methyl tosylate is quite large, it is necessary to correct for concurrent solvolysis. For equal starting concentrations a and b of ester and nucleophile, we have the following equation[14]

$$k_2 = \frac{k_1[a - (a - x)e^{k_1 t}]}{a(a - x)(e^{k_1 t} - 1)}$$

where k_2 refers to the nucleophile, k_1 refers to solvolysis, and x is the concentration of ester hydrolyzed.

In each run, six to eight determinations of both the nucleophile and hydrogen ion were made except for thiophenoxide, thiourea, and triphenylphosphine. In the case of the thiophenoxide ion, the starting concentration of the ion and the ester was about 5×10^{-3} M in the presence of 0.05 M sodium perchlorate. The rate constant for the thiophenoxide ion was calculated on the assumption that the reaction between thiophenol and methoxide ion goes to completion (30% excess of thiophenol was usually added). A small error is introduced in this way. This rate constant was further checked by titration with 0.01 M HCl.

Runs with thiourea and triphenylphosphine were performed with starting concentrations in the range from 2×10^{-2} to 4×10^{-2} M. Owing to the high extinction coefficients of these nucleophiles, aliquots had to be diluted for 25 to 50 times before measurements on the spectrophotometer could be performed. For the other ionic reagents, the starting concentrations were in the range 0.04–0.055 M. Ionic strength was made up to 0.05 with sodium perchlorate in solutions in which the concentration of nucleophile was less than 0.04 M. Good agreement with the theoretical rate equation was found for all cases up to 70% reaction except for chloride ion. For this slow reaction, an upward drift of k_2 was noted after some 20% reaction. The drift is small and its is estimated that the rate constant is good to 10% for this case, compared to 2% for the other nucleophiles. All systems were studied at least in duplicate.

Registry No.—Triethylamine, 121-44-8; triphenylphosphine, 603-35-0; thiourea, 62-56-6; thiophenoxide, 13133-62-5.

Acknowledgment.—This work was supported in part by the National Science Foundation under Grant GP6341-X.

41

Failure of the Principle of Hard and Soft Acids and Bases to Explain the Amount of Cyclization of Various Hex-5-enyl Derivatives during Acetolysis

By WALTER S. TRAHANOVSKY* and MICHAEL P. DOYLE

(*Department of Chemistry, Iowa State University, Ames, Iowa* 50010)

THE principle of hard and soft acids and bases (H.S.A.B. principle) has recently been applied to organic compounds.[1] One important type of organic reaction to which Pearson and Songstad apply the H.S.A.B. principle is the S_N2 reaction. These authors contend that if one considers the transition state for an S_N2 reaction as an acid–base complex, then symbiotic effects should operate and

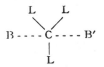

transition states with several soft or several hard ligands on the central carbon should be more stable than transition states with a mixture of soft and hard ligands. Thus, if B is an entering nucleophile and B′ the leaving group, faster rates are expected for S_N2 reactions when B and B′ are both soft or both hard.

Solvolysis of hex-5-enyl derivatives in acetic acid can be thought of as two competing substitution reactions: (a) nucleophilic attack by acetic acid which leads to hex-5-enyl acetate and (b) nucleophilic attack by the olefin which leads to the cyclic products, 1-methylcyclopentene, cyclohexene, and cyclohexyl acetate.[2] Since the olefin is clearly a soft base and acetic acid a hard base,[1] the H.S.A.B. principle predicts that hex-5-enyl iodide or bromide

should lead to *more* cyclization than hex-5-enyl *p*-nitrobenzenesulphonate or toluene-*p*-sulphonate. The results in the Table are diametrically opposed to this prediction. Hex-5-enyl *p*-nitrobenzenesulphonate and toluene-*p*-sulphonate give 41% and 30% cyclic materials, respectively after complete reaction whereas hex-5-enyl iodide and bromide give 20% and 15% cyclic materials respectively after solvolysis of all but 16% of the halides. Even if all of the remaining halide were converted into cyclic materials, which is highly unlikely, the yields of cyclic materials from the halides would not greatly exceed the yields of cyclic materials from the arenesulphonates.

Less cyclization with hex-5-enyl iodide or bromide is easily rationalized on the basis of the suggestion by DePuy and Bishop[3] confirmed experimentally by Hoffmann.[4] Hoffmann found that the ratio of rate constants for substitution reactions of toluene-*p*-sulphonates and bromides can vary from 0·36 to 5000 depending on the particular reaction. If the nucleophile is powerful and the substrate does not tend to ionize, then k_{OTs}/k_{Br} is small. Thus, less cyclization with hex-5-enyl iodide and bromide indicates that acetic acid is a stronger nucleophile than the olefin and therefore the transition state leading to direct displacement has less carbon-leaving group bond breaking than the transition state for cyclization.

Pearson and Songstad cited several collections of data to support the correlation of rates of S_N2

TABLE

Ratios of cyclic to open products from acetolysis of various hex-5-enyl derivatives at 100°[a]

Substrate	Reaction time (hr.)	Unchanged starting material, %[b]	% Cyclic products[b] % Open product
Hex-5-enyl Iodide	48	75	0·20[c]
	672	16	0·40
Hex-5-enyl Bromide	96	48	0·19[c]
	396	15	0·26
Hex-5-enyl p-nitrobenzenesulphonate[d]	50	0	0·83
Hex-5-enyl toluene-p-sulphonate[e]	72	0	0·51

[a] See ref. 2 for experimental details. [RX] = 0·1 M, [urea] = 0·2 M. [b] Actual yields of starting material and products were determined by gas-liquid partition chromatography. Recovery was 85—90% for all runs. [c] The (% Cyclic/% Open) product ratios steadily increased with extent of reaction. [d] Average of four runs. [e] Average of three runs.

reactions by the H.S.A.B. principle.[1] Most of these data have been explained in other ways by the original workers and usually contain exceptions when interpreted by the H.S.A.B. principle. Indeed, Pearson and Songstad reinterpreted Hoffman's data by the H.S.A.B. principle which can readily be done since Hoffmann's better nucleophiles were also the softer ones.

In conclusion, none of the data cited by Pearson and Songstad[1] rigorously support the H.S.A.B. principle when applied to S_N2 reactions and failure of this principle to predict the present results clearly indicates its inadequacies and the need for more detailed considerations of S_N2 reactions.

One of us (M. P. D.) thanks the U.S. Public Health Service for a Fellowship.

(*Received, August 14th, 1967*; Com. 858.)

[1] R. G. Pearson and J. Songstad, *J. Amer. Chem. Soc.*, 9167, **89**, 1827.
[2] W. S. Trahanovsky and M. P. Doyle, *J. Amer. Chem. Soc.*, 1967, in the press.
[3] C. H. DePuy and C. A. Bishop, *J. Amer. Chem. Soc.*, 1960, **82**, 2532.
[4] H. M. R. Hoffmann, *J. Chem. Soc.*, 1965, 6752, 6753.

42

Nucleophilic Reactivity Constants toward Methyl Iodide and *trans*-[Pt(py)₂Cl₂]

Ralph G. Pearson, Harold Sobel,[1] **and Jon Songstad**[2]

Contribution from the Department of Chemistry, Northwestern University, Evanston, Illinois. Received August 11, 1967

Abstract: The rates of reaction of a number of nucleophiles with methyl iodide and *trans*-[Pt(py)₂Cl₂] have been measured in methyl alcohol at 25°. Relative nucleophilic reactivity parameters, n_{CH_3I} and n_{Pt}, have been calculated. It was not found possible to correlate these numbers with each other or with other extra-kinetic data. Equations in the literature for predicting nucleophilic reactivity have only a limited range of usefulness.

The subject of nucleophilic reactivity continues to be of great interest. A number of recent reviews and discussions have appeared.[3] There are also several semiempirical equations which attempt to predict and

(1) Predoctoral fellow, National Institutes of Health, 1966–1968.
(2) Supported by the Royal Norwegian Council for Scientific and Industrial Research.

(3) J. O. Edwards and R. G. Pearson, *J. Am. Chem. Soc.*, **84**, 16 (1962); R. F. Hudson, *Chimia* (Aarau), **16**, 173 (1962); J. F. Bunnett, *Ann. Rev. Phys. Chem.*, **14**, 271 (1963); J. Miller, *J. Am. Chem. Soc.*, **85**, 1628 (1963).

correlate rate data of reactions presumed to occur by bimolecular nucleophilic displacement (SN2) mechanisms.[4] In particular, it has been of interest to try to relate rate constants to extra-kinetic properties such as redox potentials.[5]

A practical difficulty has been the lack of extensive data for a large number of nucleophiles reacting with

(4) These equations are reviewed by K. M. Ibne-Rasa, *J. Chem. Educ.*, **44**, 89 (1967).
(5) J. O. Edwards, *J. Am. Chem. Soc.*, **76**, 1540 (1954).

several substrates in a fixed environment. A limited amount of data is available for reactions of methyl bromide and several other organic substrates in water.[6] In this work we report the second-order rate constants for a large number of nucleophiles reacting with methyl iodide and *trans*-[Pt(py)$_2$Cl$_2$] in methanol at 25°.

$$CH_3I + Y^- \xrightarrow{k_2} CH_3Y + I^- \tag{1}$$

$$Pt(py)_2Cl_2 + Y^- \xrightarrow{k_2} Pt(py)_2ClY + Cl^-$$

$$Pt(py)_2ClY + Y^- \xrightarrow{fast} Pt(py)_2Y_2 + Cl^- \tag{2}$$

The solvent was selected because of the solubility characteristics of the substrates and the nucleophiles. In addition a great deal of data has already been reported in this solvent for the platinum complex.[7] It is well known that the solvent plays a major role, not only in the magnitude of rate constants, but also in orders of nucleophilic reactivity.[8] Nevertheless, thanks to the work of Parker[5] and others, the probable effect of various solvents on nucleophiles of different types can be estimated. This means that rate data in other solvents can often be converted to approximate rate data in methanol for comparison.

Methyl iodide is selected as an example of a moderately soft electrophilic center and Pt(py)$_2$Cl$_2$ as an example of a very soft electrophilic center.[9] The characteristic of a soft electrophilic center is that it reacts rapidly with highly polarizable, or soft, nucleophiles. Proton basicity of the nucleophile is usually not important. Proton transfers are examples of reactions of hard electrophilic centers. In this case the best nucleophiles are hard bases such as OH$^-$ and F$^-$. Soft bases will be effective only if they are also strong bases toward the proton, such as S^{2-}. For proton transfers, a Brønsted relationship usually exists between the rate constants and the pK_a of a series of bases.[10]

The pK_a values of a number of acids are known in methanol.[11] Again it is possible to roughly estimate the pK_a in methanol from a known pK_a in water. Charge on the conjugate acid is very important. Thus

$$HB \rightleftharpoons B^- + H^+ \tag{3}$$

$$HB^+ \rightleftharpoons B + H^+ \tag{4}$$

The pK_a for reaction 3 usually increases by about four units in going from water to methanol, whereas reaction 4 is much less affected. In other words, it is more difficult to form ions from neutral molecules in methanol than in water.

The data obtained in this paper and other data from the literature are converted into nucleophilic reactivity constants, n_{CH_3I} and n_{Pt}, which are defined as logarithms of the ratios of the second-order rate constants of the nucleophile divided by the second-order rate constant for solvolysis in methanol.[6,7]

Thus, for a nucleophile Y

$$n_{Pt} = \log (k_Y/k_{CH_3OH})$$

$$n_{CH_3I} = \log (k_Y/k_{CH_3OH})$$

These n_{Pt} values differ from those in ref 7 in that the first-order rate constant for solvolysis has now been divided by 26 moles/l. to obtain k_S. This agrees with the recent definition of $n^0{}_{Pt}$ by Belluco, *et al.*[12] However, since Swain and Scott used this same convention in their original work[6] defining n_{CH_3Br}, we have dropped the superscript zero. The n values calculated in this work are at 25°, whereas earlier[7] n_{Pt} values are at 30°. We will assume that the change with temperature is small over this range.

Experimental Section

Preparation and Purification of Reagents. Baker Analyzed methyl iodide was washed with potassium carbonate solution and then several times with water. The methyl iodide was then carefully dried over magnesium sulfate and distilled from silver wool. The purified compound was stored over silver wool under nitrogen in a flask that was painted black. This purification procedure was repeated approximately once a month.

trans-[Pt(py)$_2$Cl$_2$] was prepared according to the method of Kauffman.[13]

Baker Analyzed methanol was refluxed for at least 5 hr with magnesium methoxide and distilled under nitrogen free from moisture and carbon dioxide. The first 15% of distillate was discarded. The purified product was stored under nitrogen and used for the methyl iodide experiments. Oxygen-free water was added to the methanol when inorganic salts were insufficiently soluble in pure methanol. The purified methanol was also used for those reactions of the platinum complex with oxygen-sensitive or water-sensitive nucleophiles, but for most reactions of this complex, Baker Analyzed methanol was used without further purification.

Inorganic salts and thiourea were recrystallized from water and dried *in vacuo* if not of analytical grade.

Amines of the highest purity available were purified by recrystallizing their quaternized salts. The amines were then distilled twice from sodium hydroxide pellets through a 12-plate column with controlled temperature. All other liquid nucleophiles were purified in this way; a central cut of distillate was used for the kinetic measurements.

Anhydrous hydroxylamine was made according to the method of Bissot, *et al.*[14] Anhydrous hydrazine was prepared according to the method of Beckman;[15] the final product contained about 1% ammonia and less than 0.5% water, as determined by titration with HCl and KIO$_3$. Anhydrous ammonia was used without further purification, as was cylinder carbon monoxide. Imidazole was Eastman product frozen out from water–alcohol mixtures twice and then recrystallized from benzene.

The nucleophile ((C$_2$H$_5$)$_2$N)$_3$P was prepared according to the method of Burgada.[16] The distilled product was washed with ice-cold water, dissolved in pentane, dried, and finally distilled twice, bp 62–63° (2 mm). Because of its rapid reaction with methanol, this compound was studied kinetically in acetonitrile solution. Several other neutral nucleophiles were studied in both methanol

(6) C. S. Swain and C. B. Scott, *J. Am. Chem. Soc.*, **75**, 141 (1953).
(7) U. Belluco, L. Cattalini, F. Basolo, R. G. Pearson, and A. Turco, *ibid.*, **87**, 241 (1965).
(8) (a) A. J. Parker, *Quart. Rev.* (London), **14**, 163 (1962); (b) A. J. Parker, *et al.*, *J. Am. Chem. Soc.*, **88**, 1911 (1966); (c) *J. Chem. Soc., Sect. B*, 152 (1962).
(9) R. G. Pearson, *J. Am. Chem. Soc.*, **85**, 3533 (1963); *Science*, **151**, 172 (1966).
(10) However, negative deviations will occur for soft bases compared to hard bases of the same pK_a: M. Eigen, *Angew. Chem. Intern. Ed. Engl.*, **3**, 1 (1964).
(11) See ref 8b and E. Grunwald and E. Price, *J. Am. Chem. Soc.*, **86**, 4517 (1964), for data and earlier references.

(12) U. Belluco, M. Martelli, and A. Orio, *Inorg. Chem.*, **5**, 582 (1966).
(13) G. G. Kauffman, *Inorg. Syn.*, **7**, 251 (1963).
(14) T. C. Bissot, *et al.*, *J. Am. Chem. Soc.*, **79**, 796 (1957).
(15) R. R. Wenner and A. O. Beckman, *ibid.*, **54**, 2787 (1932).
(16) R. Burgada, *Ann. Chim.*, **8**, 347 (1963).

and acetonitrile solutions to set up reactivity ratios for interconversion of data. The conductivity method was always used in acetonitrile.

Alkylarsines and -phosphines were made from the appropriate Grignard reagents and phosphorus or arsenic trichloride. After the HCl adducts in 6 M HCl were washed several times with ether to remove by-products, the compounds were finally liberated with sodium hydroxide pellets, distilled under nitrogen or, if necessary, *in vacuo*. Commercial triphenylphosphine and triphenylarsine were recrystallized twice from methanol.

The sulfides were commercial products distilled over metallic sodium in nitrogen. The selenides were made by the method of Bird and Challenger;[17] dimethyl selenide was distilled under nitrogen after being dried over sodium. Dibenzyl selenide was recrystallized twice from lukewarm absolute methanol prior to use.

Baker Analyzed phenol was used without further purification. Eastman thiophenol and selenophenol were distilled in a nitrogen atmosphere prior to use. Trimethyl phosphite in ether solution was treated with sodium metal, filtered, and then distilled. Triphenylsilanol was recrystallized twice from benzene.

Tetraethylammonium trichlorostannite was prepared according to the method of Parshall[18] by adding together equimolar quantities of tetraethylammonium chloride and stannous chloride in 0.5 M hydrochloric acid solution. The compound was recrystallized twice from absolute ethanol and washed with cold absolute ethanol, in which the product is nearly insoluble. All operations were performed under nitrogen. The concentration of tin(II) was found to be from 97 to 98% of the expected concentration by titration with thiosulfate after the addition of excess potassium triiodide. This salt was used in the methyl iodide experiments. For the experiments with the platinum complex, where excess chloride ions would not complicate the kinetics, the trichlorostannite ion was prepared by adding concentrated hydrochloric acid to methanol solutions of stannous chloride.

Kinetics. All rate constant determinations were performed at 25°. The rates of many of the reactions of methyl iodide and of the platinum complex with neutral nucleophiles were measured by following the change in electrical conductivity at 1000 cps with an Industrial Instruments Model RC16B1 conductivity meter. The electrodes of the conductivity cells were of platinum; those cells used for the reactions of methyl iodide were painted black to avoid photoinitiated side reactions. For the experiments with methyl iodide, the conductometric method was used only for reactions under pseudo-first-order conditions; the concentrations of nucleophiles ranged from 0.04 to 0.15 M and were at least 80 times greater than the concentration of methyl iodide. For each nucleophile, at least three different concentrations were used. The measurements were performed under nitrogen when necessary.

In the experiments with the Pt(II) complex, the concentration of the platinum compound was always 5×10^{-5} M and the concentration of nucleophile ranged from 1×10^{-3} to 1×10^{-1} M, depending upon reactivity, except for the case of the phosphines, where the concentration of the platinum complex was 2.5×10^{-5} M and the phosphine concentration was 5×10^{-5} M. The rates of reaction of *trans*-[Pt(py)$_2$Cl$_2$] with all of the ionic nucleophiles and with some of the neutral nucleophiles were measured by following the change with time of optical density at some selected wavelength in the ultraviolet region by means of a Cary 14 spectrophotometer.

Second-order conditions were used for those reactions of methyl iodide that were too slow for pseudo-first-order conditions, and for reactions with ionic nucleophiles. The rate constants for nonionic nucleophiles obtained from pseudo-first-order kinetic runs were checked by second-order kinetics, and the rate constants were calculated from the first 5 to 10% reacted. The discrepancy was never greater than 3%. The concentrations of nucleophiles and methyl iodide were in the range of 0.02–0.08 M, depending upon the reactivity. After appropriate time intervals, 5- or 10-ml aliquots were withdrawn from the reaction vessel and added to 30 ml of water and 15 ml of benzene, ice-cold if necessary, in a separatory funnel. The organic layer was washed repeatedly with water, and the combined water extracts were finally washed with low-boiling petroleum ether to remove traces of benzene, which was found to make the end point in the following titration diffuse.

The determination of iodide ion was performed with silver nitrate and eosin indicator. The accuracy of this method was better than 0.2% with 0.05 M silver nitrate titrant and was approximately 0.5%

for 0.01 M silver nitrate. The loss of iodide ions during the removal of unreacted methyl iodide was found to be less than 0.5% and was therefore neglected. In each run, five to nine determinations of iodide ion were performed. The second-order rate constants were calculated from the first 70% reacted.

The rate constants for hydroxylamine, ammonia, and hydrazine were calculated from the first 10 to 20% reacted in the presence of 10 to 20% of the corresponding hydrochlorides in a concentration of 0.05 M. In the case of hydrazine a very small amount of water, 1% by volume, was added because of the low solubility of hydrazine hydrochloride in methanol.

The reaction between cyanide ion and methyl iodide was followed by determining cyanide ion by the Liebig–Deniges method.[19] Thiocyanate ions in the presence of iodide ions were determined by potentiometric titration with silver nitrate, using a Radiometer titration assembly with a Type PK499 mercury–mercurous sulfate electrode.

The reaction between sulfite ion and methyl iodide was followed by adding aliquots of the reaction mixture to excess iodine solution and titrating back with thiosulfate. In all runs sodium methoxide was added in concentrations exceeding that of the sulfite ion. All operations were performed under nitrogen. The reaction with thiosulfate was followed in the same way. Both reactions were run in an 80% methanol–20% water mixture (v/v) because of the low solubility of these salts in methanol.

The reaction between trimethyl phosphite and methyl iodide was followed by measuring the formation of the product, dimethyl methylphosphonate, by its phosphoryl absorbance[20] using a Perkin-Elmer Model 337 infrared spectrophotometer and barium fluoride cells. Because of difficulties in getting a stable base line, this value is not very accurate. Extreme care was taken during these runs to exclude moisture. Rates are initial values only.

The reaction between tetraethylammonium trichlorostannite and methyl iodide in the presence of HCl was followed by determination of liberated iodide ions. As the rate of formation of iodide ions was only a little higher than that from pure solvolysis, the rate constants will necessarily be rather inaccurate.

The rates of formation of chloride ions from *trans*-[Pt(py)$_2$Cl$_2$] and of iodide ions from methyl iodide from the reaction of methanol saturated with carbon monoxide were found to be similar to the rate of pure solvolysis. (The concentration of carbon monoxide in a saturated methanolic solution at 25° is 8.0×10^{-3} M.[21]) Allyl alcohol gave a similar result. The same result was obtained for the reaction of methyl iodide with triphenylstibine and for the reaction of Pt(py)$_2$Cl$_2$ with 0.04 M fluoride ion.

Solutions of the sodium salts of dimethyl phosphite, triphenylsilanol, and phenol were made by adding a 30 to 100% excess of the acids to solutions of sodium methoxide of known molarity. Solu-

Table I

Product		Calcd, %	Found, %
[Pt(py)$_2$(P(C$_6$H$_5$)$_3$)$_2$](OTs)$_2$	Pt	16.0	15.6
	C	59.0	58.0
	H	4.45	4.51
	N	2.29	2.19
[Pt(py)$_2$(S(benz)$_2$)$_2$](OTs)$_2$	Pt	17.3	17.7
	C	55.4	54.2
	H	4.65	4.71
	N	2.49	2.48
[Pt(py)$_2$(Se(benz)$_2$)$_2$](OTs)$_2$	Pt	16.0	16.4
	C	51.3	49.8
	H	4.30	4.19
	N	2.30	2.53
[Pt(py)$_2$(imid)$_2$]Cl$_2$	C	34.30	33.3
	H	3.24	3.32
	N	14.99	14.9
[Pt(py)$_2$(As(C$_6$H$_5$)$_3$)$_2$](OTs)$_2$	Pt	14.9	14.9
	C	55.1	53.4
	H	4.16	4.04
	N	2.14	1.97

(17) M. L. Bird and F. Challenger, *J. Chem. Soc.*, 571 (1942).
(18) G. W. Parshall, private communication.

(19) I. M. Kolthoff and E. B. Sandell, "Textbook of Quantitative Analysis," 3rd ed, The Macmillan Co., New York, N. Y., 1952, p 546.
(20) G. Aksnes and D. Aksnes, *Acta Chem. Scand.*, **18**, 38 (1964).
(21) G. Just, *Z. Physik. Chem.*, **37**, 342 (1901).

Pearson, Sobel, Songstad / *Nucleophilic Reactivity toward CH$_3$I and trans-[Pt(py)$_2$Cl$_2$]*

Table II. Rate Constants in Methanol at 25°

	Nucleophile	10^3k_2, M^{-1} sec^{-1} trans-[Pt(py)$_2$Cl$_2$]	CH$_3$I		Nucleophile	10^3k_2, M^{-1} sec^{-1} trans-[Pt(py)$_2$Cl$_2$]	CH$_3$I
1.	CH$_3$OH	0.00027	0.00000013[b]	30.	SeCN$^-$	5,150[a]	9.13[k]
2.	CH$_3$O$^-$	Very slow[a]	0.251[c]	31.	C$_6$H$_5$S$^-$	6,000[a]	1070
3.	F$^-$	Very slow	0.00005[d]	32.	SC(NH$_2$)$_2$	6,000[a]	2.41
4.	Cl$^-$	0.45[a]	0.0030[e]	33.	S$_2$O$_3$$^{2-}$	9,000[a]	114
5.	NH$_3$	0.47[a]	0.041	34.	(C$_2$H$_5$)$_3$As	14,100	1.03
6.	C$_5$H$_5$N	0.55[a]	0.022[f]	35.	(C$_6$H$_5$)$_3$P	249,000	1.29
7.	NO$_2$$^-$	0.68[a]	0.029	36.	(C$_4$H$_9$)$_3$P	272,000	63
8.	(C$_6$H$_5$CH$_2$)$_2$S	0.80	0.009	37.	(C$_2$H$_5$)$_3$P	290,000	66
9.	N$_3$$^-$	1.55[a]	0.078[e]	38.	CH$_3$COO$^-$	Very slow[o]	0.0027[l]
10.	NH$_2$OH	2.9[a]	0.50	39.	C$_6$H$_5$COO$^-$	...	0.002[l]
11.	H$_2$NNH$_2$	2.93[a]	0.51	40.	C$_6$H$_5$O$^-$	...	0.073[l]
12.	C$_6$H$_5$SH	5.7[a]	0.064	41.	CO	Very slow	Very slow
13.	Br$^-$	3.7[a]	0.0798[g]	42.	C$_6$H$_5$NH$_2$	0.43	0.052
14.	(C$_2$H$_5$)$_2$S	9.85	0.0282[h]	43.	C$_6$H$_5$N(CH$_3$)$_2$	...	0.0562[m]
15.	(CH$_3$)$_2$S	21.9	0.0452	44.	2,6-Dimethylpyridine	...	0.00042[f]
16.	(CH$_3$O)$_2$PO$^-$	30.4	1.28	45.	α-Picoline	0.05	0.0065[f]
17.	(CH$_2$)$_5$S	30.7	0.0342	46.	Pyrrolidine	...	2.2
18.	(CH$_2$)$_4$S	40.9	0.0587	47.	Piperidine	0.38	2.6
19.	SnCl$_3$$^-$	70	0.0009	48.	N,N-Dimethyl- cyclohexylamine	...	0.692[n]
20.	(C$_6$H$_5$CH$_2$)$_2$Se	101	0.022[i]	49.	(C$_2$H$_5$)$_2$NH	...	1.2[p]
21.	I$^-$	107[a]	3.42[j]	50.	(C$_2$H$_5$)$_3$N	Very slow	0.595[f]
22.	(CH$_3$)$_2$Se	148	0.268	51.	C$_6$H$_5$Se$^-$	...	7000
23.	SCN$^-$	180[a]	0.574	52.	(C$_6$H$_5$)$_3$SiO$^-$	...	0.19
24.	SO$_3$$^{2-}$	250[a]	44.5	53.	C$_6$H$_5$SO$_2$NCl$^-$	...	8.22[q]
25.	C$_6$H$_{11}$NC	640	Very slow	54.	C$_6$H$_5$SO$_2$NH$^-$	...	0.18[q]
26.	(C$_6$H$_5$)$_3$Sb	1,810	Very slow	55.	Phthalimide anion	...	0.37[q]
27.	(C$_6$H$_5$)$_3$As	2,320	0.0075	56.	[(C$_2$H$_5$)$_2$N]$_3$P	10[r]	45.0[r]
28.	CN$^-$	4,000	0.645	57.	Imidazole	0.74	0.012
29.	(CH$_3$O)$_3$P	4,890	0.02	58.	Allyl alcohol	Very slow	Very slow

[a] U. Belluco, L. Cattalini, F. Basolo, R. G. Pearson, and A. Turco, *J. Am. Chem. Soc.*, **87**, 241 (1965). These data are at 30°. [b] The first-order rate constant for solvolysis is 3.3×10^{-9}. A. J. Parker, *J. Chem. Soc.*, 1328 (1961), gives the value 4×10^{-9}, calculated from rate at 59.8°. [c] I. K. Alet and B. D. England, *J. Chem. Soc.*, 5259 (1961), give the value 2.55×10^{-4} sec^{-1}. [d] This value is calculated by using a relation between reactivities in methanol and water given by R. H. Bathgate and E. A. Moelwyn-Hughes, *ibid.*, 2642 (1959). [e] A J. Parker, *ibid.*, 1328 (1961) (25.1°). [f] N. Tokura and Y. Kondo, *Bull. Chem. Soc. Japan*, **37**, 133 (1964), give data from which the value 2.13×10^{-5} was calculated for pyridine. The values in the table for triethylamine, α-picoline, and 2,6-dimethylpyridine are from this same reference. [g] E. A. Moelwyn-Hughes, *J. Chem. Soc.*, 838 (1952). [h] The value 2.97×10^{-5} was calculated from the data of Y. K. Syrkin and I. T. Gladishev, *Acta Physicochim. USSR*, **2**, 291 (1935). [i] Since the product (C$_6$H$_5$CH$_2$)$_2$SeCH$_3$$^+$ is not stable, the rate constant was first estimated from the first 5% reacted in a pseudo-first-order run, assuming that the equivalent conductance for this ion is the same as that for the corresponding sulfonium ion. Then the rate constant was calculated from the first 5% reacted in a second-order run by titration of the liberated iodide ions. The agreement between the two runs was 3%. [j] P. Beronius, *Acta Chem. Scand.*, **14**, 1151 (1961). [k] A. J. Parker, *J. Chem. Soc.*, 4398 (1961). [l] D. Cook, I. P. Evans, E. C. E. Ko, and A. J. Parker, *ibid.*, *Sect. B*, 404 (1966). [m] D. P. Evans, *ibid.*, 422 (1944). This reference gives rate constants for other anilines. [n] E. R. A. Peeling and B. D. Stone, *Chem. Ind.* (London), 1625 (1959). [o] S. P. Tanner, F. Basolo, and R. G. Pearson, *Inorg. Chem.*, **6**, 1089 (1967). [p] From results in acetonitrile and a rate ratio $k_{CH_3CN}/k_{CH_3OH} = 122$ found for piperidine. [q] J. H. Beale, Ph.D. Thesis, Brown University, 1966. [r] From acetonitrile data and rate ratios of $k_{CH_3CN}/k_{CH_3OH} = 4.3$ found for (C$_6$H$_5$)$_3$P reacting with CH$_3$I, and $k_{CH_3CN}/k_{CH_3OH} = 2$ for pyridine reacting with Pt(py)$_2$Cl$_2$.

tions of sodium thiophenoxide and selenophenoxide were made similarly. The reaction rates of these two nucleophiles were determined by ultraviolet spectrophotometry. Cell compartments and solutions were carefully flushed with nitrogen. From more concentrated runs the desired sulfide and selenide were obtained in greater than 90% yield.

The reaction of methyl iodide with sodium triphenylsilanolate was followed by titrating the liberated iodide ions. The expected product ((C$_6$H$_5$)$_3$SiOCH$_3$) was obtained in greater than 80% yield, showing that the formation of [(C$_6$H$_5$)$_3$Si]$_2$O can be neglected. This is consistent with the findings of Grubb, who found the formation of [(C$_6$H$_5$)$_3$Si]$_2$O to be a slow reaction in methanol.[22]

The reactions of the amines with Pt(py)$_2$Cl$_2$ are complicated by some formation of *trans*-Pt(py)$_2$(OCH$_3$)$_2$. This causes an error in rates determined by the conductometric method. Accordingly rates for pyridine, piperidine, and ammonia were also checked by ultraviolet in the 320–340-mμ region. The compounds *trans*-[Pt(py)$_2$(pip)$_2$]Cl$_2$, *trans*-[Pt(py)$_2$(NH$_3$)$_2$]Cl$_2$, and [Pt(py)$_4$]Cl$_2$ were synthesized and their spectra checked in the region of interest. The species [Pt(py)$_2$(OCH$_3$)$_2$] (*trans*?) was made from *trans*-[Pt(py)$_2$I$_2$] and NaOCH$_3$, but not isolated. Its spectrum showed ϵ 4500 at 290 mμ, and ϵ 3920 at 300 mμ. The reactions of CH$_3$I in buffered solu-

tions of amines are not much affected because the amine rate is much greater than the solvolysis rate, unlike the platinum case.

In a number of other cases the expected products of reaction of the platinum complex with the nucleophiles were synthesized by larger scale runs in methanol using equivalent concentrations. Silver tosylate was added to precipitate silver chloride. After filtration, the filtrate was evaporated to dryness. The product was washed with water and with ether and then recrystallized from ether–methanol. Analyses are given in Table I.

In the case of ((C$_2$H$_5$)$_2$N)$_3$P reacting with [Pt(py)$_2$Cl$_2$], a crystalline product was not obtained. A glassy, hygroscopic material was formed whose infrared spectrum suggested that it was the expected [Pt(py)$_2$(P(NR$_2$)$_3$)$_2$]$^{2+}$. In the case of CH$_3$I the product was assumed to be ((C$_2$H$_5$)$_2$N)$_3$PCH$_3$$^+$ by analogy with other reactions of alkyl halides.[23]

The products in the case of SnCl$_3$$^-$ as a nucleophile were not isolable. From the literature they were expected to be CH$_3$SnCl$_3$[24] and *trans*-[Pt(py)$_2$(SnCl$_3$)$_2$].[25]

(22) W. T. Grubb, *J. Am. Chem. Soc.*, **76**, 3408 (1954).

(23) H. Nöth and H. J. Vetter, *Ber.*, **94**, 1505 (1961).
(24) A. Tschakirian, *et al.*, *Compt. Rend.*, **202**, 138 (1936); R. W. Leeper, *Iowa State Coll. J. Sci.*, **18**, 57 (1943).
(25) R. D. Cramer, *et al.*, *J. Am. Chem. Soc.*, **85**, 1691 (1963); A. G. Davies, *et al.*, *ibid.*, **85**, 1692 (1963).

Results

Whereas the rates of reaction of methyl iodide usually follow the simple second-order rate law

$$\text{rate} = k_2[\text{CH}_3\text{I}][\text{Y}] \qquad (5)$$

the rates for *trans*-[Pt(py)$_2$Cl$_2$] follow the two-term rate law that is well known for substitution reactions of platinum(II) complexes.[26]

$$\text{rate} = k_1[\text{complex}] + k_2[\text{complex}][\text{Y}] \qquad (6)$$

The rate constant k_1 is for the solvent path, and k_2 is for the direct reagent path. For the substitution reactions with *trans*-[Pt(py)$_2$Cl$_2$], an excess of nucleophile was used, and the pseudo-first-order rate constants k_{obsd} were determined. The experimental k_{obsd} is related to k_1 and k_2 by

$$k_{\text{obsd}} = k_1 + k_2[\text{Y}] \qquad (7)$$

For each reaction, the values of k_{obsd} were determined at five or more different concentrations of nucleophile, Y. Linear plots of k_{obsd} *vs.* Y were obtained; the extrapolated value of k_{obsd} at zero Y is k_1, and the slope is k_2. These values are given in Table II, along with the values of k_2 for the reactions of methyl iodide. A number of values from the literature are also included. Some of these are only estimated, but they are so marked. Table III gives values of n_{Pt} and $n_{\text{CH}_3\text{I}}$ along with values of E_n, pK_a in H$_2$O, and $pK_{\text{CH}_3\text{Hg}^+}$. The latter quantity refers to the equilibrium, in aqueous solution at 25°

$$\text{CH}_3\text{HgY} \rightleftharpoons \text{CH}_3\text{Hg}^+ + \text{Y}^- \qquad (8)$$

Since the methylmercury cation is a very soft Lewis acid, it might be a more suitable reference than the proton, which is hard.

Discussion

Figure 1 shows that there is little correlation between n_{Pt} and $n_{\text{CH}_3\text{I}}$. This is not too surprising since Pt(II) is a much softer electrophilic center than is carbon in methyl iodide. It can be seen that, in general, the most polarizable reagents lie above the line in Figure 1; that is, they react relatively faster with Pt(II). The straight line in Figure 1 has no significance except that of dividing the nucleophiles into roughly two groups. The least polarizable reagents then lie below the line, reacting relatively slower with Pt(II). Above the line are found bases in which the donor atom is P, As, Se, C, and Sn. Below the line are O and N donor atom bases, as well as F$^-$.

Surprisingly, I$^-$ is below the line. However, there is a marked tendency for anionic bases to lie below the line in general. Presumably this is explained by postulating that, in the transition state, the carbon atom of CH$_3$I has developed a greater positive charge than has platinum in Pt(py)$_2$Cl$_2$. This is not unreasonable in view of the probable mechanisms of the two substitution

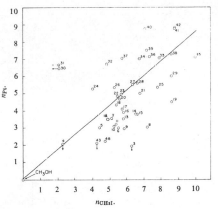

Figure 1. Plot of n_{Pt} against $n_{\text{CH}_3\text{I}}$. The numbers refer to the nucleophiles as listed in Table III.

reactions. That is, for CH$_3$I there is substantial bond breaking in the transition state, whereas for Pt(II) there is very little bond breaking, a five-coordinated intermediate being formed.[26,27]

The mechanism for Pt(II) substitution resembles more closely that for aromatic substitution, rather than aliphatic substitution. Miller[28] has developed a theory for predicting rates of aromatic substitutions. A typical reactivity sequence for nucleophiles in aromatic substitution is found experimentally and theoretically to be C$_6$H$_5$S$^-$ $\gg$ CH$_3$O$^-$ > N$_3^-$ $\gg$ SCN$^-$ > I$^-$ > Br$^-$ > Cl$^-$ > F$^-$. This is very poor with respect to CH$_3$O$^-$ and N$_3^-$ as far as Pt(II) substitution is concerned.

Unfortunately, it is rather difficult to use Miller's method to predict reactivities for other nucleophiles since bond energies, ionization potentials, and solvation energies are needed, as well as reorganization energies on forming the transition state. The theory is quite similar to that of Hudson,[29] who uses the same properties but in a more flexible way (more parameters that can be varied). We have made a serious effort to estimate $n_{\text{CH}_3\text{I}}$ values using Hudson's method for about 15 of the nucleophiles listed in Table III, where bond energies, ionization potentials, and solvation energies were available or could be estimated. Even though three entirely adjustable parameters were allowed, the results were extremely poor. Lack of bond-energy data prevented the application of Hudson's method to the calculation of n_{Pt}, but there is little reason to believe that it would be more successful.

It can be seen from Figure 2 that there is no correlation between n_{Pt} and the strength of the nucleophile as a proton base. The figures show pK_a in water, but using the known or estimated pK_a values in methanol would do little to improve the correlation. As explained

(26) C. H. Langford and H. B. Gray, "Ligand Substitution Processes," W. A. Benjamin, Inc., New York, N. Y., 1965, Chapter 2.

(27) See F. Basolo and R. G. Pearson, "Mechanisms of Inorganic Reactions," 2nd ed, John Wiley and Sons, Inc., New York, N. Y., 1967, Chapter 5.

(28) J. Miller, *J. Am. Chem. Soc.*, **85**, 1628 (1963); D. L. Hill, K. C. Ho, and J. Miller, *J. Chem. Soc.*, Sect. B, 299 (1966).

(29) R. F. Hudson, *Chimia* (Aarau), **16**, 173 (1962); "Structure and Mechanism in Organo-Phosphorus Chemistry," Academic Press Inc., New York, N. Y., 1965, Chapter 4.

Table III. Properties of Various Nucleophiles

	Nucleophile	n_{Pt}	n_{CH_3I}	$E_n{}^a$	pK_a	$pK_{CH_3Hg^+}$[h]
1.	CH₃OH	0.00	0.00	0	-1.7[a]	-1.74
2.	CO	<2.0	<2.0			
3.	CH₃O⁻	<2.4	6.29	1.65	15.7[a]	9.42
4.	F⁻	<2.2	~2.7	-0.27	3.45[b]	1.50
5.	Cl⁻	3.04	4.37	1.24	(-5.7)[a]	5.25
6.	NH₃	3.07	5.50	1.36	9.25[b]	7.60
7.	Imidazole	3.44	4.97		7.10[c]	
8.	Piperidine	3.13	7.30		11.21[c]	
9.	Aniline	3.16	5.70	(1.78)	4.58[c]	
10.	Pyridine	3.19	5.23	(1.20)	5.23[c]	4.8
11.	NO₂⁻	3.22	5.35	1.73	3.37[a]	(2.5)
12.	(C₆H₅CH₂)₂S	3.43	4.84			
13.	N₃⁻	3.58	5.78	(1.58)	4.74[a]	6.0[i]
14.	NH₂OH	3.85	6.60		5.82[c]	
15.	NH₂NH₂	3.86	6.61		7.93[c]	
16.	C₆H₅SH	4.15	5.70			
17.	Br⁻	4.18	5.79	1.51	(-7.7)[a]	6.62
18.	(C₂H₅)₂S	4.52	5.34			
19.	((C₂H₅)₂N)₃P	4.54	8.54			
20.	(CH₃)₂S	4.87	5.54		-5.3[d]	
21.	(CH₃O)₂PO⁻	5.01	7.00			
22.	(CH₂)₄S	5.02	5.42			
23.	(CH₂)₅S	5.14	5.66		-4.8[d]	
24.	SnCl₃⁻	5.44	~3.84			
25.	I⁻	5.46	7.42	2.06	(-10.7)[a]	8.60
26.	(C₆H₅CH₂)₂Se	5.53	5.23			
27.	(CH₃)₂Se	5.70	6.32			
28.	SCN⁻	5.75	6.70	1.83	(-0.7)[a]	6.05
29.	SO₃²⁻	5.79	8.53	2.57	7.26[a]	8.11
30.	C₆H₁₁NC	6.34	<2.0			
31.	(C₆H₅)₃Sb	6.79	<2.0			
32.	(C₆H₅)₃As	6.89	4.77			
33.	SeCN⁻	7.11	7.85			
34.	CN⁻	7.14	6.70	2.79	9.3[b]	14.1
35.	C₆H₅S⁻	7.17	9.92	(2.9)	6.52[e]	
36.	SC(NH₂)₂	7.17	7.27	2.18	-0.96[f]	(7)
37.	(CH₂O)₃P	7.23	~5.2			
38.	S₂O₃²⁻	7.34	8.95	2.52	1.9[a]	10.90
39.	(C₂H₅)₃As	7.68	6.90		<2.6[f]	
40.	(C₆H₅)₃P	8.93	7.00		2.73[g]	
41.	(n-C₄H₉)₃P	8.96	8.69		8.43[g]	
42.	(C₂H₅)₃P	8.99	8.72		8.69[g]	15.0
43.	CH₃COO⁻	<2.0	4.3	(0.95)	4.75[f]	
44.	C₆H₅COO⁻	...	4.5		4.19[f]	
45.	C₆H₅O⁻	...	5.75	(1.46)	9.89[f]	
46.	C₆H₅N(CH₃)₂	...	5.64		5.06[f]	
47.	2,6-DMP[o]	...	3.51			
48.	α-Picoline	~2.2	4.7		6.48[j]	
49.	Pyrrolidine	...	7.23		11.27[j]	
50.	N,N-DHA[o]	...	6.73			
51.	(C₂H₅)₃N	...	6.66		10.70[c]	
52.	(C₂H₅)₂NH	...	~7.0		11.0[c]	
53.	C₆H₅Se⁻	...	~10.7			
54.	(C₆H₅)₃SiO⁻	...	6.2			
55.	C₆H₅SO₂NCl⁻	...	6.8		(3.0)[n]	
56.	C₆H₅SO₂NH⁻	...	5.1		(8.5)[n]	
57.	Phthalimide anion	...	5.4		7.4[n]	
58.	HS⁻	...	(8)[k]	2.10	7.8[a]	(16)
59.	SO₄²⁻	...	(3.5)[k]	0.59	2.0[a]	(1.5)
60.	NO₃⁻	...	(1.5)[k]	(0.29)	-1.3[a]	
61.	(C₆H₅)₃Ge⁻	...	(12)[l]			
62.	(C₆H₅)₃Sn⁻	...	(11.5)[l]			
63.	(C₆H₅)₃Pb⁻	...	(8)[l]			
64.	Re(CO)₅⁻	...	(8)[l]			
65.	Mn(CO)₅⁻	...	(5.5)[l]			
66.	Co(CO)₄⁻	...	(3.5)[l]			
67.	HO₂⁻	...	(7.8)[m]			

[a] J. O. Edwards, *J. Am. Chem. Soc.*, **76**, 1540 (1954); data in water. [b] "Handbook of Chemistry and Physics," 45th ed, Chemical Rubber Publishing Co., Cleveland, Ohio, 1964, pp D 76–78. [c] D. D. Perrin, Ed., "Dissociation Constants," Butterworth & Co., Ltd., London, 1965. [d] E. M. Arnett, *Progr. Phys. Org. Chem.*, **1**, 324 (1963). [e] M. M. Kreevoy, *et al.*, *J. Am. Chem. Soc.*, **82**, 4899 (1960). [f] K. Issleib and H. Bruchlos, *Z. Anorg. Allgem. Chem.*, **316**, 1 (1962). [g] C. A. Streuli, *Anal. Chem.*, **32**, 985 (1960); A. I. Bokanov, B. A. Korolev, and B. I. Stepanov, *Zh. Obshch. Khim.*, **34**, 1879 (1965), give 2.61 for (C₆H₅)₃P and 8.86 for (C₂H₅)₃P. [h] Dissociation constant for CH₃HgY → CH₃-Hg⁺ + Y⁻: data at 20–25° from M. Schellenberg and G. Schwarzenbach, Proceedings of the 7th International Conference on Coordination Chemistry, Stockholm, 1962, p 158; *Helv. Chim. Acta*, **48**, 28 (1965); R. B. Simpson, *J. Am. Chem. Soc.*, **83**, 4711 (1961). Figures in parentheses are estimates based on formation constants of Ag(I); data in water. [i] T. R. Musgrave and R. N. Keller, *Inorg. Chem.*, **4**, 1793 (1965). [j] H. K. Hall, Jr., *J. Am. Chem. Soc.*, **79**, 5441 (1957). [k] Estimated from data in ref 6. [l] Estimated from relative data in diglyme reported by R. E. Dessy, R. L. Pohl, and R. B. King, *J. Am. Chem. Soc.*, **88**, 5121 (1966). [m] Estimated from rates of OH⁻ and O₂H⁻ with benzyl bromide: R. G. Pearson and D. N. Edgington, *ibid.*, **84**, 4607 (1962). [n] J. H. Beale, Ph.D. Thesis, Brown University, 1966. Values corrected to water. [o] 2,6-DMP = 2,6-dimethylpyridine; N,N-DHA = N,N-dimethylcyclohexylamine.

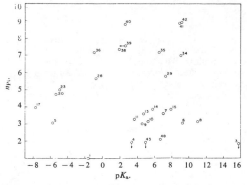

Figure 2. Plot of n_{Pt} against pK_a. The numbers refer to the nucleophiles as listed in Table III.

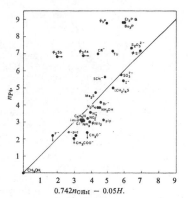

Figure 3. Plot of n_{Pt} against right-hand side of eq 13. If the equation were obeyed, all points would be on the line.

earlier, in methanol the pK_a values for the anionic bases would be increased by four units or so, and the pK_a of the neutral bases would be little changed. A plot of n_{CH_3I} against pK_a looks about like Figure 2.

There are a number of indications that proton basicity has more effect on rates with methyl iodide than with the platinum complex, however. For example, methoxide ion, which is very basic, is a good nucleophile toward methyl iodide but a very poor reagent for *trans*-[Pt-(py)₂Cl₂]. Triphenylphosphine, which is less basic than triethyl- and tributylphosphine, reacts more slowly with methyl iodide than do the other two phosphines, but its reaction rate with the platinum complex is almost identical with the rates of the two alkylphosphines. If the donor atom is held constant, the rate constant toward methyl iodide generally increases markedly with increasing base strength. This is not true for the platinum complex, as shown by the amines, for a second example. Thus proton basicity plays a smaller role in influencing reactivity toward platinum(II) than it does toward methyl iodide.

On the other hand, softness seems to be much more important for Pt, but it is difficult to find a suitable quantitative measure. One possible measure is E_n, the electrode potential of the oxidation

$$2Y_{aq}^- \rightleftharpoons Y_{2,aq} + 2e_{aq}^- \tag{9}$$

normalized to $E_n = 0$ for Y = H₂O.

The Edwards equation

$$\log (k_Y/k_S) = \alpha E_n + \beta H \tag{10}$$

has been quite successful in correlating a very wide range of data, including both kinetic and thermodynamic results from organic and inorganic systems.[5] Davis, in particular, has shown that eq 10 is quantitatively good for both nucleophilic displacements on tetrahedral carbon and divalent sulfur.[30] Since H in eq 10 is pK_a + 1.74, it is suggested that the equation be called the oxibase scale.[31]

A major disadvantage is that E_n values are determinable for very few reagents. Of those shown in Table III, the ones in parentheses are calculated from rate

(30) R. E. Davis, "Survey of Progress in Chemistry," A. Scott, Ed., Academic Press Inc., New York, N. Y., 1964; see also O. Foss, *Acta Chem. Scand.*, **4**, 404, 866 (1950).
(31) R. E. Davis, *J. Am. Chem. Soc.*, **87**, 3010 (1965).

data, assuming eq 10 to be valid. Sometimes misleading conclusions can be drawn from considering a limited amount of data. For example, there is a reasonably good correlation between the E_n values of Table III and the n_{CH_3I} values. Only cyanide ion and fluoride ion are badly off the line. Equation 10 becomes

$$n_{CH_3I} = 3.43E_n \tag{11}$$

with β equal to zero. In the same way there is a fairly good fit to the equation

$$n_{Pt} = 2.54E_n - 0.05H \tag{12}$$

though methoxide ion is badly off, if the E_n value of OH⁻ is used.

It is obvious that if both (11) and (12) were valid, one could write

$$n_{Pt} = \left(\frac{2.54}{3.43}\right)n_{CH_3I} - 0.05H \tag{13}$$

Figure 3 shows a plot of n_{Pt} against the right-hand side of (13). All points should be on the line if the equation were obeyed. The nature of the deviations is such that the softer nucleophiles, such as the phosphines, are above the line and the harder nucleophiles, such as CH₃O⁻ and CH₃COO⁻, are below the line. No combination of n_{Pt}, n_{CH_3I}, and either E_n or H works much better than eq 13.

An attempt was also made to correlate the n values with p$K_{CH_3Hg^+}$ or some combination of p$K_{CH_3Hg^+}$, H, or E_n. A very rough correlation does indeed exist between either of the n values and p$K_{CH_3Hg^+}$. However, it is far from satisfactory, nor is it possible to greatly improve it by a consideration of H or E_n as well.

It is, of course, possible to use the n_{Pt} values to predict the rates of reaction of other platinum complexes with various nucleophiles.[7] However, even in this restricted usage, substantial deviations can occur.[32] In the same way it is possible to use n_{CH_3I}, or the original n_{CH_3Br} values (in water) of Swain and Scott,[6] to predict rates of nucleophilic displacements on several organic substrates. However, the range of usefulness is limited.[4] For example, reactions at carbonyl carbon in esters and acyl halides usually do not correlate

(32) L. Cattalini, A. Orio, and M. Nicolini, *ibid.*, **88**, 5734 (1966); U. Belluco, M. Graziani, and P. Rigo, *Inorg. Chem.*, **5**, 1123 (1966).

Pearson, Sobel, Songstad | Nucleophilic Reactivity toward CH₃I and trans-[Pt(py)₂Cl₂]

well with n_{CH_3Br} or n_{CH_3I}. This electrophilic center is harder, resembling the proton. As a result a Brønsted relationship is approximately obeyed.[33] If a restriction is made to a single donor atom, such as nitrogen, then the Swain–Scott correlation does extend over a wider range of substrates.[34] However, difficulties appear with either the Swain–Scott correlation, or the Brønsted relation, if several donor atoms are included.[35] The α effect, added reactivity due to the presence of an unshared pair of electrons on the atom α to the donor atom, is also a complicating feature. Thus some substrates show a large α effect; others show no α effect.[36] It is obvious that it would be difficult to include such behavior in any equation with either one or two parameters to describe the nucleophile. Steric factors are also not completely taken care of in the Edwards or Swain–Scott equations.

A plot of n_{CH_3I} against n_{CH_3Br} is linear, with some deviations. The equation is[37]

$$n_{CH_3I} = 1.4 n_{CH_3Br} \tag{14}$$

(33) W. P. Jencks and J. Carriulo, *J. Am. Chem. Soc.*, **82**, 1778 (1960); M. Green and R. F. Hudson, *J. Chem. Soc.*, 1055 (1962); M. J. Gregory and T. C. Bruice, *J. Am. Chem. Soc.*, **89**, 2121 (1967).

(34) H. K. Hall, Jr., *J. Org. Chem.*, **29**, 3539 (1964).

(35) See, for example, T. C. Bruice and R. Lapinski, *J. Am. Chem. Soc.*, **80**, 2266 (1958); E. Thorsteinson and F. Basolo, *ibid.*, **88**, 3929 (1966).

(36) T. C. Bruice, *et al.*, *ibid.*, **89**, 2106 (1967); M. J. Gregory and T. C. Bruice, *ibid.*, **89**, 2327 (1967).

(37) R. G. Pearson and J. Songstad, *ibid.*, **89**, 1827 (1967).

Equation 14 shows an increasing relative reactivity toward methyl iodide of the more reactive, softer nucleophiles. This appears to be a consequence of the symbiotic effect.[37] A transition state containing a soft entering group and a soft leaving group has an extra stabilization. Since iodide ion is softer than bromide ion, the slope in eq 14 is greater than unity.

Such symbiosis between the entering and leaving group seems to be a general phenomenon in both organic and inorganic chemistry.[38] For example, the coefficient α in eq 10 increases steadily as the softness of the leaving group X in CH_3X increases. There is a linear relationship between α and E_n of the leaving group, which enables α values to be estimated *a priori* for certain cases.[31]

Our final conclusion is that at present it is not possible to predict quantitatively the rates of nucleophilic displacement reactions when a number of substrates of widely varying properties are considered. One can make reasonable estimates of what are apt to be good nucleophiles or poor nucleophiles. The present quantitative relationships for nucleophilic reactivity have limited ranges of application, though the Swain–Scott and Edwards equations can be quite useful.

Acknowledgment. This work was supported by the U. S. Atomic Energy Commission under Grant No. At(11-1)-1087.

(38) R. G. Pearson and J. Songstad, *J. Org. Chem.*, **32**, 2899 (1967).

Copyright 1967 by Verlag-Chemie, GmbH, Weinheim

ANGEWANDTE CHEMIE
International Edition

VOLUME 6 · NUMBER 12
DECEMBER 1967
PAGES 1013–1126

43

Nucleophilic Reactivity of the Thiophosphoryl Group

BY H. TEICHMANN AND G. HILGETAG [*]

The difference in the nucleophilic reactivities of the $>P–S$ and $>P=O$ groups can be satisfactorily explained on the basis of Pearson's acid-base concept [21]. The thiophosphoryl sulfur is a typical "soft base", and reacts preferentially with sub-group B metals, halogens, and sp^3 hybridized carbon, whereas it is largely inert to "hard acids" such as protons, carbonyl carbon, and tetrahedral phosphorus. Since a nucleophilic attack by the thiophosphoryl sulfur leads to a decrease in the charge density of the phosphorus atom, the latter (or the α-carbon atom in O-alkyl esters) becomes more open to nucleophilic attack. This reaction principle forms the basis of sulfur exchange, dealkylation, isomerization, and the Pistschimuka and Michalski reactions.

I. Introduction

Neutral phosphoryl and thiophosphoryl compounds *(1)* can react both as nucleophiles and as electrophiles. Apart from the reactivity of the substituents A, B, and

$$\begin{array}{l} A \\ B\!-\!P\!=\!X \quad (a)\!: X = O \\ C \qquad\quad (b)\!: X = S \\ (1) \end{array}$$

C (*e.g.* nucleophilic: PO-activated olefination [1]; electrophilic: dealkylation [2]), one can generally distinguish an electrophilic center at the phosphorus atom *(2)* and a nucleophilic center at the oxygen or sulfur atom *(3)*.

$$\begin{array}{l} \quad\quad X \\ N\!:^{\ominus}\!\!\rightarrow\!\!P\!\!\curvearrowleft\!C \;\rightarrow\; B\!-\!P\!=\!X \;+\; :C^{\ominus} \\ A \; B \qquad\qquad A \\ \qquad\qquad\qquad N \\ (2) \end{array}$$

e.g. A,B = alkyl, alkoxy;
C = Cl; X = O,S; N = OH

$$\begin{array}{l} A \\ B\!-\!P\!=\!X \; R\!\!\curvearrowleft\!\!Z \;\rightarrow\; B\!-\!P\!-\!X\!-\!Y \;+\; :Z^{\ominus} \\ C \qquad\qquad\qquad C \\ (3) \end{array}$$

e.g. A,B,C = alkyl; X = S;
RZ = CH₃I

Special attention has been paid in the past to the electrophilic center at the phosphorus atom, since this is the point of attack in such important reactions as hydrolysis [3] and phosphorylation [4]. The present article is confined to reactions in which the thiophosphoryl sulfur functions as a nucleophile. These are the reactions in which phosphoryl and thiophosphoryl compounds exhibit the greatest differences in behavior.

(1a) and *(1b)* differ:

1. in the electronegativities of oxygen (3.5) and of sulfur (2.5);

2. in the dissociation energy, which is 40–60 kcal/mole lower for the P=S bond (about 90 kcal/mole) than for the P=O bond (125–155 kcal/mole) [5];

3. in the extent of π bonding, which is lower for the P–S bond than for the P=O bond. This is shown by the force constants of the bonds [6], the difference in the shortening of the bond lengths [7], the difference between the dissociation

[*] Dr. H. Teichmann and Prof. Dr. G. Hilgetag
Institut für Organische Chemie
der Deutschen Akademie der Wissenschaften
An der Rudower Chaussee
DDR 1199 Berlin-Adlershof (Germany)

[1] *L. Horner, H. Hoffmann, W. Klink, H. Ertel,* and *V. G. Toscano,* Chem. Ber. *95*, 581 (1962).

[2] *G. Hilgetag* and *H. Teichmann,* Angew. Chem. *77*, 1001 (1965); Angew. Chem. internat. Edit. *4*, 914 (1965).

[3] *J. R. Cox* and *O. B. Ramsay,* Chem. Reviews *64*, 317 (1964).

[4] *D. M. Brown,* Advances org. Chem. *3*, 75 (1963); *V. M. Clark, D. W. Hutchinson, A. J. Kirby,* and *S. G. Warren,* Angew. Chem. *76*, 704 (1964).

[5] *S. B. Hartley, W. S. Holmes, J. K. Jacques, M. F. Mole,* and *J. C. McCoubrey,* Quart. Rev. *17*, 204 (1963).

[6] *H. Siebert,* Z. anorg. allg. Chem. *275*, 210 (1954).

[7] *J. R. Van Wazer,* J. Amer. chem. Soc. *78*, 5709 (1956).

Angew. Chem. internat. Edit. / Vol. 6 (1967) / No. 12

1013

399

energies of the single and double PX bonds [8], the difference in electronegativity between P and X [9], and the ^{31}P–NMR signals, which always occur at lower fields for *(1b)* than for *(1a)* [10, 11].

The P=O bond may be regarded as a semipolar bond on which is superimposed a p_π–d_π bond due to back-donation of p electrons from the oxygen atom to empty d orbitals of the phosphorus atom. Since the overlap of the p and d orbitals is greatest near the oxygen atom, the p_π–d_π bond of the P=O group is much more strongly polar than the p_π–p_π bonds usually encountered in carbon chemistry [12]. The p_π–d_π interaction, and hence the P=O bond order, increases with increasing electronegativity of the substituents A, B, and C; this is clearly shown by the position of the PO frequency, which shows a linear relationship with the sum of the electronegativities of the substituents A, B, and C [13, 14]. Calculations of the P=O bond order yield different values according to the methods used, but the results always show the same trend.

Table 1. Bond order N of the P=O bond as found (A) from the force constants [6], (B) by an MO calculation [15]. Σx = sum of the electronegativities of the substituents.

	F$_3$PO	Cl$_3$PO	Br$_3$PO	(CH$_3$)$_3$PO
Σx	12	9	8.4	6.0
N (A)	2.34	2.09	2.02	1.7
N (B)	2.98	2.19	1.96	1.0

A similar situation is to be expected for the P=S bond, the π-bond component in this case being smaller. The bonding again depends on the electronegativity of the substituents [14, 16, 17], but this dependence is less pronounced in the P=S frequency, owing to the mechanical coupling between the P=S and P(A,B,C) stretching vibrations [14, 17, 18]. The similarity of the P=O and P=S bonds is also evidenced by the fact that the frequency shifts on complex formation with acceptor molecules (cf. Section II) take place in the same direction.

Differences between *(1a)* and *(1b)* are also found:

4. in the availability of empty 3d orbitals of the thiophosphoryl sulfur, which may explain the greater stability of some thiophosphoryl complexes [19, 20].

The experimental data on the nucleophilic reactivities of phosphoryl oxygen and of thiophosphoryl sulfur

can be largely explained by *Pearson*'s concept [21], according to which oxygen is a "hard base" and so reacts preferentially with "hard acids" (protons, A metals, carbonyl carbon, phosphoryl phosphorus), whereas sulfur is a "soft base" and reacts mainly with "soft acids" (B metals, tetrahedral carbon, halogens). This is illustrated by the following examples.

1. Phosphine oxides form stable salts with inorganic and organic acids [22], and can be titrated acidimetrically [23]. The only known salt of a phosphine sulfide is evidently a hydrobromide (an oil) [24]; attempted resolutions of chiralic phosphine oxides [25], but not of phosphine sulfides [26], by salt formation with optically active acids have been successful.

Similarly, the hydrogen bonding ability is strong in phosphoryl compounds but much weaker in thiophosphoryl compounds [27]; the resulting lower solubility in water and greater resistance to hydrolysis were deciding factors in the development of the thiophosphate insecticides [28].

2. Phosphoryl compounds are extremely useful for the extraction of many metal salts from aqueous solution [29], whereas the corresponding thiophosphoryl compounds have much poorer extraction properties [30]. The only ions that are extracted particularly well by thiophosphoryl compounds [32] are those of the B metals (mainly group I, II, and VIII transition elements that are rich in d electrons and form readily polarizable cations [31]).

3. Trialkyl phosphates *(4)* react with diester chlorides of phosphoric and thiophosphoric acids *(5)* to form pyrophosphates or pyrothiophosphates *(6)*; trialkyl phosphorothionates give the same reaction only in so far as they are converted into thiol compounds [33]. The phosphoryl oxygen is thus able to carry out a nucleophilic attack on the chloro(thio)phosphate [34],

[8] *L. C. Chernick, J. B. Pedley,* and *H. A. Skinner,* J. chem. Soc. (London) *1957*, 1851.

[9] *J. Goubeau,* Angew. Chem. *69*, 77 (1957); *78*, 565 (1966); Angew. Chem. internat. Edit. *5*, 567 (1966).

[10] *J. R. Van Wazer, C. F. Callis, J. N. Shoolery,* and *R. C. Jones,* J. Amer. chem. Soc. *78*, 5715 (1956).

[11] *E. Fluck*: Die kernmagnetische Resonanz und ihre Anwendung in der anorganischen Chemie. Springer-Verlag, Berlin-Göttingen-Heidelberg 1963, p. 258 ff.

[12] *C. L. Chernick* and *H. A. Skinner,* J. chem. Soc. (London) *1956*, 1401; *D. P. Craig, A. Maccoll, R. S. Nyholm, L. E. Orgel,* and *L. E. Sutton,* J. chem. Soc. (London) *1954*, 332.

[13] *J. V. Bell, J. Heisler, H. Tannenbaum,* and *J. Goldenson,* J. Amer. chem. Soc. *76*, 5185 (1954).

[14] *Chen Wen-Ju,* Acta chim. sinica *31*, 29 (1965); Chem. Abstr. *63*, 4128 (1965).

[15] *E. L. Wagner,* J. Amer. chem. Soc. *85*, 161 (1963).

[16] *H. Siebert*: Anwendung der Schwingungsspektroskopie in der anorganischen Chemie. Springer-Verlag, Berlin-Heidelberg-New York 1966, p. 137.

[17] *A. Müller, H.-G. Horn,* and *O. Glemser,* Z. Naturforsch. *20b*, 1150 (1965).

[18] *F. N. Hooge* and *P. J. Christen,* Recueil Trav. chim. Pays-Bas *77*, 911 (1958).

[19] *W. Tefteller* and *R. A. Zingaro,* Inorg. Chem. *5*, 2151 (1966).

[20] *W. van der Veer* and *F. Jellinek,* Recueil Trav. chim. Pays-Bas *85*, 842 (1966).

[21] *R. G. Pearson,* J. Amer. chem. Soc. *85*, 3533 (1963).

[22] *R. H. Pickard* and *J. Kenyon,* J. chem. Soc. (London) *89*, 262 (1906).

[23] *D. C. Wimer,* Analytic. Chem. *30*, 2060 (1958).

[24] *R. A. Zingaro* and *R. E. McGlothlin,* J. org. Chemistry *26*, 5205 (1961).

[25] *J. Meisenheimer* and *L. Lichtenstadt,* Ber. dtsch. chem. Ges. *44*, 356 (1911); *J. Meisenheimer, J. Casper, M. Höring, W. Lauter, L. Lichtenstadt,* and *W. Samuel,* Liebigs Ann. Chem. *449*, 213 (1926).

[26] *W. C. Davies* and *F. G. Mann,* J. chem. Soc. (London) *1944*, 276.

[27] *Th. Gramstad* and *W. J. Fuglevik,* Acta chem. scand. *16*, 2368 (1962).

[28] *G. Schrader*: Die Entwicklung neuer Insektizide auf Grundlage organischer Fluor- und Phosphor-Verbindungen. 2nd Edit. Verlag Chemie, Weinheim 1952, p. 42.

[29] *R. M. Diamond* and *D. G. Tuck,* Progr. Inorg. Chem. *2*, 109 (1960); *Y. Marcus,* Chem. Reviews *63*, 147 (1963).

[30] *W. W. Wendlandt* and *J. M. Bryant,* Science (Washington) *123*, 1121 (1956).

[31] *St. Ahrland, J. Chatt,* and *N. R. Davies,* Quart. Rev. *12*, 265 (1958); *St. Ahrland,* Chem. Engng. News *43*, No. 22, p. 93 (1965).

[32] *T. H. Handley,* Talanta (London) *12*, 893 (1965); *D. E. Elliott* and *Ch. V. Banks,* Anal. Chim. Acta (Amsterdam) *33*, 237 (1965).

[33] *C. Stölzer* and *A. Simon,* Naturwissenschaften *46*, 377 (1959); Chem. Ber. *96*, 881, 896 (1963).

[34] *G. M. Kosolapoff,* Science (New York) *108*, 485 (1948); *M. Baudler* and *W. Giese,* Z. anorg. allg. Chem. *290*, 258 (1957).

1014

400

which is impossible in the case of the thiophosphoryl sulfur.

(4), X = O
(5), X = S

$$\xrightarrow{-\,RCl}\quad RO-\underset{\underset{OR}{|}}{\overset{\overset{O}{\|}}{P}}-O-\underset{\underset{OR}{|}}{\overset{\overset{X}{\|}}{P}}-OR$$

(6)

4. Whereas phosphine sulfides can be alkylated even with alkyl halides to form alkylthiophosphonium salts [35], the analogous reactions of phosphine oxides take place only with stronger alkylating agents [25, 36]. O→S transalkylation is a characteristic of many esters of phosphorothionic and phosphonothionic acids (cf. Section V 2b).

5. The "soft acids" ICl, IBr, and I_2 [21] form more stable adducts with thiophosphoryl compounds than with phosphoryl compounds [27, 37]. Even the less "soft" and more strongly oxidizing halogens Br_2 and Cl_2 give characteristic reactions with thiophosphoryl compounds (cf. Section III).

These examples already indicate a classification of the types of electrophiles to be discussed according to metal salts, halogenating agents, carbonyl and phosphoryl compounds, and alkylating agents.

II. Reactions with Metal Salts [*]

The many crystalline addition products of phosphine oxides and phosphinic, phosphonic, and phosphoric acid derivatives, as well as the solvates present in solution in extractions with phosphoryl compounds, result from a donor-acceptor interaction between phosphoryl oxygen and metal. Even in systems such as $OPCl_3/SbCl_5$, in which chloride ion transfer according to (a) (solvent system concept) was at first regarded as more likely than adduct formation at the oxygen atom (b) (coordination model), (b) has proved to be more significant [38-41].

$$OPCl_2^{\ominus}\ SbCl_6^{\ominus}\ \xleftarrow{(a)}\ OPCl_3 + SbCl_5\ \xrightarrow{(b)}\ Cl_5Sb\leftarrow OPCl_3$$

Apart from X-ray structure analysis [41], P=O-metal coordination can be deduced in particular from IR measurements. The formation of a coordinate bond from the oxygen to the metal weakens the back-donation of electrons from the oxygen to the phosphorus, and so reduces the P=O bond order; this leads to a strong bathochromic shift of the P=O frequency [42,43]. If the simultaneous decrease in the electron density on the phosphorus cannot be compensated by other substituents, this also results in a downfield shift of the ^{31}P resonance line [44,45].

Until recently, the only thiophosphoryl addition compounds known were a few metal salt adducts. However, several research groups have now described new compounds that seem to call for a revision of the earlier view that the donor character of the thiophosphoryl sulfur is insignificant. Nevertheless, the adducts of the phosphoryl compounds are generally much more stable than those of the thiophosphoryl compounds except for a few adducts of phosphorothionates [46,47] and phosphine sulfides [20], mainly with Ag and Hg(II) salts. A plausible explanation of this is offered by *Ahrland*'s view [48] that the coordinate bond between a "soft base" and a typical B metal can be strengthened by back bonding by d electrons of the metal to the ligand.

It has recently been shown that the P=S frequency also undergoes an appreciable bathochromic shift as a result of metal-sulfur coordination [20,49-51].

1. Thiophosphoryl Halides

Whereas *e.g.* very many phosphorus oxychloride adducts are known [41], the only thiophosphoryl halide adducts obtained so far are evidently [53-55]

[35] *A. Hantzsch* and *H. Hibbert*, Ber. dtsch. chem. Ges. *40*, 1508 (1907).

[36] *A. Schmidpeter, B. Wolf,* and *K. Düll*, Angew. Chem. *77*, 737 (1965); Angew. Chem. internat. Edit. *4*, 712 (1965).

[37] *R. A. Zingaro, R. E. McGlothlin,* and *E. A. Meyers*, J. phys. Chem. *66*, 2579 (1962).

[*] Including the covalent halides.

[38] *V. Gutmann*, J. physic. Chem. *63*, 378 (1959); Österr. Chemiker-Ztg. *62*, 326 (1961).

[39] Cf. discussion in *R. S. Drago* and *K. F. Purcell*, Progr. inorg. Chem. *6*, 271 (1964).

[40] *M. Baaz* and *V. Gutmann*, Mh. Chem. *90*, 426 (1959); *V. Gutmann* and *M. Baaz*, Z. anorg. allg. Chem. *298*, 121 (1959).

[41] *I. Lindqvist*: Inorganic Adduct Molecules of Oxo-Compounds. Springer-Verlag, Berlin-Göttingen-Heidelberg 1963.

[42] *J. C. Sheldon* and *S. Y. Tyree*, J. Amer. chem. Soc. *80*, 4775 (1958).

[43] *F. A. Cotton. R. D. Barnes,* and *E. Bannister*, J. chem. Soc. (London) *1960*, 2199.

[44] *J. L. Burdett* and *L. L. Burger*, Canad. J. Chem. *44*, 111 (1966).

[45] *M. Becke-Goehring* and *A. Slawisch*, Z. anorg. allg. Chem. *346*, 295 (1966).

[46] *G. Hilgetag, K.-H. Schwarz, H. Teichmann,* and *G. Lehmann*, Chem. Ber. *93*, 2687 (1960).

[47] *G. Hilgetag, H. Teichmann,* and *M. Krüger*, Chem. Ber. *98*, 864 (1965).

[48] *St. Ahrland, J. Chatt, N. R. Davies,* and *A. A. Williams*, J. chem. Soc. (London) *1958*, 276.

[49] *J. Philip* and *C. Curran*, 147th Meeting Amer. chem. Soc. *1964*, Abstracts, p. 28L.

[50] *H. Teichmann*, Angew. Chem. *77*, 809 (1965); Angew. Chem. internat. Edit. *4*, 785 (1965).

[51] *D. W. Meek* and *P. Nicpon*, J. Amer. chem. Soc. *87*, 4951 (1965).

[52] *R. Ch. Paul, K. Ch. Malhotra,* and *G. Singh*, J. Indian chem. Soc. *37*, 105 (1960).

[52a] *W. van der Veer*, Dissertation, University of Groningen 1965.

[53] *R. Gut* and *G. Schwarzenbach*, Helv. chim. Acta *42*, 2156 (1959).

[54] *F. Jellinek* and *W. van der Veer*, 7th Internat. Conf. on Coordination Chemistry *1962*, Abstracts, p. 230.

[55] *E. W. Wartenberg* and *J. Goubeau*, Z. anorg. allg. Chem. *329*, 269 (1964); *H. S. Booth* and *J. H. Walkup*, J. Amer. chem. Soc. *65*, 2334 (1943); *L. A. Niselson*, Ž. neorg. Chim. *5*, 1634 (1960).

Angew. Chem. internat. Edit. / Vol. 6 (1967) / No. 12

1015

401

SPCl₃·AlCl₃ [52], SPBr₃·AlBr₃ [52a], SPCl₃·SbCl₅ [52-54], and SPBr₃·SbCl₅ [54]. SPBr₃·SbCl₅ decomposes at room temperature, and SPCl₃·SbCl₅ at temperatures above 140 °C [54]. The low stability and the results of calorimetric measurements [56] indicate a much weaker donor character for the sulfur in the thiophosphoryl halides than for the oxygen in OPCl₃.

According to ^{31}P–NMR and conductivity measurements, SPCl₃·AlCl₃ appears to exist in solution as the chloro complex SPCl₂$^{\oplus}$ AlCl₄$^{\ominus}$ [57]. However, X-ray structure investigations on comparable OPCl₃ adducts, which dissociate similarly in solution [40], have revealed unmistakable P–O-metal coordination [41]; no corresponding investigations on thiophosphoryl compounds have yet been reported.

2. Thiophosphoryl Amides

Adducts of thiophosphoryl tripiperidide and tripyrrolidide with PdCl₂, CdBr₂, and CdI₂ [49] and of thiophosphoryl trismethylamide, triscyclohexylamide, trimorpholide, and tripiperidide with SnCl₄ and SnBr₄ [58] have been obtained.

Metal-nitrogen coordination can be discounted for these compounds on the grounds of their IR spectra (the CdHal₂ adducts show bathochromic shifts of 20–40 cm⁻¹ for the P=S frequency [49]; the P=S frequency of the triscyclohexylamide changes from 652 cm⁻¹ to 610 cm⁻¹ in the SnCl₄ adduct and to 607 cm⁻¹ in the SnBr₄ adduct [58]).

3. Phosphine Sulfides and Diphosphine Disulfides

A compound of triethylphosphine sulfide with PtCl₂ was mentioned as early as 1857 [59]. With two exceptions, however, the phosphine sulfide adducts listed in Table 2 were obtained only in the last three years. Other attempted preparations have been unsuccessful [20,49,61,62].

The frequency shifts are of roughly the same order of magnitude as those for phosphine oxide adducts, confirming that the bonding is similar. However, no information about the stability of the complex can be deduced from the magnitude of $\Delta\nu_{P=S}$. Thus some of the adducts of the trialkylphosphine sulfides can be sublimed without decomposition [50], whereas those of triphenylphosphine sulfide generally decompose on heating [20]. The SnHal₄ adducts of the trialkylphosphine sulfides undergo complete ligand exchange with trialkyl phosphates or trialkyl phosphorothiolates:

[R'₃PS]₂SnHal₄ + 2 (RO)₃PO → 2 R'₃PS + [(RO)₃PO]₂SnHal₄

[56] M. Zackrisson, Acta chem. scand. 15, 1785 (1961).

[57] L. Maier, Z. anorg. allg. Chem. 345, 29 (1966).

[58] H. Teichmann, unpublished.

[59] A. Cahours and A. W. Hofmann, Ann. chem. Pharm. 104, 1 (1857).

[60] L. Malatesta, Gazz. chim. ital. 77, 518 (1947).

[61] E. Bannister and F. A. Cotton, J. chem. Soc. (London) 1960, 1959.

[62] K. Issleib and H. Reinhold, Z. anorg. allg. Chem. 314, 113 (1962); B. W. Fitzsimmons, P. Gans, B. C. Smith, and M. A. Wassef, Chem. and Ind. 1965, 1698.

Table 2. Addition compounds of tertiary phosphine sulfides. Δνp=S = difference between the P=S frequencies in the free phosphine sulfide and in the adduct.

Compound	M.p. (°C)	Δνp=S (cm⁻¹)	Ref.
2 (CH₃)₃PS·SnCl₄	196–199	−36	[50]
2 (CH₃)₃PS·SnBr₄	182–184	−37	[50]
2 (CH₃)₃PS·ZnI₂	—	−30	[51]
2 (CH₃)₃PS·CdI₂	—	−27	[51]
2 (CH₃)₃PS·HgCl₂	—	−37	[51]
4 (CH₃)₃PS·CuClO₄	—	—	[63]
2 CH₃(C₂H₅)₂PS·3 HgBr₂	93	—	[64]
(C₂H₅)₃PS·2 HgCl₂	—	—	[60]
2 (C₂H₅)₃PS·SnCl₄	194–196	−8	[50]
2 (C₂H₅)₃PS·SnBr₄	182–184	−13, −32	[50]
2 (n-C₃H₇)₃PS·SnCl₄	166–166.5	−25, −38	[50]
2 (n-C₃H₇)₃PS·SnBr₄	176–177	−30, −43	[50]
2 (n-C₄H₉)₃PS·SnCl₄	93–95	—	[50]
2 (n-C₄H₉)₃PS·SnBr₄	101–102	−48, −55	[50]
2 (CH₃)₂C₆H₅PS·SnCl₄	174–177	−44	[50]
2 (CH₃)₂C₆H₅PS·SnBr₄	155–157	−40	[50]
2 (C₆H₅)₃PS·PdCl₂	236	—	[61]
(C₆H₅)₃PS·AuCl₃	~140	—	[65]
(C₆H₅)₃PS·AuCl	>130	—	[65]
4 (C₆H₅)₃PS·Hg(ClO₄)₂	206–210	−52	[66]
2 (C₆H₅)₃PS·HgBr₂	—	−77 [*]	[49]
(C₆H₅)₃PS·HgCl₂	225–226	−43	[20]
(C₆H₅)₃PS·AlBr₃	93–95	−56	[20]
2 (C₆H₅)₃PS·SnCl₄	163–166	−55	[20]
2 (C₆H₅)₃PS·TiCl₄	88–90	−102	[20]
(C₆H₅)₃PS·2 TiCl₄	85–88	−57	[20]
(C₆H₅)₃PS·TiBr₄	240–250	−102	[20]
(C₆H₅)₃PS·SbCl₅	72–73	−96	[20]
(C₆H₅)₃PS·NbCl₅	128–130	−62	[20]
(C₆H₅)₃PS·TaCl₄	202–205	−62	[20]
2 (C₆H₅)₃PS·AgNO₃	160	−32	[67]
(C₆H₅)₃PS·AgNO₃	82–83	−51	[67]

[*] This value is probably too high, since νP=S in (C₆H₅)₃PS was given incorrectly.

Unlike the very unstable SnCl₄ adduct of triphenylphosphine sulfide, they are not decomposed by trialkyl phosphorothionates or by diethyl ether. The relative donor strengths toward tin(IV) halides thus decrease in the order (RO)₃PO, (RO)₂(RS)PO > R₃PS > (RO)₃PS, R₂O > Ar₃PS [68].

Phosphine sulfides reduce Cu(II) to Cu(I) [63], and, to some extent, Fe(III) to Fe(II) [20]; Hg₂(ClO₄)₂ disproportionates on reaction with triphenylphosphine sulfide [66]. The resulting ions are always "softer" acids than the starting materials, and are therefore better suited for complex formation with the phosphine sulfide.

The following compounds have been prepared from diphosphine disulfides: 2 R₂P(S)P(S)R₂·CuClO₄ (R = CH₃, C₂H₅); R₂P(S)P(S)R₂·CuCl (R = CH₃) [51]; and R₂P(S)P(S)R₂·SnHal₄ (R = CH₃, C₂H₅, n-C₃H₇, n-C₄H₉; Hal = Cl, Br) [50,58]. The Cu(I) adducts are formed from diphosphine disulfides and Cu(II) salts in ethanol. A chelate structure (7) was deduced from

[63] P. E. Nicpon and D. W. Meek, 151th Meeting Amer. Chem. Soc. 1966, Abstracts H-63.

[64] L. Maier: Topics in Phosphorus Chemistry, Vol. 2, Interscience Publ., New York-London-Sydney 1965, p. 116.

[65] J. M. Keen, J. chem. Soc. (London) 1965, 5751.

[66] R. A. Potts and A. L. Allred, Inorg. Chem. 5, 1066 (1966).

[67] H. Teichmann, I. Schwandt, and G. Hilgetag, unpublished.

[68] H. Teichmann, lecture to the 3rd All-Union Conference on the Chemistry and Application of Organophosphorus Compounds, Moscow, October 27th, 1965.

1016

Angew. Chem. internat. Edit. / Vol. 6 (1967) / No. 12

conductivity measurements [51]. The SnHal$_4$ compounds, some of which are extremely sparingly soluble, appear to have a different structure.

$$R_2P{=}S \overset{\oplus}{\underset{S}{\diagdown}} S{=}PR_2 \atop R_2P{-}S \overset{}{\diagdown} S{-}PR_2 \quad X^{\ominus}$$

(7): X = ClO$_4$, CuCl$_2$

4. Phosphorothionates, Phosphonothionates, and Phosphinothionates

The reactions of these compounds clearly illustrate the difference in the behavior of "hard" and "soft" acids. While the metal-sulfur coordination is retained in the reaction products with soft acids, the A metals tend to loosen the metal-sulfur bond.

a) Reactions with "Soft" Acids

Solid adducts of triesters of phosphorothionic acid have been obtained with HgCl$_2$ [47,69-71], AuCl$_3$, PtCl$_4$ [69], AgNO$_3$ [46,69], AgNO$_2$ and AgBF$_4$ [46]. The formation of crystalline compounds of phosphorodithioates with Ag and Hg salts had been observed by *Carius* [72] as long ago as 1861. Phosphonothionates and phosphinothionates also give crystalline AgNO$_3$ adducts [73]. No corresponding compounds of esters of phosphoric acid are known; for example, triphenyl phosphate, unlike triphenyl phosphorothionate, does not react with AgNO$_3$ [46].

The partial positive charge induced on the phosphorus atom by the sulfur-metal coordination makes either the phosphorus atom itself or (in alkyl esters) the α-carbon atom of an ester residue more open to nucleophilic attack. Consequently, the adducts of trialkyl phosphorothionates with B-metal salts are generally not very stable. Thus the HgCl$_2$ adducts *(8)* lose alkyl chloride, sometimes even during preparation, but in any case when heated [69,74]:

(RO)$_3$PS·2 HgCl$_2$ → (RO)$_2$P(O)SHgCl·HgCl$_2$ + RCl
    (8)   *(9)*

A series of dealkylation products *(9)* has been described [69]. Dialkyl phosphorochloridothioates and phosphorofluoridothioates react in the same way with mercury(II) chloride or acetate [75]. In dialkyl acyl phosphorothionates *(10)*, the acyl group is particularly readily removed, so that these compounds can

[69] *P. Pistschimuka*, Ber. dtsch. chem. Ges. *41*, 3854 (1908); J. prakt. Chem. [2] *84*, 746 (1911); J. russ. physic.-chem. Soc. *44*, 1406 (1912), especially p. 1448 f.

[70] *W. G. Emmett* and *H. O. Jones*, J. chem. Soc. (London) *99*, 713 (1911).

[71] *Hu Ping-Fang*, *Li Shou-Cheng*, and *Cheng Wan-Yi*, Acta chim. sin. *22*, 49 (1956); Chem. Abstr. *52*, 6156 (1958).

[72] *L. Carius*, Ann. Chem. Pharm. *119*, 289 (1861).

[73] *G. Hilgetag*, *P. Gregorzewski*, and *H. Teichmann*, unpublished.

[74] *F. Feher* and *A. Blümcke*, Chem. Ber. *90*, 1934 (1957).

[75] *K.-H. Lohs* and *A. Donner*, personal communication.

$$RO{-}\underset{\underset{OR}{|}}{\overset{\overset{S{\rightarrow}HgCl_2}{\parallel}}{P}}{-}O\overset{\overset{\curvearrowright}{}}{{}}\underset{H{\rightharpoonup}OC_2H_5}{COR'} \longrightarrow (RO)_2P(O)SHgCl + R'COOC_2H_5 + HCl.$$

(10)

be determined alkalimetrically after addition of alcoholic HgCl$_2$ solution [76].

A similar reaction is used for the determination of the insecticide Delnav *(11)* [77]:

(11)

H_2O

(CH$_2$OH)$_2$ + (CHO)$_2$ + 2 HCl

However, this reaction is not characteristic of thiophosphates, but is observed with mercaptals in general [78].

The very large increase in electrophilic reactivity of the phosphorus atom as a result of sulfur coordination is particularly obvious in the transesterification reactions of trialkyl phosphorothionates in the presence of HgCl$_2$, which proceed even at room temperature [79]:

$$(C_2H_5O)_3PS \xrightarrow[HgCl_2]{CH_3OH} (CH_3O)_3PS\cdot2\,HgCl_2$$

The Cu-ion catalysed hydrolysis of *p*-nitrophenyl phosphorothionates and phosphonothionates [80] and the Ag-ion catalysed formation of 4-t-butylmethylenecyclohexane from dimethyl 1-hydroxy-4-t-butylcyclohexylmethylphosphonothionate [81] probably take place for the same reason.

Another unusual reaction is the smooth cleavage of triaryl phosphorothionates by silver nitrate [46,82], which is best explained by an attack on the phosphorus atom by the nitrate ion.

$$2\,(C_6H_5O)_3PS\cdot AgNO_3 \xrightarrow[-(C_6H_5O)_3PS]{} \underset{(13)}{\overset{C_6H_5O}{\underset{C_6H_5O}{\overset{C_6H_5O}{{}}}}P\overset{SAg}{\underset{\underset{NO_2}{O}}{{}}}} \longrightarrow$$

(12)   *(13)*

→ (C$_6$H$_5$O)$_2$POSAg + HOC$_6$H$_4$NO$_2$

On the other hand, the corresponding reactions of the trialkyl phosphorothionates, which also proceed *via* addition compounds *(14)* [46], are evidently true dealkylations.

[76] *M. I. Kabachnik*, *T. A. Mastryukova*, *N. P. Rodionova*, and *E. M. Popow*, Ž. obšč. Chim. *26*, 120 (1956).

[77] *C. L. Dunn*, J. agric. Food Chem. *6*, 203 (1958).

[78] *T. A. Mastryukova et al.*, Izvest. Akad. Nauk SSSR, Ser. chim. *1956*, 443; Chem. Abstr. *50*, 16662 (1956).

[79] *R. Donner*, *Kh. Lohs*, and *H. Holzhäuser*, Mber. dtsch. Akad. Wiss. Berlin *8*, 240 (1966); Z. anorg. allg. Chem. *347*, 156 (1966).

[80] *J. A. A. Ketelaar*, *H. R. Gersmann*, and *M. M. Beck*, Nature (London) *177*, 392 (1956).

[81] *E. J. Corey* and *G. T. Kwiatkowski*, J. Amer. chem. Soc. *88*, 5654 (1966).

[82] *G. Hilgetag*, *G. Lehmann*, *A. Martini*, *G. Schramm*, and *H. Teichmann*, J. prakt. Chem. [4] *8*, 207 (1959).

403

$$\underset{(14)\quad NO_3^\ominus}{\underset{C_2H_5O}{\overset{C_2H_5O}{\diagup}}P\underset{O\uparrow C_2H_5}{\overset{S\rightarrow Ag^\oplus}{\diagdown}}} \longrightarrow (C_2H_5O)_2POSAg + C_2H_5ONO_2.$$

The reactions of methyl diphenyl and of phenyl dimethyl phosphorothionates with $AgNO_3$ should otherwise yield nitrophenol in addition to methyl nitrate, but they do not [83].

b) Reactions with "Hard Acids"

Triesters of phosphorothionic acid react with metal halides of the type used as Friedel-Crafts catalysts, such as $SbCl_5$ [84,85], $SnCl_4$, $SnBr_4$ [86], $TiCl_4$, $AlCl_3$, and $FeCl_3$ [47], the ease with which the reaction takes place increasing with the alkylating power of the ester. However, there is no sulfur-metal coordination in any of the crystalline products isolated. The partial positive charge induced on the phosphorus atom by adduct formation facilitates the heterolysis of an O-alkyl bond, but this, together with the highly nucleophilic character of the sulfur toward tetrahedral carbon and the tendency of the A metals to coordinate with "harder" bases, leads to the formation of thiol compounds. For example, trimethyl phosphorothionate and $SbCl_5$ give a 1:1 adduct (15) at $-80\,^\circ C$, which changes quantitatively, even below room temperature, into a quasiphosphonium hexachloroantimonate (16) and an unstable dealkylation product (17) [84].

$$2\ (CH_3O)_3PS\rightarrow SbCl_5 \longrightarrow \underset{(16)}{\underset{OCH_3}{\overset{OCH_3}{CH_3O-\overset{\oplus}{P}-SCH_3}}}\ SbCl_6^\ominus$$

$$\underset{(15)}{}$$

$$+$$

$$\underset{(17)}{(CH_3O)_2P(S)OSbCl_4}$$

Trimethyl phosphate, on the other hand, gives a stable adduct $(CH_3O)_3PO\rightarrow SbCl_5$ [87]. The extremely powerful alkylating properties of this adduct, as well as those of (15), can be used e.g. in a simple preparation of trialkyloxonium salts (18) [88].

$$(RO)_3PX\rightarrow SbCl_5 + R_2O\rightarrow SbCl_5 \longrightarrow$$
$$R_3\overset{\oplus}{O}\ SbCl_6^\ominus + (RO)_2P(X)OSbCl_4$$
$$X = O, S \qquad\qquad (18)$$

Since quasiphosphonium ions of the type (16) can be easily dealkylated, even by weak nucleophiles [89], to yield trialkyl phosphorothiolates, and since A metals generally favor coordination with oxygen rather than with chloride ions, reactions of phosphorothionates with most Lewis acids result in formation of phosphorothiolate adducts [47,85,86].

[83] G. Hilgetag, I. Schwandt, and H. Teichmann, unpublished.
[84] H. Teichmann and G. Hilgetag, Chem. Ber. 96, 1454 (1963).
[85] G. Hilgetag and H. Teichmann, Mber. dtsch. Akad. Wiss. Berlin 6, 439 (1964).
[86] H. Teichmann and G. Hilgetag, Chem. Ber. 98, 856 (1965).
[87] G. Hilgetag, H. Teichmann, and M. Jatkowski, unpublished.
[88] G. Hilgetag and H. Teichmann, Chem. Ber. 96, 1446 (1963).
[89] G. Hilgetag and H. Teichmann, Chem. Ber. 96, 1465 (1963).

$$[(RO)_3PS]_2SnCl_4 \longrightarrow (RO)_3\overset{\oplus}{P}SR\ Cl_4Sn\leftarrow OP(OR)_2S^\ominus \longrightarrow$$
$$[(RO)_2(RS)PO]_2SnCl_4$$

(For further details of this reaction, see Section V 2b.) A few phosphorothiolate adducts were prepared in this way some decades ago, but were wrongly regarded as phosphorothionate adducts [69] (starting from the isomeric thiolates identical adducts are obtained [47]).

The relationship between this type of isomerization and the alkylating power is clearly shown in the series

$$(CH_3O)_3PS > (CH_3O)_2CH_3PS > (CH_3O)(CH_3)_2PS.$$

Phosphonothionates are also isomerized (though much more slowly than phosphorothionate) into phosphonothiolate adducts by tin(IV) halides, but with phosphinothionates the reaction stops with the formation of the thiono ester adduct. This is the first case in which the thiono and thiolo esters give isomeric adducts instead of identical adducts [68]. In both isomeric series, strong bathochromic shifts of the P=O and P=S frequence indicate the bonding site [58] (cf. Table 3).

Table 3. Isomeric adducts of thiophosphinates.

Compound	M.p. (°C)	νP=O (cm^{-1})	νP=S (cm^{-1})
$(CH_3O)(CH_3)_2PS$	—	—	585
$[(CH_3O)(CH_3)_2PS]_2SnCl_4$	121—122	—	538
$[(CH_3O)(CH_3)_2PS]_2SnBr_4$	105—107	—	540
$(CH_3S)(CH_3)_2PO$	40	1190	—
$[(CH_3S)(CH_3)_2PO]_2SnCl_4$	127.5—129	1085—1125	—
$[(CH_3S)(CH_3)_2PO]_2SnBr_4$	116—118	1075—1105	—

III. Reactions with Halogenating Agents

The most important electrophiles in this class are the elemental halogens and sulfuryl chloride. Two types of reactions can be distinguished here even more clearly than in the reactions with metal salts. Thus if the phosphorus atom carries alkoxy (or OH or SH) groups, the secondary attack by a nucleophile is directed toward the alkyl residue (or toward the proton); otherwise the nucleophile attacks the phosphorus atom, with replacement of sulfur by halogen (or, particularly with thionyl chloride, by oxygen).

Halogens oxidize thiophosphoryl compounds in the presence of water to form phosphoryl compounds and sulfate; the oxidizing agents in this reaction are hypohalite ions, which are characterized by their high nucleophilicity toward tetrahedral phosphorus. In spite of their preparative [90] and analytical importance [91], these and similar oxidations will not be discussed here.

1. Phosphine Sulfides and Diphosphine Disulfides

Triarylphosphine sulfides form crystalline adducts with ICl, IBr, and I_2 [19,37,92,93]. These adducts are more stable than those of the corresponding phosphine

[90] H. O. Fallscheer and J. W. Cook, J. Assoc. off. agric. Chemists 39, 691 (1956).
[91] K. Groves, J. agric. Food Chem. 6, 30 (1958).
[92] R. A. Zingaro and E. A. Meyers, Inorg. Chem. 1, 771 (1962); R. A. Zingaro and R. M. Hedges, J. physic. Chem. 65, 1132 (1961).
[93] R. A. Zingaro, Inorg. Chem. 2, 192 (1963).

oxides [19,37]. IR spectra [19,23] and X-ray structure investigations [19] indicate sulfur-halogen coordination and rule out the possibility of interaction between the halogen and the aromatic nuclei.

Table 4. Addition compounds R₂R'PS. Hal₂·ΔνP=S = difference between the P=S frequencies in the free phosphine sulfide and in the adduct.

R	R'	Hal₂	M.p. (°C)	ΔνP=S (cm⁻¹)	Ref.
C_6H_5	C_6H_5	ICl	124	−38	[92,93]
C_6H_5	C_6H_5	IBr	151.5	—	[92,93]
1-Naphthyl	C_6H_5	IBr	137—140	−29	[19]
C_6H_5	C_6H_5	I₂ [*]	140	−37	[92,93]
C_6H_5	1-Naphthyl	IBr	155—156	−20	[19]
C_6H_5	1-Naphthyl	I₂	157—160	−12	[19]
1-Naphthyl	C_6H_5	I₂	117—120	−23	[19]
1-Naphthyl	1-Naphthyl	I₂	174—176	−32	[19]
C_6H_5	4-Methyl-1-naphthyl	I₂	161—166	−10	[19]
4-Methyl-1-naphthyl	C_6H_5	I₂	177—180	−13	[19]
4-Methyl-1-naphthyl	4-Methyl-1-naphthyl	I₂	233—245	−17	[19]

[*] This compound has the composition $2(C_6H_5)_3PS·3\ I_2$.

The reactions with Cl₂ and Br₂ do not stop with the formation of the adducts, but lead, by substitution at the phosphorus atom (which has acquired a partial positive charge), to halogenophosphoranes and sulfur halides. For example, diarylthiophosphinyl chlorides give diaryltrichlorophosphoranes [94]; the analogous chlorination of SPCl₃ to PCl₅ has been known for a long time [95].

The unusually easy cleavage of tetraalkyldiphosphine disulfides *(19)* by Cl₂ [96,97], Br₂ [96–98], or SO₂Cl₂ [97–99] to dialkylthiophosphinyl halides *(20)* can be explained in a similar manner [100]:

$$R_2\overset{\overset{\text{S}}{\|}}{P}-\overset{\overset{\text{S}\ \overset{\curvearrowright}{\text{Br-Br}}}{}}{PR_2} \longrightarrow R_2\overset{\overset{\text{S}}{\|}}{\overset{\oplus}{P}}\underset{\text{Br}^\ominus}{\overset{\text{SBr}}{PR_2}} \longrightarrow 2\ R_2P(S)Br.$$

(19)　　　　　　　　　　　　　　　*(20)*

In agreement with this view, dithio-*(21a)* [101] and monothiohypophosphates *(21b)* [102] are cleaved by SO₂Cl₂ in a strongly exothermic reaction even below 0 °C, whereas the sulfur-free compounds *(21c)* must be heated under reflux [103].

$$(RO)_2\overset{\overset{\text{X}}{\|}}{P}-\overset{\overset{\text{Z}}{\|}}{P}(OR)_2 + SO_2Cl_2 \longrightarrow (RO)_2P(X)Cl$$

(21)　　　　　　　　　　　　　　　　+

$$(RO)_2P(Z)Cl + SO_2.$$

(a): X = Z = S; *(b)*: X = O, Z = S; *(c)*: X = Z = O

Since dialkylthiophosphinyl halides themselves react by the same principle with halogenating agents, the reaction of tetraalkyldiphosphine disulfides with an excess of Cl₂ or Br₂ yields dialkyltrihalogenophosphoranes [96], and the reaction with an excess of SO₂Cl₂ yields dialkylphosphinyl chlorides [98,104]. When SOCl₂ is used instead of SO₂Cl₂, the intermediate thiophosphinyl chlorides, though detectable [105], can no longer be isolated [105–107].

Thiophosphonyl dihalides [107,108] and phosphine sulfides [107–110] also form the corresponding phosphoryl compounds with SOCl₂. The determining factor in this reaction is the electron density, not at the phosphorus, but at the sulfur atom, as is shown by the marked decrease in reactivity in the order $(CH_3)_2P(S)Cl > (C_6H_5)CH_3P(S)Cl > C_6H_5P(S)Cl_2 > P(S)Cl_3$ [107]. A similar substituent effect is also observed in the reactions with SbF₃: phosphine sulfides and diphosphine disulfides give di- and trifluorophosphoranes respectively [111], trialkyl phosphorothionates do not react [111], and thiophosphonyl dichlorides [112] and thiophosphoramidic dichlorides [113] simply undergo halogen exchange.

2. Derivatives of Thiophosphoric, Thiophosphonic, and Thiophosphinic Acids

a) Esters, Ester Chlorides, and Ester Amides

Addition compounds of trialkyl phosphorothionates with I₂ have been detected in solution; they are more stable than the corresponding trialkyl phosphate adducts [27]. Trialkyl phosphorothionates react with Cl₂ and Br₂ even below 0 °C, with elimination of alkyl halide and formation of bis(alkoxy)phosphorylsulfenyl chlorides *(22)* (Michalski reaction) [114,115].

$$(RO)_3P=S\overset{\curvearrowright}{\ Cl-}Cl \longrightarrow \underset{\underset{Cl^\ominus}{}}{RO}\overset{\oplus}{\underset{O-R}{P}}\overset{SCl}{} \longrightarrow RCl + \underset{RO}{\overset{RO}{P}}\overset{SCl}{\underset{O}{}}$$

(22)

[94] W. G. Craig and W. A. Higgins, US-Pat. 2 727 073 (Dec. 13th, 1955), Lubrizol Corp.; Chem. Abstr. 50, 9445 (1956).

[95] Chevrier, C. R. hebd. Séances Acad. Sci. 68, 1174 (1869).

[96] W. Kuchen, H. Buchwald, K. Strolenberg, and J. Metten Liebigs Ann. Chem. 652, 28 (1962).

[97] R. Cölln and G. Schrader, German Publ. Pat. Appl. 1 054 453 (Feb. 12th, 1958) Farbenfabriken Bayer A.-G.; Chem. Zbl. 1959, 12 696.

[98] L. Maier, Chem. Ber. 94, 3051 (1961).

[99] H. Schlör and G. Schrader, German Publ. Pat. App. 1 067 021 (May 31st, 1958) Farbenfabr. Bayer A.-G.; Chem. Zbl. 1960, 5319.

[100] B. Miller, Topics in Phosphorus Chemistry, Vol. 2, Interscience Publ., New York-London-Sydney 1965, p. 142.

[101] L. Almasi and L. Paskucz, Chem. Ber. 96, 2024 (1963).

[102] J. Michalski, W. Stec, and A. Zwierzak, Bull. Acad. polon. Sci., Sér. Sci. chim. 13, 677 (1965).

[103] J. Michalski and T. Modro, Chem. and Ind. 1960, 1570.

[104] R. Cölln and G. Schrader, German Publ. Pat. Appl. 1 056 606 (May 14th, 1958) Farbenfabr. Bayer A.-G.; Chem. Zbl. 1959, 17 390.

[105] L. Maier, Angew. Chem. 71, 575 (1959); Chem. Ber. 94, 3056 (1961).

[106] H. J. Harwood and K. A. Pollart, US-Pat. 3 104 259 (July 20th, 1959) Monsanto Chem. Co.; Chem. Zbl. 1965, 38-2479.

[107] K. A. Pollart and H. J. Harwood, J. org. Chemistry 27, 4444 (1962).

[108] H. J. Harwood and K. A. Pollart, US-Pat. 3 082 256 (July 20th, 1959) Monsanto Chem. Co.; Chem. Zbl. 1964, 46-2222.

[109] H. J. Harwood and K. A. Pollart, J. org. Chemistry 28, 3430 (1963).

[110] L. Maier, Helv. chim. Acta 47, 120 (1964).

[111] R. Schmutzler, Inorg. Chem. 3, 421 (1964).

[112] R. Schmutzler, J. inorg. nucl. Chem. 25, 335 (1963); H. L. Boter and A. J. J. Ooms, Recueil Trav. chim. Pays-Bas 85, 21 (1966).

[113] G. Schrader, unpublished; cited in: Methoden der organ. Chemie (Houben-Weyl), 4th Edit., Georg Thieme-Verlag, Stuttgart 1964, Vol. XII/2, p. 753.

[114] J. Michalski and A. Skowronska, Chem. and Ind. 1958, 1199.

[115] J. Michalski, B. Pliszka-Krawiecka, and A. Skowronska, Roczniki Chem. 37, 1479 (1963).

Sulfuryl chloride is a better reagent than elemental halogen for the preparation of *(22)* [114–116]. Dialkyl ester amides, dialkyl aryl, and diaryl alkyl esters of thiophosphoric acid generally react in the same way [115]. However, conversion decreases rapidly in the order $(C_2H_5O)_3PS$ > $(C_6H_5O)(C_2H_5O)_2PS$ > $(4-O_2NC_6H_4O)(C_2H_5O)PS$ > $(4-O_2NC_6H_4O)_2(C_2H_5O)PS$ (the ethyl bis-(4-nitrophenyl) ester does not react at all), again showing the effect of the substituents on the nucleophilic strength of the thiono sulfur [115]; the reverse order applies in the ease of dealkylation by nucleophilic attack at the α-carbon atom [2].

Since compounds *(22)*, like alkanesulfenyl chlorides, add to multiple bond systems [114–117], the Michalski reaction is of great preparative interest.

Dialkyl thiophosphoryl chlorides *(23)* also enter into this reaction [118].

$$(C_2H_5O)_2P(S)Cl + SO_2Cl_2 \rightarrow \underset{(23)}{} \quad \begin{array}{c} C_2H_5O \\ P \\ ClS \quad Cl \end{array} \; O + C_2H_5Cl$$
$$+ \; SO_2.$$

On the other hand, a different reaction course is reported [119] for esters of phosphonochloridothionic acids *(24)*, the nucleophilic attack in the second step taking place, not on the carbon, but on the phosphorus atom.

$$\underset{(24)}{\begin{array}{c} C_2H_5 \quad S \\ P \\ C_2H_5O \quad Cl \end{array}} + SO_2Cl_2 \rightarrow \begin{array}{c} C_2H_5 \quad SCl \\ P \\ C_2H_5O \quad O \end{array} + SOCl_2.$$

This type of reaction is necessarily also encountered with triphenyl phosphorothionate, which gives tris-(phenoxy)dichlorophosphorane (or ionic products having the same overall composition) with Cl_2 [115] or SO_2Cl_2 [115, 120].

The formation of methylphosphonochloridates from dialkyl phosphonothionates and thionyl chloride [121] may also proceed *via* a sulfenyl chloride, which splits off sulfur when heated.

The reaction of trialkyl phosphorothionates with alkanesulfenyl chlorides *(25)* [122] also follows the scheme of the Michalski reaction.

$$(CH_3O)_3PS + CH_3SCl \rightarrow (CH_3O)_3\overset{\oplus}{P}SSCH_3 \; Cl^{\ominus} \rightarrow$$
$$(25)$$

$$\begin{array}{c} CH_3O \\ P \\ CH_3O \quad SSCH_3 \end{array} O + CH_3Cl$$

b) Thioacids and Thioacid Anhydrides

Dialkyl phosphorothioic acids and their salts, like the triesters, react with Cl_2 [123] or SO_2Cl_2 [123, 124] to form sulfenyl chlorides *(22)*. However, the overall

$$(RO)_2P(S)OH + Cl_2 \; \rightarrow \; (RO)_2P(O)SCl + HCl$$
$$(22)$$

equation conceals the fact that *(22)*, as an extremely strong electrophile, immediately reacts with the dialkyl phosphorothioate, which is present in excess, to form the disulfide *(26)*, which is then chlorinated with cleavage to form *(22)* [123].

$$(RO)_2P(S)OH \; + \; (22) \; \xrightarrow[-HCl]{} \; (RO)_2P(O)SSP(O)(OR)_2$$
$$(26) \; \xrightarrow{+Cl_2} \; 2\,(22)$$

The alkylalkoxyphosphorylsulfenyl chlorides *(27a)*, which can be prepared in a similar manner, decompose readily with deposition of sulfur [125, 126], while the dialkyl- [125] and diarylphosphorylsulfenyl chlorides [127] *(27b)* decompose during attempts to prepare them.

$$\begin{array}{c} R \quad O \\ P \\ A \quad SCl \end{array} \rightarrow \begin{array}{c} R \quad O \\ P \\ A \quad Cl \end{array} + S$$
$$(27) \; (a): A = RO; \; (b): A = R$$

The chlorination of dialkyl phosphorodithioates *(28)*, which is used for the preparation of dialkyl phosphorochloridothioates *(29)*, also proceeds through the stages sulfenyl chloride, disulfide, and sulfenyl chloride [71, 128, 129]; the bis(alkoxy)thiophosphoryl-

$$(RO)_2PS_2H \; \xrightarrow[-HCl]{+Cl_2} \; \begin{array}{c} RO \quad S \\ P \\ RO \quad SCl \end{array} \rightarrow \begin{array}{c} RO \quad S \\ P \\ RO \quad Cl \end{array} + S$$
$$(28) \qquad\qquad (29)$$

sulfenyl chlorides *(28)* have themselves only recently become available by another method [129, 130].

The chlorination of diaryldithiophosphinic acids yields diaryltrichlorophosphoranes [94, 131], and that of aryldithiophosphonic anhydrides *(30)* leads *via* isolable thiophosphonyl dichlorides *(31)* to tetrachlorophosphoranes *(32)* [132].

[116] W. H. Mueller, R. M. Rubin, and P. E. Butler, J. org. Chemistry 31, 3537 (1966).

[117] J. Michalski, B. Borecka, and S. Musierowicz, Bull. Acad. polon. Sci., Sér. Sci. chim. 6, 159 (1958); B. Borecka, T. Kapecka, and J. Michalski, Roczniki Chem. 36, 87 (1962); J. Michalski and S. Musierowicz, ibid. 36, 1655 (1962).

[118] G. C. Vegter, US-Pat. 3081329 (Jan. 26th, 1960) Shell Oil Comp.; Chem. Zbl. 1965, 1-2370.

[119] Yu. G. Gololobov and V. V. Semidetko, Ž. obšč. Chim. 36, 930 (1966).

[120] A. C. Poshkus, J. E. Herweh, and L. F. Hass, J. Amer. chem. Soc. 80, 5022 (1958).

[121] Z. Pelchowicz, J. chem. Soc. (London) 1961, 238.

[122] J. B. Douglass and W. J. Evers, J. org. Chemistry 29, 419 (1964).

[123] B. Lenard-Borecka and J. Michalski, Roczniki Chem. 31, 1167 (1957).

[124] J. Michalski and B. Lenard, Roczniki Chem. 30, 655 (1956).

[125] C. Borecki, J. Michalski, and S. Musierowicz, J. chem. Soc. (London) 1958, 4081.

[126] J. Michalski and A. Ratajczak, Chem. and Ind. 1959, 539; Roczniki Chem. 36, 911 (1962).

[127] W. G. Craig, US-Pat. 2724726 (Nov. 22nd, 1955) Lubrizol Corp.; Chem. Abstr. 50, 10129 (1956).

[128] L. Malatesta, Gazz. chim. ital. 81, 596 (1951).

[129] L. Almasi and L. Paskucz, Chem. Ber. 98, 3546 (1965).

[130] L. Almasi and A. Hantz, Chem. Ber. 97, 661 (1964); L. Almasi and L. Paskucz, ibid. 98, 613 (1965).

[131] W. A. Higgins, P. W. Vogel, and W. G. Craig, J. Amer. chem. Soc. 77, 1864 (1955); C. Stuebe, W. M. LeSuer, and G. R. Norman, ibid. 77, 3526 (1955).

[132] H. Z. Lecher, R. A. Greenwood, K. C. Whitehouse, and T. H. Chao, J. Amer. chem. Soc. 78, 5018 (1956); H. Z. Lecher and R. A. Greenwood, US-Pat. 2870204 (Jan. 20th, 1959) Amer. Cyanamid Co.; Chem. Abstr. 53, 11306 (1959).

1020

Angew. Chem. internat. Edit. / Vol. 6 (1967) / No. 12

406

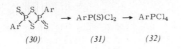

(30)   *(31)*   *(32)*

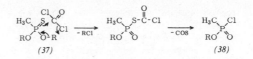

(37)   *(38)*

IV. Reactions with Carbonyl and Phosphoryl Compounds

As has been mentioned, the thiophosphoryl sulfur should not be very reactive toward the "hard" carbonyl carbon and phosphoryl phosphorus. Accordingly, only a few reactions with particularly reactive carbonyl groups have been reported, and most of these are not very characteristic.

The replacement of carbonyl oxygen by sulfur [133], which has been used in preparative chemistry, and the more recently discovered, analogous replacement of phosphoryl oxygen by sulfur [134, 135] with the aid of phosphorus(v) sulfide are of considerable importance. This reaction is only mentioned in passing, since it is not confined to thiophosphoryl sulfur, but even takes place with phosphorus sulfides, such as P₄S₃ [136], which only contain P–S–P bonded sulfur.

1. Reactions with Carbonyl Compounds

The relatively smooth reaction of trialkyl phosphates with acyl chlorides to form dialkyl acyl phosphates [137] has evidently been achieved in the thiophosphate series only for a few diester monoamides and monoester diamides *(33)* with chloroacetyl chloride *(34)* [138].

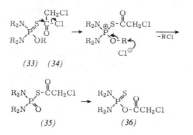

(33)   *(34)*

(35)   *(36)*

S-Acyl thiophosphates *(35)* rearrange to the more stable O-acyl derivatives *(37)* [76].

Dialkyl methylphosphonothioates *(37)*, like their oxygen analogues, react with phosgene [139] or oxalyl chloride [121] to form esters of phosphonochloridic acids *(38)*.

Contrary to earlier reports, triphenylphosphine sulfide and phenyl isocyanate do not form phenyl isothiocyanate [140] but diphenylcarbodiimide [110].

The phosphine sulfide, like esters of thiophosphonic acids [141], is effective even in catalytic quantities. Since neither carbon oxysulfide nor triphenylphosphine oxide could be detected in the reaction products [110] the reaction appears to proceed by a course different from that followed by the formation of carbodiimide in the presence of phosphine oxide [142].

Triphenylphosphine sulfide is desulfurized by diphenylketene to form the phosphine oxide [140], and trimethyl thionophosphate is desulfurized to trimethyl phosphate by chloral [143]. Both reactions are formally similar to the Wittig olefination.

$(C_6H_5)_3PS + (C_6H_5)_2C=C=O \rightarrow$
$$(C_6H_5)_3PO + 1/n[(C_6H_5)_2C=C=S]_n$$

$(CH_3O)_3PS + Cl_3CCHO \rightarrow (CH_3O)_3PO + 1/n[Cl_3CCHS]_n$

2. Reactions with Phosphoryl Compounds

Thiophosphoryl compounds exchange their sulfur for phosphoryl oxygen at 150–200 °C, the exchange taking place more readily as the sum of the electronegativities of the substituents on the thiophosphoryl group increases and as that of the substituents on the phosphoryl group decreases. The reaction rate is directly proportional to this electronegativity difference [144], and so corresponds to the substituent effects in the formation of pyrophosphate from triesters and diester chlorides of phosphoric acid [133].

Thiophosphoryl chloride is therefore the most suitable reagent for the conversion of phosphonyl [135, 144] and phosphinyl chlorides [144, 145] into their thio analogues. Under similar conditions, phosphine oxides are converted into phosphine sulfides by salts or O-alkyl esters of dimethylthiophosphinic acid [146]. O-Methyl dimethylthiophosphinate reacts with its S-methyl isomer to give esters of the dithiophosphinic acid, among other products [146], and small quantities of O,O-diphenyl S-methyl dithiophosphate are formed during the thermal isomerization of methyl diphenyl thiophosphate [147].

[133] A. Kekulé, Ann. Chem. Pharm. 90, 309 (1854).

[134] M. I. Kabachnik and N. N. Godovikov, Doklady Akad. Nauk SSSR 110, 217 (1956); N. N. Godovikov and M. I. Kabachnik, Ž. obšč. Chim. 31, 1628 (1961).

[135] E. Uhing, K. Rattenbury, and A. D. F. Toy, J. Amer. chem. Soc. 83, 2299 (1961).

[136] K. H. Rattenbury, US-Pat. 2993929 (July 25th, 1961) Victor Chemical Works; Chem. Abstr. 56, 505 (1962).

[137] Y. Nishizawa, M. Nakagawa, and T. Mizutani, Botyu-Kagaku (Sci. Insect Control) 26, 4 (1961); Chem. Zbl. 1964, 8-0817.

[138] G. A. Saul, J. W. Baker, and K. L. Godfrey, US-Pat. 2983595 (May 9th, 1961) Monsanto Chem. Co.; Chem. Abstr. 55, 19122 (1961); P. C. Hamm and G. A. Saul, US-Pat. 3020141 (March 3rd, 1958) Monsanto Chem. Co.; Chem. Abstr. 57, 2075 (1962).

[139] J. I. G. Cadogan, J. chem. Soc. (London) 1961, 3067.

[140] H. Staudinger, G. Rathsam, and F. Kjelsberg, Helv. chim. Acta 3, 853 (1920).

[141] J. J. Monagle, J. org. Chemistry 27, 3851 (1962).

[142] J. J. Monagle and J. V. Mengenhauser, J. org. Chemistry 31, 2321 (1966).

[143] H. Sohr and K.-H. Lohs, Z. Chem. 7, 153 (1967).

[144] L. C. D. Groeneweghe and J. H. Payne, J. Amer. chem. Soc. 83, 1811 (1961).

[145] L. C. D. Groeneweghe, US-Pat. 3206442 (March 9th, 1962) Monsanto Chem. Co.; Chem. Abstr. 63, 15007 (1965).

[146] H. Teichmann, P. Gregorzewski, and G. Hilgetag, unpublished.

[147] H. Teichmann and G. Hilgetag, unpublished.

The observed substituent effects require a strong positive charge at the phosphorus atom of the thiophosphoryl compound and a high electron density at the phosphoryl oxygen of the phosphoryl compound. It may therefore be concluded that the exchange reaction is initiated by a nucleophilic attack on the phosphorus atom of the thiophosphoryl compound by the phosphoryl oxygen (similar to the pyrophosphate synthesis mentioned above). However, the sulfurization of phosphine oxides by the thiophosphinate ion can hardly be understood in this way. Similarly, the formation of diphenyldithiophosphinate *(40)* from diphenylphosphinothioic amide *(39)* and chloride [148] can be explained only by the nucleophilic reactivity of the sulfur.

(39)   *(40)*

$(C_6H_5)_2PS_2^{\ominus} + {}^1/n \; [(C_6H_5)_2PN]_n$

V. Reactions with Alkylating Agents

The primary products of the action of alkylating agents on thiophosphoryl compounds are alkylthiophosphonium salts. However, these can be isolated unchanged only if they have no alkoxy groups attached to the phosphorus atom. Special conditions are required for the preparation of alkoxyalkylthiophosphonium salts [84, 89]. Alkylations of compounds such as $(RO)_2P(S)OH$ [149], $(RO)_2P(S)NHSO_2Ar$ [150], or $(RO)_2P(S)NHCOAr$ [151], which undergo tautomeric changes, will not be dealt with here.

1. Thiophosphoric, Thiophosphonic, and Thiophosphinic Amides; Phosphine Sulfides

Triamides of thiophosphoric acid are alkylated to onium salts *(41a)* by alkyl halides, in some cases even at room temperature [152, 153]; in contrast to *N*-alkylamides, *N*-arylamides do not react even under more rigorous conditions [152]. The compounds *(41b)*–*(41d)* are formed in a similar manner from thiophosphonic [153] and thiophosphinic amides [36, 148] and from trialkylphosphine sulfides [35, 154, 155]. When the products are heated under vacuum, alkyl halides are

[148] *A. Schmidpeter* and *H. Groeger*, Z. anorg. allg. Chem. *345*, 106 (1966).

[149] *M. I. Kabachnik, S. T. Joffe*, and *T. A. Mastryukova*, Ž. obšč. Chim. *25*, 684 (1955).

[150] *L. Almasi* and *A. Hantz*, Rev. Chim. (Bucuresti) *9*, 433 (1964); Chem. Abstr. *62*, 6417 (1965).

[151] *L. Almasi* and *L. Paskucz*, Chem. Ber. *99*, 3293 (1966).

[152] *H. Tolkmith*, J. Amer. chem. Soc. *85*, 3246 (1963); *H. Tolkmith*, US-Pat. 3074993 (April 2nd, 1962) Dow Chem. Co.; Chem. Abstr. *59*, 1542 (1963).

[153] *A. J. Burn* and *J. I. G. Cadogan*, J. chem. Soc. (London) *1961*, 5532.

[154] *W. Steinkopf* and *R. Bessaritsch*, J. prakt. Chem. [2] *109*, 230 (1925).

[155] *L. Horner* and *A. Winkler*, Tetrahedron Letters *1964*, 175.

sometimes split off again [153]. The salts *(41d)* obtained from aromatic phosphine sulfides are difficult to prepare with alkyl halides [26, 156]; they are best isolated as the fluoroborates or the like [36].

	(a): A, B, C = R^1R^2N
	(b): A, B = R^1R^2N; C = R^3
	(c): A = R^1R^2N; B, C = R^3
	(d): A, B, C = R^1

(41)

2. *O*-Alkyl Thiophosphates, Thiophosphonates, and Thiophosphinates

a) Reactions with Alkyl Halides

Since alkoxyalkylthiophosphonium salts *(42)* are cleaved at the *O*-alkyl bond by nucleophiles, the action of alkyl halides on alkyl esters of the thionophosphoric, thionophosphonic, and thionophosphinic acid series leads to thiol esters *(43)* (Pistschimuka reaction).

(42)   *(43)*

Suitable starting materials are trialkyl thionophosphates [69, 153, 157–159], dialkyl aryl thionophosphates [158], trialkyl dithiophosphates [69, 153, 157], trialkyl trithiophosphates [69], dialkyl thiophosphoric amides and alkyl thiophosphoric diamides [153, 157], and esters of thiophosphonic [153, 157, 158, 160] and thiophosphinic acids [158, 161].

The Pistschimuka reaction is analogous to the Michalski reaction (Section III, 2), but requires much more rigorous conditions (heating for many hours at 100–150 °C); its preparative value is frequently limited by poor yields and by side reactions such as the formation of trialkylsulfonium salts. However, these disadvantages can be overcome to a large extent by the use of strongly polar solvents, which considerably increase the reaction rate [153, 157]. The substituents A and B favor the reaction in the order Cl < RS < RO < R < R_2N [153] by increasing the nucleophilic strength of the sulfur atom.

b) Reactions without the Addition of Alkylating Agents

The role of the alkyl halide in the Pistschimuka reaction can be taken over by the thiophosphoryl com-

[156] *F. G. Mann* and *J. Watson*, J. org. Chemistry *13*, 502 (1948).

[157] *A. J. Burn, J. I. G. Cadogan*, and *A. B. Foster*, Chem. and Ind. *1961*, 591.

[158] *H. Maier-Bode* and *G. Kötz*, German. Publ. Pat. Appl. 1014107 (Dec. 28th, 1954); Chem. Zbl. *1958*, 11075.

[159] *R. A. McIvor, G. D. McCarthy*, and *G. A. Grant*, Canad. J. Chem. *34*, 1819 (1956); *Ch. Walling* and *R. Rabinowitz*, J. Amer. chem. Soc. *81*, 1243 (1959); *W. A. Sheppard*, J. org. Chemistry *26*, 1460 (1961).

[160] *M. I. Kabachnik* and *T. A. Mastryukova*, Izvest. Akad. Nauk SSSR, Ser. Chim. *1953*, 163; *M. I. Kabachnik, T. A. Mastryukova*, and *N. I. Kurochkin*, ibid. *1956*, 193.

[161] *T. A. Mastrykova, T. A. Melentyeva*, and *M. I. Kabachnik*, Ž. obšč. Chim. *35*, 1196 (1965).

1022

Angew. Chem. internat. Edit. / Vol. 6 (1967) / No. 12

408

pound itself, since the esters mentioned all have alkylating properties. Trialkyl [70, 153, 162–164], dialkyl aryl [164, 165–168], and alkyl diaryl thionophosphates [164, 165, 169, 170], dialkyl thionophosphoryl halides [33, 170], alkyl thionophosphoryl dihalides [170, 171], O,O,S-trialkyl dithiophosphates [172], and alkyl esters of phosphonothionic [173] and phosphinothionic acids [173] can therefore be thermally isomerized into the thiol esters. However, the conditions are often even more severe than those of the Pistschimuka reaction, so that the formation of the thiol esters is frequently followed by secondary reactions (cf. the reviews [2, 174]).

Solvents having a high dielectric constant again favor this reaction [153, 163]. On the other hand, the substituent effect is opposite to that observed in the Pistschimuka reaction; the isomerization depends on the alkylating power of the ester, which is increased by electron-attracting substituents at the phosphorus atom [2, 174]).

The thiophosphoryl sulfur is alkylated, not only by the thiono ester introduced, but also by the resulting thiol ester, provided that it still contains further P—O-alkyl groups; in this case the isomerization is autocatalytic [163]. Mixtures of thiono esters give mixed thiol esters [163].

$(RO)_3PS + (R'O)_3PS \rightarrow$

$(RO)_2P(O)(SR) + (R'O)_2P(O)(SR') + (RO)_2P(O)(SR') +$
$+ (R'O)_2P(O)(SR).$

In contrast to the thermal isomerization, the isomerization induced by Lewis acids (cf. Section II, 4b) often proceeds almost quantitatively at or below room temperature. Since the thiol ester adducts formed can be broken down into their components, thiono esters that are unsuitable for thermal isomerization because of the instability of the thiol isomers can be converted in high yields into the thiol esters with the aid of Lewis acids [47, 58, 86]. The substituent effects are of the same nature as in the thermal isomerization. The rate of

isomerization also depends on the acceptor strength of the Lewis acid, and decreases rapidly e.g. in the order $SnCl_4 > SnBr_4 > SnI_4$ [58, 86].

The thermal isomerization takes place more readily in allyl esters (44) than in alkyl esters; the reaction undoubtedly proceeds by a different mechanism in this case, since the products obtained are those of a Claisen rearrangement [175].

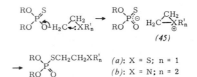

Finally, a further considerable increase in the rate of isomerization is observed in phosphorothionates containing a β-alkylthio [176–178] or dialkylamino group [179–182] in an ester residue. The same is true of β-alkylthioalkyl phosphoramidothionates [183] and phosphonothionates [178, 184]. In these cases the reaction is no longer induced by the nucleophilicity of the thiophosphoryl sulfur; instead, the neighboring group effect of the nucleophilic center in the β position leads to the ready release of a cyclic sulfonium (45a) or ammonium cation (45b), which S-alkylates the remaining thiophosphate ion in the usual manner.

The intermediate formation of the thiiranium [177] and aziridinium ions [179–181] has been demonstrated in various ways.

Received: March 6th, 1967   [A 604 IE]
German version: Angew. Chem. 79, 1077 (1967)
Translated by Express Translation Service, London

[162] G. Hilgetag, G. Schramm, and H. Teichmann, J. prakt. Chem. [4] 8, 73 (1959).

[163] H. Teichmann, unpublished.

[164] G. Hilgetag, G. Schramm, and H. Teichmann, Angew. Chem. 69, 205 (1957).

[165] G. Hilgetag and G. Schramm, unpublished.

[166] Z. M. Bakanova, J. A. Mandelbaum, and N. N. Melnikov, Ž. obšč. Chim. 26, 2575 (1956); H. L. Morrill, US-Pat. 2601219 (July 17th, 1948) Monsanto Chem. Co.; Chem. Abstr. 46, 8322 (1952); V. Tichy, Chem. Zvesti 9, 3 (1955); Chem. Zbl. 1959, 7747.

[167] J. B. McPherson and G. A. Johnson, J. agric. Food Chem. 4, 42 (1956).

[168] R. L. Metcalf and R. B. March, J. econ. Entomol. 46, 288 (1953).

[169] G. Schrader and R. Gönnert, German Pat. 949230 (April 3rd, 1955) Farbenfabr. Bayer A.-G.; Chem. Zbl. 1957, 3078.

[170] G. Hilgetag, G. Lehmann, and W. Feldheim, J. prakt. Chem. [4] 12, 1 (1960); W. Feldheim, Diploma Thesis, Jena 1954.

[171] N. N. Godovikov and M. I. Kabachnik, Ž. obšč. Chim. 31, 1628 (1961); E. M. Popov and N. E. Medenikova, ibid. 32, 3080 (1962).

[172] G. Hilgetag, H. Teichmann, and L. Nguyen, unpublished.

[173] G. Hilgetag, P. Gregorzewski, and H. Teichmann, unpublished.

[174] H. Teichmann and G. Lehmann, S.-B. dtsch. Akad. Wiss. Berlin, Kl. Chem., Geol. Biol. 1962, No. 5.

[175] A. N. Pudovik and J. M. Aladsheva, Ž. obšč. Chim. 30, 2617 (1960).

[176] A. Henglein and G. Schrader, Z. Naturforsch. 10b, 12 (1955); N. Muller and J. Goldenson, J. Amer. chem. Soc. 78, 5182 (1956); J. A. Mandelbaum, N. N. Melnikov, and V. I. Lomakina, Ž. obšč. Chim. 26, 2581 (1956); W. Dedek, Atompraxis 10, No. 4, 1 (1964); A. H. Ford-Moore and G. W. Wood, Brit. Pat. 851590 (March 30th, 1954) Ministry of Supply; Chem. Abstr. 55, 10316 (1961).

[177] T. R. Fukuto and R. L. Metcalf, J. Amer. chem. Soc. 76, 5103 (1954).

[178] F. W. Hoffmann and T. R. Moore, J. Amer. chem. Soc. 80, 1150 (1958).

[179] T. R. Fukuto and E. M. Stafford, J. Amer. chem. Soc. 79, 6083 (1957).

[180] G. Hilgetag and H. Cierpka, unpublished; Diploma Thesis H. Cierpka, Humboldt-Univ. Berlin 1958.

[181] A. Calderbank and R. Ghosh, J. chem. Soc. (London) 1960, 637.

[182] R. Ghosh and J. F. Newman, Chem. and Ind. 1955, 118; L.-E. Tammelin, Acta chem. scand. 11, 1738 (1957); G. Hilgetag and W. Hartmann, unpublished; R. Ghosh, Brit. Pat. 738839 (Nov. 9th, 1952) I.C.I. Ltd.; Chem. Abstr. 50, 13983 (1956).

[183] M. I. Kabachnik et al., Ž. obšč. Chim. 29, 2182 (1959).

[184] F. W. Hoffmann, J. W. King, and H. O. Michel, J. Amer. chem. Soc. 83, 706 (1961); M. I. Kabachnik and T. A. Medved, Izvest. Akad. Nauk SSSR, Ser. Chim. 1961, 604.

[Reprinted from the Journal of the American Chemical Society, **90**, 4076 (1968).]

The Relative Nucleophilicity of Some Common Nucleophiles toward Sulfinyl Sulfur. The Nucleophile-Catalyzed Hydrolysis of Aryl Sulfinyl Sulfones[1]

John L. Kice and Giancarlo Guaraldi

Contribution from the Department of Chemistry, Oregon State University, Corvallis, Oregon 97331. Received December 1, 1967

Abstract: The relative reactivity of seven common nucleophiles in a displacement reaction at sulfinyl sulfur (eq 8) may be determined from kinetic data on their catalysis of the hydrolysis of an aryl sulfinyl sulfone (IIa) in aqueous dioxane. These data for sulfinyl sulfur (Table IV) are compared with analogous data for substitutions at sulfenyl sulfur (eq 11), peroxide oxygen (eq 9), sp^3 carbon (eq 10), and sulfonyl sulfur (eq 12). This comparison reveals the following interesting points. (1) The substitution at sulfenyl sulfur shows a completely different pattern of reactivity ($F^- > AcO^- \gg Cl^-$) than the one at sulfinyl sulfur ($Cl^- > AcO^- > F^-$). (2) In the substitutions at both sulfenyl sulfur and peroxide oxygen, nucleophiles such as iodide or thiocyanate show a considerably greater reactivity compared to chloride ion than they do in the one at sulfinyl sulfur. (3) The relative reactivities of the various nucleophiles in the substitutions at sulfinyl sulfur and sp^3 carbon are very similar. These facts are discussed with reference to the theory of hard and soft acids and bases (HSAB). The conclusions are (a) that sulfonyl sulfur is a much harder and sulfenyl sulfur a significantly softer electrophilic center than sulfinyl sulfur, and (b) that sulfinyl sulfur is a medium soft electrophilic center analogous to sp^3 carbon.

E quation 1 is a generalized representation of a nucleophilic substitution reaction. In a protic

$$Nu^- + SX \longrightarrow NuS + X^- \qquad (1)$$

solvent the relative reactivity of a series of nucleophiles in such a reaction depends greatly on the nature of the center in the substrate SX which is being attacked by the nucleophile.[2] Nucleophiles which are "hard" bases,[3] *i.e.*, of low polarizability and high proton basicity, show up to particular advantage in substitutions involving attack on such centers as carbonyl carbon[4] or tetracoordinate phosphorus.[5] On the other hand, nucleophiles which are "soft" bases,[3] *i.e.*, of high polarizability and low proton basicity, do particularly well in substitutions involving centers such as divalent oxygen[6] or Pt^{2+}.[7] Edwards and Pearson,[8] in what has become a classic paper, attempted to evaluate and explain the relative importance of basicity, polarizability, and other effects in determining the reactivity of a nucleophile toward these and other centers in protic

solvents. Recently, Pearson and Songstad[9] have shown how the various data and conclusions can also be easily understood within the framework of the theory of hard and soft acids and bases (HSAB).

In their paper, Edwards and Pearson[8] made some predictions about the reactivity patterns that might be observed for nucleophiles reacting with different sulfur centers, but they pointed out that unfortunately no quantitative data were yet available for any of these centers. In fact for sulfinyl sulfur, $-S(=O)-$, there were not even any qualitative data. Since there has been much interest in recent years in reactions involving substitution at the sulfinyl sulfur of sulfoxides, sulfinate esters, and related compounds,[10-16] it would seem that having quantitative data for sulfinyl sulfur would be of considerable value. Furthermore, comparison of this data with suitable data for substitutions at other sulfur

(1) This research supported by the Directorate of Chemical Sciences, Air Force Office of Scientific Research, Grant AF-AFOSR-106-65. Preliminary communication: J. L. Kice and G. Guaraldi, *Tetrahedron Letters*, 6135 (1966).

(2) J. O. Edwards, "Inorganic Reaction Mechanisms," W. J. Benjamin, Inc., New York, N. Y., 1965, pp 51–72.

(3) R. G. Pearson, *J. Am. Chem. Soc.*, **85**, 3533 (1963); *Science*, **151**, 172 (1966).

(4) W. P. Jencks and J. Carriuolo, *J. Am. Chem. Soc.*, **82**, 1778 (1960).

(5) Reference 2, pp 39–63, 177–180.

(6) J. O. Edwards, "Peroxide Reaction Mechanisms," J. O. Edwards, Ed., Interscience Division, John Wiley and Sons, Inc., New York, N. Y., 1962, pp 67–106.

(7) U. Belluco, M. Martelli, and A. Orio, *Inorg. Chem.*, **5**, 592 (1966)

(8) J. O. Edwards and R. G. Pearson, *J. Am. Chem. Soc.*, **84**, 16 (1962).

(9) R. G. Pearson and J. Songstad, *ibid.*, **89**, 1827 (1967).

(10) K. Mislow, T. Simons, J. T. Melillo, and A. L. Ternay, *ibid.*, **86**, 1452 (1964).

(11) C. R. Johnson, *ibid.*, **85**, 1020 (1963); C. R. Johnson and D. McCants, Jr., *ibid.*, **87**, 5404 (1965); C. R. Johnson and W. G. Phillips, *Tetrahedron Letters*, 2101 (1965).

(12) P. Bickart, M. Axelrod, J. Jacobus, and K. Mislow, *J. Am. Chem. Soc.*, **89**, 697 (1967); K. Mislow, M. M. Green, P. Laur, J. T. Melillo, T. Simmons, and A. L. Ternay, *ibid.*, **87**, 1958 (1965); K. K. Anderson, *Tetrahedron Letters*, 93 (1962).

(13) J. Day and D. J. Cram, *J. Am. Chem. Soc.*, **87**, 4398 (1965).

(14) D. Landini, F. Montanari, H. Hogeveen, and C. Maccagnani, *Tetrahedron Letters*, 2691 (1964); G. Modena, G. Scorrano, D. Landini, and F. Montanari, *ibid.*, 3309 (1966); J. H. Krueger, *Inorg. Chem.*, **5**, 132 (1966).

. (15) T. Higuchi and K. H. Gensch, *J. Am. Chem. Soc.*, **88**, 5486 (1966); T. Higuchi, I. H. Pitman, and K. H. Gensch, *ibid.*, **88**, 5676 (1966).

(16) S. Allenmark, *Acta Chem. Scand.*, **15**, 928 (1961); **17**, 2711, 2715 (1963); **19**, 1, 1667, 2075 (1965).

410

centers would then allow a test of some of the Edwards and Pearson[8] predictions.

In earlier work,[17] we have shown that the hydrolysis of aryl sulfinyl sulfones II (eq 2) in aqueous dioxane

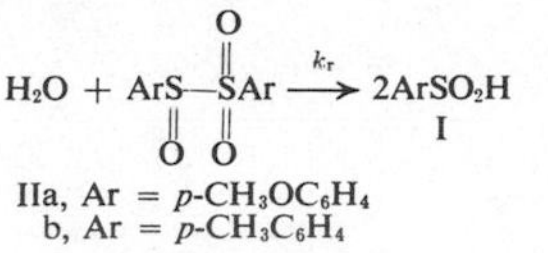

$$H_2O + ArS\overset{O}{\underset{O}{\overset{\|}{\underset{\|}{S}}}}Ar \xrightarrow{k_r} 2ArSO_2H \quad (2)$$
$$I$$

IIa, Ar = p-CH$_3$OC$_6$H$_4$
b, Ar = p-CH$_3$C$_6$H$_4$

can be markedly catalyzed by the addition of small amounts of various nucleophiles. The mechanism of this nucleophile-catalyzed hydrolysis of II is as shown in Chart I. From kinetic measurements of the catalytic

Chart I. Mechanism of Nucleophile-Catalyzed Hydrolysis of Aryl Sulfinyl Sulfones

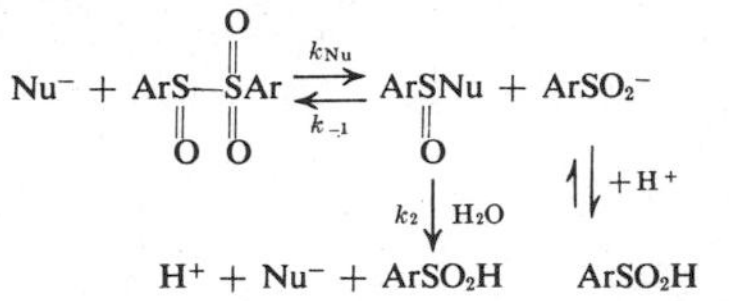

$$Nu^- + ArS\overset{O}{\underset{O}{\overset{\|}{\underset{\|}{S}}}}Ar \underset{k_{-1}}{\overset{k_{Nu}}{\rightleftharpoons}} Ar\overset{O}{\underset{O}{\overset{\|}{\underset{\|}{S}}}}Nu + ArSO_2^- \quad (3a)$$

$$\downarrow k_2 \Big| H_2O \qquad \Big\uparrow +H^+$$

$$H^+ + Nu^- + ArSO_2H \qquad ArSO_2H \quad (3b)$$

effect of a series of nucleophiles under appropriate conditions, one can extract for each nucleophile a value of k_{Nu}, the rate constant for nucleophilic attack of Nu$^-$ on the sulfinyl sulfur of II. These k_{Nu} values provide the type of quantitative data on nucleophilic reactivity in a substitution at sulfinyl sulfur that is desired.

Results

Previous studies[17] using the three halide ions I$^-$, Br$^-$, and Cl$^-$ as catalysts for the hydrolysis of IIa and IIb in 60% dioxane containing 0.01–0.80 M HClO$_4$ have shown that under these conditions the kinetics of the nucleophile-catalyzed hydrolysis are such that

$$k_r - k_r^0 = k_{Nu}(Nu^-) + k_{Nu}'(Nu^-)(H^+) \quad (4)$$

where k_r = experimental first-order rate constant for hydrolysis of II in the presence of Nu$^-$ and k_r^0 = experimental first-order rate constant for hydrolysis of II in the absence of Nu$^-$ under otherwise identical conditions. The relative magnitudes of k_{Nu} and k_{Nu}' in all cases were such that at (H$^+$) < 0.10 M the k_{Nu} term was the almost exclusive contributor to the nucleophile-catalyzed rate. All of the solutions used in this previous study[17] contained enough perchloric acid ((H$^+$) > 0.01 M) so that the ArSO$_2^-$ formed in eq 3a was immediately and completely converted to the sulfinic acid, ArSO$_2$H. Because of this the reaction ArSO$_2^-$ + ArS(O)Nu → II + Nu$^-$, i.e., the reaction having the rate constant k_{-1}, played no role under these conditions. This was confirmed[17] by the fact that runs to which a significant amount of ArSO$_2^-$ was added initially showed the same rate as those without any added sulfinate.

This procedure of avoiding the potential complications introduced by the k_{-1} reaction via the expedient of operating in a medium of sufficient acidity to ensure complete protonation of all ArSO$_2^-$ produced in eq 3a is fine for hydrolyses catalyzed by nucleophiles such as I$^-$, Br$^-$, or Cl$^-$, which are such weak bases that they will not be protonated themselves by a medium of this

(17) J. L. Kice and G. Guaraldi, *J. Am. Chem. Soc.*, **89**, 4113 (1967).

acidity, but it is, of course, totally unsuited to any study of the catalysis by such nucleophiles as AcO$^-$ or F$^-$, which are more basic than ArSO$_2^-$ itself. Furthermore, since preliminary experiments showed that the ArSO$_2^-$ + ArS(O)Nu reaction is indeed important under conditions where the pH of the medium is such that a significant fraction of the ArSO$_2^-$ remains unprotonated, one must take it into account when dealing with kinetic data obtained under these conditions.

For situations where the k_{-1} reaction is important $(k_r - k_r^0)$ will be given by[18]

$$k_r - k_r^0 = \frac{k_{Nu}(Nu^-)}{\left(1 + \dfrac{k_{-1}(ArSO_2^-)}{k_2}\right)} \quad (5a)$$

This can be written in a more useful form for our purposes as

$$\frac{1}{k_r - k_r^0} = \frac{1}{k_{Nu}(Nu^-)}\left[1 + \frac{k_{-1}(ArSO_2^-)}{k_2}\right] \quad (5b)$$

Examination of eq 5b suggests how one may obtain accurate values of k_{Nu} under conditions where the k_{-1} reaction is important. Specifically, one carries out a series of runs at constant (Nu$^-$) but with varying concentrations of initially added ArSO$_2^-$, the amount of ArSO$_2^-$ added being always considerably larger than the total amount of the same species that will be formed by the hydrolysis of II. These runs are carried out in a buffer whose pH is such that all ArSO$_2^-$ will remain unprotonated. One then plots $1/(k_r - k_r^0)$ for these runs vs. (ArSO$_2^-$). The intercept of this plot will be $1/k_{Nu}(Nu^-)$. Since the slope of the plot is $k_{-1}/k_2 \cdot k_{Nu}(Nu^-)$, one can also obtain k_{-1}/k_2, which provides information about the behavior of the reactive sulfinyl intermediate ArS(O)Nu.

Catalysis of the Hydrolysis of IIa by Acetate Ion. The rate of hydrolysis of IIa was determined in three different acetate buffers in the presence of varying amounts of added sodium p-methoxybenzenesulfinate. The concentration of acetate ion in all three buffers was the same, but the buffer ratio of (HOAc)/(OAc$^-$) varied from 10 to 0.1. The results are summarized in Table I. Figure 1 shows a plot of the data for all three

Table I. Rate of Hydrolysis of IIa in 60% Dioxane in Acetate–Acetic Acid Buffers[a]

(AcO$^-$) × 10^3, M	(HOAc)/ (AcO$^-$)	(ArSO$_2^-$) × 10^3, M	k_r × 10^3, sec^{-1}	[1/(k_r − k_r^0)] × 10^{-2} [b]
1.03	10	2.00	5.6	2.8
		1.37	6.4	2.3
		0.72	7.6	1.8
		0.60	8.3	1.6
	1.0	2.20	5.3	3.0
		1.55	6.2	2.4
		0.71	8.1	1.6
		0.46	9.1	1.4
	0.1	1.47	6.4	2.3
		0.67	8.3	1.6
		0.42	9.5	1.3

[a] All runs at 21.4°; (IIa)$_0$, 5.0–7.0 × 10^{-5} M. [b] k_r^0 equals 2.0 × 10^{-3} sec^{-1}.[17]

(18) Since the k_{-1} reaction becomes important only at very low (H$^+$), where $k_{Nu}'(Nu^-)(H^+)$ (the acid- and nucleophile-catalyzed term in eq 4) is vanishingly small compared to $k_{Nu}(Nu^-)$, we can neglect any contribution to $(k_r - k_r^0)$ from the $k_{Nu}'(Nu^-)(H^+)$ term.

Kice, Guaraldi | Hydrolysis of Aryl Sulfinyl Sulfones

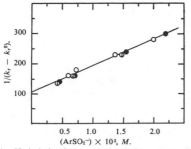

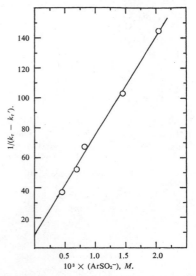

Figure 1. Hydrolysis of IIa in acetate buffers. Data of Table I plotted according to eq 5b: $\bigcirc$, (HOAc)/(AcO$^-$) = 10; $\bullet$, (HOAc)/(AcO$^-$) = 1.0; $\oplus$, (HOAc)/(AcO$^-$) = 0.1.

buffers according to eq 5b. From Figure 1 one sees that the results for all three acetate buffers are nicely correlated by a single line. This proves that it is acetate ion, whose concentration is the same in all three buffers, and not hydroxide ion, whose concentration changes by 100-fold, which is responsible for the catalysis of the hydrolysis of IIa which is being observed.

The fact that added ArSO$_2^-$ depresses the rate of the acetate-catalyzed hydrolysis rules out the otherwise plausible possibility that AcO$^-$ exerts its catalytic effect, not as shown in Chart I (Nu$^-$ = OAc$^-$), but rather by acting as a general base (eq 6). Were the role of acetate

$$AcO^- + H_2O + ArS\overset{\overset{\displaystyle O}{\|}}{\underset{\underset{\displaystyle O}{\|}}{S}}Ar \longrightarrow \left[\overset{\displaystyle Ar}{\underset{\underset{\displaystyle H \; O}{|}}{AcO\text{-}\text{-}H\text{-}\text{-}O\text{-}\text{-}\overset{\delta-}{S}\text{-}\text{-}SO_2Ar}} \right] \longrightarrow$$

transition state

$$AcOH + ArSO_2H + ArSO_2^- \quad (6)$$

as shown in eq 6, one could not explain why added sulfinate ion retards the reaction. Such retardation is, however, readily understood in terms of the mechanism in Chart I.

From the intercept of Figure 1, k_{OAc} is calculated to be 9.0 M^{-1} sec^{-1}. From the slope one estimates k_{-1}/k_2 as 8.6 × 10^2.

Catalysis of the Hydrolysis of IIa by Fluoride Ion. If a second nucleophile capable of catalyzing the hydrolysis of II is added to an acetate buffer, k_r under such conditions will be given by

$$k_r = k_r^0 + \frac{k_{OAc}(AcO^-)}{\left[1 + \dfrac{k_{-1}(ArSO_2^-)}{k_2} \right]} + \\ \frac{k_{Nu}(Nu^-)}{\left[1 + \dfrac{k_{-1}'(ArSO_2^-)}{k_2'} \right]} \quad (7a)$$

Let us define the rate of hydrolysis of IIa in the presence of acetate alone under these conditions as k_r'. This is equal to

$$k_r' = k_r^0 + \frac{k_{OAc}(AcO^-)}{\left[1 + \dfrac{k_{-1}(ArSO_2^-)}{k_2} \right]} \quad (7a')$$

and can be calculated from the data in Table I. Equation 7a can then be rewritten as

$$\frac{1}{k_r - k_r'} = \frac{1}{k_{Nu}(Nu^-)}\left[1 + \frac{k_{-1}'(ArSO_2^-)}{k_2'} \right] \quad (7b)$$

Figure 2. Hydrolysis of IIa in a 1:1 HOAc–AcO$^-$ buffer in the presence of 9.7 × 10^{-3} M chloride ion. Data of Table II plotted according to eq 7b.

This indicates that k_{Nu} for the second nucleophile should be obtainable from the intercept of a plot of $1/(k_r - k_r')$ vs. (ArSO$_2^-$) for a series of runs at a constant concentration of added Nu$^-$ but varying sulfinate ion concentration.

To test the validity of this procedure a series of experiments was carried out with chloride ion as the added nucleophile, since k_{Nu} for chloride was already accurately known.[17] The data are shown in the first part of Table II and are plotted according to eq 7b in

Table II. Catalysis of the Hydrolysis of IIa by Nucleophiles in an Acetate Buffer in 60% Dioxane[a]

Nu$^-$, concn, M	(ArSO$_2^-$) × 10^3, M	k_r × 10^3, sec^{-1}	k_r' × 10^3, sec^{-1} [b]	(1/(k_r − k_r')) × 10^{-2}
Cl$^-$, 0.97 × 10^{-2}	2.04	12.4	5.5	1.45
	1.45	16.0	6.3	1.03
	0.82	22.6	7.6	0.67
	0.69	27.1	8.0	0.52
	0.46	35.7	8.8	0.37
F$^-$, 1.07 × 10^{-2}	2.15	19.3	5.4	0.72
	1.52	23.5	6.1	0.58
	0.89	30.0	7.4	0.44
	0.76	34.0	7.8	0.38
	0.53	38.7	8.6	0.33
SCN$^-$, 0.11 × 10^{-2}	2.19	11.6	5.3	1.60
	1.58	14.1	6.0	1.23
	0.92	19.8	7.3	0.80
	0.84	24.2	7.6	0.60
	0.60	28.9	8.4	0.49

[a] All runs at 21.4° in a 1:1 acetate–acetic acid buffer containing 1.03 × 10^{-3} M acetate ion; (IIa)$_0$ equal to 4.0–7.0 × 10^{-5} M. [b] k_r' calculated from eq 7a' using k_{OAc} and k_{-1}/k_2 as determined from Figure 1.

Figure 2. From the intercept of Figure 2 k_{Nu} for Cl$^-$ is calculated to be 12 M^{-1} sec^{-1}, which is exactly the same value as that obtained from experiments under

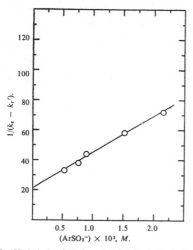

Figure 3. Hydrolysis of IIa in a 1:1 HOAc–AcO⁻ buffer in the presence of 1.07×10^{-2} M fluoride ion. Data of Table II plotted according to eq 7b.

Figure 4. Hydrolysis of IIa in the presence of thiourea. Data of Table III plotted according to eq 4.

conditions where the k_{-1} reaction is not important. We conclude that accurate values of k_{Nu} can be obtained by the procedure just described.

A similar set of runs with fluoride ion as the added nucleophile is shown in the second part of Table II, and the data are plotted in Figure 3. From the plot k_{Nu} for F⁻ is calculated to be 4.4 M^{-1} sec⁻¹ and k_{-1}'/k_2' is 1.1×10^3.

Catalysis of the Hydrolysis of IIa by Thiocyanate Ion. A set of runs was carried out in the 1:1 HOAc–AcO⁻ buffer using thiocyanate ion as the added nucleophile. These are shown in the last part of Table II. A plot of these data according to eq 7b gives an intercept from which k_{Nu} for SCN⁻ is estimated to be 1.5×10^2 M^{-1} sec⁻¹ and a slope from which k_{-1}'/k_2' is calculated to be 1.2×10^4. Because in this instance the intercept of the plot is rather small compared to the slope, we felt that k_{Nu} could not be determined as accurately by this procedure as in the previous cases discussed.

There has been considerable uncertainty in the past regarding the pK_a of thiocyanic acid, HSCN. Recent work[19] suggested, however, that it was probably a strong enough acid so that SCN⁻ would not be significantly protonated at perchloric acid concentrations ($\sim$0.01 M) known[17] to be sufficient to suppress completely the k_{-1} reaction. Studies on the nucleophile- and acid-catalyzed racemization of optically active phenyl benzenethiolsulfinate reported in an accompanying paper[20] confirmed that this was so. Consequently several runs were made with SCN⁻ as catalyst for the hydrolysis of IIa under such conditions. These are shown in the first part of Table III. They indicate the correct value of k_{Nu} for SCN⁻ is 1.7×10^2 M^{-1} sec⁻¹, or slightly higher than the value estimated from the data in Table II.

(19) T. D. B. Morgan, G. Stedman, and P. A. E. Whincup, *J. Chem. Soc.*, 4813 (1965).
(20) J. L. Kice and G. B. Large, *J. Am. Chem. Soc.*, **90**, 4069 (1968).

Table III. Catalysis of the Hydrolysis of IIa by Nucleophiles in Acidic 60% Dioxane[a]

Nucleophile	(Nu) $\times 10^4$, M	(HClO₄), M	$k_r \times 10^2$ sec⁻¹	$(k_r - k_r^0)/(Nu)$[b]
SCN⁻	0.84	0.010	1.62	1.7×10^2
	1.21	0.013	2.2	1.7×10^2
	1.67	0.01	3.1	1.7×10^2
Thiourea	0.126	0.10	6.8	5.2×10^3
		0.20	9.4	7.3×10^3
	0.0252	0.20	2.1	7.5×10^3
		0.30	2.5	9.1×10^3
		0.50	3.4	12.9×10^3

[a] All runs at 21.4°. Initial concentration of IIa, 0.7–1.1×10^{-4} M. [b] k_r^0 equals rate of hydrolysis of IIa in the absence of the nucleophile under otherwise identical conditions. See ref 17 for values used.

Catalysis of the Hydrolysis of IIa by Thiourea. Catalysis of the hydrolysis of IIa by thiourea was studied under conditions ((HClO₄) = 0.10–0.50 M) similar to those employed in our earlier study[17] of catalysis by iodide, bromide, or chloride. The results are shown in the second part of Table III. A plot of $(k_r - k_r^0)/(Nu)$ vs. (HClO₄) is linear (Figure 4), as would be expected from eq 4. For thiourea k_{Nu} is 3.5×10^3 M^{-1} sec⁻¹ and k_{Nu}', the rate constant for the acid- and thiourea-catalyzed hydrolysis, is 1.9×10^4 M^{-2} sec⁻¹.

Discussion

Relative Reactivity of Nucleophiles toward Sulfinyl Sulfur. Values of the rate constant k_{Nu} for reaction 8 in 60% dioxane for all nucleophiles studied are given

$$Nu^- + ArS\text{—}\underset{\underset{O}{\overset{\displaystyle \| }{}}}{\overset{\overset{O}{\overset{\displaystyle \| }{}}}{S}}Ar \xrightarrow{k_{Nu}} ArSNu + ArSO_2^- \quad (8)$$

$$Ar = p\text{-}CH_3OC_6H_4$$

in Table IV. (The data for chloride, bromide, and iodide ion are taken from an earlier publication.[17]) Also of use in our further discussion are the values of

Kice, Guaraldi | Hydrolysis of Aryl Sulfinyl Sulfones

Table IV. Nucleophilic Reactivity toward the Sulfinyl Sulfur of IIa

Nucleophile	k_{Nu} (eq 8), M^{-1} sec^{-1a}	k_{Nu}/k_{Cl}
F$^-$	4.4	0.37
AcO$^-$	9.0	0.75
Cl^{-b}	12	(1.0)
Br^{-b}	65	5.4
SCN$^-$	1.7×10^2	14
I^{-b}	1.0×10^3	83
Thiourea	3.5×10^3	2.9×10^2

[a] All data are at 21.4° in 60% dioxane (v/v) as solvent. [b] Reference 17.

k_{Nu}/k_{Cl} shown in the last column of Table IV. These can be compared with analogous data for substitutions at other types of centers. Such data are shown in Table V for displacements at (1) peroxide oxygen[6]

Table V. Relative Nucleophilicity of Some Common Nucleophiles in Various Substitution Reactions

Nucleophile	k_{Nu}/k_{Cl} Subst at peroxide oxygen (eq 9)^a	Subst at sp^3 carbon (eq 10)^b	Subst at sulfenyl sulfur (eq 11)^c
F$^-$	...	0.10	...
AcO$^-$	d	0.48	
Cl$^-$	(1.0)	(1.0)	(1.0)
Br$^-$	2.8×10^2	7.0	35
SCN$^-$	5.0×10^2	54	5.4×10^3
I$^-$	2.0×10^5	1.0×10^2	1.4×10^4
Thiourea	e	2.3×10^2	...

[a] Solvent, water; ref 6. [b] Solvent, water; ref 8 and 21. [c] Solvent, 60% dioxane (v/v); ref 20. [d] Too slow to measure. [e] Too fast to measure.

(eq 9), (2) sp^3 carbon[8,21] (eq 10), and (3) sulfenyl sulfur[20] (eq 11)   This comparison reveals that $k_{Nu}/$

$$Nu^- + HOOH_2^+ \longrightarrow NuOH + H_2O \qquad (9)$$

$$Nu^- + CH_3Br \longrightarrow CH_3Nu + Br^- \qquad (10)$$

$$Nu^- + \underset{\underset{OH}{|}}{PhSSPh} \longrightarrow PhSNu + PhSOH \qquad (11)$$

k_{Cl} changes with nucleophile in much the same way for the substitutions at sulfinyl sulfur and sp^3 carbon. On the other hand, in the substitutions at sulfenyl sulfur or peroxide oxygen the nucleophiles iodide, thiocyanate, and thiourea are much more reactive relative to chloride than they are in eq 8 and 10. Work currently in progress in this laboratory[22] indicates that the substitution at sulfonyl sulfur in eq 12 shows yet a third type of behavior, in that nucleophiles such as Cl$^-$ or Br$^-$ are quite unreactive compared to such species as F$^-$ or OAc$^-$.

$$Nu^- + \underset{\underset{O}{\overset{O}{\|}}}{Ar\overset{\overset{O}{\|}}{S}}{-}SAr \longrightarrow Ar\underset{\underset{O}{\overset{O}{\|}}}{S}{-}Nu + ArSO_2^- \qquad (12)$$

Nucleophiles such as iodide, thiocyanate, and thiourea are considered by HSAB[3,9] to be much "softer" bases than chloride ion, while those such as fluoride or acetate are considered to be much "harder"

bases.   According to HSAB[9] soft nucleophiles react especially well with soft electrophilic centers and hard nucleophiles react especially readily with hard electrophilic centers.   The fact that nucleophile reactivity in the substitution at sulfonyl sulfur (eq 12)[22] follows the pattern F$^-$ > AcO$^-$ >> Cl$^-$ while for the substitution at sulfinyl sulfur (eq 8) one finds Cl$^-$ > AcO$^-$ > F$^-$ means that sulfinyl sulfur is a much softer electrophilic center than sulfonyl sulfur.   This is in accord with expectations since the sulfinyl group possesses an unshared pair of outer-shell electrons on sulfur whereas sulfonyl sulfur does not; sulfinyl sulfur also has a lower positive charge on sulfur.   Both these factors should made it a softer electrophilic center.[9]

The different response of k_{Nu}/k_{Cl} to changes in Nu$^-$ for substitution at sulfinyl (eq 8) and sulfenyl (eq 11) sulfur is discussed in detail in an accompanying paper.[20]   There it is concluded that the results indicate that sulfenyl sulfur is a significantly softer electrophilic center than sulfinyl sulfur, and that sulfenyl sulfur is actually quite a soft electrophilic center.   The series –SX, –S(O)X, and –SO$_2$X would thus seem to span quite a range as far as hardness or softness of the sulfur atom as an electrophilic center is concerned, with sulfenyl sulfur occupying an intermediate position between the quite soft sulfenyl center and the quite hard sulfonyl center.

The reasonably close similarity in the k_{Nu}/k_{Cl} values for substitution at >S(O) (eq 8) and sp^3 carbon (eq 10) suggests that the two centers are about alike as far as softness is concerned.   Since the carbon of a methyl group has been classed[9] as a "medium soft" electrophilic center, this would suggest that the same term should be applied to sulfinyl sulfur.

Two factors, however, prevent one from being completely dogmatic about this conclusion.   The first is that the data for eq 10 were obtained in water as solvent, while those for eq 8 were obtained in 60% dioxane.   Although we feel that both media are sufficiently aqueous solvents so that there should be no sizable solvent effect on k_{Nu}/k_{Cl} of the type encountered[23] when one compares data for a given substitution reaction in protic vs. aprotic solvents, there could be a small solvent effect, and, if there were, it would be such as to make the values of k_{Nu}/k_{Cl} for eq 8 in Table IV somewhat smaller than they would be in water for those nucleophiles softer than Cl$^-$, and somewhat larger than they would be in water for those nucleophiles harder than Cl$^-$.   The net effect would be that a comparison of the data in Tables IV and V would make sp^3 carbon appear somewhat softer relative to >S(O) than is actually the case.

The second potential source of some uncertainty is that the leaving groups in the two reactions are not the same.   Pearson and Songstad[9] have noted that having a soft base as the leaving group in a substitution can apparently enhance the reactivity of a given center toward soft nucleophiles, while having a hard base as the leaving group can make it more reactive toward hard nucleophiles.   They call this a symbiotic effect. If Br$^-$ is actually a significantly harder base than ArSO$_2^-$, the comparison of k_{Nu}/k_{Cl} for eq 8 and 10 could tend to make sp^3 carbon appear somewhat harder relative to sulfinyl sulfur than is actually the case.

(21) C. G. Swain and C. B. Scott, *J. Am. Chem. Soc.*, **75**, 141 (1953).
(22) G. J. Kasperek, unpublished results.

(23) A. J. Parker, *J. Chem. Soc.*, 1328, 4398 (1961).

One notes that any solvent effect and any symbiotic effect would tend to compensate each other. For this reason, plus the fact that we doubt that either is very large, we believe one can probably conclude with reasonable certainty that sulfinyl sulfur represents a medium soft electrophilic center analogous to sp[3] carbon.

An interesting aspect of the data for eq 8 in Table IV is that fluoride ion is not a great deal less reactive than chloride ion ($k_F/k_{Cl} = 0.37$). Mislow, et al.,[10] have reported that, whereas aryl alkyl sulfoxides racemize readily in aqueous dioxane containing 4 M HCl, they are unaffected by aqueous dioxane containing a similar concentration of HF. The present results suggest that the failure of HF to racemize sulfoxides is not due to any lack of reactivity of fluoride ion toward tricoordinate sulfur. Presumably it must therefore be due to the much lower acidity of the hydrofluoric acid solution.

Experimental Section

Preparation and Purification of Materials. The preparation or purification of most of the reagents has already been described.[17] Sodium fluoride, potassium thiocyanate, sodium acetate, and thiourea were all Analytical Reagent grade and were in general further purified by recrystallization before use.

Procedure for Kinetic Runs. The same procedure outlined in an earlier paper[17] was followed in all cases.

Acknowledgment. We appreciate a number of stimulating discussions with Dr. B. Saville regarding HSAB.

45

The Relative Nucleophilicity of Some Common Nucleophiles toward Sulfonyl Sulfur. The Nucleophile-Catalyzed Hydrolysis and Other Nucleophilic Substitution Reactions of Aryl α-Disulfones[1a]

John L. Kice, George J. Kasperek,[1b] and Dean Patterson[1c]

Contribution from the Department of Chemistry, Oregon State University, Corvallis, Oregon 97331. Received April 21, 1969

Abstract: The relative reactivity of nine common nucleophiles in a displacement reaction at sulfonyl sulfur (eq 2) in 60% dioxane has been determined from either kinetic data on their catalysis of the hydrolysis of aryl α-disulfones (2) or, in the case of primary and secondary amines and azide ion, from direct measurement of their rate of reaction with 2. These data for sulfonyl sulfur (Table VII) are compared in Table VIII with data for some of these same nucleophiles in an exactly analogous displacement at sulfinyl sulfur (eq 1). The substitution at sulfonyl sulfur shows an entirely different pattern of nucleophile reactivity ($F^- \gg AcO^- \gg Cl^- > Br^- > H_2O$) than the one at sulfinyl sulfur ($Br^- > Cl^- \cong AcO^- > F^- \gg H_2O$). Interpreted in terms of the theory of hard and soft acids and bases (HSAB) these results indicate that sulfonyl sulfur is a much harder electrophilic center than sulfinyl sulfur, exactly as HSAB would have predicted it should be. Comparison of the data for sulfonyl sulfur with analogous data on nucleophilic reactivity in a substitution at another hard electrophilic center, carbonyl carbon (Table IX), reveals that the order of reactivity of the various nucleophiles toward sulfonyl sulfur ($RNH_2 > N_3^- > F^- > NO_2^- > AcO^-$) is about the same as toward carbonyl carbon.

In protic solvents the relative reactivity of a group of nucleophiles in a substitution reaction can change quite markedly with a change in the nature of the electrophilic center at which the substitution takes place.[2] In general, nucleophiles which are of low polarizability and high electronegativity, so-called "hard" bases,[3] enjoy an advantage over other nucleophiles in substitutions at centers such as carbonyl carbon[4] or tetracoordinate phosphorus.[5] Nucleophiles which are of high polarizability and low electronegativity, so-called "soft" bases,[3] react particularly readily in substitutions involving centers such as Pt^{II}[6] or peroxide oxygen.[7] A thoughtful

and thorough analysis of these effects was first given by Edwards and Pearson.[8] More recently, Pearson and Songstad[9] have shown that the data can also be nicely rationalized using the concepts of the theory of hard and soft acids and bases (HSAB).

In an earlier study[10] the relative reactivity of a series of common nucleophiles toward sulfinyl sulfur was determined by measurement of the rates of a series of nucleophilic substitutions involving aryl sulfinyl sulfones (eq 1). The present paper shows that similar data for nucleophilic substitution at sulfonyl sulfur can be obtained from measurement of the rates of analogous substitutions of aryl α-disulfones (eq 2). Since eq 1 and 2 involve the same leaving group ($ArSO_2$) and have been studied in the same solvent (60% dioxane), comparison of the results for the two systems allows one to evaluate in a completely unequivocal manner what effect a change in substitution site from sulfinyl to sulfonyl sulfur has on the relative reactivity of various nucleophiles.

(1) (a) This research supported by the National Science Foundation, Grant GP-6952; (b) NDEA Fellow, 1966–1969; (c) NSF Summer Undergraduate Research Participant, 1968.
(2) J. O. Edwards, "Inorganic Reaction Mechanisms," W. A. Benjamin, Inc., New York, N. Y., 1965, pp 51–72.
(3) R. G. Pearson, *J. Am. Chem. Soc.*, 85, 3533 (1963); *Science*, 151, 172 (1966).
(4) W. P. Jencks and J. Carruiolo, *J. Am. Chem. Soc.*, 82, 1778 (1960).
(5) Reference 2, pp 59–63, 177–180.
(6) R. G. Pearson, H. Sobel, and J. Songstad, *J. Am. Chem. Soc.*, 90, 319 (1968).
(7) J. O. Edwards, "Peroxide Reaction Mechanisms," J. O. Edwards, Ed., Interscience Publishers, New York, N. Y., 1962, pp 67–106.

(8) J. O. Edwards and R. G. Pearson, *J. Am. Chem. Soc.*, 84, 16 (1962).
(9) R. G. Pearson and J. Songstad, *ibid.*, 89, 1827 (1967).
(10) J. L. Kice and G. Guaraldi, *ibid.*, 90, 4076 (1968).

416

$$Nu^- + \underset{\underset{O}{\overset{\overset{O}{\|}}{ArS}}}{\overset{}{}}\text{—}SAr \xrightarrow{k_{Nu}^{SO}} ArSNu + \underset{\overset{}{O}}{\overset{\overset{O}{\|}}{ArSO_2^-}} \quad (1)$$

$$Nu^- + \underset{\underset{O}{\overset{\overset{O\ \ O}{\|\ \ \|}}{ArS}}}{\overset{}{}}\text{—}\underset{\overset{}{O}}{\overset{}{S}}Ar \xrightarrow{k_{Nu}^{SO_2}} \underset{\overset{}{O}}{\overset{\overset{O}{\|}}{ArSNu}} + ArSO_2^- \quad (2)$$

$$\mathbf{2}$$

HSAB theory[9] would predict that sulfonyl sulfur should be a significantly harder electrophilic center than sulfinyl sulfur. Expressed in terms of the concepts used by Edwards and Pearson[8] this means that the basicity and electronegativity, rather than the polarizability, of a series of nucleophiles should be paramount in determining their order of relative reactivity toward sulfonyl sulfur, in contrast to the situation with sulfinyl sulfur[10] where polarizability, electronegativity, and basicity are all three important. The objective of the present work was to investigate the extent to which this prediction was in fact borne out by experiment by (1) suitable comparison of the results for sulfinyl and sulfonyl sulfur from eq 1 and 2 and by (2) comparing the behavior of sulfonyl sulfur with that of other centers, like carbonyl carbon, where basicity and electronegativity are thought to be the principal factors determining relative reactivity of a series of nucleophiles in substitutions at that center.

Results

The procedure for obtaining data on the reactivity of nucleophiles in the reaction shown in eq 2 varied with the nucleophile involved. Thus, with some nucleophiles, such as primary or secondary amines or azide ion, the product $ArSO_2Nu$ was not readily hydrolyzed further in 60% dioxane, and kinetic measurements on such systems gave the rate of conversion of the α-disulfone to $ArSO_2Nu$. On the other hand, with other nucleophiles, such as acetate or fluoride ion, the product $ArSO_2Nu$ underwent rapid hydrolysis to the sulfonic acid. In such systems what one can study kinetically is the catalysis of the hydrolysis of α-disulfones (eq 3) by the nucleophile.

$$\underset{\underset{O}{\overset{\overset{O\ \ O}{\|\ \ \|}}{ArS}}}{\overset{}{}}\text{—}\underset{\overset{}{O}}{\overset{}{S}}Ar + H_2O \xrightarrow{k_h} ArSO_3H + ArSO_2H \quad (3)$$

Catalysis of the Hydrolysis of α-Disulfones by Added Nucleophiles. The hydrolyses of the α-disulfones were followed spectrophotometrically using the same procedure described in an accompanying paper.[11] Plots of $\log (A - A_\infty/A_0 - A_\infty)$ vs. time showed excellent linearity in every case.

Catalysis by Acetate Ion. The rate of hydrolysis of phenyl α-disulfone (**2a**, Ar = C_6H_5) was measured in a variety of acetate buffers in 60% dioxane at four different temperatures. The results are summarized in Table I. The following points are worth noting. (1) A series of runs at 80.3° at different buffer ratios ranging from (HOAc)/(AcO⁻) = 10 to (HOAc)/(AcO⁻) = 0.10, but with a fixed (AcO⁻) concentration, show that there is no

(11) J. L. Kice and G. J. Kasperek, *J. Am. Chem. Soc.*, **91**, 5510 (1969).

Table I. Catalysis of the Hydrolysis of Phenyl α-Disulfone by Acetate Ion in 60% Dioxane[a]

Temp, °C	(AcO⁻) $\times 10^3$, M	(HOAc) $\times 10^3$, M	$k_h \times 10^4$, sec⁻¹	k_{OAc},[b] M⁻¹ sec⁻¹
90.8	0.43	0.43	3.8	0.34
	0.72	0.72	4.6	
	1.08	1.08	5.7	
	1.44	1.44	7.1	
	1.80	1.80	8.0	
	2.17	2.17	9.0	
80.3	4.0	40.0	7.6	0.18
	4.0	4.0	7.7	
	4.0	0.40	8.2	
	1.0	1.0	2.9	
	1.0	10.0	3.1	
	2.0	2.0	4.2	
	3.0	3.0	5.5	
	5.0	5.0	9.2	
67.9	3.6	3.6	3.2	0.076
	7.2	7.2	6.0	
	10.8	10.8	8.4	
54.9	0.00	0.00	0.19	0.033
	3.6	3.6	1.6	
	5.8	5.8	2.4	
	7.2	7.2	2.7	
	9.4	9.4	3.4	
	13.0	13.0	4.6	
	14.4	14.4	5.1	
	18.0	18.0	6.1	
	100	100	33.5	
	100	100	30.8 (D₂O)	$(k_{OAc}^{H_2O}/k_{OAc}^{D_2O}) = 1.1$

[a] Initial concentration of **2a**, 5×10^{-5} M. [b] Evaluated from slope of plot of k_h vs. (AcO⁻).

significant dependence of k_h on the buffer ratio. This demonstrates that it is acetate ion, not hydroxide ion, which is responsible for all the increase in rate of hydrolysis which is observed. (2) Plots of k_h vs. (AcO⁻) are linear in all cases, showing that the rate of the acetate-catalyzed reaction is proportional to the first power of acetate ion concentration. (3) The slopes of such plots are equal to k_{OAc}, the actual second-order rate constant for the acetate ion catalyzed hydrolysis. (4) The last two runs in the table show that k_{OAc} exhibits only a very small solvent isotope effect ($k_{OAc}^{H_2O}/k_{OAc}^{D_2O}$) = 1.1.

From a plot of $\log k_{OAc}$ vs. $1/T$ one calculates the following activation parameters for the acetate-catalyzed hydrolysis: E_a = 15.6 kcal/mole; $\Delta S^{\ddagger}$ = −19.8 eu.

Catalysis by Fluoride Ion. The rate of hydrolysis of **2a** was measured at four different temperatures spanning a 60° temperature range in 1:1 HOAc–AcO⁻ buffers containing varying amounts of added fluoride ion. The results are summarized in the first part of Table II. Figure 1 shows a plot of k_h vs. (F⁻) for the data at 80.3°. Equally good linear relationships between k_h and (F⁻) were obtained at the other three temperatures.

The values of k_F for **2a** calculated from these plots are given in the last column of the table. From a plot of $\log k_F$ vs. $1/T$ one estimates E_a = 14.4 kcal/mole and $\Delta S^{\ddagger}$ = −15.9 eu for the fluoride-catalyzed reaction of **2a**.

Effect of Aryl Group Structure on k_F. The variation of k_F with changes in the structure of the aryl group of the α-disulfone was investigated at 21.3° using **2a**, **2b** (Ar = p-CH₃C₆H₄), **2c** (Ar = p-ClC₆H₄), and **2d** (Ar = p-CH₃OC₆H₄). The data for **2b–d** are shown in the last three sections of Table II. A plot of the log k_F values for **2a–d** vs. σ for the *para* substituent of the aryl group gives a good straight line with a slope, ρ, of +5.0. (One

Kice, Kasperek, Patterson / Nucleophilic Substitution Reactions of Aryl α-Disulfones

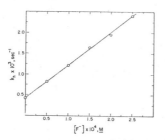

Figure 1. Fluoride ion catalyzed hydrolysis of **2a** in 60% dioxane at 80.3°. Plot of k_h vs. (F⁻) for a series of runs in a 1:1 HOAc–AcO⁻ buffer 0.002 M in AcO⁻.

will recall that the spontaneous hydrolysis of **2**[11] also shows a large positive ρ value, +3.5.)

Catalysis by Chloride and Bromide Ion. In contrast to the marked catalysis of the hydrolysis of **2a** by very small concentrations ($\sim 10^{-4} M$) of added fluoride ion, even quite large concentrations of chloride or bromide ion have only a small effect on the hydrolysis rate. This can be seen from Table III which summarizes the data obtained with these additives at 80.3°.

Table II. Catalysis of the Hydrolysis of Aryl α-Disulfones by Fluoride Ion in 60% Dioxane[a]

ArSO₂SO₂Ar, Ar =	Temp, °C	(AcO⁻) in 1:1 HOAc–AcO⁻, buffer, M	(F⁻) × 10⁴, M	k_h × 10³ sec⁻¹	k_F, M^{-1} sec⁻¹ [b]
C₆H₅	80.3	0.002	0.50	0.82	
			1.0	1.21	7.8
			1.5	1.64	
			2.0	1.95	
			2.5	2.4	
	54.9	0.002	2.0	0.43	
			3.0	0.60	1.8
			4.0	0.79	
			5.0	0.99	
	36.0	0.020	4.0	0.27	
			12.0	0.65	
			20.0	0.94	0.39
			28.0	1.20	
			40.0	1.68	
	21.3	0.020	8.0	0.13	
			16.0	0.23	
			24.0	0.34	0.126
			32.0	0.43	
			40.0	0.54	
			50.0	0.66	
			50.0[c]	0.64	
p-ClC₆H₄	21.3	0.020	3.2	0.95	
			6.4	1.33	
			9.6	1.65	1.20
			12.8	2.16	
			16.0	2.45	
p-CH₃C₆H₄	21.3	0.020	20	0.053	
			40	0.092	
			60	0.13	0.018
			80	0.17	
			100	0.19	
p-CH₃OC₆H₄	21.3	0.020	16	0.0073	
			40	0.016	
			60	0.022	0.0034
			80	0.029	
			100	0.034	

[a] Initial concentration of **2**, 5 × 10⁻⁵ M. [b] Evaluated from slope of plot of k_h vs. (F⁻); see Figure 1 for an example. [c] Benzenesulfinic acid, 0.001 M, added initially.

Table III. Hydrolysis of Phenyl α-Disulfone in the Presence of Added Chloride or Bromide in 60% Dioxane

Temp, °C	(HClO₄), M	(LiClO₄), M	(Cl⁻), M	(Br⁻), M	k_h × 10⁴ sec⁻¹	k_{Cl} or k_{Br}, M^{-1} sec⁻¹ [a]
80.3	0.01	0.40	0.00		1.13	
		0.30	0.10		1.42	
		0.20	0.20		1.72	0.00030
		0.10	0.30		2.02	
		0.00	0.40		2.36	
80.3	0.01	0.20		0.20	1.47	
		0.10		0.30	1.63	0.00017
		0.00		0.40	1.80	

[a] Calculated from slope of a plot of k_h vs. (X⁻).

From a plot of k_h vs. (Cl⁻) one finds that k_{Cl} has a value of only 3 × 10⁻⁴ M^{-1} sec⁻¹ at 80.3°, or only about 4/100,000 the rate constant for the fluoride ion catalyzed reaction at the same temperature.

A similar plot of the data for bromide ion yields a value for k_{Br} of 1.7 × 10⁻⁴ M^{-1} sec⁻¹ for the same reaction conditions.

Catalysis by Nitrite Ion. This anion, which is intermediate between acetate and fluoride in its catalytic effectiveness, was studied in the same manner as for fluoride ion. The results are given in Table IV. From plots of k_h vs. (NO₂⁻) the values of k_{NO_2} shown in the last column of the table were obtained. A plot of log k_{NO_2} vs. 1/T yields E_a = 12.9 kcal/mole and $\Delta S^{\ddagger}$ = −24.6 eu for the activation parameters of the nitrite ion catalyzed hydrolysis of **2a**.

Table IV. Catalysis of the Hydrolysis of Aryl α-Disulfones by Nitrite Ion in 60% Dioxane[a]

ArSO₂SO₂Ar, Ar =	Temp, °C	(AcO⁻) in 1:1 HOAc–AcO⁻ buffer, M	(NO₂⁻) × 10³ M	k_h × 10⁴ sec⁻¹	k_{NO_2}, M^{-1} sec⁻¹ [b]
C₆H₅	80.3	0.002	0.50	8.7	
			1.0	12.3	0.79
			1.5	16.5	
	67.9	0.002	0.30	4.3	
			0.60	5.1	
			0.90	6.5	0.47
			1.2	8.4	
			1.5	9.2	
	54.9	0.002	0.60	2.3	
			0.90	2.9	0.20
			1.2	3.5	
			1.5	4.0	
	39.7	0.002	0.60	0.84	
			0.90	1.10	
			1.2	1.32	0.082
			1.5	1.55	
p-CH₃C₆H₄	80.3	0.002	0.00	1.24	
			0.50	2.28	0.20
			1.0	3.3	
			1.5	4.2	

[a] Initial concentration of **2**, 5 × 10⁻⁵ M. [b] Calculated from slope of plot of k_h vs. (NO₂⁻) at each temperature.

A series of runs with **2b** at 80.3° (last part of Table IV) showed that k_{NO_2} for that compound was only 0.26k_{NO_2} for **2a** under the same conditions, thereby further demonstrating what we have already seen from the fluoride-catalyzed reactions, i.e., that the rate constants are very sensitive to changes in the structure of the aryl group.

Reaction of α-Disulfones with Sodium Azide. Reaction of sodium azide ($0.0125 M$) with **2b** ($0.005 M$) in 60% dioxane gave a 90% yield of p-toluenesulfonyl azide (**3b**). The kinetics of this reaction can be followed

$$N_3^- + ArS\underset{\underset{O}{\parallel}}{\overset{\overset{O}{\parallel}}{S}}Ar \xrightarrow{k_{N_3}} ArS\underset{\underset{O}{\parallel}}{\overset{\overset{O}{\parallel}}{}}N_3 + ArSO_2^- \quad (4)$$
$$\underset{3}{}$$

spectrophotometrically by the same type of procedure used to follow the hydrolysis of **2b**, although a slightly smaller over-all change in optical density is involved. The results of such a kinetic study are given in Table V. A large excess of azide ion over **2b** was employed so that the disappearance of **2b** followed first-order kinetics. A plot of k_1, the experimental first-order rate constant, $vs.$ (N_3^-) was linear. From its slope k_{N_3} for **2b** is found to be 0.115 M^{-1} sec^{-1} at 21.3°.

Table V. Reaction of Azide Ion with p-Tolyl α-Disulfone in 60% Dioxane at 21.3° ^{a}

(2b)$_0$ × 10^5, M	(N$_3^-$) × 10^2, M	k_1 × 10^3, sec^{-1} ^{b}	k_{N_3}, M^{-1} sec^{-1} ^{c}
7.0	2.0	2.24	
	1.6	1.96	
	1.2	1.41	0.115
	0.80	0.97	
	0.40	0.47	

^{a} Initial concentration of **2b**, 7×10^{-5} M. Ionic strength maintained at 0.04 in all runs by addition of lithium perchlorate. ^{b} Experimental first-order rate constant for disappearance of **2b**. ^{c} Evaluated from slope of a plot of log k_1 $vs.$ (N$_3^-$).

Because of absorption by azide ion at the wavelength maximum for **2a** the kinetics of its reaction with azide could not be followed.

Reaction of Primary and Secondary Amines with α-Disulfones. Primary and secondary amines are known[12] to react with aryl α-disulfones to give sulfonamides (eq 5). In neutral or weakly alkaline 60% dioxane the sulfonamides do not themselves hydrolyze at an appreciable rate.

$$2R_2NH + ArS\underset{\underset{O}{\parallel}}{\overset{\overset{O}{\parallel}}{S}}\underset{\underset{O}{\parallel}}{\overset{\overset{O}{\parallel}}{S}}Ar \longrightarrow$$
$$\underset{2}{}$$
$$ArS\underset{\underset{O}{\parallel}}{\overset{\overset{O}{\parallel}}{}}NR_2 + R_2NH_2^+ + ArSO_2^- \quad (5)$$

Although one might think that hydroxide ion, formed from

$$R_2NH + H_2O \rightleftharpoons R_2NH_2^+ + OH^-$$

would be able to compete with the amine for the α-disulfone, it turns out that primary and secondary n-alkylamines are sufficiently reactive that in a 1:1 amine–alkylammonium salt buffer containing 0.01 M or more

(12) H. J. Backer, *Rec. Trav. Chim.*, **70**, 254 (1951).

amine the reaction with the amine (eq 5) is the only process of importance.

The kinetics of eq 5 can be followed spectrophotometrically in the same general way as the hydrolysis of **2**. All of the α-disulfone–amine reactions were studied under conditions where the amine was present in at least 100-fold stoichiometric excess over **2**, so that the disappearance of the α-disulfone followed first-order kinetics. The results are summarized in Table VI. Plots of the experimental first-order rate constant for the disappearance of **2b**, k_1, $vs.$ [amine] are linear in all cases. The slope of the plot of k_1 $vs.$ [amine] is in each case equal to k_{Nu}, the second-order rate constant for eq 2 with Nu$^-$ = amine. These rate constants are shown in the last column of Table VI.

Table VI. Reaction of Primary and Secondary Amines with p-Tolyl α-Disulfone in 60% Dioxane at 21.3° ^{a}

Amine	(2b)$_0$ × 10^5, M	(Amine) × 10^2, M^b	(Amine-H$^+$) × 10^2, M	k_1 × 10^3, sec^{-1c}	k_{amine}, M^{-1} sec^{-1d}
n-BuNH$_2$	5.0	2.4	2.4	79	
		2.0	2.0	67	
		1.6	1.6	48	3.3
		1.2	1.2	37	
		0.80	0.80	26	
Et$_2$NH	5.0	4.0	4.0	10.2	
		3.6	3.6	8.6	
		2.8	2.8	6.9	0.24
		2.0	2.0	5.1	
		0.80	0.80	2.2	
i-BuNH$_2$	5.0	3.2	3.2	51	
		2.8	2.8	45	1.6
		2.0	2.0	33	
		0.8	0.8	13.2	

^{a} Ionic strength maintained at 0.04 in all runs by addition of lithium perchlorate. ^{b} Solutions prepared by adding calculated amount of HCl to solution of the amine. ^{c} Experimental first-order rate constant for disappearance of **2b**. ^{d} Evaluated from slope of plot of k_1 $vs.$ (amine).

Discussion

Mechanism of the Nucleophile-Catalyzed Hydrolysis of α-Disulfones. In principle catalysis of the hydrolysis of **2** by added nucleophiles could be due either to nucleophilic catalysis (eq 6) or to general base catalysis

$$Nu^- + ArS\underset{\underset{O}{\parallel}}{\overset{\overset{O}{\parallel}}{S}}\underset{\underset{O}{\parallel}}{\overset{\overset{O}{\parallel}}{S}}Ar \xrightarrow[\text{rate determining}]{k_{Nu}}$$

$$ArS\underset{\underset{O}{\parallel}}{\overset{\overset{O}{\parallel}}{}}Nu + ArSO_2^- \quad (6)$$

$$\downarrow \text{H}_2\text{O, fast}$$

$$ArSO_3H + H^+ + Nu^-$$

by the nucleophile of the attack of water on **2** (eq 7). Measurement of the solvent isotope effect associated

$$Nu^- + H_2O + ArS\underset{\underset{O}{\parallel}}{\overset{\overset{O}{\parallel}}{S}}\underset{\underset{O}{\parallel}}{\overset{\overset{O}{\parallel}}{S}}Ar \xrightarrow{k_{gb}}$$

$$NuH + ArSO_3H + ArSO_2^- \quad (7)$$

Kice, Kasperek, Patterson / *Nucleophilic Substitution Reactions of Aryl α-Disulfones* with the acetate ion catalyzed hydrolysis revealed that $(k_{H_2O}^{OAc}/k_{D_2O}^{OAc})$ was only 1.1. This is entirely consistent with the nucleophilic catalysis mechanism in eq 6 and totally inconsistent with eq 7, since all previous examples where acetate functions as a general base catalyst in the manner shown in eq 7 have been associated with much larger solvent isotope effects, ranging from 1.7 to 2.7.[13]

Since, even though they are both weaker bases than acetate ion, fluoride and nitrite are both much better catalysts for the hydrolysis of 2, it seems mandatory that if acetate is acting as a nucleophilic catalyst they must also be functioning in that capacity. Chloride and bromide ion have never been observed to act as general base catalysts for hydrolysis reactions, and so their weak catalytic effect on the present reaction must also be due to nucleophilic catalysis.

We thus conclude that in every case the catalysis of the hydrolysis of 2 by the various nucleophiles studied in the present work involves nucleophilic catalysis according to the mechanism shown in eq 6. Furthermore, since initial addition of $ArSO_2^-$ in large excess over that which is formed by hydrolysis of 2 (see runs with 2a at 21.3° in Table II for an example) does not lead to any decrease in the rate, the attack of the nucleophiles on 2 must be the rate-determining step of the nucleophile-catalyzed hydrolyses. As a result the second-order rate constants (k_{OAc}, k_F, etc,) for the different nucleophile-catalyzed hydrolyses are in each instance the rate constant ($k_{Nu}^{SO_2}$ of eq 2) for the attack of the nucleophile on the sulfonyl group of 2.

For those nucleophiles, like azide ion or the primary and secondary amines, which form stable products $ArSO_2Nu$ on reaction with 2, the second-order rate constants (k_{N_3}, k_{amine}) for their reactions with the α-disulfone are obviously also equal to $k_{Nu}^{SO_2}$ for those nucleophiles.

Relative Reactivity of Nucleophiles toward Sulfonyl Sulfur. Values of $k_{Nu}^{SO_2}$ for reaction 2 (Ar = C_6H_5) in 60% dioxane at 21.3° for all nucleophiles studied are given in Table VII. (The values for the primary and

Table VII. Reactivity of Nucleophiles toward Phenyl α-Disulfone

Nucleophile	$k_{Nu}^{SO_2}$ (eq 2), M^{-1} sec^{-1} [a]	$(k_{Nu}^{SO_2}/k_{OAc}^{SO_2})$
n-BuNH$_2$	13	5.9×10^3
i-BuNH$_2$	6.4	2.9×10^3
Et$_2$NH	0.96	4.4×10^2
N$_3^-$	0.72	3.3×10^2
F$^-$	0.13	59
NO$_2^-$	0.022	10
AcO$^-$	0.0022	(1.0)
Cl$^-$	3.5×10^{-6} [b]	0.0016[c]
Br$^-$	2.0×10^{-6} [b]	0.0009[c]

[a] All data are at 21.3° in 60% dioxane as solvent. [b] Calculated from data at 80.3° assuming $k_{Cl}^{SO_2}$ and $k_{Br}^{SO_2}$ show the same dependence on temperature as $k_{OAc}^{SO_2}$. [c] Measured at 80.3°.

secondary amines are calculated from the rate constants for the reactions of these nucleophiles with 2b (Ar = p-CH$_3$C$_6$H$_4$) assuming that $k_{Nu}^{SO_2}$ shows the same dependence on Ar as does the spontaneous hydrolysis of

(13) For a compilation, see S. L. Johnson, "Advances in Physical Organic Chemistry," Vol. 5, V. Gold, Ed., Academic Press, New York, N. Y., 1967, p 281.

2,[11] another reaction involving attack of an uncharged nucleophile on 2. In the case of azide ion we assumed that k_{N_3} for 2a would be greater than that measured for 2b by the same factor by which k_F for these two α-disulfones differs.) One sees that the spread in reactivity between the most reactive nucleophile studied, n-butylamine, and the least reactive, bromide ion, is about 10^7. Also noteworthy is the fact that fluoride ion is over 10^4 more reactive toward sulfonyl sulfur than either chloride or bromide ions. Azide ion is even more reactive than fluoride ion by a factor of 6.

For comparison with data on the relative reactivity of various nucleophiles toward other centers, it is advantageous to express the data for sulfonyl sulfur in terms of the relative reactivity of each nucleophile compared to that of acetate ion as the standard. These values of $(k_{Nu}^{SO_2}/k_{OAc}^{SO_2})$ are given in the last column of Table VII.

Comparison of the Reactivity of Nucleophiles toward Sulfonyl and Sulfinyl Sulfur. The reactivity (k_{Nu}^{SO}) of a series of nucleophiles in a nucleophilic substitution at the sulfinyl group of p-anisyl p-methoxybenzenesulfinyl sulfone (eq 1, Ar = p-CH$_3$OC$_6$H$_4$) was measured in an earlier study.[10] Since eq 1 and 2 have been studied in the same solvent (60% dioxane) and both involve an arenesulfinate ion ($ArSO_2^-$) as the leaving group, comparison of $(k_{Nu}^{SO}/k_{OAc}^{SO})$ for eq 1 with $(k_{Nu}^{SO_2}/k_{OAc}^{SO_2})$ for the same nucleophiles in eq 2 provides an unambiguous indicator of the change in relative reactivity of the various nucleophiles caused by a change in substitution site from sulfinyl to sulfonyl sulfur. The relevant data for those nucleophiles for which data are available for both eq 1 and eq 2 are shown in Table VIII. Inspection of this table reveals that *the change*

Table VIII. Relative Nucleophilicity toward Sulfinyl vs. Sulfonyl Sulfur[a]

Nucleophile	Sulfinyl sulfur $(k_{Nu}^{SO}/k_{OAc}^{SO})$ in eq 1	Sulfonyl sulfur $(k_{Nu}^{SO_2}/k_{OAc}^{SO_2})$ in eq 2
Br$^-$	7.2	0.0009
Cl$^-$	1.3	0.0016
OAc$^-$	(1.0)	(1.0)
F$^-$	0.49	59
H$_2$O	1.1×10^{-5} [b]	3.0×10^{-5} [b]

[a] All data are for 60% dioxane as solvent. For eq 1 Ar = p-MeOC$_6$H$_4$: for eq 2, Ar = C_6H_5. [b] $k_{H_2O}^{SO}$ and $k_{H_2O}^{SO_2}$ are the rates of spontaneous hydrolysis of 1 (ref 10) and 2 (ref 11), respectively.

in substitution site from sulfinyl to sulfonyl sulfur leads to an almost complete reversal in the reactivity order of the various nucleophiles. Toward sulfinyl sulfur the relative reactivity of the nucleophiles is in the order Br$^-$ > Cl$^-$ $\cong$ AcO$^-$ > F$^-$ >>> H$_2$O, an order both qualitatively and quantitatively similar to that found for S$_N$2 substitutions at sp^3 carbon.[14] On the other hand, toward sulfonyl sulfur the relative reactivity of the same nucleophiles is in the order F$^-$ $\gg$ OAc$^-$ $\gg$ Cl$^-$ > Br$^-$ > H$_2$O, fluoride and acetate being now both much more reactive than chloride or bromide. Note, however, that with the two oxygen nucleophiles, acetate and water, $(k_{H_2O}^{SO}/k_{OAc}^{SO})$ is about the same as $(k_{H_2O}^{SO_2}/k_{OAc}^{SO_2})$, even though in both instances acetate is much more reactive than water.

(14) See ref 10 for a full discussion of this point.

In the terminology of HSAB[3] nucleophiles like fluoride ion, acetate ion, or water are considerably harder bases than chloride or bromide ions. The application of HSAB to nucleophilic substitution reactions[9] leads to the conclusion that hard nucleophiles should react particularly readily with hard electrophilic centers and soft nucleophiles with soft electrophilic centers. *The HSAB interpretation of the results in Table VIII would accordingly be that sulfonyl sulfur represents a much harder electrophilic center than sulfinyl sulfur.*

Is this in accord with what HSAB theory would have predicted in advance about these two centers? According to HSAB, two of the principal factors which tend to make an electrophilic center hard are (1) the absence of unshared pairs of easily excited outer shell electrons on the atom being attacked and (2) a high positive charge on that atom.[15] Sulfinyl sulfur has an unshared pair of outer shell electrons on sulfur; sulfonyl sulfur does not. Because of the dipolar character of an $>S^{\delta+}=O^{\delta-}$ bond sulfonyl sulfur also has a significantly higher positive charge on sulfur than does sulfinyl sulfur.

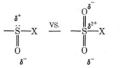

fur. HSAB would have predicted that the combination of these two effects should make sulfonyl sulfur a much harder electrophilic center than sulfinyl sulfur, and this is just what we have found experimentally.

Comparison of the Reactivity of Nucleophiles toward Sulfonyl Sulfur and Carbonyl Carbon. Thanks to the work of Jencks and Gilchrist,[16] data are available on the reactivity of a wide range of nucleophiles toward both 2,4-dinitrophenyl acetate (eq 8) and 1-acetoxy-4-methoxypyridinium perchlorate, 4 (eq 9). With all nucleophiles studied the rate-determining step in eq 9 is

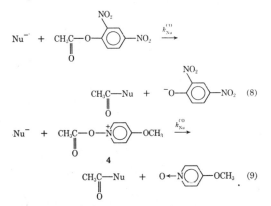

without doubt the attack of the nucleophile on the carbonyl carbon of 4, *not* the breakdown of the tetrahedral intermediate 5. The same is probably true for most nucleophiles in eq 8. Jencks and Gilchrist's[16] data

(15) The other principal factor, the size of the atom, is not important here because for both sulfinyl and sulfonyl sulfur we are dealing with attack at sulfur.
(16) W. P. Jencks and M. Gilchrist, *J. Am. Chem. Soc.*, **90**, 2622 (1968).

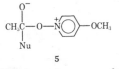

5

therefore provide an accurate measure of the relative reactivity of the various nucleophiles toward carbonyl carbon, another hard electrophilic center.

Values of $(k_{Nu}^{CO}/k_{OAc}^{CO})$ for both eq 8 and 9 are shown in Table IX for all those nucleophiles studied by

Table IX. Relative Nucleophilicity toward Sulfonyl Sulfur *vs.* Carbonyl Carbon

Nucleophile	Sulfonyl sulfur[a] $(k_{Nu}^{SO_2}/k_{OAc}^{SO_2})$ in eq 2	Carbonyl carbon[b] $(k_{Nu}^{CO}/k_{OAc}^{CO})$ in eq 8	$(k_{Nu}^{CO}/k_{OAc}^{CO})$ in eq 9
n-BuNH$_2$	5.9×10^3	2.1×10^5[c]	3.1×10^3[c]
N$_3^-$	3.3×10^2	1.7×10^3	3.5×10^2
F$^-$	59	5.6	3.3
NO$_2^-$	10	15	7.7
AcO$^-$	(1.0)	(1.0)	(1.0)
H$_2$O	3×10^{-5}[d]	3.5×10^{-4}	2×10^{-4}

[a] Solvent, 60% dioxane; temperature, 21.3°; Ar = C$_6$H$_5$. [b] Solvent, water; temperature, 25°. [c] Data are for n-PrNH$_2$ rather than for n-BuNH$_2$. [d] $k_{H_2O}^{SO_2}$ is the spontaneous rate of hydrolysis of 2 (ref 11)

Jencks and Gilchrist[16] for which $(k_{Nu}^{SO_2}/k_{OAc}^{SO_2})$ is also available for eq 2. One sees that the pattern of relative reactivities is roughly the same for sulfonyl sulfur and carbonyl carbon. Thus for sulfonyl sulfur the relative reactivity of the various nucleophiles is in the order n-BuNH$_2 \gg$ N$_3^- >$ F$^- >$ NO$_2^- >$ AcO$^- \gg$ H$_2$O, while for carbonyl carbon in eq 8 or 9 it is n-BuNH$_2 \gg$ N$_3^- \gg$ NO$_2^- >$ F$^- >$ AcO$^- \gg$ H$_2$O. The only difference is the reversal in the order of relative reactivities of fluoride and nitrite, the former being more reactive toward sulfonyl sulfur but less reactive toward carbonyl carbon. Since it is likely that fluoride ion is a harder base than nitrite or acetate ion, this difference *perhaps* suggests that sulfonyl sulfur is a somewhat harder electrophilic center than carbonyl carbon. Because of the difference in the nature of the leaving group in eq 2 as compared to eq 8 and 9 and the fact that the latter two reactions were studied in water rather than 60 %dioxane, this conclusion must be considered to be strictly tentative, however.

One other aspect of the results deserves comment. Azide ion is usually classified by HSAB as a base of hardness comparable to chloride ion. As such its high reactivity toward hard electrophilic centers like sulfonyl sulfur and carbonyl carbon is surprising. Perhaps a partial answer to this dilemma is provided if we remember that azide ion is a much stronger base than chloride ion, and in terms of the oxibase scale equation

$$\log (k/k_0) = \alpha E + \beta H$$

which Davis[17] has proposed to correlate nucleophilic reactivity in various substitutions, the βH term ($H = pK_a$ of NuH + 1.74) should be very important in deter-

(17) R. E. Davis, "Organosulfur Chemistry," M. J. Janssen, Ed., Interscience Publishers, New York, N. Y., 1967, pp 311–328; R. E. Davis, "Survey of Progress in Chemistry," Vol. 2, A. F. Scott, Ed., Academic Press, New York, N. Y., 1964, pp 189–238.

mining the rates of substitutions at hard electrophilic centers. Since H for N_3^- is 6.46 while that for Cl^- is about -3 one can see how azide ion could be much more reactive in substitutions at this type of center than chloride ion.

Experimental Section

Preparation and Purification of Materials. The preparation and purification of the various α-disulfones have been described in an accompanying paper,[11] as was the purification of the dioxane used as solvent. Lithium bromide, sodium bromide, and sodium fluoride were all reagent grade and were recrystallized from distilled water and dried before using. Sodium acetate, sodium nitrite, acetic acid, and sodium azide were all reagent grade and were used without further purification. The various amines were of the highest purity obtainable from Matheson Coleman and Bell and were further purified by distillation from barium oxide before use.

Reaction of 2b with Sodium Azide. p-Tolyl α-disulfone (**2b**), 0.62 g (2.00 mmoles), and sodium azide, 0.33 g (5.0 mmoles), were dissolved in 400 ml of 60% dioxane (v/v) and the solution was heated for 5 hr at 60°. The solution was then evaporated to dryness under reduced pressure. The residue was treated with a mixture of water and ether. The ether layer was washed first with dilute sulfuric acid, then with dilute sodium hydroxide, and finally with water. The ether layer was then dried over anhydrous magnesium sulfate and the ether was evaporated under reduced pressure. The residue crystallized upon being cooled and scratched with a stirring rod: yield, 0.35 g (90%) of p-toluenesulfonyl azide; mp 22° (lit.[18] mp 22°); infrared spectrum identical with that of a known sample prepared by the method of Curtius.[18]

Procedure for Kinetic Runs on the Catalysis of the Hydrolysis of 2 by Added Nucleophiles. The exact procedure depended on the temperature at which the run was to be carried out. For those runs carried out at temperatures *higher* than 21.3° the procedure was as follows. A standard solution of the α-disulfone was prepared in dioxane and the proper volume of this solution was pipetted into the reaction flask of the same apparatus used in an accompanying paper[11] to study the uncatalyzed hydrolysis of **2**. The proper volumes of standard aqueous solutions of the other reagents required were then pipetted into the same reaction vessel, and the solutions were thoroughly mixed. From this point on, the procedure was the same as that used for following the uncatalyzed hydrolysis.[11]

For the runs at 21.3° the apparatus employed was that used by Kice and Guaraldi[19] in studying the hydrolysis of **1** in 60% dioxane. To make a run, 3 ml of a standard solution of the α-disulfone in freshly distilled dioxane was placed in chamber A of this apparatus, and 2 ml of an aqueous solution containing all the remaining reagents was placed in chamber B. The apparatus was then immersed in the constant-temperature bath. After 5 min the two solutions were rapidly mixed, and the resulting solution was transferred to chamber C of the apparatus, a 1-cm spectrophotometer cell. The apparatus was then immediately placed in a thermostated cell holder inside a Cary Model 15 spectrophotometer, and the progress of the hydrolysis of **2** was monitored spectrophotometrically at the same wavelengths used[11] for the uncatalyzed hydrolysis.

Procedure for Kinetic Studies of the Reaction of Azide with 2b. The general procedure was the same as for the runs at 21.3° with the catalyzed hydrolysis. However, because of the absorption of sodium azide in the 240-mμ range, the kinetics of the disappearance of **2b** were followed at 275 mμ, rather than at 258 mμ. At 275 mμ **2b** still has an appreciable absorption but azide does not.

Procedure for Kinetic Studies of the Reaction of Amines with 2b. The general procedure was exactly the same as that used for studying the other reactions of **2b** at 21.3°. The disappearance of the α-disulfone was followed spectrophotometrically at 270 mμ. The 1:1 RNH_2-RNH_3^+ buffer solutions were prepared by adding the calculated amount of standard hydrochloric acid solution to a standard solution of the amine in water.

---

(18) T. Curtius and G. Kraemer, *J. Prakt. Chem.*, **125**, 323 (1930).

(19) J. L. Kice and G. Guaraldi, *J. Am. Chem. Soc.*, **89**, 4113 (1967).

Part VII

Editor's Comments on Papers 46–49

This last section contains several applications of HSAB which are neither organic nor inorganic. Two of the papers by Barclay discuss the adsorption of ions on metallic electrodes. As expected, bulk metal surfaces are soft Lewis acids and also soft Lewis bases.

The paper by Ugo discusses homogeneous catalysis in similar terms, the metal ion acting like a soft Lewis acid in some cases and as a hard Lewis acid in other cases. Bulk metals in heterogeneous catalysis act as soft Lewis acids or bases. Besides metals, the other great class of heterogeneous catalysts are the metal oxides. These are typically hard Lewis acids and bases, e.g., Al_2O_3. However, going from aluminum to a transition metal ion, such as chromium, should make for a softer Lewis acid, with a corresponding change in behavior. The paper by Burwell *et al.* discusses the properties of chromia in catalysis from this viewpoint.

ELECTROANALYTICAL CHEMISTRY AND INTERFACIAL ELECTROCHEMISTRY
Elsevier Sequoia S.A., Lausanne – Printed in The Netherlands

Chemical softness and specific adsorption at electrodes

Donald J. Barclay

The principle of soft and hard acids and bases (SHAB) has been shown to be useful in rationalizing a wide variety of chemical phenomena[1]. No unequivocal definition of hardness and softness has been given but it is generally true that hardness is associated with low polarizability and isolated electronic ground states, and that softness implies high polarizability and low-lying electronic states[2]. The SHAB principle states that hard acids prefer to coordinate with hard bases, and soft acids coordinate preferentially with soft bases. Various tabulations of soft and hard acids and bases have been given[3-6]. Hard interactions are normally ionic. Soft interactions tend to be covalent.

A chemical phenomenon which is not well understood is the specific adsorption of ions at the electrode–electrolyte interface[7]. In terms of Grahame's model of the double-layer at this interface, ions which are contained in the inner layer are said to be specifically adsorbed (see ref. 8 for a more complete definition of specific adsorption). GRAHAME assumed that specific adsorption involved covalent bonding between the adsorbed anion and the (mercury) electrode[9]. This interpretation was rejected by LEVINE et al.[10] who proposed image energy as the origin of specific adsorption, and by BOCKRIS et al.[11] who suggested that the degree and type of ionic solvation were predominant in determining whether an ion would be specifically adsorbed. The latter authors demonstrated that image energy was insufficient to explain the magnitude of adsorption of most ions.

As specific adsorption resembles to some extent adsorption in heterogeneous catalysis (for example, similar isotherms are used in discussing both types of adsorption) and as the latter phenomenon has been examined in terms of the SHAB principle[3], it seemed possible that specific adsorption may be amenable to explanation, or at least rationalization, using this principle.

PEARSON[3] has shown that the adsorptive characteristics of the surface of a bulk metal are consistent with the metal surface having the properties of a soft acid (metals in the zero-valent state are soft) and of a soft base ("conduction electrons are non-innocent ligands *par excellence*"[2]). That is, soft acids and bases are adsorbed from the gaseous phase whereas hard species are not. It does not seem unreasonable to consider a metallic electrode in the same manner.

If an electrode at the point of zero charge (p.z.c., *i.e.*, when the electronic charge on the electrode surface is zero) can be considered to be soft then, if specific adsorption is a soft interaction, it would be predicted that soft anions and soft cations would be adsorbed.

Anions are more strongly adsorbed than cations on mercury (at which electrode most adsorption studies have been made). While there is no quantitative relationship between the potential at the p.z.c. (E_{pzc}) and the amount of adsorbed anion, it is found in solutions of simple salts, and in the absence of specific adsorption of cations (Li^+, Na^+ and K^+ are negligibly adsorbed), that E_{pzc} shifts cathodic with increasing specific adsorption of the anion. E_{pzc} for 0.1 M solutions of numerous salts is given in Table 1, and applying PEARSON's classification of softness of anions it can

TABLE 1

RELATIONSHIP BETWEEN E_{pzc} AND SOFTNESS OF ANIONS

Anion	$-E_{pzc}$ (mV vs. SCE)	PEARSON's classification of anion	Anion	$-E_{pzc}$ (mV vs. SCE)	PEARSON's classification of anion
S^{2-}	880	Soft	Cl^-	461	Hard
I^-	693	Soft	OAc^-	456	Hard
CN^-	645	Soft	NO_2^-	450	Intermediate*
CNS^-	589	Soft	HCO_3^-	440	Hard
Br^-	535	Intermediate	CO_3^{2-}	440	Hard
N_3^-	509	Intermediate	SO_4^{2-}	438	Hard
NO_3^-	478	Hard	F^-	437	Hard
ClO_4^-	470	Hard			

These results were obtained in the presence of a non-specifically adsorbed cation. Salt concns. were 0.1 M except in the case of sulfide (0.5 M). This concn. difference would not affect the given order.
The data on N_3^- and NO_2^- were obtained by J. LAWRENCE, E. GONZALEZ AND R. PARSONS (private communication). Other E_{pzc} were taken from *Handbook of Analytical Chemistry*, edited by L. MEITES, McGraw-Hill, Inc., New York, 1963, chap. 5.
* Another classification by F. BASOLO AND R. G. PEARSON (*Mechanisms of Inorganic Reactions* J. Wiley and Sons, Inc., New York 1967, p. 140) suggests that NO_2^- is hard.

be seen that strongly adsorbed anions are soft, and that weakly adsorbed anions are hard. This table also reproduces very closely the order of softness given by PEARSON[3].

There has been much debate on the existence or non-existence of specific adsorption of cations but it is now generally accepted that the ions, NR_4^+ (where R is an alkyl radical), Tl^+ and Cs^+, are specifically adsorbed on mercury. These are soft cations. The softest metal ions (Hg^{2+}, Au^+, Tl^{3+} and Ag^+ according to a recent classification[6]) are reduced at potentials more anodic than that at which they might be expected to adsorb. Hard cations are not detectably adsorbed.

On the basis of this apparent correlation between softness and adsorbability we suggest that specific adsorption is a soft interaction (at least in the region of the p.z.c.) and therefore involves the formation of a covalent bond between the electrode and the adsorbed ion. BOCKRIS *et al.*[11] criticized GRAHAME's concept of covalency

J. Electroanal. Chem., 19 (1968) 318–321

on, principally, the grounds that in the halide series the order of specific adsorption on mercury was the reverse of the Hg–halide bond energy in the gaseous phase. This reversal can be brought about by solvation effects. In covalent interactions, desolvation of the reacting species, prior to bond formation, often presents a substantial energy barrier and it is therefore not valid to compare gaseous bond energies and stability in solution without consideration of desolvation energies.

KLOPMAN[6] has recently suggested that a scale of ionic softness in solution can be developed by taking, for a given ion, the sum of its desolvation energy and orbital energy. Softness in a base is characterized by low desolvation and high orbital energy, and in an acid by low desolvation and low orbital energy. On the basis of this description, soft interactions are essentially covalent bonding between empty and filled orbitals of comparative energies. In the halide series, the orbital energy *increases* in the order $F^- < Cl^- < Br^- < I^-$, while the desolvation energy *decreases* in the same order, predicting the softness of the halides (and on the present hypothesis, their adsorbability) to decrease in the order, $I^- > Br^- > Cl^- > F^-$, in agreement with experiment.

BOCKRIS *et al.*[11] indicated that a relationship existed between the extent of hydration of an ion and its adsorbability, (the less hydrated ions were strongly adsorbed) which agrees with what is being said here inasmuch as soft ions are often, though not invariably, weakly hydrated. However, these authors go further and propose that low hydration is a *necessary* condition for the specific adsorption of an ion. More recently, DEVANATHAN AND TILAK[12] have augmented this and suggested that for specific adsorption, "ease of hydration of the ion is a necessary condition, and a covalent ion–metal bond is the sufficient condition."

In direct contradiction to this view of low hydration being a necessary condition for specific adsorption to occur, ARMSTRONG *et al.*[13] have recently reported very strong specific adsorption of the sulfide ion on mercury (see *Note**). The sulfide ion, which is strongly hydrated[14], is very soft which supports the suggestion that the factor that determines the adsorbability of an ion is not its degree of solvation but its softness.

The anion PF_6^- though weakly hydrated[15], would be expected to be hard, as the coordination number of fluorine seldom exceeds 1. Recent measurements in aqueous[16] and non-aqueous[16,17] solvents have shown this ion to be only slightly adsorbed, again suggesting a softness criteria for specific adsorption. It seems reasonable to predict that the smaller BF_4^- would be even less adsorbed than PF_6^-.

It is to be expected that the extent (and possibly the order) of adsorbability of ions will vary with the solvent, as an increase in the interaction between the electrode and the solvent will tend to decrease ion adsorption, and the softness of an ion is determined in part by the interaction of the ion with the solvent. KLOPMAN[6] has indicated that the order of softness can change radically with decreasing dielectric constant of the solvent.

When the electrode is positively charged it would be predicted that its acidic character would be enhanced and its basic character reduced. Anion adsorption should then increase, and cation adsorption decrease with increasing positive charge, as is

Note: The adsorption data for sulfide ion was obtained in a medium of $0.5\ M$ $Na_2S + 1\ M$ $NaHCO_3$ where the concentration of free sulfide ion is only $1.5 \cdot 10^{-5}\ M$. In view of this, and of the large concentration of the soft bisulfide ion, it is probable that this ion is co-adsorbed with sulfide.

observed (the converse behavior would of course occur when the electrode is negatively charged). At the extremes of positive polarization, the electrode would become increasingly harder (as an acid) and it is then feasible that soft anions could be desorbed and hard anions be adsorbed. In fact, PAYNE[18] has proposed that extrapolation of his data for the adsorption of the nitrate ion on mercury from nitrate–fluoride mixtures indicates that at high positive charges nitrate desorbs, and that either water or fluoride is adsorbed. As this does not appear to occur in the absence of fluoride, it is probable that fluoride is adsorbed at high positive charges. This, on the present hypothesis, is reasonable as fluoride (or water) is much harder than nitrate and the SHAB principle predicts that hard interactions are to be preferred to hard–soft interactions.

Acknowledgement

This work was supported by the National Science Foundation.

*Gates and Crellin Laboratories of Chemistry**   DONALD J. BARCLAY
California Institute of Technology
Pasadena, Calif. 91109 (U.S.A.)

1 (a) 1st Symposium on *Hard and Soft Acids and Bases*, Cologny, Geneva, 1965; (b) 2nd Symposium on *Hard and Soft Acids and Bases*, Northern Polytechnic, London, 1967.
2 C. F. K. JØRGENSEN. *Structure and Bonding,* 1 (1966) 234.
3 (a) R. G. PEARSON, *J. Am. Chem. Soc.*, 85 (1963) 3533.
   (b) R. G. PEARSON, *Chem. in Britain*, 2 (1967) 103.
4 R. S. DRAGO AND B. B. WAYLAND, *J. Am. Chem. Soc.*, 87 (1965) 3571.
5 A. YINGST AND D. H. MCDANIEL, *Inorg. Chem.*, 6 (1967) 1067.
6 G. KLOPMAN, *J. Am. Chem. Soc.*, 90 (1968) 223.
7 P. DELAHAY, *Double Layer and Electrode Kinetics*, Interscience Publishers, New York, 1965.
8 D. M. MOHILNER, *Electroanalytical Chemistry*, Vol. I, Marcel Dekker, New York, 1966.
9 D. C. GRAHAME, *Chem. Rev.*, 41 (1947) 441.
10 S. LEVINE, G. M. BELL AND D. CALVERT, *Can. J. Chem.*, 40 (1962) 518.
11 J. O'M. BOCKRIS, M. A. V. DEVANATHAN AND K. MULLER, *Proc. Roy. Soc., London*, 274A (1963) 55.
12 M. A. V. DEVANATHAN AND B. V. K. S. R. A. TILAK, *Chem. Rev.*, 65 (1965) 635.
13 R. D. ARMSTRONG, D. F. PORTER AND H. R. THIRSK, *J. Electroanal. Chem.*, 16 (1968) 219.
14 B. E. CONWAY AND J. O'M. BOCKRIS, *Modern Aspects of Electrochemistry*, No. 1., Butterworths, London, 1954, p. 85.
15 E. R. NIGHTINGALE JR., *Chemical Physics of Ionic Solutions*, John Wiley and Sons, Inc., New York, 1966, chap. 7, p. 90 *et seq.*
16 R. PAYNE, *J. Am. Chem. Soc.*, 89 (1967) 489.
17 J. LAWRENCE AND R. PARSONS, *Trans. Faraday Soc.*, 64 (1968) 751.
18 R. PAYNE, *J. Electrochem. Soc.*, 113 (1966) 999.

Received May 23rd, 1968

---

* Contribution No. 3695.

J. Electroanal. Chem., 19 (1968) 318–321

CCA-669   541.132:541.183
                                          Original Scientific Paper

Structural Factors Involved in Ionic Adsorption

D. J. Barclay* and J. Čaja**

*Department of Chemistry, University of Glasgow, Glasgow, W. 2., Scotland and Electrochemistry Laboratory, »Ruđer Bošković« Institute, Zagreb, Croatia, Yugoslavia

Received April 9, 1971

This paper deals with the adsorption of simple inorganic anions on metallic electrodes. Thesis was made that adsorption may be regarded as a co-ordination reaction similar to complex formation in solution.

Pearson's Principle of Soft and Hard Acids and Bases (SHAB) was applied. It suggests that adsorption is a typical soft co-ordination reaction (*i. e.* metal surfaces are soft acceptors) in which an electron pair on the anion is shared with a vacant orbital on a surface metal atom. Thus soft anions and electrodes with high φ_M or E. A. adsorb strongest.

Variations in adsorption order of anions and electrodes can be understood in terms of the symmetry of surface and anion orbitals and by the change in the balance between solvation and bonding energies which can occur on electrodes of different bonding propensity.

INTRODUCTION

This paper deals with the adsorption of simple inorganic anions on metallic electrodes. The thesis will be developed that adsorption of this nature may be regarded as a co-ordination reaction similar to complex formation in solution. It is anticipated that this approach will have greater success in rationalising adsorption phenomena than had previous hypotheses based on the formation of a normal covalent bond or on the supposition that the principal limitation to adsorption is the extent of solvation of the anion. The distinction between a coordinate and normal covalent bond is that the former may be regarded, in solutions of high dielectric constant, as resulting from the heterolytic combination of ions while the latter results from the homolytic combination of atoms.

Like co-ordination reactions between cations and ligands, all that may be achieved at present is an understanding of the *relative* variations in adsorption behaviour of anions and electrodes rather than quantification of the adsorption free energy (ΔG_{ADS}.) in terms of a theoretical model. Thus it is required to elucidate the properties which lead to the order of adsorptivity of anions at the potential of zero charge, E_Z:

$$S^{2-} > I^- > CN^- > NCS^- > Br^- > S_2O_3^{2-} \sim N_3^-$$
$$> NO_3^- > ClO_4^- > Cl^- > NO_2^- > CO_3^{2-} > SO_4^{2-} > F^-$$

** Present address: »Croatia« Research Laboratory, Koturaška 69, 41000 Zagreb, Croatia, Yugoslavia

and of electrodes[1]

Au, Ag, Cu $>$ Pt $>$ Hg $>$ Sb, Bi $>$ Tl, In, Ga

Softness and Adsorptivity

It has previously been shown that soft anions are the most strongly adsorbed on mercury[2]. As the order, unlike the degree, of adsorption of anions given above does not vary significantly with the electrode material — some exceptions will be discussed — this relationship between softness and adsorptivity can be generalised for all metallic electrodes. Because Pearson[3] developed his Principle of Soft and Hard Acids and Bases (SHAB) mainly to rationalise the co-ordination properties of ions in solution this correlation suggests that adsorption is a typical soft co-ordination reaction (*i. e.* metal surfaces are soft acceptors) in which an electron pair on the anion is shared with a vacant orbital on a surface metal atom. While no quantitative explanation has been given which can account in detail for the SHAB principle, a treatment due to Klopman[4] indicates that soft interactions occur as a result of a balance between the energy spent in partial desolvation of the reacting particles and the energy gained in formation of a co-ordinate covalent bond. Soft interactions, and hence adsorption, are favoured by weak solvation *and* by the energies of the frontier orbitals on the donor and acceptor being nearly equal. The problem then is to determine the relative value of these properties as a function of electrode and anion.

Variation in Adsorption with Electrode Material

The approximately linear relationship between E_Z in the absence of specific adsorption and the work function, φ_M, *in vacuo* suggests that the surface potential due to adsorbed water is sensibly constant on most metals[5], indicating that the variation in adsorption with electrode material does not arise from the difference in desolvation energies of electrodes. This conclusion is borne out to some extent by the results of calculations on the dispersion interaction between adsorbed water molecules and electrodes[6]. The relative adsorption properties of electrodes may then be regarded as being determined by the electronic properties of the surface metal atoms.

The strongest co-ordinate covalent bond occurs when the energies of the donor and acceptor orbitals on the anion and electrode respectively are equal and a bond is only formed when these orbitals have the same symmetry properties. This is simply expressed in an equation suggested by Basolo and Pearson[7] for evaluation of the stabilisation of soft-soft interaction:

$$\Delta E = H_A - H_M + (\Delta H^2 + 4 \beta^2)^{1/2} \qquad (1)$$

Where the negative quantities H_A, H_M are Coulomb Integrals on the anion and electrode, ΔH is the difference between them, and β is the Exchange Integral. Maximum stabilisation obtains if $H_A = H_M$ when

$$\Delta E_{MAX} = 2 \beta \qquad (2)$$

where β may be expressed as[8]:

$$\beta = - 2 SH_{AV} \qquad (3)$$

S being the Overlap Integral and H_{AV} the average value of the Coulomb Integrals. S, and hence ΔE, is nonzero only for orbitals of the same symmetry.

H_M essentially expresses the capability of surface atoms to accept electrons in a bond with the anion and may be taken as roughly proportional to their electron affinity, E. A., with atoms of the highest E. A. forming the strongest bond. As previously pointed out the E. A. of surface atoms may be approximated either by φ_M of the metal or by the E. A. of isolated metal atoms[9]. These respectively take the extreme views of regarding the surface atoms as having the properties of the bulk or as atoms free from the influence of the crystal. The true state of surface atoms will differ from either of these extremes though there is accumulated evidence, specially from gas-phase adsorption studies, that adsorption properties may be related to the chemical behaviour of individual metal atoms[10].

Table I lists values of φ_M of selected metals[11] together with the E. A. of the isolated metal atoms[12]. The strongest adsorbing electrodes have the highest value of these parameters as would be expected if they represent the relative value of H_M and a co-ordinate bond is formed on adsorption. The experimental evidence available is not, in general, of sufficient quality to allow us to determine what factors contribute to the increased adsorptivity of certain

TABLE I

Work Functions and Electron Affinities for Selected Metals

M	φ_M (eV)	E. A. (eV)
Cu	4.5	2.0 (1.1)[a]
Ag	4.7	2.2 (1.0)[a]
Au	4.8	2.7 (2.2)[a]
Pt	5.4	1.6
Zn	3.7	0.7
Cd	4.0	0.6
Hg	4.5	1.0
Ga	3.8	(—0.1)[b]
In	—	(0.0)[b]

[a] — Taken from 12 (b)
[b] — H. O. P r i t c h a r d and H. A. S k i n n e r, *Chem. Revs.*, 55 (1955) 745. For comparison, Ag had an E. A. of 1.0 eV on this compilation.

metals. However, the lack of the Esin-Markov effect in the adsorption of chloride and bromide on platinum[13] may be interpreted as indicating greater electron transfer in the Pt—Cl and Pt—Br adsorption bonds than in the corresponding mercury systems where the Esin-Markov effect is observed. The latter effect is the shift in E_Z resulting from specific adsorption and will be greatest when no electron-sharing occurs between the adsorbed ion and the electrode.

It is assumed that the surface orbitals have the same symmetry as the directional orbitals of the bulk metal atoms. Thus, for transition metals *dsp* hybrids occur and empty and filled orbitals with σ and π symmetry project at the metal surface allowing σ and π donor and acceptor bonds to be formed. Non-transitional metals will evolve *sp* hybrids and σ donor and acceptor properties should predominate. The greater adsorptivity of anions over cations suggest that the acceptor mode of surface orbitals is energetically more favourable. However, the greater solvation of cations is also associated with their generally weak adsorption.

This distinction in terms of symmetry between the types of orbitals emerging at transition and non-transition metal surfaces immediately allows an explanation for the reversal of the degree of adsorption of certain anions on mercury and platinum. For example, cyanide is more strongly adsorbed than iodide on platinum[14] while the reverse is true on mercury. The cyanide ion is a good π-electron acceptor as well as being σ donor and hence extra stabilisation is obtained on metals, such as platinum, which are π-donors through filled t_{2g} levels (see Fig. 1). The back-bonding capabilities of iodide are much less than cyanide.

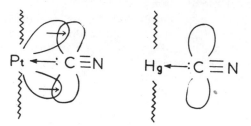

Fig. 1. Illustration of difference in bonding mode of cyanide to platinum and mercury. The filled t_{2g} orbitals on platinum can overlap with vacant orbitals on cyanide. Arrows indicate direction of electron transfer.

The relatively weak adsorption of the nitrite ion on mercury, even though it is classified as being intermediately soft, may be due in part to its softness being determined by its acceptor capabilities. It should be noted that orbitals of varying symmetry will emerge from different crystallographic faces of the electrode[10] and that π-bonding in the adsorption of acceptor anions should be demostrable by measuring their adsorption as a function of crystal orientation.

Variation in Adsorption with Anion

Unlike metal surfaces, where the degree of solvation is regarded as being invariant with the nature of the electrode, the relative co-ordination — or in the present context, adsorptivity — properties of anions are dependent on the solvation as well as the orbital energy. Generally speaking, the solvation contribution to ΔG_{ADS} will favour the adsorption of weakly solvated anions although strong adsorption of highly solvated anions (e. g. S^{2-} and $S_2O_3^{2-}$) sometimes occur. In the latter instances the stabilisation brought about by the covalent bond more than compensates for the loss in solvation energy. On the other hand weakly solvated anions such as ClO_4^- and PF_6^- show little tendency to form covalent bond and do not adsorb strongly.

H_A in equation (1) is related to the 1st and 2nd ionisation potentials (I. P.)[15] of the adsorbing anion and most often anions with low I. P.'s will be best matched in energy with the relatively high energy acceptor orbitals at the metal surface. This facet is illustrated in Fig. 2 where the adsorption stabilisation obtained with an easily ionisable anion is contrasted with that for a difficult to ionise anion.

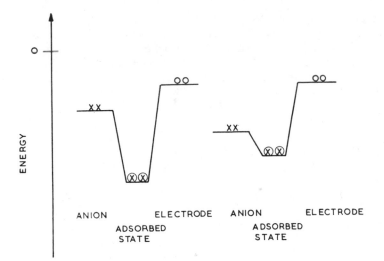

It is not yet feasible to predict on the basis of ionic parameters how the balance between solvation and orbital energies will influence the adsorptivity of anions. At this point the SHAB principle becomes most useful. In Table II the extreme combinations of solvation and energy and orbital

TABLE II

Adsorption of Anions as a Function of Solvation and Orbital Energy

Solvation	LOW	HIGH	HIGH	HIGH	LOW
Orbital Energy	HIGH	LOW	HIGH	HIGH	LOW
Example	I^-	F^-	OH^-	$S_2O_3^{2-}$	ClO_4^-
Pearson's Classification	SOFT	HARD	HARD	SOFT	HARD
Adsorptivity	HIGH	LOW	LOW	HIGH	LOW

energy ($\sim H_A$) are given with examples of the adsorption behaviour of ions with these properties. Pearson's classification, determined empirically from the co-ordination trends of ions in solution best describes the result of the interplay between solvation and bonding tendencies. This interplay can result in a variation in the order of adsorptivity of ions on different electrodes. For example, on platinum chloride adsorbs to a greater extent than perchlorate[16] (at E_Z) while the reverse is true on mercury. This may be explained as follows. The adsorption of perchlorate, with little or no bonding capabilities, is determined by its weak solvation and is unlikely to vary with electrode, whereas chloride, though much more strongly solvated, has good bonding ability and its adsorption will be enhanced on electrodes such as platinum.

The relationship between softness and adsorbability allows speculation on the orientation of ambidentate anions at electrode surfaces. Thus the most probable positions of CN^-, NO_2^-, SCN^- and $S_2O_3^{2-}$ on mercury are as shown in Fig. 3. The angular Hg—S—C bond is a general phenomena of sulphur-bonded SCN^-. This configuration for adsorbed thiocyanate is necessary to explain the adsorption of zinc (II) from thiocyanate media[17] and allows a rationalization of the multiple ligand-bridging which occurs in the oxidation of Cr (II) in the presence of NCS^- [18]. The adsorption of cadmium (II) from

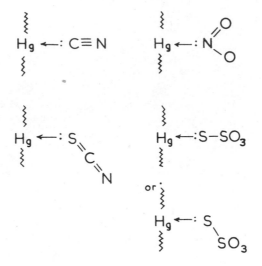

Fig. 3. Orientation of ambidentate ligands at a mercury electrode. The bonding atom is the one which normally co-ordinates to soft cations.

thiosulphate media is understooood on the basis of the proposed bonding of thiosulphate to the electrode[19]. The assignment for NO_2^- is rather more tentative than for the other anions as its weak adsorptivity on mercury suggests that the interaction may be non-specific.

CONCLUSION

The proposal that adsorption of anions on metal electrodes be regarded as a co-ordination reaction allows the gross variations in adsorptive behaviour to be rationalised. Thus soft anions and electrodes with high φ_M or E. A. adsorb strongest. Variations in adsorption order of anions and electrodes can be understood in terms of the symmetry of surface and anion orbitals and by the change in the balance between solvation and bonding energies which can occur on electrodes of markedly different bonding propensity. On some electrodes (e. g. Ga) the relatively strong adsorption of water may have a significant influence on the extent of anion adsorption[20]. Reasonable assignments can be made of the orientation of adsorbed anions at electrode surfaces. A complete description of adsorption would have to include the influences of image and electron-repulsion forces. While the former do not vary greatly with the anion if no chemical bond is formed[21], the reduction in charge on the adsorbed anion resulting from electron transfer infers that the image energy gained when covalent bonding occurs will be less than for purely ionic adsorption. The correct model would maximise the covalent and image energy contribution. Electron-repulsion forces generally have a small influence on relative bond-energies. However, in the present context where ΔG_{ADS} results from differences in fairly large numbers it would be desirable to quantify them. It would not be correct to include an attractive polarisation term in an equation which also included covalent stabilisation.

Acknowledgements. Support for this work to one of us (D.J.B.) by S.R.C., London is gratefully acknowledged. Many of the ideas expressed here were developed while both authors held research positions in Prof. F.C. Anson's group at the California Institute of Technology, Pasadena, Calif., USA.

REFERENCES

1. T. N. Andersen, J. L. Anderson, and H. Eyring, *J. Phys. Chem.* **73** (1969) 3562.
2. D. J. Barclay, *J. Electronal. Chem.* **19** (1968) 318.
3. R. G. Pearson, *J. Am. Chem. Soc.* **85** (1963) 3533.
4. G. Klopman, *J. Am. Chem. Soc.* **90** (1968) 223.
5. R. S. Perkins and T. N. Andersen in *Modern Aspects of Electrochemistry,* J. O'M. Bockris and B. E. Conway, (Eds.), Vol. 5 Plenum Press, 1969, p. 203.
6. J. O'M. Bockris and D. A. J. Swinkels, *J. Electrochem. Soc.* **111** (1964) 736.
7. F. Basolo and R. G. Pearson, *Mechanisms of Inorganic Reactions,* John Wiley and Sons, New York, 1970, p. 114.
8. M. Wolfsberg and L. Helmholtz, *J. Chem. Phys.* **20** (1952) 837.
9. D. J. Barclay, *J. Electroanal. Chem.* **28** (1970) 443.
10. G. C. Bond, Discussions Faraday Soc. **41** (1966) 200.
11. *Handbook of Chemistry and Physics,* Chemical Rubber Co., 50th ed. 1969—70, pE-87.
12. a. R. S. Nyholm, *Proc. Chem. Soc.* **1961**, 273.
    b. idem, **3rd** *Congress on Catalysis,* Vol. 1, North Holland Publishing Co., Amsterdam, 1965, p. 25.
13. A. Frumkin, O. Petry, A. Kossaya, V. Entina and V. Topolev, *J. Electroanal. Chem.* **16** (1968) 175; **17** (1968) 244.
14. K. Schwabe, *Electrochim. Acta* **6** (1962) 223.
15. R. G. Pearson and H. B. Gray, *Inorg. Chem.* **2** (1962) 358.
16. Ying-Chech Chiu and M. A. Genshaw, *J. Phys. Chem.* **75** (1969) 3571.
17. D. J. Barclay and F. C. Anson, *J. Electroanal. Chem.* **71** (1970) 28.
18. D. J. Barclay, E. Passeron, and F. C. Anson, *Inorg. Chem.* **9** (1970) 1024.
19. D. J. Barclay and F. C. Anson, *J. Electrochem. Soc.* **116** (1969) 438.
20. V. A. Kir'yanov, V. S. Krylov, and N. B. Grigor'ev, *Sov. Electrochem.* **4** (1968) 361.
21. T. N. Andersen and J. O'M. Bockris, *Electrochim. Acta* **9** (1964) 347.

IZVOD

Strukturalni faktori pri jonskoj adsorpciji

D. J. Barclay i J. Čaja

U ovom radu obrađuje se adsorpcija jednostavnih anorganskih aniona na metalnim elektrodama. Pretpostavljeno je da se adsorpcija može promatrati koordinacijskom reakcijom, slično kao što je reakcija stvaranja kompleksa u otopini.

Primijenjen je Pearsonov princip slabih i jakih kiselina i baza. Na osnovu toga adsorpcija se može razumjeti kao slaba koordinacijska reakcija (tj. metalne površine su slabi akceptori), prema kojoj anion dijeli elektronski par s praznom orbitalom atoma na površini metala. Zato se slabi anioni i elektrode s visokim φ_M i E. A. vrlo snažno adsorbiraju.

Simetrija orbitala aniona i površine te promjene u jakosti solvatacije i energije veze uzrokuju razliku u adsorbiranju serije aniona i elektroda.

UNIVERSITY OF GLASGOW
    GLASGOW, SCOTLAND
              i
INSTITUT »RUĐER BOŠKOVIĆ«
        ZAGREB

Primljeno 9. travnja 1971.

Reprinted from *Estratto della Rivista* "La Chimica e l'Industria"
Copyright 1969 by Soc. P. Az. Editrice di Chimica, Milano

General Features of Homogeneous Catalysis with Transition Metals (*)

48

R. Ugo

After a brief introduction, the homogeneous catalysis is classified as « hard » and « soft ». Only the « soft » catalysis of π unsaturated hydrocarbons with transition metal complexes is discussed in detail.
Four catalytic steps are recognised as very important: the coordinative unsaturation, the metal ion activation, the ligand activation by coordination and the template ligand reactions.
These points are discussed in terms of transition metals organometallic chemistry.
An analysis of some homogeneous catalytic mechanisms shows that they can be easily explained by applying in the right order the above four reaction steps.
It is emphasized that many theoretical concepts, developed in homogeneous catalysis, can be easily applied to other fields such as heterogeneous catalysis or biochemistry.

Homogeneous catalysis by transition metal complexes has stimulated in the last twenty years the interest of many groups both in academic and industrial circles.

This interest arises either from the discovery of a variety of new, and often unusual, catalytic reactions or from the attempt to use the results obtained in the field of homogeneous catalysis for the resolution of problems rising from closely related work in heterogeneous catalysis, organic syntheses in solution and biochemistry.

The present rapid expansion of the homogeneous field can be in fact explained both by the success of some important petrochemical applications of a certain economical dimension such as acetaldehyde or vinylacetate synthesis from ethylene or olefin hydroformylation and by the attempt of rationalising the many aspects of catalysis as a sequence of a few simple and theoretically understandable chemical steps.

The second point is not, in my opinion, less important, indeed the subject of catalysis is still widely regarded as black art and its practitioners as alchemists, if not indeed magicians.

There is unfortunately more than a grain of truth in this view, in fact catalysis in its classical aspects, that is heterogenous and biological catalysis, is a subject of great experimental complexities and theoretical difficulties. It is probably for this reason that in some academic circles there was a certain sympathy with the aphorism ascribed to OSTWALD that « No one who wishes to be taken seriously should study catalysis ». Until few years ago, the investigation of a catalytic process was carried out by empirical steps or at best by inspired guesswork towards the goal of obtaining more active and selective catalysts.

However this situation is now slowly changing and the goal of a type of research is not only that of obtaining a good catalyst but also of understanding why a certain catalyst is a good one. However

as more experimental work is carried out and more results obtained more areas of theoretical ignorance are revealed. This is the evergrowing chaos [1].

It thus appeared necessary to many researchers to study some theoretical aspects and to develop slowly the theoretical concepts which should rationalize and systemize some of the many facts and pseudo-facts which constitute the collective unconscious of the catalyst practitioner. This means to create an emerging order.

Homogeneous catalysis is undoubtedly one of the more powerful methods of obtaining theoretical information and of studying complex catalytic reactions in detail. In fact it offers less experimental difficulties and a better knowledge of the catalytic intermediates than for instance heterogeneous catalysis.

This fact, on the other hand, has been well known for a long time, indeed the results obtained by the investigation of the acid-base catalysis in solution [2] have largely been used in the theoretical interpretation of catalytic mechanisms which take place on acid or basic surfaces [3] or in some steps of biological processes [4].

The scope of the present paper is that of demonstrating that the chemical and theoretical knowledge of homogeneous catalysis by transition metal ions is at a good stage, and that the concepts developed in this particular field can be usefully extended to many other aspects of catalysis.

A Classification of Organic Homogeneous Catalytic Processes.

In table 1 is given a rough classification of the many aspects of homogeneous organic catalysis. In fact this paper will deal only with catalysis on organic substrates.

To a first approximation homogeneous catalysis can be divided into « hard » and « soft » [5], because in any case the first step of a catalytic process is an electronic exchange or pertubation which should correspond to an acid-base interaction between the catalyst and the organic substrate.

(*) Paper presented at the 20th International Congress « Chemistry Days » on Industrial Catalysis, Milan, 19-21 May 1969.

Table 1.

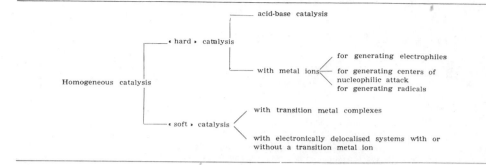

« Hard » catalysis means that the catalyst is a « hard » centre such as H^+ or Zn^{2+}; the action of this type of catalyst is that of polarising, through an electrostatic interaction or the formation of a labile complex, the charge of a certain bond (acid-base catalysis or catalysis with metal ions for generating electrophiles or centers of nucleophilic attack) or that of favouring an electron transfer, through the formation of labile complexes, in order to generate radicals ([6]). In table 2 are reported some examples.

The metal ions which are used in « hard » catalysis are generally in a high oxidation state; their catalytic properties are directly related to their polarisability (which must be low) and to their polarizing power (which must be high). These properties are related to the well known FAJANS parameters ([7]) such as ionic radius and charge

However the catalytic generation of electrophiles and centers of nucleophilic attack is typical not only of transition metal ions such as Cu^{2+}, Ni^{2+},

Mn^{2+}, VO^{2+}, etc. or $FeCl_3$, $CrCl_3$ but also of many non transition metal ions such as Zn^{2+}, Be^{2+}, Hg^{2+}, Cd^{2+}, Sb^{3+} or BF_3, $SbCl_3$ $AlCl_3$ etc.

Indeed if we consider some sophisticated catalysts such as enzymes, we can see that sometimes they use indifferently transition (Mn^{2+}) or non transition (Zn^{2+}, Mg^{2+}) metal ions.

The metal ion catalysis to form radicals is really on the border line between « hardness » and « softness ». In fact this kind of catalysis is typical of transition metal ions, since whether such a reaction will occur at a measurable rate will be determined by the ease of electron transfer from the substrate to the metallic species, that is on a suitable metal's redox potential.

« Soft » catalysis corresponds to a good electron exchange between the metal and the substrate. In this case factors such as electrostatic interactions or polarisation are less important than true electron exchanges via covalent interactions.

Table 2 - Examples of «hard» catalysis with metal ions.

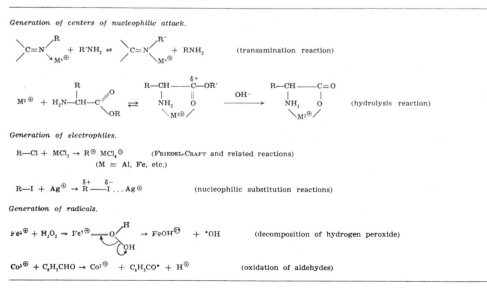

For this reason « soft » catalysis involves « soft » transition metal complexes and « soft » organic or inorganic molecules such as π unsaturated hydrocarbons, carbon monoxide, cyanide ion etc.

Transition metals also have the ability to easily supply or accept additional electrons which corresponds to variable valence and coordination number. Besides, reducing the activation energy of organic reactions, transition metal complexes assist syntheses by organising the reactants through coordination bonds which are not strong enough to seriously bar their combination. Thus these syntheses proceed through, and are understandable in terms of the chemistry of transition metal organometallics ([8]). A similar type of catalysis, which however is limited to activation of small molecules such as oxygen or nitrogen or carbon dioxide, can be achieved by largely delocalised π-systems.

Typical examples can be found in biochemistry by considering metal-emines or metal-corrins.

In this article only « soft » catalysis by simple transition metal complexes on π unsaturated hydrocarbons will be treated, this being the most important aspect from the industrial point of view. Indeed petrochemical catalytic reactions are generally carried out with this kind of transition metal complexes.

The Steps of the « Soft » Catalytic Activity of Transition Metal Ions.

It became evident during the many investigations on the mechanism of some reactions catalysed by « soft » transition metal complexes that four points (or steps) are the most important in defining the path of the catalytic activity of a transition metal complex.

The purpose of this article will be that of describing and discussing in detail the four points and of demonstrating that many « soft » homogeneous catalytic reactions can be easily explained by applying in the right order these four reaction steps.

1) COORDINATIVE UNSATURATION. The importance of « coordinative unsaturation » ([9]) in transition metal complexes is becoming recognised as a major factor influencing homogeneous behaviour. This fact is related to the presence of « vacant sites » of coordination which resemble the concept of « active centres » in heterogeneous systems. The coordinative unsaturation can be due to the low coordination number of the metal ion or to the lability of a certain number of the ligands, bound to the metal ion, which dissociate in solution giving rise to a coordinatively unsaturated species.

2) METAL ION ACTIVATION. The process of catalytic activation of a metal complex ion takes place through the formation of particularly reactive metal-hydrogen or metal-carbon bonds. This is obtained generally by the action of a co-catalyst.

3) LIGAND ACTIVATION BY COORDINATION. The coordination of a π unsaturated molecule, which takes place on the « vacant sites » of the metal ion, causes a significant change in the electronic structure and bonding of that molecule. It follows that changes in chemical reactivity can occur in the ligand and these may lead either to « activation » or « deactivation ». Indeed « activation » or « deactivation » depends on the type of catalytic reaction which is considered.

4) TEMPLATE LIGAND REACTIONS. When two ligands are coordinated in adjacent positions on the coordinative sphere of a metal ion they greatly enhance the possibilities of reaction between them, particularly in a definite stereochemical way. The metal ion may act as a template for ligand reactions.

We will discuss now in detail these four points.

Coordinative Unsaturation.

The majority of « soft » catalytic reactions, which will be discussed in this article, involves as catalysts coordination compounds, generally with « soft » ligands such as carbon monoxide, cyanide, tertiary phosphines etc., of metals near the end of each transition series notably the platinum group. It is important to point out that the electron configurations of the metals in catalytic complexes are generally in the range d^8 to d^{10} with the greatest number with d^8 configuration. The catalytic complexes occur generally with the metal in a low oxidation state and they are, because of the presence of « soft » ligands, of the spin-paired or low-spin type corresponding to a sufficiently large ligand field splitting.

The stable coordination numbers of low spin complexes range from eight to two and exhibit a systematic inverse dependence on the number of d electrons of the metal atom as shown in table 3.

Table 3.

Coordination number	Complexes	Electron configuration
8-7	$[Mo(V)(CN)_8]^{3-}$, $[Mo(IV)(CN)_8]^{4-}$	d^1, d^2
6	$[M(CN)_6]^{3-}$ $[M = Cr(III), Mn(III), Fe(III), Co(III)]$	d^3, d^4, d^5, d^6
	$M(CO)_6$ $[M — V(0), Cr(0), Mo(0), W(0)]$	d^5, d^6
5	$[Co(II)(CN)_5]^{3-}$, $[Ni(II)(CN)_5]^{3-}$, $[Mn(— I)(CO)_5]^-$	d^7 d^8
	$[Fe(0)(CO)_5]$	
4 (square planar)	$[Ni(II)(CN)_4]^{2-}$, $[Rh(I)(CO)_2Cl]_2$, $[Ir(I)(CO)(PPh_3)_2Cl]$	d^8
4 (tetrahedral)	$[Ni(0)(CO)_4]$, $[Co(— I)(NO)(CO)_3]$, $[Cu(I)(PPh_3)_4]^+$	d^{10}
	$[Cu(I)(CN)_4]^{3-}$	
	$[Ni(II)(PPh_3)_2Cl_2]$, $[Ni(I)(PPh_3)_3Cl]$	d^8, d^9
3	$Pt(0)(PPh_3)_3$, $[Ag(I)(PPh_3)_2Cl]$	d^{10}
2	$Pt(0)(PPh_3)_2$, $Ag(I)(PPh_3)Cl$, $[M(CN)_2]^-$	d^{10}
	$[M = Cu(I), Ag(I), Au(I)]$	

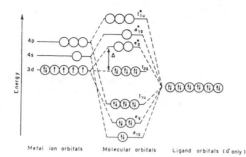

Fig. 1 - Molecular/orbitals in an octahedral complex.

The inverse dependence has its origin in the fact that the higher the coordination number the fewer the d electrons which can be accomodated in stable (bonding or nearly non bonding) orbitals of the metal ion [10]. In the case of an octahedral complex for example (fig. 1) the three stable t_{2g} orbitals (non-σ bonding or possibly slightly π-bonding in the case of π-acceptor ligands such as CO or CN^-) can accomodate up to 6 d electrons.

Any additional electrons are forced to occupy the $e_g^{\bullet}$ orbitals which are strongly anti-bonding. This explains why, despite the different electron configuration, metal ions with electron configuration from d^3 to d^6 prefer hexacoordination while an electron configuration with more than 6 d electrons leads to the destabilisation of the coordination number 6 in favour of a lower coordination number which permits the d electrons to be accommodated in stable orbitals [10]. A typical example of this rule is here reported:

$$[Co(CN)_6]^{3-} \xrightarrow{e^-} [Co(CN)_6]^{4-} \rightleftarrows [Co(CN)_5]^{3-} + CN^-$$
d^6 stable $\qquad$ d^7 unstable $\qquad$ d^7 stable

Analogously while pentacoordination can be stable for a d^8 configuration, it is not stable if more electrons are added:

$$Fe(CO)_5 \xrightarrow{2e^-} [Fe(CO)_5]^{2-} \rightleftarrows [Fe(CO)_4]^{2-} + CO$$
d^8 stable $\qquad$ d^{10} unstable $\qquad$ d^{10} stable

The formation of coordinatively unsaturated species corresponding to low coordination numbers is thus related to a configuration with many d electrons and to a high electron density on the metal. Both properties are satisfied in low oxidation state complexes.

The effective metal electron density is very important in defining more subtle electronic balances such as in the case of the coordination number of the same metal in the same electron configuration with ligands having different σ-π properties [10].

When there are in the coordination sphere ligands having good π acceptor properties the coordination number is generally higher than when these ligands are not present. Some examples are reported in table 4.

Generally speaking the higher the metal electron density the lower is the corresponding coordination number. It follows that in metal complexes of the same metal in the same electron configuration the formation of « vacant sites » is facilitated by ligands such as tertiary phosphines, isonitriles or tin(II)trichloride ($SnCl_3^-$) which are enough « soft » to stabilise a low oxidation state, but which are not able to delocalise strongly the metal electron density via a good π-back donation.

It is important to point out here that, at the moment, the conditions which favour the formation of stable coordinatively unsaturated species have been discussed.

Unstable species with « vacant sites » can however be formed in an appreciable concentration in cases which should not be favourable to coordinative unsaturation.

Of course more drastic conditions are required in order to reach an energetically excited state; the source of energy can be heat or radiation. Some examples are here reported:

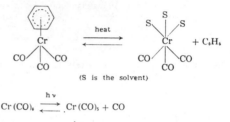

(S is the solvent)

$$Cr(CO)_6 \xrightarrow{h\nu} Cr(CO)_5 + CO$$

$$Ru(CO)_3(PPh_3)_2 \xrightarrow{h\nu} Ru(CO)_2(PPh_3)_2 + CO$$

It follows that for creating « vacant sites » of coordination it is necessary to vary on the electron configuration and effective electron density of the central metal ion or on the reaction conditions.

Table 4.

Electron configuration	Complex	Strongly π acceptor ligand
d^8	Rh(CO)(PPh$_3$)$_2$Cl (very slight dissociation)	CO
	Rh(PPh$_3$)$_3$Cl [high dissociation to Rh(PPh$_3$)$_2$Cl]	—
d^8	Ni(PPh$_3$)$_2$(CN)$_2$ [they give place to stable pentacoordinated Ni(PPh$_3$)$_3$(CN)$_2$ species]	CN$^-$
	Ni(PPh$_3$)$_2$Cl$_2$ [they do not give place to stable pentacoordinated Ni(PPh$_3$)$_3$Cl$_2$ species]	—
d^{10}	Ni(Co)$_2$(PPh$_3$)$_2$ (very slight dissociation)	CO
	Ni(PPh$_3$)$_4$ [high dissociation to Ni(PPh$_3$)$_3$ and Ni(PPh$_3$)$_2$]	—

Metal ion activation.

The creation of « vacant sites » corresponds, from a certain point of view, to an activation of the metal ion. Indeed a « vacant site » is a coordination centre free and available to the π unsaturated hydrocarbon which can in this way interact with the metal electron density. For instance this process corresponds very closely to chemisorption on metal surfaces.

However the creation of « vacant sites » is not generally enough for catalytic activity, It is necessary in fact, and we will see later the reason, to have on the transition metal some very reactive bonds such as metal-hydrogen or σ metal-carbon bonds.

Both hydrogen and carbon atoms, when bound to a transition metal, have peculiar properties.

The transition metal hydrides have for instance proton N.M.R. absorptions in a higher range (+ 5 to + 18 p.p.m. relative to tetramethylsilane) than any other non transition metal hydride [11]. This fact suggests that hydrogen is highly shielded by d electronic density of the metal. The hydrogen atom, although covalently bound to the transition metal, has a certain negative charge and can behave as a nucleophilic center or sometimes as atomic hydrogen that is like a radical. The σ alkyltransition metal bond is generally thermodynamically and kinetically unstable. In fact the dissociation energy of the alkyl carbon-transition metal σ bond appears to be much less than the carbon-metal σ bond of non transition metals. The reasons for the instability are:

a) the very small « covalent » energy of the metal-carbon bond

b) the relatively small differences in electronegativities between the transition metal and the carbon atom, which account for the small « ionic resonance » energy contribution.

In fact, following the so-called MULLIKEN Magic Formula [12], the bond dissociation energy can be given, in first approximation, by two terms: the covalent bond energy and the ionic resonance energy.

JAFFÈ and DOAK [13] have evaluated the stability of alkyl carbon-metal bonds of a variety of metals; they have pointed out that the covalent energy for carbon-metal bonds of transition metals is appreciably smaller (perhaps one half) than the corresponding values for other elements, but also the ionic resonance energy is appreciably smaller (perhaps one third) than that of alkyl-alkali or alkyl-alkaline earth metal bonds. Of course the stability of this type of bond can be increased by using perfluoroalkyl radicals (large ionic resonance energy contribution) or vinyl or aryl groups (p_π-d_π contributions to the bond, different electronegativity of an sp^2 carbon atom with respect to a sp^3 carbon atom etc.).

The above discussion is concerned solely with some factors affecting the thermodynamic stability, it is however appropriate to consider also factors affecting kinetic stability, although frequently it is difficult to separate thermodynamic and kinetic contributions to stability. If we consider the transition

Fig. 2

metal-sp^3 carbon bond in a simple molecular orbital scheme (fig. 2) we will see that the bonding electrons are in a σ molecular orbital which is mainly formed by the sp^3 hybrid carbon atomic orbital.

This means that the polarity of the bond is $M^{\delta+}$—$C^{\delta-}$; moreover the transition metal, unlike other metals, is characterised by partly filled non bonding d orbitals which are close, in energy, to sp^3 hybrid carbon orbitals.

It would be possible thus to promote d electrons in the anti-bonding σ orbital of the metal carbon bond (ΔE_2) or an electron from the σ bonding metal-carbon orbital into an empty d orbital on the metal (ΔE_1).

Both type of electronic excitation would weaken the metal carbon bond and cleavage would lead to a true carbanion (ΔE_2) or radical (ΔE_1) which are both very reactive.

It is also possible to regulate such an electronic situation and to maximise the electronic promotion energy (which corresponds to a kinetically and thermodynamically more stable metal-carbon bond). This can be achieved by having around the transition metal π acceptor ligands (such as CO, CN$^-$, etc.) which (as shown in fig. 3) stabilize the d non-bonding electrons with a π bond and lead to increased values of ΔE_1 and ΔE_2.

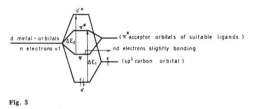

Fig. 3

Of course the electronic balance for making more or less stable transition metal-carbon bonds is very delicate but the scope of this work is not that of discussing in detail the many aspects of the electronic situation in transition metal complexes.

The most important conclusion of the above metal-hydrogen or metal-carbon bonds discussion is that they are anionic centers which can lead easily to hydrides and carbanions or radicals.

Metal ion activation by these bonds can be obtained in different ways using many different cocatalysts.

In first approximation three types of co-catalytic reactions are generally used, that is:

a) Hydrogen abstraction from an activated carbon-hydrogen bond of metal coordinated organic molecules.

b) Oxidative additions reactions to the metal ion of molecules such as hydrogen, acids, alkyl or acyl halides.

c) Reduction or alkylation of the metal ion *via* non transition metal hydrides (such as BH_4^-, AlH_4^- or N_2H_4) or non transition metal alkyls (such as AlR_3, RMgX or LiR).

Hydrogen abstraction from an activated carbon-hydrogen bond is one of the most interesting reactions; indeed the carbon-hydrogen bond is ubiquitous in organic chemistry and the unusual ability of transition metal catalysts to assist in its breaking and making is particularly useful.

Many different organic molecules can be used as hydrogen source and some examples are here reported:

$$Pt(PPh_3)_2Cl(OC_2H_5) \rightarrow PtH(Cl)(PPh_3)_2 + CH_3CHO$$

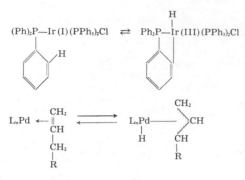

$$Pt(C_2H_5)(Cl)(PPh_3)_2 \rightleftarrows PtH(Cl)(PPh_3)_2 + C_2H_4$$

The first example reports the hydrogen abstraction from an oxygenated molecule which could be an alcohol, as in the example, or an ether, a ketone, an aldehyde or a derivative of formic acid such as formamides or esters; the other examples show abstraction from a phenyl ring, from the allylic position of an olefin and from a σ organic radical. The factors facilitating the abstraction are not yet clear so that general rules can not be used. Generally a high reaction temperature and a low oxidation state of the transition metal favour the process. The oxidative additions [4] to a transition metal complex is in my opinion, the most important activation process.

They can be generalised by the folloing reactions:

$$M^m L_n + HX \rightleftarrows M^{(m+2)}L_n(H)X$$

$$(X = H, Cl, HC=CR_2, C\equiv CR \text{ etc.})$$

$$M^m L_n + RX \rightleftarrows M^{(m+2)}L_n(R)X$$

(R = alkyl or acyl or aroyl radicals;
X = halogen or pseudo halogen)

L is a ligand, m is the oxidation state of the metal.

In this type of reaction the transition metal increases formally of two units its oxidation state whilst the added molecule is dissociated. It is interesting to point out that both hydrogen abstraction and oxidative addition correspond to a well known type of chemisorption, that is dissociative chemisorption.

The examples reported in table 5 should make clear the parallelism of the two situations.

The factors facilitating oxidative additions are enough well known [14], in fact the energy variation in an oxidative addition is given, in first approximation, by:

$$M^m L_n + XY \rightleftarrows M^{(m+2)}(X)(Y)L_n$$

$$\Delta E = E_{M-X} + E_{M-Y} - (D_{X-Y} + P_{m \rightarrow m+2})$$

The E_{M-X} and E_{M-Y} terms, that is the energies of the new bonds, generally can not be evaluated easily as they depend on both the metal and on the nature of X and Y, but they seem to be less relevant, in the total balance, than D_{X-Y} (the energy of dissociation of X—Y) or $P_{m \rightarrow m+2}$ (the promotion energy of the metal from the valence state corresponding to formal oxidation state m to the valence state corresponding to the oxidation state $m + 2$).

Table 5.

Homogeneous phase	Heterogeneous phase
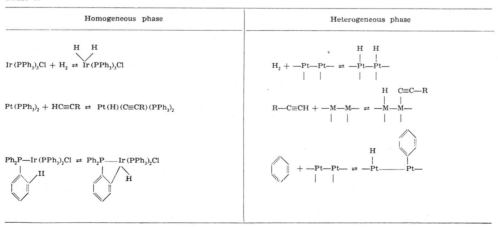	

Considering the same molecule XY the promotion energy becomes the most important factor which must be minimized.

It is well known that a low promotion energy can be obtained with metals in low oxidation state and with a large electron density localised on the metal itself, that is with metal complexes having « basic properties ».

The third method of activating the metal ion is perhaps the most useful one and, being a classical method, does not need any further analysis. The general aspects of this type of reactions are summarized in the following examples:

$$ML_nX_y \xrightarrow{\text{AlR}_3 \text{ or LiR} \atop \text{or RMgX}} ML_nX_{(y-1)}R_l$$

$$ML_nX_y \xrightarrow{(CH_3)_2CHMgBr} ML_nX_{y-1}CH(CH_3)_2 \rightarrow$$

$$\rightarrow ML_nX_{y-1}(H) + CH_3-CH=CH_2$$

Ligand Activation by Coordination.

As already briefly pointed out the coordination of an unsaturated hydrocarbon is associated with a process of electronic exchange which should affect the reactivity of the hydrocarbon.

We are thus discussing the step generally considered as the most important of the catalytic process: the activation *via* coordination. The first point which will be analysed is the following: how the hydrocarbon reactivity affected by coordination to a transition metal? Some reviews have been written on the subject of reactions of coordinated ligands, but only the classification proposed recently by WHITE ([15]) for the electrophilic and nucleophilic attack on organometallic compounds will be used here.

In fact, following WHITE, the classes of reactions can be designated as being of the type $E_{n/m}$ and

Table 6 - Examples of electrophilic and nucleophilic reactions.

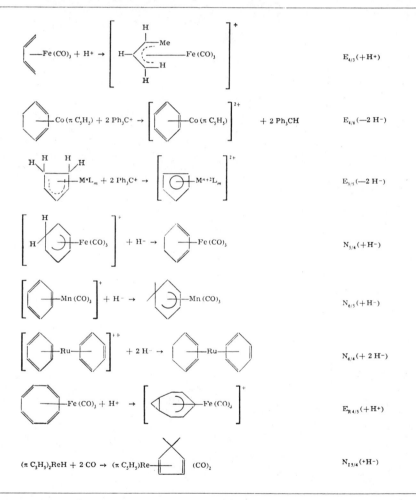

443

$N_{n/m}$ ($\pm$ X) respectively where n and m denote the number of C atoms involved in bonding to the transition metal before and after the reaction, and X denotes the moiety added to or removed from the organic ligand in the reaction.

The examples reported in table 6 should make this convention clear.

Suffixes to E or N will be used to indicate rearrangement of the ligand during the reaction (suffix R) and the occurrence of intramolecular attack (suffix I). Two examples are reported in table 6.

Electrophilic or nucleophilic attack can both bring about either an increase or decrease of the number of carbon atoms involved in bonding to the metal but the most common reactions, which are really important in homogeneous catalysis, show a change of one carbon atom in magnitude. Having clarified this classification we wish to examine the electrophilic and nucleophilic reactivity of a π hydrocarbon (taking as example the olefins) coordinated to a transition metal.

If we consider changes in the infrared absorption of olefins upon coordination, they indicate that the reactivity of the double bond should be enhanced; the C=C stretching frequency is in fact lowered at lower wavelength. This may be attributed at first sight to an appreciable reduction of electron density which, like the electronegative group of α,β-unsaturated carbonyl compounds, renders the olefin susceptible to nucleophilic attack, an attack which could be further assisted by transfer of olefin electron density to the transition metal.

However the olefin, in π bonding to a transition metal, does not act only as σ donor but also as π acceptor, following the qualitative, symmetry-based view of the bonding put forward by DEWAR, CHATT and DUNCANSON (see fig. 4).

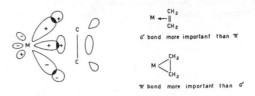

Fig. 4 - The Dewar, Chatt, Duncanson model.

It can happen that, in some cases, the π back-donation from the metal to the olefin is larger than the σ donation from the olefin to the metal; in this case the C=C bond order (related to the stretching frequency) decreases, the antibonding orbitals being occupied, but the olefin acts as a carbanion because it accepts electrons from the metal.

In this case the olefin should be susceptible to electrophilic attack. The two extreme, different situations can be reached by changing the properties and nature of the transition metal; such changes can be dictated by very simple rules.

1) Metal complexes with the metal in a low oxidation state (0, + 1) are susceptible only to electrophilic attack (as a large part of the electron density is localised on the metal and on the ligands), while metal complexes with the metal ion in a high oxidation state (+ 2, + 3) are susceptible, for the opposite reason, only to nucleophilic attack. The examples reported in table 7 should make clear these statements.

2) There are however many other factors such as the coordination number, the nature of the ligands and so on which make the effect of the

Table 7 - Effect of the oxidation state.

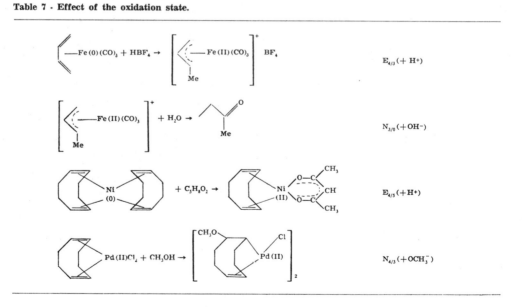

444

Table 8 · Effect of the charge.

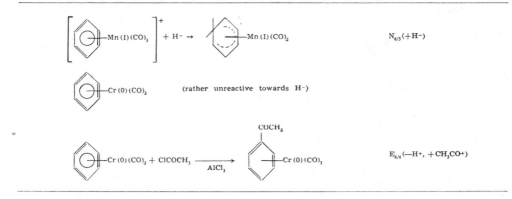

$N_{6/5}(+H^-)$

(rather unreactive towards H^-)

$E_{6/6}(-H^+, +CH_3CO^+)$

---

oxidation state rather blurred ([15]). The effect of the charge is more sharply defined: cationic complexes are susceptible only to nucleophilic attack whereas anionic complexes are only susceptible to electrophilic attack.

The effect of the charge can be easily seen looking at some isoelectronic compounds (table 8).

3) The substituents on the organic moiety may induce a different reactivity. For instance the presence of electron withdrawing substituents may lead to inhibition of electrophilic attack

$$(\pi\, C_5H_5)Fe(CO)_2CH_2CH_3 + HCl \rightarrow$$
$$\rightarrow (\pi\, C_5H_5)Fe(CO)_2Cl + C_2H_6\ E_{1/0}(+H^+)$$
$$(\pi\, C_5H_5)Fe(CO)_2CF_2CF_2H\ or$$
$$(\pi\, C_5H_5)Fe(CO)_2CH_2CH_2CN\ are\ stable\ to\ acids.$$

In other cases they may induce susceptibility to nucleophilic attack as in tricarbonyl(perfluorocyclohexa-1,3-diene)iron(O) ([16]).

In conclusion coordination can favour nucleophilic or electrophilic reactivity with the following rules, corresponding to extreme positions:

Increased reactivity to nucleophiles: high oxidation state of the metal, positive charge of the metal complex;

Increased reactivity to electrophiles: low oxidation state of the metal, negative charge of the metal complex.

Template Ligand Reactions.

The most important reaction which is facilitated by transition metal ion, acting as a template, is the so-called *cis* « insertion reaction ».

Such reaction is illustrated in fig. 5:

Fig. 5 · Cis insertion reaction.

These reactions are generally reversible and one of the reactant, X in our case, is an unsaturated molecule such as an olefin, acetylene or carbon monoxide. The bond M—Y must be labile; it is generally the metal-hydrogen or metal-carbon formed during the metal ion activation process.

Three types of *cis* «insertion» reaction are very important:

1) The olefin or acetylene, metal-hydrogen insertion

2) The olefin or acetylene, metal-carbon insertion

3) The carbon monoxide, metal-carbon insertion.

The term olefin or carbon monoxide insertion implies olefin or carbon monoxide mobility, but it is misleading since it was demonstrated that olefin or carbon monoxide do not move, but only hydrogen or alkyl groups move to the new position ([17]) (fig. 6).

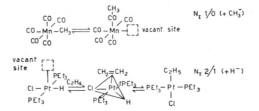

$N_I\ 1/0\ (+CH_3^-)$

$N_I\ 2/1\ (+H^-)$

Fig. 6 · Mechanism of the cis insertion reaction.

These reactions can be described, using the WHITE's classification, as intramolecular nucleophilic attacks on coordinated carbon monoxide or coordinated olefin.

The nucleophilic species being localised, as already shown, in the metal-carbon or metal-hydrogen bonds.

However these reactions have not been investigated in detail except in few cases so that following some Authors [18] in some cases the insertion reaction can be considered a radical reaction more than a nucleophilic reaction.

The « insertion » of moieties other than olefins and carbon monoxide into metal-alkyl and related bonds probably also involves an internal nucleophilic attack: as do the sulphur dioxide and cyanide insertions. A less important template reaction is related to the cyclobutanetion [14] or cyclobutadiene stabilisation by metal complexes (fig. 7):

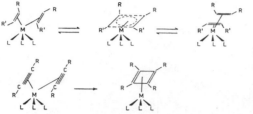

Fig. 7 - Cyclobutanetion reaction.

In this case the metal, through its electronic density, favours a process which is thermally forbidden on the ground of WOODWARD-HOFFMANN symmetry rules [19].

Such reactions have been invoked to explain the alkyne cyclic trimerisation or tetramerisation [20]; however recent investigations in this field [21] seem to demonstrate that the cyclisation takes place as follows (fig. 8):

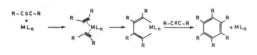

Fig. 8 - Trimerisation of alkynes.

Some Mechanisms of Homogeneous Catalytic Reactions.

We will now describe the mechanism of few catalytic reactions of olefins or alkynes, in the attempt to demonstrate that the catalytic cycles can be easily explained using the concepts reported above. Of course the many catalysts so far described and studied will be not treated here in detail but only the general aspects of the catalytic cycles will be analyzed.

Hydrogenation of Unsaturated Hydrocarbons.

In all the really homogeneous hydrogenating catalytic systems reported in the literature [9] it is observed that the transition metal complex interacts with molecular hydrogen to form a new complex containing at least one metal-hydrogen bond as the reactive intermediate. This corresponds to « metal activation ». This intermediate must be able to activate the olefin so it must be « coordinatively unsaturated » (see scheme 1).

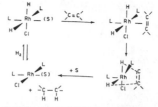

Scheme 1 - Catalytic cycle of hydrogenation (S is the solvent).

Once coordinated the olefin can interact with the hydride group by an « insertion reaction » [$N_{12/1}$ (+ H⁻) reaction].

The catalyst can be regenerated by interaction of the transition metal alkyl species with molecular hydrogen or with another hydridic ligand [$E_{1/0}$ (H⁺) reaction], either *via* an intramolecular (as in the example) or intermolecular mechanism.

Alkynes are reduced with a similar mechanism, the only difference being that an intermediate transition metal vinyl species is formed in place of a metal alkyl species.

$$L_nM-\overset{H}{\underset{CH_2}{\overset{|}{\underset{\|}{CH_2}}}} \rightleftarrows L_nM-CH_2CH_3$$

$$L_nM-\overset{H}{\underset{CH}{\overset{|}{\underset{\||}{CH}}}} \rightleftarrows L_nM-CH=CH_2$$

A similar mechanism has been proposed for the hydrogenation by metals in heterogeneous phase. Indeed the chemisorption of hydrogen gas and of unsaturated hydrocarbons is very similar to the hydrogen activation and the olefin coordination of the homogeneous process.

Hydroformylation of Olefins.

This important industrial process for the production of aldehydes and alcohols, involves the catalytic formation of an aldehyde from carbon monoxide, hydrogen and the olefin:

$$R-CH=CH_2 + CO + H_2 \rightarrow R-CH_2-CH_2CHO$$

Many carbonyl complexes can be used as catalysts but cobalt carbonyl is actually still the most used industrially.

The catalytic cycle using cobalt carbonyl is given here as a general example (scheme 2).

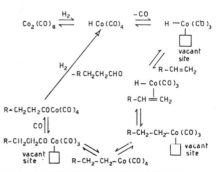

Scheme 2 - Catalytic cycle of olefin hydroformilation.

The first step is the « metal complex activation » by molecular hydrogen from the hydridotetracarbonylcobalt (I) $HCo(CO)_4$ species. In order to generate a vacant site the temperature must be rather high; in these conditions the $HCo(CO)_4$ species dissociates losing a carbon monoxide to form the coordinatively unsaturated complex $HCo(CO)_3$. An olefin molecule can be thus activated by the metal by coordination and inserted into the metal-hydrogen bond [$N_{I2/1}$ (+ H$^-$) type reaction]. The new metal-alkyl intermediate is coordinatively unsaturated and readily takes up carbon monoxide. A coordinated carbonyl group then inserts into the metal-alkyl bond forming an unsaturated cobalt acyl derivative [$N_{II/1}$ (+ RCH_2CH_2)]. This complex takes up a further carbon monoxide. The resulting species is attacked by molecular hydrogen or by a second mole of $HCo(CO)_4$ to give the aldehyde as product.

Similar basic steps can be envisaged in all the so-called carbonylation reactions to form acids, esters or acyl halides.

The only difference consists in the fact that acyl group reacts with nucleophilic species X$^-$ such as Cl$^-$, OCH$_3^-$, with a $N_{1/0}$ (+ X$^-$) reaction

$$L_nM\!-\!COR \xrightarrow{\text{X}^-} [L_nM]^- + RCOX$$
$$\downarrow \text{H}^+$$
$$L_nMH$$

The catalytic complex is reactivated by reaction with the proton to form a metal-hydrogen bond.

Dimerisation and Polymerisation of Olefins.

Olefin dimerisation is the first step of the olefin polymerisation process. Indeed while some catalysts, mainly of the noble metals group, are able to only dimerise olefins, other catalysts, the so-called ZIEGLER-NATTA catalysts, polymerise the same olefins.

However the basic steps of both processes can be considered rather similar.

In the following example (scheme 3) a catalytic cycle for ethylene dimerisation by a rhodium complex (^{22}) is reported:

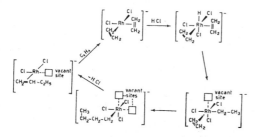

Scheme 3 - Catalytic cycle of rhodium catalysed ethylene dimerisation.

In this case the metal ion activation takes place with an oxidative addition of hydrogen chloride to a rhodium (I) complex $[RhCl_2(C_2H_4)_2]^-$. The sum of hydrogen chloride and the subsequent olefin insertion into the metal-hydrogen bond can be also considered as an electrophilic attack $E_{2/1}$ (+ H$^+$) to the olefin; such a reaction is facilitated in fact by the low oxidation state of the metal and by the negative charge of the ion. The next step involves the insertion of an adjacent coordinated ethylene into the metal-alkyl bond to form a longer-chain alkyl derivative.

This is an internal nucleophilic reaction, such a reaction is to be expected because now the rhodium atom has a higher oxidation state. A reverse insertion reaction takes subsequently place yielding a Rh(I) π-bonded butene complex; the starting complex is regenerated by the excess of ethylene.

A further insertion of ethylene into the σ butyl derivative is not possible in this case because the intermediate butyl derivative is not sufficiently stable, this is probably the reason why polymerisation does not occur with noble metal complexes.

ZIEGLER-NATTA catalysts, which undergo low pressure polymerisation reactions, are generally complex but recent studies indicate that the transition metal is the active polymerisation site.

The aluminium-alkyl, or other similar alkylating agents, functions mainly as co-catalyst for activating the transition metal ion. A mechanism proposed (^{23}) to explain ethylene polymerisation by Ti^{3+} ion is given here (scheme 4):

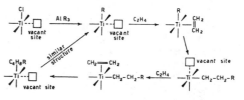

Scheme 4 - Catalytic cycle of Ziegler-Natta polymerisation.

After the metal ion alkylation, ethylene is activated by coordination to the transition metal.

As the metal is in a rather high oxidation state the nucleophilic attack of the alkyl group on the olefin is facilitated by the coordination and a cis insertion with the neighbouring metal-alkyl bond occurs. This process continues until chain termination occurs.

It is interesting to point out that the many reported co-dimerisation of olefins with dienes or dienes dimerisation and trimerisation always have as the most important step the olefin insertion into a σ metal-alkyl or a rather mobile $\sigma - \pi$ metal-alkyl-bond, that is a $N_{II/1}$ type reaction.

Olefin Oxidation and Acetylene Hydration.

Palladium (II) salts are used in to the stoichiometric oxidation of ethylene to acetaldehyde or vinyl-acetate with palladium deposition. The presence of cupric salts in solution under these conditions prevents the formation of palladium metal, oxidizing it immediately back to palladium (II).

The cuprous salts, formed by this reaction, can be reoxidized to cupric salts by stream of air.

This scheme is the basis of the « WACKER » process for homogeneous oxidation of olefins (scheme 5):

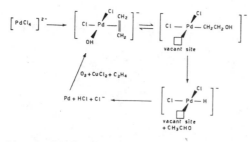

Scheme 5 - Catalytic cycle of ethylene oxidation.

The first step consists in palladium ion activation by hydrolysis of a Pd-Cl bond and formation of a Pd-OH bond. The coordination of the olefin to palladium (II) ion favours the nucleophilic attack $N_{12/1}$ (+ OH$^-$) on the olefin itself. It is not yet clear if this attack is really an insertion reaction that is an internal nucleophilic attack; however an external nucleophilic attack should be very difficult due to the anionic nature of the ion. A similar attack can be obtained with other nucleophiles such as CH_3COO^-, OCH_3^-, amines etc.

The subsequent step is a hydrogen abstraction or hydride shift to palladium; such a reaction is not new, indeed it was already found in the olefin isomerisation and dimerisation process (see later). It can be considered an $E_{11/0}$ (— H$^-$) reaction.

Other metals in high oxidation state such as Ir (III), Rh (III) and Ru (III) oxidize olefins probably with the same mechanism.

The catalytic hydration of acetylene by Ru (III) chloride ([25]) is very similar to olefin oxidation by palladium (II) salts (scheme 6):

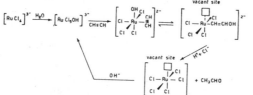

Scheme 6 - Catalytic cycle of ruthenium catalysed acetylene hydration.

Also in this case there is an internal nucleophilic attack of the OH$^-$ ion corresponding to an acetylene insertion reaction. However the subsequent step is an electrophilic attack $E_{1/0}$ (+ H$^+$) on the vinyl group in place of a hydrogen abstraction $E_{1/0}$ (— H$^-$).

Olefin Isomerisation.

Transition metals complexes easily isomerize olefins, generally they need co-catalytic activation ([22]) (with hydrogen or hydrogen chloride or ethanol and

so on) to form a metal-hydrogen bond and an unsaturated or kinetically labile complex (scheme 7).

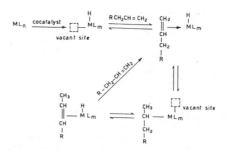

Scheme 7 - Catalytic cycle of cocatalysed olefin isomerisation.

In this case, after the metal ion activation, the reaction is a sequence of insertion $N_{12/1}$ (+ H$^-$) and abstraction $E_{12/1}$ (— H$^-$) reactions. In some cases the isomerisation seems to proceed without requiring a co-catalyst, in these cases in fact the olefin itself is the source of hydridic hydrogen.

The mechanism proposed by some Authors is quite distinct from the more general mechanism involving an intermediate alkyl (scheme 8).

Scheme 8 - Catalytic cycle of olefin isomerisation.

The first step in fact involves a hydrogen abstraction with the formation of a labile π-allyl hydride (this reaction is $E_{12/3}$ (— H$^-$) type reaction). The subsequent rearrangement corresponds to an allyl insertion into the metal-hydrogen bond [$N_{12/3}$ (+ H$^-$) type reaction].

A similar sequence has been proposed by KETLEY and coworkers ([26]) to explain ethylene dimerisation by Pd (II) salts (scheme 9).

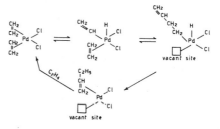

Scheme 9 - Catalytic cycle of palladium catalysed ethylene dimerisation.

1330

After a very unusual vinyl hydrogen abstraction $[E_{n/1} (-H^-)]$, an olefin insertion into the metal-carbon bond of the σ vinyl $[N_{f2/1} (+C_2H_3^-)]$ a hydrogen shift from the metal to the alkyl group $[N_{11/2} (+H^-)]$ concludes the catalytic cycle.

It is interesting to point out that the mechanism proposed for the olefin isomerisation or dimerisation on metal surfaces [27] is completely similar either in the cases in which a co-catalyst is used (hydrogen usually) or when the reaction is carried out without co-catalyst.

Conclusion.

In conclusion, although much work remains to be done to clarify many homogeneous catalytic reactions, certain general principles underlying the great catalytic versability of coordination compounds of the transition metals are already apparent.

These general principles, just analysed above in detail, give a rational interpretation of some experimental facts such as the ability of transition metal complexes to stabilize molecular fragments and reaction intermediates which have been proposed for a longtime in heterogeneous catalysis.

It is for this reason that the interpretation of mechanisms of heterogeneous catalytic reactions has received recently a great help from the studies in homogeneous phase, indeed similarities between the mechanisms of homogeneous and heterogeneous are becoming increasingly apparent.

There is also reason to hope and anticipate, although the difficulty of this goal should not be understimated, that the detailed understanding, that is emerging from studies of relatively simple homogeneous systems, will make a meaningful contribution to the understanding of some enzymic systems.

It is also expected (or, at least, hoped) in the near future that through increased understanding, homogeneous catalysts may be designed whose specificity and efficiency will approach those of enzymes.

At the moment however the understanding is limited to the basic principles, while it is known that the catalytic reactivity is controlled by many factors such as the oxidation state of the metal itself, the nature of the solvent and mainly by the auxiliary ligands in the catalytic metal-complex.

For instance, the effect of the auxiliary ligands, is very important being involved in several ways. First they may be electron conductors for the oxidation or reduction of the transition metal, second they can assist the mobility of alkyl or hydride ligands, finally, by induction and resonance, they help control the electronic density of the transition metal at the level required for catalysis.

All these aspects are not at the moment completely understood, however, once the effects of these factors are fully understood, homogeneous catalytic systems can be designed for particular uses.

A great deal of work is being done at the moment in this direction. However not only problems of understanding remain to be solved, but also of application. Among these we can note the homogeneous activation of saturated hydrocarbons and of molecular nitrogen or of oxygen (in a certain type of reactions).

The scope of all these researches is certainly of great industrial interest, being that of producing, with a rational approach, soluble catalysts that are more efficient and, in particular, more selective than those heterogeneous systems at present in use.

Istituto di Chimica generale dell'Università, Centro del C.N.R., Milano.

R. Ugo

References

[1] G. C. BOND, *Chem. Eng. 63*, 3 (1968).
[2] R. P. BELL: « The Proton in Chemistry ». Cornell University Press, Ithaca, N. Y. 1959.
[3] J. M. THOMAS, W. J. THOMAS: « Introduction to the Principles of Heterogeneous Catalysis ». Academic Press, New York 1967.
[4] H. R. MAHLER, E. H. CORDES: « Biological Chemistry ». Harper Int. Ed., New York 1966.
[5] R. G. PEARSON, *J. Am. Chem. Soc. 85*, 3533 (1963).
[6] M. M. JONES: « Ligand Reactivity and Catalysis ». Academic Press, New York 1968.
[7] K. FAJANS: « Structure and Bonding ». Vol. 3, Springer-Verlag, Berlino 1967.
[8] M. L. H. GREEN: « Organometallic Compounds ». Vol. 2, Methuen, London 1968.
[9] J. A. OSBORNE, *Endeavour 27*, 144 (1967).
[10] L. E. ORGEL: « An Introduction to Transition-Metal Chemistry, Ligand Field Theory ». Methuen, London 1961.
[11] J. CHATT, *Proc. Chem. Soc.*, 318 (1962).
[12] R. S. MULLIKEN, *J. Phys. Chem. 56*, 295 (1952).
[13] H. H. JAFFÉ, G. O. DOAK, *J. Chem. Phys. 21*, 196 (1953).
[14] S. CARRÀ, R. UGO, *Inorg. Chim. Acta Rev. 1*, 49 (1967).
[15] D. A. WHITE, *Organometall. Chem. Rev. A 3*, 497 (1968).
[16] G. W. PARSHALL, G. WILKINSON, *J. Chem. Soc.*, 1132 (1962).
[17] K. NOAK, F. CALDERAZZO, *J. Organometall. Chem. 10*, 101 (1967).
[18] J. HALPERN, LAI-YONG WONG, *J. Am. Chem. Soc. 90*, 6665 (1968).
[19] R. HOFFMANN, R. B. WOODWARD, *J. Am. Chem. Soc. 87*, 2048 (1965).
[20] G. N. SCHRAUZER, P. GLOCKNER, S. FICHLER, *Angew. Chem. (Int. Ed.) 3*, 185 (1964).
[21] J. P. COLLMAN, *Accounts of Chem. Res. 1*, 136 (1968).
[22] R. CRAMER, *Accounts of Chem. Res. 1*, 186 (1968).
[23] P. COSSEE, *J. Catalysis 3*, 80 (1964).
[24] A. AGUILÒ: « Adv. Organometall. Chem. ». Vol. 5, Academic Press, New York 1967.
[25] J. HALPERN, B. R. JAMES, A. L. W. KEMP, *J. Am. Chem. Soc. 83*, 4097 (1961).
[26] A. D. KETLEY, L. P. FISHER, A. J. BERLIN, C. R. MORGAN, E. H. GORMAN, T. R. STEADMAN, *Inorg. Chem. 6*, 657 (1967).
[27] S. CARRÀ, V. RAGAINI, *J. Catalysis 10*, 230 (1968).

Ricevuto l'8 ottobre 1969.

Tip. Stefano Pinelli   via Farneti 8   Milano

Aspetti generali della catalisi eterogenea con metalli di transizione

Dopo una breve introduzione, la catalisi omogenea è stata classificata come « dura » e « molle ». E' stata discussa nei dettagli solo la catalisi « molle » di idrocarburi π insaturi con complessi di metalli di transizione.

Sono stati riconosciuti come molto importanti quattro passaggi catalitici: insaturazione coordinativa, attivazione dello ione del metallo, attivazione del legante per coordinazione e reazioni del modello legante. Questi punti sono discussi in termini di chimica organometallica dei metalli di transizione. Un'analisi di alcuni meccanismi di catalisi omogenea ha dimostrato che questi possono essere agevolmente spiegati considerando nel giusto ordine i citati quattro passaggi di reazione.

Si rileva il fatto che molti concetti teorici, sviluppati nella catalisi omogenea, possono essere facilmente applicati in altri campi, come la catalisi eterogenea e la biochimica.

R. Ugo

General Features of Homogeneous Catalysis with Transition Metals

After a brief introduction, the homogeneous catalysis is classified as « hard » and « soft ». Only the « soft » catalysis of π unsaturated hydrocarbons with transition metal complexes is discussed in detail.

Four catalytic steps are recognised as very important: the coordinative unsaturation, the metal ion activation, the ligand activation by coordination and the template ligand reactions.

These points are discussed in terms of transition metals organometallic chemistry.

An analysis of some homogeneous catalytic mechanisms shows that they can be easily explained by applying in the right order the above four reaction steps.

It is emphasized that many theoretical concepts, developed in homogeneous catalysis, can be easily applied to other fields such as heterogeneous catalysis or biochemistry.

R. Ugo

Copyright 1969 by Akademische Verlagsgesellschaft, Frankfurt

Sonderdruck: „Zeitschrift für Physikalische Chemie Neue Folge", 64 (1969) 18–25

Herausgegeben von G. Briegleb, Th. Förster, G. Schmid, G.-M. Schwab, E. Wicke

49

Adsorptive and Catalytic Properties of Chromia

By

ROBERT L. BURWELL, JR., JOHN F. READ, KATHLEEN C. TAYLOR
and GARY L. HALLER

Ipatieff Catalytic Laboratory, Department of Chemistry, Northwestern
University, Evanston, Illinois, U.S.A.

Dedicated to Professor Dr. Georg-Maria Schwab on his 70th birthday

With 2 figures

(Received September 30, 1968)

AKADEMISCHE VERLAGSGESELLSCHAFT
FRANKFURT AM MAIN
1969

Abstract

Chemisorptions of ammonia, carbon monoxide, carbon dioxide and oxygen and the rate of hydrogenation of 1-hexene have been measured on amorphous chromia as functions of activation temperature between 25 and 400°C. These quantities have also been measured on microcrystalline α-Cr_2O_3 prepared by heating chromia in hydrogen at about 425°. An attempt is made to interpret the data in terms of simple coordinative adsorption at coordinatively unsaturated surface (cus) Cr^{3+} (oxygen and carbon monoxide chemisorption); of heterolytic dissociative adsorption at pair sites, Cr^{3+} (cus) and O^{2-} (cus) (hydrogenation of 1-hexene); and of ligand displacement adsorption (ammonia chemisorption).

A greater specification of the nature of the unsaturation in surface sites has been a major objective in heterogeneous catalysis. Treatment of the unsaturation as *coordinative unsaturation*[1,2] appears to be useful for certain catalysts. One might expect that it would be possible to characterize the surfaces of oxides as composed of an ensemble of separate sites more readily than the surfaces of metals owing to the maze of orbitals present at the surfaces of metals. After preparation in the presence of moisture, many metallic oxide catalysts are covered

---

[1] R. L. BURWELL, JR., and C. J. LONER, Proceedings of the Third International Congress on Catalysis, North-Holland Publishing Co., Amsterdam, 1965, Vol. II. p. 804.

[2] R. L. BURWELL, JR., Chem. Engng. News, Aug. 22 (1966) 56.

by a layer of hydroxide ions which condense during preliminary activation to generate coordinatively unsaturated surface (cus) ions.

$$OH^- \ OH^- \ OH^- \quad OH^- \qquad O^{2-}$$
$$\rightarrow \qquad\qquad\qquad + H_2O(g) \qquad (1)$$
$$M^{n+} \ M^{n+} \ M^{n+} \qquad M^{n+} \ M^{n+} \ M^{n+}.$$

Active sites, if (cus) ions, should add suitable ligands from the gas phase, *i.e.*, chemisorb suitable gases. It has been well recognized that the lone M^{n+} ion at the right of Eq. (1) is coordinatively unsaturated but it has been less well recognized that the O^{2-} ion is also coordinatively unsaturated.

As one increases the temperature of activation, one would expect the concentration of (cus) ions to increase monotonically but the nature of the sites might change with increasing temperature.

The catalytic activity of chromia (by which we mean some hydrated chromic oxide) for isotopic exchange between deuterium and alkanes[3] and benzene[1] and for hydrogenation of ethylene[4] increases steadily as one increases the activation temperature. If chemisorption behaves similarly, the ratio of chemisorption to catalytic activity *vs.* activation temperature should provide much more information about interrelations than usual studies on a catalyst activated at only one temperature. Interpretation should be facilitated for chromia by the possibility of close interrelations with coordination chemistry and by mechanistic interrelations with homogeneous catalysis[2,5].

We have employed chromia gels precipitated from an aqueous solution of chromic nitrate by the urea method[1]. The gel may be considered to be a condensation polymer of $Cr(H_2O)_3(OH)_3$. As dried at $100\,^\circ C$, the "formula" is about $Cr_2O_3 \cdot 4H_2O$. Exposure to higher temperatures leads to a progressive loss of water by various bulk condensation processes and, to a much smaller extent, by Eq. (1).

Let us consider what kinds of adsorption might occur at the Cr^{3+} (cus) and O^{2-}·(cus) ions generated by Eq. (1).

----

[3] R. L. BURWELL, JR., A. B. LITTLEWOOD, M. CARDEW, G. PASS and C. T. H. STODDART, J. Amer. chem. Soc. 82 (1960) 6272.

[4] Y. I. PECHERSKAYA, V. B. KAZANSKII and V. V. VOEVODSKII, Kinetics Catalysis (USSR) 3 (1962) 90.

[5] J. HALPERN, Chem. Engng. News, Oct. 31 (1966) 68.

2*

1. Simple coordinative adsorption with an increase in the coordination number of Cr^{3+} (cus).

$$Cr^{3+} + O(CH_3)_2 = Cr^{3+}O(CH_3)_2. \tag{2}$$

Carbon monoxide could similarly chemisorb to form a surface carbonyl, $Cr^{3+}CO$, with additional binding from a $d—p$ π-bond.

The ion, Cr^{3+} is hard[6]. Dimethyl ether is hard but carbon monoxide and ethylene are soft. Thus, one would expect binding of dimethyl ether to be stronger than that of carbon monoxide and ethylene, whereas, on the surfaces of transition metals, the reverse should be true.

2. Simple adsorption at basic sites like O^{2-} (cus).

$$OH^- + CO_2 = HCO_3^- \text{ or } O^{2-} + CO_2 = CO_3^{2-}. \tag{3}$$

3. Heterolytic dissociative adsorption at pair sites.

$$H_2(g)$$
$$H^-$$
$$Cr^{3+} O^{2-} = Cr^{3+} OH^-. \tag{4}$$

Methane might adsorb similarly to form $Cr^{3+}CH_3^- + OH^-$. We use the convention of coordination chemistry in writing hydride and methyl carbanion but one would expect considerable charge neutralization. The reverse of Eq. (1) would represent heterolytic adsorption of water. In this case, all interactions are hard-hard whereas H^- and CH_3^- result in soft-hard interactions. One would expect water to adsorb more strongly as is indeed the case.

4. Reductive adsorption.

$$H_2(g) + 2O^{2-} \text{ (cus)} + 2Cr^{3+} \text{ (cus)} = 2OH^- + 2Cr^{2+} \text{ (cus?)}. \tag{5}$$

This might occur for *bulk* Cr^{3+} ions at high temperatures but not at temperatures of interest in this paper. Cr^{2+} has very poor crystal field stabilization in the octahedral coordination of a chromia lattice.

5. Ligand displacement adsorption.

$$O^{2-} Cr^{3+} + B(g) = O^{2-} + Cr^{3+} B \tag{6}$$

$$O^{2-} Cr^{3+} + HB(g) = OH^- + Cr^{3+} B^-. \tag{7}$$

---

[6] F. Basolo and R. G. Pearson, Mechanisms of Inorganic Reactions. 2nd ed., John Wiley and Sons, Inc., New York, N. Y., p. 23, 113.

The last equation with HB being H_2O corresponds to the reverse of the bulk condensation processes which lead to water loss during the heating of hydrated chromia gel. Surface coordinative unsaturation is not involved but surface mobility is needed as B or B^- displaces O^2 from the coordination sphere of Cr^{3+}.

The ideas of 3. have been used in a mechanism proposed for hydrogenation of olefins[3], as exemplified by ethylene.

$$H_2(g) \qquad\qquad\qquad\quad CH_3 \qquad C_2H_6(g)$$
$$\qquad\qquad\qquad\qquad\qquad\qquad |$$
$$\qquad H^- \qquad\qquad\qquad\quad -CH_2$$

$$Cr^{3+}\ O^{2-} \rightarrow Cr^{3+}\ OH^- \xrightarrow{\ CH_2=CH_2\ } Cr^{3+}\ OH^- \rightarrow Cr^{3+}\ O^{2-}. \qquad (8)$$

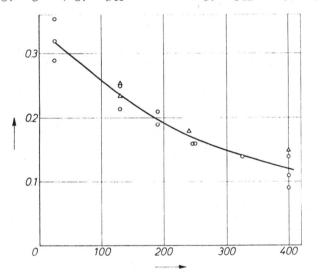

Fig. 1. Adsorption of ammonia on chromia at 25 °C as a function of the temperature of activation, ($\triangle$) chromia originally dried in air at 110 °C, ($\bigcirc$) chromia originally dried in air at 25 °C. A new sample of chromia was used for each point. x-axis: Activation temperature, °C; y-axis: NH_3/Cr^{3+}

Reverse of the last step leads to isotopic exchange between deuterium and alkanes.

Figure 1 shows that the adsorption of ammonia at 25 °C decreases slowly with activation temperature. The chromia was heated to the activation temperature in flowing, pure hydrogen and held there for two hours. A gas chromatographic technic was used for measurement of adsorption. In a run after activation at 130 °C, 28% of the adsorbed ammonia desorbed in flowing hydrogen in 3.5 hours at 25 °C, a further

12%, in two hours at 95° and all by 400°. At 25°, hydrogen sulfide adsorbed to $H_2S/Cr^{3+} = 0.12$ on a chromia activated to 130° and none was released by 400°. Such large adsorptions must represent some degree of bulk conversion and we suggest that they involve heterolytic ligand displacement adsorption. Some adsorbed ammonia can be held by hydrogen bonding but this is unlikely for hydrogen sulfide. We have no evidence that ligand substitution adsorption is involved in any specific catalytic process but it may join other types of adsorption

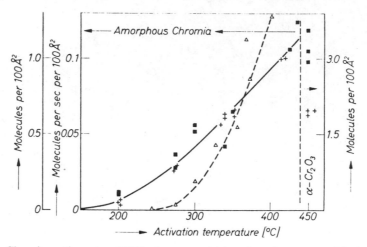

Fig. 2. Chemisorption at −78°C of oxygen (+) and carbon monoxide (■) and rate of hydrogenation of 1-hexene at 64°C (△) as functions of the temperature of activation

in processes in which one needs to bind two ligands to one five-coordinated surface Cr^{3+}. The process would be similar to that of loss of a solvent molecule or of a halide ion which is involved in some cases of homogeneous catalysis by transition metal complexes[5].

Figure 2 shows that chemisorption at −78°C of carbon monoxide and of oxygen appears after activation to about 175° and increases steadily with temperature of activation. Except for the lowest temperatures of activation, adsorptions of the two gases are equal on amorphous chromias. On α-Cr_2O_3, only 65% as much oxygen adsorbs as carbon monoxide in agreement with MacIver and Tobin[7].

Since there are potentially about 10 Cr^{3+} ions per 100 Å² on the surface of chromias[7], degrees of dehydroxylation of amorphous chromias are rather low but of α-Cr_2O_3, rather large. We suggest that

[7] D. S. MacIver and H. H. Tobin, J. physic. Chem. 64 (1960) 451.

most surface sites are five-coordinate Cr^{3+} and that carbon monoxide and oxygen adsorb on amorphous chromia to form

$$Cr^{3+}CO \quad \text{and} \quad Cr^{4+}O_2^-$$

but, that on $\alpha\text{-}Cr_2O_3$, sites are often close enough together so that oxygen often adsorbs by oxidative chemisorption as

$$O—O^{2-}$$

$$Cr^{4+} \quad Cr^{4+}.$$

Chemisorptions of oxygen and carbon monoxide were measured with a Cahn recording microbalance in a continuous flow of highly purified helium (diffusion through fused silica, liquid nitrogen trap) or hydrogen (diffusion through silver-palladium alloy)[8].

Activations involved heating in flowing hydrogen at the rate of $50°$ per hour to the activation temperature followed by an arrest for twelve hours. Helium was substituted for hydrogen during the arrest or at $300°C$ in activations above $300°$. Activation in hydrogen at about $400°$ gives microcrystalline $\alpha\text{-}Cr_2O_3$ but, after heating at $450°$ in helium or nitrogen, the material is still amorphous to x-ray[8,9].

Chemisorption of carbon monoxide is taken as gas not desorbed in flowing helium at $-78°C$. Much chemisorbed carbon monoxide is weakly held and 75% is liberated upon warming to $25°$ in flowing helium. Heating to near the original activation temperature restores the original chemisorptive capacity. The points shown for $300°$ in Fig. 2 are for a catalyst originally activated at $350°$ and might more appropriately be listed at that temperature. At $25°$, the LANGMUIR-like adsorption isotherm of carbon monoxide saturates at a value which agrees almost exactly with adsorption irreversible at $-78°$.

The definition of oxygen chemisorption was taken to be the same as that of carbon monoxide, not because the chemisorption is weak but because at $25°$ there is a slow secondary adsorption. In four hours at $100°$, a sample activated at $350°$ adsorbed four times as much oxygen as that shown in Fig. 2. Oxidation of Cr^{3+} (cus) must expose further Cr^{3+}. This process is presumably related to the effect of oxygen in promoting crystallization of amorphous chromia at about $350°$[9]. Heating in helium does not regenerate chromia covered

[8] R. L. BURWELL, JR., K. C. TAYLOR and G. L. HALLER, J. physic. Chem. **71** (1967) 4580.

[9] J. D. CARRUTHERS and K. S. W. SING, Chem. and Ind. (1967) 1919.

with oxygen but heating in hydrogen does. At about 100° slow reductive adsorption of hydrogen occurs until $H_2/O_2(ads) \sim 2$. Water desorbs at higher temperatures.

Chemisorption of carbon dioxide at 25°C follows neither the pattern of ammonia nor that of oxygen and carbon monoxide. Coverage rises slowly from about 0.65 molecules per 100 Å^2 after activation at 200° to about 1.1 after 437°. However, activation above 100° seems to be necessary for detectable chemisorption. Coverage on α-Cr_2O_3 is about 1.8. It appears probable that the nature of the chemisorption changes with increasing activation temperature.

Figure 2 shows that activity for the hydrogenation of 1-hexene at 64°C develops with activation temperature more rapidly than chemisorptive capacity for oxygen and carbon monoxide. Thus, between 275 and 350°C, chemisorption increases by a factor of 2.5, hydrogenation activity, by 18. Activations were in flowing hydrogen except that helium was used above 375° for catalysts we wished to keep amorphous. In studies at an activation temperature of 300°, activation *in vacuo* with introduction of hydrogen only at 64° led to nearly the same activity as that in hydrogen. Rates are presented as molecules hydrogenated initially per sec. per 100 Å^2. Rates on microcrystalline α-Cr_2O_3 were 0.42 to 0.58.

In view of the mechanism of Eq. (8) one would expect reactivity to involve not only a contribution from the exact nature and locations of the ions surrounding Cr^{3+} (cus) but also from the basicity of O^{2-} (cus). Thus, we believe that, whereas the adsorption of carbon monoxide diagnoses the total concentration of Cr^{3+} (cus), hydrogenation largely reflects that of more active pair sites generated mainly above 300°C.

Hydrogenation was measured in a flow apparatus at space velocities chosen to keep conversions under 40%. The gas feed was deuterium or hydrogen at a mole ratio to 1-hexene of 5 and a total pressure of 750 Torr. Further details of hydrogenation and accompanying isomerization and isotopic exchange as well as results for other catalytic reactions will be reserved for Vol. 20 of *Advances in Catalysis*.

The porosity of our amorphous chromias is mostly in micropores[7,8]. The average pore diameter of our microcrystalline α-Cr_2O_3 is about 25 Å[7]. Surface areas in m^2/g were: after activation at 200°, 303; 350°, 280; 425°, 209; α-Cr_2O_3, about 80. Mass transport limitations may affect the rates of the more rapid catalytic reactions.

There has been considerable discussion with regard to whether sites on chromia are Cr^{3+} or Cr^{2+} [10]. Our results indicate that sites on amorphous chromia are Cr^{3+}. Very probably, formation of any substantial concentration of Cr^{2+} labilizes the lattice for conversion to $\alpha\text{-}Cr_2O_3$. However, on our $\alpha\text{-}Cr_2O_3$, Cr^{2+} may well be involved.

Acknowledgment

This work was supported by the Air Force Office of Scientific Research and the Petroleum Research Fund. G.L.H. was a National Science Foundation Predoctoral Fellow, 1964 to 1966.

[10] L. L. VAN REIJEN, W. M. H. SACHTLER, P. COSSEE and D. M. BROUWER, Proceeding of the Third International Congress on Catalysis, North-Holland Publishing Co., Amsterdam, 1965, Vol. II, p. 829.

Author Citation Index

254, 256, 267, 277, 294, 295, 300,
309, 310, 311, 322, 323, 325, 329,
332, 333, 334, 342, 344, 346, 351,
355, 364, 365, 366, 377, 382, 383,
387, 388, 390, 391, 392, 395, 396,
398, 400, 410, 416, 428, 435, 449, 454

Pecherskaya, Y. I., 453
Pecile, C., 134, 300, 307
Peddle, G. J. D., 373
Pedley, J. B., 400
Pelchowicz, Z., 406
Pelletie, W. M., 387
Pengelly, R. M., 158
Penneman, R. A., 294
Peri, J. B., 66
Perkins, R. S., 435
Perrin, D. D., 219, 396
Person, W. B., 278
Persson, H., 190
Perumareddi, J. R., 52
Peter, S., 371
Peterson, H., Jr., 294
Petrov, K. A., 375
Petrovich, J. P., 351, 352
Petry, O., 435
Pfluger, C. E., 294
Pfluger, H. L., 14
Philip, J., 401
Phillips, C. S. G., 120, 158, 191
Phillips, D. J., 333
Phillips, W. G., 410
Philpot, J. St. L., 13
Pickard, R. H., 400
Pierson, R. G., 322
Pimentel, G. C., 275
Pinching, G. D., 190
Piper, T. S., 294, 350
Pistschimuka, P., 403
Pitman, I. H., 410
Pitt, B. M., 45
Pitzer, K. S., 42, 57, 66, 84, 190, 191, 219,
234, 257
Planc, R. A., 305
Platt, J. R., 42
Pliszka-Krawiecka, B., 405
Poë, A. J., 59, 120, 300, 309, 310
Pohl, R. L., 396
Pollart, K. A., 405
Pöllmann, P., 322
Popov, E. M., 403, 409
Porter, D. F., 428
Porter, G., 56

Porter, M. R., 375
Posey, F. A., 190, 316
Poshkus, A. C., 406
Poth, M. A., 18
Potts, R. A., 402
Poulsen, I., 191
Powell, D. L., 56
Powers, J. W., 354
Powers, R. M., 190
Prater, B. E., 322
Pratt, J. M., 149, 316
Preisler, P. W., 12
Preston, F. J., 322
Prestt, B., 44
Prewitt, C. T., 333
Price, E., 392
Price, H. J., 344
Proper, R., 316, 335
Prue, J. E., 35, 84, 219
Pryor, W. A., 55, 351
Pu, L. S., 322
Pudovik, A. N., 409
Purcell, K. F., 84, 267, 401
Purdie, D., 300

Rabinowitch, E., 16
Rabinowitz, R., 408
Radda, G. K., 255
Radtke, D. D., 159
Ragaini, V., 449
Ralf, E., 372
Ramsey, B., 352
Ramsey, O. B., 399
Ransil, B. J., 229, 233
Rao, K. N., 331
Ratajczak, A., 406
Rathsam, G., 407
Rattenbury, K. H., 407
Raymond, K. N., 84, 294
Reed, R. I., 322
Rees, C. W., 322
Reid, F. J., 149
Reinen, D., 219
Reinheimer, J. D., 57
Reinhold, H., 402
Reinsalu, V. P., 322
Reishus, J. W., 326
Remy, H., 134
Reuter, B., 372
Reynolds, W. L., 346
Rich, A., 87

Subject Index

Acetylacetonate complexes, 282–284, 296–299

Adsorption
  as acid–base process, 56, 59, 63, 424, 425–428, 429–436, 451–459
  structural factors involved in, 429–436

Alpha effect, 3, 40, 43–44, 383

Alumia, acid–base properties of, 76, 424

Ambident ions, 77, 131, 141, 255, 264, 282, 358, 384
  defined, 363–364
  in platinum complexes, 296–299
  and stabilization of metal complexes, 285–295
  in thiophosphoryl group, 399–409

Anti-symbiosis, 51, 282–283, 309–310

Biological systems, 66, 117, 372, 438, 441

Bond energies, estimation of, 86–87, 99, 182, 220–222, 223–231, 232–239, 244

Brønsted acids, defined, 2–3, 67, 72, 156–157

Brønsted bases, defined, 2, 72, 156–157

Brønsted relationship, 10, 44, 74, 323, 383, 392

Carbene, as Lewis acid, 70, 284, 381

Catalysis, 5, 59, 70, 156, 283, 311–317, 334–340, 358, 365–376, 424
  of aryl α-disulfones, 416–422
  of aryl sufinyl sulfones, 410–415
  of chromia, 451–459
  with transition metals, 437–450

Charge-controlled reactions, 80, 255–266

Charge-transfer complexes, 56, 68, 73

Chelate effect, 173, 213–215

Chromia
  acid–base properties, 424, 451–459
  catalysis of, 451–459

Class (a) and (b), defined, 4, 21–24, 28

Coordinative unsaturation, 439, 452

Dialectric constant, role of, 114, 162, 178, 199–200, 206, 212, 257

Drago–Wayland equation, 71, 122, 157, 243, 250, 254, 256, 267–279

Edwards equation, 10, 12–19, 41, 74, 242, 254, 326, 379, 398

Electronegativity, 6, 7, 54, 77, 154, 326, 359, 399
  and carbon, 379
  defined, 136
  and donor atoms, 128–129
  and Madelung energy, 228–231
  and Pauling's equation, 220–222, 223–231
  size effects, 232–239
  and softness, 260

Electronegativity equalization, 226

Electrophilic substitution, 358, 366–367, 377, 443
  on pyridine oxide, 264–265

Enthalpy, 156, 195–196, 209, 214, 223
  of adduct formation, 267–279
  of atomization, 238
  of reaction, 225–226, 228

Entropy, 195–197, 212–218
  of solvation, 359

Free radicals, acid–base properties of, 56, 70, 73, 347–356, 366

Frontier-controlled reactions, 80, 255–266

HSAB tables, 50, 64, 68, 75

Hydrogen molecule, cleavage of, 370, 442, 454

Innocent ligands, 5, 135–148

Intrinsic strength, 69, 73, 76, 223, 377